Lecture Notes in Computer Science 1685

Edited by G. Goos, J. Hartmanis and J. van Leeuwen

W0107471

Springer-Verlag Berlin Heidelberg GmbH

Patrick Amestoy   Philippe Berger
Michel Daydé   Iain Duff   Valérie Frayssé
Luc Giraud   Daniel Ruiz   (Eds.)

Euro-Par'99 Parallel Processing

5th International Euro-Par Conference
Toulouse, France, August 31 – September 3, 1999
Proceedings

 Springer

Series Editors

Gerhard Goos, Karlsruhe University, Germany
Juris Hartmanis, Cornell University, NY, USA
Jan van Leeuwen, Utrecht University, The Netherlands

Volume Editors

Patrick Amestoy
Philippe Berger
Michel Daydé
Daniel Ruiz
ENSEEIHT, 2, Rue Camichel, F-31071 Toulouse Cedex 7, France
E-mail: {amestoy,berger,dayde,ruiz}@enseeiht.fr

Iain Duff
Valérie Fraysse
Luc Giraud
CERFACS, 42, Av. Gaspard Coriolis, F-31057 Toulouse Cedex 1, France
E-mail: {duff,fraysse,giraud}@cerfacs.fr

Cataloging-in-Publication data applied for

Die Deutsche Bibliothek - CIP-Einheitsaufnahme

Parallel processing : proceedings / Euro-Par '99, 5th International
Euro-Par Conference, Toulouse, France, August 31 - September 3,
1999. Patrick Amestoy ... (ed.). [ACM ; IFIR]. - Berlin ; Heidelberg ;
New York ; Barcelona ; Hong Kong ; London ; Milan ; Paris ;
Singapore ; Tokyo : Springer, 1999
  (Lecture notes in computer science ; Vol. 1685)
  ISBN 978-3-540-66443-7

CR Subject Classification (1998): C.1-4, D.1-4, F.1-2, G.1-2, E.1, H.2

ISSN 0302-9743

ISBN 978-3-540-66443-7   ISBN 978-3-540-48311-3 (eBook)
DOI 10.1007/978-3-540-48311-3

This work is subject to copyright. All rights are reserved, whether the whole or part of the material is concerned, specifically the rights of translation, reprinting, re-use of illustrations, recitation, broadcasting, re-
production on microfilms or in any other way, and storage in data banks. Duplication of this publication or
parts thereof is permitted only under the provisions of the German Copyright Law of September 9, 1965, in
its current version, and permission for use must always be obtained from Springer-Verlag Berlin Heidelberg
GmbH. Violations are liable for prosecution under the German Copyright Law.

© Springer-Verlag Berlin Heidelberg 1999
Originally published by Springer-Verlag Berlin Heidelberg New York in 1999

Typesetting: Camera-ready by author
SPIN: 10704397   06/3142 – 5 4 3 2 1 0   Printed on acid-free paper

Preface

Euro-Par is an international conference dedicated to the promotion and advancement of all aspects of parallel computing. The major themes can be divided into the broad categories of hardware, software, algorithms and applications for parallel computing. The objective of Euro-Par is to provide a forum within which to promote the development of parallel computing both as an industrial technique and an academic discipline, extending the frontier of both the state of the art and the state of the practice. This is particularly important at a time when parallel computing is undergoing strong and sustained development and experiencing real industrial take-up. The main audience for and participants in Euro-Par are seen as researchers in academic departments, government laboratories and industrial organisations. Euro-Par's objective is to become the primary choice of such professionals for the presentation of new results in their specific areas. Euro-Par is also interested in applications which demonstrate the effectiveness of the main Euro-Par themes.

There is now a permanent Web site for the series `http://brahms.fmi.uni-passau.de/cl/europar` where the history of the conference is described. Euro-Par is now sponsored by the Association of Computer Machinery and the International Federation of Information Processing.

Euro-Par'99

The format of Euro-Par'99 follows that of the past four conferences and consists of a number of topics each individually monitored by a committee of four. There were originally 23 topics for this year's conference. The call for papers attracted 343 submissions of which 188 were accepted. Of the papers accepted, 4 were judged as distinguished, 111 as regular and 73 as short papers. Distinguished papers are allowed 12 pages in the proceedings and 30 minutes for presentation, regular papers are allowed 8 pages and 30 minutes for presentation, short papers are allowed 4 pages and 15 minutes for presentation. Distinguished papers are indicated in the table of contents with a superscribed asterisk after the title. Four extra pages could be purchased for distinguished papers, two for regular papers and one for short paper. There were on average 3.5 reviews per paper. Submissions were received from 44 countries, 34 of which are represented at the conference. The principal contributors by country are France with 39 papers, the U.S.A. with 22, Germany with 19, and the U.K., Italy and Spain with 13 papers each. This year's conference, Euro-Par'99, features new topics such as fault avoidance and fault removal in real-time systems, instruction-level parallelism and uniprocessor architecture, and global environment modelling.

The Web site for the conference is `http://www.enseeiht.fr/europar99/`.

Acknowledgments

Knowing the quality of the past Euro-Par conferences makes the task of organising one daunting indeed and we have many people to thank. Ron Perrott, Christian Lengauer, Luc Bougé, Jeff Reeve and David Pritchard have given us the benefit of their experience and helped us generously throughout the past 18 months. The topic structure of the conference means that we must depend on the goodwill and enthusiasm of all the 90 programme committee members listed below. Their professionalism makes this the most academically rigorous conference in the field worldwide. The programme committee meeting at Toulouse in April was well attended and, thanks to sound preparation by everyone and Ron Perrott's guidance, resulted in a coherent, well-structured conference. The smooth running of the organisation of the conference can be attributed to a few individuals. Firstly the software for the submission and refereeing of papers that we inherited from Lyon via Passau was significantly enhanced by Daniel Ruiz. This attracted many compliments from those who benefited. Valérie Frayssé and Luc Giraud spent copious hours checking, printing and correcting papers. Finally, Brigitte Yzel, secretary to the conference and the CERFACS Parallel Algorithms Project, has been invaluable in monitoring the conference organisation and seeing to the myriad of tasks that invariably arise.

June 1999

Patrick Amestoy
Philippe Berger
Michel Daydé
Iain Duff
Valérie Frayssé
Luc Giraud
Daniel Ruiz

Euro-Par Steering Committee

Chair
Ron Perrott · Queen's University Belfast, UK
Vice Chair
Christian Lengauer · University of Passau, Germany
Committee
Makoto Amamiya · Kyushu University, Japan
Luc Bougé · ENS Lyon, France
Helmar Burkhart · University of Basel, Switzerland
Pierrick Fillon-Ashida · European Commission, Belgium
Paul Feautrier · University of Versailles, France
Ian Foster · Argonne National Laboratory, USA
Jeff Reeve · University of Southampton, UK
Henk Sips · Technical University Delft, The Netherlands
Paul Spirakis · CTI, Greece
Marian Vajtersic · Slovak Academy, Slovakia
Marco Vanneschi · University of Pisa, Italy
Jens Volkert · Johannes Kepler University Linz, Austria
Emilio Zapata · University of Malaga, Spain

Euro-Par'99 Local Organisation

The conference has been jointly organised by CERFACS (Centre Européen de Recherche et de Formation Avancée en Calcul Scientifique) and ENSEEIHT-IRIT (Ecole Nationale Supérieure d'Electronique, d'Electrotechnique, d'Informatique et d'Hydraulique de Toulouse - Institut de Recherche en Informatique de Toulouse).

General Chair
Iain Duff · CERFACS and RAL
Vice Chair
Michel Daydé · ENSEEIHT-IRIT
Committee
Patrick Amestoy · ENSEEIHT-IRIT
Philippe Berger · ENSEEIHT-IRIT
Valérie Frayssé · CERFACS
Luc Giraud · CERFACS
Informatics Chair
Daniel Ruiz · ENSEEIHT-IRIT
Secretary
Brigitte Yzel · CERFACS

Euro-Par'99 Programme Committee

Topic 01: Support Tools and Environments

Global Chair
Ian Foster   Argonne Nat. Lab., USA
Local Chair
Frédéric Desprez   LIP-INRIA, Lyon, France
Vice Chairs
Jean-Marie Garcia   LAAS, Toulouse, France
Thomas Ludwig   TU Munich, Germany

Topic 02: Performance Evaluation and Prediction

Global Chair
Aad van der Steen   Utrecht Univ., The Netherlands
Local Chair
Jean-Marc Vincent   IMAG, Grenoble, France
Vice Chairs
Erich Strohmaier   Univ. of Tennessee at Knoxville, USA
Jerzy Wasniewski   Uni-C, Denmark

Topic 03: Scheduling and Load Balancing

Global Chair
Jean-Marc Geib   LIFL, Lille, France
Local Chair
Jean Roman   LaBRI, Bordeaux, France
Vice Chairs
Bruce Hendrickson   SNL, Albuquerque, USA
Pierre Manneback   PIP, Mons, Belgium

Topic 04: Compilers for High Performance Systems

Global Chair
Barbara Chapman   University of Southampton, UK
Local Chair
François Bodin   INRIA-IRISA, Rennes, France
Vice Chairs
Louis Féraud   IRIT, Toulouse, France
Chris Lengauer   FMI, University of Passau, Germany

Topic 05: Parallel and Distributed Databases

Global Chair
Burkhard Freitag   University of Passau, Germany
Local Chair
Kader Hameurlain   IRIT, Toulouse, France
Vice Chairs
László Böszörmenyi   University of Klagenfurt, Austria
Waqar Hasan   DB WIZARD, CA, USA

Topics 06 + 20: Fault Avoidance and Fault Removal in Real-Time Systems

Global Chairs
Ravi Iyer   Univ. of Illinois, USA
Tomasz Szmuc   St. Staszic TU, Krakow, Poland
Local Chairs
Gilles Motet   INSA, Toulouse, France
David Powell   LAAS-CNRS, Toulouse, France
Vice Chairs
Wolfgang Halang   FernUniversität, Hagen, Germany
Andras Pataricza   TU Budapest, Hungary
Joao Gabriel Silva   University of Coimbra, Portugal
Janusz Zalewski   Univ. of Central Florida, Orlando, USA

Topic 07: Theory and Models for Parallel Computation

Global Chair
Michel Cosnard   LORIA-INRIA, Nancy, France
Local Chair
Afonso Ferreira   CNRS-I3S-INRIA, Sophia Antipolis, France
Vice Chairs
Sajal Das   University of North Texas, USA
Frank Dehne   Carleton University, Canada

Topic 08: High-Performance Computing and Applications

Global Chair
Horst Simon   NERSC, L. Berkeley Lab., CA, USA
Local Chair
Michael Rudgyard   Oxford Univ. and CERFACS, England
Vice Chairs
Wolfgang Gentzsch   GENIAS Software, Germany
Jesus Labarta   CEPBA, Spain

Topic 09: Parallel Computer Architecture

Global Chair
Chris Jesshope   Massey University, New Zealand
Local Chair
Daniel Litaize   Université Paul Sabatier, France
Vice Chairs
Karl Dieter Reinartz   University of Erlangen, Germany
Per Stenstrom   Chalmers University, Sweden

Topic 10: Distributed Systems and Algorithms

Global Chair
André Schiper   EPFL, Lausanne, Switzerland
Local Chair
Gérard Padiou   ENSEEIHT-IRIT, Toulouse, France
Vice Chairs
Roberto Baldoni   University of Rome, Italy
Jerzy Brzezinski   Fac. of Electrical Eng., Poznan, Poland
Friedemann Mattern   TU-Darmstadt, Germany
Luis Rodrigues   Faculty of Sciences, Univ. of Lisbon, Portugal

Topic 11: Parallel Programming: Models, Methods and Languages

Global Chair
Luc Bougé   ENSL, Lyon, France
Local Chair
Mamoun Filali   IRIT, Toulouse, France
Vice Chairs
Bill McColl   Oxford Univ. Computing Lab., UK
Henk J. Sips   TU Delft, The Netherlands

Topic 12: Architectures and Algorithms for Vision and other Senses

Global Chair
Virginio Cantoni   University of Pavia, Italy
Local Chair
Alain Ayache   ENSEEIHT-IRIT, Toulouse, France
Vice Chairs
Concettina Guerra   Purdue Univ., USA, and Padoa Univ., Italy
Pieter Jonke   TU Delft, The Netherlands

Topics 13+19: Numerical Algorithms for Linear and Nonlinear Algebra

Global Chairs
Jack Dongarra   Univ. of Tennessee and ORNL, USA
Bernard Philippe   INRIA-IRISA, Rennes, France
Local Chairs
Patrick Amestoy   ENSEEIHT-IRIT, Toulouse, France
Valérie Frayssé   CERFACS, Toulouse, France
Vice Chairs
Françoise Chaitin-Chatelin   Univ. Toulouse I and CERFACS, France
Donato Trigiante   University of Florence, Italy
Marian Vajtersic   Slovak Academy, Bratislava
Henk van der Vorst   Utrecht University, The Netherlands

Topic 14: Emerging Topics in Advanced Computing in Europe

Global Chair
Renato Campo   European Commission, DG XIII, Belgium
Local Chair
Luc Giraud   CERFACS, Toulouse, France
Vice Chair
Pierrick Fillon-Ashida   European Commission, DG XIII, Belgium
Rizos Sakellariou   CRPC, Rice University, Houston, USA

Topic 15: Routing and Communication in Interconnection Networks

Global Chair
Ernst W. Mayr   TU Munich, Germany
Local Chair
Pierre Fraigniaud   LRI, Orsay, France
Vice Chairs
Bruce Maggs   Carnegie Mellon Univ., Pittsburg, USA
Jop Sibeyn   MPI, Saarbrücken, Germany

Topic 16: Instruction-Level Parallelism and Uniprocessor Architecture

Global Chair
Mateo Valero   CEPBA, Polytech. Univ. of Catalunya, Spain
Local Chair
Pascal Sainrat   IRIT, Toulouse, France
Vice Chairs
D. K. Arvind   Edinburgh Univ., Computer Science Dept., UK
Stamatis Vassiliadis   TU Delft, The Netherlands

Topic 17: Concurrent and Distributed Programming with Objects

Global Chair
Akinori Yonezawa   University of Tokyo, Japan
Local Chair
Patrick Sallé   ENSEEIHT-IRIT, Toulouse, France
Vice Chairs
Joe Armstrong   Ericsson, Sweden
Vasco T. Vasconcelos   University of Lisbon, Portugal
Marc Pantel   ENSEEIHT-IRIT, Toulouse, France

Topic 18: Global Environment Modelling

Global Chair
David Burridge   ECMWF, Reading, UK
Local Chair
Michel Déqué   Météo-France, Toulouse, France
Vice Chairs
Daniel Cariolle   Météo-France, Toulouse, France
Jean-Pascal van Ypersele   Louvain la Neuve Inst. of Astrophysics,
                             Belgium

Topic 22: High-Performance Data Mining and Knowledge Discovery

Global Chair
David Skillicorn   Queens University, Kingston, Canada
Local Chair
Domenico Talia   ISI-CNR, Rende, Italy
Vice Chairs
Vipin Kumar   Univ. of Minnesota, USA
Hannu Toivonen   RNI, University of Helsinki, Finland

Topic 23: Symbolic Computation

Global Chair
John Fitch   University of Bath, UK
Local Chair
Mike Dewar   NAG Ltd, Oxford, UK
Vice Chairs
Erich Kaltofen   North Carolina State Univ., Raleigh, USA
Anthony D. Kennedy   Edinburgh University, Scotland

Euro-Par'99 Referees

(excluding members of the programme and organisation committees)

Adelantado, Martin
Akl, Selim
Albanesi, Maria Grazia
Alvarez-Hamelin, I.
Amir, Yair
Amodio, Pierluigi
Ancourt, Corinne
Arbab, Farhad
Arbenz, Peter
Arnold, Marnix
Aronson, Leon
Asenjo, Rafael
Avresky, D.
Aydt, Ruth
Ayguade, Eduard
Aykanat, Cevdet
Baden, Scott
Bagchi, Saurabh
Bai, Z.
Baker, Mark
Bal, Henri
Baquero, Carlos
Barreiro, Nuno
Bartoli, Alberto
Baudon, Olivier
Becchetti, Luca
Béchennec, Jean-Luc
Becka, Martin
Beemster, Marcel
Bellegarde, Françoise
Bellosa, Frank
Benzi, Michele
Beraldi, Roberto
Bernon, Carole
Berthomé, Pascal
Bétourné, Claude
Biancardi, Alberto
Birkett, Nick
Bodeveix, Jean-Paul
Boichat, Romain
Bolychevsky, Alex
Bos, André

Boukerche, Azzedine
Boulet, Pierre
Bourzoufi, Hafid
Braconnier, Thierry
Bradford, Russell
Brandes, Thomas
Braunl, Thomas
Broggi, Alberto
Broom, Bradley
Browne, Shirley
Brugnano, Luigi
Brun, Olivier
Bungartz, Hans-Joachim
Burkhart, Helmar
Buvry, Max
Caires, Luis
Calamoneri, Tiziana
Cameron, Stephen
Caniaux, Guy
Cappello, Franck
Carissimi, Alexandre
Caromel, Denis
Carter, Larry
Cassé, Hugues
Castro, Miguel
Catthoor, Francky
Cavalheiro, Gerson
Chartier, Philippe
Charvillat, Vincent
Chassin de
Kergommeaux, Jacques
Chauhan, Arun
Chaumette, Serge
Cheng, Ben-Chung
Chesneaux, Jean-Marie
Chich, Thierry
Chmielewski, Rafal
Cilio, Andréa
Cinque, Luigi
Ciuffoletti, Augusto
Clocksin, William
Coelho, Fabien

Cole, Murray
Conter, Jean
Corbal, Jesus
Correia, Miguel
Costa, Antonio
Costa, Ernesto
Cotofana, Sorin
Coudert, David
Coulette, Bernard
Counilh, Marie-Christine
Courtier, Philippe
Cristal, Adrian
Crouzet, Yves
Crégut, Xavier
Cubero-Castan, Michel
Cucchiara, Rita
Cung, Van-Dat
Dackland, Krister
Daoudi, El Mostafa
Dechering, Paul
Defago, Xavier
Dekeyser, Jean-Luc
Delmas, Olivier
Delord, Xavier
Denissen, Will
Denneulin, Yves
Di, Vito
Di Martino, Beniamino
Di Stefano, Luigi
Doblas, Francisco Javier
Dominique, Orban
Doreille, Mathias
Dörfel, Mathias
Douville, Hervé
Drira, Khalil
Duato, José
Duchien, Laurence
Durand, Bruno
Eigenmann, Rudolf
Eijkhout, Victor
Eisenbeis, Christine
Eldershaw, Craig

Ellmenreich, Nils
Endo, Toshio
Erhel, Jocelyne
Etiemble, Daniel
Evans., D.J.
Even, Shimon
Fabbri, Alessandro
Faber, Peter
Farcy, Alexandre
Feautrier, Paul
Fent, Alfred
Ferragina, Paolo
Ferreira, Maria
Fidge, Colin
Fimmel, Dirk
Fischer, Claude
Fischer, Markus
Fleury, Eric
Fonlupt, Cyril
Fraboul, Christian
Fradet, Pascal
Fragopoulou, Paraskevi
Frietman, Edward E.E.
Fünfrocken, Stefan
Galilee, François
Gallopoulos, Efstratios
Gamba, Paolo
Gandriau, Marcel
Gaspar, Graca
Gatlin, Kang Su
Gauchard, David
Gautama, Hasyim
Gavaghan, David
Gavoille, Cyril
Geist, Al
Geleyn, Jean-François
van Gemund, Arjan
Gengler, Marc
Genius, Daniela
Gerndt, Michael
Getov, Vladimir
Girau, Bernard
Goldman, Alfredo
Gonzalez, Antonio
Gonzalez, José

Gonzalez-Escribano, A.
Goossens, Bernard
Grammatikakis, Miltos
Gratton, Serge
Greiner, Alain
Griebl, Martin
Grislin-Le Strugeon, E.
Gropp, William
Guedes, Paulo
Guivarch, Ronan
Guo, Katherine
Gupta, Anshul
Gustedt, Jens
Guyennet, Hervé
Hains, Gaetan
Hains, Gaétan
Halleux, Jean-Pierre
Han, Eui-Hong
Haquin, Thierry
Harald, Kosch
Hart, William
Hatcher, Phil
Hegland, Markus
Helary, Jean-Michel
Henty, David
Herley, Kieran
Herrmann, Christoph
Higham, Nicholas J.
Hitz, Markus
Hoisie, Adolfy
Houzet, Dominique
Hruz, Tomas
Hurfin, Michel
Ierotheou, Cos
Izu, Cruz
Jabouille, Patrick
Jeannot, Emmanuel
Jégou, Yvon
Jin, Guohua
Jonker, Pieter
Joshi, Mahesh
Jourdan, Stephan
Juanole, Guy
Juurlink, Ben
Kågström, Bo

Kakugawa, Hirotsugu
Kalbarczyk, Zbigniew
Kale, Laxmikant
Katayama, Takuya
Kaufman, Linda
Kehr, Roger
Keleher, Peter
Kelly, Paul
Kesselman, Carl
Killijian, Marc-Olivier
Kindermann, Stephan
Kiper, Ayse
Klasing, Ralf
Knijnenburg, Peter
Knoop, Jens
Koelbel, Charles
Kolda, Tamara
Kosch, Harald
Kossmann, Donald
Koster, Jacko
Kredel, Heinz
Kruse, Hans-Guenther
Kuchen, Herbert
Kucherov, Gregory
Kurmann, Christian
Kvasnicka, Dieter
de Laat, Cees
Ladagnous, Philippe
Lahlou, Chams
Langlois, Philippe
Larriba-Pey, Josep L.
Laure, Erwin
Lavenier, Dominique
LeCun, Bertrand
Lecomber, David
Lecomber, david
Lecussan, Bernard
Leese, Robert
Legall, Françoise
Leuschel, Michael
L'Excellent, Jean-Yves
Li, Xiaoye S.
Lindermeier, Markus
van Lohuizen, Marcel P.
Lombardi, Luca

Lombardo, Maria-Paola
Lopes, Cristina
Lopes, Luis
Lopez, David
Loucks, Wayne
Luksch, Peter
Lusk, Ewing
Lutton, Evelyne
Llosa, Josep
Manolopoulos, Yannis
Marchetti, Carlo
Marchetti-Spaccamela, A.
Marcuello, Pedro
Marquet, Pascal
Marquet, Philippe
Martel, Ivàn
Marthon, Philippe
Martin, Bruno
Maslennikov, Oleg
Masuhara, Hidehiko
Mauran, Philippe
Maurice, Pierre
Mautor, Thierry
Mechelli, Marco
Méhaut, Jean-François
Mehofer, Eduard
Mehrotra, Piyush
Meister, Gerd
Melideo, Giovanna
Merlin, John
Mery, Dominique
Meszaros, Tamas
Meurant, Gérard
Mevel, Yann
Meyer, Ulrich
Michael, W.
Michaud, Pierre
Miller, Bart
Miller, Quentin
Moe, Randi
Monnier, Nicolas
Montanvert, Annick
Monteil, Thierry
Moreira, Ana
Morère, Pierre

Morin, Christine
Morris, John
Morvan, Franck
Mostefaoui, Achour
Mounie, Gregory
Mucci, Philip
Muller, Jean-Michel
Mutka, Matt
Mzoughi, Abdelaziz
Nash, Jonathan
Nemeth, Zsolt
Neves, Nuno
Nichitiu, Codrin
Nicole, Denis
Norman, Arthur
Novillo, Diego
O'Boyle, Michael
Ogihara, Mitsunori
Ogston, Elizabeth
Oksa, Gabriel
Oliveira, Rui
Ortega, Daniel
Owezarski, Philippe
Oyama, Yoshihiro
Oztekin, Bilgehan U.
Paillassa, Béatrice
Pantel, Marc
Paprzycki, Marcin
Paugam-Moisy, Héléne
Pazat, Jean-Louis
Pedone, Fernando
Pellégrini, Franois
Pereira, Carlos
Perez, Christian
Perrin, Guy-René
Perry, Nigel
Petitet, Antoine
Petiton, Serge
Petrosino, Alfredo
Phillips, Cynthia
Pialot, Laurent
Piccardi, Massimo
Pietracaprina, Andréa
Pingali, Keshav
Pitt-Francis, Joe

Pizzuti, Clara
Plaks, Toomas
Planton, Serge
Porto, Stella
Pozo, Roldan
Prakash, Ravi
Pritchard, David
Prylli, Loïc
Pryor, Rich
Pucci, Geppino
Puglisi, Chiara
Quaglia, Francesco
Quinton, Patrice
Quéinnec, Philippe
Rackl, Guenther
Radulescu, Andrei
Raghavan, Padma
Rajopadhye, Sanjay
Ramirez, Alex
Rapine, Christophe
Rastello, Fabrice
Rau-Chaplin, Andrew
Rauber, Thomas
Ravara, Antnio
Raynal, Michel
van Reeuwijk, C.
Renaudin, Marc
Rescigno, Adele
Revol, Nathalie
Rezgui, Abdelmounaam
Richter, Harald
Risset, Tanguy
Roberto, Vito
Roch, Jean-Louis
Rochange, Christine
Roos, Steven
Rose, Kristoffer
Royer, Jean-François
Rubi, Franck
Rychlik, Bohuslav
Sagnol, David
Sajkowski, Michal
Sanchez, Jesus
Sanders, Peter
Sanjeevan, K.

Santos, Eunice
Saouter, Yannick
Savoia, Giancarlo
Schaerf, Andréa
Schreiber, Robert S.
Schreiner, Wolfgang
Schulz, Martin
Schuster, Assaf
Schwertner-Charao, A.
Seznec, André
Siegle, Markus
Sieh, Volkmar
Silva, Luis Moura
Silva, Mario
Slomka, Frank
Smith, Jim
Smith, Warren
Sobaniec, Cezary
Spezzano, Giandomenico
Stanca, Marian

Stanton, Jonathan
Stathis, Pyrrhos
Stein, Benhur
Steinbach, Michael
Stricker, Thomas M.
Strout, Michelle Mills
Stunkel, Craig
Subhlok, Jaspal
Sumii, Eijiro
Surlaker, Kapil
Swarbrick, Ian
Swierstra, Doaitse
Sybein, Jop
Szychowiak, Michal
Talbi, El ghazali
Talcott, Carolyn
Taylor, Valerie
Temam, Olivier
Thakur, Rajeev
Thiesse, Bernard

Thuné, Michael
Tiskin, Alexandre
Toetenel, Hans
Torng, Eric
Tortorici, Adèle
Trehel, Michel
Trichina, Eléna
Trystram, Denis
Tsikas, Themos
Tuecke, Steve
Tůma, Miroslav
Tvrdik, Pavel
Ubeda, Stephane
Ueberhuber, C. W.
Ugarte, Asier
Vakalis, Ignatis
Zomaya, Albert
Zory, Julien
Zwiers, Job

Table of Contents

Invited Talks

Some Parallel Algorithms for Integer Factorisation 1
  Richard P. Brent

MERCATOR, the Mission 23
  Philippe Courtier

Adaptive Scheduling for Task Farming with Grid Middleware 30
  Henri Casanova, MyungHo Kim, James S. Plank, Jack J. Dongarra

Applying Human Factors to the Design of Performance Tools 44
  Cherri M. Pancake

Building the Teraflops/Petabytes Production Supercomputing Center 61
  Horst D. Simon, William T.C. Kramer, Robert F. Lucas

A Coming of Age for Beowulf-Class Computing 78
  Thomas Sterling, Daniel Savarese

Topic 01
Support Tools and Environments 89
  Frédéric Desprez

Systematic Debugging of Parallel Programs in DIWIDE Based on
Collective Breakpoints and Macrosteps 90
  P. Kacsuk, R. Lovas, J. Kovács

Project Workspaces for Parallel Computing - The TRAPPER Approach .. 98
  Dino Ahr, Andreas Bäcker

PVMbuilder - A Tool for Parallel Programming 108
  Jan B. Pedersen, Alan Wagner

Message-Passing Specification in a *CORBA* Environment 113
  T. Es-sqalli, E. Fleury, E. Dillon, J. Guyard

Using Preemptive Thread Migration to Load-Balance Data-Parallel
Applications 117
  Gabriel Antoniu, Christian Perez

FITS—A Light-Weight Integrated Programming Environment 125
  B. Chapman, F. Bodin, L. Hill, J. Merlin, G. Viland, F. Wollenweber

INTERLACE: An Interoperation and Linking Architecture for
Computational Engines 135
    Matthew J. Sottile, Allen D. Malony

Multi-protocol Communications and High Speed Networks 139
    Benoît Planquelle, Jean-François Méhaut, Nathalie Revol

An Online Algorithm for Dimension-Bound Analysis 144
    Paul A.S. Ward

Correction of Monitor Intrusion for Testing Nondeterministic
MPI-Programs 154
    D. Kranzlmüller, J. Chassin de Kergommeaux, Ch. Schaubschläger

Improving the Performance of Distributed Shared Memory Environments
on Grid Multiprocessors 159
    Dimitris Dimitrelos, Constantine Halatsis

Topic 02
Performance Evaluation and Prediction 163

    Jean-Marc Vincent

Performance Analysis of Wormhole Switching with Adaptive Routing in a
Two-Dimensional Torus 165
    M. Colajanni, B. Ciciani, F. Quaglia

Message Passing Evaluation and Analysis on Cray T3E and SGI Origin
2000 Systems 173
    M. Prieto, D. Espadas, I.M. Llorente, F. Tirado

Performance Evaluation and Modeling of the Fujitsu AP3000
Message-Passing Libraries 183
    Juan Touriño, Ramón Doallo

Improving Communication Support for Parallel Applications 188
    Joerg Cordsen, Marco Dimas Gubitoso

A Performance Estimator for Parallel Programs 193
    Jeff Reeve

Min-Cut Methods for Mapping Dataflow Graphs 203
    Volker Elling, Karsten Schwan

Influence of Variable Time Operations in Static Instruction Scheduling 213
    Patricia Borensztejn, Cristina Barrado, Jesus Labarta

Evaluation of LH*LH for a Multicomputer Architecture* 217
    Andy D. Pimentel, Louis O. Hertzberger

Set Associative Cache Behavior Optimization 229
    Ramón Doallo, Basilio B. Fraguela, Emilio L. Zapata

A Performance Study of Modern Web Server Applications 239
    Ramesh Radhakrishnan, Lizy Kurian John

An Evaluation of High Performance Fortran Compilers Using the
HPFBench Benchmark Suite 248
    Guohua Jin, Y. Charlie Hu

Performance Evaluation of Object Oriented Middleware 258
    László Böszörményi, Andreas Wickner, Harald Wolf

PopSPY: A PowerPC Instrumentation Tool for Multiprocessor Simulation. 262
    C. Limousin, A. Vartanian, J-L. Béchennec

Performance Evaluation and Benchmarking of Native Signal Processing ... 266
    Deependra Talla, Lizy Kurian John

Topic 03
Scheduling and Load Balancing **271**
    Jean-Marc Geib, Bruce Hendrickson, Pierre Manneback, Jean Roman

A Polynomial-Time Branching Procedure for the Multiprocessor
Scheduling Problem 272
    Ricardo C. Corrêa, Afonso Ferreira

Optimal and Alternating-Direction Load Balancing Schemes 280
    Robert Elsässer, Andreas Frommer, Burkhard Monien, Robert Preis

Process Mapping Given by Processor and Network Dynamic Load
Prediction 291
    Jean-Marie Garcia, David Gauchard, Thierry Monteil, Olivier Brun

Ordering Unsymmetric Matrices into Bordered Block Diagonal Form for
Parallel Processing 295
    Y.F. Hu, K.C.F. Maguire, R.J. Blake

Dynamic Load Balancing for Ocean Circulation Model with Adaptive
Meshing 303
    Eric Blayo, Laurent Debreu, Grégory Mounié, Denis Trystram

DRAMA: A Library for Parallel Dynamic Load Balancing of Finite
Element Applications 313
    *Bart Maerten, Dirk Roose, Achim Basermann, Jochen Fingberg,
    Guy Lonsdale*

Job Scheduling in a Multi-layer Vision System 317
    M. Fikret Ercan, Ceyda Oğuz, Yu-Fai Fung

A New Algorithm for Multi-objective Graph Partitioning 322
Kirk Schloegel, George Karypis, Vipin Kumar

Scheduling Iterative Programs onto LogPMachine 332
Welf Löwe, Wolf Zimmermann

Scheduling Arbitrary Task Graphs on LogP Machines 340
Cristina Boeres, Aline Nascimento, Vinod E.F. Rebello

Scheduling with Communication Delays and On-Line Disturbances 350
Aziz Moukrim, Eric Sanlaville, Frédéric Guinand

Scheduling User-Level Threads on Distributed Shared-Memory
Multiprocessors 358
Eleftherios D. Polychronopoulos, Theodore S. Papatheodorou

Using Duplication for the Multiprocessor Scheduling Problem with
Hierarchical Communications 369
Evripidis Bampis, Rodolphe Giroudeau, Jean-Claude König

Topic 04
Compilers for High Performance Systems 373
Barbara Chapman

Storage Mapping Optimization for Parallel Programs 375
Albert Cohen, Vincent Lefebvre

Array SSA for Explicitly Parallel Programs 383
Jean-François Collard

Parallel Data-Flow Analysis of Explicitly Parallel Programs 391
Jens Knoop

Localization of Data Transfer in Processor Arrays 401
Dirk Fimmel, Renate Merker

Scheduling Structured Systems 409
Jason B. Crop, Doran K. Wilde

Compiling Data Parallel Tasks for Coordinated Execution 413
Erwin Laure, Matthew Haines, Piyush Mehrotra, Hans Zima

Flexible Data Distribution in PGHPF 418
Mark Leair, Douglas Miles, Vincent Schuster, Michael Wolfe

On Automatic Parallelization of Irregular Reductions on Scalable Shared
Memory Systems.. 422
E. Gutiérrez, O. Plata, E.L. Zapata

I/O-Conscious Tiling Strategy for Disk-Resident Data 430
  Mahmut Kandemir, Alok Choudhary, J. Ramanujam

Post-Scheduling Optimization of Parallel Programs 440
  Stephen Shafer, Kanad Ghose

Piecewise Execution of Nested Parallel Programs - A Thread-Based
Approach 445
  W. Pfannenstiel

Topic 05
Parallel and Distributed Databases 449
  Burkhard Freitag, Kader Hameurlain

Distributed Database Checkpointing 450
  Roberto Baldoni, Francesco Quaglia, Michel Raynal

A Generalized Transaction Theory for Database and Non-database Tasks . 459
  Armin Feßler, Hans-Jörg Schek

On Disk Allocation of Intermediate Query Results in Parallel Database
Systems 469
  Holger Märtens

Highly Concurrent Locking in Shared Memory Database Systems 477
  Christian Jacobi, Cédric Lichtenau

Parallel Processing of Multiple Text Queries on Hypercube Interconnection
Networks 482
  Basilis Mamalis, Paul Spirakis, Basil Tampakas

Topic 06 + 20
Fault Avoidance and Fault Removal in Real-Time Systems &
Fault-Tolerant Computing 487
  Gilles Motet, David Powell

Quality of Service Management in Distributed Asynchronous Real-Time
Systems 489
  Binoy Ravindran

Multiprocessor Scheduling of Real-Time Tasks with Resource
Requirements 497
  Costas Mourlas

Designing Multiprocessor/Distributed Real-Time Systems Using the
ASSERTS Toolkit 505
  *Kanad Ghose, Sudhir Aggarwal, Abhrajit Ghosh, David Goldman,
  Peter Sulatycke, Pavel Vasek, David R. Vogel*

UML Framework for the Design of Real-Time Robot Controllers 511
  L. Carroll, B. Tondu, C. Baron, J.C. Geffroy

Software Implemented Fault Tolerance in Hypercube 515
  D.R. Avresky, S. Geoghegan

Managing Fault Tolerance Transparently Using CORBA Services 519
  René Meier, Paddy Nixon

Topic 07
Theory and Models for Parallel Computation **523**
  Michel Cosnard

Parallel Algorithms for Grounded Range Search and Applications 525
  Michael G. Lamoureux, Andrew Rau-Chaplin

Multi-level Cooperative Search: A New Paradigm for Combinatorial
Optimization and an Application to Graph Partitioning 533
  Michel Toulouse, Krishnaiyan Thulasiraman, Fred Glover

A Quantitative Measure of Portability with Application to
Bandwidth-Latency Models for Parallel Computing 543
  Gianfranco Bilardi, Andrea Pietracaprina, Geppino Pucci

A Cost Model for Asynchronous and Structured Message Passing 552
  Emmanuel Melin, Bruno Raffin, Xavier Rebeuf, Bernard Virot

A Parallel Simulation of Cellular Automata by Spatial Machines 557
  Bruno Martin

Topic 08
High-Performance Computing and Applications **561**
  Wolfgang Gentzsch

Null Messages Cancellation Through Load Balancing in Distributed
Simulations 562
  Azzedine Boukerche, Sajal K. Das

Efficient Load-Balancing and Communication Overlap in Parallel
Shear-Warp Algorithm on a Cluster of PCs 570
  Frédérique Chaussumier, Frédéric Desprez, Michel Loi

A Hierarchical Approach for Parallelization of a Global Optimization
Method for Protein Structure Prediction 578
  S. Crivelli, T. Head-Gordon, R. Byrd, E. Eskow, R. Schnabel

Parallelization of a Compositional Simulator with a Galerkin Coarse/Fine
Method 586
  Geir Åge Øye, Hilde Reme

Some Investigations of Domain Decomposition Techniques in Parallel
CFD 595
    F. Chalot, G. Chevalier, Q.V. Dinh, L. Giraud

A Parallel Ocean Model for High Resolution Studies 603
    Marc Guyon, Gurvan Madec, François-Xavier Roux, Maurice Imbard

Nonoverlapping Domain Decomposition Applied to a Computational Fluid
Mechanics Code 608
    Paulo B. Vasconcelos, Filomena D. d'Almeida

A PC Cluster with Application-Quality MPI 613
    M. Gołębiewski, A. Basermann, M. Baum, R. Hempel, H. Ritzdorf,
    J.L. Träff

Using Network of Workstations to Support a Web-Based Visualization
Service 624
    Wilfrid Lefer, Jean-Marc Pierson

High-Speed LANs: New Environments for Parallel and Distributed
Applications 633
    Patrick Geoffray, Laurent Lefèvre, CongDuc Pham, Loïc Prylli,
    Olivier Reymann, Bernard Tourancheau, Roland Westrelin

Consequences of Modern Hardware Design for Numerical Simulations and
Their Realization in FEAST 643
    Ch. Becker, S. Kilian, S. Turek, the FEAST Group

A Structured SADT Approach to the Support of a Parallel Adaptive 3D
CFD Code 651
    Jonathan Nash, Martin Berzins, Paul Selwood

A Parallel Algorithm for 3D Geometry Transformations in OpenGL 659
    J. Sébot Julien, A. Vartanian, J-L. Béchennec, N. Drach-Temam

Parallel Implementation in a Industrial Framework of Statistical
Tolerancing Analysis in Microelectronics 663
    Salvatore Rinaudo, Francesco Moschella, Marcello A. Anile

Interaction Between Data Parallel Compilation and Data Transfer and
Storage Cost Minimization for Multimedia Applications 668
    Chidamber Kulkarni, Koen Danckaert, Francky Catthoor, Manish Gupta

Parallel Numerical Simulation of a Marine Host-Parasite System 677
    Michel Langlais, Guillaume Latu, Jean Roman, Patrick Silan

Parallel Methods of Training for Multilayer Neural Network 686
    El Mostafa Daoudi, El Miloud Jaâra

Partitioning of Vector-Topological Data for Parallel GIS Operations:
Assessment and Performance Analysis 691
   Terence M. Sloan, Michael J. Mineter, Steve Dowers,
   Connor Mulholland, Gordon Darling, Bruce M. Gittings

Topic 09
Parallel Computer Architecture - What Is Its Future? 695
   Chris Jesshope

The Algebraic Path Problem Revisited 698
   Sanjay Rajopadhye, Claude Tadonki, Tanguy Risset

Vector ISA Extension for Sparse Matrix-Vector Multiplication 708
   Stamatis Vassiliadis, Sorin Cotofana, Pyrrhos Stathis

A Study of a Simultaneous Multithreaded Processor Implementation 716
   Dominik Madoń, Eduardo Sánchez, Stefan Monnier

The MorphoSys Parallel Reconfigurable System 727
   Guangming Lu, Hartej Singh, Ming-hau Lee, Nader Bagherzadeh,
   Fadi Kurdahi, Eliseu M.C. Filho

A Graph-Oriented Task Manager for Small Multiprocessor Systems 735
   Xavier Verians, Jean-Didier Legat, Jean-Jacques Quisquater,
   Benoit Macq

Implementing Snoop-Coherence Protocol for Future SMP Architectures ... 745
   Wissam Hlayhel, Jacques Collet, Laurent Fesquet

An Adaptive Limited Pointers Directory Scheme for Cache Coherence of
Scalable Multiprocessors 753
   Cheol Ho Park, Jong Hyuk Choi, Kyu Ho Park, Daeyeon Park

Two Schemes to Improve the Performance of a *Sort-Last* 3D Parallel
Rendering Machine with Texture Caches 757
   Alexis Vartanian, Jean-Luc Béchennec, Nathalie Drach-Temam

ManArray Processor Interconnection Network: An Introduction 761
   Gerald G. Pechanek, Stamatis Vassiliadis, Nikos Pitsianis

Topic 10
Distributed Systems and Algorithms 767
   Gérard Padiou, André Schiper

A Cooperation Service for CORBA Objects. From the Model to the
Applications 769
   Khalil Drira, Frédéric Gouëzec, Michel Diaz

Symphony: Managing Virtual Servers in the Global Village 777
  Roy Friedman, Assaf Schuster, Ayal Itzkovitz, Eli Biham, Erez Hadad,
  Vladislav Kalinovsky, Sergey Kleyman, Roman Vitenberg

Épidaure: A *Java* Distributed Tool for Building DAI Applications 785
  Djamel Fezzani, Jocelyn Desbiens

A Client/Broker/Server Substrate with $50\mu s$ Round-Trip Overhead 790
  Olivier Richard, Franck Cappello

Universal Constructs in Distributed Computations 795
  Ajay D. Kshemkalyani, Mukesh Singhal

Illustrating the Use of Vector Clocks in Property Detection: An Example
and a Counter-Example 806
  Michel Raynal

A Node Count-Independent Logical Clock for Scaling Lazy Release
Consistency Protocol 815
  Luciana Bezerra Arantes, Bertil Folliot, Pierre Sens

Mutual Exclusion Between Neighboring Nodes in an Arbitrary System
Graph Tree That Stabilizes Using Read/Write Atomicity 823
  Gheorghe Antonoiu, Pradip K. Srimani

Topic 11
Parallel Programming: Models, Methods and Languages 831

  Luc Bougé, Bill McColl, Mamoun Filali, Henk Sips

Exploiting Advanced Task Parallelism in High Performance Fortran via a
Task Library* 833
  Thomas Brandes

A Run-Time System for Dynamic Grain Packing 845
  João Luís Sobral, Alberto José Proença

Optimising Skeletal-Stream Parallelism on a BSP Computer 853
  Andrea Zavanella

Parallel Programming by Transformation 858
  Noel Winstanley

Condensed Graphs: A Multi-level, Parallel, Intermediate Representation . . 866
  John P. Morrison, Niall J. Dalton

A Skeleton for Parallel Dynamic Programming 877
  D. Morales, F. Almeida, F. Garcia, J. Gonzalez, J. Roda, C. Rodriguez

Programming Effort vs. Performance with a Hybrid Programming Model
for Distributed Memory Parallel Architectures 888
Andreas Rodman, Mats Brorsson

DAOS — Scalable And-Or Parallelism 899
*Luís Fernando Castro, Vítor Santos Costa, Cláudio F.R. Geyer,
Fernando Silva, Patrícia Kayser Vargas, Manuel E. Correia*

Write Detection in Home-Based Software DSMs 909
Weiwu Hu, Weisong Shi, Zhimin Tang

D'Caml: Native Support for Distributed ML Programming in
Heterogeneous Environment 914
Ken Wakita, Takashi Asano, Masataka Sassa

ParBlocks - A New Methodology for Specifying Concurrent Method
Executions in Opus 925
Erwin Laure

Static Parallelization of Functional Programs: Elimination of Higher-Order
Functions & Optimized Inlining 930
*Christoph A. Herrmann, Jan Laitenberger, Christian Lengauer,
Christian Schaller*

A Library to Implement Neural Networks on MIMD Machines 935
Yann Boniface, Frédéric Alexandre, Stéphane Vialle

Topic 12
Architectures and Algorithms for Vision and Other Senses 939

Alain Ayache, Virginio Cantoni, Concettina Guerra, Pieter Jonker

LUX: An Heterogeneous Function Composition Parallel Computer for
Graphics 940
Stéphane Mancini, Renaud Pacalet

A Parallel Accelerator Architecture for Multimedia Video Compression ... 950
Bertil Schmidt, Manfred Schimmler

A Parallel Architecture for Stereoscopic Processing 961
Milton Romero, Bruno Ciciani

A Robust Neural Network Based Object Recognition System and Its SIMD
Implementation 969
Alfredo Petrosino, Giuseppe Salvi

Multimedia Extensions and Sub-word Parallelism in Image Processing:
Preliminary Results 977
Marco Ferretti, Davide Rizzo

Vanishing Point Detection in the Hough Transform Space 987
Andrea Matessi, Luca Lombardi

Parallel Structure in an Integrated Speech-Recognition Network 995
M. Fleury, A.C. Downton, A.F. Clark

3D Optoelectronic Fix Point Unit and Its Advantages Processing 3D
Data 1005
B. Kasche, D. Fey, T. Höhn, W. Erhard

Parallel Wavelet Transforms on Multiprocessors 1013
Manfred Feil, Rade Kutil, Andreas Uhl

Vector Quantization-Fractal Image Coding Algorithm Based on Delaunay
Triangulation 1018
Zahia Brahimi, Karima Ait Saadi, Noria Baraka

Topic 13+19
Numerical Algorithms for Linear and Nonlinear Algebra 1023

mpC + ScaLAPACK = Efficient Solving Linear Algebra Problems on
Heterogeneous Networks 1024
Alexey Kalinov, Alexey Lastovetsky

Parallel Subdomain-Based Preconditioner for the Schur Complement 1032
Luiz M. Carvalho, Luc Giraud

A Preconditioner for Improved Fermion Actions 1040
*Wolfgang Bietenholz, Norbert Eicker, Andreas Frommer,
Thomas Lippert, Björn Medeke, Klaus Schilling*

Application of a Class of Preconditioners to Large Scale Linear
Programming Problems 1044
Venansius Baryamureeba, Trond Steihaug, Yin Zhang

Estimating Computer Performance for Parallel Sparse QR Factorisation . 1049
David J. Miron, Patrick M. Lenders

A Mapping and Scheduling Algorithm for Parallel Sparse Fan-In Numerical
Factorization 1059
Pascal Hénon, Pierre Ramet, Jean Roman

Scheduling of Algorithms Based on Elimination Trees on NUMA Systems 1068
María J. Martín, Inmaculada Pardines, Francisco F. Rivera

Block-Striped Partitioning and Neville Elimination 1073
P. Alonso, R. Cortina, J. Ranilla

A Comparison of Parallel Solvers for Diagonally Dominant and General
Narrow-Banded Linear Systems II 1078
 Peter Arbenz, Andrew Cleary, Jack Dongarra, Markus Hegland

Using Pentangular Factorizations for the Reduction to Banded Form 1088
 B. Großer, B. Lang

Experience with a Recursive Perturbation Based Algorithm for Symmetric
Indefinite Linear Systems 1096
 Anshul Gupta, Fred Gustavson, Alexander Karaivanov,
 Jerzy Wasniewski, Plamen Yalamov

Parallel Cyclic Wavefront Algorithms for Solving Semidefinite Lyapunov
Equations 1104
 José M. Claver, Vicente Hernández, Enrique S. Quintana-Ortí

Parallel Constrained Optimization via Distribution of Variables 1112
 Claudia A. Sagastizábal, Mikhail V. Solodov

Solving Stable Stein Equations on Distributed Memory Computers 1120
 Peter Benner, Enrique S. Quintana-Ortí, Gregorio Quintana-Ortí

Convergence Acceleration for the Euler Equations Using a Parallel
Semi-Toeplitz Preconditioner 1124
 Andreas Kähäri, Samuel Sundberg

A Stable and Efficient Parallel Block Gram-Schmidt Algorithm 1128
 Denis Vanderstraeten

On the Extension of the Code GAM for Parallel Computing 1136
 Felice Iavernaro, Francesca Mazzia

PAMIHR. A Parallel FORTRAN Program for Multidimensional
Quadrature on Distributed Memory Architectures 1144
 G. Laccetti, M. Lapegna

Stability Issues of the Wang's Partitioning Algorithm for Banded and
Tridiagonal Linear Systems 1149
 Velisar Pavlov, Plamen Yalamov

Topic 14
Emerging Topics in Advanced Computing in Europe 1153
 Renato Campo, Luc Giraud

The HPF+ Project: Supporting HPF for Advanced Industrial
Applications 1155
 Siegfried Benkner, Guy Lonsdale, Hans Zima

TIRAN: Flexible and Portable Fault Tolerance Solutions for Cost Effective
Dependable Applications 1166
   O. Botti, V. De Florio, G. Deconinck, F. Cassinari, S. Donatelli,
   A. Bobbio, A. Klein, H. Kufner, R. Lauwereins, E. Thurner,
   E. Verhulst

OCEANS – Optimising Compilers for Embedded Applications 1171
   Michel Barreteau, François Bodin, Zbigniew Chamski,
   Henri-Pierre Charles, Christine Eisenbeis, John Gurd,
   Jan Hoogerbrugge, Ping Hu, William Jalby, Toru Kisuki,
   Peter M.W. Knijnenburg, Paul van der Mark, Andy Nisbet,
   Michael F.P. O'Boyle, Erven Rohou, André Seznec,
   Elena A. Stöhr, Menno Treffers, Harry A.G. Wijshoff

Cray T3E Performances of a Parallel Code for a Stochastic Dynamic Assets
and Liabilities Management Model 1176
   G. Zanghirati, F. Cocco, F. Taddei, G. Paruolo

Parametric Simulation of Multi-body Systems on Networks of
Heterogeneous Computers 1187
   Javier G. Izaguirre, José M. Jiménez, Unai Martín, Bruno Thomas,
   Alberto Larzábal, Luis M. Matey

Parallel Data Mining in the HYPERBANK Project 1195
   S. Fotis, J. A. Keane, R. I. Scott

High Performance Computing for Optimum Design of Multi-body
Systems 1199
   José M. Jiménez, Nassouh A. Chehayeb, Javier G. Izaguirre,
   Beidi Hamma, Yan Thiaudière

Topic 15
Routing and Communication in Interconnection Networks 1203

Optimizing Message Delivery in Asynchronous Distributed Applications . 1204
   Girindra D. Sharma, Nael B. Abu-Ghazaleh,
   Umesh Kumar V. Rajasekaran, Philip A. Wilsey

Circuit-Switched Broadcasting in Multi-port Multi-dimensional Torus
Networks* 1209
   San-Yuan Wang, Yu-Chee Tseng, Sze-Yao Ni, Jang-Ping Sheu

Impact of the Head-of-Line Blocking on Parallel Computer Networks:
Hardware to Applications 1222
   V. Puente, J.A. Gregorio, C. Izu, R. Beivide

Interval Routing on Layered Cross Product of Trees and Cycles 1231
   R. Královič, B. Rovan, P. Ružička

Topic 16

Instruction-Level Parallelism and Uniprocessor Architecture....1241
Pascal Sainrat, Mateo Valero

Design Considerations of High Performance Data Cache with Prefetching .1243
Chi-Hung Chi, Jun-Li Yuan

Annotated Memory References: A Mechanism for Informed Cache
Management 1251
Alvin R. Lebeck, David R. Raymond, Chia-Lin Yang,
Mithuna S. Thottethodi

Understanding and Improving Register Assignment.....................1255
Cindy Norris, James B. Fenwick, Jr.

Compiler-Directed Reordering of Data by Cyclic Graph Coloring........1260
Daniela Genius, Sylvain Lelait

Code Cloning Tracing: A "Pay per Trace" Approach 1265
Thierry Lafage, André Seznec, Erven Rohou, François Bodin

Execution-Based Scheduling for VLIW Architectures 1269
Kemal Ebcioğlu, Erik R. Altman, Sumedh Sathaye, Michael Gschwind

Decoupling Recovery Mechanism for Data Speculation from Dynamic
Instruction Scheduling Structure 1281
Toshinori Sato

Implementation of Hybrid Context Based Value Predictors Using Value
Sequence Classification 1291
Luis Piñuel, Rafael A. Moreno, Francisco Tirado

Heterogeneous Clustered Processors: Organization and Design 1296
Francesco Pessolano

An Architecture Framework for Introducing Predicated Execution into
Embedded Microprocessors..1301
Daniel A. Connors, Jean-Michel Puiatti, David I. August,
Kevin M. Crozier, Wen-mei W. Hwu

Multi-stage Cascaded Prediction 1312
Karel Driesen, Urs Hölzle

Mispredicted Path Cache Effects 1322
Jonathan Combs, Candice Bechem Combs, John Paul Shen

Topic 17

Concurrent and Distributed Programming with Objects 1333
Patrick Sallé, Marc Pantel

Non-regular Process Types 1334
  Franz Puntigam

Decision Procedure for Temporal Logic of Concurrent Objects 1344
  Jean-Paul Bahsoun, Rami El-Baïda, Hugues-Olivier Yar

Aliasing Models for Object Migration............................... 1353
  Uwe Nestmann, Hans Hüttel, Josva Kleist, Massimo Merro

Dynamic Extension of CORBA Servers 1369
  Marco Catunda, Noemi Rodriguez, Roberto Ierusalimschy

On the Concurrent Object Model of UML........................... 1377
  Iulian Ober, Ileana Stan

Object Oriented Design for Reusable Parallel Linear Algebra Software ... 1385
  Eric Noulard, Nahid Emad

Topic 18
Global Environment Modelling 1393
  Michel Déqué

The Parallelization of the Princeton Ocean Model 1395
  L.A. Boukas, N.Th. Mimikou, N.M. Missirlis, G.L. Mellor,
  A. Lascaratos, G. Korres

Modular Fortran 90 Implementation of a Parallel Atmospheric General
Circulation Model.. 1403
  William Sawyer, Lawrence Takacs, Andrea Molod, Robert Lucchesi

Implementation of the Limited-Area Numerical Weather Prediction Model
Aladin in Distributed Memory 1411
  Claude Fischer, Jean-François Estrade, Jure Jerman

Parallelization of the French Meteorological Mesoscale Model MésoNH ... 1417
  Patrick Jabouille, Ronan Guivarch, Philippe Kloos, Didier Gazen,
  Nicolas Gicquel, Luc Giraud, Nicole Asencio, Veronique Ducrocq,
  Juan Escobar, Jean-Luc Redelsperger, Joël Stein, Jean-Pierre Pinty

The PALM Project: MPMD Paradigm for an Oceanic Data Assimilation
Software 1423
  A. Fouilloux, A. Piacentini

A Parallel Distributed Fast 3D Poisson Solver for Méso-NH 1431
  Luc Giraud, Ronan Guivarch, Joël Stein

Porting a Limited Area Numerical Weather Forecasting Model on a
Scalable Shared Memory Parallel Computer 1435
  Roberto Ansaloni, Paolo Malfetti, Tiziana Paccagnella

Topic 22
High-Performance Data Mining and Knowledge Discovery **1439**
David Skillicorn, Domenico Talia

Mining of Association Rules in Very Large Databases: A Structured
Parallel Approach 1441
P. Becuzzi, M. Coppola, M. Vanneschi

Parallel k/h-Means Clustering for Large Data Sets 1451
Kilian Stoffel, Abdelkader Belkoniene

Performance Analysis for Parallel Generalized Association Rule Mining on
a Large Scale PC Cluster 1455
Takahiko Shintani, Masato Oguchi, Masaru Kitsuregawa

Inducing Load Balancing and Efficient Data Distribution Prior to
Association Rule Discovery in a Parallel Environment 1460
Anna M. Manning, John A. Keane

Topic 23
Symbolic Computation **1465**
Mike Dewar

Parallelism in ALDOR — The Communication Library Π^{it} for Parallel,
Distributed Computation 1466
Thierry Gautier, Niklaus Mannhart

A Library for Parallel Modular Arithmetic 1476
David Power, Russell Bradford

Performance Evaluation of Or-Parallel Logic Programming Systems on
Distributed Shared-Memory Architectures............................ 1484
Vanusa Menditi Calegario, Inês de Castro Dutra

A Parallel Symbolic Computation Environment: Structures and
Mechanics 1492
Mantŝika Matooane, Arthur Norman

Index of Authors 1497

ManArray Processor Interconnection Network: An Introduction

Gerald G. Pechanek [1], Stamatis Vassiliadis [2], Nikos Pitsianis [1,3]

[1] Billions of Operations Per Second, (BOPS) Inc.,
Chapel Hill, NC, USA gpechanek@bops.com
[2] Delft University of Technology, Department of Electrical Engineering
Delft, The Netherlands stamatis@einstein.et.tudelft.nl
[3] Duke University, Department of Computer Science
Chapel Hill, NC, USA nikos@BOPS.com

Abstract. The present paper introduces the new interconnection network of the BOPS ManArray family of available core products. To form a ManArray network, the processing elements are completely connected within clusters and communicate with members of only two other clusters thereby reducing signal fan-out and wiring density. With this simple network, single-step communications between a hypercube and its compliment node, single-step transpose operations, and a diameter of 2 are achieved.

1 Introduction

As chip densities continue to improve, demand increases for the low cost integration of high performance parallel processing systems. For example audio, video, and communication signal processing are but three areas requiring very high performance that are in demand for low cost consumer products.

High performance multi-processor array systems, such as the mesh, torus, and hypercube [1, 2, 3, 4] have some characteristics, we feel, that limits their use for System-On-a-Chip products. Consequently, the ManArray family of processor cores was developed for high volume commercial applications with improved network connectivity, a simple implementation scheme, and a simple programming model.

This paper introduces the ManArray processor platform as a strong contender to be a ubiquitous, high-volume signal processor in commercial applications.

2 Background

In this section we briefly describe some characteristics of conventional networks, which we felt were too limiting for the intended products.

A crossbar switch interconnection network is generally known to be expensive to implement since for N processors it has a cost of $O(N^2)$. Even on single chip systems, with relatively small N, this wiring, fan-out, and logic delay limits its acceptability for a pervasive scalable approach to single chip multiprocessing.

The torus has limited connectivity between processing elements (PEs), which can cause high communication latency effects. Even though new approaches to arrays have been proposed in [5, 6, 7, 8], due to their irregularity of PE combinations there are problems in the implementation and with the generality of connections.

P. Amestoy et al. (Eds.): Euro-Par'99, LNCS 1685, pp. 761-765, 1999.
© Springer-Verlag Berlin Heidelberg 1999

The hypercube [9, 10] reduces the communication latency from O(n) on a nxn torus to O(logn), the distance between two binary complement nodes. But even O(logn) can represent a high latency on large networks. Reducing the longest path between complement PEs has been deemed difficult and costly.

In the next section we present the ManArray network, which alleviates the previously stated concerns by improving the connectivity among the PEs with low implementation expense and low interconnection wiring requirements.

3   ManArray Network

The ManArray network achieves the goals of providing higher connectivity than a mesh, torus, or hypercube network, a simple switch implementation for multiple array sizes, and a simple programming model.

First we explain how we create the ManArray organization of PEs. Consider by way of example, a 2D 4 x 4 torus and the corresponding embedded 4D hypercube, written as a 4 x 4 table with the hypercube node labels, see Fig. 1A. In Fig. 1A, the $PE_{i,j}$ cluster nodes are labeled in Gray-code as follows: $PE_{G(i).G(j)}$ where G(x) is the Gray code of x. Along the rows and the columns of this table, the distance between adjacent elements is one. If columns 2, 3 and 4 are rotated one position up, then the distance of the corresponding elements between the first and the second column becomes two. Repeating the same rotation with columns 3 and 4 and then column 4, the distance between elements of a column with the corresponding elements of the adjacent columns is two. The resulting 4D ManArray table is shown in Fig. 1B.

It is important to note that each row of the table contains a grouping of 4 nodes, including two pairs of diametrically opposite hypercube nodes. In higher dimensional tori, and thus hypercubes, the grouping of diametrically opposite nodes is achieved by the same rotation along each new dimension except the last one.

PE-0,0	PE-0,1	PE-0,2	PE-0,3
0000	0001	0011	0010
PE-1,0	PE-1,1	PE-1,2	PE-1,3
0100	0101	0111	0110
PE-2,0	PE-2,1	PE-2,2	PE-2,3
1100	1101	1111	1110
PE-3,0	PE-3,1	PE-3,2	PE-3,3
1000	1001	1011	1010

Figure 1A

PE-0,0	PE-1,1	PE-2,2	PE-3,3
0000	0101	1111	1010
PE-1,0	PE-2,1	PE-3,2	Pe-0,3
0100	1101	1011	0010
PE-2,0	PE-3,1	PE-0,2	PE-1,3
1100	1001	0011	0110
PE-3,0	PE-0,1	PE-1,2	PE-2,3
1000	0001	0111	1110

Figure 1B

Using these groupings, it can be shown that the complexity of the ManArray network is small although its connectivity is high. To demonstrate this, we show how this organization of PEs is interconnected with a simple cluster switch network.

A 4x4 ManArray with torus and hypercube node Ids, in Fig. 2, consists of four 2x2 clusters. The cluster-switch for the upper left hand 2x2 is shown partitioned into four groups, each consisting of a 4-input and a 3-input multiplexer. Each of these groups is associated with a particular PE and this has been indicated with the dotted line arrows. For example, $PE_{0,0}$ is associated with the "A" group multiplexers a1 and a2. The circled multiplexers are controlled by their associated PE.

The 4x4 ManArray includes connection paths that connect hypercube complements as shown in Fig. 2. For example, PE 0111 ($PE_{1,2}$) can communicate with PE 1000 ($PE_{3,0}$) as well as the other members of its cluster. The longest path between hypercube complement PEs, 4 steps for a 4D hypercube, is reduced to 1 step in the ManArray network. The improved connectivity and simplicity of the ManArray network supports single-cycle communications and efficient algorithms.

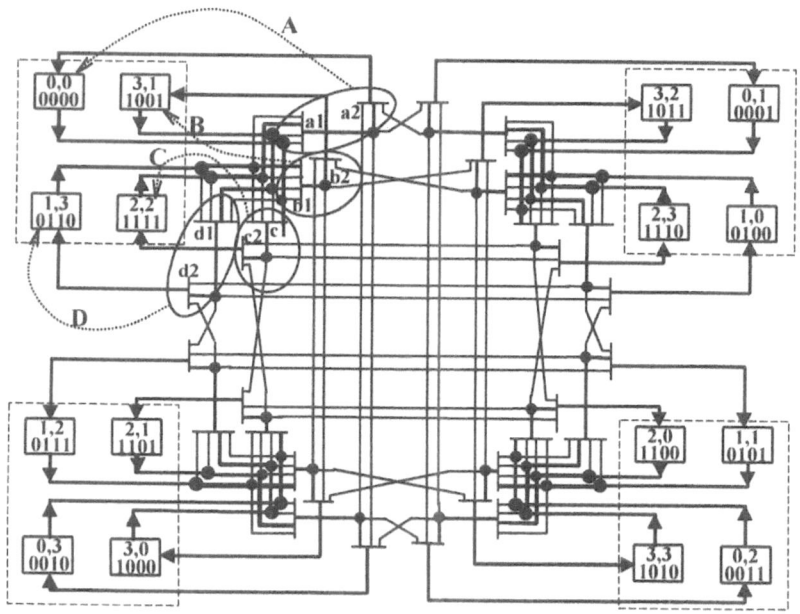

Figure 2 4x4 ManArray highlighting cluster switch control

4   ManArray Network Properties

In this section, we discuss some of the properties of the ManArray network within the problem domain of single-chip parallel processors. For the purposes of this paper we constrain our discussion to network sizes that can be implemented on a single chip. With the advancement of technology though, the number of PEs in a ManArray processor scale with the technology allowing larger array sizes to be developed for future products.

The network diameter is the largest distance between any pair of nodes and captures the worst case number of steps required for node-to-node communication. The smaller the diameter, the fewer steps needed to communicate between far away nodes. Small network diameters are desirable. As the table below shows, the network diameter of a d-dimensional hypercube is d, and with the addition of the complementary node connections it becomes, $\lceil d/2 \rceil$. Note that only the edges connecting complementary nodes are accounted for in the middle column.

For this introductory paper, we add the third column labeled "ManArray Network", which indicates the number of edges contained in the structure as well as the constant network diameter of 2 for current ManArray single chip implementations. In this paper, we show by way of example the connectivity of a 4x4 ManArray (Figure 2). Each 4-PE cluster contains 12 uni-directional edges or 6 bi-directional edges, if you exclude the self-connecting edges. With four clusters, this amounts to 24 bi-directional edges.   In ManArray, any PE in a cluster can communicate with any PE in an adjacent cluster. Consequently, there are 16 bi-directional edges between any cluster. The ManArray needs only 8 uni-directional connections between the clusters since that is the maximum number of paths that can be connected between 8 PEs at any one time. By sharing these 8 uni-directional links appropriately with the multiplexers used in the cluster-switches, the 16 bi-directional path combinations can be created. The total number of edges in a 4x4 ManArray is $24 + 4*16 = 88$ corresponding to d=4 and k=2 in the table.

	Hypercube*	Hypercube+ compliment edges*	ManArray Network
Nodes	2^d	2^d	2^d
Edges	$d2^{d-1}$	$(d+1)2^{d-1}$	$2^{2k-1}((4*3^{k-1})-1)$; for d=2k $2^{2k}((8*3^{k-1})-1)$; for d=2k+1
Diameter	d	$\lceil d/2 \rceil$	2

* A hypercube and a hypercube with complementary edges are proper subgraphs of the ManArray.

Where the upper bound on "k" and "d" depends upon the chosen process technology and the processor cycle time requirements.

With the full number of ManArray edges provided as shown in the third column above, the network diameter is reduced to a constant diameter of 2 for all *d*, within the design constraints of the process technology.

5   ManArray Processing

Generally speaking, ManArray combines PEs in clusters that also contain a Sequence Processor (SP), uniquely merged into the PE array, and a cluster-switch, interconnecting the PEs. The SP provides program control, contains the instruction and data address generation units, and dispatches instructions to the processor array.

Each PE contains five execution units (a multiply accumulate unit, an arithmetic logic unit, a data select unit (DSU), a load unit, and a store unit, supporting various 8/16/32/64-bit packed-data types) a 32x32-bit reconfigurable register file, a VLIW-Instruction-Memory unit, and local data memory. The DSU supports shifts, rotates, and single-cycle PE-to-PE communications across the ManArray network. With the indirect VLIW (iVLIW) architecture, the communications operations can be overlapped with the compute operations, thereby providing *zero-latency* data transfers between PEs. The load and store units provide independent data paths between the local memory in each PE in the array. This allows very high memory bandwidth support for compute-intensive algorithms.

6 Conclusions

Using the ManArray network, BOPS has implemented an advanced, scalable family of DSP cores for emerging applications such as broadband communications, digital video, digital audio, imaging, and graphics.   The BOPS ManArray (Hardware, Software, and Programming Environment) is the culmination of a thorough examination of DSP requirements, dozens of innovative ground-breaking patents, and hundreds of man-years of development effort.   The ManArray elegantly provides three basic levels of parallelism (indirectVLIW, packed-data, and multi-processing), all independent of each other and available to the compiler or programmer on an as-needed basis.   These features are combined in a way which allows a 2x2 ManArray processor to produce a radix 4 distributed 256 point FFT in 425 cycles using complex numbers of 32 bits (16 bits for real and imaginary parts) and an 8x8 IDCT in 34 cycles that meets IEEE standards.   And finally, because these emerging markets are primarily System-On-Chip markets, BOPS is providing the ManArray as licensable IP in the form of Cores, SW, and Programming Tools.

REFERENCES

1. R. Cypher and J.L.C. Sanz, "SIMD Architectures and Algorithms for Image Processing and Computer_Vision," IEEE Transactions on Acoustics, Speech and Signal Processing, Vol. 37, No. 12, pp. 2158-2174, Dec. 1989.

2. K.E. Batcher, "Design of a Massively Parallel Processor," IEEE Transactions on Computers, Vol. C-29 No. 9, pp. 836-840, Sept. 1980.

3. "Multiprocessor FFTs", P. N. Swarztrauber, Parallel Computing 5, pp. 197-210, Elzevier Science Publishers B. V. (North-Holland) 1987.

4. Ian Foster, "Designing and Building Parallel Programs", 1995 Addison-Wesley Publishing Company, Inc., pp.83-135.

5. "M.F.A.S.T.: A Single Chip Highly Parallel Image Processing Architecture", G. G. Pechanek, M. Stojancic, S. Vassiliadis, and C. J. Glossner, Proceedings of the IEEE 1995 International Conference on Image Processing, pp. 69-72, Oct. 22-25, 1995 Washington,D.C.

6. "A Massively Parallel Diagonal Fold Array Processor", G.G. Pechanek et al., 1993 International Conference on Application Specific Array Processors, pp. 140-143, Oct. 25-27, 1993, Venice, Italy.

7."Digital Neural Emulators Using Tree Accumulation and Communication Structures", G. G. Pechanek, S. Vassiliadis, J. G. Delgado-Frias, IEEE Transactions on Neural Networks Vol. 3, No. 6, pp. 934-950, Nov. 1992.

8. "Multiple Fold Clustered Processor Torus Array", G.G. Pechanek, et. al., Proceedings Fifth NASA Symposium on VLSI Design, pp. 8.4.1-11, Nov. 4-5, 1993, University of New Mexico, Albuquerque, New Mexico.

9. Robert Cypher and Jorge L.C. Sanz, "The SIMD Model of Parallel Computation", 1994 Springer-Verlag, New York, pp. 61-68.

10. F. Thomas Leighton, "Introduction To Parallel Algorithms and Architectures: Arrays, Trees, Hypercubes," 1992 Morgan Kaufman Publishers, Inc., San Mateo, CA, pp. 389-404,.

Topic 10
Distributed Systems and Algorithms

Gérard Padiou and André Schiper

Co-chairmen

We received 24 papers representing a wide spectrum of research areas in distributed computing: simulation, distributed shared memories, protocols (including mobile computing), middlewares based on CORBA or Java and, of course, distributed algorithms. The variety of submissions shows the importance and the diversity of the research activities in the field of distributed computing. Two comments about the importance of this research:

- first, basic research on distributed systems modelling remains necessary. The usual model for specifying and describing distributed algorithms remains primitive and there is still no agreement on a more abstract formal model;
- second, the area of applications in distributed computing is continuously increasing. These new applications (for instance multimedia) raise new challenges for distributed system designers. This leads to the development of middlewares requiring a strong effort of normalization.

Of the 24 submissions, the Program Committee accepted 6 regular papers and 2 short papers, which have been grouped in two sessions: one devoted to *Distributed algorithms*, and the other to *Distributed systems and middleware*.

Four regular papers are presented in the *Distributed algorithms* session. The paper by A. Kshemkalyani and M. Singhal presents new abstractions to reason about distributed computations. Identifying good abstractions has always been an important contribution of the research in distributed systems. Two papers are related to vector clocks, a classical mechanism in the context of distributed algorithms. The paper by M. Raynal shows on two examples first the usefulness and then the limits of vector clocks. The paper by L. Arantes et al. proposes an optimization related to vector clocks in the context of implementing lazy release consistency. Finally, the paper by G. Antonoiu and P.K. Srimani proposes a mutual exclusion algorithm based on self-stabilization, a paradigm for designing robust distributed algorithms.

Two regular and two short papers are presented in the *Distributed systems and middleware* session. The paper by K. Drira et al. opens the session with an example of a middleware dedicated to multimedia services. The authors propose a multi-level structuring approach to solve cooperative interactions among CORBA objects. The Symphony environment described in the paper by R. Friedman et al. tackles the problem of dynamic reconfiguration of Internet services and provides an infrastructure for adapting servers according to the load. The Epidaure system (paper by D. Fezzani and J. Desbiens) uses the Java technology to provide a tool for building Distributed Artificial Intelligence applications. The

P. Amestoy et al. (Eds.): Euro-Par'99, LNCS 1685, pp. 767–768, 1999.
© Springer-Verlag Berlin Heidelberg 1999

last paper written by O. Richard and F. Cappello considers local networks and describes a high performance implementation of client/server interaction.

We hope that these two sessions will lead to interesting and fruitful discussions.

A Cooperation Service for CORBA Objects.
From the Model to the Applications*

Khalil Drira[1], Frédéric Gouëzec[1], and Michel Diaz[1]

LAAS-CNRS, 7 Av. Colonel Roche 31077 Toulouse Cedex 04 FRANCE
drira@laas.fr,
http://www.laas.fr/~khalil

Abstract. This paper introduces a CORBA-based cooperation service
for distributed applications and its underlying design model distinguishing coordination and communication. The communication service provides a group communication facility. Coordination is more sophisticated
and addresses dynamic architecture evolution and shared work space access management functions. A formal development technique based on
graph grammars is defined and used to implement this service.

Keywords: cooperation, coordination, distributed applications, CORBA

1 Introduction

Coordination and communication constitute two basic issues for cooperative distributed applications. Coordination functions include shared work spaces access
management (SWSAM) in collaborative activities support applications; floor
control in multimedia conferencing [2]; concurrency control in concurrent programming [1] and in multi-database systems [6].

The emergence of Internet-based collaborative activities with increasing complexity and with strong scalability and evolutivity requirements demonstrated
the necessity of Dynamic Architecture Evolution (DAE) for collaborative activities support systems. This issue constitutes an active research area since the last
decade and many works related to architecture description languages have been
conducted. The most relevant for comparison with our work is the new approach
based on graph grammars proposed recently by Le Mtayer [7] for verification of
DAE rules correctness w.r.t. a required architecture style.

We propose, here, graphs to describe coordination without separation between coordination functions dealing with architecture and those dealing with
access management. We propose a graph grammar-based technique to describe
both SWSAM and DAE rules. For this purpose, the classical graph structure
is extended by considering active (runnable) information in graph nodes. We
obtain the so-called *coordination graph* defined as a graph whose nodes may

---

* This work has been partially supported by The TELECOMMUNICATION Program
of CNRS under grant   TL-97028.

P. Amestoy et al. (Eds.): Euro-Par'99, LNCS 1685, pp. 769–776, 1999.
© Springer-Verlag Berlin Heidelberg 1999

be associated with user-defined behaviors implemented here using CORBA objects. Our approach differs from the one of Le Mtayer [7], by its application to object-oriented development and by considering simultaneously the DAE and the SWSAM functions. In order to validate this approach, we have implemented a graph grammar system allowing this technique to be used for objects activation and deactivation in CORBA-based applications.

Group communication is also an important support that underlies the exchange of coordination primitives and cooperative information [10] between group entities. Group communication and coordination functions for group support constitute a set of generic functions underlying the so-called *cooperation service* proposed in this paper and designed following the CCC model (which stands for Communication and Coordination support for group Cooperation), a structured model designed to support formal development of object-oriented and cooperative distributed applications.

This paper is organized into five sections. Section 2 presents the model on which is based the CORBA service for cooperation which is presented in section 3 and used in section 4 to implement SWSAM for shared work space with tree-like structure. In conclusion, we outline our main contributions and introduce future work.

2   The CCC Model

As depicted by Fig. 2, in the CCC model, a distributed cooperative application is composed of distributed objects that cooperate according to a set of cooperation rules using a coordination and a communication medium.

The cooperation rules specify **what** information is needed for the activity achievement and **how** to produce and transform information shared by the cooperating objects supporting this activity. This includes scheduling and execution of relevant coordination and communication rules.

The coordination rules specify **when** access to, sharing of, production and transformation of information are possible or required; and **how** architecture should be modified to enable this. This includes access to the coordination medium to generate coordination directives and execution of communication rules to diffuse these directives.

The communication rules describe **where** and **how** collaborative information and coordination directives are diffused.

2.1 The Cooperation Rules

The cooperation evolves according to a set of events that may be internally generated by the cooperating objects (such as alarms and notifications) or by their environment (such as events coming from graphical interfaces interacting with a user). In both cases, an "*action*" is associated with the occuring event and implies the request to apply one or more coordination or communication rules according to a given scheduling algorithm.

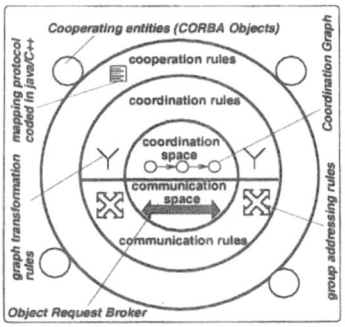

Fig. 1. The CCC model and its application to CORBA communication model

2.2 The Coordination Medium

The coordination medium is composed of the *coordination graph* and a set of *transformation rules* that operate on this graph.

The coordination graph is a typed graph whose nodes are associated with the behaviors of cooperating objects. Other attributes may be added to graph nodes and they will be associated with user-defined parameters. The field "behavior" associated with graph nodes allows a dynamic architecture to be represented by the graph and allows such a dynamic architecture to be managed using objects activation/deactivation actions associated with the transformation rules.

The coordination rules are described as a set of transformation rules over the coordination graph. A graph transformation rule is again a graph where we distinguish three fragments: the *"remove"* and the *"preserve"* fragments (called the required fragment in the sequel) and the *"insert"* fragment. Rules may have parameters. A *guard* may also be used in rules. It describes boolean conditions relating the rule parameters and the variables appearing in the rule graph. Graph transformation operates as follows: (1) A subgraph isomorphic to the required fragments is searched. (2) If any found, the subgraph matching the *"remove"* fragment of the rule is removed from the original graph. A graph isomorphic to the insertion fragment is then embedded into the previously transformed graph.

A simple graph transformation rule and its application to a graph are shown in Fig. 2.(a,c) and Fig. 2.(b) respectively. In the transformation "$G \longrightarrow G'_1$", the required fragment is mapped on the original graph by associating:

node $\_x$ (resp. $\_y$) of the rule with node A (resp. B) of the graph

edge $\_x \longrightarrow \_y$ of the rule with edge A $\longrightarrow$ B of the graph.

In a coordination graph description, we should be able to classify cooperating objects, e.g. to distinguish a cooperating object handling text from another handling images. For these reasons, we use a typed version of graph transformation techniques: nodes have types that productions must preserve. The applicability of a rule is conditioned by typed graph morphism laws without distinguishing runnable nodes. The application of a rule leads to both graph transformation

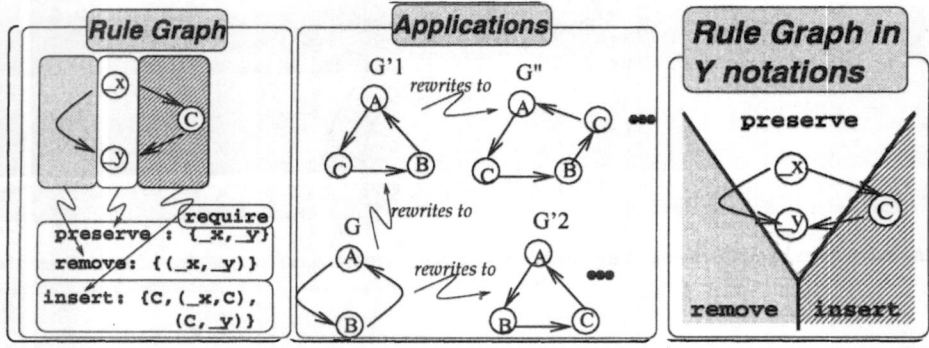

Fig. 2. (a): a transformation rule. (b): an initial graph and some applications (c) the Y notation [4]

and DAE management actions. In the case of the example of Fig. 2, a cooperating object whose behavior is specified by the node C is activated on each connected site every time the rule is applied.

2.3 The Communication Medium

The communication medium involves a set of communication rules operating on a communication space.

The communication rules allow the cooperating objects to address a set of partners (other cooperating objects) that may be located on the same machine or on distributed machines.

The group address is defined as a pair $\langle siteAddress, ObjectsGroupsAddress\rangle$.

Where *siteAddress* is a group address defined according to the Set-Based Format which distinguishes the three following addresses:

- *All* : every site is in the scope of this group address;
- *AllOf(S)*: every site with identifier in S is in the scope of this group address;
- *AllBut(S)*: in this case, all the sites are in the scope of this group address except those with identifiers in S; with S being a set of site addresses.

And where *ObjectsGroupsAddress* has one of the following formats:

- *"AllObjects"* meaning that all objects of the addressed site are concerned.
- *TheseObjects(A)* with A being a set of *ObjectAddresses*; and an *ObjectAddress* being a pair whose first element is defined according to the Set-Based Format, and whose second element is the *"type"* of the addressed objects. An *"ObjectAddress"* may then correspond to one of the three following formats:
  - Type-based addressing: $(All, Atype)$: all objects of type *"Atype"* are concerned.
  - Explicit addressing: $(AllOf(S), ?)$ every object with identifier in the set S is concerned regardless of type information.
  - Hybrid addressing: $(AllBut(S), Atype)$ addresses all objects of type *"Atype"* excepting objects with identifiers in the set S.

The communication space is delegated to the Object Request Broker [8].

3 Implementation for CORBA-Based Applications

The main CORBA service is a factory that allows different cooperation servers to be created. Each cooperation server is composed of two CORBA objects: the coordination server and the communication server. This package has been developed using the IDL/Java mapping and OrbixWeb. The coordination server implements the graph rewrite system and the architecture management module as illustrated by Fig. 3. Transformation rules are provided by the application programmer in a Java class that implements a predefined Java interface. Cooperating objects behavior may be implemented using the IDL/Java or IDL/C++ mappings. We also use CorbaScript, a CORBA scripting language [9], to easily administrate the servers.

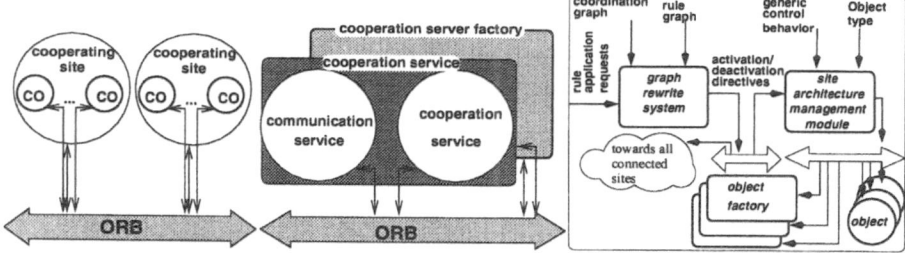

Fig. 3. (a) Application components description where CO stands for Cooperating Object; (b) Functional architecture.

The architecture of this service allows the programmer to develop a standalone application or an applet executed using WWW browsers. The communication being relayed by the group communication server, this architecture is in conformance with the safety requirements of web browsers stating that an applet cannot establish a connection except with the Web server it was downloaded from. For this kind of applications, the cooperation server must run on the same machine as the web server.

Global Application Architecture:

The application is viewed as a collection of distributed cooperating objects that run on distributed sites. A site may be distributed on different machines (different IP addresses). It is also possible to group objects of different sites on the same machine. This possible due to the utilization of object factories to activate objects as depicted by Fig. 3 (right part).

When no coordination action is involved, cooperating objects may directly communicate with their peers, without using the coordination medium. Cooperating Objects use the coordination medium when communication is not sufficient.

4   The Tree-Like Model for SWSAM Coordination Service

Several coordination services can be defined for SWSAM depending on how the shared work space is organized. The organization may correspond to a physical ordering of information such as character precedence in a text file. It may also correspond to logical dependencies between information compounds.

In a recent work [3], we have developed a SWSAM model for linearly organized shared work spaces. This model applies well to textual document sharing but may not be used for non-linear organization. We develop here a more general model that overcomes this limitation. The model manages shared work spaces that are structured as a tree where intermediate nodes represent folders and where leaves represent documents (left part of Fig. 4). As depicted by Fig. 4 the documents can be available or locked (in use). Different types of nodes are used to represent this state in the coordination graph that represents the shared work space state (right part of Fig. 4).

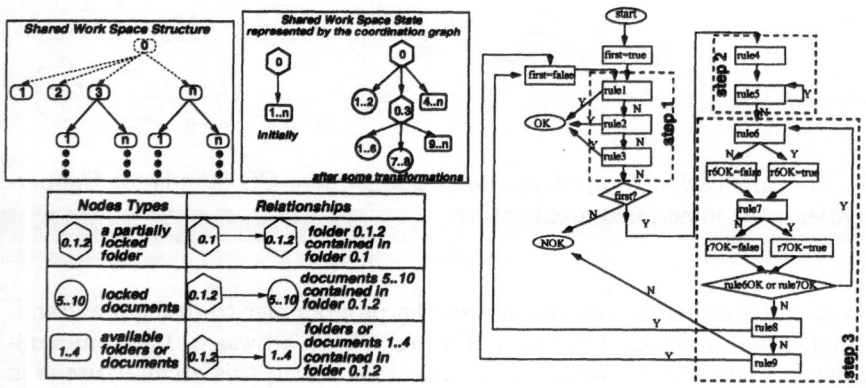

Fig. 4. (left) structure and state of the shared work space; (right) cooperation rules described as an algorithm that maps a request-for-lock to applications of coordination rules.

The coordination rules (formally depicted by Fig. 5) operate as follows: the request to lock a document or a folder will lock this document or this folder as well as all the tree of all the folders and the documents it contains. Besides all the chain of the containing folders will become partially locked. They cannot be requested for lock before the release of all their (partially or totally) locked successors (represented respectively by 9 and #); only available successors (i.e. unlocked folders and documents represented by ☐) of partially locked folders may be totally locked. Fig. 4 (left) describes the node types and the relationships. Nodes types that are not significant for coordination are not described in Fig. 4. Namely, the "label typed node" ▷ is never present in the coordination graph.

It is only used for intermediate transformations and at the end of the rules applications it is definitely removed from the graph.

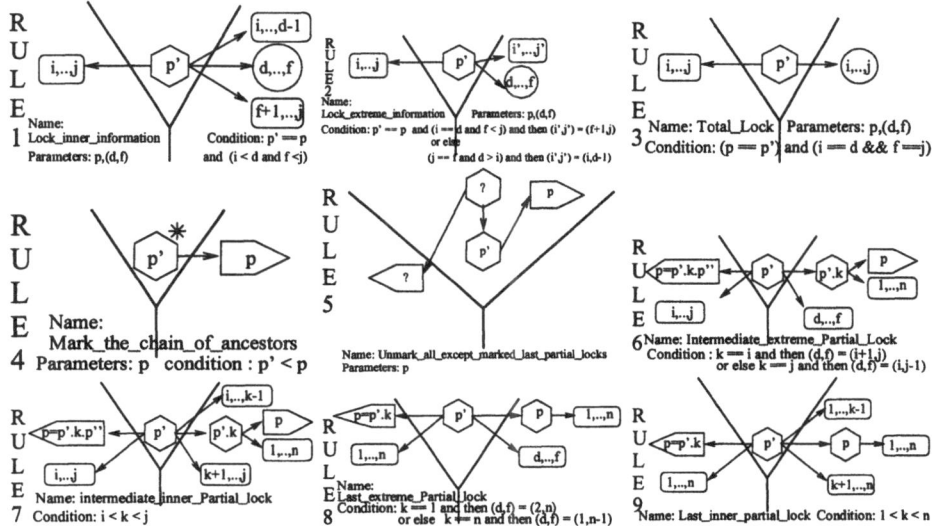

Fig. 5. Coordination rules for Tree-like organized shared work space

As described by the algorithm of Fig. 4, the cooperation rules consist in applying coordination rules until a valid coordination graph (i.e. a graph which does not contain any "label typed node") is obtained. They act as follows:

(step 1) creation of the locked list of documents (or folders) if the request for lock concerns an available list of documents (or folders) $[p.d..p.f]$ whose immediate containing folder (p' such that $p' == p$) is already partially locked (rules 1,2,3), then the appropriate available node is split into one (rule 3), two (rule 2) or three (rule 1) nodes.

(step 2) identification of the last created containing folder This step is composed of two sub-steps:

– by applying rule 4 mark all the chain of already created containing folders by adding the relationship ($folder \rightarrow label$). The star symbol (*) means that this rule is applied to ALL the ancestors, and not only to one of them as it would be the case without the star.

– by applying several times rule 5 (while this application is possible) unmark all ancestors but not the last one.

(step 3) creation of the chain of partially locked containing folders This step is composed of two sub-steps:

– creation of the intermediate list of partially locked containing folders: starting from this last folder, create all the intermediate partially locked folders and mark them to allow the creation of intermediates in order to continue from each newly created folder: rule 6 and rule 7.

– creation of the last partially locked containing folders After this third step, step 1 is applied once again.

5   Conclusion

This paper proposed a cooperation service including group communication facilities inherited from CORBA and new coordination facilities using the coordination graph transformation technique. The originality of this approach comes from its abstraction power, from the friendliness of the graph structure and from its independence with respect to the cooperation paradigms underlying the model or the language the coordinated objects are implemented with. Moreover describing coordination rules using graph transformation allows formal verification to be conducted to check the correctness of the implemented architecture [7] leading to significantly improve safety and reliability of applications. The cooperation service was implemented using the Java programming language and OrbixWeb ORB. Several experiments are being conducted for the validation of the implemented services.

References

[1] P. Ciancarini and C. Hankin, editors. *Coordination Languages and Models*, volume 1061 of *Lecture Notes in Computer Science*. Springer Verlag, 1996.

[2] Hans-Peter Dommel and J.J. Garcia-Luna-Aceves. Floor Control for multimedia conferencing and collaboration. *Multimedia Systems*, 5:23–38, 1997.

[3] K. Drira, F. Gouezec, and M. Diaz. Design and implementation of coordination protocols for distributed cooperating objects. A general graph-based technique applied to CORBA. In *Third IFIP International Conference on Formal Methods for Open Object-based Distributed Systems*, Florence, Italy, February 15-18, 1999. to appear/a paraitre.

[4] H. Göttler. Attributed graph grammars for graphics. In G. Rozenberg H. Ehrig, M. Nagl, editor, *Graph Grammars and their application to Computer Science*, LNCS 153, pages 130–142, 1982.

[5] ISO/ITU. *Basic Reference Model of Open Distributed Processing*. ISO 10746-3/ITU-T X.903, 1994.

[6] K. Barkaoui and R. Benamara. Concurrency Control in Multidatabase Systems with an Advanced Transaction Model. In H. Arabnia et al., editor, *Proc. of PDPTA'99*, Las Vegas, 28 june-01 july, 1999.

[7] D. Le Métayer. Describing software architecture styles using graph grammars. *IEEE Transactions on Software Engineering*, 24(7), July 1998.

[8] Thomas J. Mowbray and Ron Zahavi. *The Essential CORBA: Systems Integration Using Distributed Objects*. OMG/Wiley (ISBN 0-471-10611-9), 1995.

[9] Christophe Gransart Philippe Merle, Jean-Francois Roos, and Jean-Marc Geib. Corbascript: A dedicated corba scripting language. In *CHEP'98 Computing in High Energy Physics*, Chicago, Illinois, USA, August 31 - September 4, 1998.

[10] E. Yavatkar and K. Lakshman. Communication support for distributed collaborative applications. *Multimedia Systems*, 2(2):74–88, August 1994.

Symphony: Managing Virtual Servers in the Global Village*

Roy Friedman, Assaf Schuster, Ayal Itzkovitz, Eli Biham, Erez Hadad,
Vladislav Kalinovsky, Sergey Kleyman, and Roman Vitenberg

Department of Computer Science
The Technion
Haifa 32000
Israel

Abstract. A *virtual server* is a server whose location in an internet is
virtual; it may move from one physical site to another, and it may span
a dynamically changing number of physical sites. In particular, during
periods of high load, it may grow to new machines, while in other times it
may shrink into a single host, and may even allow other virtual servers to
run on the same host. This paper describes the design and architecture of
Symphony, a management infrastructure for executing virtual servers in
internet settings. This design is based on combining CORBA technology
with group communication capabilities, for added reliability and fault
tolerance.

1 Introduction

The emergence of internets, and in particular the Internet, created a potential for
(literally) global resource sharing. If we examine the total load on all computing
servers in the world at any given moment, we may see the following phenomena:
While the load on a given computer may increase and decrease throughout the
day, the overall load on all computers remains more or less constant. Moreover, at
times when some servers are overloaded to the point of crushing, other machines
are sitting idle.

To solve this problem, we propose the notion of *virtual servers*. A virtual
server is a server whose identity is not bound to a fixed physical computer, but
rather is changing and evolving with time and in reaction to the load on this
server. Thus, a virtual server may move from one location in the network to
another, and the number of physical computers on which it resides may change
dynamically.

Consider, for example, an e-commerce application that shifts its location
from the US to Japan when night falls on Los Angeles, and then from Japan
to Israel when when day breaks in Tel-Aviv. In this example, being able to
shift the location of the server guarantees that at any given moment, the server
is located in proximity to its current clients. Another example is a large ray

---

* Supported by the Ministry of Science, Basic Infrastructure Fund, Project #9762.

P. Amestoy et al. (Eds.): Euro-Par'99, LNCS 1685, pp. 777–784, 1999.
© Springer-Verlag Berlin Heidelberg 1999

tracing application running on a cluster of PCs in the Technion at night, using a distributed shared memory system like Millipede [8, 9]. This application suddenly notices that there are idle workstations at the Hebrew University in Jerusalem, and then decides to shift some of its computing tasks there, in order to enjoy higher parallelism. Similarly, Web based news sites become overloaded when an important event, such as elections or an important chess game, take place. These Web sites need to be able to grow, perhaps even to distant hosts, in order to handle the load.

In this work we describe the internal architecture of *Symphony*, an infrastructure for managing and executing virtual servers in internet settings.[1] Symphony's architecture is based on the fundamental principles underlying CORBA [2]. However, it is not a "pure CORBA" design, in the sense that services and applications (virtual servers) are replicated and can communicate internally in a non-CORBA compliant fashion. This distinction is important since it enables us to integrate high-performance servers, such as distributed shared memory, for which CORBA's performance is insufficient at best. Also, our main management services, as described later in this paper, are built using the Ensemble [7] group communication technology [3]; this design allows us to build scalable management services without giving up the power of group communication.

A service based design has the benefits of scalability, interoperability, and extendibility. The design is scalable, since the location of a service in the system is virtual. A service can run anywhere from a single host to the entire set of machines, yet be accessible everywhere. Also, services present an abstraction of objects, where only their interface is known, so that different implementation of the same service can be replaced without having to recompile the system. Finally, it is always possible to add services, or modify existing ones, in order to extend the functionality of the system.

In summary, our design tries to enjoy the benefits of both worlds. We use group communication as the main building block for our management infrastructure, use a CORBA-oriented architecture, and allow replicated/distributed servers to communicate internally with their own communication stacks for high performance. (Of course, servers that whish to communicate internally using CORBA can do so, but we do not impose this on other server.) We would like to emphasize that our work *does not restrict the programming model or language* (in particular we do not restrict ourselves to Java and/or sand-box models), yet puts emphasis on security, as discussed later in this paper.

2   Architectural Overview

Our architecture follows the CORBA approach of providing services to perform common tasks needed by applications and other services in the system. We arrange our services in several *domains*, as depicted in Figure 1. Services are *logically* organized in a hierarchical structure such that services in one domain

---

[1]   A prototype of Symphony has recently been built, with an initial release planned for later this year.

tend to draw upon the functionality provided by services in lower or equal domains, but not in higher domains, as discussed below. Also, Symphony's run-time system is located below all services alongside with the the operating system.

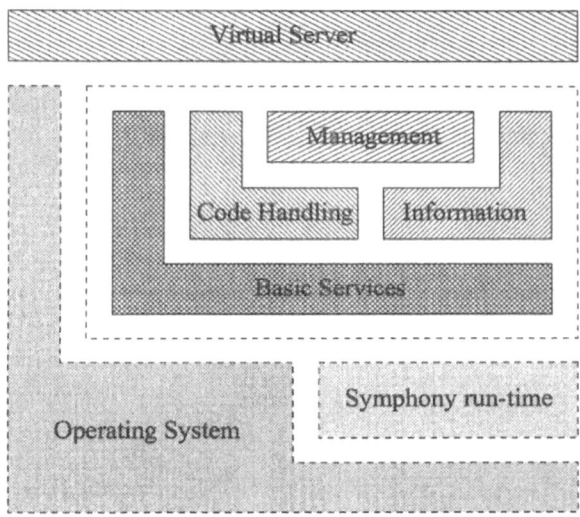

Fig. 1. Symphony's Service-Based Architecture.

Management services lie at the heart of our design. These services provide support for automatically replicating and migrating virtual servers based on the characteristics of these servers, the topology of the network, computers availability and so forth. The application can register with these services a description of its requirements and cost functions for locating "ideal" machines for it. The application is free to choose between two modes of operations: (a) The application can either allow the management services to automatically migrate and replicate itself based on the above descriptions (and restrictions), or (b) request non binding recommendations from the management services, but maintain the final word as to where and when its replicas should be (re)moved. This latter option is given for virtual servers that wish to retain control regarding the location of their replicas.

Management services draw upon the functionality provided by both *informational* services and *code-handler services*. The code-handler services do the actual task of starting an image of a code on a given machine upon demand, or evacuating machines under certain conditions. These services may receive instructions from either management services or applications, receive operational information from informational services, and utilize the functionality of the basic services as part of their normal operation.

Informational services, on the other hand, collect and report information about the system. This information includes the load on participating machines,

the topology of the system, the given environment on each of the machines, and
the location of various applications. The information gathered by informational
services can either be reported to subscribers using registered up-calls, or given
as a response to an explicit request.

Basic services are located at the bottom of the hierarchy. These services
provide the basic functionality that we view as necessary for most common
applications, as well as to other services. These include a naming services, for
registering and locating other services, a *security* service, *transaction* service,
replication service, and an *event notification* service.

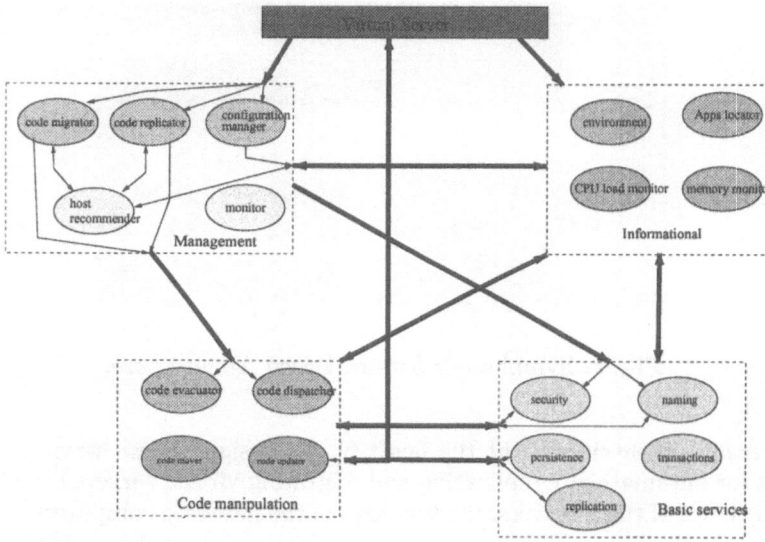

Fig. 2. Flow of information in Symphony.

Figure 2 presents a deeper look into the bowels of the system, by examining an
ideal picture of the interactions between some of the main services in the system.
In this example we see that the application (virtual server) registers itself with
the management services, mainly the *code-replicator* and *code-migrator* services.
(We discuss these services and others in the full version of this paper.) The code-
replicator and code-migrator, in return, register with the informational services
to be notified about the status of the system and to be notified about changes as
they occur. When one of the management services decides to start a replica on
a certain machine, or to evacuate the machine, it contacts the *code-dispatcher*,
or *code-evacuator* respectively, to perform the job.

Applications as well as Symphony's services may use the name service and
the security service to locate other services and to communicate securely with
them.

2.1 Communicating with Replicated Services

Most of the services we propose need to be replicated for efficiency and high-availability. Also, different services may need different levels of replication. Some services may require a replica on every hosts, e.g., the code-dispatcher, while for other services it may be sufficient to instantiate a single replica in each cluster or subnet.

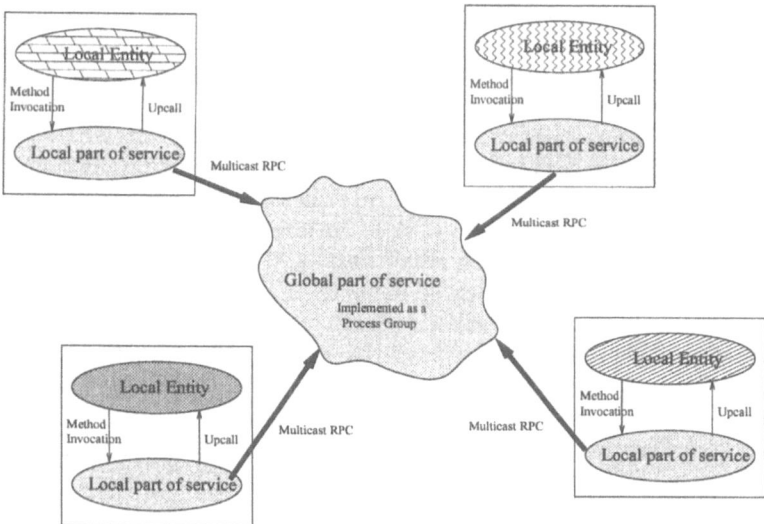

Fig. 3. Communicating with Replicated Services.

   This goal is realized by splitting each service into a local stub that has to be started on each host, and a global part that is replicated, as illustrated in Figure 3. The global part consists of several replicas that form a process group; these replicas communicate through a group communication system that also manages membership changes, i.e., discovery of both failures of existing replicas and joins by new replicas.[2] The local stub communicates directly with the application or service entity on the local host. When contacted, the stub issues a multicast RPC request to the global part of the service, and is then responsible for collecting the replies and forwarding them to the local entity that initiated the request. In particular, the local stub can be asked to deliver the first reply that arrives from any replica, deliver all replies, or deliver the most common reply. Also, invocations on the local stub may be either synchronous, deferred synchronous, or asynchronous [10].

---

[2] The Event Notification Service of CORBA can be viewed as a plausible alternative for implementing replicated services. Felber and Guerraoui explain in [5] why ENS is in fact unsuitable for this task.

3 Security in Symphony

Trust Model A trust model, as suggested by its name, defines for each entity the kind of trust it has in other entities, and how this trust is verified. In Symphony, an entity can be a host, a message, or an image of a code. Each entity *e* holds a data structure that contains the following fields: (a) list of entities to whom *e* is willing to send messages, (b) list of entities from whom *e* is willing to accept messages, (c) list of entities to whom *e* is willing to send code, (d) list of entities from whom *e* is willing to accept code, (e) list of entities to whom *e* is willing to migrate/replicate, (f) list of entities to whom *e* is willing to provide services, and (g) list of entities with whom *e* is willing to share resources. Each of these categories is defined by specifying an access control list, and can be refined using the following subcriteria: (i) type (of communication/service/code), (ii) authentication scheme, (iii) limitations on the way the communication was carried, e.g., has to go through a firewall, (iv) other conditions for the interaction, and (v) recommendations in case multiple options occur.

Thus, it is possible to specify that host *A* is willing to accept any code written by Assaf, if it is authenticated using the El Gamal scheme, and it passed through a given firewall. *A* may also be willing to accept any Java applet that was generated at the Technion, and is signed with an MD5 hash function and a known secret key. Similarly, it is possible to specify that a replica of *"Your-Fastest-Web-Search-Engine"* never runs on the same machine as a replica of *"The-Most-Complete-Web-Search-Engine"*, while some server *A* may require encryption of queries, but is willing to receive authenticated, but not encrypted, replies.

The trust model we have just described is not only used by virtual servers, but also by Symphony's internal services. In particular, some of the management and informational service may not accept certain types of data from untrusted entities. This is important in order to prevent malicious parties from creating havoc in the system. For example, if the host recommender recommends bad hosts, it may cause the system to migrate code to the wrong nodes, which will degrade the performance of the system, and make it unusable.

Message Security Securing messages can be done by authentication and encryption. Symphony supports both, and provides key management services for managing keys used in the entire system.

Securing Imported Code Execution In recent years there has been a large body of research on how to securely execute imported code. In Symphony, an exported code is accompanied with a *descriptive object*, which defines the restrictions the code is willing to run under, as well as the runtime environments it requires. Also, part of the information gathered by the informational services is the policies employed at each of the sites on the network towards such code. In particular, a code can be authenticated as coming from Assaf at the Technion, and some hosts in Jerusalem may be configured to allow full execution rights to Assaf's code, but restrict all other code to a Java-like sand-box model.

When sending code for remote execution, the code-dispatcher service consults with the appropriate information services regarding the possibility of placing the given code on the requested machine. The code is then executed on the remote site only if the site's policy allow the code to run there.

Moreover, the descriptive object accompanying the code includes the maximum amount of resources the process needs. Before starting a code image, the code-dispatcher verifies that the hosting machine is willing to provide that much resources to the imported code. These resources include memory consumption, disk space, cpu time, communication, and processes. If the answer is no, it will not be started on the given machine. If the answer is yes, it will be started, but the amount of resources it consumes will be monitored periodically; if they are exceeded, the process(es) will be stopped.

4   Related Work

Due to space limitations, here we only briefly mention some of the most related works. The full version of this paper includes a more detailed comparison of Symphony to other approaches and systems.

OSF DCE offers a service based environment for distributed computing [1]. DCE services include, e.g., a distributed time service, security, a distributed file system, and a distributed directory service. DCE aims to be a platform and vendor independent system, and supports its own RPC style communication. Symphony, on the other hand, concentrates on issues specific to virtual servers, and in particular supports services that automate the migration and replication of such virtual servers. Thus, Symphony could have been built as an extension over DCE, although we have chosen to use CORBA as our base platform.

Legion [6] is another object-oriented distributed system that provides an illusion of a single virtual machine composed of wide area collection of workstations, parallel computers, and fast massive multiprocessor machines. Class interfaces in Legion can be written in either CORBA's IDL or MDL [6]. Unlike Symphony, in which an application can choose the set of service it wises to use, Legion requires all objects to conform to its model. Each object should inherit from a predefines LegionObject and each class should be also instantiated and inherited from a base class named LegionClass. This programming model (rules) enables the Legion runtime system to efficiently support the mobility of application's objects, security in object invocation and heterogeneity. Legion deals with issues of scalability, security, fault-tolerance, and management of a large scale (sometimes called nation-wide) meta-computer. Note that, Legion address scalability by cloning objects, but does not take Symphony's approach of replicated services based on group communication.

There are several known approaches for combining CORBA and group communication, such as the ones developed in Electra and Orbix+ISIS [10]. The full version of this paper discusses them in more detail, and compare them to our approach.

User-level software-only distributed shared memory (DSM) systems, such as Millipede [8, 9], are another example of systems that need to migrate code in a cluster of workstations environment. In particular, in Millipede both code and data can migrate between machines based on their load and on access patterns to shared memory pages. Symphony complements Millipedes functionality, as Symphony provides services for replicating code, and for finding candidate machines for migrating code based on their load and communication capabilities.

Condor [4] and Utopia/LSF [11] are systems for doing batch scheduling in distributed environments. They focus on queuing many independent computation tasks and distributing them among a large network, whenever idle machines are detected. Since both these systems only handle independent computation tasks, they do not need to interact with the applications they run.

References

[1] DCE Home Page. http://www.osf.org/dce.

[2] The OMG Home Page. http://www.omg.org.

[3] K. Birman. The Process Group Approach to Reliable Distributed Computing. *Communications of the ACM*, 36(12):37–53, December 1993.

[4] D.H.J. Epema, M. Livny, R. van Dantzig, X. Evers, and J. Pruyne. A Worldwide Flock of Condors: Load Sharing Among Workstation Clusters. *Journal on Future Generations of Computer Systems*, 12, 1996.

[5] P. Felber, R. Guerraoui, and A. Schiper. Replicating objects using the corba event service. In *Proc. of the 6th IEEE Workshop on Future Trends of Distributed Computing Systems*, October 1997.

[6] A.S. Grimshaw and W.A. Wulf. The Legion Vision of a Worldwide Virtual Computer. *Communications of the ACM*, 40(1), January 1997.

[7] M. Hayden. The Ensemble Syste. Technical Report TR98-1662, Department of Computer Science, Cornell University, January 1998.

[8] A. Itzkovitz, A. Schuster, and L. Wolfovich. Supporting Multiple Programming Paradigm On Top Of A Single Virtual Parallel Machine. In *Proc. of Second International Workshop on High-Level Parallel Programming Models and Supportive Environments (HIPS'97)*, pages 25–34, April 1997. Earlier version appeared as Technion CS Technical Report LPCR #9607.

[9] A. Itzkovitz, A. Schuster, and L. Wolfovich. Thread Migration and its Applications in Distributed Shared Memory Systems. *The Journal of Systems and Software*, 1998. To appear. Also available as Technion CS Technical Report LPCR #9603.

[10] S. Landis and S. Maffeis. Building Reliable Distributed Systems with CORBA. *Theory and Practice of Object Systems*, April 1997.

[11] S. Zhou, J. Wang, X. Zheng, and P. Delisle. Utopia: A load sharing facility for large, heterogeneous distributed computing sytesm functionality. Technical Report CSRI-257, Computer Systems Research Institute, University of Toronto, 1992.

Épidaure: A *Java* Distributed Tool for Building DAI Applications

Djamel Fezzani and Jocelyn Desbiens

INRS–Télécommunications
16, place du Commerce, Verdun (Québec) Canada, H3E 1H6.
Tel. (514) 761-8666 Fax. (514) 761-8619
{fezzani, desbiens}@inrs-telecom.uquebec.ca

Abstract. This paper presents the main aspects of the combination of a *Java* implementation of the actor model, called *Épidaure*, and the *Java* expert system shell *JESS*. This association of *Épidaure* with *JESS* provides a distributed computational environment within which each *JESS* is an active independent computational entity with the ability to communicate freely with other *JESS*.
The paper will go into more detail on the communication mechanism in the *Épidaure* system and will give a short description on how the association of *Épidaure* with *JESS* is used to write DAI applications. The paper will also review some advantages and potential applications and will conclude with new perspectives.

Keywords: Actor model, communication languages, distributed artificial intelligence (DAI), *Java* language, cooperating expert systems.

1 Introduction

In recent years many scientific and engineering disciplines have relied on the results of complex, time and memory consuming programs. Several problems of interest are currently too big for conventional workstations, and this is why parallel or distributed programs must be developed.

One paradigm for overcoming the complexity barrier is to build systems of smaller more manageable components which can communicate and cooperate. This paradigm is the emerging sub-field of distributed artificial intelligence (DAI) which is concerned with systems that consist of multiple independent entities that interact in a domain [1, 2]. DAI aims to provide both principles for construction of complex systems involving multiple agents and mechanisms for coordination of independent agents' behaviors.

After having introduced motivations and the context of this paper, the authors will go into more details on how the association of *Épidaure* with *JESS* is designed and written. Firstly, we highlight some key characteristics of the actor model that influenced *Épidaure* and briefly describe the system layer that

P. Amestoy et al. (Eds.): Euro-Par'99, LNCS 1685, pp. 785–789, 1999.
© Springer-Verlag Berlin Heidelberg 1999

constitutes the foundation of distributed asynchronous interacting *JESS*. Then a short description showing how to use the association of *Épidaure* with *JESS* will be presented. Finally, we will brush a quick picture of a few potential areas of research and development that could benefit from such distributed environment and will provide our conclusion on the subject.

2   The Conceptual Framework of *Épidaure* & *JESS* Environment

2.1   The Actor Model and *Épidaure*

To achieve communication between computational entities, DAI systems rely on two approaches: the blackboard model first introduced in the HEARSAY-II system and used in several recent systems [3]; and the actor model which the original model comes from Hewitt [4] and Agha's [5] work on open systems. In the blackboard approach, knowledge sources use available information without knowing its origin and produce information without worrying about its destination: the communication between the modules is limited to reading and writing in the working memory; each module must read/write in a format acceptable to other modules [6], whereas in the actor approach, an actor communicates information directly to another actor via message-passing.

In this model, Hewitt proposed the concept of a self-contained, interactive concurrently executing object which he termed 'actor'. Within the actor model, the analogous role played by the objects of objects-oriented languages is identified by the term 'actor'. Viewed as a computational entity, an actor also comprises two parts: a script, that defines the behaviors of the actor upon receipt of a message (*e.g.*, what corresponds to the definition of an object module within object-oriented languages); and a finite set of acquaintances which are the other actors known to the actor.

Typically, *Épidaure* is a hybrid implementation of the actor model, rewritten entirely in *Java* and can be run on a distributed heterogeneous network of processors. The present version of *Épidaure*, can effectively execute over a network of personal computers (under Windows'98 or NT), and SUN SPARC or HP workstations (under UNIX).

2.2   *Épidaure* Messages

The management of all messages is done entirely by the class **Message**, which coordinates the sending of messages between agents. There are two stages involved in sending a message: the creation of an object of type **Message** followed by the actual sending of the message. The first stage is executed by the constructors of the class **Message**. This class has four constructors, two of which can be called by the user:

- Message(AgentName, MethodName, MethodArguments);
- Message(AgentNumber, MethodName, MethodArguments).

The instanciation of the class **Message** creates an object message that either knows the name or the number of the agent to which the message should be sent. The second stage is executed using overload **send** methods. There are two possible message passing cases: either both agents wishing to communicate are located in the same processor or they are on different processors.

2.3 New User Messages Related to the Association of *Épidaure* with *JESS*

In addition to the above messages that are used by any *Épidaure* agent (outside the association of *Épidaure* with *JESS*), we have implemented other messages, which are mainly used by the *JESS* agents.

The interaction among the *JESS* is asynchronous (synchronocity can also used when needed). The interchange of knowledge between these extended *JESS* can involve exchanging facts, rules, and any other *JESS* data object or functionality. We enumerate (see Table 1) some of the functions specific to our environment and describe their respective use and functionality. The list is not exhaustive, but gives an outline on some main statements which can be used directly by the user. We assume a familiarity with expert systems in general and with *JESS* in particular [7].

	How to load a file?
synopsis:	*(jsend<string-or-symbol-agent-name><method-name>(batch <file-name>))*
behavior:	The agent*<string-or-symbol-agent-name>*possesses the expertise specified in *<file-name>*
	How to activate a *JESS* agent at run-time?
synopsis:	*(jsend <string-or-symbol-agent-name> <method-name> <run>)*
behavior:	The agent *<string-or-symbol-agent-name>* is activated at run-time by the 'run' message.
	How to assert a run-time fact?
synopsis:	*(jsend <string-or-symbol-agent-name> <method-name> (rassert <string>))*
behavior:	The fact *<string>* is asserted in the agent *<string-or-symbol-agent-name>* who then executes
	How to remove a rule?
synopsis:	*(jsend <string-or-symbol-agent-name> <method-name> (undefrule <symbol-rule-name>))*
behavior:	The rule *<symbol-rule-name>* is removed from the agent *<string-or-symbol-agent-name>*. This rule will never fire again

Table 1. Conceptual framework of the *jsend* method

```
package epidaure.agent;
import java.util.*;
import java.io.*;
import jess.*;
import Epidaure.*;
  public class AgentProg extends Agent {
    public AgentProg (String Name) {
          super(Name);
  }
      public void Execute(String Args[]) {
            Rete MyJess = new Rete("jess",new NullDisplay());
            Reference JessRef = MyJess.GetReference();
            . . .
            Message msg = new Message("ExecuteCommand", new Object [] {"(run)"});
            msg.Send(JessRef);
      }
  }
```

Fig. 1. Definition of a *JESS* agent

3   Using *Épidaure* & *JESS*

Épidaure is a *Java* package that allows applications written in *Java* to communicate via message-passing. An application running under the *Épidaure* system
is composed of at least one task or virtual machine. Generally speaking, an
application will lump together many tasks, each one of these being compiled
separately and distributed over the nodes of a massively parallel computer or
a network of workstations by a load balancing algorithm. When compiled with
the *Épidaure* class library, each of these components is actually a task under
the control of the *Épidaure* run time environment which provides services to the
agents composing the tasks. Those services implement a form of uniform address
space in the distributed memory system. In each task there is also a message
server (the communication agent) that controls the message reception for agents
in the task. Each task can serve one or many agents.

We will examine herein an agent that uses *Épidaure*. This agent implements
a simplified expert system shell *JESS*, and functions within a suite of *Épidaure*
methods. The code for this agent is quite sparse, illustrating the ease with which
agents can be constructed using *Épidaure*. *Épidaure* agents typically consist of
one main execution **AgentProg** and a number of service agents. The **AgentProg**'s
main is depicted in Fig. 1.

This agent has a **main** method so that it can be started directly via the
command 'java epidaure.agent.Epidaure'. This agent is an instanciation of
the **Rete** class which contains the inference engine based on the *Rete* algorithm.
The **Rete** class is derived from the **Agent** class which indicates to *Épidaure* that
this class will represent an agent.

As shown in Fig. 1, one can invoke the *JESS* agent in **AgentProg** by just
sending it a message of type **Send**, which is a message without a return value.
The sending of a message is a two steps process: *i*) its construction, and *ii*) its
actual mailing. At the setup time of a message, *i.e.* when the class **Message**

is instanciated, the arguments are encoded and the message is uniquely identified. Once the message is setup, *Épidaure* hands it to the local communication agent and the resolution algorithm is applied. It is *Épidaure* which manages the message system and the remote methods execution [8].

4   Conclusion

In this paper we have introduced a distributed computational environment within which each *JESS* is an active independent computational entity communicating freely with other *JESS*. This environment is currently developed and tested within our software telecommunications group of the INRS. It is based on the portable programming language *Java* and implements the essential aspects of the actor model. In the same way, we have presented how to use this environment to develop DAI application to demonstrate the relative ease with which agents can be build. The potential applications of such environment are considerable.

By building this environment, we wanted to offer the future software developers a tool combining the advantages of the *Java* language with those of the open systems. Such a tool allows independent systems to cooperate in solving problems and to work in parallel on common problems. It allows also unrelated systems to share expertise. In this environment, participant systems can be distributed physically to make the best use of the available processing power.

We are presently putting the final touch to a next version which is intended to do a sound mathematical analysis of the communication costs involved in several updating algorithms. This later release will also include support for other languages and CORBA integration.

References

[1] M. Wooldridge and N. Jennings, "Intelligent agents: Theory and practice," *Knowledge Engineering Review*, vol. 2, no. 10, pp. 115–152, 1995.

[2] N. Jennings and M. Wooldridge, "Applying agent technology," *Applied Artificial Intelligence*, vol. 6, no. 9, pp. 357–370, 1995.

[3] J. Ferber, *Les systmes multi-agents : vers une intelligence collective*. Paris: InterEditions, 1997.

[4] C. Hewitt, P. Bishop, and R. Steiger, "A universal modular actor formalism for artificial intelligence," in *Proc. of the 1st IJCAI*, pp. 235–245, 1973.

[5] G. Agha, *ACTORS: A Model of Concurrent Computation in Distributed Systems*. Cambridge, Massachusetts: The MIT Press, 1986.

[6] R. Engelmore and T. Morgan, *Blackboard Systems*. Inc.: Addison-Wesley Publishing Company, 1988.

[7] E. Friedman-Hill, *"JESS*, the *Java* expert system shell," tech. rep., Distributed Computing Systems, Sandia National Laboratories, Livermore CA, Mar. 1998. (version 4.0).

[8] J. Desbiens, M. Lavoie, and S. Pouzyreff, "Une approche actorisée du mécanisme de communication d'un système à objets répartis," in *Langages et Modèles à Objets*, pp. 207–219, 1995.

A Client/Broker/Server Substrate with 50 μ s Round-Trip Overhead

Olivier Richard and Franck Cappello

LRI, Université Paris-Sud, 91405
Orsay, France
(fci)@lri.fr.

Abstract. This paper describes an environment (API and runtime) for fast remote executions in the context of NOWs using high performance networks and low cost multiprocessors. This environment is based on the usual client/broker/server architecture. It is designed and optimized for the features of local area networks. The main result of the early performance measurements is a $50\mu s$ remote execution overhead.

1 Introduction

A significant advance in networks of workstations are the availability of high performance network hardwares (ATM, Myrinet, SCI and Ethernet 1 Gbits/s) and protocols (VIA). An other advance is the progress of software technology toward the use of components. Distributed environments like NetSovle [1] and Ninf [2] already allow remote execution through the use of software components. These *Metacomputing* tools are mostly dedicated to MAN and WAN. In this note, we present a low level software mechanism (a substrate) designed for fast remote executions inside a LAN and based on high performance networks.

2 Software Architecture

The controlled remote execution substrate is based on the classical Client-Broker-Server (CBS) architecture (Figure 1). The broker provides a strong frontier between users and servers. It has the responsibility to furnish users with a unique view of servers. Users can not directly access servers. The substrate had been designed to fulfill a very low round-trip latency for remote executions. Figure 1 presents the critical path for remote executions.

Three significant points have influenced the substrate design towards a CBS dedicated to LAN. First, high performance LANs (Myrinet, SCI, Ethernet 1 Gigabits) allow to use high performance protocols (BIP, PM, Unet, Active messages, Fast Messages, VIA) because the network hardware have several key properties (almost error free, packets stay ordered, homogeneous physical support). Second, the broker and server architectures are optimized for performance. In particular objects management, heterogeneity and security are not fundamental

P. Amestoy et al. (Eds.): Euro-Par'99, LNCS 1685, pp. 790–794, 1999.
© Springer-Verlag Berlin Heidelberg 1999

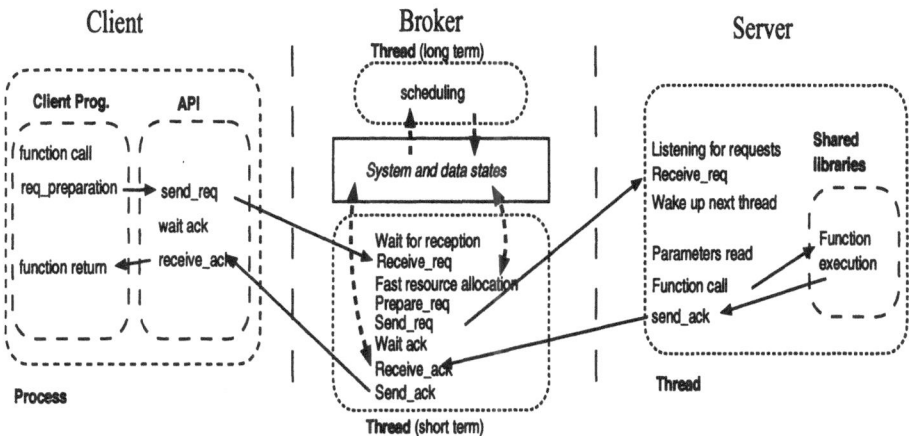

Fig. 1. Critical Path for Remote Executions.

issues for our substrate. The substrate follows a multi-thread oriented design. Servers execute only library function calls and users cannot directly launch processes on the server. Since server libraries are suppose to be fault free, there is no particular memory protection mechanism on the server. Third, the substrate uses a global naming system for the data sets on clients, broker and servers to limit the data transfers (between these entities) to the ones required by the program semantic. Function results and parameters are passed by references to avoid multiple exchange transfers between client and server when consecutive functions presents data-flow dependencies. Four client operations allow to manage data on the server side: **create, kill, put** and **get**. First operations create (destroy) a variable and return references that may be used to store and read data sets by client remote function calls. Last operations write (read) values to existing variables on the server side using references. When consecutive remote functions present dependencies, values are simply communicated between functions using references.

Clients interact with the substrate using a low level API. Users make requests to the substrate annotating their programs with directives. Directive annotations concerns the substrate operations and users functions to execute remotely. A preprocessor translates user directives in API function calls. The following program shows how annotations are used in a C program to request a remote execution of a matrix product.

```
float a[100];
//ovm create A,100   /* create a 100 entries array in the server side */
//ovm put A,a        /* send the value of the a array to A array */
//ovm req(bserv,dgemm,N,A,B,C) /* Requests remote execution of dgemm */
```

The client first requests to create an array A on the server side. A is a reference to a contiguous memory region (100 words) on the selected server. Then the value of the client local array a is transfered to the remote array A (*put* directive). Matrix element distribution (column first or row first) in memory regions is not relevant for the substrate which conserves the words order during the transfer. Finally, the client requests the remote execution of the **dgemm** function.

3   Performance Evaluations

The platform contains a Myrinet network. 2*400 Mhz Pentium II biprocessor nodes and 2 300 Mhz Pentium II biprocessor nodes are used for the experiments. The software environment includes Linux 2.0.36, BIP 0.95c [3] and Linux Pthread library. BIP raw performances on Myrinet is a latency of 5 us and a bandwidth of 1 Gbit/s. Every tests use 2*300 Mhz PII and 1*400 Mhz PII. Client and server use 300 Mhz nodes and the broker runs on a 400 Mhz node.

We measure the bandwidth and the latency for data transfers between a client and a server through the broker. The protocols used for the **Put** and **Get** transfers are implemented on top of BIP low level communication operations. **Put** operations use two protocols according to the message size. The threshold for protocol change is 240 Bytes. Figure 2 presents the communication bandwidths for both transfer types (**Put** and **Get**) and the communication bandwidth obtained with the BIP library between two nodes.

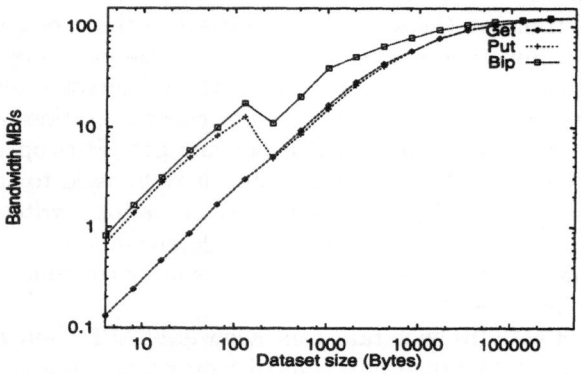

Fig. 2. Communication Bandwidth for BIP Library, **Put** and **Get** Operations

Put and **Get** operations reach the bandwidth of BIP for large data transfers ($> 1MB$). Client **Get** operations require $32\mu s$ to get small data sets from servers. Client **Put** operations have a $6\mu s$ inter-sending delay.

Client request latencies are shown in table 1.

Performance results do not consider the servers management algorithm executed by the broker. The client part of remote execution is about 40 times higher than a client local void function call.

Figure 3 presents the execution time seen by the client program for the matrix product routine (**dgemm**) of the BLAS library. We compare the local uniprocessor, the remote uniprocessor and the remote biprocessor execution times. These timings do not take into account the matrix transfers.

asynchronous remote void execution + **get**	50 μs
synchronous remote void execution	32 μs
client part of remote execution	7 μs
client local void function call	0,16 μs

Table 1. Execution Delays for Main Remote Operations of the CBS Substrate

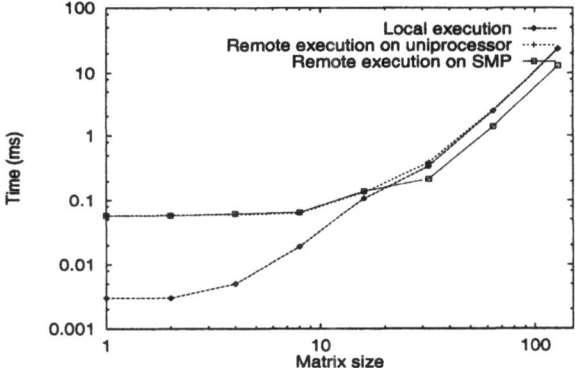

Fig. 3. Execution Times of the Matrix Products for Several Configurations.

Remote uniprocessor execution time equals the local uniprocessor execution time from 32 * 32 matrix. For higher matrix size, the parallel biprocessor implementation can be used efficiently.

Table 2 presents the speed-up of several implementations of the CG NAS NPB 2.3 serial benchmark (class W). The conj_grad function is executed remotely. We measure sequential as well as multi-threaded (2 threads) versions.

Executions	Remote uniprocessor	Local biprocessor	Remote biprocessor
Speed-up	0.96	1.34	1.32

Table 2. Speed-up from Local Uniprocessor of Several Remote Executions for the CG Class W Serial Benchmark

Remote executions of serial and multi-threaded versions reach respectively 96% and 98% the performance of local executions. The data transfers surrounding the remote execution of the conj_grad function are main limiting factors.

4 Conclusion

The CBS substrate has been designed in the perspective of very low remote execution overhead inside a LAN. Results section shows that remote execution overhead is lower than $50us$ and very short remote executions ($400us$) may be effective if they need few data transfers. By lowering the granularity of relevant remote executions, the CBS substrate enlarges the number of potential functions that are worthwhile to execute remotely and the applications domain.

References

[1] Henri Casanova et Al. NetSolve: A network-enabled server for solving computational science problems. *The International Journal of Supercomputer Applications and High Performance Computing*, 11(3):212–223, Fall 1997.

[2] A. Takefusa et Al. Multi-client LAN/WAN performance analysis of Ninf: A high performance global computing system. *SC'97; High Performance Networking and Computing: Proceedings of the Conference*, November 15–21, 1997.

[3] L. Prylli et Al. Bip: a new protocol designed for high performance networking on myrinet. *Workshop on Personal Computers based Networks Of Workstations*, 1998.

Universal Constructs in Distributed Computations

Ajay D. Kshemkalyani[1] and Mukesh Singhal[2]

[1] Dept. of EECS, University of Illinois at Chicago, Chicago, IL 60607-7053, USA
ajayk@eecs.uic.edu
[2] Dept. of CIS, The Ohio State University, Columbus, OH 43210, USA
singhal@cis.ohio-state.edu

Abstract. This paper identifies two classes of communication patterns that occur in distributed computations and explores their properties. It first examines local patterns, primarily *IO* and *OI intervals*, that occur at nodes in distributed computations. These local patterns form building blocks that are then used to define the global patterns, termed *segments* and *paths*, that occur across nodes in distributed computations. By controlling the predicates on the local patterns used to define segments and paths, various types of segments and paths can be defined. A number of key concepts and structures characterizing distributed computations are special cases of and are expressed in terms of the patterns identified.

1 Introduction

Analyzing the structure of a distributed computation helps to understand the concurrency and leads to a better design of distributed applications, algorithms, and systems. To this end, this paper identifies two classes of communication patterns that occur in every distributed computation and examines their properties. The first class of patterns consists of local patterns or intervals, primarily *IO* and *OI intervals*, that occur at processes [6]. These local patterns are specified in terms of messages received and messages sent by a process, and are distinguished by the order in which a pair of messages is sent and/or received by a process. Domain-specific predicates can be defined on how the interval at one process is related to the interval at another process. The use of such predicates on intervals at different processes allows intervals to be used as building blocks to formulate the second class of patterns, which is comprised of two global patterns, termed *segments* and *paths*. These global patterns occur across processes in a distributed computation and signify the flow of information and coupling among the events at different processes. Segments and paths generalize causal chains. While a causal chain only captures the causal relation, certain other message sequences also play a significant role in the analysis of a distributed computation. This paper generalizes segments and paths identified in [6]. By controlling the predicates on the intervals used to define segments and paths, different types of segments and paths can be defined.

Several key concepts and structures characterizing distributed computations are special cases of and can be expressed using the identified patterns. These patterns are shown to be useful in areas such as synchronous and causally ordered

P. Amestoy et al. (Eds.): Euro-Par'99, LNCS 1685, pp. 795–805, 1999.
© Springer-Verlag Berlin Heidelberg 1999

communication [4], transfer of knowledge [3], concurrency measures [5], necessary and sufficient conditions for a consistent global state [2, 9] which is useful in checkpointing and recovery [1], and distributed deadlock detection [6].

Section 2 gives the system model. Section 3 defines the local patterns and gives their properties. Section 4 defines the global patterns that occur across nodes and shows their applications. Section 5 concludes. The full paper is in [7].

2   System Model

The system is a network of N nodes (sites) with a logical channel between each pair of nodes. The nodes communicate by passing messages over the logical channels and do not share memory. We assume without loss of generality that each node in the system has one process running on it. Hence, nodes are synonymous to processes. Process execution and message transfers are asynchronous. Messages are delivered reliably but not necessarily in the order sent.

The execution of a computation at a node is modeled by three types of events: message send events, message receive events, and internal events. Let s_i^x and r_j^x denote the send and the receive events at which the message with label x is sent at node i and received at node j, respectively. The superscript and/or subscript will be omitted when it is not important. Let $dest(s_i^x)$ denote the destination of the message sent at s_i^x. An execution of a distributed computation associates with each node i a totally ordered set C_i of events. Let $C = \bigcup C_i$ be the possibly infinite set of all events. The state of a node is defined by the values of the variables associated with its computation, which are a function of the history of events executed by it at any time. A distributed computation is represented by the poset $(C, \prec)$, where $\prec$ is the causality relation on C [8].

3   Local Communication Patterns: Intervals at a Node

This section formalizes the local communication patterns that occur at nodes. The next section shows how these patterns are used as building blocks to formulate two global patterns that occur across nodes.

At the time a node i sends a message at s_i, an "outward dependency" gets established at i. At the time a node i receives a message at r_i, an "inward dependency" gets established at i. An *interval* at a node is the period between the times that two such dependencies get established [6]. There are two main types of intervals, shown in Fig. 1 (a) and (b), based on whether the inward dependency is established before the outward dependency or vice-versa. The former interval is an *IO interval* and the latter is an *OI interval*. Analogous to IO and OI intervals, *II intervals* and *OO intervals* can also be defined.

An *interval begins* at node i whenever one of the following two events occurs (see Fig. 1): (a) i receives a message from some node j, or (b) i sends a message to some node k. For the two cases (a) and (b) in which an *interval* begins, the IO or OI *interval completes* when, in case (a), i sends a message to some node k, and in case (b), i receives a message from some node j, respectively. Similar explanations hold for II and OO intervals.

The formation of an interval at a node signifies the participation of the node in global communication patterns that span across nodes. Note that intervals

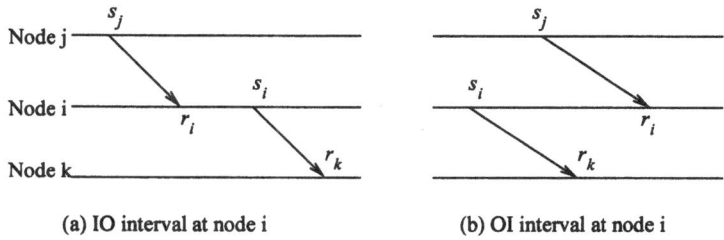

(a) IO interval at node i   (b) OI interval at node i

Fig. 1. IO and OI intervals [6].

at a node can overlap. For example, in Fig. 2, the following pairs of intervals overlap at node 3: (i) the OI interval between s_3^3 and r_3^2, the OI interval between s_3^3 and r_3^5; (ii) the IO interval between r_3^2 and s_3^6, the IO interval between r_3^5 and s_3^6; and (iii) the OI interval between s_3^3 and r_3^5, the IO interval between r_3^2 and s_3^6. There are numerous intervals at each process in the computation.

Each node has a set of application-specific semantic-defined "distinguished" events that are identified by monotonically increasing functions such as the sequence number of the events at the node. An example of a distinguished event is a checkpointing of the local state of a node. The time span from the xth to the $(x+1)$th distinguished event at node i is called the xth duration at i. IO or OI intervals of interest to an application are those that satisfy a certain application-specific relationship on the durations in which the send and receive events identifying the interval occur. Each duration x at node i is associated with a predicate $\Phi_{i,x}$ which is true only during that duration. The duration at node i in which an event e_i occurs is denoted by $D(e_i)$. A send event s_i and a receive event r_i can be related at a node i in one of the following ways:

1. $D(s_i) - D(r_i) = 0$. Events s_i and r_i belong to the same duration and identify an IO or an OI interval, based on whether $r_i \prec s_i$, or vice-versa, resp..
2. $D(s_i) - D(r_i) > 0$. In this case, events s_i and r_i identify an IO interval.
3. $D(s_i) - D(r_i) < 0$. In this case, events s_i and r_i identify an OI interval.

The semantics attached to an IO or OI interval can be of the following types (classification of global communication patterns uses these semantics):

1. No semantics is attached to the events of an OI or an IO interval. No conditions are imposed on the relation between $D(s_i)$ and $D(r_i)$.
2. The s_i and r_i events of an interval satisfy constraints on $D(s_i)$ and $D(r_i)$, the local durations to which they belong. For example, events of an IO interval are in two different local durations, but the events of an OI interval are in the same duration.
3. The events of an interval satisfy certain constraints on the durations they belong to and on the durations of the events in their causal past.

4   Global Communication Patterns: Paths and Segments

This section defines global patterns that span nodes in a computation. It then shows that several key concepts and structures characterizing distributed computations are special cases of and can be expressed using these patterns.

Domain-specific predicates can be defined on how the IO or OI interval at one node is related to the IO or OI interval at another node. The use of such predicates on IO and OI intervals at different nodes allows IO and OI intervals to be used as building blocks to formulate the global patterns: segments and paths. These global patterns occur across different nodes and signify a sequence of message exchanges such that any two adjacent messages in the sequence are related at a node by an IO or an OI interval. By controlling the predicates (or conditions) on the IO and OI intervals used to define segments and paths, different types of segments and paths can be defined.

Based on the semantics attached to the events identifying IO and OI intervals, three versions of segments and paths are presented. The first version (Sect. 4.1) is for a general computation where no restrictions are imposed and any s_i and any r_i events at a node i participate in OI and IO intervals. This version has applications in characterizing distributed computations by identifying structures like a crown, deriving concurrency measures, and analyzing knowledge transfer. In the second version (Sect. 4.2), distinguished events are assigned values of a monotonically nondecreasing function. This version has applications in characterizing global checkpoints. In the last version (Sect. 4.3), the distinguished events signify participation in a stable property. This version has applications in characterizing stable properties like distributed deadlocks.

Before defining the global patterns, we introduce some primitive predicates (conditions) on a distributed computation. The global patterns for various semantic models are defined using these conditions. In a computation, at any instant, there could have existed a sequence $\langle s_{i_1}, s_{i_2}, \ldots, s_{i_n} \rangle$ of send events on nodes $i_j \in \{i_1, i_2, \ldots, i_n\}$ satisfying a combination of the following conditions (henceforth, $i_j \in \{i_1, i_2, \ldots, i_n\}$):

(C1) Convey predicate to successor: $D(s_{i_j}) = x_{i_j}$ and $dest(s_{i_j}) = i_{j+1}$, for $1 \leq j \leq n - 1$.

(C2) Predicate conveyed from predecessor: A node i_j (except for $j = 1$) has received the message sent by i_{j-1} at $s_{i_{j-1}}$ before s_{i_j}.

(C3) No local violation of predecessor's predicate: Each node i_j (except for $j = 1$) has not invalidated the predicate $\Phi_{i_{j-1}, x_{i_{j-1}}}$ at node i_{j-1}.

(C4) No violation of predecessor's predicate: A node i_j (except for $j = 1$) has not received any message, in the causal past of which i_{j-1}'s predicate $\Phi_{i_{j-1}, x_{i_{j-1}}}$ got invalidated.

(C5) Local predicate valid in duration: Each node i_j is in its x_{i_j}th duration and $\Phi_{i_j, x_{i_j}}$ is currently true.

(C6) Duration of send event does not occur before duration of receive event: $D(s_{i_j}) \geq D(r_{i_j})$.

4.1 Segments and Paths for General Computations

In a general computation, no semantics is attached to the events of an interval and no constraints are imposed on the relation between $D(s_i)$ and $D(r_i)$.

Definition 1. A "segment" for a general computation, denoted $S_g(s_{i_1}, r_{i_{n+1}})$, is a sequence of events $\langle s_{i_1}, s_{i_2}, \ldots, s_{i_n} \rangle$ satisfying (C1) $\bigwedge$ (C2).

Every event in a segment occurs at a node that has sent a message to the node at which the successor event in the segment occurs. (Henceforth, a reference to "a node on a segment/path" will mean "a node with an event on a segment/path".) Moreover, when a node i_j sends the message at s_{i_j} (as per (C1)), the message sent at the previous event $s_{i_{j-1}}$ in the sequence has been received (as per (C2)). Therefore, a segment denotes a sequence of nodes such that the dependencies on their successor nodes in the segment are created sequentially. That is, $\forall i_j{:}1 \le j < n \ ::\ s_{i_j} \preceq s_{i_{j+1}}$. A segment thus denotes a causal chain of messages in which the events signify completed IO intervals.

For a sequence of events $\langle s_{i_1}, s_{i_2}, \ldots, s_{i_n} \rangle$ such that $dest(s_{i_j}) = i_{j+1}$ for $1 \le i \le n-1$, it may happen that $\exists j{:}\ s_{i_j} \nprec s_{i_{j+1}}$, that is, node i_{j+1} has an OI interval. A *path* is defined next to capture such a sequence of events.

Definition 2. *A "path" for a general computation, denoted $P_g(s_{i_1}, r_{i_{n+1}})$, is a sequence of events $\langle s_{i_1}, s_{i_2}, \ldots, s_{i_n} \rangle$ satisfying (C1).*

The formation of an interval at a node signifies the participation of the node in a path or a segment. In a path, successive messages are related by either an IO or an OI interval. In a segment, successive messages are related only by IO intervals. Thus, the successive events in a sequence at which outward dependencies are established satisfy a weaker causal relationship in a path than in a segment. A path may contain several segments; a segment is always a path.

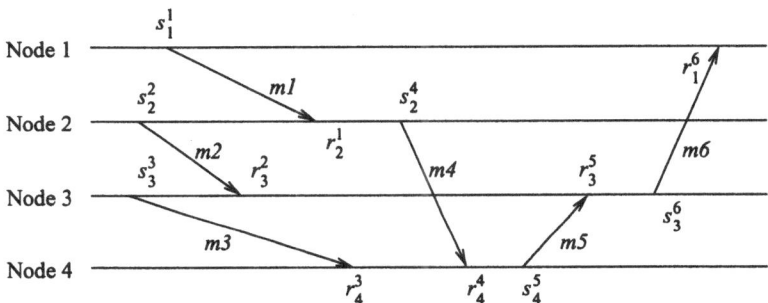

Fig. 2. An example computation.

Fig. 2 gives examples of paths and segments. Some segments are: $\langle s_1^1, s_2^4, s_4^5, s_3^6 \rangle$, $\langle s_2^2, s_3^6 \rangle$, $\langle s_3^3, s_4^5, s_3^6 \rangle$, and all subsequences of the above. By definition, each segment is a path. The following are some paths with at least one OI interval: $\langle s_1^1, s_2^2, s_3^3 \rangle$, $\langle s_1^1, s_2^1, s_3^6 \rangle$, $\langle s_2^2, s_3^3, s_4^5, s_3^6 \rangle$. Subsequences of these are also paths.

A *maximal path* is a path which cannot be extended by the addition of a send event at either end. For example, in Fig. 2, $\langle s_1^1, s_2^2, s_3^3, s_4^5, s_3^6 \rangle$ is a maximal path. The longest maximal path in the computation is $\langle s_2^2, s_3^3, s_4^5 s_3^6, s_1^1, s_2^4 \rangle$ which happens to consist of all the messages in the computation. A *maximal segment* is defined likewise. In Fig. 2, $\langle s_1^1, s_2^4, s_4^5, s_3^6 \rangle$ is a maximal segment.

Maximal segments and maximal paths are useful concepts in analyzing properties of a distributed computation. A maximal segment is a causal chain that signifies the maximum length of the serial execution of "thread of control" represented by the segment. On the other hand, a maximal path whose events are related by OI intervals at nodes provides a measure of the concurrency in a computation. The higher the number of OI intervals, the higher the concurrency in the computation. The ratio of the average of the sizes of maximal paths to the average of the sizes of maximal segments in a distributed computation is a good indicator of the concurrency in the computation [5].

We next show how segments and paths can be used to express some communication patterns that are important in analyzing distributed computations.

Realizable Synchronous Computations: The Crown Criterion. Charron-Bost et al. observed that a distributed algorithm designed to run correctly on asynchronous systems (called *A-computations*) may not run correctly on synchronous systems – an algorithm that runs on an asynchronous system may *deadlock* on a synchronous system [4].

A-computations that can be realized under synchronous communication are called *Realizable with Synchronous Communication* (RSC) computations. Formally, a computation C is RSC if there exists a non-separated linear extension of the poset $(C, \prec)$. A non-separated linear extension of $(C, \prec)$ is a linear extension of $(C, \prec)$ such that for each pair of send event s and corresponding receive event r, the interval $\{ x \in C \mid s \prec x \prec r \}$ is empty. [4] showed that RSC computations are a proper subset of causally ordered computations, which are a proper subset of FIFO computations.

Charron-Bost et al. [4] developed a criterion (called the *crown criterion*) to determine if an A-computation can be realized on a system with synchronous communication. This criterion uses a structure called *crown*, defined next.

Definition 3. *Let C be a computation. A crown of size k in C is a sequence $\langle (s^i, r^i), i \in \{ 0, ..., k\text{-}1 \} \rangle$ of pairs of corresponding send and receive events such that: $s^0 \prec r^1, s^1 \prec r^2, ... s^{k-2} \prec r^{k-1}, s^{k-1} \prec r^0$.*

Charron-Bost et al. [4] showed that a computation is RSC iff it contains no crown.

Fig. 3 shows a crown having six pairs of corresponding send and receive events (s^i, r^i), $i \in [0, 5]$. There is also a causal chain from s^i to $r^{(i+1) \bmod 6}$, for $i \in [0, 5]$. Defn. 3 specifies the constraints between s^i and $r^{(i+1) \bmod k}$, for $i \in [0, k-1]$. Each such constraint simply represents a segment $S_g(s^i, r^{(i+1) \bmod k})$. We next define crowns in terms of segments.

Definition 4. *In terms of segments, a crown of size k in a computation is a sequence $\langle (s^i, r^i), i \in \{ 0, ..., k\text{-}1 \} \rangle$ of pairs of corresponding send and receive events such that $\forall i \in [0, k-1], S_g(s^i, r^{(i+1) \bmod k})$. (To simply notation, the crown will also be expressed by just the conditions $\{ S_g(s^i, r^{(i+1) \bmod k}) : i \in [0, k-1] \}$.)*

Example 1 (Crown in Fig. 3): $CROWN = \{ S_g(s^i, r^{(i+1) \bmod k}) : i \in [0,5] \}$.
Refinement of Defn. 4: Defn. 4 expresses a crown of size k in terms of k segments. A crown of size k can generally be expressed in terms of less than k segments and paths.

A segment $S_g(s^i, r^j)$ such that events s^i and r^j lie on the same node is called a *local* segment. Note that in Fig. 3, (i) segments $S_g(s^2, r^3)$ and $S_g(s^3, r^4)$ are

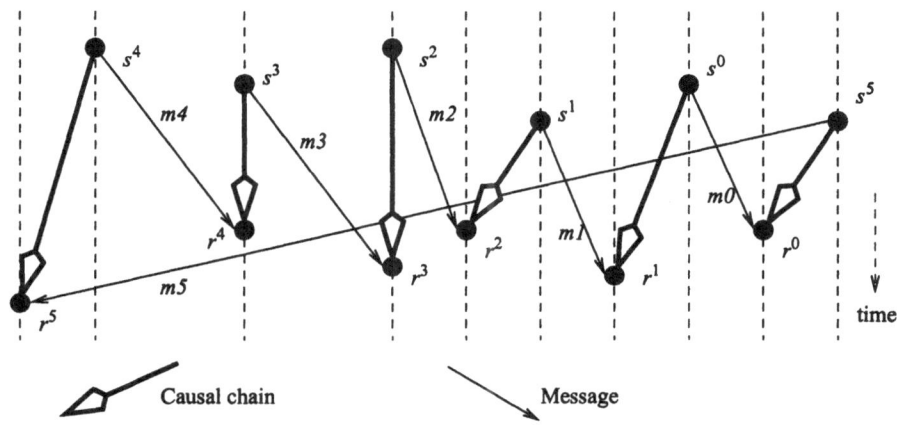

Fig. 3. A crown of size 6.

local segments and (ii) these two segments are connected by message (s^3, r^3). In this situation, segments $S_g(s^2, r^3)$ and $S_g(s^3, r^4)$ can be represented by path $\langle s^4, s^3, s^2 \rangle$. Consequently, the conditions represented by segments $S_g(s^2, r^3)$ and $S_g(s^3, r^4)$ in the expression of the crown can be equivalently stated in terms of path $\langle s^4, s^3, s^2 \rangle$ (which happens to contain only OI intervals). Given a crown, we present an algorithm that replaces clusters of local segments connected by messages, by equivalent paths. This algorithm compacts consecutive segments into paths wherever possible. (Such paths consist of OI intervals only and a cyclic path with OI intervals only is always a crown.)

A Crown-Compaction Algorithm:

1. $CR\_ALT = CROWN$;
2. Identify each maximal sequence of consecutive integers, modulo k, from x to y satisfying $\forall\, j \in [x, (y) mod\ k]$, s^j and $r^{(j+1) mod\ k}$ occur on the same node. For each such sequence, do the following:
   (a) $CR\_ALT = CR\_ALT \setminus \{S_g(s^i, r^{(i+1) mod\ k}) : i \in [x, (y) mod\ k]\}$
   (b) $CR\_ALT = CR\_ALT \bigcup \{\langle s^{(y+1) mod\ k}, s^y, s^{(y-1) mod\ k}, \ldots \ldots s^{(x+1) mod\ k}, s^x \rangle\}$

Example 1 (contd.): In the crown in Fig. 3, s^2 and r^3 lie on the same node, and s^3 and r^4 lie on the same node. As there is a range of consecutive integers $[x, y] = [2, 3]$ such that $\forall i \in [2, 3]$, s^i and $r^{(i+1) mod\ k}$ lie on the same node, segments $S_g(s^2, r^3)$ and $S_g(s^3, r^4)$ can be replaced by path $\langle s^4, s^3, s^2 \rangle$. Hence, $CR\_ALT = \{S_g(s^i, r^{(i+1) mod\ k}) : i \in \{0, 1, 4, 5\}\} \bigcup \{\langle s^4, s^3, s^2 \rangle\}$.

Thus, a crown which is an example of various structures in distributed computations can be expressed more compactly in terms of paths and segments.

Knowledge Transfer. Knowledge in distributed systems refers to the states of the nodes and is defined as temporal and spatial predicates over the variables of the nodes. Knowledge plays a significant role in the evaluation of global predicates, debugging, monitoring, establishing breakpoints, evaluating triggers, industrial process control, and controlling a distributed execution [11].

Knowledge is transferred among nodes through send and receive events [3]; the extent of knowledge dissemination is determined by the message communication pattern among nodes and is identified by the causality relation between events. A segment from event e_i to event e_j signifies the flow of knowledge of node i's state preceding event e_i to all the events following e_j. In Fig. 2, messages forming segment $\langle s_1^1, s_2^4, s_4^5 \rangle$ transfer the knowledge about the local state of node 1 just before event s_1^1 to event r_3^5. A path with an OI interval denotes a disrupted transfer of knowledge among the nodes along the path. In Fig. 2, knowledge about the local state of node 1 just before event s_1^1 is not transferred to event r_3^2. The knowledge transfer is disrupted at node 2 due to the OI interval formed by s_2^2 and r_2^1. Thus, paths and segments are useful tools to identify the extent of knowledge transfer.

4.2   Segments and Paths for Monotonically Nondecr. Functions

In distributed computations with monotonically nondecreasing functions at nodes (e.g., the local clock time at the occurrence of an event [8]), the distinguished events at a node are associated with monotonically nondecreasing values.

The definition of a segment for a monotonically nondecreasing function (Defn. 5) is the same as for a general function (Defn. 1). The definition of a path for a monotonically nondecreasing function (Defn. 6) differs from the corresponding Defn. 2 in that the events of an OI interval must belong to the same duration.

Definition 5. *A "segment" for a monotonically nondecreasing computation, denoted* $S_m(s_{i_1}, r_{i_{n+1}})$, *is a sequence of events* $\langle s_{i_1}, s_{i_2}, \ldots, s_{i_n} \rangle$ *satisfying (C1)* $\bigwedge$ *(C2).*

Definition 6. *A "path" for a monotonically nondecreasing function, denoted* $P_m(s_{i_1}, r_{i_{n+1}})$, *is a sequence of events* $\langle s_{i_1}, s_{i_2}, \ldots, s_{i_n} \rangle$ *satisfying (C1)* $\bigwedge$ *(C6).*

A *closed path* for a monotonically nondecreasing function is a path $P_m(s_{i_1}, r_{i_{n+1}})$ such that events s_{i_1} and $r_{i_{n+1}}$ occur at the same node (i.e., $i_1 = i_{n+1}$).

Necessary and Sufficient Conditions for a Global Snapshot: Zigzag Paths. Checkpointing is used in fault-tolerant computing [1], and parallel and distributed debugging [11]. Each node can take local checkpoints asynchronously; a consistent global checkpoint is constructed by chosing a local checkpoint from each node. Checkpoints are the "distinguished events" which demarcate consecutive durations at nodes. The xth duration (or xth *checkpoint interval*) at a node denotes the computation from its xth to its $x + 1$th checkpoint.

An important problem is to determine if an arbitrary set of local checkpoints belongs to a consistent global checkpoint [2]. Netzer and Xu used the zigzag path, a generalization of Lamport's causality relation [8], and showed that two local checkpoints cannot lie on a consistent global checkpoint iff a zigzag path exists between the checkpoints [9]. Let $C_{i,x}$ denote the xth local checkpoint at node i and let $e_{i,x}$ denote the event of taking $C_{i,x}$.

Definition 7. *A zigzag path exists from $C_{i,x}$ to $C_{j,y}$ iff there are messages m_1, m_2, ..., m_n $(n > 1)$ such that*

1. *m_1 is sent by node i after $C_{i,x}$*
2. *if m_k $(1 \leq k < n)$ is received at node r, then m_{k+1} is sent by r in the same or a later checkpoint interval*
3. *m_n is received by process j before $C_{j,y}$.*

In Fig. 4, $m1$, $m2$, and $m3$ form a zigzag path from checkpoint C_{11} at node 1 to checkpoint C_{32} at node 3. Likewise, $m4$, $m5$, and $m6$ form a zigzag path from checkpoint C_{12} at node 1 to checkpoint C_{42} at node 4. Note from Defn. 7 that a zigzag path is a chain of messages that are connected by OI or IO intervals at nodes. Thus, a zigzag path is nothing but a "path" (Defn. 6) and can be expressed using paths as follows:

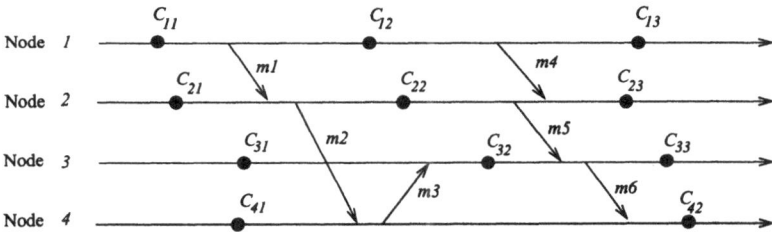

Fig. 4. Zigzag paths.

Definition 8. *A zigzag path exists from $C_{i,x}$ to $C_{j,y}$ iff $\exists\, P_m = \langle s_{i_1}, s_{i_2}, \ldots, s_{i_n} \rangle$ such that (i) $e_{i,x} \prec s_{i_1}$, and (ii) a message sent at s_{i_n} to j is received before $e_{j,y}$.*

A checkpoint is on a Z-cycle iff there is a zigzag path from the checkpoint to itself. In Fig. 4, checkpoint C_{32} lies on a Z-cycle consisting of messages $m6$ and $m3$. A checkpoint can be part of a consistent snapshot iff it is not involved in a Z-cycle. Observe that a Z-cycle is nothing but a closed path.

4.3   Segments and Paths for Stable Properties

A stable property is a property of the system state such that once it becomes true, it continues to hold unless there is external intervention [10]. Examples of such properties are deadlocks, termination of a computation, etc. In this section, segments and paths are defined for stable properties, with special emphasis on deadlocks [6]. A similar approach can be used for other stable properties.

Conditions for Deadlocks. We consider deadlocks in the request-reply model. In this model, a process sends a request and blocks until it receives a reply to its request. "Distinguished" events at a node are the events at which a node sends a request and blocks waiting for a reply. The predicate $\Phi_{i,x}$ stands for "node i is blocked on the request it sent at its xth distinguished event". This predicate becomes true at the start of the duration between two distinguished events and becomes false on the receipt of the reply at some time before the next distinguished event. In this context, a segment and a path are defined next [6].

Definition 9. *A "segment" in the request-reply model, denoted $S_d(s_{i_1}, r_{i_{n+1}})$, is a sequence of events $\langle s_{i_1}, s_{i_2}, \ldots, s_{i_n} \rangle$ satisfying the following conditions :*
(I) (C1) $\wedge$ (C2) $\wedge$ (C3) $\wedge$ (C4).   /* conditions on distinguished events. */
(II) (C5).   /* conditions when the system is observed. */

Definition 10. *A "path" in the request-reply model, denoted $P_d(s_{i_1}, r_{i_{n+1}})$, is a sequence of events $\langle s_{i_1}, s_{i_2}, \ldots, s_{i_n} \rangle$ satisfying the following conditions:*
(I) (C1) $\wedge$ ((C2) $\Longrightarrow$ ((C3) $\wedge$(C4))). /* conditions on distinguished events. */
(II) (C5).   /* conditions when the system is observed. */

Condition (C2) indicates that the request sent at $s_{i_{j-1}}$ has been received before s_{i_j}. Condition (C3) indicates that i_j has not sent back a reply to i_{j-1} and thus has not invalidated $\Phi_{i_{j-1}, x_{i_{j-1}}}$. By Condition (C4), i_j has not received a message indicating that i_{j-1} got unblocked, i.e., $\Phi_{i_{j-1}, x_{i_{j-1}}}$ got invalidated. As no node in a distributed system has instantaneous knowledge of the entire system, while declaring a segment/path, it must be ensured that the nodes that are believed to be blocked are still blocked; Condition (C5) asserts this. A detailed explanation of Defns. 9 and 10 is given in [6].

Paths in which each node is blocked waiting for a reply from its successor and the last node never receives a reply denote deadlocks.

Definition 11. *A closed path is a path $P_d(s_{i_1}, r_{i_{n+1}})$ such that events s_{i_1} and $r_{i_{n+1}}$ occur at the same node (i.e., $i_1 = i_{n+1}$).*

A closed path denotes a deadlock because no node with an event on the closed path will ever receive a reply and get unblocked. A closed path has at least one OI interval, i.e., at least two segments. Condition (C5) helps to ensure that false deadlocks are not detected.

5   Conclusion

We identified two classes of universal communication patterns in distributed computations. IO, OI, II and OO intervals are local patterns that occur at nodes, whereas paths and segments are global patterns which occur across nodes in a distributed computation and are defined in terms of the local patterns. It is seen that the communication patterns identified are universal to all distributed computations. We showed that a number of key concepts and structures characterizing distributed computations are special cases of the proposed patterns and can be expressed using these patterns. By controlling the predicates on local patterns used to define segments and paths, different types of segments and paths can be defined to address the needs of other applications also.

References

[1] B. Bhargava, S.R. Lian, Checkpointing and rollback recovery in distributed systems – an optimistic approach, *Proc. 7th IEEE SRDS*, 3-12, Oct. 1988.

[2] K. M. Chandy, L. Lamport, Distributed snapshots: Global states of a distributed system, *ACM Trans. Comput. Systems*, 3(1):63-75, 1985.

[3] K. M. Chandy, J. Misra, How processes learn, *Distributed Computing*, 1, 40-52, 1986.

[4] B. Charron-Bost, F. Mattern, G. Tel, Synchronous, asynchronous, and causally ordered communication, *Distributed Computing*, 9(4):173-191, 1996.

[5] C. J. Fidge, A simple run-time concurrency measure, In: T. Bossomaier et al. (Eds.), *The Transputer in Australasia (ATOUG-3)*, 92-41, IOS Press, 1990.

[6] A. D. Kshemkalyani, M. Singhal, On characterization and correctness of distributed deadlock detection, *Journal of Parallel and Distributed Computing*, 22(1), 44-59, July 1994. (Tech. Rep. TR-06/90-TR15, Ohio State Univ., 1990.)

[7] A. D. Kshemkalyani, M. Singhal, Universal constructs in distributed computations, *Technical Report 29.2136*, IBM, March 1996.

[8] L. Lamport, Time, clocks, and the ordering of events in a distributed system, *Communications of the ACM*, 21(7):558-565, July 1978.

[9] R. Netzer, J. Xu, Necessary and sufficient conditions for consistent global snapshots, *IEEE Trans. on Parallel and Distributed Systems*, 6(2):165-169, 1995.

[10] A. Schiper, A. Sandoz, Strong stable properties in distributed systems, *Distributed Computing*, 8:93-103, 1994.

[11] M. Spezialetti, R. Gupta, Debugging distributed programs through the detection of simultaneous events, *Proc. 14th IEEE ICDCS*, 634-641, June 1994.

Illustrating the Use of Vector Clocks in Property Detection: An Example and a Counter-Example

Michel Raynal

IRISA, Campus de Beaulieu, 5042 Rennes Cedex, France
raynal@irisa.fr

1 Introduction

Logical (scalar, vector or matrix) clocks are a powerful mechanism used by a lot of distributed algorithms. This paper is on vector clocks: it surveys their main features and is particularly focused on their power and their limitation. In that sense, this paper complements [5, 6] and may be seen as a critical and practical introduction to vector clocks.

The paper is divided into four main sections. Section 2 introduces a model for distributed executions. Section 3 describes vector clocks and their basic properties. Vector clocks are a simple mechanism that allows processes to track causality between the events they produce. Then, Section 4 and Section 5 study two problems related to causality. The first problem consists in detecting a conjunction of stable local predicates. The second problem consists in recognizing the occurrence of a very simple event pattern. It is shown that simple vector clocks are insufficient to solve the second problem which, actually, requires more sophisticated clocks, namely, vector of vector clocks. So, this paper exhibits a frontier between problems that can be solved using simple vector clocks and problems requiring more sophisticated vector clock systems.

2 Distributed Computations

2.1 Partially Ordered Set of Events

Distributed Computations A distributed program is made up of n sequential local programs which can communicate and synchronize only by exchanging messages. The execution of a local program gives rise to a sequential process. Let $P_1, \ldots, P_n$ be this finite set of processes. We assume that, at run-time, each ordered pair of communicating processes (P_i, P_j) is connected by a reliable channel. Message transmission delays are finite but unpredictable, process speeds are positive but arbitrary: the underlying computation model is asynchronous.
Partial Order of Events Execution of an internal/send/receive statement produces a corresponding internal/send/receive event. Let e_i^x be the x-th event produced by process P_i. The sequence $h_i = e_i^1 e_i^2 \ldots e_i^x \ldots$ constitutes the history of P_i. Let H be the set of events produced by a distributed computation. This set is structured as a partial order by Lamport's causal precedence relation [3], denoted "$\to$" and defined as follows:

P. Amestoy et al. (Eds.): Euro-Par'99, LNCS 1685, pp. 806–814, 1999.
© Springer-Verlag Berlin Heidelberg 1999

$$(i = j \wedge x \leq y) \text{ (local precedence)} \quad \vee$$
$$e_i^x \to e_j^y \Leftrightarrow (\exists m \; : e_i^x = send(m) \quad \wedge \; e_j^y = receive(m)) \text{ (msg prec.)} \quad \vee$$
$$(\exists \; e_k^z \; : \; e_i^x \to e_k^z \wedge e \frac{z}{k} \to e_j^y) \quad \text{(transitive closure)}.$$

The partial order $\widehat{H} = (H, \to)$ constitutes a model of the distributed computation it is associated with. Figure 1 depicts a distributed computation where events are denoted by black points. Two events e and f are *concurrent* (or *causally independent*) if $\neg(e \to f) \wedge \neg(f \to e)$. The *causal past* of event e is the partially ordered set of events f such that $f \to e$. Similarly, the *causal future* of event e is the partially ordered set of events f such that $e \to f$.

2.2 Partially Ordered Set of Local States

Local States Let σ_i^0 be the initial state of process P_i. Event e_i^x entails P_i's local state change from σ_i^{x-1} to σ_i^x: σ_i^x is the local state of P_i resulting from its partial history $h_i^x = e_i^1 \ldots e_i^x$. We say that e_i^y *belongs to* σ_i^x (denoted $e_i^y \in \sigma_i^x$) if $y \leq x$. In Figure 1 local states are represented by rectangular boxes.

Partial Order of Local States Let Σ be the set of all local states associated with a distributed computation $\widehat{H}$. Lamport's precedence relation can be extended to local states in the following way:

$$\boxed{(\sigma_i^x \to \sigma_j^y) \Leftrightarrow (e_i^{x+1} \to e_j^y) \qquad (SE)}$$

Local states not related by "$\to$" are said to be *concurrent*, (denoted $\|$). More formally: $(\sigma_i^x \| \sigma_j^y) \Leftrightarrow (\neg(\sigma_i^x \to \sigma_j^y) \wedge \neg(\sigma_j^y \to \sigma_i^x))$. In Figure 1, we have $\sigma_3^1 \to \sigma_1^4$ and $\sigma_1^3 \| \sigma_3^5$. As for events, the *causal past* of a local state σ_i^x is the partially ordered set of local states σ_k^z such that $\sigma_k^z \to \sigma_i^x$.

Consistent Global States A global state Σ is a set of n local states $(\sigma_1, \cdots, \sigma_n)$ one from each process. It is consistent if $\forall i \neq j$ we have $\sigma_i \| \sigma_j$. This means that a global state $\Sigma = (\sigma_1, \cdots, \sigma_n)$ is consistent if, for any pair of its local states (σ_i, σ_j), there is no message m such that $receive(m) \in \sigma_i \wedge send(m) \notin \sigma_j$. If such a message m exists, it is called *orphan* with respect to the pair (σ_i, σ_j). Let us again consider Figure 1. $\Sigma_1 = (\sigma_1^3, \sigma_2^4, \sigma_3^2)$ is consistent, while $\Sigma_2 = (\sigma_1^5, \sigma_2^6, \sigma_3^4)$ is not consistent. There is no orphan message with respect to Σ_1, while m_4 is orphan with respect to the pair $(\sigma_3^4, \sigma_2^6) \subset \Sigma_2$. Actually, m_4 creates the dependency $\sigma_3^4 \to \sigma_2^6$ which makes Σ_2 inconsistent.

3 Vector Clocks

As a concept, with the associated theory, vector clocks have been introduced in 1988, simultaneously and independently by Fidge [1] and by Mattern [4].

A Causality Tracking Timestamping Mechanism A vector clock system is a mechanism that associates timestamps with events (local states) in such a way that the comparison of their timestamps indicates whether the corresponding events (local states) are or not causally related (and, if they are, which one is the first). This timestamping system is implemented in the following way. Each process P_i has a vector of integers $VC_i[1..n]$ and:

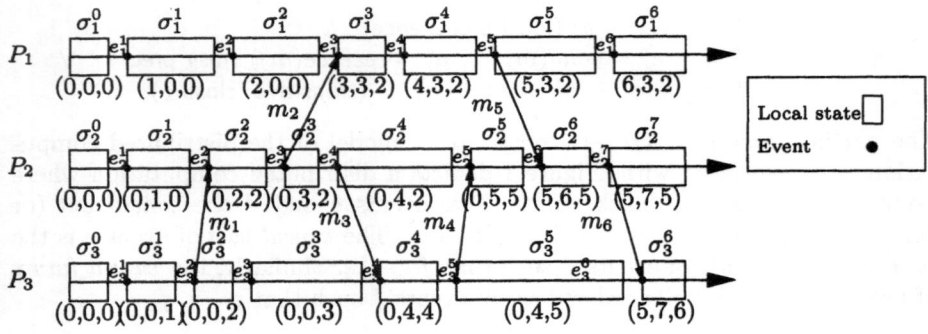

Fig. 1. Local States of a Distributed Computation

R1 Each time it produces an event, P_i increments its vector clock entry $VC_i[i]$ $(VC_i[i] := VC_i[i] + 1)$ to indicate it has progressed.

R2 When a process P_i sends a message m, it attaches to it the current value of VC_i. Let $m.VC$ denote this value.

R3 When P_i receives a message m, it updates its vector clock in the following way: $\forall x : VC_i[x] := max(VC_i[x], m.VC[x])$ (this operation is usually abbreviated as $VC_i := max(VC_i, m.VC)$).

Note that $VC_i[i]$ counts the number of events produced so far by P_i. Moreover, $VC_i[j]$ represents the number of events produced by P_j that belongs to the current causal past of P_i. When a process P_i produces an event e (and enters the associated local state σ) it associates with them a vector timestamp whose value is equal to the current value of VC_i, namely, $e.VC = \sigma.VC =$ current value of VC_i. Figure 1 shows vector timestamp values associated with events and local states. As an example we have: $e_4^2.VC = \sigma_4^2.VC = (0, 4, 2)$.

Let $e.VC$ and $f.VC$ be the vector timestamps associated with two distinct events e and f, respectively. The following property is the fundamental property associated with vector clocks [1, 2, 4][1]:

$$((e \to f) \Leftrightarrow ((\forall k : e.VC[k] \le f.VC[k]) \wedge (\exists k : e.VC[k] < f.VC[k])))$$

Let P_i be the process that has produced e. This additional information allows to simplify the previous relation that reduces to [1, 4]:

$$\boxed{(e \to f) \Leftrightarrow (e.VC[i] \le f.VC[i]) \qquad (R)}$$

Let $\sigma1$ and $\sigma2$ be local states of P_i and P_j, respectively. From the definition SE and the previous relations we get: $(\sigma1 \to \sigma2) \Leftrightarrow (\sigma1.VC[i] < \sigma2.VC[i])$.

Timestamps of Global States Relations between vector clocks and global states have been studied from a formal point of view by several authors [2, 4]. Here, we only consider some of these properties through a set of simple examples taken from Figure 1.

---

[1] $(\forall k : e.VC[k] \le f.VC[k]) \wedge (\exists k : e.VC[k] < f.VC[k])$ is denoted $e.VC < f.VC$.

Vector clocks allow to associate timestamps to global states in a very easy way. First, let us consider a local state, e.g., σ_3^4. It is timestamped $(0, 4, 4)$ and this timestamp identifies the consistent global state $(\sigma_1^0, \sigma_2^4, \sigma_3^4)$ which is the first consistent global state to which σ_3^4 belongs. More generally, the timestamp of a consistent global state Σ is defined as the component-wise maximum of the timestamps of the local states that compose it. Let us consider $\Sigma = (\sigma_1^4, \sigma_2^4, \sigma_3^4)$. Its timestamp is the component-wise maximum (denoted max) of $(4, 3, 2)$, $(0, 4, 2)$ and $(0, 4, 4)$, i.e., $(4, 4, 4)$.

In the same way if we consider two (or more) consistent global states Σ_1 and Σ_2, timestamped $\Sigma_1.VC$ and $\Sigma_2.VC$, respectively, the global state defined by the timestamp $max(\Sigma_1.VC, \Sigma_2.VC)$ is the first consistent global state (called $max(\Sigma_1, \Sigma_2)$) that includes the causal past of both Σ_1 and Σ_2. As an example, let us consider $\Sigma_1 = (\sigma_1^0, \sigma_2^2, \sigma_3^3)$ timestamped $(0, 2, 3)$, and the global state $\Sigma_2 = (\sigma_1^4, \sigma_2^4, \sigma_3^2)$ timestamped $(4, 4, 2)$. The global state $\Sigma = max(\Sigma_1, \Sigma_2)$ is a consistent global state, namely $(\sigma_1^4, \sigma_2^4, \sigma_3^3)$, timestamped $(4, 4, 3)$, that occurs after Σ_1 and after Σ_2. (Actually, from a theoretic point of view, the set of consistent global states defines a lattice [2, 4, 6]).

4   Detection of a Conjunction of Stable Local Predicates

4.1   The Problem

A predicate is *local* to a process P_i if it is only on local variables of P_i. A local predicate LP_i is *stable* if, as soon as it becomes true, it remains true forever. The notation $\sigma_i \models LP_i$ will be used to indicate that P_i's local state σ_i satisfies the local predicate LP_i. Let LP_1, LP_2, ..., LP_n be n local predicates, one per process. A consistent global state $\Sigma = (\sigma_1, \cdots, \sigma_n)$ satisfies the global predicate $LP_1 \wedge LP_2 \cdots \wedge LP_n$ (denoted $\Sigma \models (\bigwedge_i LP_i)$) if $\bigwedge_i (\sigma_i \models LP_i)$.

The problem consists in detecting on-the-fly, and without using additional control messages, the first consistent global state Σ that satisfies a conjunction of stable local predicates.

4.2   How to Solve It

Let us consider Figures 2.a and 2.b. The fact that local predicate LP_i is satisfied, is indicated by thickening process P_i's axis. In both figures, the global state $\Sigma = (\sigma_1^{y_1}, \sigma_2^{y_2}, \sigma_3^{y_3})$ (timestamped (y_1, y_2, y_3)) is the first to satisfy the conjunction of stable local predicates. Let us first examine Figure 2.a.

- When P_1 receives m_1, it learns nothing new about predicate detection.
- When P_1 receives m_3 (*i.e.*, when it enters local state $\sigma_1^{x_1}$), it can learn (by appropriately tracking causality with vector clocks) that there is a consistent global state, namely $(\sigma_1^0, \sigma_2^{x_2}, \sigma_3^0)$, that partially satisfies the global predicate $(\sigma_2^{x_2} \models LP_2)$.
- Message m_4 gives P_2 the knowledge that $\sigma_3^{x_3} \models LP_3$. So, (using vector clocks) m_5 can carry the information that the global state timestamped $(0, x_2, x_3)$ partially satisfies the global predicate (namely, $(\sigma_1^0, \sigma_2^{x_2}, \sigma_3^{x_3}) \models LP_2 \wedge LP_3$).

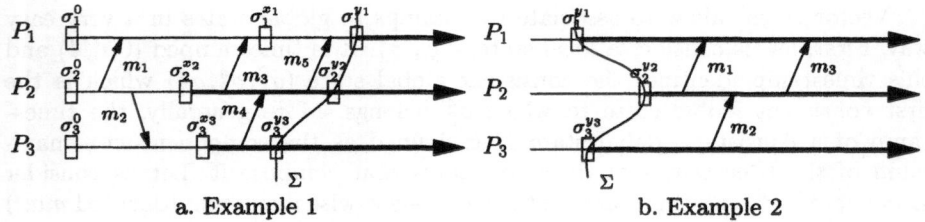

Fig. 2. Detection of $\bigwedge_i LP_i$

- So, when P_1 receives m_5 it learns this, and when LP_1 becomes true, P_1 can safely detects that Σ, timestamped (y_1, y_2, y_3), is the first consistent global state satisfying $\bigwedge_i LP_i$ ($\sigma_2^{y_2}$ and $\sigma_3^{y_3}$ being the local states that immediately follow the sendings of m_5 and m_4, respectively).

Let us now examine Figure 2.b. In that situation, due to flow of exchanged messages, P_1 can learn, when it receives m_3, that Σ, timestamped (y_1, y_2, y_3), is the first consistent global state satisfying the global predicate $LP_1 \wedge LP_2 \wedge LP_3$.

4.3   Data Structures

The previous discussion shows that we need two thinks to solve the problem: (1) Track causality to be able to identify *consistent* global states; and (2) Track which local predicates are *satisfied* and track the *first* consistent global state in which those local predicates are satisfied. Hence, each process P_i is endowed with the following data structures:

- $VC_i[1..n]$: a vector clock (initialized to $(0, 0, ..., 0)$).
- SAT_i : a set (initially empty) of process identifiers, with the following meaning: $j \in SAT_i \Leftrightarrow P_i$ knows P_j entered a local state from which LP_j is true.
- $FIRST_i$: a vector timestamp (initialized to $(0, 0, ..., 0)$) that defines the first consistent global state, known by P_i, in which all the predicates defined in SAT_i are satisfied. More precisely: $\forall j \ : (j \in SAT_i \Rightarrow (\sigma_j^{FIRST_i[j]} \models LP_j))$. Note that if $SAT_i = \{1, 2, \cdots, n\}$, then $FIRST_i$ defines the first consistent global state Σ that satisfies $\bigwedge_j LP_j$. In Figure 2.a, $FIRST_1 = VC_1 = (y_1, y_2, y_3)$ when LP_1 becomes true. In Figure 2.b, when P_1 receives m_3, it updates SAT_1 (the value of which becomes $\{1, 2, 3\}$) and $FIRST_1$ takes the value $max((y_1, y_2, 0), (0, y_2, y_3))$, i.e., $FIRST_1 = (y_1, y_2, y_3)$. In both figures, $(\sigma_1^{y_1}, \sigma_2^{y_2}, \sigma_3^{y_3}) \models (LP_1 \wedge LP_2 \wedge LP_3)$.

4.4   The Detection Protocol

The protocol executed by a process P_i is described in Figure 3. Let e and σ denote the last event produced by P_i and the corresponding local state (entered by P_i just after executing e). Their vector timestamps ($e.VC$ and $\sigma.VC$) have the same value which is equal the current value of VC_i. The protocol is composed of two procedures and three statements $S1$, $S2$ and $S3$ associated with the production by P_i of an internal event, a send event and a receive event, respectively. The local variable $done_i$ (initialized to $false$) is set to the value $true$ the first time LP_i is satisfied (then it remains true forever). The notation $VC := max(VC1, VC2)$ (statement S3) is an abbreviation for $\forall k \in 1..n : VC[k] := max(VC1[k], VC2[k])$. It defines a vector clock update. The protocol consists in:

---

procedure *detected?* **is**
    if $SAT_i = \{1, 2, \ldots, n\}$ **then** $FIRST_i$ defines the first consistent
                         global state Σ that satisfies $\bigwedge_j LP_j$ **fi**
procedure *check_LP_i* **is**
    if $(\sigma_i^x \models LP_i)$ **then** $SAT_i := SAT_i \cup \{i\}; FIRST_i := VC_i;$
                    $done_i := true;$ *detected?* **fi**

(S1) **when** P_i **produces an internal event** (e)
    $VC_i[i] := VC_i[i] + 1;$ execute e and move to $\sigma;$
    if $\neg done_i$ **then** *check_LP_i* **fi**

(S2) **when** P_i **produces a send event** (e=send m **to** P_j)
    $VC_i[i] := VC_i[i] + 1;$ move to $\sigma;$ **if** $\neg done_i$ **then** *check_LP_i* **fi**;
    $m.VC := VC_i;$ $m.SAT := SAT_i;$ $m.FIRST := FIRST_i;$
    send (m) **to** P_j % m carries $m.VC, m.SAT$ and $m.FIRST$ %

(S3) **when** P_i **produces a receive event** (e=receive (m))
    $VC_i[i] := VC_i[i] + 1;$ $VC_i := max(VC_i, m.VC);$
    move to $\sigma;$ % by delivering m to the process % **if** $\neg done_i$ **then** *check_LP_i* **fi**;
    if $\neg(m.SAT \subseteq SAT_i)$ **then** $SAT_i := SAT_i \cup m.SAT;$
                            $FIRST_i := max(FIRST_i, m.FIRST);$
                          $detected?$ **fi**

Fig. 3. Detection Protocol for $\bigwedge_i LP_i$

- The management of the vector clock VC_i (in $S1$, $S2$ and $S3$).
- The combined management of variables SAT_i and $FIRST_i$:
- When LP_i becomes true, VC_i defines the first consistent global state in which LP_i, plus the local predicates already in SAT_i, are satisfied. So $FIRST_i$ is defined as VC_i (procedure *check_LP_i*).
- When P_i sends a message m (Statement $S2$), it attaches to it the current values of SAT_i and $FIRST_i$. Those values are denoted $m.SAT$ and $m.FIRST$.
- When P_i receives a message m (Statement S3), it first executes the reception, moves to the new current state and tests if its local predicate becomes satisfied. Then, P_i checks if it learns something new about the satisfaction of local local predicates (by testing $\neg(m.SAT \subseteq SAT_i)$). If it learns something new, it computes the first consistent global state in which the local predicates of $m.SAT \cup SAT_i$ are satisfied. As indicated in the previous discussion, this global state is timestamped $max(FIRST_i, m.FIRST)$.

- When something new happens (from the point of view of local predicates detection), P_i tests the condition $SAT_i = \{1, 2, \cdots, n\}$, to know if the full global predicate, namely $\bigwedge_i LP_i$, is satisfied.

5 Limitation of Simple Vector Clocks

Vector clocks provide each process with a "counter-based view" of its causal past that displays limitations to solve some "causality"-related problems. We present here such a problem and its solution.

5.1 Recognition of a Simple Pattern

Let us consider a distributed execution that produces two types of internal events: *black* and *white* (communication events are tagged *white*). Given two blacks events s and t, the problem consists in deciding if there is another black event u such that $s \to u \ \wedge \ u \to t$? Let $black(e)$ be a predicate indicating if event e is black. More formally, given two events s and t, the problem consists in deciding if the following predicate $\mathcal{P}(s, t)$ is true:

$$\mathcal{P}(s, t) \equiv (black(s) \wedge black(t)) \wedge (\exists u \neq s, t : (black(u) \wedge (s \to u \wedge u \to t)))$$

To show that vector clocks do not allow to solve this problem let us consider Figures 4.a and 4.b. In these two executions, both event s have the same timestamp: $s.VC = (0, 0, 2)$. Similarly, both events t have also the same timestamp, namely, $t.VC = (3, 4, 2)$. But, as the reader can verify, execution (b) satisfies the pattern, while (a) does not.

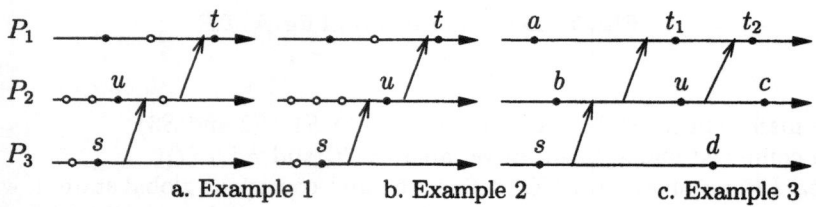

Fig. 4. Recognize a Pattern

5.2 How to Solve It

Observe that for the predicate $\mathcal{P}(s, t)$ to be true, there must be a black event in the causal past of t which has s in its causal past. This problem is related to causality, but two levels of predecessors appear in the predicate $\mathcal{P}$. Track "second order knowledge" on the past requires vector of vectors clocks.

Which Clocks? $P(s,t)$ can be decomposed into two predicates $P_1(s,u,t)$ and $P_2(s,u,t)$ in the following way: $P(s,t) \equiv (\exists u \; : \; P_1(s,u,t) \wedge P_2(s,u,t))$, where $P_1(s,u,t) \equiv (black(s) \wedge black(u) \wedge black(t))$ and $P_2(s,u,t) \equiv (s \to u \wedge u \to t)$.

P_1 indicates that only the black events are relevant for the predicate detection: so, only black events have to be tracked. This means vector clocks can be managed in the following way: (1) A process P_i increments $VC_i[i]$ only when it produces a black event; (2) The other statements associated with vector clocks are left unchanged.

Let us consider Figure 4.c, where only black events are indicated. We have $P(s,t_1) = false$, while (due to u or to t_1) $P(s,t_2) = true$. The underlying idea to solve the problem lies in associating two timestamps with each black event e:

• A vector timestamp $e.VC$ (counting only black events).
• An array of vector timestamps $e.MC[1..n]$ such that $e.MC[j]$ contains the vector timestamp of the last black event of P_j that causally precedes e ($e.MC[j]$ can be considered as a pointer from e to the last event that precedes it on P_j). When considering Figure 4.c, we have: $t_1.MC[1] = a.VC$, $t_1.MC[2] = b.VC$, $t_1.MC[3] = s.VC$, $t_2.MC[1] = t_1.VC$, $t_2.MC[2] = u.VC$ and $t_2.MC[3] = s.VC$.

Each process P_i has a vector clock $VC_i[1..n]$ and a vector of vector clocks $MC_i[1..n]$. Those variables are managed as described in Figure 5. Let us note that, in statement S3, $MC_i[k]$ and $m.MC[k]$ contain vector timestamps of two black events of P_k. It follows that one of them is greater or equal to the other: $max(MC_i[k], m.MC[k])$ is its timestamp.

(S1) **when P_i produces a black event** (e)
   $VC_i[i] := VC_i[i] + 1$; % one more black event on P_i %
   $e.VC = VC_i$; $e.MC = MC_i$;
   $MC_i[i] := VC_i$ % vector timestamp of P_i's last black event %

(S2) **when P_i executes a send event** (e=send m to P_j)
   $m.VC := VC_i$; $m.MC := MC_i$;
   send (m) **to** P_j % m carries $m.VC$ and $m.MC$ %

(S3) **when P_i executes a receive event** (e=receive(m))
   $VC_i := max(VC_i, m.VC)$; % update of the local vector clock %
   $\forall \, k : \; MC_i[k] := max(MC_i[k], m.MC[k])$
   % record vector timestamp of the last black predecessor on each P_k %

Fig. 5. Clock Management for Detecting $P(s,t)$

An Operational Pattern Detection Predicate Let us first remark that, as the protocol considers only black events, the predicate P_1 is trivially satisfied by any triple of events. So, detecting $P(s,t)$ amounts to only detect $\exists u : P_2(s,u,t)$. Given s and t with their timestamps (namely, $s.VC$ and $s.MC$ for s; $t.VC$ and $t.MC$ for t), the predicate $(\exists u \; : \; P_2(s,u,t)) \equiv (\exists u : s \to u \to t)$ can be

restated in a more operational way using vector timestamps. More precisely:
$(\exists u : s \rightarrow u \rightarrow t) \equiv (\exists u : s.VC < u.VC < t.VC)$.

If such an event u does exist, it has been produced by some process P_k and belongs to the causal past of t. Consequently, its vector timestamp is such that: $\exists k : u.VC \leq t.MC[k]$. From this observation, the previous relation translates in: $(\exists u : s \rightarrow u \rightarrow t) \equiv (\exists k : s.VC < t.MC[k] < t.VC)$.

As $\forall k$, $t.MC[k]$ is the vector timestamp of a black event in the causal past of t, we have $\forall k : t.MC[k] < t.VC$. Consequently, the pattern detection predicate simplifies and becomes: $\boxed{\mathcal{P}(s,t) \equiv (\exists k : s.VC < t.MC[k])}$.

References

[1] Fidge C.J., Timestamp in Message Passing Systems that Preserves Partial Ordering, *Proc. 11th Australian Computing Conference*, pp. 56-66, 1988.

[2] Garg V.K., *Principles of Distributed Systems*, Kluwer Acad. Press, 274 pages, 1996.

[3] Lamport L., Time, Clocks and the Ordering of Events in a Distributed System, *Communications of the ACM*, 21(7):558-565, 1978.

[4] Mattern F., Virtual Time and Global States of Distributed Systems, *Proc. "Parallel and Distributed Algorithms" Conference*, North-Holland, pp. 215-226, 1988.

[5] Raynal M. and Singhal S., Logical Time: Capturing Causality in Distributed Systems, *IEEE Computer*, 29(2):49-57, 1996.

[6] Schwarz R. and Mattern F., Detecting Causal Relationships in Distributed Computations: in Search of the Holy Grail. *Distributed Computing*, 7:149-174, 1994.

A Node Count-Independent Logical Clock for Scaling Lazy Release Consistency Protocol

Luciana Bezerra Arantes*, Bertil Folliot, and Pierre Sens

LIP6 Laboratory.
Universit Pierre et Marie Curie.
4, Place Jussieu 75252 Paris Cedex 05, France.
[Luciana.Arantes, Bertil.Folliot, Pierre.Sens]@lip6.fr

Abstract. The use of per processor vector logical clocks in lazy release consistency (LRC) protocol implementation may restrict its scalability since the size of these clocks depends on the number of nodes of the system. We propose a new logical clock, the *barrier-lock*, whose concept is based on the causality of synchronization operations. Its size is proportional to the number of synchronization variables used by the application, being not affected by the number of nodes of the system.

1 Introduction

Distributed shared memory systems (DSM) simulate a shared-memory address space on top of loosely coupled multiprocessor systems. The physical distribution of data among the nodes as well as the consistency of the shared memory is made by the DSM layer, being completely transparent for the application. One of the most efficient DSM protocol for memory consistency is the lazy release consistency (LRC) [1]. By relaxing memory consistency model, LRC reduces the number of messages and data transferred among the processors. In LRC, synchronization operations set up the ordering of memory accesses and the propagation of shared data coherence information.

This paper presents our proposal for a LRC protocol whose implementation does not depend on the number of nodes of the system. This approach can be quite interesting for platforms with a large number of nodes, i.e., large-scale DSM systems. In these systems, scalability is an important feature.

One of the limitations for scaling current LRC implementations is the fact that, for controlling shared memory updates causality, they use logical clocks to timestamp synchronization operations [6] [4]. These logical clocks consist of vector structures which have one entry for each node (process) of the system. Hence, their size is proportional to the total number of nodes. We propose a new logical clock, whose size is independent of the number of nodes. We have named it the *barrier-lock* clocks. Its concept is based on operations on locks and barriers, i.e., the basic synchronization variables provided by most LRC DSM

---

* Ph.D. scholar from CAPES (Brazil)

P. Amestoy et al. (Eds.): Euro-Par'99, LNCS 1685, pp. 815–822, 1999.
© Springer-Verlag Berlin Heidelberg 1999

systems. Its size is proportional to the number of synchronization variables used by the DSM application.

Section 2 of this paper gives an overview of the lazy release consistency protocol. Section 3 presents the *barrier-lock* logical clocks and some performance measures obtained with a first *barrier-lock* LRC DSM prototype. In section 4 some related works are discussed, while the last section summarizes the contributions of this work.

2   Lazy Release Consistency Overview

In lazy release consistency memory model, originally defined by TreadMarks [1], ordinary memory accesses are distinguished from synchronization ones. Synchronization operations are divided into acquire and release ones. The updates made by a process on its local shared data copy are propagated out to a second one only when the latter performs an acquire operation. This postponement reduces much of the communication required to make shared data consistent.

In LRC, the execution of each process is divided into intervals. A new interval begins at each acquire or release operation. Synchronization operations settle a causal ordering between intervals, which, in their turn, define the causality of shared memory updates. Intervals are partially ordered (**LRC happened-before**) as follows:

- intervals of the same process are totally ordered, i.e., if i and i' are intervals of the same process, and i occurs before i' in program order, then $i \rightarrow i'$;
- if i is an interval on process P_1 and i' is an interval of process P_2 then $i \rightarrow i'$ if i' begins with the acquire operation which is subsequent to the release operation which concluded i.
- if $i \rightarrow i'$ and $i' \rightarrow i''$, then $i \rightarrow i''$.

LRC controls partial order between intervals by using logical vector clocks, as defined by **Mattern** [6] and **Fidge** [4]. A vector clock timestamp is assigned to every interval. Each process keeps a local vector clock variable of N entries, where N is the total number of nodes (processes) of the system. A process j controls the intervals created by itself by using the jth entry of its vector clock variable. The other entries store the current knowledge that this process has of the intervals of the other processes. Thus, process Pj updates its vector clock v_j at each synchronization operation, based on the following rules:

r_1 : If it is an acquire operation and the acquirer P_j is different from the releaser Pr, whose local clock variable value is v_r, then:
$$0 \leq k \leq N - 1 : v_j[k] = max(v_j[k], v_r[k]).$$
r_2 : $v_j[j] = v_j[j] + 1.$

Hence, at a remote acquire, the acquirer P_j sends its current clock value to the releaser P_r. This one sends back all the intervals "covered" by its own local vector clock, but not by P_j's, including the identification of the pages that have been modified in each interval. Each identification is stored in a structure called *write-notice*. When receiving a *write-notice*, the acquiring process invalidates its corresponding local page (invalidate protocol). The first access to an invalid page will cause a page fault. The faulting process will then ask for all the updates that it does not have yet, applying them in the order defined by the causality of the intervals. These updates come in the form of *diffs*, a word-by-word comparison between a copy of the original page and its last version. Each *diff* is associated with a *write-notice* which in its turn is associated with an interval.

LRC DSMs usually offer to application two types of synchronization variables: **locks** and **barriers**. The first ones are used to control accesses to shared memory critical regions, while the second ones to sequence the execution of the program. Both of them can be mapped onto an acquire and/or release operations. Operations on a **lock** can be directly mapped onto acquire or release operations: obtaining a lock corresponds to an acquire operation, while granting it corresponds to a release. A **barrier** is a synchronization point, executed in parallel by all processes, where each process incorporates consistency information (*write-notices*) from all other processes. After the execution of a barrier, the local vector clocks of all processes are set to the same value. Basically, when arriving at a barrier each process performs a release operation, sending to the barrier manager process the intervals that the latter does not have. On the other hand, the departure from the barrier is seen as an acquire, since the manager process sends to all other processes the intervals that they do not have yet. Thereby, if I_{bbar} is the set of intervals that happened *before* a barrier call and I_{abar} is the set of intervals that happened *after* it, we have:

$$\forall i_j \in I_{bbar} \text{ and } \forall i_k \in I_{abar} : i_j \to i_k \qquad \textbf{(barrier property)}.$$

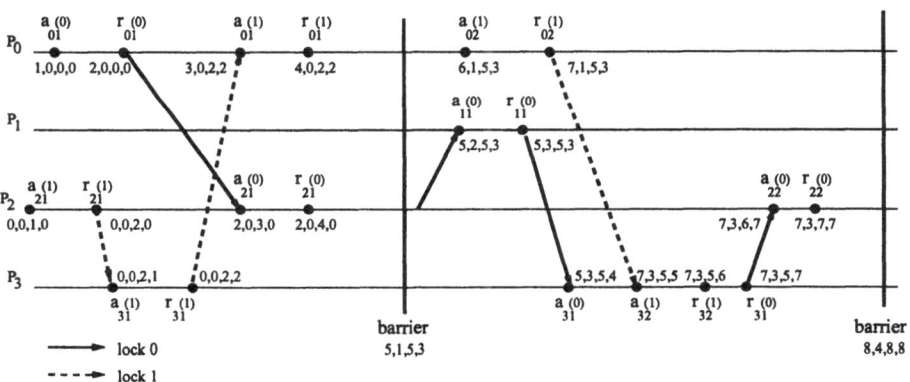

Fig. 1. LRC timing diagram based on vector clocks.

Figure 1 shows the partial ordering timing diagram corresponding to operations on 2 locks made by 4 processors. A barrier is called two times. For simplicity, shared memory operations or data structures (*write-notices*, *diffs*) are not shown. The notations $a_{ij}(l)$ and $r_{ij}(l)$ respectively represent the *jth* acquire or release operation of processor P_i on lock l. Each synchronization operation is timestamped with $v_i[0], v_i[1], v_i[2], v_i[3]$.

3   LRC Based on Barrier-Lock Clocks

Vector clocks precisely control causality of LRC synchronization operations. However, their size is proportional to the number of processes of the system, which can be quite huge in a large-scale DSM. Due to LRC transitive propagation of updates, several of them can be included in a synchronization message.

As explained in the last section, at a barrier, processes set their local vector clocks to the same value. Thus, barriers could be seen as stop points for the restarting of a new set of lock operations, i.e., a program execution could be divided into *barrier-intervals* and each *barrier-interval* would be subdivided into *lock-intervals*. A new *barrier-interval* would begin at each barrier call, while a *lock-interval* at each acquire or release operation on a lock. This is exactly the idea behind the *barrier-lock* clock.

A *barrier-lock* timestamp is represented by the tuple $(b, vl)_i$, where b_i is a barrier call counter and vl_i is a per lock vector. Similar to per processor vector clocks, *barrier-lock* ones precisely control **LRC happened-before** partial ordering. At each barrier call, the counter b_i of all processes is incremented, while their lock vector vl_i is reset. On the other hand, for controlling processes' operations on locks within a single *barrier-interval*, the concept of the *poset-diagram*, as presented by Mattern in [6], is used. This diagram shows the logical relationship of events. If $a \rightarrow b$, than it is possible to follow a *path of causality* from a to b, which is easier seen in the *poset-diagram*. The timing diagram and the *poset-diagram* are isomorphic. Figure 2 illustrates the *poset-diagram*, related to the operations on locks 0 and 1 of the first *barrier-interval* of figure 1.

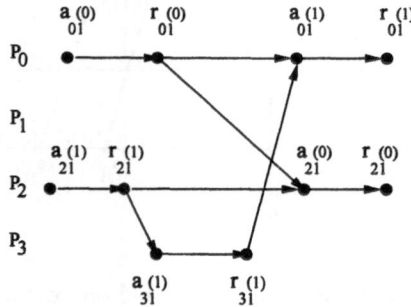

Fig. 2. Lock-intervals poset-diagram within a barrier-interval.

As locks are used to control processes' accesses to shared data critical regions, lazy release consistency protocol assures that a lock can only be held by one process at a time. Its behavior is like a token. This means that if two acquire or release operations on locks are mutually independent, they correspond to operations on different locks. It is guaranteed that two processes are not going to increment the same entry of their local variable vl concurrently. Therefore, for timestamping synchronization operations on locks, we can employ a vector variable where each of its entry is associated with a lock. The size of the *barrier-lock* clock is then $L + 1$, L being the number of locks used by the application. Very often this number is quite small (e.g. barrier-like programs). It is worth remarking that it is the "lock token" behavior that justifies why, in spite of the fact that **Charon-Bost** [3] has proved that causality can only be characterized by vector clocks with N entries, we have managed to have a clock, that precisely captures causality, but whose size can be smaller than the total number of processes (nodes) N. In other words, since locks are held in mutual exclusion, operations on them can be represented as the union of L chains (*paths of causality*). Hence, a vector of L entries is sufficient to characterize causality of synchronization operations on locks.

Let N be the number of processes (nodes) of the system and L, the number of locks used by the application. A *barrier-lock* clock timestamp of processor P_j is represented by the tuple $(b, vl)_j$, where vl has dimension L. A new timestamp is computed based on the following rules:

r_1 : If it is assigned to a new *barrier-interval*:
    $a : 0 \leq i \leq N - 1 : \{0 \leq k \leq L - 1 : vl_i[k] = 0\}$;
    $b : 0 \leq i \leq N - 1 : b_i = b_i + 1$;

r_2 : If it is assigned to a new *lock-interval*, corresponding to an acquire or release operation on lock l:
    a : in the case of an acquire operation in which the acquirer process Pj is different from the releaser Pr:
        $0 \leq k \leq L - 1 : vl_j[k] = max(vl_j[k], vl_r[k])$;
    b : $vl_j[l] = vl_j[l] + 1$.

Figure 3 shows the same figure 1, using *barrier-lock* timestamps. The notation b-vl[0],vl[1] is used for each timestamp assigned to a synchronization operation.

For comparing two timestamps, (b, vl) and (b', vl'), the following relations are defined:

$$- \ (b, vl) < (b', vl') \Leftrightarrow \begin{cases} b < b' \\ or \\ (b = b') \ and \ (vl < vl') \end{cases}$$

$$- \ (b, vl) \ \| \ (b', vl') \Leftrightarrow \begin{cases} (b = b') \ and \ (vl = vl' = 0) \\ or \\ \neg(vl < vl') \ and \ \neg(vl' < vl) \end{cases}$$

The above relations guarantee that the *barrier-lock* clocks are also strongly consistent. So, for any two intervals i and i', respectively identified by (b, vl) and (b', vl') timestamps, we have:

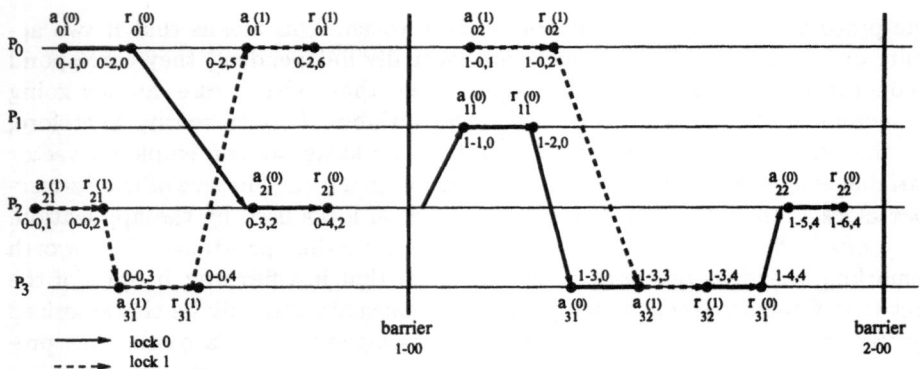

Fig. 3. LRC based on barrier-lock clocks.

$$i \rightarrow i' \Leftrightarrow (b, vl) < (b', vl')$$

Proof. If i and i' belong to different *barrier-intervals*, then $b < b'$ (rule r_1). If they belong to the same *barrier-interval* ($b = b'$), let i'', with (b, vl'') timestamp, be the latest interval that directly precedes i' and l the lock corresponding to the synchronization operation (acquire or release) which started interval i'. If i' and i'' belong to different processes then i' refers to the interval that begins with the acquire of l, which has been released with the ending of i''. Then $vl' = max(vl'', vl') + 1$ (rule $r_2.a$ and $r_2.b$). If they belong to the same process then $vl'[l] = vl''[l] + 1$ (rule $r_2.b$). Hence, in both case $vl'' < vl'$. We can apply this arguments to any pair of intervals belonging to all *paths of causality* that lead to i'. Due to the transitive dependency property of LRC intervals, we have that $vl < vl'$.

Conversely, if $b < b'$, intervals i and i' belong to different *barrier-intervals*. Then, based on the **barrier principle**, described in the previous section, $i \rightarrow i'$. On the other hand, if $b = b'$ then i and i' are *lock-intervals* within the same *barrier-interval*. Let (b, vl'') be the greatest timestamp smaller than vl' within this *barrier-interval*. If these timestamps have been created by different processes, then (b, vl') corresponds to the acquire of lock l, whose releasing operation has been identified by (b, vl''); if they belong to the same process, they correspond to totally order intervals and i'' occurred before i'. Then, based on the **LRC happened-before** partial ordering, i'' directly precedes i', which means that $i'' \rightarrow i'$. Applying this same arguments for each pair of *lock-intervals* belonging to all *paths of causality* that lead to i' and considering the transitive property of LRC protocol, we have $i \rightarrow i'$.   □

3.1   Performance

In order to verify the feasibility of the *barrier-lock* clocks, we have implemented a prototype, replacing the traditional vector clocks by the *barrier-lock* ones in TreadMarks software DSM, version 0.10.1. The tests have been made on top of 8 Sun-sparc-5 stations linked by a 100 Mbit/s Ethernet backbone with 5 well-known applications: SOR and TSP (distributed by TreadMarks [1]), IS and 3D FTT (NAS benchmark [2]) and Barnes-Hut(SPLASH benchmark [7]). The number of locks used by these applications are quite small (from 0 to 2). For showing the scalability of a LRC DSM, we have simulated a platform with a huge number of nodes by increasing the constant that specifies the number of processes in the system. We have then measured the number of bytes exchanged between the processes at synchronization operations. Figure 4 shows the ratio: *number of bytes of synchronization messages on top of barrier-lock LRC prototype / the number of bytes of synchronization messages on top of TreadMarks.*

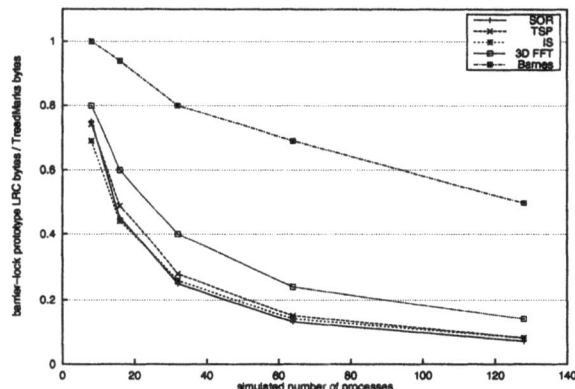

Fig. 4. Ratio of data exchanged at synchronization operations.

We can remark a considerable cut down in the amount of data exchanged at synchronization operations when the *barrier-lock* clocks are used, since all the applications use a small number of locks.

4   Related Works

TreadMarks [1] was the first DSM to implement the LRC memory model. Automatic Update Release Consistency (AURC) [10] and Home-based Lazy Release Consistency (HLRC) [10] are variations of LRC protocol. The three of them employ vector clocks to ensure the causality of synchronization operations. The last two adopt a "home-based" protocol in which updates of a page are eagerly propagated to a home node associated with the page.

Some authors have proposed a different implementation of Mattern and Fidge vector clocks. For instance, in **Singhal** and **Kshemkalyani** technique [8], a process sends to another only those entries of its vector clock that have been modified since they have last communicated to each other. Hence, the size of the message exchanged between processes is reduced. **Fowler** and **Zwaenepoel**

[5] have proposed a vector clock implementation where each process only keeps direct dependencies on others. This implies in a considerable cut in memory storage and communication overhead. However, for capturing transitive causal relations, it is necessary to recursively trace causal dependencies.

There are some clocks, as the **plausible** ones [9], that can be constructed with a constant number of entries, independently of the number of nodes of the system. However, even if they offer a high level of ordering accuracy, they do not guarantee that certain pairs of concurrent events will not be ordered. Therefore, they are not appropriate for implementing LRC protocol, as this ordering can lead to unnecessary consistency operations and remote requests.

5 Conclusions

We have presented the *barrier-lock* clocks, whose size is proportional to the number of synchronization variables used by a DSM application. They have been modeled to precisely capture causality of synchronisation operations of lazy release consistency protocol. As their size is not affected by the number of nodes of the system, these clocks are quite appropriate for scaling DSM systems that provide such protocol. The proof that *barrier-lock* clocks can precisely control causal ordering of synchronization operations was presented in section 3, while the results of section 4 validate them.

References

[1] C. Amza, A. Cox, S. Dwarkadas, P. Keleher, H. Lu, R. Rajamony and W. Zwaenepoel. TreadMarks: Shared Memory Computing on Networks of Workstations. *IEEE Computer*, 29(2):18–28, February 1996.

[2] D. Bailey, J. Barton, T. Lasinski and H. Simon. The NAS Parallel Benchmark. Technical Report 103863, NASA, July 1993.

[3] B. Charon-Bost. Concerning the Size of Logical Clocks. *Information Processing Letters*, 39:11–16, July 1991.

[4] C. Fidge. Logical Time in Distributed Computing Systems. *IEEE Computer*, 24(8):28–33, August 1991.

[5] J. Fowler and W. Zwaenepoel. Causal Distributed Breakpoints. In *the 10th International Conference on Distribute Computing Systems*, pages 131–41, 1990.

[6] F. Mattern. Virtual Time and Global States in Distributed Systems. In *Workshop on Parallel and Distributed Algorithms*, Elsevier (Holland), October 1988.

[7] P.Singh, W. Weber and A. Gupta. SPLASH: Stanford Parallel Applications for Shared-memory. Computer Architecture News, 20(1):5–44, March 1992.

[8] M. Singhal, M. and A. Kshemkalyani. An Efficient Implementation of Vector Clocks. *Information Processing Letters*, 33:47–53, August, 1992.

[9] F. Torres-Rojas, F. and M. Ahamad, Plausible Clocks: Constant Size Logical Clocks for Distributed Systems. In *the 10th International Workshop on Distributed Algorithms*,Bologna(Italy), Octobre, 1996.

[10] Y. Zhou, L. Iftode and K. Li. Performance Evaluation of Two Home-Based Lazy Release Consistency Protocols for Shared Virtual Memory Systems. In *the 2nd Symposium on Operating Systems Design and Implementation*, Octobre 1996.

Mutual Exclusion Between Neighboring Nodes in an Arbitrary System Graph Tree That Stabilizes Using Read/Write Atomicity

Gheorghe Antonoiu[1] and Pradip K. Srimani[1]

Department of Computer Science, Colorado State University, Ft. Collins, CO 80523

Abstract. Our purpose in this paper is to propose a new protocol that can ensure mutual exclusion between neighboring nodes in an arbitrary distributed system, i.e., under the given protocol no two neighboring nodes can execute their critical sections concurrently. This protocol can be used to run a serial model self stabilizing algorithm in a distributed environment that accepts as atomic operations only "send a message", "receive a message", and "update a state". Unlike the scheme in [1], our protocol does not use time-stamps (which are basically unbounded integers); our protocol is a generalization of the protocol described in [2] which was restricted to work only for tree structured distributed systems. Like the protocol in [2], our algorithm uses only bounded integers and can be easily implemented.

1 Introduction

The *serial model* for self-stabilizing distributed algorithms assumes that the reading of the states of the neighbors and the updating its own state are grouped together into one atomic operation (indivisible execution step); the model also assumes that each node can simultaneously see the current states of all its neighbors and that only one node executes an atomic step at a time. Because of the popularity of the serial model and the relative ease of its use in designing and proving correctness for new self-stabilizing algorithms, a large number of such algorithms have appeared in the literature. This serial model makes stronger assumptions about the run time environment than those normally provided by a real time message passing distributed system. On the other hand designing and proving correctness for self-stabilizing algorithms in a distributed model are relatively difficult. Thus, in order to make use of the self-stabilizing algorithms designed for a serial model, it is necessary to design lower level self-stabilizing protocols such that an algorithm developed for a serial model can be run in a distributed environment. Our purpose in this paper is to propose a new protocol that can be used to run a serial model self stabilizing algorithm in a arbitrary distributed environment (an arbitrary system graph) that uses read write atomicity, i.e., accepts, as atomic operations, "send a message", "receive a message" and "update a state". Two major contributions of this paper are: (1) unlike the scheme in [1], our protocol does not use time-stamps (which are unbounded integers); our algorithm uses only bounded integers and can be easily implemented; (2) unlike the scheme in [2] (which works only for a tree structured system),

P. Amestoy et al. (Eds.): Euro-Par'99, LNCS 1685, pp. 823-830, 1999.
© Springer-Verlag Berlin Heidelberg 1999

our protocol can be used in any arbitrary distributed system. We use the same system model as in [2]; proofs for Lemmas and Theorems are omitted for lack of space.

2 Mutual Exclusion Protocols for Neighboring Nodes in a System Graph

2.1 Protocol Using Unbounded Integers

The protocol for mutual exclusion between neighboring nodes presented in this section uses an unbounded integer C_x as a local state variable at the node x. In addition to C_x, each node x maintains a local array $LC_x[\ ]$ used to store the copies of the C variables of its neighbors; We use the notation $LC_x(y)$ to denote the component of the state vector LC_x that stores a copy of the state variable C_y of node y, $\forall y \in \mathcal{N}(x)$, where $\mathcal{N}(x)$ denotes the set of neighbors of node x in the system graph G. We assume that each node x in the system has a unique identification id_x; we define two binary relations $\lhd$ and $\rhd$ between a local variable C_x and a local vector element $LC_x(y)$.

Definition 1.
(i) $C_x \lhd LC_x(y)$ iff $((C_x < LC_x(y)) \vee ((C_x = LC_x(y)) \wedge (id_x < id_y)))$;
(ii) $C_x \rhd LC_x(y)$ iff $(C_x > LC_x(y)) \vee ((C_x = LC_x(y)) \wedge (id_x > id_y))$;

Remark 1. (1)We can similarly define $LC_x(y) \lhd LC_x(z)$ and $LC_x(y) \rhd LC_x(z)$, for all $y, z \in \mathcal{N}(x)$.
$\max_{y \in \mathcal{N}(x)}^{\lhd}(LC_x(y))$, defined as $LC_x(z)$ such that $LC_x(z) \rhd LC_x(y)_{y \in \mathcal{N}(x)-\{z\}}$, is well defined and unique; we denote this simply by $\max_{y \in \mathcal{N}(x)}(LC_x(y))$ (omitting the superscript) without ambiguity. To compute $\max_{y \in \mathcal{N}(x)}(LC_x(y))$ the node x must know the id's of its neighbors; this can be easily done if every message from a neighbor contains the id of the sender. (2) Since the node id's are unique and we use them as tie-breakers, either $C_x \rhd \max\limits_{y \in \mathcal{N}(x)}(LC_x(y))$ or $C_x \lhd \max\limits_{y \in \mathcal{N}(x)}(LC_x(y))$ is always true.

The uniform protocol (algorithm) for any node x in G is shown in Figure 1; note that the state of a node x is defined by the variable C_x and the vector LC_x; the global system state is defined by the local states of all nodes in G.

Definition 2. *A system state is legitimate iff* $\forall (x,y) \in E$, $C_x \geq LC_y(x)$.

Remark 2. Note that since the system graph G is undirected, if $(x,y) \in E \rightarrow (y,x) \in E$.

Lemma 1. *Starting from a legitimate state, the system remain in a legitimate state after an arbitrary move by any node.*

Theorem 1. *For an arbitrary system graph G in any legitimate state, no two neighboring nodes can execute their critical sections simultaneously, i.e., the mutual exclusion condition between neighboring nodes holds in a legitimate state.*

Theorem 2. *Starting from an arbitrary state, the system reaches a legitimate system state in a finite number of moves.*

$$R_x^1 \quad : \quad LC_x(n_x[1]) = C_{n_x[1]};$$
$$R_x^2 \quad : \quad LC_x(n_x[2]) = C_{n_x[2]};$$
$$\cdots \quad : \quad \cdots;$$
$$R_x^{d_x} \quad : \quad LC_x(n_x[d_x]) = C_{n_x[d_x]};$$
$$CS_x \quad : \quad \textbf{if } (\ C_x \lhd \max_{y \in \mathcal{N}(x)} (LC_x(y)) \ ) \textbf{ then} \text{ execute Critical Section;}$$
$$W_x \quad : \textbf{if } (\ C_x \lhd \max_{y \in \mathcal{N}(x)} (LC_x(y)) \ ) \textbf{ then} C_x = \max_{y \in \mathcal{N}(x)} (LC_x(y)) + 1;$$

Fig. 1. Protocol for an Arbitrary Node Using an Unbounded Integer

2.2 A Bounded Integer Protocol

The protocol presented in the previous section uses unbounded integers to store the state variables. In absence of any faults, the protocol may be used in real applications by resetting the state variables to zero and using very large integer variables to store the state variables. In a self-stabilizing system, transient faults may start the system from any *arbitrary* system state and hence however large integers we use, the state variables cannot increase monotonically. Simple straightforward extension of the protocol by using **mod** M operation (when the local state variables are written) does not work either since some nodes may starve. Consider the network of 3 nodes, "a", "b" and "c" as shown in Figure 2; an arbitrary system state is shown with corresponding C values. When $C_b = M - 2$ the node "b" can enter in the critical section; when C_b is updated, $C_b = M - 1 + 1 \ mod \ M = 0$. Thus, node "b" can move again and the cycle may repeat; the nodes "a" and "c" starve (can not enter the critical section).

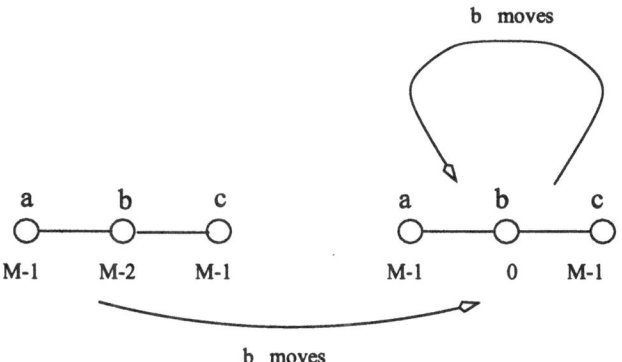

Fig. 2. An Example

Note that using a large M does not prevent the pathological behavior presented in Figure 2; however large M is, the system would not work in the desired fashion. The problem stems from two facts: (1) the system must be able to start from an *arbitrary* initial state; (2) when the state variables wrap around (monotonically increases to $M - 1$ and then goes back to 0 and start increasing again), the monotonic increasing property is destroyed which is essential to guarantee mutual exclusion. If we could force the initial values of C variables to 0, a large enough M may be acceptable for any practical purpose. The first step in our approach to solve this problem is to construct a spanning tree of the network in a self-stabilizing manner starting from an arbitrary initial state.

A Spanning Tree of the System Graph The algorithm for spanning tree construction must satisfy 3 conditions: (i) it is self-stabilizing; (ii) it uses read-write atomicity; (iii) for each node x, it determines its predecessor P_x and the set $C(x)$ of its *immediate successors* in the spanning tree. Note that since the given system graph G is arbitrary, in general $P_x \cup C(x) \subseteq \mathcal{N}(x)$. There exist several self-stabilizing algorithms for spanning tree generation [3,4], but they use composite atomicity and are not useful for our purpose. We use the self-stabilizing spanning tree generation algorithm of [5] that uses read-write atomicity in an *id*-based network. The basic idea of this algorithm is to choose the node with maximum *id* as the root node for the spanning tree; due to distinct node *id* property this node is unique. We rewrite the algorithm from [5] in Figure 3 to conform to our notations; see [5] for details and correctness proofs. Each node x maintains three local variables, RT_x for the *id* of the root node, P_x for the immediate predecessor of node x in the spanning tree x, and an integer variable D_x to store the distance of the root node from node x; in addition, each node x also maintains three local state vectors, $LRT_x[\ ]$, $LP_x[\ ]$ and $LD_x[\ ]$ respectively to store local copies of the corresponding state variables at each neighboring node. For example, $LRT_x(y)$ will store (at node x) a copy of the variable RT_y at node y where y is a neighbor of node x. Also as before, id_x denotes the unique *id* of the node x. Note that M is some positive integer greater than the number of nodes in the system graph. The pseudo-code of the algorithm that is to be executed at each node x is shown in Figure 3. Note that, the algorithm does not directly produce the set $C(x)$, the set of immediate successors of the node x in the spanning tree generated; this can be easily accomplished since in a legitimate state of the system, for any node x we find that $C(x) = \{y \in \mathcal{N}(x) \mid LP_x(y) = x\}$.

For the rest of the paper, we assume that a rooted spanning tree is already generated for the given system graph, i.e., a legitimate state for the spanning tree algorithm has been reached. Each node $x \in G$, thus, knows its predecessor P_x and the the set $C(x)$ of its immediate successors in the spanning tree. Note that a node x can decide if it is the root node of the spanning tree, i.e. a node x is the root node if $P_x = id_x$ and a node x knows if it is a leaf node, i.e., when $|C(x)| = 1$. Note that in general $\{P_x\} \cup C(x) \subseteq \mathcal{N}(x)$ for any node $x \in G$.

Remark 3. (1)We view the spanning tree T of the system graph $G = (V, E)$ to be a subgraph of G such that $T = (V, E')$, $E' \subseteq E$ and any edge $(x, y) \in E$ belongs to spanning tree T iff $y \in C(x)$. (2) Since the algorithm for spanning tree generation has stabilized, P_x and $C(x)$ for each node x are constant and exactly one node knows that it is the root; we refer to this node as node r ($P(r)$ is null).

(R_x^C) :   **for** $y \in \mathcal{N}(x)$ **do** $LRT_x(y) = RT_y$;
(R_x^P) :   **for** $y \in \mathcal{N}(x)$ **do** $LP_x(y) = P_y$;
(R_x^D) :   **for** $y \in \mathcal{N}(x)$ **do** $LD_x(y) = D_y$;
W_x :   **if** $((RT_x < id_x) \vee (P_x = id_x \wedge (RT_x \neq id_x \vee D_x \neq 0))$
               $\vee (F_x \notin (\mathcal{N}(x) \cup \{x\}) \vee D_x > M)$
         then $RT_x = id_x, P_x = id_x, D_x = 0$;
      elseif $(P_x = id_y \wedge (y \in \mathcal{N}(x)) \wedge (D_x < M) \wedge (RT_x \neq LRT_x(y)$
               $\vee D_x \neq LD_x(y) + 1))$
         then $RT_x = LRT_x(y), D_x = LD_x(y) + 1$;
      elseif $((RT_x < LRT_x(y) \wedge (y \in \mathcal{N}(x)) \wedge LD_x(y) < M)$
               $\vee (RT_x = LRT_x(y) \wedge y \in \mathcal{N}(x) \wedge LD_x(y) + 1 < M)$
         then $RT_x = LRT_x(y), F_x = id_y, D_x = LD_x(y) + 1$;

Fig. 3. Spanning Tree Algorithm with Read/Write Atomicity

A Strategy to Use Bounded Integers In our approach to design a mutual exclusion protocol with bounded integers, we take two steps: (1) we use a large bounded integer M, $M \gg n$ (n is the number of nodes in the system) and use mod M operation to increment the system variables; (2) we use a "dynamic upper bound" on the local state variable values; the local state variable C_x at node x is not allowed to exceed its own upper bound, say H_x, where the local bound variable H_x itself is dynamic. The dynamic upper bound variables at nodes move cyclically in a range between 0 and M (H_x for any node x starts at an arbitrary value, goes up to $M-1$, resets to 0 and then monotonically increases to M and so on). We show that our proposed protocol can achieve the mutual exclusion condition between neighboring nodes in a legitimate state except for a relatively short period of time when the bound variables at different nodes are being wrapped around. To avoid neighboring nodes entering the critical section during the wrap around phase of the upper bounds, we prevent nodes entering their critical sections during the time period when the upper bounds are wrapped around; we design our protocol such that the upper bound variables are wrapped around in a finite time period when they reach the maximum. Thus, we use two constants V_{max} and V_{min} such that whenever the bound variable H_x at node x is such that $H_x < V_{min}$ or $H_x > V_{max}$, node x is not allowed in its critical section. We prove later that $V_{max} = M - n$ and $V_{min} = 2n + 4$. Thus, we observe several key features of the protocol:

1. When the system starts from a legitimate state, the system state will periodically enter into illegitimate states even in absence of faults for a relatively short period of time.
2. When the system enters into periods of illegitimacy, the neighboring mutual exclusion condition is maintained by effectively prohibiting each node from entering into its critical section.
3. If we choose M to be sufficiently large (large M/n ratio), the periods of illegitimacy would be very small compared to the periods the system will be available.

The Protocol Using Bounded Integers Each node, x, maintains two local state variables, C_x, H_x, and the two corresponding local state vectors LC_x, LH_x. The variable, C_x, has a functionality similar to the C_x variable in the previous protocol with unbounded integers. The other state variable H_x is used as a "sliding upper bound" for the C_x variable. We need to define a predicate $\Psi(x)$ for a node $x \in G$.

Definition 3.

$$\Psi(x) = \left(C_x \lhd \min_{y \in \mathcal{N}(x)} LC_x(y) \right) \wedge \left(\max_{y \in \mathcal{N}(x)} LC_x(y) + 1 \leq H(x) \right)$$

where the $\max$ and the $\min$ operation is performed for the order relation $\lhd$ (see Definition 1 and Remark 1).

The complete pseudo code of the algorithm is given in Figure 4; note that the protocols for root node, internal nodes and the leaf nodes with respect to the generated spanning tree are different. As before the code at each node x is an infinite loop of atomic operations of reading the states of its neighbors (R_x), computing a predicate and entering the critical section if the predicate is true (CS_x), and then writing its own state variables (W_x). There are three constants in our protocol: M is a large integer, $M \gg n$; V_{max} and V_{min} are two constant integers whose values are to be determined later. First, we observe the following¿

Remark 4.

1. The first two lines in the program executed by each node represent actually a set of atomic actions; each line is written as a **for** loop in one statement for brevity and clarity of presentation.
2. The root node and the leaf nodes execute the same rules for updating the C variables, but they execute different rules for the H variables.
3. The atomic write action for any node can be of one of three types: (i) either the C variable is set to the value of the H variable, or (ii) the C variable is incremented, or (iii) the H variable is updated.
4. The values assumed by the variable H_x at each node x repeat a cyclic pattern from 0 to $M - 1$ and then to 0 and so on.

Note that the basic difference between this protocol and the protocol using unbounded integers is to use the variable H_x as a sliding upper bound for C_x and to increment the variable H_x in a modular fashion in the domain 0 to $M - 1$. As long as modifications of C_x remains not greater than H_x, this algorithm behaves exactly like the algorithm presented in the previous section.

Definition 4. *We say that a node executes: (i) a **type A** write action when the node sets its C variable to the value of its H variable, (ii) a **type B** write action when it increments its C variable, and (iii) a **type C** write action when its updates its H variable. A type A or a type B action by a node x does not change the H_x variable.*

Lemma 2. *Any fair execution contains infinitely many type C actions by each node, i.e., in a fair execution H_x at each node x will increase (mod M) infinitely many times.*

(R_x^c) : **for** $y \in \mathcal{N}(x)$ **do** $LC_x(y) = C_y$;
(R_x^h) : **for** $y \in C(x) \cup \{P_x\}$ **do** $LH_x(y) = H_y$;
CS_x : **if** $(C_x \lhd \min\limits_{y \in \mathcal{N}(x)} LC_x(y)) \wedge ((H_x \geq V_{min}) \wedge (H_x \leq V_{max}))$
         then Execute CS;
W_x : **if** $(C_x > H_x)$ **then** $C_x = H_x$;
         elseif $\Psi(x)$ **then** $C_x = \max\limits_{y \in \mathcal{N}(x)} LC_x(y) + 1$;
         elseif $((\forall_{y \in C(x)})\ H_x = LH_x(y))$ **then** $H_x = LH_x(P_x)$;

(i) Protocol for the Internal Node x.

(R_x^c) : **for** $y \in \mathcal{N}(x)$ **do** $LC_x(y) = C_y$;
(R_x^h) : **for** $y \in C(x)$ **do** $LH_x(y) = H_y$;
CS_x : **if** $(C_x \lhd \min\limits_{y \in \mathcal{N}(x)} LC_x(y)) \wedge ((H_x \geq V_{min}) \wedge (H_x \leq V_{max}))$
         then Execute CS;
W_x : **if** $(C_x > H_x)$ **then** $C_x = H_x$;
         elseif $\Psi(x)$ **then** $C_x = \max\limits_{y \in \mathcal{N}(x)} LC_x(y) + 1$;
         elseif $((\forall_{y \in C(x)})\ H_x = LH_x(y))$ **then** $H_x = H_x + 1 \mod M$;

(ii) Protocol for the Root Node x.

(R_x^c) : **for** $y \in \mathcal{N}(x)$ **do** $LC_x(y) = C_y$;
(R_x^h) : $LH_x(P_x) = H_{P_x}$;
CS_x : **if** $(C_x \lhd \min\limits_{y \in \mathcal{N}(x)} LC_x(y)) \wedge ((H_x \geq V_{min}) \wedge (H_x \leq V_{max}))$
         then Execute CS;
W_x : **if** $(C_x > H_x)$ **then** $C_x = H_x$;
         elseif $\Psi(x)$ **then** $C_x = \max\limits_{y \in \mathcal{N}(x)} LC_x(y) + 1$;
         else $H_x = LH_x(P_x)$;

(iii) Protocol for the Leaf Node x.

Fig. 4. Protocol for Mutual exclusion with Bounded Integers

Definition 5. *We use the notation $x \succeq y$, if $x = y$ or $x = (y + 1) \mod M$, where $x, y \in Z_M$, where Z_M is the field of integers modulo M.*

Definition 6.

1. *A system state is **c-legitimate** if $\forall\ (x, y) \in E,\ C_x \geq LC_y(x)$.*
2. *An edge $(x, y) \in T$ (y is an immediate successor of x in the spanning tree T), is **h-legitimate** if (i) $(H_x \succeq LH_y(x)) \wedge (LH_y(x) \succeq H_y) \wedge (H_y \succeq LH_x(y))$ and (ii) at most one pair of successive variables in the sequence $(H_x,\ LH_y(x),\ H_y,\ LH_x(y))$ are unequal.*
3. *A global system state is **h-legitimate** iff $\forall\ (x, y) \in T$ the edge (x, y) is h-legitimate.*
4. *A global system state is **legitimate** iff the system state is **both** c-legitimate and h-legitimate.*

Remark 5. The definition of c-legitimate state uses <u>all</u> edges of the system graph G while the definition of h-legitimate edge applies only to the directed edges of the rooted spanning tree T (Remark 3).

Theorem 3. *For an arbitrary system graph in a legitimate state, no two neighboring processes can execute their critical sections simultaneously.*

When the system starts from an arbitrary state (possibly illegitimate), we need to show that the system is brought to a legitimate state in finite time and once the system is in a legitimate state, it remains in a legitimate state until the occurrence of the next fault. From Definition 6.4, in order to be legitimate a system state must be both h-legitimate and c-legitimate. Once the system state is h-legitimate, it always remains so in all subsequent moves; starting from an arbitrary state, the system reaches a h-legitimate state in finitely many moves. Starting from a h-legitimate state, the system reaches a c-legitimate state in finite time; once the system state is c-legitimate, it remains so until $H_r = M - n$ (r is the root node of the spanning tree); the system is no longer c-legitimate when the H-variables in the system are in "wrap-around phase", but again becomes c-legitimate when $H_r \geq 3n + 1$. During this period, when the system is h-legitimate but not c-legitimate, the system state is not legitimate, but does not violate mutual exclusive entry into critical section by neighboring nodes (by effectively prohibiting each node from enetering into critical sections). Thus, for the protocol to function correctly, we choose $V_{min} = 3n + 1$ and $V_{max} = M - n$. Obviously, larger the ratio M/n is, longer the time period for system legitimacy is. We need to choose M to be a large integer compared to the number of nodes in the system, but M is a bounded integer and the protocol behaves correctly.

References

1. M. Mizuno and H. Kakugawa. A timestamp based transformation of self-stabilizing programs for distributed computing environments. In *Proceedings of the 10th International Workshop on Distributed Algorithms (WDAG'96)*, volume 304–321, 1996.
2. G. Antonoiu and P. K. Srimani. Mutual exclusion between neighboring nodes in a tree that stabilizes using read/write atomicity. In *Proceedings of Euro-Par 98; Lecture Notes in Computer Science*, volume LNCS 1470, pages 545–553, Springer Verlag, 1998.
3. S. Sur and P. K. Srimani. A self-stabilizing distributed algorithm to construct BFS spanning tress of a symmetric graph. *Parallel Processing Letters*, 2(2,3):171–180, September 1992.
4. N. S. Chen, H. P. Yu, and S. T. Huang. A self-stabilizing algorithm for constructing spanning trees. *Inf. Processing Letters*, 39(3):14–151, 1991.
5. A. Arora and M. Gouda. Distributed reset. *IEEE Trans. Comput.*, 43(9):1026–1038, September 1994.

Topic 11
Parallel Programming: Models, Methods, and Languages

Luc Bougé, Bill McColl, Mamoun Filali, and Henk Sips

Co-chairmen

For the last three years, Euro-Par has included two separate lines of workshops on parallel programming models, languages and methods.

- The first has been devoted to the design of parallel languages and their semantics: *Parallel languages, programming and semantics* (Ian Foster, 1996), *Parallel Languages* (Ron Perrott, 1997), *Parallel Programming Languages* (Henk Sips, 1998).
- The second has been devoted to the abstract models for parallel programming: *Theory and models of parallel computing* (Bill McColl, 1996), *Programming Models and Methods* (David Skillicorn, 1997; Christian Lengauer, 1998),

This year, we have decided to merge the two lines of workshops into a single topic, to emphasize the necessary collaboration between language designers and those working on theoretical models. We feel that these two fields have reached a stage where it is no longer appropriate to pursue thses issues in isolation from one another. Developing abstract programming models in the long term makes sense only if they can be "embodied" into some sort of programming language: only then can they be tested against *real* problems. Conversely, the common experience is that it is hopeless to promote yet-another-programming-language (or library) unless it is based on some sort of abstract model which helps the users in reasoning about their programs in a structured way: parallel programming in the large is far too complex to rely on intuition only.

There has already been several examples of this converging trend in the past. For instance, papers about BSP, skeletons, or data-parallel programming can be found in both lines of workshops, depending on whether the stress is put on the use of languages derived from these models, or on the study of their mathematical properties. The decision to merge the two fields in this year's Euro-Par recognizes the importance of this evolutionary trend, and aims to emphasize it in current and future research.

This decision to merge these two strands has resulted in a large number of papers being submitted: 33, one of the highest figures among this year's topics. The selection was a difficult task, as the scientific level of the submissions was high, and covered a wide range of issues related to parallel programming

P. Amestoy et al. (Eds.): Euro-Par'99, LNCS 1685, pp. 831–832, 1999.
© Springer-Verlag Berlin Heidelberg 1999

models. Which criteria may be used to identify a model as specific to parallel programming? How can one evaluate the practical usability of a new programming concept? We hope that this conference, and others which follow, will provide a place to discuss these and other important related questions.

We wish to thank Dominique Méry whose help and encouragement were crucial in making this merging possible.

Exploiting Advanced Task Parallelism in High Performance Fortran via a Task Library

Thomas Brandes

Institute for Algorithms and Scientific Computing (SCAI)
German National Research Center for Information Technology (GMD)
Schloß Birlinghoven, D-53754 St. Augustin, Germany
brandes@gmd.de

Abstract. As task parallelism has been proven to be useful for applications like real-time signal processing, branch and bound problems, and multidisciplinary applications, the new standard HPF 2.0 of the data parallel language High Performance Fortran (HPF) provides approved extensions for task parallelism that allow nested task and data parallelism. Unfortunately, these extensions allow the spawning of tasks but do not allow interaction like synchronization and communication between tasks during their execution and therefore might be too restrictive for certain application classes. E.g., they are not suitable for expressing the complex interactions among asynchronous tasks as required by multidisciplinary applications. They do not support any parallel programming style that is based on non-deterministic communication patterns.

This paper discusses the extension of the task model provided by HPF 2.0 with a task library that allows interaction between tasks during their lifetime, mainly by message passing with an user-friendly HPF binding. The same library with the same interface can also be used for single processors in the local HPF model. The task model of HPF 2.0 and the task library have been implemented in the ADAPTOR HPF compilation system that is available in the public domain. Some experimental results show the easy use of the concepts and the efficiency of the chosen approach.

Keywords: Data Parallelism, Task Parallelism, High Performance Fortran

1   Introduction

High Performance Fortran (HPF) [10] is a data parallel, high level programming language for parallel computing that might be more convenient than explicit message passing and that should allow higher productivity in software development. With HPF, programmers provide directives to specify processor and data layouts, and express data parallelism by array operations or by directives specifying independent computations.

Some users are reluctant to use HPF because many applications do not completely fit into the data parallel programming model. The applications contain

P. Amestoy et al. (Eds.): Euro-Par'99, LNCS 1685, pp. 833–844, 1999.
© Springer-Verlag Berlin Heidelberg 1999

data parallelism, but task parallelism is needed to represent the natural computation structure or to enhance performance. Many results verify that a mixed task/data parallel computation can outperform a pure data parallel version (e.g. see [7]) if the granularity of the data is not sufficient; a typical example is the pipelining of data parallel tasks for image and signal processing. Multiblock codes are more naturally programmed as interacting tasks, and applications that interact with external devices. Another important application area is the coupling of different simulation codes (multidisciplinary applications).

Some features supporting task parallelism are available as approved extensions in HPF 2.0[10]. The TASK_REGION construct provides the means to create independent coarse-grain tasks, each of which can itself execute a data-parallel or nested task-parallel computation. This kind of task parallelism has been implemented and evaluated within the public domain HPF compilation system ADAPTOR [3]. Currently, ADAPTOR is the only HPF compiler supporting the task model as defined in the new HPF 2.0 standard.

But the HPF task model does not allow interaction between the independent tasks during their execution. The introduction of a task library that allows interaction of HPF data parallel tasks during their lifetime enhances the possibilities of the current model. This paper describes the functionality of such a library and shows that it goes conform with the existing language concepts. The implementation of the task library in an existent HPF compiler should be rather straightforward like it was in the ADAPTOR compilation system.

The rest of this paper is organized as follows. Section 2 describes related work regarding the integration of task and data parallelism. The current HPF task model and its restrictions are outlined in Section 3. Section 4 introduces the HPF task library for interaction of data parallel tasks. Some experimental results in Section 5 show the applicability of the advanced tasking model.

2   Related Work

The need of task parallelism and its benefits have been discussed by many authors and at many places (e.g. in [6]). The promising possibility of integrating task parallelism within the HPF framework has attracted much attention [13, 9, 10, 4]

An integrated task and data parallel model has been implemented in the Fx compiler at Carnegie Mellon [14]. It is mainly based on the specification of subgroups and the assignment of arrays, variables and computations to the subgroups. Communication between the tasks must be visible at the coordination level specified as a TASK_REGION. The execution model is not based on message passing, programs can still be executed in the serial model. A variation of this model has become an approved extension for HPF 2.0 [10].

Kohr, Foster et al. [7] developed a coordination library based on MPI to exchange distributed data structures between different HPF programs. The data parallel language has not to be extended at all. Unfortunately, their implementation supports only coupling of different HPF programs, but not of different

HPF tasks created within a task region. But the ideas can be generalized for nested task parallelism.

Zima, Mehrotra et al. [4] propose an interaction mechanism using shared modules with access controlled by a monitor mechanism. A form of remote procedure call (RPC) is used to operate on data in the shared module. The monitor mechanism ensures mutual exclusion of concurrent RPC's to the same module. This concept enhances modularity and is particularly good for multidisciplinary applications. Due to the use of an intermediate space, it appears less well suited for fine-grained or communication-intensive applications.

Orlando and Perego [11] provide run-time support for the coordination of concurrent and communicating HPF tasks. $COLT_{HPF}$ provides suitable mechanisms for starting distinct data-parallel tasks on disjoint group of processors. It also allows the specification of the task interaction on a high level from which they generate automatically corresponding code skeletons. For the implementation of communication between the data parallel tasks they use the pitfalls algorithm [12].

3   Support of Task Parallelism in HPF

The introduction of task parallelism in High Performance Fortran is strongly connected with the ON directive that allows the user to control explicitly the distribution of computations among the processors of a parallel machine. It allows dividing processors into subgroups which is essential for task parallelism. The RESIDENT directive tells the compiler that only local data is accessed and no communication has to be generated. A code block guided by the ON and RESIDENT directive is called a *lexical task*. An execution instance of a lexical task is called an *execution task*. Every execution task is associated with a set of *active processors* on which the task is executed.

Though the ON and RESIDENT directive on their own allow task parallelism, HPF 2.0 provides the TASK_REGION construct. A task region surrounds a certain number of lexical tasks where other statements in the TASK_REGION can be used to specify data transfers and task interaction between these tasks.

3.1   The ON Directive

In the HOME clause of the ON directive, the user can specify a processor array or a processor subset or an array (template) or a subsection of an array (template). The ON directive restricts the active processor set for a computation to those processors named in the home, or to the processors that own at least one element of the specified array or template. As HPF allows the mapping of arrays to processor subsets, the exploitation of task parallelism is more convenient.

It should be noted that the ON directive only advises the compiler to use the corresponding processors to perform the ON statement or block. Certain statements cannot be executed by an arbitrary processor subset, e.g. allocation, deallocation and redistribution of arrays must include all processors involved in

```
          real, dimension (N) :: A1, A2
!hpf$     processors PROCS(8)
!hpf$     distribute A1 (block) onto PROCS(1:4)
!hpf$     distribute A2 (block) onto PROCS(5:8)
          ...
          do IT = 1, NITER
!hpf$       task_region
!hpf$       on (PROCS(1:4)), resident
              call TASK1 (A1,N)
            if (mod (IT,2) == 0) A1 = A2      ! task interaction
!hpf$       on home (A2), resident
              call TASK2 (A2,N)
!hpf$       end task_region
          end do
```

Fig. 1. Mixed task and data parallelism in High Performance Fortran.

it. But the compiler should inform the user if it overrides the user's advice. Not respecting the ON directive can suppress the task parallelism intended by the user.

3.2 The RESIDENT Directive

If a statement or a block should be executed by a processor subset, the compiler must make sure that all data is mapped onto the corresponding active processors. This data transfer can involve other processors that are not part of the active processors. Unfortunately, compilers are conservative and can also introduce synchronization or communication where it is not really necessary. The RESIDENT directive tells the compiler that only local data is accessed and no communication has to be generated. This guarantees that only the specified processors are involved and the code can be skipped definitively by the other processors. The RESIDENT directive is very useful for task parallelism where subroutines are called. It gives the compiler the important information that within the routine only resident data is accessed. This might also allow the compiler to respect the specified HOME where it was not possible before.

3.3 The TASK_REGION Construct

Though the ON and RESIDENT directive on their own allow task parallelism, HPF 2.0 provides the TASK_REGION construct. A TASK_REGION surrounds a certain number of lexical tasks. The TASK_REGION construct specifies clearly where task parallelism appears, it provides syntactical restrictions (every ON directive must be combined with the RESIDENT directive), and the user guarantees no I/O interferences between the different execution tasks. Data dependencies within a TASK_REGION can result in a serial execution of tasks. Nevertheless, parallelism might be achieved due to the outer loop around the TASK_REGION resulting in a pipelined execution of the tasks (see example in Fig. 1).

The HPF model allows nested task and data parallelism. There is no restriction that execution tasks within one TASK_REGION are executed on disjoint processor subgroups. The compiler can ignore the ON directive without changing the semantic of the program. Task interaction must be visible at the coordination level which is the code within the task region outside the lexical tasks.

The main disadvantage is the lack of any possibility for task interaction during the execution of the tasks. Communication between tasks in the TASK_REGION is deterministic and does not allow any self-scheduling.

4   The HPF Task Library

The HPF task library is intended to allow interaction between different data parallel tasks via message passing. The use of new language constructs has not been considered as it would complicate the whole development.

4.1   Problems and Design Issues

Any task interaction can only be useful if the concurrent execution of the tasks is guaranteed. All execution tasks must be mapped to disjoint processor subsets that is not mandatory for the task concept of HPF. With task interaction, an HPF program might no longer run on a serial machine. Furthermore, task interaction requires the unique identification of tasks, e.g. by a task identifier, that is used to specify the source and destination of message passing.

In the first place, it might have been useful to define the message passing routines between data parallel tasks in analogy to MPI [8], but an own HPF binding of such routines offers a lot of advantages. It allows to pass whole arrays or subsection of arrays to the routines without specifying the data type, the distribution, or the size of the data. The data must not be contiguous and the use of derived data types like in MPI is not really necessary. The routines do not require a communicator as the communicator is implicitly given by the current task nesting level (see also Section 4.4).

The task library is not intended to be realized compiler-independently but as a part of the HPF compiler. Most of the necessary functionality must be available in the HPF runtime system of an HPF compiler. As internal descriptors for arrays and their mappings and internal representations for communication schedules are far away from any standardization, the task library is most efficiently implemented by the compiler vendor itself.

The routines of the HPF_TASK_LIBRARY are available in global, local and serial HPF programs by the USE statement of Fortran. Two subroutines, HPF_TASK_INIT and HPF_TASK_EXIT support the initialization and termination of data parallel tasks. They should verify at runtime that the tasks of the current context are really mapped to disjoint processor subgroups. Furthermore, at the end it should be verified that there are no pending messages between the tasks. The two subroutines HPF_TASK_SIZE and HPF_TASK_RANK return, similar to their MPI counterparts, the size (number of data parallel tasks in the current context) and the rank of the calling task ($1 \leq rank \leq size$).

There are no mechanisms for the creation or termination of task processes at runtime. Tasks can only be defined as subtasks within the current context. Support of task migration can be an option of the runtime system but is not part of the language or library.

4.2 Spawning of HPF Tasks

Tasks can be spawned within a TASK_REGION. Their concurrent execution requires that the execution of tasks does not depend on any values computed within the tasks and that there are no dependencies at the coordination level. The user has to assert this property by the keyword INDEPENDENT. The following example shows how to invoke data parallel tasks for pipelined data parallelism as described in section 5.2.

```
!hpf$ processors PROCS (20)
      ...
!hpf$ independent task_region
!hpf$ on (PROCS(1:4)), resident              ! will be task 1
          call STAGE1 ()
!hpf$ on (PROCS(5:10)), resident             ! will be task 2
          call STAGE2 ()
!hpf$ on (PROCS(11:16)), resident            ! will be task 3
          call STAGE2 ()
!hpf$ on (PROCS(17:20)), resident            ! will be task 4
          call STAGE3 ()
!hpf$ end task_region
```

All execution tasks within the INDEPENDET TASK_REGION get a task identifier starting with 1. Processor subsets in the ON directive must not be known at compile time. Therefore the user can implement algorithms on its own to compute the processor subsets for the scheduling of his data parallel tasks.

The HPF Task Library allows also the coupling of separately compiled data parallel programs. Some parallel architectures provide the possibility of loading distinct executables on distinct nodes. Then the data parallel programs will be executed on disjoint processor subsets that are specified on an outer level. The task initialization within the HPF runtime system guarantees that all data parallel tasks know their current task context.

HPF allows the use of local routines via the EXTRINSIC mechanism. A local routine allows to write single-processor code that works only on data that is mapped to a given physical processor. The processors executing the local subprogram can be viewed as single processor tasks where task interaction can be used in exactly the same way as for data parallel tasks. The routines in the local model have the same syntax as the corresponding routines for communication between data parallel tasks.

4.3   Communication between Data Parallel Tasks

For the sending of data (scalars, arrays or array sections of any type), the task identifier of the destination task dest must be specified. For the receiving of data, the specification of the source task is optional to allow the receiving of data from any task. Every send must have a matching receive.

```
subroutine HPF_SEND (data, dest [,tag] [,order])
subroutine HPF_RECV (data [,source] [,tag])
```

The optional ORDER argument for the sending of data must be of type integer, rank one, and of size equal to the rank of DATA. Its elements must be a permutation of $(1, 2, ..., n)$, where n is the rank of the data. If the ORDER argument is available, the axes of the sending data will be permuted similar to the extended TRANSPOSE routine of HPF 2.0.

Due to the HPF/Fortran 90 binding, the routines HPF_SEND and HPF_RECV can be called with array arguments and the arguments can be named. This makes the use of the routines easier and better readable.

Figure 2 shows the communication pattern between the single processors if TASK 1 runs on a 2×2 processor subset and TASK 2 on a processor subset of three processors. The implementation of point-to-point communication between data parallel tasks results in communication between the processors of the two processor subgroups that are involved.

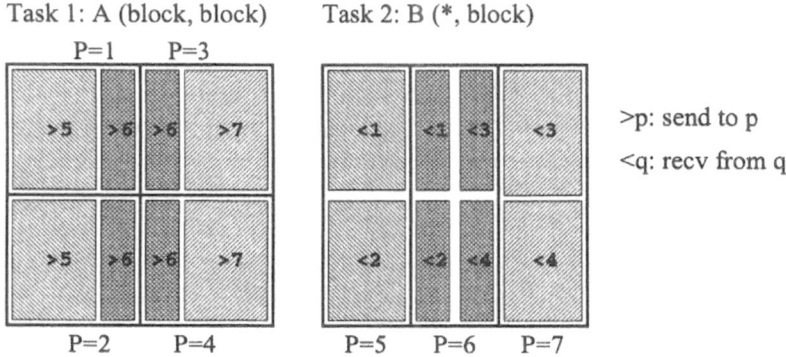

Fig. 2. Example of point-to-point communication between data parallel tasks.

In fact, the implementation of the message passing routines should be straightforward for all HPF compilers as the send and receive operation together correspond to the HPF array assignment between the two arguments. The only difference is that due to the local allocation of the arrays in the processor subset the exchange of the mapping information (*descriptor exchange*) is necessary.

Sending and receiving of distributed data must be assumed to be blocking. When executing shift operations across a chain of tasks or when two tasks are exchanging data, one needs to order the sends and receives correctly (e.g. even tasks send, then receive, odd tasks receive first, then send) so as to prevent cyclic dependencies that may lead to deadlock. When using a send-receive routine, the system takes care of these issues.

```
subroutine HPF_SEND_RECV (send_data, dest, recv_data, source
                          [,send_tag] [,recv_tag] [,order])
```

The point-to-point communication between two data parallel tasks causes a certain overhead due to the exchanging of the mapping information. This overhead as well as the computation of the schedule between the single processors of the tasks can be reduced by introduction of internal handles (*request*). The use of these routines might cause serious problems when the distribution or sizes of data has changed.

```
subroutine HPF_SEND_INIT (data, dest, request [,tag] [,order])
subroutine HPF_RECV_INIT (data, source, request [,tag])
subroutine HPF_TASK_COMM (request)
```

Collective communication like in MPI might also be useful for HPF tasks. Especially the broadcast of data and the barrier proved to be useful. It should be observed that the context of these operations is given by the current task context. A barrier synchronizes the tasks of the current context, not the processors within this task.

```
subroutine HPF_BCAST (data [,root])
subroutine HPF_BARRIER ()
```

4.4   Nested Task Parallelism

Task parallelism can be nested. Every independent TASK_REGION defines a set of data parallel tasks, in each task new subtasks can be created hierarchically.

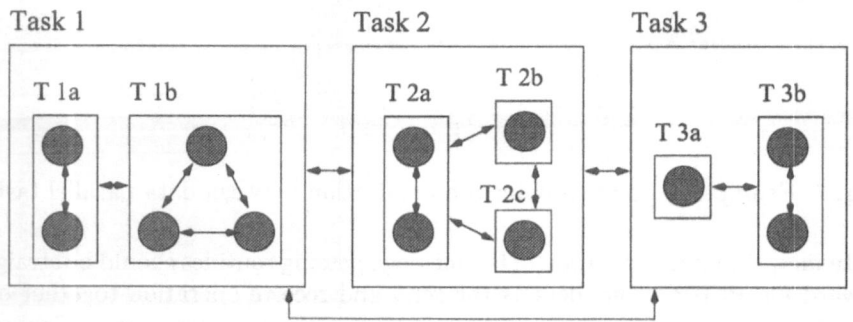

Fig. 3. Task interaction for nested task parallelism.

Task interaction is only possible between the tasks within one context. In the example of Figure 3, task communication is possible between the tasks TASK 1,

TASK 2 and TASK 3 as long as they are not spawned into subtasks. When TASK 2 is divided into the three subtasks TASK 2a, TASK 2b and TASK 2c, only communication between these subtasks is possible. Task communication between e.g. TASK 1a and Task 2a is not possible.

5   Experiments and Results

This section presents some typical patterns for using the HPF task library. More examples regarding task parallelism in HPF and more detailed results can be found in [2].

5.1   Recursive Tree Structured Algorithms

Many problems can be solved recursively by splitting the problem into two partial problems of the half size. Typical examples are the Fast Fourier Transformation (FFT) that plays a key role in many areas of computational science and engineering. Another recursive tree structured algorithm is the Barnes-Hut algorithm for N-body problems [1]. The recursive splitting can be used to split the active processors into two processor subsets via task parallelism. The two subtasks can exchange data via the task library. When the recursion ends up in a single processor, it might be more efficient to call a serial and iterative algorithm.

Table 1 shows the execution times (in seconds) of an one-dimensional FFT ($N = 2^{19}$) that has been implemented in this way. The task parallel version shows reasonable speed-ups when compared with the serial version of the FFT based on the efficient Cooley-Tukey algorithm [5]. It outperforms a data parallel version that uses indirect addressing.

	P=1	P=2	P=4	P=8	P=16	P=32
serial	1.506	n.a.	n.a.	n.a.	n.a.	n.a.
data	5.399	3.363	1.955	1.157	0.671	0.444
task	1.784	1.144	0.762	0.493	0.295	0.184

Table 1. Results for one-dimensional FFT ($N = 2^K, K = 19$) on IBM SP2.

5.2   HPF Task Farming

The HPF task library allows the realization of task farming within a pipeline of data parallel tasks. Let the pipeline have three stages. While the first and the last stage are exactly one task, there are a certain number of worker tasks for the second stage. The load is scheduled to available tasks on this second stage. A worker task sends a ready signal to the first stage when it is free for doing work. It receives the data, works on it and sends it at the end to the final stage NT in the pipeline where NT is the number of tasks. Figure 5 shows the principle of message passing between the different tasks.

```
      recursive subroutine DO_IT (A, N, NP)
      use HPF_TASK_LIBRARY
      integer :: N, NP, N2, NP2, SIZE, RANK, IP
      real, dimension(N) :: A, H
!hpf$ processors PROCS(1:NP)
!hpf$ distribute (block) onto PROCS :: A, H
      call HPF_TASK_INIT ()
      call HPF_TASK_RANK (rank=RANK)
      IP = 3 - RANK
      call HPF_SEND_RECV (send_data=A, dest=IP, recv_data=H, source=IP)
      ...
      if (NP .eq. 1) then
         call DO_IT_SERIAL (A, N); return
      end if
      NP2 = NP/2; N2   = N/2
!hpf$ independent task_region
!hpf$ on (PROCS(1:NP2)), resident
         call DO_IT (A(1:N2), N2, Np2)
!hpf$ on (PROCS(NP2+1:NP)), resident
         call DO_IT (A(N2+1:N), N2, Np2)
!hpf$ end task_region
      ...
      call HPF_TASK_EXIT ()
      end subroutine DO_IT
```

Fig. 4. Example of a recursively nested task and data parallel program.

Task farming becomes especially useful when running data parallel programs on heterogeneous architectures.

6 Conclusions

The HPF task model allows the coupling of data parallel tasks in a simple way as long as the interaction between the tasks is completely visible at the coordination level in the TASK_REGION. As all information about the mapping of the arrays is available, no exchange of descriptors or distribution information is necessary. The deterministic communication allows the program still to be run on a serial machine. This task model is relatively easy to implement in an HPF environment if the ON directive and related clauses are supported as well as the mapping of data to processor subsets.

Moving task interaction within the tasks is rather straightforward. An assignment of data from one task to data of another task becomes a send in the first and a receive in the other task. New allocated data within the tasks (e.g. local variables) can now also be exchanged. Furthermore, only task interaction within the task allows the receiving of values from any other task resulting in non-deterministic behavior. Many well established parallel programming styles like farming can be used in a data parallel framework. Task parallelism can now also be exploited for separately compiled HPF programs. The task library combines

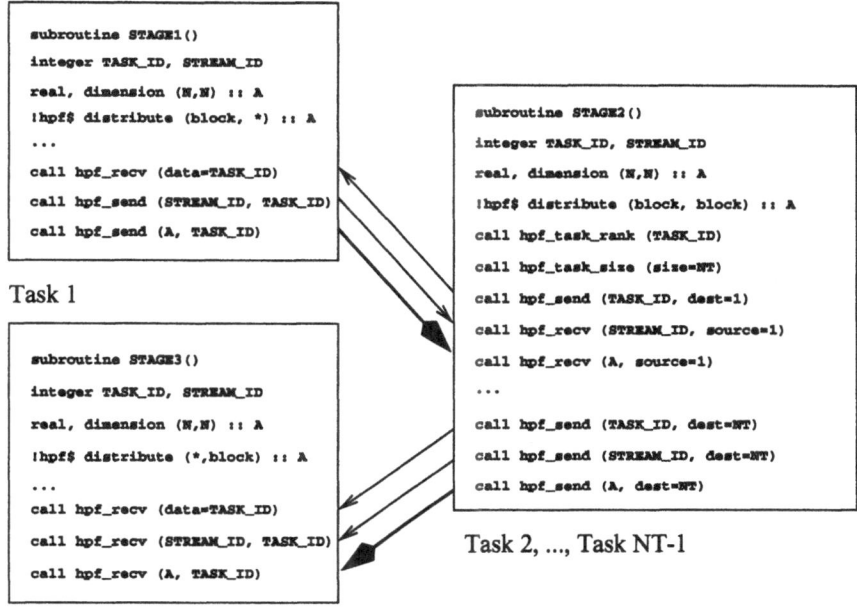

Fig. 5. Task farming for data parallel tasks.

the advantages of the HPF task model based on processor subsets and a sequential semantic with the advantages of the great flexibility when using message passing. The concept of the task library goes conform with the other extrinsic models of HPF and therefore allows the combination with other programming models, e.g. MPI.

The use of communication via send and receive in a language like HPF seems to destroy the high level intention. But data parallel computations still do not need any message passing, it is restricted only to the interaction of running data parallel tasks where it is indeed quite natural. The high level nature of HPF is taken into account by providing a high level HPF binding of the communication routines (no arguments for size, data type, distribution, context, etc.).

The ADAPTOR HPF compilation systems provides task parallelism as specified in HPF and the task library as described in this paper. This library or a more standardized library with a similar and improved functionality could be provided by any HPF compiler vendor as it can be implemented rather easily by using the HPF runtime system that has to support redistributions in any case. Experiments so far have shown that the use of task parallelism in HPF is very user-friendly, no more limited, and, most important, efficient enough to be a very attractive alternative. Its support became also necessary as ADAPTOR is intended to support nested process and thread parallelism for hierarchical systems in future.

Acknowledgements

Most thanks are due to Salvatore Orlando (Universita di Venezia) and Raffaele Perego (CNUCE, Italy) for many valuable discussions and a lot of technical hints. I am indebted to Mike Delves (NA Software, Liverpool) for the great idea to make the task library also available in the local HPF model.

References

[1] J. Barnes and P. Hut. A hierarchical $O(NlogN)$ force calculation algorithm. *Nature 4*, 324:446–449, 1986.

[2] T. Brandes. Implementation and Evaluation of Nested Task and Data Parallelism for High Performance Fortran within the ADAPTOR Compilation System. Working paper (unpublished), GMD, 1998. available via anonymous ftp as ftp.gmd.de:/GMD/adaptor/docs/tasking.ps.

[3] T. Brandes and F. Zimmermann. ADAPTOR - A Transformation Tool for HPF Programs. In K. Decker and R. Rehmann, editors, *Programming Environments for Massively Parallel Distributed Systems*, pages 91–96. Birkhäuser Verlag, Apr. 1994.

[4] B. Chapman, M. Haines, P. Mehrotra, H. Zima, and J. Van Rosendale. Opus: A Coordination Language for Multidisciplinary Applications. *Scientific Programming*, 6(4):345–361, 1997.

[5] J. W. Cooley and J. W. Tukey. An Algorithm for the machine calculation of complex Fourier series. *Mathematical Computing*, 19:297–301, 1965.

[6] P. Dinda, T. Gross, D. O'Hallaron, E. Segall, E. Stichnoth, J. Subhlok, J. Webb, and B. Yang. The CMU Task Parallel Program Suite. Technical Report CMU-CS-94-131, School of Computer Science, Carnegie Mellon University, Mar. 1994.

[7] I. Foster, D. Kohr, K. R., and A. Choudhary. Double Standards: Bringing Task Parallelism to HPF via the Message Passing Interface. In P. Pittsburgh, editor, *Supercomputing '96*, Nov. 1996.

[8] W. Groop, E. Lusk, and A. Skjellum. *Using MPI : Portable Parallel Programming with the Message-Passing Interface*. Scientific and Engineering Computation Series. The MIT Press, Cambridge, MA, 1994.

[9] T. Gross, D. O'Hallaron, and J. Subhlok. Task Parallelism in a High Performance Fortran Framework. *IEEE Parallel and Distributed Technology*, 2(2):16–26, 1994.

[10] High Performance Fortran Forum. High Performance Fortran Language Specification. Version 2.0, Department of Computer Science, Rice University, Jan. 1997.

[11] S. Orlando and R. Perego. $COLT_{HPF}$, a Coordination Layer for HPF Tasks. Technical Report Series on Computer Science CS-98-4, Universita ca' Foscari di Venezia, Mar. 1998. Paper submitted to Concurrency: Practice and Experience.

[12] S. Ramaswamy and P. Banerjee. Automatic Generation of Efficient Array Redistribution Routines for Distributed Memory Multicomputers. In *Proceedings of the Fifth Symposium on the Frontiers of Massively Parallel Computations (FRONTIERS'95)*, pages 342–394, Feb. 1995.

[13] S. Ramaswamy, S. Spatnekar, and P. Banerjee. A Framework for Exploiting Task and Data Parallelism on Distributed Memory Multicomputers. *IEEE Transaction on Parallel and Distributed Systems*, 8(11):1098–1116, Nov. 1997.

[14] J. Subhlok and B. Yang. A New Model for Integrated Nested Task and Data Parallel Programming. In *PPOPP 97*, 1997.

A Run-Time System for Dynamic Grain Packing

João Luís Sobral, Alberto José Proença

Departamento de Informática - Universidade do Minho
4709 Braga Codex - PORTUGAL
{jls, aproenca}@di.uminho.pt

Abstract. The SCOOPP (Scalable Object Oriented Parallel Programming) system is an hybrid compile and run-time system, that extracts parallelism, supports explicit parallelism and dynamically serialises parallel tasks in excess, to dynamically scale applications through a wide range of target platforms. This paper describes the run-time system of the current SCOOPP prototype - the ParC++ - and its mechanism to serialise parallelism. Low level performance results are presented, which indicate that the proposed methodology is effective and provides an high reduction in parallelism overheads. These features can improve the scalability of parallel applications with excessive parallelism.

1 Introduction

When developing parallel applications, programmers are often faced with a parallelism granularity control decision: a larger number of fine parallel grains may help to scale up the parallel application and it may also improve the load balancing; however, it requires more control overhead, and performance may degrade due to excessive communication over computation, if grains are too fine.

Static granularity control [1][2] is usually applied to fine grained tasks, whose behaviour is known at compile-time. However, parallel applications with irregular parallel tasks - where tasks behaviour is not known at compile-time - require dynamic granularity control to achieve an acceptable performance. Programmer based dynamic granularity control adds an extra burden on the programmer activity, requires a deep knowledge of both architecture and algorithm behaviours and decreases the program clarity.

The SCOOPP system [3] is an hybrid compile and run-time system, that extracts parallelism, supports explicit parallelism and dynamically serialises parallel tasks in excess at run-time to dynamically scale applications through a wide range of target platforms. This paper focus on the methodology to remove excess parallelism in the run-time system in the current SCOOPP prototype (ParC++) and it attempts to measure its impact on performance.

This work was partially supported by the SETNA-ParComp project (Scalable Environments, Tools and Numerical Algorithms in Parallel Computing), under PRAXIS XXI funding (Ref. 2/2.1/TIT/1557/95).

P. Amestoy et al. (Eds.): Euro-Par'99, LNCS 1685, pp. 845-852, 1999.
© Springer-Verlag Berlin Heidelberg 1999

Section 2 presents the SCOOPP system and Section 3 discusses the alternatives to dynamically remove excess parallelism. Section 4 and 5 present the ParC++ system and some performance results. Section 6 concludes the paper with suggestions for future work.

2   SCOOPP System Overview

Explicit parallelism is specified in SCOOPP through a special type of object: the parallel object. These objects model parallel activities and may be placed at remote processing nodes. Parallel objects communicate through methods calls, either asynchronously, when no return value is expected (i.e. the caller does not wait for method completion) or synchronously, when a value is expected. References to parallel objects may be freely copied or used as method arguments, supporting the development of parallel applications with complex inter-object communication graphs. When this feature is not used the inter-object communication graph is a tree.

Parallel objects may also create "sequential objects", taking advantage of existing code. These objects are placed in the context of the parallel object which created them and only copies of them are allowed to move between parallel objects. Method calls on these objects are always synchronous.

The SCOOPP system granularity control is accomplished in two steps: at compile-time - the compiler and/or the programmer specifies a large number of parallel objects - and at run-time - parallel objects are packed into larger grains, according to the application/target platform behaviour and based on security and performance issues.

Parallelism extraction is performed by transforming selected sequential objects into parallel objects, whereas parallelism serialisation (i.e. grain packing) is performed by transforming parallel objects into sequential ones. In this serialisation process, compiler transformed parallel objects are preferred for serialisation; parallel objects are only serialised when all compiler transformed parallel objects have been serialised (more details in [6]).

3   Removing Excess Parallelism

Conventional grain packing mechanisms [4][5] are based on fork/join parallelism. Grain-size is increased by ignoring the fork construct and executing tasks sequentially, instead of spanning a new parallel activity to execute the forked task. This mechanism to increase the grain-size is fault free when tasks have no intermediate communication, i.e., inter-tasks dependencies form direct acyclic graphs. However, parallel objects (tasks) may perform intermediate communications, which, in turn, may generate rather complex inter-tasks communication graphs, namely, on explicit parallelism. Alternative mechanisms are required to increase the grain-size and these should guarantee a correct program behaviour.

The SCOOPP system packs grains by joining several parallel objects in a single computing grain and by serialising intra-grain operations. Intra-grain method calls - between objects within the same grain - are synchronous and are usually performed directly as a normal procedure call. Asynchronous inter-grain calls are implemented through standard inter-tasks communication mechanisms.

On the SCOOPP ParC++ prototype, the development of the grain packing mechanism was guided by the following requirements:

- *correctness* - grain packed tasks should have the correct behaviour, namely, packing should not introduce deadlock;
- *reversibility* - the system should provide both packing and unpacking operations;
- *fairness* - packed tasks should have the same opportunities of execution as non packed tasks.

To ensure program *correctness* some intra-grain operations should not be serialised. Deadlock may occur with cyclic inter-grain communication - when asynchronous method calls are changed to synchronous (Fig.1a) - and in inter-grain synchronous calls when packing cross referencing objects in the same grain (Fig.1b); in this case, one of the grains must span a new thread to serve the incoming request, to avoid a deadlock.

Without grain packing, both calls in Fig.1a are executed simultaneously by two separate threads/processes. Since calls are performed asynchronously, if either object is busy, the request is queued into the object message buffer. When these grains are packed together and calls are changed to synchronous, the same thread must execute both calls. It may first execute the call from object a to object b suspending the execution of a and executing object b code. Later, when executing a call from object b to object a, deadlock will occur if object a can not be re-entered. To avoid this, each thread traces the intra-grain calls and when a cyclic call is detected the call closing the cycle is not performed directly, but through a standard inter-object communication mechanism.

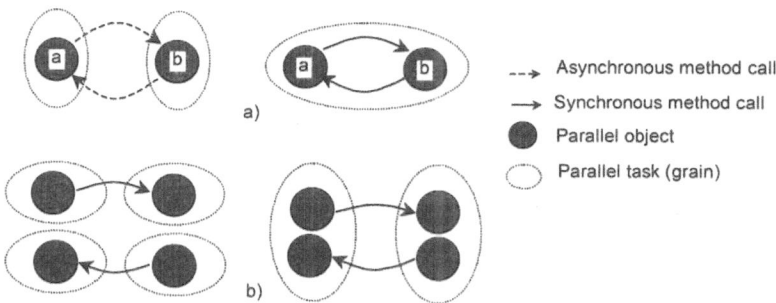

Fig. 1. Possible deadlock situations for grain packing: a) in intra-grain cyclic calls; b) in inter-grain synchronous calls

To support the *reversibility* of the packing process, packed parallel objects are not transformed into "pure" sequential objects. Instead, they remain with "parallel object" functionality, but they share a thread with other objects in the same grain. As a consequence, the management of packed objects has a small overhead when compared with sequential objects. However, reversing the packing process is simpler and faster on this approach, since packing/unpacking is just sharing/not sharing a thread among parallel objects. Moreover, remote access to packed objects is simplified since it is an intrinsic feature of parallel objects.

To support *fairness*, requests for method calls are not allowed to wait longer than a predefined time slot before being executed, reducing the deadlock impact showed in Fig.1b. If s is the time-slice allocated to each thread and n the number of threads on a node, any request to a free object will start being served in less than $(n-1)*s$, without grain packing. With grain packing, a longer wait may occur, since packed parallel objects may suffer starvation from a long running method call in the same grain, as a consequence of the intra-grain serialisation. The implemented maximum waiting for external requests is directly proportional to the number of objects on each node.

Fairness is further improved by establishing a non-fixed association among threads and computing grains, i.e. any local thread may serve a request for any local object (Fig.2). In contrast to a fixed association model, this non-fixed association gives equals execution opportunities to all the objects in a processing node; it automatically balances the size of each grain, and it automatically performs local packing/unpacking operations, since threads are associated to objects in requests arrive. However, it does require additional checks on method calls for inter-thread synchronisation.

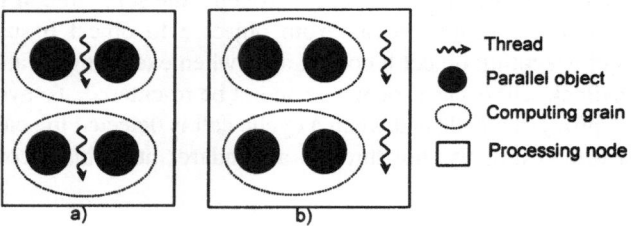

Fig. 2. Fixed a) and non-fixed b) association between threads and computing grains

4   The Run-Time System in ParC++ Prototype

The current SCOOPP ParC++ prototype, supports some extensions to C++. It includes a ParC++ pre-processor, several C++ support classes and a run-time system. The pre-processor analyses the application - by retrieving information about the declared parallel objects - and generates code for remote object creation and remote method invocation. It may also mark C++ objects to be transformed into parallel objects by the run-time system. The prototype has been tested on several distributed memory parallel architectures: a Parsytec MC-3 (with 112 Transputer based nodes), a Parsytec PowerExplorer (16 nodes, each one with one PowerPC as a computing processor and one Transputer as a communication processor) and PC clusters.

The ParC++ run-time system (RTS) is based on three object classes: proxy objects (PO), implementation objects (IO) and server objects (SO).

A PO represents a local or a remote parallel object and has the same interface as the object it represents. It transparently replaces a parallel object and forwards all method invocations to the IO, the object that implements the parallel object methods. SO are active entities (i.e. threads) that continuously receive messages from PO objects, calling the requested method on local IO and, if needed, returning the result value to the caller. A PO maintains the node address of the IO, as well as its system identification and the identification of its SO (which is shared by all SO for packed objects, under the non-fixed association model).

On inter-grains method calls the PO forwards the call to a remote SO, which activates the corresponding method on the IO (calls a in Fig.3). On intra-grain calls, the PO directly calls the corresponding method on the local IO (call b in Fig.3), since the PO is placed on the same processing node as the IO. To preserve the program correctness, when the IO is busy - when it is executing a method call that does not allow another concurrent method execution or under a cyclic call - the call is forwarded to a local SO, which will delay the request until the IO is ready.

The RTS provides run-time grain-size adaptation and load balancing through the object manager (OM). The application entry code creates one instance of this object on each processing node. The OM controls the grain-size adaptation by determining the number of SO's per node. This decision is based on run-time measurements of traffic and computing loads, currently undergoing work on related I&D projects [7].

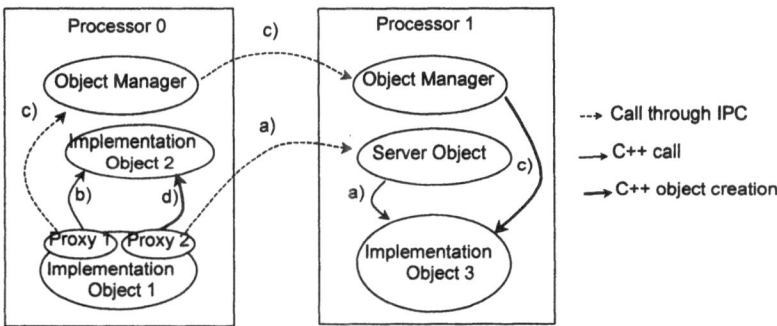

Fig. 3. Inter-grains a) and intra-grain b) method calls; RTS c) and d) direct object creation

On receiving a request to create a parallel object, the RTS creates a new local PO. The first task of the newly created PO is to request the creation of the IO. When parallelism is not being removed, the OM selects a processing node to create a new IO (according to the current load distribution policy), selects or creates the associated SO and returns their identifier to the PO (calls c in Fig.3). When the RTS is removing excess parallelism, the PO directly creates the parallel object, by locally creating the

IO (call $\underline{d}$ in Fig.3) and notifying the RTS. The PO always destroys a local IO; non local objects are destroyed by the RTS, upon a request from the PO.

The RTS may temporarily increase the number of SO's in a node to prevent deadlock and improve fairness. This happens when external requests are waiting longer than a predefined time. Additional SO's may be destroyed when the requests are completed.

5   Performance Evaluation

One of the main goals of SCOOPP is dynamic scalability. To measure its impact on performance, a low level evaluation was performed. The evaluation runs measures the execution times of different size tasks (related to granularity) and the overhead time to manage the parallel objects on 2 parallel systems. These benchmarks systems had no application objects, and both used point-to-point Transputer based communications: a PowerExplorer with Parix 1.3 and a MC-3 with Parix 1.2 (both providing a clock precision of 1μs). The measured values were taken on the ParC++ system with support classes version 1.14 and the pre-processor version 1.61.

The main overheads in OO applications include the object creation, the method calls and the object destruction. These operations usually correspond to object space allocation and initialization, procedure calls and object space de-allocation. On parallel OO applications, the additional overheads to manage parallel objects in SCOOPP include the RTS creation of PO, SO and IO (on object creation), the proxy request for remote method execution (on method calls) and the RTS deletion of PO, SO and IO (on object destruction).

The evaluation runs aim to measure the ratio between the objects management overhead and the execution time of the object's computational task (the object grain-size). Fig.4 shows 4 sets of measured times, required to create, activate and destroy one parallel object on a 2 node system:
- with no granularity control (non packed objects), with remote placement;
- with granularity control (packed objects) and remotely packed;
- packed into another local grain;
- included in the source grain.

The first set has no RTS optimisation, while the other three show different approaches to remove excess of parallelism.

An overhead of 100% in Fig.4 means that the time required to manage a parallel object equals the effective execution time of the required task, i.e., it is an indication of a crossover value to make the decision to pack objects, under these benchmark conditions. When adding more computing nodes or computational tasks, all the curves move up in Fig.4.

For an overhead of 100%, Fig.4 shows that remotely packed objects improves the overhead ratio by supporting efficient finer parallelism granularity, mainly due to object creation time: the server object creation (and its process) is not required for non-local packed objects. This improvement is strongly dependent on the process management overheads (process creation, scheduling and destruction).

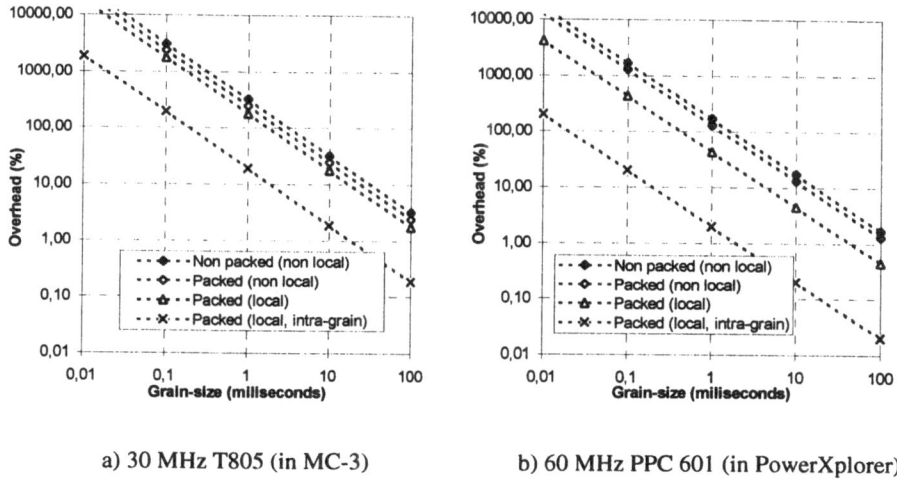

a) 30 MHz T805 (in MC-3)   b) 60 MHz PPC 601 (in PowerXplorer)

Fig. 4. Parallelism overhead as a function of the grain-size

Further improvement is obtained when these packed objects are also locally placed, due to a swap from non-local to local IPC communication, and depends on communication latency and bandwidth.

A significant improvement can be seen when parallel objects are packed into the source grain, since the RTS operations on packed objects are replaced by the proxy intervention, where IPC is swapped by a procedure call.

When the evaluation runs moves from MC-3 to the PowerXplorer, the improvements in parallelism granularity – i.e., the finest supported grain-size for a given overhead ratio – are from one to two orders of magnitude, only due to a faster processor at the computing nodes, with the same communication processor and bandwidth. This suggests that with faster processors, the overhead ratio is further improved. However, if faster communication lines are used, it can be expected no change on local packing and a significant improvement on remotely packed objects.

Other low level results have been presented in [8] and application oriented performance results have been presented in [3], which obtained a reduction on an image processing application's execution time down to 79% on the MC-3 and 56% on the PowerXplorer.

Conclusions and Future Work

A dynamic grain packing mechanism largely decreases the minimum cost of an programmer/pre-processor specified parallel task, which allows the expression of parallel OO applications in a natural way. Grain packing should encourage the programmer and/or the pre-processor to detect and specify all the parallelism opportunities and let the RTS to remove the excess parallelism.

Evaluation of the current ParC++ prototype showed that the specification of parallel objects through a new keyword – which required a pre-processor – also allows a natural implementation of implicit parallelism extraction by the same pre-processor. This feature, together with the use of proxy concept to implement remote parallel objects, provides a powerful tool for a transparent dynamic (run-time) grain-size adaptation. The results obtained so far show that serialising parallel activities (by increasing the grain-size) provides an high reduction in the parallelism overheads.

Further work is currently being performed to complete the dynamic grain packing mechanism. It includes the development of techniques to determine: (i) when to perform the grain packing/unpacking based on the system load, (ii) which objects should be packed/unpacked, and (iii) on which grains.

Current R&D work also includes further evaluation of the prototype with case studies from virtual environment generation.

References

1. Kruatrachue, B., Lewis, T.: Grain Size Determination for Parallel Processing, IEEE Software, Vol. 5(1), January (1988)
2. Gresoulis, A., Yang., T.: On the Granularity and Clustering of Direct Acyclic Graphs, IEEE Transactions on Parallel and Distributed Systems, Vol. 4(6), June (1993)
3. Sobral, J., Proença., A.: Dynamic Grain-Size Adaptation on Object-Oriented Parallel Programming - The SCOOPP Approach, Proc. of the 2nd Merged IPPS/SPDP 1999, Puerto Rico, April (1999)
4. Mohr, E., Kranz, A., Halstead, R.: Lazy Task Creation: A Technique for Increasing the Granularity of Parallel Programs, IEEE Transactions on Parallel and Distributed Processing, v2(3), July (1991)
5. Lopez, P., Hermenegildo, M., Debray, S.: A Methodology for Granularity Based Control of Parallelism in Logic Programs, Journal of Symbolic Computation, Vol. 22, (1998)
6. Sobral, J., Proença, A.: ParC++: A Simple Extension of C++ to Parallel Systems, Proceedings of the 6th Euromicro Workshop on Parallel and Distributed Applications (PDP'98), Madrid, Spain, January (1998)
7. Santos, L., Chalmers, A., Proença, A.: A messages density monitoring strategy for distributed memory parallel system, 2nd Int. Conf. on Software for Multiprocessors and Supercomputers: theory, practice and experience, Moscow, September (1994)
8. Sobral, J., Proença, A.: Overheads on the dynamical removal of excess of parallelism on OO irregular applications, 1st Work. Parallel Computing for Irregular Appl., 5th Int. Symp. HPC Arch.(HPCA-5), www-apache.imag.fr/manifestations/PCIA, Orlando, January (1999)

Optimising Skeletal-Stream Parallelism on a BSP Computer

Andrea Zavanella*

Dipartimento di Informatica
Università di Pisa Italy
zavanell@di.unipi.it
http://www.di.unipi.it/~zavanell

Abstract. Stream parallelism allows parallel programs to exploit the potential of executing different parts of the computation on distinct input data items. Stream parallelism can also exploit the concurrent evaluation of the same function on different input items. These techniques are usually named "pipelining" and "farming out". The P^3L language includes two stream parallel skeletons: the Pipe and the Farm constructors. The paper presents a methodology for efficient implementation of the P^3L Pipe and Farm on a BSP computer. The methodology provides a set of analytical models to predict the constructors performance using the BSP cost model. Therefore a set of optimisation rules to decide the optimal degree of parallelism and the optimal size for input tasks (grain) are derived. A prototype has been validated on a Cluster of PC and on a Cray T3D computer.

1 Introduction

Parallel computers can exploit an input stream to execute in parallel different parts of code while data items flow through the PEs. Generally the optimisation of a pipeline computation requires the programmer to deal with the bottleneck problem (i.e. a stage is particularly slower than the others). The goal of such a decomposition is to minimise the service time of the module, which is now limited by the maximum of the stages service times, while the latency per item is normally greater than in sequential monolithic versions. A different solution to exploit the input stream arises when a function f can be computed, over a single item, independently from the evaluation of the others. In this case computation can be "farmed out" to a group of PEs (generally named "workers"). The goal of farming the computation of the function is mainly to minimise the service time of the module. Implementing a "farm" strategy charges programmers of decisions as the degree of parallelism (i.e to decide the number of workers), the scheduling policy to adopt and the size of the "task" to schedule. In this article we propose a strategy to efficiently implement the two decomposition schemes presented

---

* This Work has been partially supported by the italian M.U.R.S.T. in the framework of the Project MOSAICO and by the UE with a TMR (Marie Curie) grant

P. Amestoy et al. (Eds.): Euro-Par'99, LNCS 1685, pp. 853–857, 1999.
© Springer-Verlag Berlin Heidelberg 1999

above on a generic BSP computer [8, 6]. The strategy is proposed with the goal in mind of designing P^3L as a portable skeleton language [3, 1]. In particular the paper provides a set of optimisation rules for a portable implementation of the P^3L constructors Pipe and Farm.

2 P^3L and the BSP Computer

A suitable approach to reconcile performance and abstraction in parallel programming is provided by the skeletons methodology [2, 4, 7]. In skeletal programming the parallelism is introduced using a limited set of parallel patterns which are automatically implemented by the support. The P^3L language allows several parallel constructors to be composed and nested in order to write fully structured parallel programs. The parallel constructors included by the language are: Pipe, Farm, Loop, Map, Reduce, Scan and Comp.

The BSP computer is a parallel abstract machine [8] composed by: *a set of memory-processor couples, an interconnection structure,* and a *synchronisation device* A BSP computation is a sequence of so-called *supersteps*. Each superstep includes three phases: *local computation phase*, a *interprocessor communication phase* and a *global synchronisation phase*. In the BSP model the differences between parallel machines are reflected by a small set of *machine-dependent parameters*: p is the number of available processors; g_∞ is the asymptotic communication cost for very large messages expressed in time steps; $N_{1/2}$ is the size of the message that produces half the optimal bandwidth, in other words $g(N_{1/2}) = 2g_\infty$ and a message of size h can be routed in $g(h) * h$ time steps; l number of time steps for barrier synchronisation; s is the number of time steps per second. Assuming that W is the maximum number of operations executed by one processor during the local computation phase and h is the maximum amount of data moved by one processor in the interprocessor communication phase: the cost of a superstep is given by $T_{sstep} = W + g_\infty h(1 + \frac{N_{1/2}}{h}) + l$. The issues for efficient implementations of the BSP model have been addressed on several papers [6] and the framework has already been used to derive optimisation rules for data parallel programs and data parallel skeletons [5, 9].

3   Implementing P^3L Pipe on the BSP Computer

A cost-efficient implementation for the Pipe constructor is derived using the following variables: the inter-arrival time (per item) is T_e; a set of functions $f_i : T_i \longrightarrow T_{i+1}$; the size of T_i is d_i; the time to compute f_i is t_i A simple scheme to implement a pipeline on a BSP computer exploits a double buffering technique which implies that during each superstep each PE performs three operations: a) computing the function on the previously received item; b) receiving the next input item; c) sending the result to the next stage. The service time of Pipe using tasks of k items is $T_{serv}^{pipe}(k) = Max(kT_e, T_{serv}^{stag}(k))$, where:

$$T_{serv}^{stag}(k) = Max_{[0 < i < z-1]} T_{serv}^i(k) \tag{1}$$

$$T_{serv}^i(k) = (t_i k + l + g_\infty k(d_i + d_{i+1}) + 2g_\infty N_{1/2})$$ (2)

Since balance between the external time and service times of the stages is convenient we propose two optimisation strategies:

- *adjusting the grain*: the value k of the number of items per task can be tuned. Assuming that $T = Max_{[0<i<z-1]}(t_i + g_\infty(d_i + d_{i+1})$ we consider the two cases: $T < T_e$ and $T \geq T_e$. In the first case the value of k balancing the service time of the Pipe with the interarrival time is given by: $k = \frac{l + 2g_\infty N_{1/2}}{T_e - T}$ In the second case the value $k = 1$ makes the service time of the Pipe minimal.
- *merging fast stages*: this technique provides an implementation which uses *less resources* of the starting one and which has the same service time. We can apply this technique when $\exists j : S(j) + S(j+1) < Max_{[0<j<z-1]}S(i)$ where: $S_{(i)} = f_i + g_\infty(d_i + d_{i+1})$. Under this assumption we can merge the stage j and $j+1$ of the pipe in a sequential stage which does not increase the service time.

4   Implementing P^3L Farm on the BSP Computer

Assuming that the interarrival time of the input stream is T_e and the time to compute f on a single item is t_f. In the case $t_f >> T_e$ and f cannot be further decomposed we can attempt to speed-up our computation using a farming technique. The template proposed uses a synchronous pipeline scheme of three types of processes: emitter, workers (n) and collector. The service times of emitter-collector and workers per item are:

$$T_{serv}^{em-col}(n, k) = 2g_\infty d_0 + \frac{g_\infty N_{1/2}}{k} + \frac{g_\infty N_{1/2} + l}{nk}$$ (3)

$$T_{serv}^{work}(n, k) = \frac{2g_\infty d_0 + t_f}{n} + \frac{2g_\infty N_{1/2} + l}{nk}$$ (4)

The service time in Eq. 3 introduces two optimisation cases, when: $2g_\infty d_0 \leq T_e$ both the Eq. 3-4 can be decreased to the interarrival time T_e choosing a couple of values: $(\hat{n}, \hat{k})$ such that be $\mathcal{P}$ the property:

$$\mathcal{P}(n, k) \equiv T_{serv}^{em}(n, k) < T_e \ \& \ T_{serv}^{work}(n, k) < T_e$$ (5)

Then $P(\hat{n}, \hat{k})$ and: $\forall(n, k) \neq (\hat{n}, \hat{k}) P(n, k) \Rightarrow n > \hat{n}$ From Eq. 3-4 using a practical optimisation rule:

$$\hat{n} = Max(|\frac{g_\infty N_{1/2}}{(T_e - 2g_\infty d_0) + g_\infty N_{1/2}}|, |\frac{2g_\infty d_0 + t_f}{T_e}| + 1)$$ (6)

$$\hat{k} = Max(1, |\frac{2g_\infty N_{1/2} + l}{T_e}|)$$ (7)

When: $2g_\infty d_0 > T_e$ we use a different rule to compute a value of k giving an approximation of the minimum for Eq 3 such that: $T_{serv}^{em} < (2g_\infty d_0)(1 + 1/prec)$. The precision parameter $prec$ can be tuned as a compiling option.

$$\hat{k} = \frac{prec(g_\infty N_{1/2} + l)}{2g_\infty d_0} \tag{8}$$

Fixed the value of $\hat{k}$ we can derive $\hat{n}$:

$$\hat{n} = \frac{prec(2g_\infty d_0 + t_f + 2g_\infty d_0)}{(prec + 1)2(g_\infty d_0)} \approx 1 + |\frac{4g_\infty d_0 + t_f}{2g_\infty d_0}| \tag{9}$$

5   Experiments

The optimisation rules of Section 4 have been tested on two different parallel architectures: a Cluster of 10 PC with PentiumII 266 Mhz processors running Linux connected by a 100Mbit Fast Ethernet technology (Backus), and a Cray T3D with 512 Alpha processors with clock rate of 150 Mhz connected by a three dimensional torus having a 300 Mbyte/s bandwidth per link. A group of tests have been executed using the C+BSP-lib and using a stream of items (C-type: double). The predictions of the model prove to be accurate and we see that the values choosen by the model for $\hat{n}$ and $\hat{k}$ are extremely close to the optimal ones.

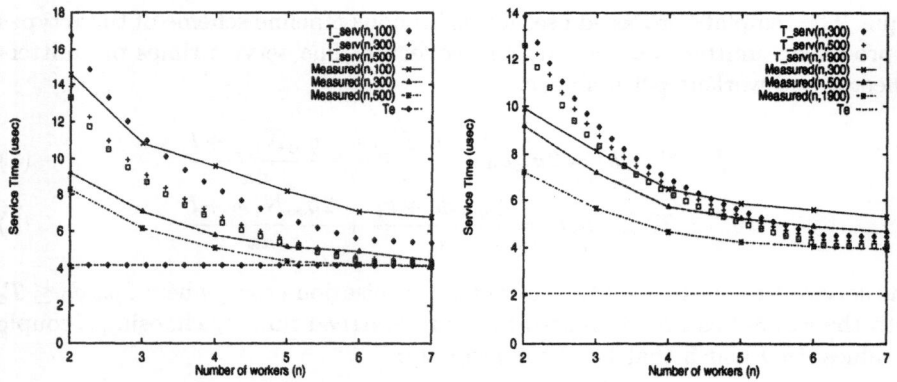

Fig. 1. Predicted and Measured times: Test Backus

6   Conclusions and Related Works

A methodology for an optimised implementation of Stream Parallelism on a BSP computer has been proposed and validated. The model provides accuracy

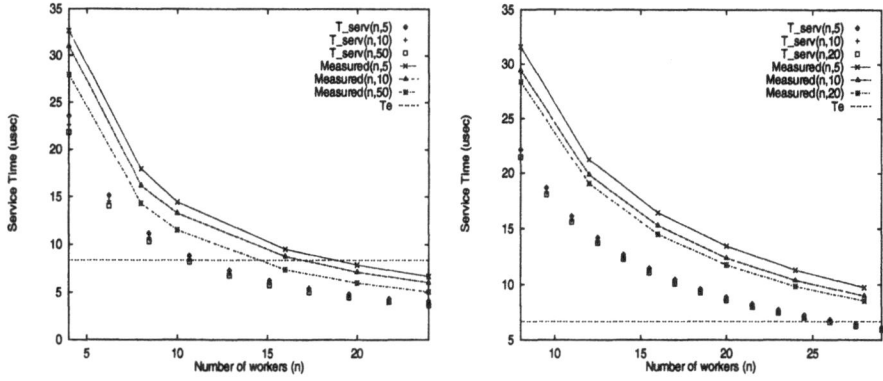

Fig. 2. Predicted and Measured times: Test T3D

enough to automatically decide the degree of parallelism and grain of data. This work also shows how a skeleton languages like the P^3L can be implemented achieving both performance and portability. The methodology together with other recent works [9, 5] is a further step towards the design of a complete compiling technology having as its first goal the performance portability.

References

[1] S. Ciarpaglini, M. Danelutto, L. Folchi, C. Manconi, and S. Pelagatti. ANACLETO: a Template-based p3l Compiler. In *Proceedings of the Seventh Parallel Computing Workshop (PCW '97)*, Australian National University, Canberra, August 1997.

[2] M. Cole. *Algorithmic Skeletons: Structured Management of Parallel Computation.* The MIT Press, Cambridge, Massachusetts, 1989.

[3] M. Danelutto, R. Di Meglio, S. Orlando, S. Pelagatti, and M. Vanneschi. A Methodology for the Development and the Support of Massively Parallel Programs. In D.B. Skillicorn and D. Talia, editors, *Programming Languages for Parallel Processing.* IEEE Computer Society Press, 1994.

[4] J. Darlington, M. Ghanem, and H. W. To. Structured Parallel Programming. In *Proceedings of the Working Conference MPPM '93*, Berlin, September 1993. GMD FIRST.

[5] D.B.Skillicorn, M. Danelutto, S. Pelagatti, and A. Zavanella. Optimising Data-Parallel Programs Using the BSP Model. In *Proceeding of EUROPAR98*, LNCS. Springer, 1998.

[6] D.B. Skillicorn, J.M.D. Hill, and W.F. McColl. Questions and answers about BSP. Technical Report PRG-TR-15-96, Oxford University Computing Laboratory, 1996.

[7] S.Pelagatti. *Structured Development of Parallel Programs.* Taylor & Francis, 1997.

[8] L. G. Valiant. A bridging model for parallel computation. *Communications of the ACM*, 33(8):103, August 1990.

[9] A. Zavanella and S. Pelagatti. Using BSP to Optimize Data-Distribution in Skeleton Programs. In *Proceedings of HPCN99*, number 1593 in LNCS, pages 613–622, April 1999.

Parallel Programming by Transformation

Noel Winstanley

Department of Computing Science
University of Glasgow

Abstract. This paper presents a system to produce efficient implementations of parallel array-based algorithms from high-level specifications. It is structured as a transformation through a series of progressively more detailed representations. This allows the use of high-level programming features without losing the fine control of low-level languages. During the transformation process, parallel implementation decisions are introduced. Finally, a representation is reached which can be translated to C+MPI.

1 Introduction

The *Abstract Parallel Machine* (APM) methodology [OR97] is used to structure complex parallel algorithm derivations by defining parallel operations at an appropriate level of abstraction. An APM contains a set of computation sites which cooperate to implement a set of parallel operations. These operations are defined by recursive equations over the input, output and state of the sites. APM parallel operations are usually specified in the lazy functional language Haskell [PHA+97].

The APMs are organised into a directed acyclic graph, where the child nodes are APMs which implement (at some lower level of abstraction) the operations provided by the the parent APM. Program derivation is by correctness-preserving transformation by equational reasoning within and between APMs in the hierarchy.

Note that the aim is not to parallelise arbitrary functional programs: rather a functional language is being used in a systematic way to model and transform imperative parallel algorithms.

Although the specifications are executable and informative, they omit many of the details required when producing a final imperative implementation. This is due to the use of high-level language features such as laziness and higher order functions, and a programming style based on garbage-collected lists rather than statically allocated arrays.

This paper presents a system to bridge the gap between the abstraction of an APM algorithm specification and the detail required for efficient implementation. We define a series of languages, where each one introduces more implementation details. The languages have APM definitions for their parallel constructs, but enforce a more imperative programming style than the arbitrary Haskell code used in higher-level specifications. This means that the languages fit into the

P. Amestoy et al. (Eds.): Euro-Par'99, LNCS 1685, pp. 858–865, 1999.
© Springer-Verlag Berlin Heidelberg 1999

APM hierarchy, but are more amenable to translation to C+MPI, our target language.

Figure 1 illustrates the language sequence, where each language is represented as an oval. The first language in the sequence (*Sequential*) allows the expression of array-based algorithms; the following languages in turn: identify potential parallelism; introduce data distribution; and communication details, until at the end of the sequence a form is reached that can be translated to C+MPI.

Single Assignment C(SAC)[Sch94] – a functional variant of C with a system of extensible high-level array operations – is used as an intermediate stage in the translation to C. This simplifies the translation process, as SAC shares some of the features of the transformation languages. Furthermore, the SAC compiler is heavily optimising, producing code comparable with hand-written FORTRAN [Sch98].

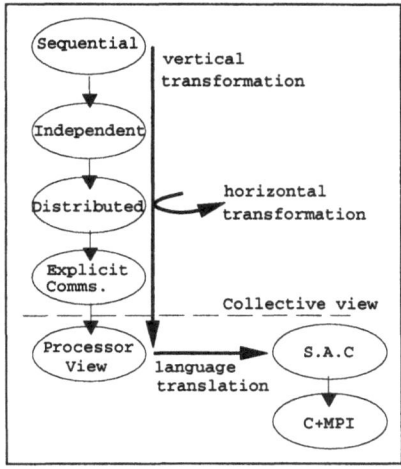

In common with the APM methodology, derivation proceeds by transformation from one language to the next. The transformation is guided by a set of axioms. Some axioms equate constructs within a single language (*horizontal transformations*) and are used to introduce op-

Fig. 1. Language sequence

timisations; others (*vertical transformations*) relate constructs in adjacent languages in the sequence: these introduce greater implementation detail. The series of transformations applied to a program can be used for correctness proofs and for retargetting – the transformations can be unrolled to a common point and then new machine-specific details introduced.

The languages are implemented in Haskell as libraries of combinators – i.e. as *Domain Specific Embedded Languages* [Hud98]. Doing this means that the benefits of the host language are inherited: strong typing of a rich type language; the infrastructure of compilers and tools; and a semantics that ensures the validity of equational reasoning. Importantly, the host language provides a common semantic base with which to relate constructs of the different transformation languages and more abstract APM specifications.

The next section introduces the first language in the transformation sequence. The following languages of the sequence are described in section 3, where the transformation process is informally illustrated with a small example.

2   The Sequential Language

The first language of the sequence forms a common core of algorithmic constructs to which later languages add constructs to express parallelism. The main datatype is the array, supported by a set of whole-array combinators such as those found in APL[Ive62] or SAC. User-defined structures and unions are also permitted. The update of variables is disallowed: this preserves referential transparency which simplifies the axiomatisation and use of transformations.

To ease the transformation process we must control language features that do not sit will with an imperative language, such as first-classness, laziness, higher-order functions and arbitrary recursion.

Monads [Wad92] are a standard technique for structuring computations in Haskell. The transformation languages can be said to be *monad-bound* – there is no way for the programmer to escape from the sequencing that the monad defines. This gives a fixed execution order and prevents computations relying on laziness for termination.

In Haskell all types are first-class – any type can be a parameter to another function or stored in a data structure. In the transformation languages, as with typical imperative languages, there is a discrimination between the type of procedures and the values that may be passed to or returned by them. A further subset of types may be stored in data structures. Extensible records [GJ96] and type classes [HHPW96] are used to encode these constraints

$$\textbf{for } (i = 0; i < B; i{+}{+})$$
$$arr[i] \ = \ f(arr[i]); \qquad (1)$$

$$\textbf{seqVect . for } (0, \ (< B), \ (+1))$$
$$(\lambda i \rightarrow \textbf{do } v \leftarrow arr\,!\,i \qquad (2)$$
$$\qquad f\,v)$$

$$\textbf{map } (\lambda(i, v) \rightarrow f\,v)\,arr \qquad (3)$$

Fig. 2. Mapping an array

into the type system of the host language.

In functional languages, control is described using recursion, either coded explicitly or packaged as higher-order functions. As both are problematic to translate, the transformation languages have iteration constructs similar to those found in imperative languages.

Compared to the large number of recursion combinators used by a functional language, imperative languages have a small set of iterative constructs. These are sufficient because of the use of mutable state – a for loop can be equivalent to a map, fold or scan depending on how variables are updated in the loop body.

As update is prohibited in the transformation languages, anything altered within a loop body must be explicitly propagated to the next iteration. This causes the set of loop constructs required to drastically expand. Although such explicit description of dependencies between loop iterations is useful for parallelisation, it is inconvenient for the programmer and makes a language unwieldy.

We circumvent this by expressing a loop construct as the composition of two or three combinators: one that generates a sequence of computations, possibly one which manipulates this sequence, and one that executes the sequence and return a result.

This is best illustrated by example: each blocks of code in Fig. 2 performs a map over an array. The C code (1) applies f to each element of *arr*. The equivalent code in our sequential language (2), uses the **for** loop generator to create a sequence of computations. The loop body is a lambda-abstraction over the loop index i. The loop executor **seqVect** executes a sequence of computations, and produces a vector of their results. Due to the functional nature of the language a new array must be created rather than updating the old array in place.

Since mapping over arrays is such a common operation, a whole array combinator **map** is provided. Equivalent code using this construct is given in (3).

Figure 3 gives another example of iteration constructs. The C code in (4) folds an array by summing the elements using an accumulating variable. A direct translation to our sequential language would produce (5). Here, the **forAccum** loop generator constructs a sequence of computations where the result of one computation is passed as a parameter to the next. The **final** loop combinator executes the sequence and returns the result of the final computation in it.

$$sum = 0;$$
$$\textbf{for } (i = 0; i < B; i{+}{+}) \qquad (4)$$
$$sum {+} {=} arr[i];$$

$$\textbf{final}$$
$$. \, \textbf{forAccum} \, (0, (< B), (+1)) \, 0 \qquad (5)$$
$$(\lambda i \, sum \rightarrow sum + arr\,!\,i)$$

$$\textbf{foldSeq} \, (\lambda a \, b \rightarrow \textbf{return} \, (a + b)) \, 0$$
$$. \, \textbf{for} \, (0, (< B), (+1)) \qquad (6)$$
$$(\lambda i \rightarrow arr\,!\,i)$$

$$\textbf{fold} \, (\lambda a \, b \rightarrow \textbf{return} \, (a + b)) \, 0 \, arr \qquad (7)$$

Fig. 3. Folding an array

Alternatively, this could be expressed using the **foldSeq** loop executor (6) which combines the results of a sequence of computations using an auxiliary function. We can now substitute a **for** loop generator for the **forAccum**. This implementation had the advantage of removing the data dependency between one loop iteration and the next, introducing the possibility of parallelising the execution of loop bodies. As folding an array occurs frequently, a whole-array combinator is provided. The equivalent code for **fold** is shown in (7).

3   The Language Sequence

The language introduced in the previous section allows the description algorithms, but not their parallel behaviour.

Programs expressed in this language can be viewed as executing sequentially, or as specifying the behaviour of a parallel algorithm without committing to implementation details.

This section introduces the other languages in the transformation sequence. The program fragment in Fig. 4 will be used to illustrate how each language in the sequence introduces extra detail. The **gen** generates a new array of bounds bnd where each element is defined by f – a

$$a \leftarrow \textbf{gen} \, bnd \, f$$
$$b \leftarrow \textbf{fold} \, op \, v \, a$$
$$c \leftarrow \textbf{map} \, g \, a$$

Fig. 4. Sequential program

function from array index to element value. The array a produced is involved in two subsequent computations – a fold and a map. For conciseness, this example contains only array combinators. However, each languages has equivalent constructs for loop combinators.

During derivation, a program spends much time 'between' languages – where some code has already been vertically transformed into the next language while the remainder is still expressed in constructs of the previous language. It would be constricting to force the programmer to transform the entire program from one language to the next before the program had a defined semantics. To solve this problem, adjacent languages in the transformation sequence can be freely mixed within a program. As every transformation produces an executable intermediate program incremental development and testing is much easier.

3.1   Independent Computation

The second language in the sequence identifies potential parallelism: the machine model is an idealised parallel machine with infinite processors and no communication cost. The language introduces variants of the constructs provided by the original sequential language; these distribute computation over the idealised machine and perform implicit communication to return results to the main coordination program.

The computation performed by such a construct represents one *macro-step* in the SPMD programming methodology. Once all computations in a macro-step have completed, a result is returned to the the main program which redistributes it to the constructs which comprise the next macro-step.

As much of program as possible is transformed to use these macro-step constructs. In many algorithms there are sequential parts that cannot be parallelised; these are moved into a sequential macro-step containing one processor. The main program is now a description of the communication patterns between macro-steps, with all computation taking place on (unnamed) processors.

$$a \quad \leftarrow \textbf{indepGen } bnd \ f$$
$$(b, c) \leftarrow \textbf{onProc } (\textbf{fold } op \ v \ a)$$
$$\mathord{<}|\mathord{>} \textbf{indepMap } g \ a$$

Fig. 5. Identifying parallelism

The code in Fig. 5 is equivalent to the previous program fragment but has been vertically transformed to the second language in the sequence by introducing parallel constructs.

The **gen** has been substituted by a **indepGen** construct. This produces an array where each element is computed independently on a separate processor. As there are no data dependencies between the fold and map, they may be computed in parallel – the $\mathord{<}|\mathord{>}$ operator expresses this and returns a tuple of the results. The fold is sequential and so is placed on a single processor by using the **onProc** keyword, while the map can be parallelised using the **indepMap** construct. Therefore the above code has two macro-steps, with an implicit redistribution of a taking place so that the fold can be computed on a single processor.

3.2 Distributed Computation

In the previous language parallelism was introduced and the program partitioned into macro-steps. Transformation of the program to the third language maps the unbounded algorithm onto a machine with a finite number of processors.

This language has a set of parallel constructs similar to those in the previous language. However, most take an extra parameter: a data distribution that describes the distribution of the computation performed by the construct. These data distributions are an adaptation of *parameterised distribution functions* [RR95] which Rünger and Rauber use to describe the distribution of array elements amongst processors; they are capable of expressing irregular, block, cyclic and block-cyclic distributions.

Before any parallel computation can take place, the processor groups and topologies used must be defined. Figure 6 shows the block of code transformed to the third language, with processor group and virtual topology definitions added at the start of the program. The **getMainGroup** construct returns a sorted vector of the machine's processor identifiers. A new group (ag) is declared, containing all processors of the main group apart from processor 1 – which is used to compute the **fold**. A two-dimensional processor topology tp of dimensions (x, y) is then defined; this maps from two-dimensional coordinates to the processors identifiers in ag.

$$
\begin{aligned}
mg &\leftarrow \textbf{getMainGroup} \\
ag &\leftarrow \textbf{drop}\, 1\ mg \\
tp &\leftarrow \textbf{mkTopol}\, (x, y)\ ag \\
&\ \ \vdots \\
a &\leftarrow \textbf{distGen}\, (block\ tp)\ bnd\ f \\
(b, c) &\leftarrow \textbf{on}\, 1\ (\textbf{fold}\ op\ v\ a) \\
&\quad <|> \textbf{distMap}\, (block\ tp)\ g\ a
\end{aligned}
$$

Fig. 6. Adding data distribution

The array is now created using **distGen** and is distributed in a block-wise manner over tp. The **distMap** produces a new array with the same distribution as a. Meanwhile, the fold is performed on processor 1.

3.3 Explicit Communication

Due to scoping the results of one macro-step are available for use in later macro-steps. Therefore values resident on a processor may be accessed by a computation executing on a different processor. This non-local access means that a hidden communication is being performed. In the language implementations, array values maintain a record of the distribution of their elements. Operations that access the elements of distributed arrays check that the element is present on the local processor. If a non-local access is attempted, the program continues, but issues a warning.

The next stage of the transformation makes these hidden communications explicit by adding communication primitives to the program. Communications such as broadcast or scatter are represented as permutation functions, mapping from one data distribution to another. Applying a communication primitive to a distributed array returns a copy of the array that has the new distribution.

In Fig. 7 a **gather** collective communication has been added to the running example. This satisfies the non-local accesses of a on processor 1 by communicating to that processor all elements of the array resident on processors in ag. The other parameter to the communication, **selAll**, is a *se-*

$$
\begin{aligned}
a &\leftarrow \textbf{distGen}\,(block\ tp)\ bnd\ f \\
a' &\leftarrow \textbf{gather}\ 1\ ag\ \textbf{selAll}\ a \\
(b, c) &\leftarrow \textbf{on}\ 1\ (\textbf{fold}\ op\ v\ a') \\
&\quad\ \ \texttt{<|>}\ \textbf{distMap}\,(block\ tp)\ g\ a
\end{aligned}
$$

Fig. 7. Explicit communication

lector – a predicate on indexes that indicates which array elements to communicate.

In this case, the entire array is communicated. More subtle patterns can be expressed by using subsets of the group of processors the array is resident on or by using a different selector. There are a set of pre-defined selectors and logical connectives: using these, complex strides and halo communications can be expressed concisely.

3.4 Per-Processor View

Conventional sequential languages such as C present a *per-processor view* of the parallel machine; they describe the computation performed on each processor, but the behaviour of the entire machine is hard to ascertain. In contrast, the transformation languages presented so far give a *collective view* of the parallel machine – a program describes how the entire machine performs a computation. We believe that this has advantages over the per-processor view, as all parallelisation and distribution information is represented within the program.

The final stage in the derivation transforms the collective view program to a transformation language providing a per-processor view similar to C.

The transformation involves substituting the collective language constructs for equivalent processor-view constructs which test on the processor identifier so that only computations distributed to the current processor are executed. Likewise, the collective view communications primitives are replaced with their processor-view equivalents – i.e. MPI library calls.

Optimisations can now be performed, such as bundling similar communications together, reordering computations and partially evaluating parameters to constructs. The program is now in a form that can be straightforwardly translated to SAC+MPI which will then compile into a heavily optimised C+MPI program.

4 Conclusions & Further Work

In this paper we present a system for producing implementations from high-level parallel algorithm derivations by transformation through a series of languages.

The languages progressively introduce more implementation details until a form is reached that can be translated to C+MPI via SAC.

The languages produced are restrictive in some ways; due to their compilation method they must have an imperative flavour, but must still preserve referential transparency. Case studies in progress, taking APM specifications as their starting point, will show how troublesome this actually is.

Performing transformations by hand is tedious and error-prone. Some of the transformations are automatable: we plan to write tools to support these. However we do not aim to build a parallelising compiler – there is no requirement to automate every stage. Stages that require human insight will be supported by interactive tools.

Acknowledgements. I wish to thank my supervisor John O'Donnell, Joy Goodman, Richard Reid, Meurig Sage & Keith Sibson for helpful discussions and feedback.

References

[GJ96]   Benedict R Gaster and Mark P Jones. A polymorphic type system for extensible records and variants. Technical Report NOTTCS-TR-96-3, Department of Computer Science,Univerity of Nottingham, November 1996. http://www.cd.nott.ac.uk/Department/Techreports/96-3.html.

[HHPW96]   Cordelia V. Hall, Kevin Hammond, Simon L. Peyton Jones, and Philip L. Wadler. Type classes in Haskell. *ACM Transactions on Programming Languages and Systems*, 18(2):109–138, March 1996.

[Hud98]   Paul Hudak. Modular domain specific languages and tools. In *Fifth International Conference on Software Reuse*, 1998.

[Ive62]   K E Iverson. *A Programming Language*. Wiley, New York, 1962.

[OR97]   John O'Donnell and Gudula Rünger. A methodology for deriving parallel programs with a family of abstract parallel machines. In *Third International EuroPar Conference*, pages 662–669, 1997.

[PHA+97]   John Peterson[editor], Kevin Hammond[editor], Lennart Augustsson, et al. Haskell 1.4, A non-strict, purely functional language. Report YALEU / DCS / RR-1106, Department of Computer Science, Yale University, April 1997.

[RR95]   Thomas Rauber and Gudula Rünger. Parallel numerical algorithms with data distribution types. Technical Report 07-95, University of Saabrücken, 1995.

[Sch94]   Sven-Bodo Scholz. Single Assignment C – functional programming using imperative style. In *IFL '94*. University of East Anglia, Norwich, UK, 1994.

[Sch98]   Sven-Bodo Scholz. A case study: Effects of WITH-loop-folding on the NAS benchmark MG in SAC. In *IFL'98*. University College, London, UK., 1998.

[Wad92]   Philip Wadler. The essence of functional programming (invited talk). In *Conference record of the Nineteenth Annual ACM SIGPLAN-SIGACT Symposium on Principles of Programming Languages: papers presented at the symposium, Albuquerque, New Mexico, January 19–22, 1992*, pages 1–14, New York, NY, USA, 1992. ACM Press.

Condensed Graphs: A Multi-level, Parallel, Intermediate Representation*

John P. Morrison and Niall J. Dalton

National University of Ireland, Cork. Ireland

Abstract. Condensed graphs are proposed as an intermediate representation for functional and imperative languages. This representation may be executed on a variety of architectures, implementing a multilevel Condensed Graphs abstract machine. This machine incorporates characteristics and feedback information of its underlying architecture and guides dynamic topological transformations of the representation so as to optimize execution. These transformations can add or remove parallelism and change evaluation orders.

Simulated executions of the intermediate representation, utilizing varying evaluation orders, on the abstract machine are presented.

1 Introduction

The design and implementation of Higher Order Typed (or HOT) languages including Haskell, ML, and Java have generated much interest due to their power of expression and strong typing. Although these languages represent differing paradigms, a number of efforts to construct common intermediate formats, compiler infrastructures and runtime systems for efficient sequential implementation have been made. One such effort, FLINT [15], aims to provide fine-grain interoperability although it has not considered lazy versus eager evaluation in the intermediate format in contrast to Peyton-Jones et al. [4].

Non-HOT languages such as C and FORTRAN already have an efficient sequential implementation. Research efforts like SUIF [16] also employ a common intermediate format and concentrate on parallelizing and statically transforming these languages. Static optimizations are limited. They cannot exploit architectures and resources unknown at compile time and this leads to portability problems.

Platform independence can be achieved by defining an abstract machine and implementing it on different architectures. The level at which the abstract machine is placed will determine the extent and the type of the possible optimizations: placing it too low omits contextual program information, placing it too high obscures specific machine details. To date, these approaches appear to have only been applied to sequential programs and architectures. Abstract machines can be used to encapsulate various architectures to make their analysis more

---

* This work is supported by the Irish National Software Directorate

tractable by providing cost models and primitives for communication and synchronization [7]. Rather than using an abstract machine for code execution, the bulk synchronous parallel (BSP) computer model[8] provides the programmer with an abstraction of all machines in terms of four performance parameters[9]. In this way it views all architectures as BSP machines. BSP may be implemented as a library on top of standard programming systems[11, 12], and in novel languages such as GPL [10] which provide the programmer with access to three of these parameters. In doing this, the programmer still has control over, and presumably responsibility for, synchronization and resource management. The advantage is that only a single architecture needs to be considered and source code portability can be achieved.

The Condensed Graph Abstract Machine (CGAM) will be parameterized with various machine characteristics typically common to all levels in an underlying multilevel architecture. These characteristics can represent static and dynamic information. Static parameters represent BSP type information whereas dynamic parameters such as network load, memory usage and processor contention may be crucial for efficient implementation in a multiprogramming environment. The same parameters exist at each level of the multilevel machine albeit with different values. Each set of these values can be used to define a corresponding level in the CGAM.

Each CGAM will contain a common set of graph transformation algorithms. Specific algorithms are invoked at each level in the CGAM based on parameter values and on the initial graph topology whose representation includes the amount of parallelism and the way the problem is partitioned. Problem partitioning is represented by embedding graphs in nodes to form a hierarchical structure.

The CGAM may be used as a dynamic runtime system or a static compiler infrastructure.

2   The Condensed Graph Model

While being conceptually as simple as classical dataflow schemes [14, 2], the Condensed Graphs ($\mathcal{CG}$) model [1] is far more general and powerful. It can be described concisely, although not completely, by comparison. Classical dataflow is based on data dependency graphs in which nodes represent operators, and edges are data paths conveying simple data values between them. Data arrive at *operand ports* of nodes along input edges and so trigger the execution of the associated operator (in dataflow parlance, they cause the node to *fire*). During execution these data are consumed and a resultant datum is produced on the node's outgoing edges, acting as input to successor nodes. Operand sets are used as the basis of the firing rules in data-driven systems. These rules may be *strict* or *non-strict*. A strict firing rule requires a complete operand set to exist before a node can fire; a non-strict firing rule triggers execution as soon as a specific proper subset of the operand set is formed. The latter rule gives rise to more parallelism but also can result in overhead due to remaining packet garbage (RPG).

Like classical dataflow, the CG model is graph-based and uses the flow of entities on arcs to trigger execution. In contrast, CGs are directed acyclic graphs in which every node contains, not only operand ports, but also an operator and a destination port. Arcs incident on these respective ports carry other CGs representing operands, operators, and destinations. Condensed Graphs are so called because their nodes may be condensations, or abstractions, of other CGs. (Condensation is a concept used by graph theoreticians for exposing meta-level information from a graph by partitioning its vertex set, defining each subset of the partition to be a node in the condensation, and by connecting those nodes according to a well-defined rule [13].) Condensed Graphs can thus be represented by a single node (called a *condensed node*) in a graph at a higher level of abstraction. The number of possible abstraction levels derivable from a specific graph depends on the number of nodes in that graph and the partitions chosen for each condensation. Each graph in this sequence of condensations represents the same information at a different level of abstraction. It is possible to navigate between these abstraction levels, moving from the specific to the abstract through condensation, and from the abstract to the specific through a complimentary process called *evaporation*.

The basis of the CG firing rule is the presence of a CG in every port of a node. That is, a CG representing an operand is associated with every operand port, an operator CG with the operator port, and a destination CG with the destination port. This way, the three essential ingredients of an instruction are brought together (these ingredients are also present in the dataflow model; only there, the operator and destination are statically part of the graph).

A condensed node, a node representing a datum, and a multinode CG can all be operands. A node represents a datum with the value on the *operator* port of the node. Data are then considered as zero-arity operators. Datum nodes represent graphs which cannot be evaluated further and so are said to be in *normal form*. Condensed node operands represent unevaluated expressions, they cannot be fired since they lack a destination. Similarly, multinode CG operands represent partially evaluated expressions. The processing of condensed node and multinode operands is discussed below.

Any CG may represent an operator. It may be a condensed node, a node whose operator port is associated with a machine primitive (or a sequence of machine primitives), or it may be a multinode CG.

The present representation of a destination in the CG model is as a node whose own destination port is associated with one or more port identifications. The expressiveness of the CG model can be increased by allowing any CG to be a destination but this is not considered further here. Fig. 1 illustrates the congregation of instruction elements at a node and the resultant rewriting that takes place.

When a CG is associated with every port of a node it can be fired. Even though the CG firing rule takes accounts of the presence of operands, operators

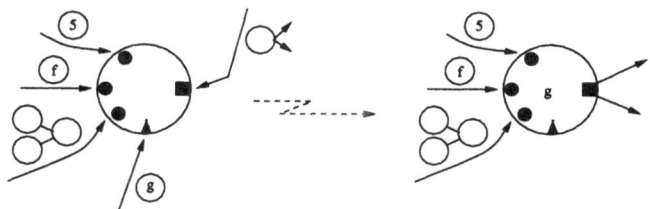

Fig. 1. *CGs congregating at a node to form an instruction*

and destinations, it is conceptually as simple as the dataflow rule. Requiring that the node contain a *CG* in every port before firing prevents the production of RPG. As outlined below, this does not preclude the use of non-strict operators or limit parallelism.

A *grafting* process is employed to ensure that operands are in the appropriate form for the operator: non-strict operators will readily accept condensed or multinode *CGs* as input to their non-strict operands. Strict operators require all operands to be data. Operator strictness can be used to determine the strictness of operand ports: a strict port must contain a datum *CG* before execution can proceed, a non-strict port may contain any *CG*. If, by computation, a condensed or multinode *CG* attempts to flow to a strict operand port, the *grafting* process intervenes to construct a destination *CG* representing that strict port, and sents it to the operand.

The grafting process thus facilitates the evaluation of the operand by supplying it with a destination and, in a well constructed graph, the subsequent evaluation of that operand will result in the production of a *CG* in the appropriate form for the operator. The grafting process, in conjunction with port strictness, ensures that operands are only evaluated when needed. An inverse process called *stemming* removed destinations from a node to prevent it from firing.

Strict operands are consumed in an instruction execution but non-strict operands may be either consumed or propagated. The *CG* operators can be divided into two categories: those that are "value-transforming" and those that only move *CGs* from one node to another (in a well-defined manner). Value-transforming operators are intimately connected with the underlying machine and can range from simple arithmetic operations to the invocation of sequential subroutines and may even include specialized operations like matrix multiplication. In contrast, *CG* moving instructions are few in number and are architecture independent. Two interesting examples are the condensed node operator and the *filter* node. Filter node have three operand ports: a Boolean, a *then*, and an *else*. Of these, only the Boolean is strict. Depending on the computed value of the Boolean, the node fires to send either the *then CG* or the *else CG* to its destination. In the process, the other operand is consumed and disappears from the computation. This action can greatly reduce the amount of work that needs to be performed in a computation if the consumed operands represent an unevaluated or partially evaluated expression. All condensed node operators are non-strict in all operands and fire to propagate all their operands to appropriate destinations in their associated graph. This action may result in condensed node operands (representing unevaluated expressions) being copied to many different parts of the computation. If one of these copies is evaluated by grafting, the graph corresponding to the condensed operand will be invoked to produce a result. This result is held local to the graph and returned in response to the grafting of the other copies. This mechanism is reminiscent of parallel graph reduction [3] but is not restricted to a purely lazy framework.

$\mathcal{CG}$s which evaluate their operands and operator in parallel can easily be constructed by introducing *spec* (speculation) nodes to act as destinations for each operand. The spec node has a single operand port which is strict. The multinode $\mathcal{CG}$ operand containing the spec node is treated by non-strict operand ports in the same way as every other $\mathcal{CG}$, however, if it is associated with a strict port, the spec node's operand is simply transferred to that port. If that operand already had fully evaluated it could be used directly in the strict port, otherwise, it is grafted onto the strict port as described above. This is illustrated in Fig. 2.

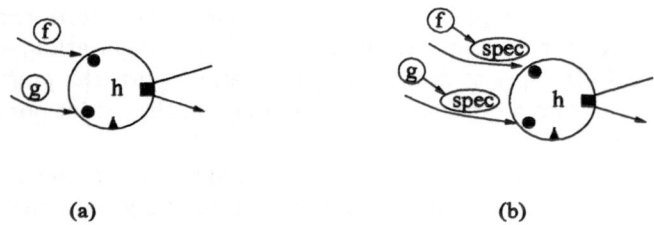

(a)   (b)

Fig. 2. *Increasing parallelism by speculatively evaluating operands*

Stored structures can be naturally represented in the $\mathcal{CG}$ model. $\mathcal{CG}$s are associated with the operand ports of a node which initially contain no operator or destination. These operands structures can then be fetched by sending appropriate *fetch* operators and destinations to these nodes. These fetch operators also form part of the $\mathcal{CG}$ moving operators and so are machine independent.

The power of the operators in the $\mathcal{CG}$ model can be greatly enhanced by, associating with each, specific *deconstruction semantics*. These specify if $\mathcal{CG}$s can be removed from the ports of a node after firing. In general, every node will be deconstructed to remove its destination after firing, this renders a node incomplete and prevents it from being erroneously refired. The deconstruction semantics of the fetch operator cause the operator and the destination to be removed after firing. This leaves the stored structure in its original state ready for a subsequent fetch. Fig. 3 illustrates the process of fetching a stored structure.

The $\mathcal{CG}$ moving instructions are extended to include assignment which, with appropriate deconstruction semantics, implement mutable structures. Side-effect computations are thus encapsulated in graphs representing mutable, stored struc-

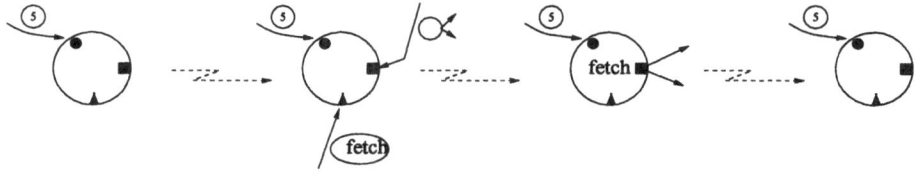

Fig. 3. *Sequence of events showing the fetching of a stored structure and subsequent node deconstruction*

tures. Condensed nodes representing these structures may flow on graphs to ensure that the side-effects are carried out in the correct order. Those parts of the computation not affected by the side-effects can proceed as normal, in parallel. This can be likened to the monadic approach for introducing side-effects into pure, non-strict languages: where the monad structure[5] is used to encapsulate the side-effects and ensure correct sequencing[6].

By statically constructing a $\mathcal{CG}$ to contain operators and destinations, the flow of operand $\mathcal{CG}$s sequence the computation in a dataflow manner. Similarly, constructing a $\mathcal{CG}$ to statically contain operands and operators, the flow of destination $\mathcal{CG}$s will drive the computation in a demand-driven manner. Finally, by constructing $\mathcal{CG}$s to statically contain operands and destinations, the flow of operators will result in a control-driven evaluation. This latter evaluation order, in conjunction with side-effects, is used to implement imperative semantics. The power of the $\mathcal{CG}$ models result from being able to exploit all of these evaluation strategies in the same computation, and dynamically move between them, using one single, uniform, formalism.

3 Executing Condensed Graphs

To illustrate the versatility of Condensed Graphs, the function

$$\Psi(n) = \begin{cases} 1 & , n = 0,1 \\ \Psi(n/2) + \Psi(n/2) & , n \text{ even} \\ 2 \times \Psi(n/3) & , n \bmod 3 = 0 \\ 2 \times \Psi(n-1) & , \text{otherwise} \end{cases}$$

is simulated with the following assumptions:

1. All instructions take the same amount of time to execute.
2. All fireable instructions are executed in the same time step.

The initial expression of Ψ is given in Fig. 4. This graph is basically data-driven, however, node ifel-2 is stemmed to ensure termination. This eager expression of Ψ consumes memory quickly and gives rise to much parallelism. It should be noted that the latter half of the computation is relatively sequential due to the late firing of a chain of ifel-2 nodes which are dynamically constructed between graph instances.

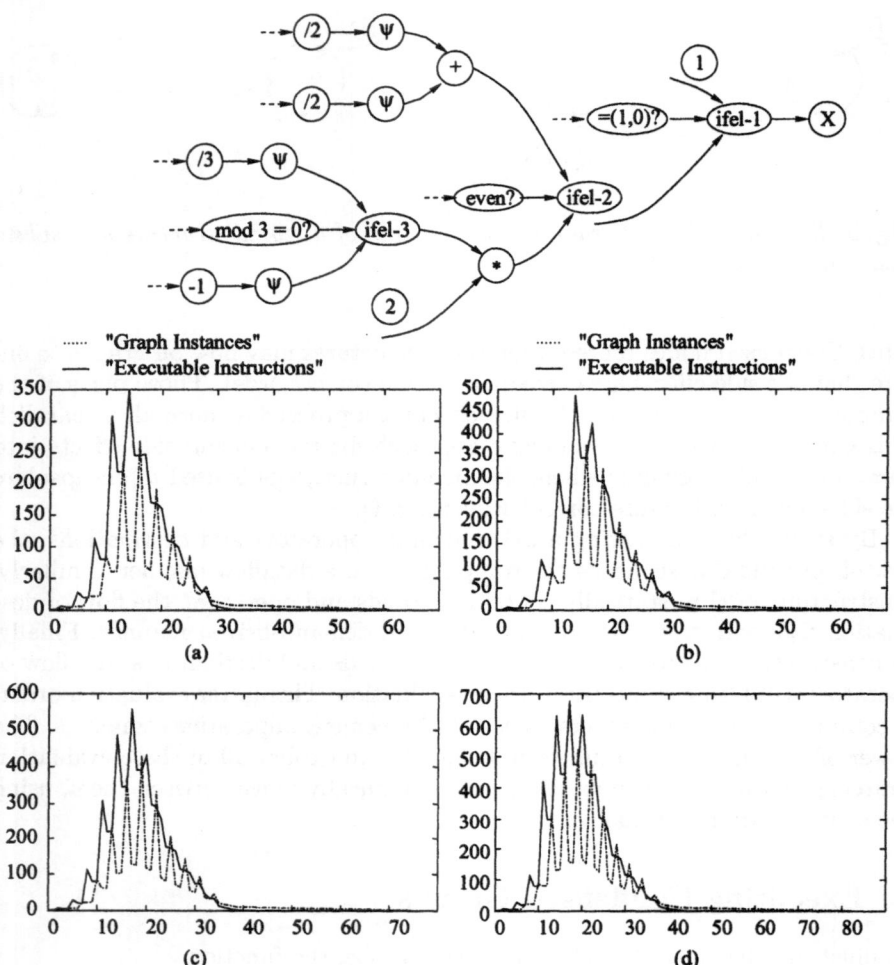

Fig. 4. *Profile of eager evaluation of Ψ with n = 9 in (a), n = 10 in (b), n = 11 in (c) and n = 12 in (d). The resource usage and parallelism increases in proportion to the input value.*

Fig. 5 illustrates a lazy expression of Ψ. The spikes in the corresponding profiles indicate the demand-driven nature of the execution. Each spike corresponds to the evaluation of a part of the equation required at that time. Consequently, less resources are consumed and less time is also required since the path to the solution is being directed from the start. In contrast, Fig. 4 expanded all possible paths before ultimately picking the correct one. These two extremes are typical of traditional eager and lazy evaluations.

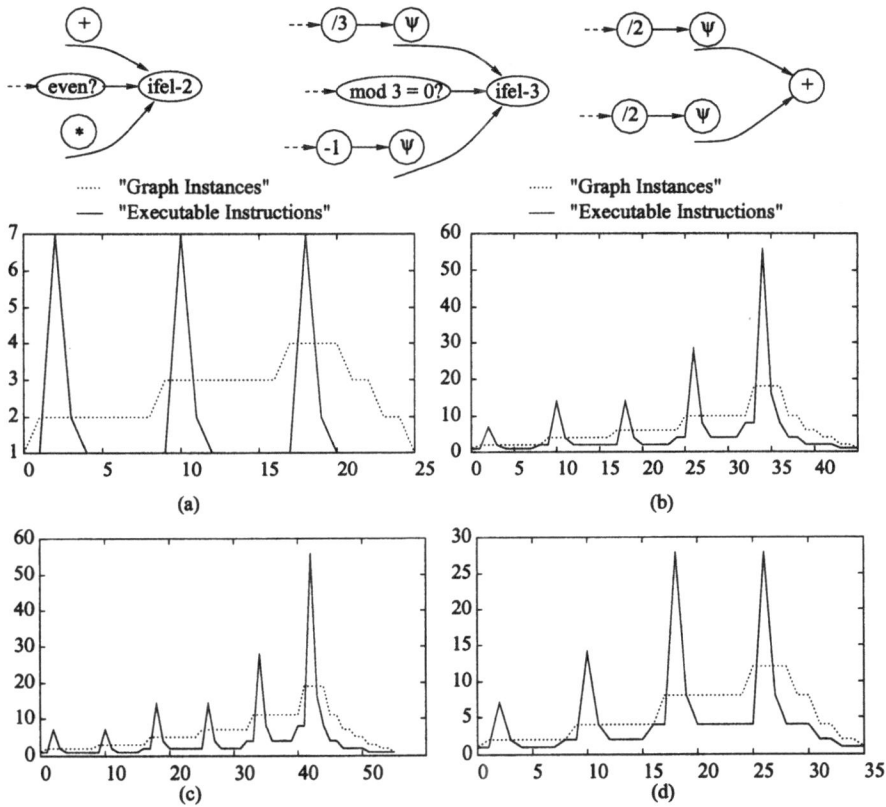

Fig. 5. *Profile of lazy evaluation of* Ψ *with* $n = 9$ *in (a),* $n = 10$ *in (b),* $n = 11$ *in (c) and* $n = 12$ *in (d). The appropriate changes to Fig. 4 are indicated.*

The power of the $\mathcal{CG}$ model lies in its ability to combine these strategies in one formalism. Fig. 6 illustrates such a combination. Stemming the plus (+) and multiplication (*) nodes facilitates the earlier firing of the ifel-2 node and so eliminates the sequential chain characteristic of Fig. 4. Thus, while all paths are still being expanded, the lazy evaluation is simultaneously directing the path choice. As a result the execution time is less than both the eager and lazy versions, even though the resource requirements are similar to the eager version.

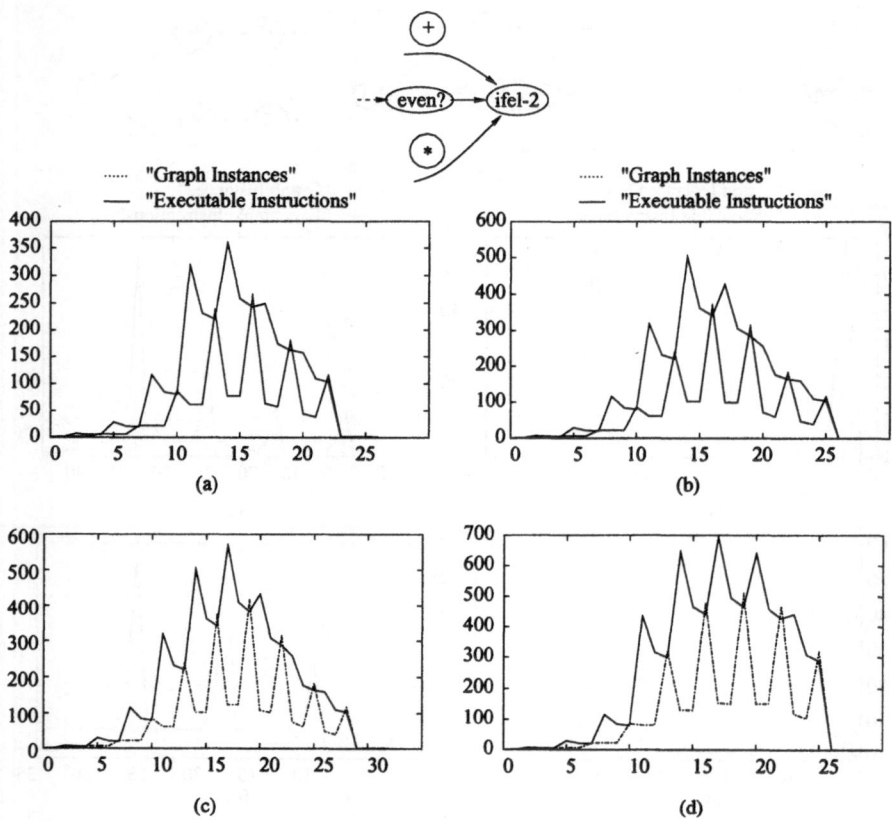

Fig. 6. *(a)* $n = 9$, *(b)* $n = 10$, *(c)* $n = 11$ and *(d)* $n = 12$. *Mixing evaluation orders eliminates the sequential chain of Fig. 4. The appropriate change to Fig. 4 is indicated.*

Runtime profiling could be used to identify parts of a computation which are rarely required. By making those parts lazy the common case can be made to execute as quickly as before, but with less resources. This is illustrated in Fig. 7 where the common cases execute as fast as in Fig. 6 but with significantly less resources. The rarer case (when n is a power of 3) takes longer to execute since the required executions were delayed, nonetheless the resource requirements are still small.

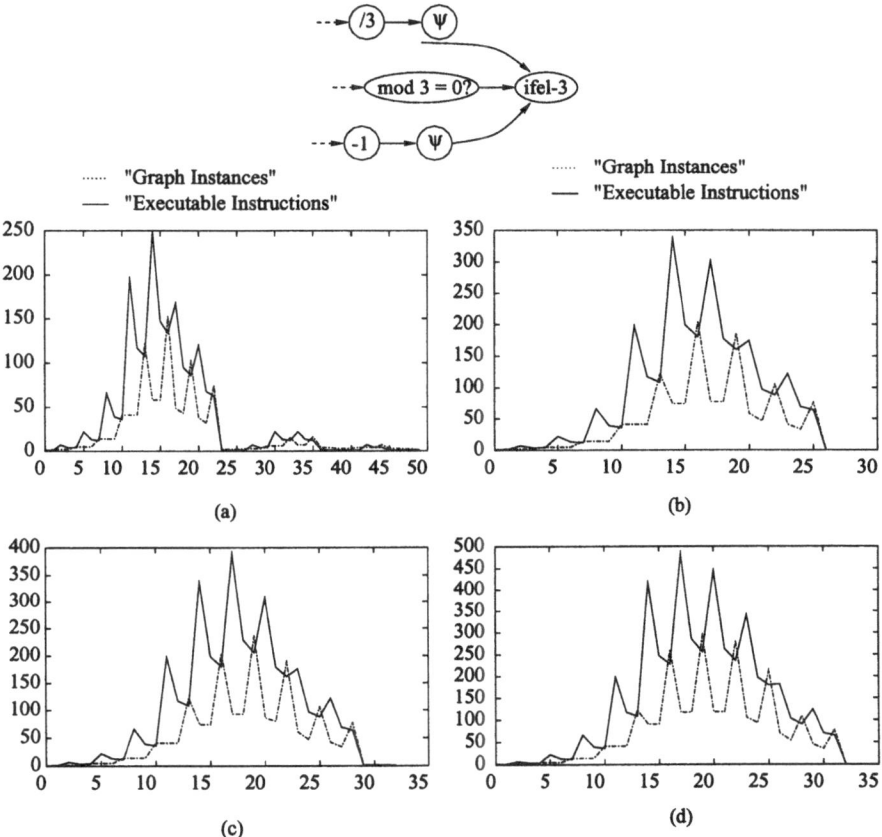

Fig. 7. *(a)* $n = 9$, *(b)* $n = 10$, *(c)* $n = 11$ and *(d)* $n = 12$. *The rare case in which n is a power of 3 is made lazy, reducing resource usage but sometimes lengthening execution time.*

4   Final Remarks

An intermediate representation based on Condensed Graphs provides a single, uniform formalism for expressing different evaluation orders and these orders can be changed using topological transformations known as stemming and grafting. Condensation and evaporation processes are used to navigate through various abstraction levels of a program facilitating optimizations at each of these levels. By mixing evaluation orders it has been shown that resource usage and execution time can be optimized.

We think that Condensed Graphs are expressiveness enough to act as a single intermediate representation for imperative and functional languages. Information required for different types of optimizations is available at different abstractions levels within the Condensed Graph. Therefore, the space of Condensed Graphs is closed under the application of our optimization algorithms. These algorithms can be applied either statically as part of a compiler or dynamically as part of a runtime system.

References

[1] John P. Morrison Condensed Graphs: Unifying Availability-Driven, Coercion-Driven and Control-Driven Computing. ISBN: 90-386-0478-5. Technische Universiteit Eindhoven, October 1996.

[2] Arvind and Kim P. Gostelow A Computer Capable of Exchanging Processors for Time In *Information Processing 77 Proceedings of IFIP Congress 77* Pages 849-853, Toronto, Canada, August 1977.

[3] Rinus Plasmeijer and Marko van Eekelen *Functional Programming and Parallel Graph Reduction* ISBN: 0-201-41663-8 Addison-Wesley Publishers Ltd.

[4] Simon Peyton Jones, John Launchbury, Mark Shields, Andrew Tolmac Bridging the gulf: a common intermediate language for ML and Haskell *Principles of Programming Languages 1998*

[5] Eugenio Moggi Computational lambda-calculus and monads. *Tech. Report ECS-LFCS-88-66, Edinburgh Univ., 1988*

[6] Simon Peyton Jones and Philip Wadler Imperative functional programming *20th ACM Symposium on Principles of Programming Languages (POPL'93), Charleston, Jan 1993, pp71-84*

[7] S. Fortune and J. Wyllie Parallelism in Random Access Machines *Proceedings of the 10th ACM Synposium on the Theory of Computing (1978) 114-118*

[8] Leslie G. Valiant A bridging model for parallel computation Communications of the ACM, August 1990, v33 n8 p103

[9] M. Goudreau, K. Lang, S. B. Rao, T Suel and T. Tsantilas. Towards Efficiency and Portability: Programming with the BSP Model *In the Proceedings of the 8th Annual ACM Symposium on Parallel Algorithms and Architectures, June 1996*

[10] W. F. McColl, and Q. Miller The GPL Language: Reference Manual *Technical Report (ESPRIT GEPPCOM Project), Oxford University Computing Laboratory October 1995*

[11] A. Geist, A. Beguelin, J. Dongarra, W. Jiang, R. Manchek, and V. Sunderam *PVM: Parallel Virtual Machine - A Users' Guide and Tutorial for Networked Parallel Computing. MIT Press, Cambridge, MA, 1994*

[12] W. Gropp, E. Lusk and A. Skjellum *Using MPI: Portable Parallel Programming with the Message-Passing Interface. MIT Press, Cambridge, MA, 1994*

[13] Frank Harary, Robert Norman and Dorwin Cartwright *Structural Models: An Introduction to the Theory of Directed Graphs John Wiley and Sons,1969*

[14] J.R. Gurd, C.C. Kirkham, and Ian Watson. The Manchester Prototype Data Flow Computer. *Communications of The ACM*, 28(1):34–52, January 1985.

[15] Zhong Shao Typed Common Intermediate Format *997 USENIX Conference on Domain-Specific Languages October 1997, Santa Barbara, USA*

[16] M. W. Hall, J. M. Anderson, S. P. Amarasinghe, B. R. Murphy, S.-W. Liao, E. Bugnion and M. S. Lam Maximizing Multiprocessor Performance with the SUIF Compiler *IEEE Computer, December 1996*

A Skeleton for Parallel Dynamic Programming

D. Morales, F. Almeida, F. Garcia, J. Gonzalez, J. Roda, and C. Rodriguez

Centro Superior de Informática
Dpto. E.I.O. y Computación
Universidad de La Laguna, Tenerife, Spain
falmeida@ull.es

Abstract. The development of skeleton tools constitutes an alternative to cover the gap between current parallel architectures and sequential programmers. Its contruction involves formal models, paradigms and methologies. Based in the automata theory we have developed a formal model for Parallel Dynamic Programming over pipeline networks. This model makes up a paradigm which is the core of skeleton tools oriented to the Dynamic Programming Technique. Following the methodology coerced by the model, we present a tool that provides the user with the ability to obtain parallel programs adapted to the parallel architecture. The efficiency is contrasted on three current parallel platforms: Cray T3E, IBM SP2 and SG Origin 2000.

1 Introduction

Dynamic programming (DP) is an important problem-solving technique that has widely been used in various fields such as control theory, operations research, biology and computer science. Sequential computation for dynamic programming has been studied extensively. General discrete DP programs viewed as multistage finite automatons with a certain superimposed cost structure were presented in [7]. Parallel dynamic programming algorithms for specific problems and specific recurrence relations have been presented for different architectures [2, 4, 8, 6]. However, few effort has been done for a general parallel dynamic programming formal model. The only proposal for a general parallel DP approach was due to Wah et al. in 1985 [10]. They suggest parallelizations of certain classes of DP formulations using parallel matrix product algorithms. However, the resolution of DP problems through matrix products leads, in general, to inefficient solutions. Neither this approach nor any other, lead to a general tool or skeleton for DP.

Following Ibaraki's [7] discrete DP approach, we extend the sequential model for DP to a parallel model. We propose a general parallel dynamic programming algorithm for pipeline and ring networks for multistage automatons. The parallel algorithm presented constitutes a paradigm to solve DP problems and provides the conceptual framework for the development of skeleton tools. Based in this paradigm, we present a skeletton oriented tool for the implementation of Multistage DP algorithms. The level of this skeleton tool not only simplifies the writing of parallel programs but also the development of the sequential case.

P. Amestoy et al. (Eds.): Euro-Par'99, LNCS 1685, pp. 877–887, 1999.
© Springer-Verlag Berlin Heidelberg 1999

The tool presented derives a parallel DP program from the user specification of a sequential code. The tool manages the optimal mapping of the stages of the associated DP automata in the actual processors, the generation of the corresponding messages and the handle of the involved data structures.

To study the performance of our skeleton, experiments were carried out in three different architectures: a distributed memory computer, the IBM-SP2; a distributed-shared memory computer, the Origin 2000 and a distributed memory with shared address space computer, the Cray T3E. The SP2 from IBM has 44 processors, 12 GB main memory, 431 GB disk storage and 18.55 Gflop/s theoretical peak. The Origin 2000 from Silicon Graphics is a 64 R10000 processors (196 MHz) machine with 8 GB main memory, 288 GB disk and 25.08 Gflop/s theoretical peak. The Cray T3E has 32 DEC 211164 processors. Each processor has 128 Mb of memory and reaches 600 Mflops.

Section 2 introduces a formal approach for DP based in the automata theory. The model is illustrated with the Single Resource Allocation Problem. Section 3 presents the class of Non Decreasing Finite Automata, for which an optimal pipeline algorithm is delivered. In section 4, we describe tools for the automatic parallelization of DP. In section 5, we will show the computational results. Finally, section 6 presents the conclusions.

2   The Dynamic Programming Technique

Dynamic Programming is an important technique for solving combinatorial optimization problems. It is based on the principle of optimality introduced by R. Bellman [3]. Our goal is to solve the optimization problem:

$$minimize \; (maximize) \; f(x)$$
$$subject \; x \; \epsilon \; S$$
(1)

in which the set of constraints S can be modeled as a language, that is, $S \subseteq \Sigma^*$. The set Σ is a finite set called the set of decisions. The elements of Σ^* which are elements of the Kleene closure of Σ, are referred or called policies. A policy $x \; \epsilon \; \Sigma^*$ is said to be feasible if $x \; \epsilon \; S$.

The Dynamic Programming technique works by associating to the optimization problem an automaton with costs [7]:

Definition: An automaton with costs is a triple $\prod = (M, \; \mu, \; \xi_0)$, where:

1) M is a finite automaton $M = (Q, \; \Sigma, \; q_0, \; \lambda, \; Q_F)$ in which:
   Q denotes the set of states,
   Σ is the set of decisions,
   $q_0 \; \epsilon \; Q$ is the initial state,
   $\lambda : Q \times \Sigma \to Q$, is the transition function and
   $Q_F \subset Q$ is the set of final states.

The transition function (λ) can be extended to the set of policies; $\lambda : Q \times \Sigma^* \to Q$ by defining $\lambda(q, \epsilon) = q$ and $\lambda(q, xa) = \lambda(\lambda(q, x), a)$, where $q \; \epsilon \; Q$,

$x \in \Sigma^*$, and $a \in \Sigma$. The notation $\lambda(x) \cong \lambda(q_0, x)$ for all $x \in \Sigma^*$ and $\lambda(q) = \{q' \in Q$ such that exists $a \in \Sigma : \lambda(q, a) = q'\}$ will be used.

2) $\mu : \Re \times Q \times \Sigma \rightarrow \Re$ is the cost function, where $\Re$ denotes the set of real numbers. If decision $a \in \Sigma$ is applied to state $q \in Q$, and the accumulated cost is ξ, the automaton changes into the new state $\lambda(q, a)$ with new cost given by $\mu(\xi, q, a)$. The cost function μ can be extended to $\Re \times Q \times \Sigma^* \rightarrow \Re$ by defining $\mu(\xi, q, xa) = \mu(\mu(\xi, q, x), \lambda(q, x), a)$ for $\xi \in \Re$, $q \in Q$, $x \in \Sigma^*$, $a \in \Sigma$ and $\mu(\xi, q, \epsilon) = \xi$. We will use notation $\mu(x)$ to denote the cost of policy x starting in q_0, i.e.: $\mu(x) \cong \mu(\xi_0, q_0, x)$.

3) $\xi_0 \in \Re$ is the setup cost of the initial state q_0.

Definition: A finite automaton with costs $\prod = (M, \mu, \xi_0)$ represents the optimization problem (1) if and only if it satisfies:

1) The language accepted by the automaton M is the set S of feasible solutions, i.e.: $L(M) = \{x \in \Sigma^* / \lambda(x) \in Q_F\} = S$.

2) For all feasible policy $x \in S$, $\mu(x) = f(x)$.

Definition: A feasible policy $x \in S$ is said to be optimal if and only if $\mu(x) \leq \mu(y)$ for any other feasible policy $y \in S$.

Definition: For any state $q \in Q$ we define $G(q)$ as the optimal value of any policy leading from the start state q_0 to state q, i.e., $G(q) \cong \min\{\mu(x) / x \in \Sigma^*, \lambda(x) = q\}$ when there exists an $x \in \Sigma^*$ such that $\lambda(x) = q$, and $G(q) = +\infty$ otherwise.

Let be G^* the optimum of the values for the feasible policies, that is $G^* = \min\{G(q) / q \in Q_F\}$. To solve the optimization problem (1) is equivalent [7] to finding the optimal value G^*, i. e., $G^* = \min\{f(x) / x \in S\}$. The optimal value $G(q)$ can be obtained from the optimal values of the states from wich q is reacheable through some decision a, i.e.,

for all $q \neq q_0$, $G(q) = \min\{\mu(G(q'), q', a) / q' \in Q, a \in \Sigma, \lambda(q', a) = q\}$ and $G(q_0) = \xi_0$.

Definition: Let M be a finite automaton, M is a multistage automaton if and only if, there is a partition of the set Q, $Q = \{q_0\} \cup Q_1 \cup ... \cup Q_n \cup Q_F \cup \{q_d\}$ such that $Q_i \cap Q_j = \emptyset$ when $i \neq j$, $q_d \in Q - Q_F$ is called the dead state, and
$\lambda(q_0, a) \in Q_1 \cup \{q_d\}$, for all $a \in \Sigma$
$\lambda(q, a) \in Q_{i+1} \cup \{q_d\}$, for all $q \in Q_i$, $a \in \Sigma$, $i = 1, 2, .., n - 1$.
$\lambda(q_d, a) = q_d$, for all $a \in \Sigma$.
$\lambda(q, a) \in Q_F \cup \{q_d\}$, for all $q \in Q_n$ and for all $a \in \Sigma$.

The sequential dynamic programming algorithm in figure 1 solves the optimization problem (1) for multistage automatons with costs. The algorithm assumes the existence a FIFO queue where the values needed to compute the next STAGE will be stored. In figure 1, functions INSERT and REMOVE introduces and eliminates data from the queue respectively. The algorithm runs in time $O(|\ Q\ | \cdot |\sum| \cdot O(\mu))$. Following this methodology, our tool will create to the user the illusion of being writting and executing sequential programs while the code generated will be parallel code. The user omits the loop in line 2 of figure 1 and the tool will automatically substitute the calls to INSERT and REMOVE by the classical message passing primitives to send and receive data. In the parallel case, the queue will disappear at all.

```
1:Begin
2:   for STAGE = 0 to n do
3:   begin
4:     if (STAGE == 0) /* First STAGE */
5:     begin
6:         G(q0) = ξ0;
7:         INSERT(G(q0));
8:     end
9:     else /* General STAGE */
10:        for qSTAGE-1 i ∈ QSTAGE-1 do
11:        begin
12:           REMOVE (G(qSTAGE-1 i));
13:           for a ∈ ∑ such that λ(qSTAGE-1 i, a) = qSTAGE i do
14:              G(qSTAGE i) = min{G(qSTAGE i), μ(G(qSTAGE-1 i), qSTAGE-1 i, a)};
15:           if (STAGE != n) /* last STAGE do not INSERT in queue */
16:              INSERT (G(qSTAGE i));
17:           for qSTAGE j = λ(qSTAGE-1 i, a) with a ∈ ∑ do
18:              G(qSTAGE j) = min{G(qSTAGE j), μ(G(qSTAGE-1 i), qSTAGE-1 i, a)};
19:        end;
20: end;
21:End.
```

Fig. 1. Sequential Dynamic Programming Algorithm

2.1 An Example: The Single Resource Allocation Problem (RAP)

The Single Resource Allocation Problem can be stated as follows [7]:

$$maximize\ z\ = \sum_{j=1}^{N} f_j(x_j)$$
$$subject\ to \sum_{j=1}^{N} x_j = M \tag{2}$$

Namely, it is required to allocate M units of an indivisible resource to N activities so that the sum of the effectiveness measured by $f_j(x_j)$ is maximized.

In order to solve this problem by dynamic programming, we introduce the following states q_{km} and their values $G(.)$.

q_{00}: The initial state representing the null allocation.

q_{km}: The state representing k-dimensional allocation vectors $(x_1, x_2, ..., x_k)$ of nonnegative integers such that $\sum_{j=1}^{k} x_j = m$, for $k = 1, 2, ..., N$, $m = 0, 1, ..., M$.

q_{NM}: The final state representing the allocation vectors satisfying (2).

$G(q)$: The maximum of $\sum_{j=1}^{k} f_j(x_j)$ for the allocation vectors represented by state q.

The transition function can be defined as: $\lambda(q_{km}, x) = q_{k+1, m+x}$ $x = 0, ..., M - m$

By definition, $G(q_{NM})$ is the optimum objective value we want to compute. It is equal to the length of a longest path from the source q_{00} to the sink q_{NM} where each edge $(q_{k-1 m}, q_{km+x_k})$ has a length $f_k(x_k)$. The value $G(q_{km})$ can be recursively computed by

$G(q_{00}) = 0$

$G(q_{1m}) = f_1(m), m = 0, ..., M$

$G(q_{km}) = \max_{0 \le x_k \le m} \{G(q_{k-1\ m-x_k}) + f_k(x_k)\}, k = 2, 3, ..., N, m = 0, 1, ..., M.$

3   Parallelization of the Dynamic Programming Technique

Definition: A multistage automaton M is called a non decreasing automaton if, and only if, there is a total order for each stage, $Q_k = \{q_{k1}, ..., q_{km}\}$, such that for every q_{ki} on stage Q_k, the condition $\lambda(q_{ki}) \subset [q_{k+1i}, q_{k+1r}]$, where $r = |Q_{k+1}|$ is satisfied.

```
1:Begin
2:    for q_{k-1i} ε Q_{k-1} do
3:    begin
4:        ReceiveFromLeft (G(q_{k-1i}));
5:        for a ε ∑ such that λ(q_{k-1i}, a) = q_{ki} do
6:            G(q_{ki}) = min{G(q_{ki}), μ(G(q_{k-1i}), q_{k-1i}, a)};
7:        SendToRight (G(q_{ki}));
8:        for q_{kj} = λ(q_{k-1i}, a) with a ε∑ do
9:            G(q_{kj}) = min{G(q_{kj}), μ(G(q_{k-1i}), q_{k-1i}, a)};
10:   end;
11:END
```

Fig. 2. A Pipeline Algorithm Parallelizing on the Stages for non decreasing automatons (PAPS). Code for processor k.

Along the rest of this paragraph we will consider the automaton M in $\prod = (M, \mu, \xi_0)$ to be a non decreasing automaton.

Since we are dealing with a multistage automaton, the problem can be solved using a pipeline with as many processors as stages. The computation of optimal values in stage k, $G(q)$, $q \ \epsilon \ Q_k$ is allocated to processor k (Algorithm PAPS of figure 2). During the *ith* iteration processor k receives from processor $k-1$ the optimal value $G(q_{k-1})$. As soon as $G(q_{k-1})$ is received, the values $G(q)$ for those q such that there exists $a \epsilon \Sigma$ with $\lambda(q_{k-1}, a) = q$ are updated. Property 1 of a non decreasing automaton grants that all the optimal values from which state q_{ki} depends on have been received, and so the optimal value $G(q_{ki})$ is completely computed and can be sent to processor $k+1$. The algorithm stops when processor n computes the optimal values for the final states $q_f \epsilon Q_F$. Processor n holds the optimal solution G^*. The conditions for the optimality of this algorithm can be found in [1].

4   Tools for Parallel Dynamic Programming

We aim to the building of an skeleton for parallel DP based on the exposed methodology. The project is divided in two stages: 1) The construction of a tool (called *La Laguna Pipeline*, *llp*) providing the programmer with a virtually infinite pipeline. The programmer just has to take control of the behavior of a general cell of the pipe. *llp* simulates the total amount of processors and maps them onto the physical ones using a mix of block and cyclic policies adapted to the grain g of the target architecture. 2) The developement of the parallel dynamic programming skeleton (called *La Laguna Parallel DP* or *llpdp*), built on top of the *llp*.

```
void solver_1 (int N) {
   int  result;
   if (NAME == 0) f_0(); /* First stage */
   else if (NAME == (N - 1)) {
      result = f_n-1(); /* Last stage */
      GPRINTF("\n% d:Result = % d",NAME,result);
   }
   else f(); /* A general stage of the pipeline */
   BROADCAST(BCAST, "variable", count, offset);
}
```

Fig. 3. First and Last Processors run different code.

4.1   The Basic Tool: Giving Support for Pipeline Paralellism

La Laguna Pipeline, *llp*, is a general purpose tool for the pipeline programming paradigm, and it is not specifically conceived for DP. *llp* is conceived as a macro based library and its portability is guaranteed by the wide number of message passing libraries supported (MPI, PVM, InmosC). Figures 3, 4 and 5 illustrate the easiness of use of applying a general pipeline scheme. In figure 3 funcion solver_1(), constitutes a virtually infinite pipeline to be executed. The code for the first and last processors of the pipe must be differentiated from the remaining processors. The variable NAME identifies the virtual processor in the pipeline. The functions f_i can use the macros *IN* and *OUT* to establish the corresponding communications with the left and right neighbors. *llp*, provides parallel Input/Output routines as the *GPRINTF* call in figure 3, allowing processors to read and write from shared input and output streams. The macro *BROADCAST* update values of global data distributed around the processors previously declared as *SHARED* (line 8 in figure 4). As in figure 4, several pipelines can be iteratively executed. A call to the macro *PIPE* has as arguments the generic routine and the size of the pipeline, *solver_1()* and N respectively for the first case. This macro introduces the routine in a loop simulating the virtual processors assigned to the actual processor according to the mapping policy.

```
1: #include "llp.h"
2: ....
3: int  main (PARAMETERS) {
4:    INITPIPE;
5:    Type variable[MAX];
6:    if (ARGC != 3)  abort();
7:    N = atoi(ARGV(1)); M = atoi(ARGV(2));
8:    SHARED(&variable, "variable", Type);
9:    PIPE(solver_1(N), N);
10:   PIPE(solver_2(M), M);
11:   EXITPIPE;
12: }
```

Fig. 4. Function main(): expanding PIPEs of N and M virtual processors.

4.2 Implementing Cyclic and Blocking-Cyclic Mapping Policies

The algorithm in figure 2, requires a variable number of processors, usually proportional to the number of stages or the number of states per stage. It does not consider the fact that for most cases the number of processors required exceeds the number of available processors. Therefore, it is necessary to assign the processes between the available processors. This is the problem of finding an efficient mapping of the virtual pipeline on the actual parallel machine. Fortes and Moldovan propose in [9] to partition the set of processes in sets $B(i)$ of the same cardinal as the number p of processors, in such a way that process $q \in B(i)$ if and only if $q/p = i$. Process q is processed by processor $q \bmod p$. All processes in the same set $B(i)$ are processed in parallel. Implementation can be easily achieved on a one way ring topology where the first and last processors are connected through a buffering process. This approach leads to a pure cyclic mapping. Better results depending on the grain g of the underlying platform may be achieved if this cyclic mapping is combined with block mapping. On a block-cyclic mapping, process q is assigned to processor $(q/g) \bmod p$. The implementation of a pure cyclic mapping policy for a pipeline can be easily obtained over a ring as a particular case of the macro PIPE described in figure 5 taking $g = 1$. Every processor calculates the number of STAGES to compute, (n/p) (line 2), and then goes into a loop running the function f for the appropiated first virtual processor NAME (line 4).

The classical technique to obtain block-cyclic mapping consisting of g sequential executions of the f function is not a good approximation to the problem. The delay introduced by each processor produces a parallel algorithm as slow as the sequential one. To obtain an efficient implementation, processors must start to work as soon as possible and they must be fed with data when needed. To fulfil this conditions we implement the block-cyclic mapping using the Unix standard library "*setjmp.h*". This library allows unconditional jumps to variable labels. To execute function f in pipeline with grain g, automatically g copies of the local data of function f are expanded, one for each one of the processes to be executed in one iteration of the loop. The code to be executed is the same for each process: the function f. Therefore, only the Program Counter of every process need to be stored. Macros IN and OUT differentiate between internal and external communications. External comunications use the communication primitives provided by PVM/MPI. Internal comunications produce the context switch between processes using the functions *setjmp()* and *longjmp()* of the library "*setjmp.h*". The context switch just consist of updating the NAME of the process to get the Program Counter and the address of the local variables for this process. Since a context switch occurs only when needed and it is implemented very efficiently, the overhead introduced by the tool in a block-cyclic partition is minimum.

```
1: #define PIPE(f, n) {
2:     int bands = numbands(n, g); /* The number of STAGES to compute */
3:     for (i = 0; i < bands; i++) {
           /* Physical Processor pname calculates the */
           /* NAME of the first process to simulate */
4:         NAME = g * (pname + i * numproc);
5:         if (NAME ≤ (n - 1)) f;
6:     }
7: }
```

Fig. 5. The Macro PIPE.

4.3   The Dynamic Programming Skeleton, llpdp

The *llpdp* dynamic programming skeleton takes advantage of the fact that the behavior of a pipeline is similar to a FIFO queue. This way, the *llpdp* user is asked to write the sequential code for a generic automaton stage using a FIFO queue provided by the *llpdp* system. Instead of speaking of processor NAMEs, the *llpdp* uses STAGES and IN and OUTS are substituted by INSERTs and REMOVEs. This simply renaming and semantic change means that the code generated is similar that the generated by the *llp* tool. The parallel code generated is optimal when the automaton is Non Decreasing and the user inserts the states following the natural automaton order. The efficiency achieved by the skeleton is similar to the obtained by the *llp* programming system. The *llpdp* code in figure 6 expresses the DP formulation of the Single Resource Allocation problem. Note the similarity of this code with the sequential one of figure 1.

```
void   solve_RAP () {
  LOCAL_VAR;
    int m, j, x, G[M + 1];
  BEGIN;
  if (STAGE == 0) /* First STAGE */
    for( m = 0; m ≤ M; m++) {
    G[m] = 0;
    INSERT(G[m]);
    }
  else { /* General STAGE */
      for( m = 0; m ≤ M; m++) G[m] = 0;
      for (m = 0; m ≤ M; m++) {
      REMOVE(&x);
      G[m] = max(G[m], x + f(STAGE-1, 0));
      if (STAGE != N) INSERT(G[m], 1, sizeof(int));
      for (j = m + 1; j ≤ M; j++) G[j] = max(G[j], x + f(STAGE - 1, j - m));
      } /* for m ... */
      } /* else ... */
  END;
} /* solve_RAP */
```

Fig. 6. Simple Pipeline Algorithm for the RAP. Every processor runs the same Code.

5   Computational Results

Table 1 shows the times of the sequential algorithm in the three machines. Table 2 presents the times of the algorithm implemented using the tool. The executions were performed varying the grain from 1 to 20 and the number of processors ranging from 2 to 8. The results obtained for each machine appear in two columns. The column labeled *Time* indicates the time in seconds, while the labeled *S* keeps the speedup. The introduction of mechanisms to manage the grain implies a slight loss of performance. It can be detected comparing the results with grain 1 and grain 2. However, an increasing of the grain generally carries an increase of the performance.

Table 1. Sequential time. Problem of size N = 400, M = 1000

Cray T3E	IBM SP2	Origin 2000
27.46	88.29	28.47

Table 2. Times and speedups for the RAP. Problem of size N = 400, M = 1000

		Cray T3E		IBM SP2		Origin 2000	
Grain	Proc.	Time	S	Time	S	Time	S
1	2	27.03	1.0	70.49	1.3	35.50	0.8
1	4	14.00	2.0	36,71	2.4	19.59	1.5
1	8	7.51	3.7	18.63	4.7	16.53	1.7
2	2	36.63	0.7	69.87	1.3	44.60	0.6
2	4	18.56	1.5	35.05	2.5	22.35	1.3
2	8	9.47	2.9	17.71	5.0	11.22	2.5
5	2	36.06	0.8	68.13	1.3	43.07	0.7
5	4	18.26	1.5	34.24	2.6	21.57	1.3
5	8	9.33	2.9	17.20	5.1	10.87	2.6
10	2	36.24	0.8	67.72	1.3	42.87	0.7
10	4	18.31	1.5	33.95	2.6	21.47	1.3
10	8	9.25	3.0	17.07	5.2	10.77	2.6
20	2	36.16	0.8	67.52	1.3	42.59	0.7
20	4	18.25	1.5	33.86	2.6	21.35	1.3
20	8	7.41	3.7	13.62	6.5	8.57	3.3

6   Conclusions

A general procedure that can be applied to the whole class of dynamic programming problems have been presented. The proposed scheme leads to efficient implementations of both parallel and sequential dynamic programs, as confirm the computational experiments presented in this paper. Based in the proposed algorithms, a general parallel tool for Dynamic Programming has been built. This tool comes to fill a remarkable absence of such a kind of tool both in the sequential and parallel fields.

Acknowledgments

We wish to thank to CEPBA, CESCA and CIEMAT for allowing us the access to their machines.

References

[1] Almeida F.,Models in Parallel Dynamic Programming for the Discrete Case. PHDegree. DEIOC. Unniversidad de La Laguna. October 1996.

[2] Andonov, Raimbault, Quinton. *Dynamic Programming Parallel Implementations for the Knapsack Problem.* Technical Report 740. IRISA. June 1993.

[3] Bellman R.. *Dynamic Programming.* Princeton U. P.. 1957.

[4] Chen G., Jang J.. *An Improved Parallel Algorithm for 0/1 Knapsack Problems.* Parallel Computing. 18. 811-821. 1992.

[5] Edmonds P., Chu E., George A.. *Dynamic Programming on a Shared Memory Multiprocessor.* Parallel Computing. 19, 9-22. 1993.

[6] Gibbons A., Rytter W., *Efficient Parallel Algorithms.* Cambridge University Press. 1988.

[7] Ibaraki T.. *Enumerative Approaches to Combinatorial Optimization, Part II.* Annals of Operations Research. Volume 11, N. 1-4, 1988.

[8] Kindervater G., Trienekens H.. *Experiments with Parallel Algorithms for Combinatorial Problems.* European Journal of Operational Research. 33, 65-81. 1988.

[9] Moldovan D., Fortes J.. *Partitioning and Mapping Algorithms into fixed size Systolic arrays.* IEEE Trans. Comput. C-35 (1), 1-12. 1986.

[10] Wah B., Li G., Fen C.. *Multiprocessing of Combinatorial Search Problems.* Computer. Vol. 18. 6, 93-108. 1985.

Programming Effort vs. Performance with a Hybrid Programming Model for Distributed Memory Parallel Architectures

Andreas Rodman and Mats Brorsson

Department of Information Technology, Lund University
P.O. Box 118, S-221 00 LUND, Sweden
email: Andreas.Rodman@it.lth.se

Abstract. We investigate here the programming effort and performance of a programming model which is a hybrid between shared memory and message-passing. This model permits an easy implementation in shared memory, while still being able to benefit from performance advantages of message-passing for performance critical tasks. We have integrated message-passing with a software DSM system, and evaluated the programming effort and performance with three different applications and various degree of message-passing in the applications.

In two of the applications we found that only a small fraction of the source code lines responsible for interprocess communication were performance critical and it was therefore easy to convert only those to message-passing primitives and still approach the performance of pure message-passing.

1   Introduction

Parallel computing research has so far mainly focused on how to achieve high performance. The progress in this important area continues, although at a slower pace, and it is time to also look at an often neglected research topic related to the programming effort needed to achieve this high performance. Even though parallel machines have existed on the market for quite some time now, the commercial success has been limited and we believe that much of this can be attributed to the lack of convenient and effective parallel programming models. In this paper we advocate the use of a hybrid parallel programming model that combines the convenience of shared memory with the effectiveness of message-passing for distributed memory parallel architectures.

Recently, the OpenMP initiative has led to a standard for parallel programs using a shared memory model [12]. This model is well suited to parallel architectures where shared memory is supported in hardware but in distributed memory architectures it is necessary to use a software distributed shared memory (software DSM) system which normally results in lower performance. The alternative is to use message-passing, which is effective [10, 11], but more cumbersome.

We have explored and measured the programming effort vs. performance when we combine a shared memory programming model and a standard message-passing model. Our hypothesis is that there is locality in the code where inter-process communication takes place. We believe that with small and uncomplicated modifications in a shared memory algorithm we can change the bulk of the implicit communication in

P. Amestoy et al. (Eds.): Euro-Par'99, LNCS 1685, pp. 888-898, 1999.
© Springer-Verlag Berlin Heidelberg 1999

the shared memory model to explicit, and thus more effective, communication with message-passing.

We show that this is indeed true with results from experiments on an IBM SP-2. We have used three different parallel applications that are written with a shared memory model and have gradually transformed each application into a message-passing program. Unfortunately, the performance increase that we achieve is paid with an increased programming effort. We propose a method to measure the programming effort needed when we replace implicit shared memory communication with explicit message-passing and correlates the result with the changes in performance. Even though it is difficult to measure the programming effort objectively as this most often is a subjective task, we have created a yardstick consisting of both an objective and a subjective rating.

In the next chapter we discuss our methodology after which we go through the results of our experiments in section 3. Before concluding the paper we discuss some of the related research in section 4.

2   Methodology

2.1   Estimation of Programming Effort

As mentioned before, this study focuses on the programming effort of a hybrid programming model. This is unfortunately a difficult thing to measure and often very subjective, which means that it depends on the experience and preference of the actual programmer. To establish some kind of objective programmability metric, we keep track of the number of source code lines that needs to be changed, deleted or added and compare this to the gained performance.

To reinforce the evaluation of programming effort we have also included a subjective grading of each separate modification of an application on a scale from 1 to 3, where 1 means simple and 3 difficult. All grades are thoroughly motivated, but it is still a subjective metric and should only be seen as a complement to the objective method of counting the number of changed source code lines.

We base our evaluation on shared memory versions of the programs. To ensure fairness in the study, we neither change the algorithms, nor the communication patterns, only the method of communication. To be able to assess the impact of changing shared memory communication to message-passing primitives we have developed five versions of each program. One with communication purely through shared memory, one with only message-passing and three hybrid versions with increasing amount of message-passing.

2.2   Experimental Environment

All our experiments are based on TreadMarks [1], a software DSM system. The evaluation is done on three parallel applications: Barnes-Hut and Raytrace that originates from the SPLASH-2 benchmark suite [14], and 3DFFT from the NAS parallel benchmarks [2]. All applications are measured using an IBM SP2 with 160 MHz POWER2 processors. The interconnection switch transfer rate is 110 Mbyte/s.

We have extended the Treadmarks programming interface with some simple MPI-like [3] functions to create a hybrid shared memory/message-passing programming model. These message-passing functions are *non-blocking send*, *blocking receive*, *broadcast* and *allgather*. Even though the MPI-standard has a much larger variety of communication methods, these few functions still provide us with a good evaluation base, since most of the communication patterns can be fairly effectively implemented using these few building blocks. Let us now go over to present the results of the evaluation of programming effort vs. performance using our hybrid programming model

3   Experimental Results

3.1   Barnes-Hut

This application simulates the gravitational interaction between a set of bodies in a three-dimensional space. The simulation is divided into time steps, where each step computes the net force on every body and updates its position and velocity in space. We have classified the communication in the Barnes-Hut algorithm into 3 modification groups, A-C, shown in tab. 1. We also note the number of source code lines that needs to be modified in order to move from shared memory to message-passing, and a subjective estimate of the effort to make these changes.

Table 1. Effort to transform shared memory communication into message-passing in Barnes-Hut.

Modification Group	Description	Number of modified source code lines	Effort (1-3)
A	Broadcast of tree structure	2	1
B	Broadcast and return of body structure	35	2
C	misc.: start-up, global reduction etc.	77	1

Group A handles the communication of the tree structure that is used to approximate large sets of bodies as a single mass to improve performance. In the shared memory version, this tree is allocated in shared memory and the master node reads all bodies to build a new tree in beginning of each time step. The nodes then read the tree using the shared memory layer to calculate new positions and velocities for the bodies. It is thus a single-producer, all-consumer communication pattern. The conversion from shared memory to message-passing is done by broadcasting the tree structure after the tree construction at the beginning of each time step. The effort of modification A is small since it is very easy to implement and only needs a single broadcast.

The bodies are allocated in large shared vector in the shared memory version. At the beginning of each time step the master node reads all bodies and builds the tree. The bodies are then divided among the nodes and each node updates the position and velocity of its assigned bodies. Modification B is the conversion of this communication into message-passing. After the master node has calculated the tree it broadcasts the entire set of bodies to all nodes. Then each node updates its assigned bodies. At the end of the time step the nodes pack all modified bodies into a large packet and return them to the master node so it can perform the next tree calculation. Creating the packing procedure is quite simple but not obvious. Therefore modification B is graded 2.

Finally, group C is the distribution of start-up data and collection of global time step information. The distribution of start-up data was replaced by some simple broadcast statements in the message-passing version. In the shared memory version the global time step information was gathered by the master node at the end of each time step using a shared vector where each node has its own element. The master node then reads the entire vector and merges the result. This communication was replaced by a send of the local time step information from all nodes to the master node at the end of each time step. Even though all this requires a lot of modifications, the conversion is straightforward and hence we have given it the grade 1.

We have combined these 3 modifications according to tab. 2 which leads to five different versions of Barnes-Hut. Barnes-Hut-SM is the pure shared memory version, Barnes-Hut-MP the pure message-passing version and the versions whose name ends with H1, H2 and H3 are the three hybrid versions. The highest programming effort is the maximum subjective rating of 1-3 of all the groups included.

Table 2. The different versions of Barnes-Hut.

Program name	Groups	Total number of source code lines changed in program	Highest programming effort (1-3)
Barnes-Hut-SM	-	0	none
Barnes-Hut-H1	A	2	1
Barnes-Hut-H2	B	35	2
Barnes-Hut-H3	A,B	37	2
Barnes-Hut-MP	A,B,C	114	2

In our experiment we have simulated a set of 65536 bodies for 6 time steps. The sequential execution time is 241 seconds. Fig. 1 shows the relative speedups obtained in our experiments. As expected, the speedup increases with the substitution of shared memory communication for message-passing directives.

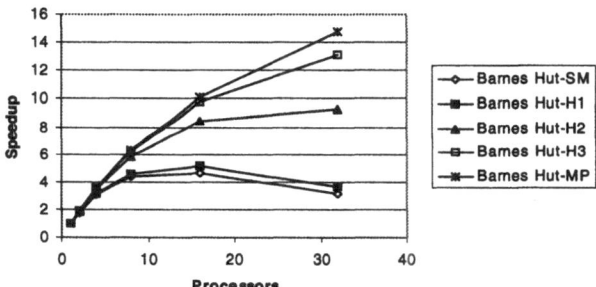

Fig. 1. The speedup of the different hybrid versions of the Barnes-Hut algorithm.

The amount of communicated data, as can be seen in fig. 2, grows when message-passing instructions are substituted for shared memory communication and this might seem a bit strange since it also results in a higher speedup. However, the amount of data does not lead to any problem on the IBM SP2 because of the high performance switch which has a high data transfer rate.

The major reason for the speedup when we increase the amount of communication through message-passing is of course the latency in the request-reply function involved when using a software DSM. When we are executing the MP version on 32 processors, only 558 send statements and 22 broadcasts are invoked, while the software DSM version transmits a total of 2618032 messages.

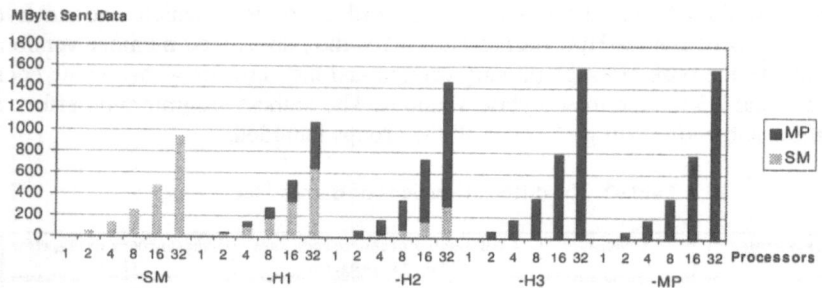

Fig. 2. The amount of communicated data through the shared memory layer (SM) and the message passing layer (MP) for Barnes-Hut.

3.2   3DFFT

This program uses a three-dimensional fast Fourier transformation algorithm for large matrices, and is used to solve partial differential equations. The communication in 3DFFT has been divided into three modification groups as shown in tab. 3.

Table 3. Effort to transform shared memory communication to message-passing in 3DFFT.

Modification Group	Description	Number of modified source code lines	Effort (1-3)
A	Checksum calculation	13	1
B	Matrix transpose operation	77	3
C	Start-up	11	1

Group A handles a checksum calculation done for each iteration. In the shared memory version this is done by a global vector where each node enters its result in its own element. The master node reads the entire vector and calculates the result. This communication is with message-passing implemented as a simple send of each vector element from every node to the master node. It is a simple task to convert this into message-passing and is therefore given a grading of 1.

The 3DFFT algorithm is implemented as three separate FFT:s in each dimension. Two of the three FFT:s can be done locally for each iteration, but the third FFT requires data from all the other nodes. This communication is done by a global transpose operation and is marked as group B. The transpose operation requires that each node sends a piece of its calculated result to all the other nodes, which means that this is an all-to-all communication. In the shared memory version, the entire matrix is allocated in shared memory and the communication occurs when each node transposes its part of the matrix. The message-passing version sends a message from every node to every other node in order to transpose the matrix. It is quite difficult to grasp the communication pattern of this operation and it is therefore given a grading of 3.

Group C consists of distribution at start-up of shared memory variables in the shared memory version. In the message-passing version the communication was replaced by some simple broadcasts. The programming effort was estimated to 1 since the conversion was very easy. These groups were then combined into the program versions seen in tab. 4. Our input data consists of a 64×64×64 matrix and we are running 64 iterations. A sequential execution takes 55 seconds.

Table 4. The different versions of 3DFFT.

Program name	Groups	Total number of source code lines changed in program	Highest programming effort (1-3)
3DFFT-SM	-	0	none
3DFFT-H1	A	13	1
3DFFT-H2	B	77	3
3DFFT-H3	A,B	90	3
3DFFT-MP	A,B,C	101	3

Fig. 3 shows the relative speedup for the five versions of 3DFFT. A major performance impact can first be seen for version H2. As stated earlier, this version handles the matrix transpose operation by message-passing instead of shared memory and most of the communication in the program occurs in this operation. Hence, all other versions have little impact on both the number of source code lines needed and on the actual speedup. The modification to handle the transpose operation in message-passing is relatively difficult and in this case results in 77 lines of code that needs to be modified.

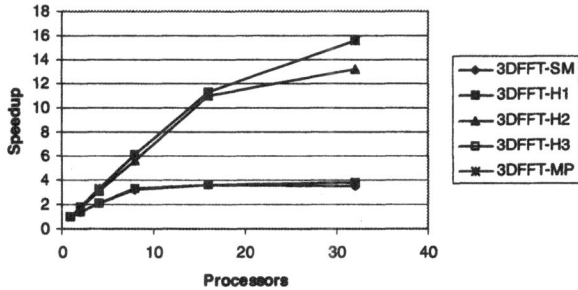

Fig. 3. The speedup of the different hybrid versions of the 3DFFT algorithm.

In contrast to Barnes-Hut, TreadMarks transmits a lot more data than needed over the network which together with the high number of messages is the major reason for the poor performance of the shared memory version of 3DFFT (see fig 4). The cause for the excessive amount of communication is false sharing due to the large coherence granularity. For 8 nodes the amount of data is about the same, but there is still a significant gap between the speedup for the SM and MP version. Again we have the request-reply dilemma for the shared memory version as discussed earlier for Barnes-Hut. The push technique used in message-passing has a much lower overall latency and the MP version thus performs better.

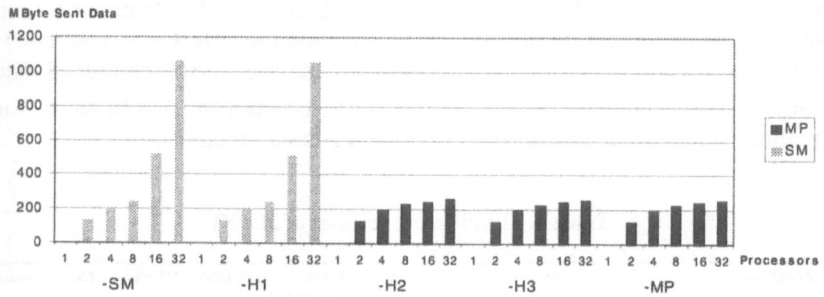

Fig. 4. The amount of communicated data through the shared memory layer (SM) and the message passing layer (MP) for 3DFFT.

3DFFT is different compared to Barnes-Hut since we needed a relatively large and difficult modification of the source code in order to get any appreciable performance benefit. In this case, the programming effort required to get a higher performance is similar to pure message-passing since there are few incremental steps.

3.3   Raytrace

The raytrace application is a standard image rendering program of a three-dimensional scene. The communication in Raytrace has been divided into four groups (see tab. 5).

Table 5. Effort to transform shared memory communication into message-passing in Raytrace.

Modification Group	Description	Number of modified source code lines	Effort (1-3)
A	Broadcast of model	9	1
B	Gathering of image result	27	2
C	Removal of distributed work pool	11	2
D	Start-up	75	1

Group A is the distribution of the three-dimensional scene. In the shared memory version the scene is placed in a large global structure and is then read by the nodes during execution. This means that if a node has a ray that hits a part of the scene not previously read, the software DSM will request the page from master node. The effect is that every node reads a large part of the scene by requesting it page by page from the master node. In the message-passing version this communication has been replaced by a broadcast of the entire scene before the calculation begins. This is only a few lines of simple code and is given the subjective grade of 1.

Group B handles the rendered image output. Since each node has rendered a small part of the picture, a merging reduction operation was done in message-passing so each node can add their result to the final picture. The merging process mainly consists of a send from all nodes to the master node containing their rendered part. The master node then copies each part into the correct position of the picture. The shared memory program has the resulting image in a shared array and the merging process is automatically done through the consistency protocol of the DSM. The conversion required a programming effort of 2.

There are large variations in the amount of work required to render each part of the picture. To handle this possible work imbalance, a distributed work pool exists in the shared memory version. With modification of group C we have performed a slight deviation from the original algorithm and removed the global work pool and thereby changed the communication structure. The shared memory version of Raytrace uses task stealing between the individual work pools, to dynamically balance the workload. The individual pools are protected by locks and the communication pattern has a very dynamic behavior. It is difficult to implement the work pool using message-passing without adding a task server process and therefore we have chosen to change the dynamic work pool into a static partitioning and to measure the resulting work imbalance. This will unfortunately not be an entirely fair comparison, but it will still give us some idea of the programming effort versus performance gain using a message-passing model. In other words the work pool has been removed and thereby the communication. In the end we have given this solution the grade 2 since it is mostly a removal of code.

Group D is in the shared memory version a set of variable distributions at start-up. These were substituted with a set of broadcasts in the message-passing version. Since the modification is very simple the effort was graded to 1. The groups were then combined into 5 versions seen in tab. 6. During evaluation, we raytraced a 256×256 pixels image of the car scene that is supplied with the program. The evaluation image took 20 seconds to raytrace sequentially.

Table 6. The different versions of Raytrace.

Program name	Groups	Total number of source code lines changed in program	Highest programming effort (1-3)
Raytrace-SM	-	0	none
Raytrace-H1	A	9	1
Raytrace-H2	A,B	36	2
Raytrace-H3	A,B,C	47	2
Raytrace-MP	A,B,C,D	122	2

The relative speedup for the versions are shown in fig. 5. The speedup is small for all the versions due to the low computation-to-communication ratio in this example which depends on both the three-dimensional scene and on the size of the produced image.

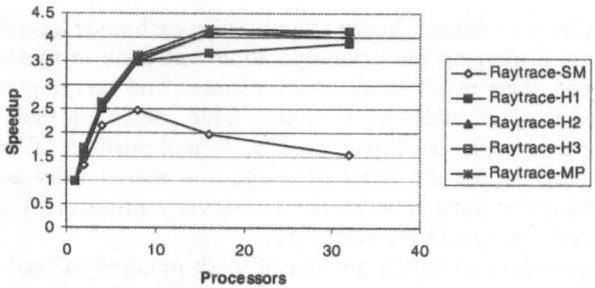

Fig. 5. The speedup of the different hybrid versions of the Raytrace algorithm.

Almost all improvements are achieved when we transform the SM version into H1 which only needed 9 lines of changed code. Even though the H1 version sends far too much data, which can be seen in fig. 6, the push technique of message-passing will again become the favorable method. Hence a lot of speedup was gained with a minimal programming effort

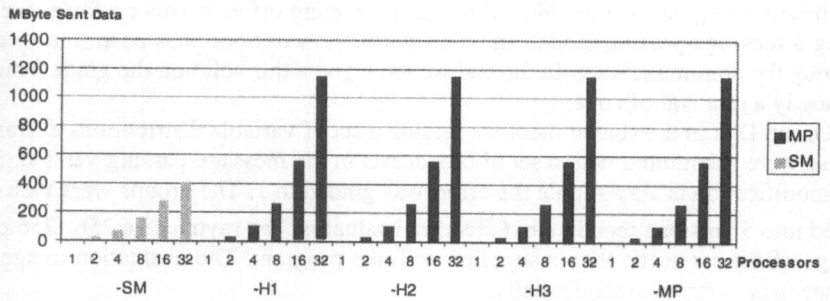

Fig. 6. : The amount of communicated data through the shared memory layer (SM) and the message passing layer (MP) for Raytrace.

H3 and MP with static workload balance displayed a higher speedup than all other versions. We measured the maximum imbalance and found that for 32 nodes and an execution time of 4.9 seconds we got a maximum work imbalance of only 1.7 seconds. So no processor spent more than 35% of its execution waiting for the other processors. This could, however, be much worse, and we can only conclude that we were lucky to get such a good static partitioning after all. If we were to change the scene or the granularity, then we would probably arrive at a much higher workload imbalance. The H3 and MP versions without the work pool, lack the robustness of the other versions.

The Raytrace example has a lot to gain from being implemented in a hybrid model when considering both programmability and performance. We can increase the performance immensely with only minimal programming effort without sacrificing the robustness of the dynamic workload balancing.

4   Related Work

The integration of shared memory and message-passing has, unfortunately, not been studied extensively. The groups working on Stanford Flash [6] and MIT Alewife [9] have argued for hardware support for both programming models for the same reasons that we do here. They have however only studied the hybrid programming technique from a performance aspect, and not investigated the programming effort.

There have also been a few attempts to integrate message-passing with a software DSM. Koch et al. proposed a system called CarlOS [8], which uses message driven consistency. The work was however strictly from a performance view, and no measurements were done on the code of the algorithms. Cordsen et al. [4] also examined a hybrid programming model based on software DSM. Again there were no actual recording or discussion of the impact the hybrid model had on the code. Recently the Brazos [13] group has started to implement a programming environment supporting both OpenMP using a software DSM and a complete MPI library. As previous no study was made on programming effort but only on performance.

Attempts have been made to supply the shared memory system with the same type of communication pattern information as given by explicit massage-passing statements and thereby improving the performance. Dwarkadas et al. [5] investigated the use of compiler analyzing to extract the communication pattern. Larus et al. proposed a programming model called CICO [7] that used a collection of code annotations to supply the system with the appropriate communication pattern information.

5   Conclusion

The hybrid model between shared memory and message-passing improves the performance for a software DSM at an acceptable programmability price. The core of an algorithm usually gains from being implemented in message-passing, but the core often only involves a minority of the entire code handling communication. Therefore the hybrid model definitely has a future since a substantial part of the program is non performance critical and suited for being implemented in shared memory while the performance critical core gains from a message-passing model. The hybrid model also enables an incremental parallelization from shared memory to message-passing. We would therefore recommend to extend an existing MPI library with software DSM support to construct a more programmer friendly model.

Acknowledgments

The research in this paper was supported with computing resources by the Swedish council for High Performance Computing (HPDR) and Center for High Performance Computing (PDC), Royal Institute of Technology.

References

1. C. Amza, et al. *TreadMarks: Shared Memory Computing on Networks of Workstations*, IEEE Computer, Vol. 29, no. 2, pp. 18-28, February 1996.
2. D. Bailey. et al. *The NAS Parallel Benchmarks 2.0*, Report NAS-95-020, Nasa Ames Research Center, Moffett Field, Ca, 94035, USA. December, 1995.

3. L. Clarke, I. Glendinning and R. Hempel. *The MPI Message Passing Interface Standard*, The MPI Forum, March 1994
4. J. Cordsen and W. Schröder-Preiskcha. On the Coexistance of Shared-Memory and Message-Passing in the programming of Parallel Applications. In *Proc. of High-Performance Computing and Networking: International Conference and Exhibition*, Vienna, Austria, April 1997.
5. S. Dwarkadas, et al. *Combining Compile-Time and Run-Time Support for Efficient Software Distributed Shared Memory*. Proceedings of the IEEE, pp. 476-486, Vol 87, Number 3, March 1999.
6. J. Heinlein, et al. Integration of Message Passing and Shared Memory in the Stanford FLASH Multiprocessor, in *Proc. of the 6th Int. Conf. on Support for Programming Languages and Operating Systems*, pp. 38-50, October 1994.
7. M. D. Hill, et al. *Cooperative Shared Memory: Software and Hardware for Scalable Multiprocessors*, ACM Transactions on Computer Systems, November 1993
8. P. T. Koch, R. J. Fowler. Carlsberg: A Distributed Execution Environment Providing Coherent Shared Memory and Integrated Message-passing, in *Proc. of Nordic Workshop on Programming Environment Research*, pp 279-293, Lund, Sweden, June 1994.
9. D. Kranz, et al. Integrating Message-Passing and Shared-Memory: Early Experiences, in *Proc. of the Fourth Symposium on Principles and Practices of Parallel Programming*, pp. 54-63, May 1993.
10. H. Lu, S. Dwarkadas, A. L. Cox, and W. Zwaenepoel. *Quantifying the Performance Differences Between PVM and TreadMarks*, Journal of Parallel and Distributed Computing, Vol. 43, No. 2, pp. 65-78, June 1997.
11. J. M. MacLaren and J. M. Bull. Lessons Learned when Comparing Shared Memory and Message-passing Codes on Three Modern Parallel Architectures, in *Proc. of High-Performance Computing and Networking: International Conference and Exhibition*, Amsterdam, The Netherlands, April 1998, pp. 337-346.
12. OpenMP consortium, *OpenMP: A Proposed Standard API for Shared Memory Programming*, White paper, http://www.openmp.org.
13. E. Speight, H. Abdel-Schafi and J.K. Bennet. An Integrated Shared-Memory / Message Passing API for Cluster-Based Multicomputing. in *Proc. of the Second International Conference on Parallel and Distributed Computing and Networks* (PDCN), December 1998.
14. S. C. Woo, et al. The SPLASH-2 Programs: Characterization and Methodological Considerations. in *Proc. of the 22nd International Symposium on Computer Architecture*, pp. 24-36, Santa Margherita Ligure, Italy, June 1995.

DAOS — Scalable And-Or Parallelism

Luís Fernando Castro[1], Vítor Santos Costa[2], Cláudio F.R. Geyer[1],
Fernando Silva[2], Patrícia Kayser Vargas[1], and Manuel E. Correia[2]

[1] Universidade Federal do Rio Grande do Sul, Porto Alegre - RS, Brasil
{lfcastro,geyer,kayser}@inf.ufrgs.br
[2] Universidade do Porto, Porto, Portugal
{vsc,fds,mcc}@ncc.up.pt

Abstract. This paper presents DAOS, a model for exploitation of And-
and Or-parallelism in logic programs. DAOS assumes a physically distributed memory environment and a logically shared address space. Exploiting both major forms of implicit parallelism should serve a broadest
range of applications. Besides, a model that uses a distributed memory
environment provides scalability and can be implemented over a computer network. However, distributed implementations of logic programs
have to deal with communication overhead and inherent complexity of
distributed memory managent. DAOS overcomes those problems through
the use of a distributed shared memory layer to provide single-writer,
multiple-readers sharing for the main execution stacks combined with
explicit message passing for work distribution and management.
Keywords: Parallel Logic Programming, And/Or Model, Scheduling,
Distributed Shared Memory

1 Introduction

Logic programs are amenable to the exploitation of two major forms of *implicit*
parallelism: or- and and-parallelism. *Or-parallelism* (ORP) aims at exploring
different alternatives to a goal in parallel and arises naturally in search problems.
And-parallelism (ANDP) consists in the parallel execution of two or more goals
that cooperate in determining the solutions to a query. One important form of
and-parallelism is independent and-parallelism (IAP), where parallel goals do not
share variables. This form of parallelism arises in divide-and-conquer algorithms.
Another type is called dependent and-parallelism (DAP), that allows goals to
share variables. This is common in consumer-producer applications. Parallel logic
programming systems (PLPs) should support both ANDP and ORP due to serve
the broadest range of applications and to become popular. In practice, exploiting
just one form of implicit parallelism requires sophisticated system design and
exploiting two distinct forms of parallelism is even harder.

Shared memory PLPs have been the most successful and widespread so far.
There are several examples of systems supporting full Prolog. Aurora [8] and
Muse [1] are well-known ORP systems while &-Prolog, DASWAM [13], and
&-ACE [9] are IAP systems that have been used to parallelise sizeable appli-
cations. Andorra-I [12] is a further example that supports both determinate

P. Amestoy et al. (Eds.): Euro-Par'99, LNCS 1685, pp. 899–908, 1999.
© Springer-Verlag Berlin Heidelberg 1999

and-parallelism and or-parallelism. Several distributed memory PLPs have been proposed. Some support pure IAP [15], or pure ORP [2].

The differences between shared and distributed memory machines have become less significant in the last few years due to distributed shared memory systems (DSMs) that gives a shared memory abstraction. Work on PLPs for hardware DSMs has given interesting results. Dorpp ORP system [14] achieved good performance of a DSM machine. More recently, Santos Costa et al [10] had analyzed Andorra-I system on a DASH-like simulated machine. Their numbers confirm that most read cache misses result from scheduling and from accessing the code area. Misses to the execution stacks (most part eviction or cold misses) varied from 8% on a ORP parallel application to 30% on an ANDP application. Only the ORP application has significant sharing misses for an execution stack, the choice-point stack (60%), because this stack is also used for scheduling. We argue that the previous analysis suggests a new approach to distributed PLPs. First, most large data structures in these systems are built in the execution stacks. We should take the best advantage of caches to reduce network traffic. In contrast, the previous studies show high rates of sharing misses in scheduler data-structures. This suggests using explicit messages for scheduling.

DAOS: Distributing And/Or in Scalable machines is the first PLP model for distributed systems that supports these innovations. Its main contributions are:

- binding data representation: it is both simple to implement on PLPs and it naturally adapts to DSM techniques allowing the use of previously designed synchronization and scheduling algorithms;
- combined DSM and message passing techniques in the same framework: DAOS innovates over shared memory PLPs by explicitly addressing the distribution problems inherent to scalable machines.

Next, some considerations about And/Or exploitation are presented. Then, the DAOS model is presented in Section3. Section 4 analyze how workers can be implemented in DAOS. Finally, there are some conclusions.

2   Exploiting And/Or Parallelism

ORP and IAP are arguably two of the more interesting forms of parallelism available in Prolog programs. Several methods have been proposed to exploit And/Or parallelism in PLPs. This section will present the three main techniques to deal with IAP. Consider the following and-parallel query:

```
?- a(X) & b(Y).
```

where & represents a parallel conjunction. Figure 1 shows that both goals have several solutions. Besides running the goals a(X) and b(Y) in parallel, one needs to combine the solutions. One alternative is to use a special node, that maintains pointers to each solution for each goal. Solutions to the conjunction are obtained by calculating the cross-product between values for X and for Y. This approach is known as *reuse* as presented in Figure 1(a).

Recomputation-based models are based on having an Or search tree, such that and-parallel goals correspond to building bits of the search tree in advance. Thus, when one starts a(X) and b(X) in parallel, the solution to b(X) is a continuation branch for a(X), and is thus *associated with a specific solution for a(X)*, as shown in Figure 1(b). These models are named recomputation-based because to exploit several alternatives for a(X), the solution for b(X) has to be recomputed for each alternative. Recomputation avoids cross-product node implementation overheads and symplifies Full Prolog semantic support. Reuse saves some effort in recomputing goal but Prolog programmers usually try to reduce search space [6].

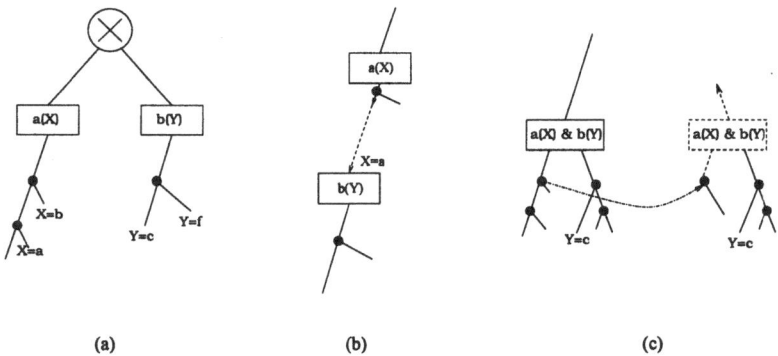

(a)   (b)   (c)

Fig. 1. (a) Reuse (b) Recomputation (c) C-Tree

The C-tree is shown in Figure 1(c). In the C-tree [6], whenever a worker picks an alternative from another process, it creates a copy of all parallel conjunctions found along the way and restarts work to the right. Traditionally, the C-tree has been associated with a notion of *team*, i.e, a group of workers. Normally, IAP work is explored inside a team and ORP work is executed between teams. DAOS gives freedom to system designer to decide if he or she wants to use or not teams. The use of teams has the advantage of simplifying scheduling allowing (re)use of IAP schedulers within teams and ORP scheduler between teams. This organization also simplifies the solution propagation in IAP through the possibility of use multicast inside a group.

3   DAOS: Distributed And-Or in Scalable System

DAOS aims at two goals: improve efficiency over traditional distributed systems, and preserve Prolog semantics. It is a fundamental point in DAOS establish which data areas should be *private* to a worker, and which ones should be *virtually shared*. There two opposite approachs: (a) all stacks must be private as in distributed PLPs, or (b) all stacks must be shared through a DSM layer. The later option seems interesting because we could use a previous implementation to shared memory systems. However, this may be inefficient due to scheduling data structures. DAOS presents a intermediate solution: the major data-structures area will be logically shared and the work management areas will be private.

3.1   A Shared Address Space

How to implement the virtually shared address space is one of the key aspects of DAOS. This shared space must contains the major data structures used in Prolog, such as all Prolog variables and compound terms. Or-parallelism exploitation in a shared memory space normally is done using a binding array (BA) based approach as in Aurora [8] and Andorra-I [12]. The original BA was designed for ORP and keeps a private slot for every *variable* in the current search-tree. This slot stores all conditional bindings made by a worker, instead of writing on the shared space. Accesses to other memory areas are read-only This gives a important single-writer, multiple-reader pattern for the shared memory.

Unfortunately, the original BA design is not suitable to IAP, because the number of cells for each and-goal is not know beforehand. Management of slots between workers running in and-parallel becomes highly complex [6]. In DAOS, we propose to use the Sparse Binding Array (SBA) [11] to manage bindings to shared variables. The SBA addresses the memory management problems in traditional BAs by *shadowing* the whole shared stacks, that is, *every cell* in a shared stack has a private "shadow" in a worker or team's SBA. SBA was designed to organize workers in teams. Each team should share the same choice-points and thus the same SBA. This approach is not a good one to DAOS since the SBA is write-intensive. So, we propose a different SBA solution: *Each worker will have a private SBA. SBAs are synchronized, through the trail, both when sharing ORP and IAP work.*

3.2   Sharing Work in DAOS

In this section we present the shared and private areas in the Prolog execution environment. We follow the Warren Abstract Machine (WAM) [16] organisation as found in most current PLPs, where each worker has a set of stacks. *Heap* and *Environment Stack* support forward execution. *Control Stack* is used both for backtracking and for parallelism. *Trail* supports backtracking. *Sparse Binding Array* (SBA) provides access to private bindings. Last, *Goal Stack* has goals to be exploited in IAP.

The Forward Execution Stacks The two largest stacks store terms and logical variables. Both have data structures that are virtually shared: The *Environment stack* stores environments, corresponding to the activation records of traditional imperative languages. The *Global stack* or *Heap* accommodates compound terms and the remaining logical variables.

False sharing may happen in two situations [14] First, workers can make previous work *public*, and then continue execution on the same stack. In this case, their next updates to the stack might be sent to the sharing workers. Such sharing updates can be treated by relaxed consistency techniques [3] which ensures that new updates from the worker will only be sent at synchronisation points. A second source of false sharing is backtracking. In general, when all alternatives have been exploited, both global and environment stack space may be reclaimed, and next reutilised to store new work. Updates to this space may then be sent to the workers which had originally shared the work, unless one uses an invalidate-based protocol.

The Control Stack This stack stores Choice-points. Choice-point is a structure that includes pointers to the stacks and to the goal's arguments before creating the choice-point plus pointers to the available alternatives. Some ORP systems also store scheduling informations in choice-points while IAP systems extend Control stack to include *parcall frames* that describe the conjunctions to be executed in parallel. ORP systems had used Control stack to manage the available work. For instance, in the `chat-80` ORP-only benchmark running under Aurora, about a third of the sharing misses originated from Control stack (the rest originated from the scheduler data structures). Although a similar study is not available for IAP systems, parcall-frames are expected to also be a source of sharing misses.

The Trail Trail Stack stores all *conditional* bindings. In a sequential execution this is required to undo bindings to variables when backtracking. In a BA-based system, the Trail is a fundamental data-structure, as it is used to synchronise bindings between different branches. Since conditional bindings *may not* be placed directly on the stacks, the alternative is to store these bindings in the Trail. When workers fetch work, they read the bindings from the Trail and stored them in the SBA. So, the first operation to be performed in DAOS when sharing work is installing all conditional bindings in the SBA. This requires access to corresponding Trail section. Deciding whether to keep Trail under DSM or to use explicit distribution is a fundamental open question:

- All chunks of the Trail must be present in a worker *before* it can start work. This argues for sending the required trail segments immediately when we share choice-points or goals and thus for a fully-distributed solution.
- The Trail tends to be a relatively small stack. After an initial delay, one may take advantage of the sharing facilities in a DSM system to actually have the Trail segments before they are asked.
- IAP programs tend to require much larger stacks than ORP programs, as they perform much less backtracking, but they do tend to perform less conditional bindings. Trail segments are expected to grow larger.

A final decision will depend on several factors, such as the efficiency of the DSM system and of the message passing implementation. In the IDAOS implementation, as discussed next, we have decided to initially follow a fully distributed implementation because trail copying can be naturally integrated with Control-stack copying.

4   IDAOS: A DAOS Implementation

We have so far discussed the DAOS model. We now concentrate on the design of our DAOS prototype, IDAOS (an Implementation of DAOS). In our design, each IDAOS processor or *worker* consists of three modules that are implemented as threads. The *Engine* is responsible for the execution of Prolog code. Most of the execution time will be spent in this code which should have performance close to good sequential implementations. We base our work on Diaz and Codognet's wamcc [4] which has performance close to the best Prolog implementations currently available. The *Work-dispatcher* module controls the exportation of both and- and or-work. This module and the Engine have exclusive access to the Control Stack, Trail, and Goal stack. The *Memory-Manager* module controls the major execution stacks, namely the Environment stack and the Heap through a page-based software DSM.

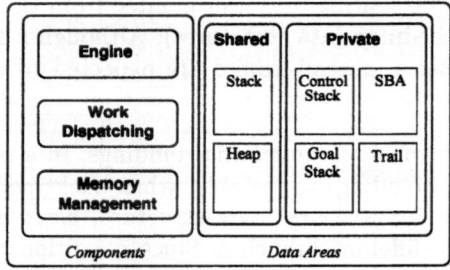

Fig. 2. Worker Organisation

IDAOS uses MPI to implement message passing and the commercial software TreadMarks [3] for DSM. Having both a message passing and DSM mechanisms creates an interesting problem: both Treadmarks and MPI want to initialise the distributed processors. One solution is to give management to TreadMarks and use dynamic process functionality from current MPI implementations (that will be standardised in MPI-2).

Next main issues in the implementation of the IDAOS Engine, and in section 4.2 issues about implementation of Trail and Control stacks are presented.

4.1   And/Or Support in Engine Module

As we have explained, the Engine thread implements an SBA based abstract machine. It combines the use of SBA [5] to deal with or-parallelism, with mechanisms based on Hermenegildo's RAP-WAM [7] to deal with and-parallelism. We had to perform major changes to the wamcc to support both IAP and ORP. Regarding ORP, the major changes in wamcc are as follows:

- conditional bindings must be stored and read from the SBA. We have found that this has an impact of from 19% up to 38% on the execution time on a Pentium-II 333MHz PC with wamcc-2.2, depending on the benchmark.
- in ORP the trail is used to both install and deinstall bindings. Thus, each trail entry receives an extra field containing the new value to be marked. We found the overhead to vary from 2% up to 10% on the same machine.
- Last, choice-points must include new fields, to support sharing or-work.

Supporting IAP requires supporting an extensive set of new data-structures[7]: the parcall-frame (representing a parallel conjunction), a goal-slot (representing a sub-goal in the conjunction), a goal-stack, and markers (to perform stack management). The data structures must handle three forms of execution: forward execution; inside backtracking, that is backtracking within the parallel conjunction; and outside backtracking, that is, backtracking caused by a goal in the continuation that failed. Last, the compiler must also support the sequential conjunction. Our design tries to follow closely Hermenegildo's, and more recent implementations such as Shen's DASWAM [13], Pontelli and Gupta's &-ACE [9], and Correia's SBA [5]. We discuss in detail some issues specifically important for IDAOS.

First, in IDAOS there is not a shared and-tree of goals, rooted at the parcall-frame. Goals are instead sent to an external worker, that executes in its stacks. The parcall-frame therefore must contain information on which processors are executing a goal, not just direct links.

Second, on receiving a goal, the receiver will not have direct access to the parent's parcall-frame, as it is stored in the sender. This is a problem because traditional implementations of IAP use markers to mark the new space being allocated. These markers are linked to the goal slot, and from there to the parcall frame, resulting in an involved chain of pointers in shared memory [1]. In IDAOS a new goal is started with a *starting choice-point (SCP)*. The SCP marks the stacks allocated for this goal, and thus fulfills the task of the markers. The only alternative in SCP is the code to be executed when backtracking out of the task. When completing the task, we install a *final choice-point (FCP)*. The FCP stores the final values for the stacks, points to the SCP, and its alternative is the code to be executed when backtracking to the task. A further advantage of using choice-points is that all memory management is now performed through choice-points, simplifying integration with ORP. Memory allocation in this model will be based in segments to implement the so-called cactus stack. The idea is that all segments, for ORP or for IAP, start with a choice-point and can be allocated and recovered using the same techniques.

---

[1] This could be a good argument for storing the Control stack in the DSM, but this process is closely related to scheduling and, thus, should be made explicit.

4.2   Trail and Control Management

A key data-structure in IDAOS is the Trail. It is used both to propagate conditional bindings during exportation of and- and or-work, and to return conditional bindings performed during remote execution of and-goals:

- Whenever a team imports or-work, workers have to move up or down in the search tree. This is performed by copying the trail from the exporter and installing the bindings in their SBAs.
- When a worker receives an and-goal to execute, it starts its execution in its own memory address space. Any conditional bindings, or any bindings to variables created prior to the and-parallel conjunction , is stored in the SBA, and trailed. At the end of execution of the goal, the worker returns its trail to the exporter.

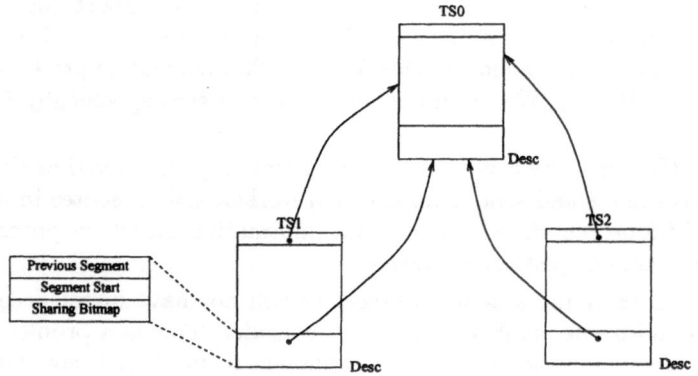

Fig. 3. Trail Segments

To support And/Or parallelism the trail needs to be segmented. The trail is physically divided into *trail segments (TSs)* which corresponds to a contiguous computation. In PLPs the Trail forms a tree: each TS is a children from the one it got work. Each TS starts with a descriptor, followed by a sequence of bindings, and terminates with a special parent pointer, that points to where previous bindings are stored. Figure 3 shows a situation where trail segment TS0 corresponds to several choicepoints generated in a row. TS1 and TS2 correspond to new segments that are rooted in TS0, but the computation for TS3 starts from an older choicepoint than TS1 (we assume the Trail grows downwards). Each trail segment descriptor contains a pointer to the start of the segment, a direct pointer to the ancestor node, and a bitmap indicating which nodes already have this segment. Copying the trail implies start copying from the segment of the goal or choice-point being exported and then follow to the root until a segment that has already been sent is found. MPI's buffering can be used to send all the TSs in a single message.

Note that in this case we are implementing our own DSM mechanism. So, coherence problem must be treated. There are several solutions to this problem: keep the bitmaps or the whole Trail in a DSM area; use broadcast for trail segments; or simply ignore the problem and accept duplicated broadcasts of TSs. Lack of space prevents us from discussing the issues here in detail, but in IDAOS we are using the ostrich algorithm. Performance evaluation will then tell us if more sophisticated solutions are required.

Other important stack in IDAOS is the Control Stack. Our principle in designing DAOS was to use scheduling to avoid unnecessary pressure over the DSM subsystem. To reduce sharing in this area, one solution would be to split the Control stack into a choice-point stack that would be under the DSM, and in a Control stack that would be managed by the message-passing system. For simplicity reasons, we will favour a simpler solution: maintaining the stack as fully distributed. The owner of a parcall-frame or of a choice-point to be the one that can directly access the data-structure, and that accesses from other workers will be performed through the communication protocol.

5   Conclusions

We have proposed a scheme for Distributed And/Or execution in Scalable systems, DAOS. The model innovates by taking advantage of recent work in DSM technology to obtain efficient sharing, and efficiently supporting both IAP and ORP in a distributed environment. We have found that the DSM mechanism is quite effective in simplifying our design allowing us to focus on the issues that we believe have a major impact in performance. Work in the IDAOS prototype is progressing at Universidade Federal do Rio Grande do Sul (UFRGS), Federal do Rio de Janeiro (UFRJ) and Porto (UP). Our target is a network of workstations connected by a fast network, such as Myrinet or Fast Ethernet. Both the UFRGS and the UP groups have access to such networks. Changes required to support IAP and ORP have already been included to the base system wamcc, and work will next move on to experimenting with the distributed platform. We expect that after implementing the message mechanism on top of MPI most of the work will move on to scheduler design, as it is traditional in parallel logic programming systems.

Acknowledgments

We would like to acknowledge Inês de Castro Dutra, Ricardo Bianchini, Gopal Gupta, Enrico Pontelli, Cristiano Costa, and Kish Shen for their contribution and influence. This work has been partially supported by the CNPq/ProTem-CC project Appelo and by funds granted to LIACC through the Programa de Financiamento Plurianual, Fundação para a Ciência e Tecnologia and Programa PRAXIS.

References

[1] K. A. M. Ali and R. Karlsson. The Muse Or-parallel Prolog Model and its Performance. In *Proceedings of the North American Conference on Logic Programming*, pages 757–776. MIT Press, October 1990.

[2] J. Briat, M. Favre, C. Geyer, and J. Chassin. Scheduling of or-parallel Prolog on a scaleable, reconfigurable, distributed-memory multiprocessor. In *Proceedings of Parallel Architecture and Languages Europe*. Springer Verlag, 1991.

[3] C. Amza et al. TreadMarks: Shared memory computing on networks of workstations. *IEEE Computer*, 19(2):18–28, February 1996.

[4] P. Codognet and D. Diaz. wamcc: Compiling Prolog to C. In *12th International Conference on Logic Programming*. The MIT Press, 1995.

[5] M. E. Correia, F. M. A. Silva, and V. Santos Costa. The SBA: Exploiting orthogonality in OR-AND Parallel Systems. In *Proceedings of the 1997 International Logic Programming Symposium*, October 1997.

[6] G. Gupta, M. Hermenegildo, and V. Santos Costa. And-Or Parallel Prolog: A Recomputation based Approach. *New Generation Computing*, 11(3,4):770–782, 1993.

[7] M. V. Hermenegildo. An Abstract Machine for Restricted And-Parallel Execution of Logic Programs. In E. Shapiro, editor, *Third International Conference on Logic Programming, London*, pages 25–39. Springer-Verlag, July 1986.

[8] E. Lusk, R. Butler, T. Disz, R. Olson, R. Overbeek, R. Stevens, D. H. D. Warren, A. Calderwood, P. Szeredi, S. Haridi, P. Brand, M. Carlsson, A. Ciepelewski, and B. Hausman. The Aurora or-parallel Prolog system. In *International Conference on Fifth Generation Computer Systems 1988*, pages 819–830. ICOT, Tokyo, Japan, Nov. 1988.

[9] E. Pontelli, G. Gupta, M. Hermenegildo, M. Carro, and D. Tang. Efficient Implementation of And-Parallel Logic Programming Systems. *Computer Languages*, 22(2/3), 1996.

[10] V. Santos Costa, R. Bianchini, and I. C. Dutra. Parallel Logic Programming Systems on Scalable Multiprocessors. In *Proceedings of the 2nd International Symposium on Parallel Symbolic Computation, PASCO'97*, pages 58–67, July 1997.

[11] V. Santos Costa, M. E. Correia, and F. Silva. Performance of Sparse Binding Arrays for Or-Parallelism. In *Proceedings of the VIII Brazilian Symposium on Computer Architecture and High Performance Processing – SBAC-PAD*, August 1996.

[12] V. Santos Costa, D. H. D. Warren, and R. Yang. Andorra-I: A Parallel Prolog System that Transparently Exploits both And- and Or-Parallelism. In *Third ACM SIGPLAN Symposium on Principles & Practice of Parallel Programming PPOPP*, pages 83–93. ACM press, April 1991. SIGPLAN Notices vol 26(7), July 1991.

[13] K. Shen. Initial Results from the Parallel Implementation of DASWAM. In M. Maher, editor, *Proceedings of the 1996 Joint International Conference and Symposium on Logic Programming*. The MIT Press, 1996.

[14] F. M. A. Silva. *An Implementation of Or-Parallel Prolog on a Distributed Shared Memory Architecture*. PhD thesis, Dept. of Computer Science, Univ. of Manchester, September 1993.

[15] A. R. Verden and H. Glaser. Independent And-Parallel Prolog for Distributed Memory Architectures. Technical report, Department of Electronics and Computer Science, University of Southampton, Apr. 1990.

[16] D. H. D. Warren. An Abstract Prolog Instruction Set. Technical Note 309, SRI International, 1983.

Write Detection in Home-Based Software DSMs*

Weiwu Hu, Weisong Shi, and Zhimin Tang

Institute of Computing Technology
Chinese Academy of Sciences, Beijing 100080
{hww,wsshi,tang}@water.chpc.ict.ac.cn

Abstract. Write detection is essential in multiple writer protocols to identify writes to shared pages so that these writes can be correctly propagated. This paper studies different write detection schemes in a home-based software DSM system called JIAJIA. It compares the performance of three write detection schemes: the traditional virtual memory page fault write detection scheme which write-protects both home and cached pages at the beginning of an interval, a cache only write detection scheme which does not detect writes to home pages but invalidates all cached pages at the beginning of an interval, and an API write detection scheme which requires the programmer or pre-complier to explicitly records writes in program. Evaluation with some well-known DSM benchmarks reveals that tradeoffs of different write detection schemes vary with data (home) distribution and memory reference patterns of applications.

1 Introduction

Most recent software DSM systems employ the multiple writer cache coherence protocol alleviate the impact of false sharing caused by large granularity of coherence in software DSM systems. The multiple writer protocol needs to detect writes to shared memory so that the protocol can be activated to correctly propagate the writes. Write detection of home-based software DSMs is more complex than that in homeless software DSMs because detecting of writes to home pages has to be taken into account as well. This paper studies different write detection schemes with a software DSM system called JIAJIA[2]. These write detection schemes include: (1) The traditional virtual memory write detection (VM-WD) scheme which identifies writes to shared pages through page faults and twins. With this scheme, both home pages and cached pages are write-protected at the beginning of an interval to detect writes to shared pages in that interval. (2) A cache only write detection (CO-WD) scheme which does not detect writes to home pages. Only writes to cached pages are detected to generate *diffs* which are sent to their home at the end of an interval. This scheme does not write-protect home pages but invalidates all cached pages at the beginning of an interval previously. (3) An API write detection (API-WD) scheme which requires the

---

* The work of this paper is supported partly by National Climbing Program of China and National Natural Science Foundation of China (Grant No. 69703002).

P. Amestoy et al. (Eds.): Euro-Par'99, LNCS 1685, pp. 909–913, 1999.
© Springer-Verlag Berlin Heidelberg 1999

programmer or pre-complier to explicitly records writes in program. Instead of detecting writes by the software DSM system automatically, this scheme provides a function jia_wtnt() in the API to record writes to the shared locations.

Evaluation with some widely accepted DSM benchmarks indicate that, the tradeoffs of different write detection schemes vary with data distribution and memory reference patterns of applications, performance can be significantly improved if proper write detection is adopted in accordance with the data distribution and memory reference pattern of the application.

2   The JIAJIA Software DSM System

JIAJIA characterizes itself with a new lock-based cache coherence protocol and novel memory organization scheme which combines the physical memories of multiple workstations to form a large shared address space. Like many research or commercial systems such as TreadMarks[4], JIAJIA is implemented entirely as a user-level library and currently runs on many mainstream Unix platforms and Windows NT platform. One important characteristic of JIAJIA is its supporting of home migration scheme, which is the first home-based software DSM system that implements this idea. Multiple writer technique is employed to alleviate false sharing.

3   Write Detection in JIAJIA

3.1   Virtual Memory Write Detection (VM-WD)

With the VM-WD scheme, both home and cached shared pages are initially write-protected at the beginning of an interval. A SIGSEGV signal is delivered when a processor first writes to a shared page in the interval.

For a write fault on a cached page, a *twin* of the page is created and a write notice is recorded for this page in the SIGSEGV handler. Write protection on the shared page is then removed so that further writes to this page can occur without page faults. At the ending of the interval, a word-by-word comparison is performed between the written page and its twin to produce *diff* about this page. Write notices and *diff*s are then sent to the associated lock and home respectively.

If the SIGSEGV signal is caused by a write fault on a read-only page, then a write notice is recorded for this page and write protection on the shared page is removed. At the ending of the interval, write notices about home pages are sent to the associated lock too.

3.2   Cache Only Write Detection (CO-WD)

The above VM-WD scheme detects writes through page faults and entails additional runtime overhead on the protocol. Our previous experiments with JIAJIA shows that, write-protecting home pages causes significant overheads for applications with large shared data set and good data distribution so that most writes hit in the home.

The CO-WD scheme reduces the overhead of home pages write detection at the cost of some extra cache miss. In the CO-WD scheme, all cached shared pages are conservatively assumed to be obsolete and are invalidated at the beginning of an interval. No write detection about home pages is required in CO-WD because the purpose of detecting write notices of home pages is to maintain coherence through invalidating associated cached pages, and the CO-WD scheme has already invalidated all cached pages when starting an interval. Only writes to cached pages are detected to generate *diffs* which are sent to their home at the end of an interval. *Diffs* are generated through comparing the dirty page with its twin as in the VM-WD scheme.

3.3   API Write Detection (API-WD)

The API write detection scheme is similar to the *dirtybit* approach[7] except that it depends on the programmer or pre-complier instead of the complier to record writes.

In the API-WD scheme, each node of JIAJIA maintains a dirty bit for each home page and a dirty bit vector for each cached page. Setting the dirty bit of a home page indicates that the page is modified, while setting the ith bit of a cached page's dirty bit vector indicates that the ith words of the page is modified. A function jia_wtnt(addr,len) is provided to record writes to the shared locations from addr to addr+len. The recorded shared region [addr,addr+len) can across page boundary. The programmer (or a pre-complier) bears the responsibility of inserting a jia_wtnt(&a,len) after every write to shared variable a. The dirty bit of each home page and the dirty bit vector of each cached page is reset at the beginning of an interval. At the ending of an interval, the dirty bits and dirty bit vectors are checked to determine whether a page is modified and which part of a cached page is modified in the interval.

4   Performance Evaluation and Analysis

The evaluation is done in the Dawning-1000A parallel machine developed by the National Center of Intelligent Computing Systems. The machine has eight nodes each with a 200MHz PowerPC 604 processor and 256MB memory. These nodes are connected through a 100Mbps switch Ethernet.

We port some widely accepted DSM benchmarks to evaluate the effect of home migration in JIAJIA. This paper shows the results of seven applications, include LU from SPLASH2[6], EP and IS from NAS Parallel Benchmarks[1], and TSP and SOR from Rice University[5].

Table 1 shows characteristics and execution results of the benchmarks. It can been seen from Table 1 that, both CO-WD and API-WD outperform VM-WD significantly in LU and SOR, while VM-WD slightly outperforms CO-WD and

Table 1. Characteristics and Execution Results of Benchmarks

Appl.	Size	Shared Mem	Seq. Time	8-proc. Time			SEGV #			Remote Accesses		
				VM	CO	API	VM	CO	API	VM	CO	API
LU	2048	32MB	84.86	25.11	24.64	24.92	32663	12072	12072	18135	20032	18135
EP	2^{24}	4KB	49.69	6.25	6.27	6.26	22	21	21	14	14	14
IS	2^{24}	4KB	30.10	4.79	4.76	4.70	230	210	210	140	140	140
SOR	2048	16MB	6.97	2.14	1.20	1.13	41200	280	280	280	280	280
TSP	19 cities	788KB	258.33	37.48	38.06	38.71	5880	6179	5711	4733	5218	4775

API-WD in TSP. The difference among three write detection schemes is trivial in EP and IS.

In LU and SOR, matrices are distributed across processors in a way that each processor only writes to its home part of the matrices in the computing. Since the computation of an iteration is synchronized with barriers and passing a barrier causes all shared pages to be write-protected in VM-WD, page faults occur for writing all home pages in an iteration. The CO-WD and the API-WD scheme, on the other hand, does not write protect shared pages on a barrier, and writing to home pages of a processor can process smoothly without any intervention. Table 1 shows that the CO-WD and API-WD scheme causes much less page faults than the VM-WD scheme in LU and SOR.

In TSP, the number of shared pages is not large and all shared pages are allocated at host 0. As a result, the advantage of CO-WD and API-WD over VM-WD in keeping home pages writable on a synchronization point is not significant. Since CO-WD invalidate all cached pages on an acquire, it has largest number of page faults among three write detection schemes. On the other hand, TSP is an application with the tight sharing memory reference pattern, a cached page is usually invalidated by the coherence protocol on an acquire because it has been written by other processors, the disadvantage of VM-WD over API-WD in write-protecting writable cached pages on synchronization points is also not significant, while the extra overhead of executing jia_wtnt() in API-WD dominates and VM-WD slightly outperforms API-WD.

5   Conclusions and Future Work

It can be seen from the above evaluation and analysis that write detection constitutes a significant overhead for home-based software DSMs. Evaluation results show that the tradeoffs of different write detection schemes vary with data distribution and memory reference patterns of applications, performance can be significantly improved if proper write detection is adopted in accordance with the data distribution and memory reference pattern of the application.

References

[1] D. Bailey, J. Barton, T. Lasinski, and H. Simon, "The NAS Parallel Benchmarks", *Technical Report 103863*, NASA, July 1993.

[2] W. Hu, W. Shi, and Z. Tang, "JIAJIA:An SVM System Based on A New Cache Coherence Protocol", in *Proceedings of the 7th High Performance Computing and Networking Europe*, pp. 463-272, April 1999.

[3] L. Iftode, J. Singh and K. Li, "Scope Consistency: A Bridge Between Release Consistency and Entry Consistency", in *Proceedings of the 8th Annual ACM Symposium on Parallel Algorithms and Architectures*, June 1996.

[4] P. Keleher, S. Dwarkadas, A. Cox, and W. Zwaenepoel, "TreadMarks Distributed Shared Memory on Standard Workstations and Operating Systems", in *Proceedings of the 1994 Winter Usenix Conference*, pp. 115–131, January 1994.

[5] H. Lu, S. Dwarkadas, A. Cox, and W. Zwaenepoel, "Quantifying the Performance Differences Between PVM and TreadMarks", *Journal of Parallel and Distributed Computing*, Vol. 43, No. 2, pp. 65–78, June 1997.

[6] S. Woo, M. Ohara, E. Torrie, J. Singh, and A. Gupta, "The SPLASH-2 Programs: Characterization and Methodological Considerations", in *Proceedings of the 22th Annual Symposium on Computer Architecture*, pp. 24–36, 1995.

[7] M. Zekauskas, W. Sawdon, and B. Bershad, "Software Write Detection for a Distributed Shared Memory", in *Proceedings of the first International Symposium on Operating System Design and Implementation*, pp. 87–100, November 1994.

D'Caml: Native Support for Distributed ML Programming in Heterogeneous Environment

Ken Wakita, Takashi Asano, and Masataka Sassa

Department of Mathematical and Computing Sciences,
Tokyo Institute of Technology, Tokyo 152-8552, Japan
{wakita,asano,sassa}@is.titech.ac.jp

Abstract. Distributed Caml (D'Caml) is a distributed implementation of Caml, a dialect of ML. The compiler produces native code for diverse execution platforms. The distributed shared memory allows transmission and sharing of arbitrary ML objects including higher-order functions, exceptions, and mutable objects without affecting the sequential semantics of ML. The distributed garbage collector reclaims unused distributed data-structures. Examples demonstrate expressivity of higher-order distributed programming using Distributed Caml. The paper presents the design, implementation, and preliminary performance results of the system.

1 Introduction

Due to significant advances in network and platform technologies, exchange of digital information on the Internet is replacing the traditional methods of information exchange and has increased the demand for high-quality distributed services and applications for various use. However development of distributed applications remains hardest in software development and thus it is getting more difficult to meet the rapidly growing demand for high-quality Internet software. Obstacles in development of distributed software, among others, are (i) hardware and software heterogeneity of computers that participate in distributed computing, (ii) data conversion required for inter-node communication, and (iii) branches of knowledge required for system software development such as OS system calls, message passing libraries, theories of parallel/distributed/real-time computing, fault tolerance, security, and variety of libraries for system programming.

Distributed Caml (D'Caml) is a distributed programming language and system being developed by the authors. Our research goal is to offer a programming system which alleviates above mentioned obstacles, hereby invite novice programmer to practical distributed programming.

A D'Caml application is a collection of cooperative distributed processes that execute on a heterogeneous workstation cluster. Here, "heterogeneity" refers to difference in hardware resources (instruction set, CPU architecture, memory hierarchy, accessible hardware devices, etc.) and software configuration (operating system, versions of libraries, etc.). For instance, our development platform comprises Sun SPARC Stations (Ultra SPARC/Solaris), Digital workstations (Alpha/Digital UNIX and Linux), and PCs (Pentium/Linux and Solaris) connected via Ethernet.

P. Amestoy et al. (Eds.): Euro-Par'99, LNCS 1685, pp. 914–924, 1999.
© Springer-Verlag Berlin Heidelberg 1999

The distributed shared memory (DSM) substrate of the D'Caml runtime system entirely hides the underlying heterogeneous, distributed environment and provides the programmer for virtually shared, single address space abstraction. This functionality enables transmission and sharing of arbitrary ML objects including functional closures (functions with binding of free variables).

Remote closure invocation is a programming mechanism in D'Caml that allows a thread to invoke an arbitrary function as a new thread in a remote process. Integration of distributed shared memory and remote closure invocation offers the programmer a programming style similar to single-node multi-thread programming like Java, Concurrent ML [13], and Objective Caml [9].

Remote closure invocation differs from standard RPC mechanisms, as found in Ada and Java + RMI, in several important respects. It supports dynamically created functions (closures). When a closure is passed, not only the code pointer but also its free lexical references are transmitted to the remote site. Secondly, these variables are virtually shared between the sender and the receiver so that sharing semantics is maintained consistent with the sequential semantics. On the other hand, in most RPC system, variables are simply copied and thus violating sharing semantics for them (i.e., side-effect made on the variable is invisible by remote processes). Thirdly, unhandled exception is forwarded to and can be caught by the thread that issued remote invocation. Finally, there is no constraint on the kinds of objects that can be transmitted during remote closure invocation.

The rest of the paper is organized as follows. Section 2 describes the language and the system of D'Caml. Section 3 shows the implementation scheme of the system and Section 4 presents its preliminary performance results. Section 5 compares our work with others and concludes this paper.

2 The Distributed Caml System

An application produced by the D'Caml system executes as a collection of cooperative distributed processes running on a (possibly heterogeneous) workstation cluster.

D'Caml supports two kinds of execution platforms . The D'Caml/MPI configuration utilizes the standard message passing interface [7] for low-level communication. In this configuration, an application starts as a process which invokes all other cooperative processes in the cluster. We call the first process that invokes others the *host process*. A distributed application created with D'Caml/MPI configuration is static in the sense that it does not support dynamic creation of processes after the application startup time.[1] The D'Caml/TCP configuration supports dynamic process creation. In this configuration, an application starts as a single-process application which can spawn a new process at remote computer arbitrarily.

The D'Caml language is a superset of Objective Caml (O'Caml) [9] developed at INRIA. In addition to functionality supported by O'Caml, D'Caml offers distributed shared memory, types and functions for distributed computing as summarized in Figure 1), and distributed garbage collector.

---

[1] The lack of dynamic process creation is due to the specification of MPI 1. MPI 2 proposal suggests inclusion of this facility.

```
type node
val current_node: unit -> node
val is_host: unit -> bool
val get_nodes: unit -> node array      (* only for MPI *)
val start_node: string -> string -> string -> node (* only for TCP *)
val spawn: node -> (unit -> 'a) -> unit
val rcall: node -> (unit -> 'a) -> 'a
```

Fig. 1. Some of the primitive functions defined in the *Dcaml* module

The node type is an abstraction of the distributed processes which encapsulates heterogeneity of the underlying hardware and software resources. An instance of the node type identifies one of the distributed processes that participate in the distributed computing. The function *current_node* gives the node that corresponds to the process where this function is executed. The function *is_host* tells if the current process is the host process.

In the D'Caml/MPI configuration the *get_nodes* function gives an array of the participant nodes In the D'Caml/TCP configuration, the *start_node* function is used to spawn a new D'Caml process at a remote site. Three parameters specify the network address, the working directory, and path to the D'Caml executable code.

All the D'Caml processes execute the same Caml program in a SPMD manner. However typical D'Caml application initiates execution only at the host process which serves as the master process using the *is_host* primitive:

> **if** *Dcaml.is_host* ( ) **then** *print_string* (*Unix.gethostname* ( ))

Two functions, *spawn* and *rcall*, achieves remote closure invocation. They take a node and a closure, send the closure to the remote node, and invoke the closure as a new thread in the remote address space of the node. The *spawn* primitive is asynchronous: its execution immediately finishes and the remote thread runs independently with the current one. On the other hand, execution of *rcall* synchronizes with the termination of the remote thread and takes its return value as its own return. In this way, *rcall* behaves like a remote procedure call. The following program collects host names for all the nodes in an array *hostnames* using the *rcall* primitive.

> **if** *is_host* ( ) **then**
>   **let** *nodes* = *Dcaml.get_nodes* ( ) **in**
>   **let** *hostnames* = *Array.make* (*Array.length nodes*) " " **in**
>   *Array.iteri*
>   (**fun** *i node* ->
>   *Dcaml.rcall node*
>   (**fun** ( ) -> *hostnames* . (*i*) <- *Unix.gethostname* ( )))
>   *nodes*

This program is interesting in two aspects. The lexical scope of the closure being *rcalled* contains free references to local names, *hostnames* and *i*. As mentioned earlier, these free references are sent as part of the closure and become accessible by the *rcall*ed closure to a remote node. Secondly, values referenced by these free references are shared between distributed nodes. Therefore side-effect made on them by one node is observable by the others. In the above example, the host node can retrieve hostnames assigned by remote nodes.

Inter-node communication using remote closure invocation is advantageous in comparison with standard message passing because the programmer is released from implementing with communication protocols. D'Caml offers basic communication protocols as functions (e.g., *spawn* and *rcall*). Higher order functional programming allows us to build higher-level communication protocols from lower-level primitives. For example, let us consider the *future* primitive as found in Multilisp [8]. The future primitive takes a node and a function and executes the function at the designated node. Execution of the future primitive immediately finishes and returns a handle for the return value which can be used to retrieve the return value in an arbitrary future time.

```
let future node f =    (* node -> (unit -> 'a) -> (unit -> 'a) *)
   let rbox = ref None in
   let sem = Mutex.create () in
   Mutex.lock sem;
   Dcaml.spawn node
      (fun () -> rbox := Some (f ()); Mutex.unlock sem);
      (fun () -> Mutex.lock sem; match !rbox with Some v -> v);;

let touch = future some_node a_closure
in ...
      let result = touch () in ...
```

In this implementation, the function *f* is *spawn*ed to the designated node and its result is stored in a reply box. The handle to the reply box is represented by a function which synchronizes with the execution of *f* using a semaphore.

The last example in this section demonstrates how standard parallelizing techniques are expressed in D'Caml. The program presented in Figure 2 paints a graphical image of Mandelbrot's fractal set defined over the complex number space. Because each pixel color can be computed independently from others, this computation is an inherently parallel computation. The program achieves parallelism in a *master-workers style*: single master process divides the entire computation into smaller sub-computations and feed them to multiple worker processes. In the program, computation for a two-dimensional matrix is divided into many sub-computations that calculates pixel colors in a row. The host node serves as the master and all other become workers. For each worker, the host node starts a master thread which continuously feeds the corresponding worker node with a *new_job* until entire computation finishes.

3 Implementation

In this section, we describe the implementation scheme of the D'Caml system. We briefly overview its software organization in the next subsection. Subsequent subsections describe the implementation scheme in detail.

```
exception Done;;
let size = 500;;
let image = new_image size size and next = ref 0;;
let mandelbrot i j = ...;;
let compute_row i =
  let row = Array.create size 0 in
  Array.iteri (fun j point -> point <- mandelbrot i j) row;;
let new_job () =
  let i = !next in
  if i < size then (incr next; fun () -> compute_row i)
  else raise Done;;
let master node =
  try while true do paint image (Dcaml.rcall node (new_job ()))
    done
  with Done -> ();;
if Dcaml.is_host () then
  Array.iter (fun node ->
    Dcaml.spawn (Dcaml.current_node ()) (fun () -> master node)
  (Dcaml.get_nodes ())
```

Fig. 2. Master-workers style parallel computation of the Mandelbrot set

3.1 Software Architecture

Figure 3 illustrates the software architecture of the Distributed Caml system. A D'Caml application runs on a heterogeneous workstation cluster. Given a Caml program, the D'Caml compiler produces native code for all architectures involved in the cluster.

D'Caml DSM is a software-only implementation of single address space that spans over heterogeneous, physically distributed address spaces. Major difficulties implementation of a DSM for a functional language, namely transmission of functional closures (Subsection 3.2) and dealing with shared mutable objects (Subsection 3.3). Low-level message transmission is established by per-node thread that we call *message handler* (Subsection 3.6).

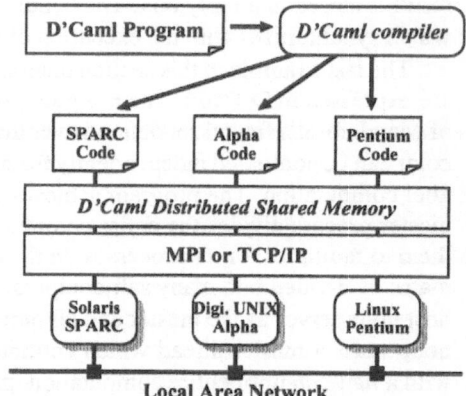

Fig. 3. Software architecture of the Distributed Caml system

3.2   Marshaling

To transmit Caml objects over the network we need a way to convert their internal data representation to and from their architecture-independent forms, namely *marshaling* and *unmarshaling*.

With most implementations of functional languages, each runtime data is attached with GC tag which gives its structural information, size and sort of object (e.g., constant, arrays, or pointer aggregates, etc). This gives sufficient information for marshaling most heap allocated Caml objects. In fact, the built-in marshaling mechanism of O'Caml traces the data structure using this information.

The marshaling mechanism of O'Caml, however, does not deal with distributed and heterogeneous environment because it is assumed that the marshaled data will be loaded back by the same process at later time. D'Caml extends the marshaling mechanism of O'Caml by support for distributed data structures and ability to safely transmit closures. Distributed data structures are expressed using remote pointers as described in Subsection 3.3. A difficulty in passing a closure over the network is the representation of code that implements the closure. This issue is discussed in Subsection 3.4.

3.3   Remote Pointers

In order to express distributed data structures, D'Caml DSM incorporates the notion of *remote pointers* by which a data can remotely reference an object allocated in a remote address space. Figure 4 illustrates the representation of a remote pointer. A remote pointer points to an entry in a statically allocated *import table* whose entry consists of the node identifier n_1 and entry index j of the referenced object. The pair of n_1 and j identifies an entry in the export table of the remote node, which contains a regular pointer pointing the remotely referenced object.

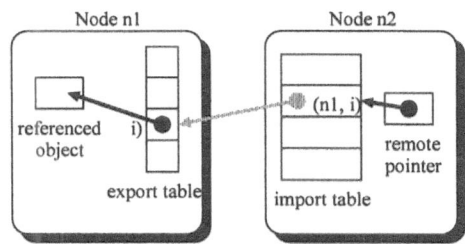

Fig. 4. Implementation of remote pointers.

The export table is contained in the GC root set of the per-node local garbage collector so that remotely referenced object is not collected even if it is not referenced locally.

Import table achieves fast dereference of regular pointers. Dereference of a possibly remote pointer requires to determine if it is a remote pointer (*pointer testing*). In D'Caml, pointer testing is issued for each access to *possibly remote* pointer. Because import table is statically allocated each pointer testing is as cheaply done as single address comparison.

Now, we discuss the data transmission scheme and how remote pointers are introduced there. Types of data transmission in D'Caml falls in three cases: (1) Remote closure invocation transmits a closure which may contain various Caml objects in its defining environment. (2) Dereference of a remote pointer requires the remotely referenced object to be transfered. (3) Update of remotely referenced object requires a value to be sent and stored in the remote object.

When a node transmits a D'Caml object to other node, it replicates the object in the remote node substituting all regular pointers referencing mutable objects with remote pointers. This scheme guarantees that mutable objects are not replicated but referenced through remote pointers.

D'Caml compiler takes advantage of the fact that immutable objects are always pointed to by regular pointers and removes pointer testing for them. Therefore D'Caml program coded in mostly functional style does not suffer from pointer testing overhead. The type system of ML, which covers object mutability, offers the compiler precise information and enables this optimization.

3.4 Closure Passing

A closure is a function defined in a local scope. The value assignment environment for the variables in the lexical scope is called the defining environment for the closure. The defining environment is a set of value assignments to variables that occur free in the function definition.

In modern implementation of functional programming languages, runtime closure representation comprises the code fragment that implements the function and an array of values assigned to the free variables. In native implementation, the code is actually represented by the address of the entry point of the code fragment of the function implementation which we call *entry address*.

In order to transmit closures across the heterogeneous network, we need to interpret an entry address used in one address space into another entry address which correctly points to the corresponding code fragment in the remote address space.

The D'Caml compiler assumes that all the D'Caml processes are running native code that is produced from the same Caml program. Given this, the D'Caml compiler and linker assigns for each code fragment a unique identifier (*entry point identifier*) and the runtime system translates entry addresses to and from entry point identifiers.

Figure 5 sketches our remote closure passing scheme. The D'Caml compiler processes module source programs. The code generator of D'Caml assigns an architecture-independent, module-wide unique number for each entry point in the produced code. The module identifier assigned by the statup-time linker and the entry point number produced by the D'Caml compiler together serve as network-wide, unique identifier of the code fragment. Figure 5 depicts passing of the following closure defined in a module identified by m from node n_1 to another node n_2:

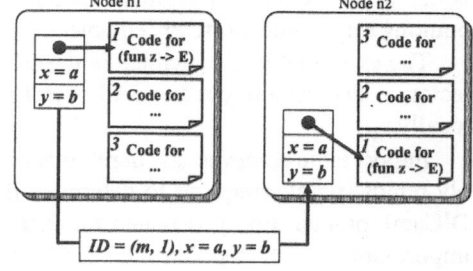

Fig. 5. Remote closure passing scheme

let $x = a$ **and** $y = b$ **in** (**fun** z -> E)

A variant of this address translation technique is used to implement remote exception handling and transmission of atomic objects.

3.5 Garbage Collection

The D'Caml system supports automatic reclamation of unreferenced objects by coordination of per-node local garbage collectors and a distributed garbage collector. The local garbage collector is a slightly modified version of O'Caml garbage collector. It deallocates unreferenced heap-allocated objects. As mentioned before, the GC root set includes the export table, to inform the local garbage collector of remote references.

The distributed garbage collector deallocates unused entries in the export tables which point to heap-allocated objects. If these objects are not locally referenced, they will be deallocated at the next attempt of local garbage collection. The distributed garbage collection mechanism is based on a variant of reference counting algorithm called *weighted reference counting (WRC)* [3].

3.6 Message Handler

Each D'Caml process executes a thread called the *message handler* which sends outgoing messages and dispatches incoming requests. Types of messages include requests for remote closure invocation, requests for dereference and update of remote pointer, and reply for remote dereference, and messages used by the runtime system such as the distributed garbage collector.

In the D'Caml/MPI configuration, the message handler waits for the message by polling every few milli-seconds. This active polling consumes noticeable CPU time and reduces system's response time for a communication-intensive applications. This design is due to thread unawareness of the underlying MPI implementation. This overhead is eliminated in the D'Caml/TCP configuration.

4 Evaluation

This section presents some of preliminary evaluation results of the D'Caml/MPI.

Speed-up: The first result shown in Figure 6-(a) is obtained from running the program shown in Figure 2 for 500 × 500 matrix on varying number of nodes in the cluster. The cluster consists of 15 nodes of SPARC Station 20 running Solaris 2.5.1 connected with 100 Base-T Ethernet. The result is the average of 10 runs of the program ranging the number of clients for 1, 3, 6, 9, and 12 nodes. The result does not say much but it suggests D'Caml effectively utilizes inherent parallelism in a small scale workstation cluster.

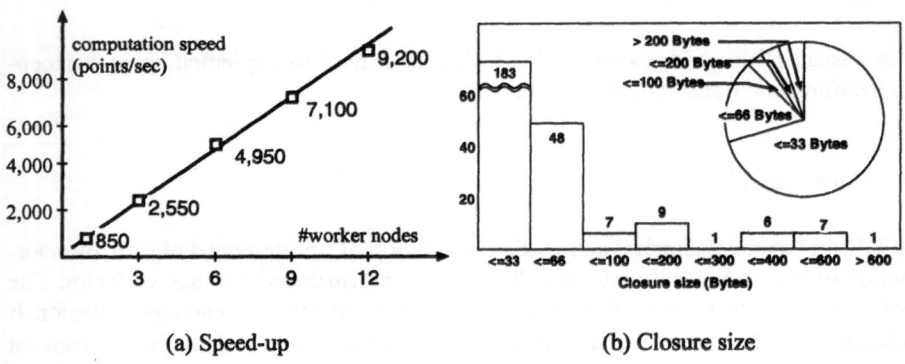

(a) Speed-up   (b) Closure size

Fig. 6. Evaluation results

Size of marshaled closures: A closure involves all its free references to other closures, which in turn involves other closures, and so on. The reader may fear that the external representation of closures can be huge making sending over the network impractical. It turned out, however, the extern-ed closures are practically small enough. Figure 6-(b) is the result we obtained from measuring the size for each marshaled string of 262 functions defined in the standard library. About 90% of them occupy less than 100 bytes and all of them fit in an Ether packet whose size is about 1,000 bytes. The reason for their smaller size against our intuition is that highly optimizing closure conversion [1] mechanism of the O'Caml compiler excludes top-level defined functions from closure representation. We have found aggressive use of this technique can further be applied to reduce extern-ed closure size. This technique reduces the largest closure (683 bytes) to 150 bytes and others less than 100 bytes. Currently we have not adopted this optimization because the changes required for closure representation introduces small invocation overhead.

5   Related Work and Conclusions

There are several projects working on distributed functional programming languages. ParaML [2] is a distributed ML based on SPMD computation model and Distributed ML [6] allows channel based inter-node communication. Both of them do not consider heterogeneous environment.

Facile [14] is a mobile programming language that integrates a concurrent calculus with ML. It allows passing ML modules across the heterogeneous network and execute them native by using dynamic compiler and linker. It does not support distributed address space and is incapable to send closures across the network.

The dML [11] incorporates dynamic types in the ML type system and proposes a theoretical foundation of inter-node communication between independent ML programs.

Kali Scheme [5] and NeXeme [10] are distributed dialects of Scheme and Obliq [4] is a mobile programming language. They offer single address space abstractions (or in Obliq's terminology *distributed scope*) and support heterogeneous environment. Kali Scheme and Obliq provides automatic coherence protocol for distributed shared mutable objects but the NeXeme leave this programmers' responsibility allowing development of relaxed coherence protocols. All of them are executed by bytecode interpreters.

An alternative to native implementation like D'Caml is the combination of bytecode compiler and just-in-time technology. Attempts have been made to target functional languages for JVM (e.g, Kawa scheme) but currently JIT fails to accelerate such implementations as efficient as standard implementation. As for distributed implementation, inability to modify the VM leads to significant degradation of execution efficiency due to pointer-testing overhead.

Distributed dialects of Scheme [12, 5] typically prohibit assignment to variables but incorporate mutable data structures called boxes, similar to `ref` type in ML. The purpose of boxes is for optimization of DSM and its coherence protocol [12]. In ML, the type system precisely tells mutability information and thus we can apply the same optimization without affecting the syntax and semantics of the language.

Distributed Caml offers a single address space for native code distributed application that executes on a heterogeneous network cluster. Integration of higher-order programming and communication based on remote closure invocation achieves a flexible and extensible environment for distributed programming. Two important limitation of our system are (1) communication between independently developed Caml programs and (2) the support of mobility. These issues require future work.

Acknowledgment The authors thank anonymous Euro-Par '99 reviewers for giving us precious comments. Tetsu Ushijima implemented earlier version of D'Caml/MPI. The work is funded by Research Institute of Software Engineering, Japan.

References

[1] A. Appel. *Compiling with Continuations*, chapter 10. Cambridge UP, 1992.

[2] P. Bailey et al. Supporting coarse and fine grain parallelism in an extention of ML. In *CONPAR '94, LNCS 854*, pages 693–704, 1994.

[3] D. Bevan. Distributed garbage collection using reference counting. In *Parallel Architectures and Languages, LNCS 258*, pages 176–187, 1987.

[4] L. Cardelli. A language with distributed scope. In *POPL '95*, pages 286–298, 1995.

[5] H. Cejtin, S. Jagannathan, and R. Kelsey. Higher-order distributed objects. *TOPLAS*, 17(5):704–739, 1995.

[6] R. Cooper and C. Krumvieda. Distributed programming with asynchronous ordered channels in distributed ML. In *Workshop on ML and Applications*, 1992.

[7] Message Passing Interface Forum. A Message Passing Interface Standard. Technical Report CS-94-230, Dept. CS, Univ. Tennessee, 1994.

[8] R. Halstead. Multilisp: A language for concurrent symbolic computation. *ACM Trans. on Programming Languages and Systems*, pages 501–538, October 1985.

[9] X. Leroy. *Objective Caml*, 1997. Available at http://pauillac.inria.fr/ocaml/.

[10] L. Moreau et al. NeXeme: A distributed Scheme based on Nexus. In *EuroPar '97, LNCS 1300*, pages 581–590, 1997.

[11] A. Ohori and K. Kato. Semantics for communication primitives in a polymorphic language. In *POPL'93*, pages 99–112, 1993.

[12] C. Queinnec. DMEROON: a distributed class-based causally-coherent data model. In *Parallel Symbolic Computing: Languages, Systems, and Applications, LNCS 1068*, 1995.

[13] J. Reppy. CML: A higher-order concurrent language. In *PLDI '91*, pages 293–306, 1991.

[14] B. Thomsen et al. Facile antigua release – programming guide. Technical Report ECRC-93-20, ECRC, 1993.

ParBlocks - A New Methodology for Specifying Concurrent Method Executions in Opus*

Erwin Laure

Institute for Software Technology and Parallel Systems
University of Vienna
erwin@par.univie.ac.at

Abstract. Many applications make use of hybrid programming models intermixing task and data parallelism in order to exploit modern architectures more efficiently. However, unbalanced computational load or idle times due to tasks that are blocked in I/O or waiting on results from other tasks can cause significant performance problems. Fortunately, such idle times can be overlapped with useful computation in many cases. In this paper we propose a simple, yet powerful methodology for specifying intra-object parallelism and synchronization in the context of the coordination language Opus.

1 Introduction

We recently introduced the coordination Language Opus [2] which allows a high level management of data parallel tasks. Its central concept is the *shared abstraction (SDA)*, which generalizes Fortran 90/HPF modules using an object-based approach and imposing monitor semantics. SDAs can be internally data parallel while task parallelism is exploited between different SDAs. SDAs communicate with one another via synchronous or asynchronous method invocation; arguments are passed with copy-in/copy-out semantics.

The monitor semantics of SDAs ensure a consistent state of the SDA data at the expense of potential parallelism losses. In fact, there may well be multiple method executions safely active within an SDA object. Weakening the monitor semantics of SDAs has the benefit of introducing an additional level of parallelism which can be exploited on systems with shared address space; but also on systems with distributed memory idle times, due to communication or synchronization with other tasks, can be overlapped with useful computation, thus reaching a better utilization of the available computation nodes.

Allowing concurrent executions of multiple methods within an SDA poses a number of difficulties (see [7] for a detailed discussion of intra-object concurrency) among which the most important one is how to specify potential parallelism and needed synchronization among methods. Compiler analysis can be used for detecting some potential for intra-SDA parallelism. However, a compiler

---

* The work described in this paper was partially supported by the Special Research Program SFB F011 "AURORA" of the Austrian Science Fund.

P. Amestoy et al. (Eds.): Euro-Par'99, LNCS 1685, pp. 925–929, 1999.
© Springer-Verlag Berlin Heidelberg 1999

is generally not able to detect all cases and therefore some support from the user is needed in order to exploit intra-SDA parallelism to some greater extent. This is also the case in other approaches described below:

Java [3] allows all methods of an object to execute in parallel unless explicitly synchronized. However, the `synchronized` attribute is ill suited for expressing partial concurrency and synchronization that is based upon the state of an object. Similarly, *OpenMP* [8] allows all methods to be invoked in parallel from work-sharing constructs, unless they are explicitly synchronized. On the contrary, *Fortran 95* [4] allows parallel invocation only for `pure` procedures that are free from certain side effects. *Path Expressions* [1] are an elegant means of specifying synchronization between processes by describing how a process is allowed to execute in relation to others, irrespective of their invocation order. With the help of Path Expressions complex synchronization patterns can be specified, however, it is not possible to specify synchronization which depends on the state of a process.

In this paper we propose a compiler directive called *ParBlock* which can be used for specifying potential parallelism and necessary synchronization among methods in a simple and intuitive way. Synchronization can be specified statically or dynamically.

2   The Opus Approach to Intra-SDA-Parallelism

Due to the specific properties of SDAs (SDAs are kind of "active" objects which are triggered by other objects) we identify a set of properties the specification mechanism for parallelism/synchronization has to fulfill:

Parallelism/synchronization information should be *encapsulated* within an SDA, since SDAs can be accessed by a set of tasks which are not necessarily aware of each other. Hence, it is necessary that a consistent internal state is guaranteed by an SDA itself rather than by synchronizing the accessing tasks. Synchronization should be possible in a *static* (i.e., independent of an SDA internal state) and and *dynamic* (i.e., state dependent) way. Although static synchronization can be seen as a special case of dynamic synchronization, having means for specifying synchronization statically allows a more efficient implementation. All parallelism/synchronization information should be specified on the highest possible level. Finally, the user should only be compelled to specify as much synchronization as necessary. Consequently, exclusive access to an SDA is still the default property of a method.

2.1   ParSets

Opus already provides some support for dynamic synchronization on the method level: the *condition clauses*. Condition clauses can be used to guard the execution of a method with a side-effect free logical condition. However, with this feature only synchronization that depends on the state of the SDA can be specified. It is not possible to synchronize two method executions independently of the internal data of the SDA. Hence, new means for specifying *static* parallelism and synchronization, in particular pairwise interference freedom among methods, are required. We propose that every method should be annotated with a set of method names representing all the methods with which its execution can safely overlap. This set is called *ParSet*. Note that ParSets are symmetric but not transitive. By default, the ParSet of a method is empty and thus the method has exclusive access to the SDA.

ParSets and condition clauses can be used in conjunction: while ParSets *statically* specify potential parallelism, condition clauses can be used to synchronize method executions in a *dynamic* way. The execution order of methods is derived implicitly from both, the parallelism specification and condition clauses, since before launching the execution of a method it is necessary to check if (1) the method is allowed to execute in parallel with all other methods currently being executed, and (2) its condition clause is satisfied. Obviously, both checks have to form an atomic action.

The direct specification of ParSets for every method is a cumbersome task and specifying ParSets in a consistent way is not trivial. Hence, we need higher level constructs for specifying static parallelism/synchronization.

In Section 1 we discussed *Path Expressions* which can be used to specify process parallelism at a high level. Such a technique could also be applied to Opus, however, Path Expressions explicitly specify the execution order of methods, irrespective of the invocation order. The direct specification of the execution order, however, is unwanted, since non-deterministic executions are deliberately enabled in Opus; condition clauses can be used for imposing specific execution orders, instead.

2.2   ParBlocks

Instead of specifying ParSets for every method we propose a new compiler directive called *ParBlock* for the static specification of parallelism/synchronization. ParSets can be derived from ParBlocks by the compiler. ParBlocks borrow from Path Expressions in that they allow the specification of parallel and mutual exclusive method executions, but without fixing the execution order of methods. The body of the ParBlock directive is a list of method names where all comma separated methods can execute in parallel while semi-colon separated methods need to execute mutually exclusive. We refer to a comma separated list as *par-section* and to a semi-colon separated one as *sync-section*. Both sections can be arbitrarily nested (using parenthesis) thus allowing complex synchronization patterns. In addition, multiple ParBlocks can be specified for an SDA. However, no method name may occur more than once in a given ParBlock nor in more than one ParBlock to avoid inconsistent declarations.

The syntax of the ParBlock directive is similar to HPF directives: the **PARBLOCK** keyword, which is preceded by the Opus compiler-directive-origin !OC$ is followed by arbitrarily nested par- or sync-sections.

Although ParBlocks have enough expressiveness for inter-method parallelism, it is not possible to specify an overlapping of different instances of the same method. This can be accomplished by giving the method the F95 "pure" attribute. Moreover, pure methods can safely run mutually in parallel. Therefore, the compiler will generate an additional ParBlock containing all pure methods which is consistent with the F95/HPF standard.

Summarizing the above, Opus provides a set of features for specifying intra-SDA parallelism and synchronization, both dependent and independent of the SDA's internal state:

- **condition clauses:** for specifying dynamic synchronization based upon the internal state of the SDA,
- **ParBlocks:** for specifying parallelism as well as synchronization independently of the SDA's internal state, and
- **pure attributes:** for specifying potential parallelism according to the Fortran 95 standard.

The following example illustrates the use and expressiveness of ParBlocks:

Example 1. Consider an SDA with 6 methods, a, b, c, d, e, and f. All methods are allowed to execute in parallel but with the restrictions that (1) method b and c cannot execute concurrently and (2) method d cannot execute concurrently with neither e nor f. A sync-section is used for restriction (1): (b;c). For restriction (2) we need to nest a sync-section with a par-section. Let's first specify that method e and f can run in parallel: (e,f). Now we extend this expression specifying the synchronization of d: (d;(e,f)). We have now specified all the necessary synchronization and can put everything in a par-section. The resulting ParBlock for our example is:
(a,(b;c),(d;(e,f))). In an Opus program the required directive would look like: !OC$ PARBLOCK(a,(b;c),(d;(e,f))).   ∎

2.3   Implementation

The Opus compiler parses the ParBlock-directives of an SDA and constructs a ParSet for each method of an SDA (see [5] for a detailed description of the algorithms).

These ParSets are used at runtime to check if a method can start executing in parallel with other methods. In particular, a new method can start its execution if and only if the set of the currently executing methods is a subset of its ParSet and its condition clause is satisfied as well.

The Opus implementation design must also be modified to facilitate overlapping method execution. In particular, we described in [6] that an SDA is compiled to an *active object* consisting of two threads: a *server thread* responsible for retrieving incoming request and an *execution thread* which executes the methods. To allow concurrent method executions within an SDA, an SDA object requires a set of execution threads instead of only one.

3 Conclusions

In this paper we introduced a simple, yet expressive method for specifying intra-SDA parallelism. In [5] we present the applicability of our approach to a set of synchronization problems and discuss its benefits wrt. other approaches, like the Java multithreading model.

The proposed methodology is currently being implemented in our Opus compilation and runtime framework.

References

[1] R.H. Campbell. *Path Expressions: A technique for specifying process synchronization*. PhD thesis, The University of Newcastle Upon Tyne, 1976.

[2] B. Chapman, M. Haines, P. Mehrotra, J. Van Rosendale, and H. Zima. OPUS: A Coordination Language for Multidisciplinary Applications. *Scientific Programming*, 6/9:345–362, Winter 1997.

[3] J. Gosling, B. Joy, and G. Steele. *The Java Language Specification*. Addison-Wesely, 1996.

[4] ISO. Fortran 95 Standard. ISO/IEC 1539 :1997.

[5] E. Laure. ParBlocks - A new Methodology for Specifying Concurrent Method Executions in Opus. Technical Report TR99-05, Institute for Software Technology and Parallel Systems, University of Vienna, 1999.

[6] E. Laure, M. Haines, P. Mehrotra, and H. Zima. On the Implementation of the Opus Coordination Language. *Concurreny: Practice and Experience*, to appear 1999.

[7] B. Meyer. *Object-Oriented Software Construction*. Prentice Hall, 1997.

[8] *OpenMP C and C++ Application Program Interface Version 1.0*. http://www.openmp.org/, October 1998.

Static Parallelization of Functional Programs: Elimination of Higher-Order Functions & Optimized Inlining

Christoph A. Herrmann, Jan Laitenberger, Christian Lengauer, and
Christian Schaller

Fakultät für Mathematik und Informatik,
Universität Passau, Germany
{herrmann,lengauer}@fmi.uni-passau.de
http://www.fmi.uni-passau.de/~lengauer

Abstract. Functional programs have long been recognized as attractive
subjects of an implicit static parallelization because functional programming excludes artificial dependences, which would restrict parallelism.
One central concept which makes functional programming a powerful
paradigm is the higher-order function, which can have functions appearing in its arguments or result. We present an automatic method of eliminating higher-order functions, which is based on earlier work by Bell,
Bellegarde and Hook [2]. The number of auxiliary functions added in the
process is subsequently minimized by inlining transformations.

Keywords: functional programming, Haskell, higher-order function, inlining, parallelization, skeletons.

1 Introduction

We report on first experiences with a new compiler for a functional language,
called $\mathcal{HDC}$, which supports the use of skeletons, i.e., higher-order functions
which have customized parallel implementations, collected in a skeleton library.
The overall idea is to equip $\mathcal{HDC}$ with implicit, high-quality parallelism through
these skeleton implementations.

The compiler consists of quite a number of phases; we concentrate on two
here. One performs higher-order elimination (HOE), the other inlining. More
complete information on the compiler is available elsewhere [6].

The following section sketches the different phases of our parallelizing compiler. Section 3 describes the HOE algorithm proposed in the literature and our
modifications to it. Section 4 comments on the quality of the generated first-order
program. Section 5 concludes.

2 The $\mathcal{HDC}$ Compiler

The $\mathcal{HDC}$ compiler [6, 9] translates a subset of Haskell [3] into an imperative language – at present, C with MPI calls. The main difference to Haskell is that $\mathcal{HDC}$

P. Amestoy et al. (Eds.): Euro-Par'99, LNCS 1685, pp. 930–934, 1999.
© Springer-Verlag Berlin Heidelberg 1999

is strict, in order to facilitate a compile-time parallelization. (However, invisibly to the programmer, strictness is partly eliminated by inlining transformations.)

The compiler is based on the principle of compilation by transformation, which has already been used successfully in the Glasgow Haskell compiler **GHC** [10], and consists of the following phases [6]:

1. scanning/parsing, using the tool happy
2. desugaring
3. list comprehension simplification
4. lambda lifting, let elimination [8]
5. simplification of list comprehensions
6. type checking
7. monomorphization
8. elimination of functional arguments (HOE)
9. elimination of mutual recursion (optional)
10. case elimination
11. generation of intermediate DAG code
12. tuple elimination
13. optimization cycle (optional)
    - inline expansion
    - rule-based DAG optimizations
    - size inference [5]
14. abstract code generation
15. automatic parallelization (optional)
16. code generation

3 Higher-Order Elimination (HOE)

The program subject to HOE must be well-typed according to the Hindley-Milner rules. It must also be closed, i.e., all functions cited must be available to the HOE procedure for a global analysis and transformation. The result of the HOE is an equivalent first-order functional program, which is also well-typed.

We are applying HOE in $\mathcal{HDC}$ because we want to avoid having to deal with higher-orderness in our target C code.

We base our work on a previous HOE algorithm for a more general setting [2]. We were able to simplify this algorithm significantly for our purposes. Most importantly, in order to simplify the generation of the target C code, our input to the HOE algorithm is monomorphic. The large amount of functions, which are introduced in the HOE and in the prerequisite phase of desugaring, is subsequently reduced substantially by the inlining transformations.

There is also a source translation from ML to Ada [13], which is based on the same general algorithm.

The general HOE algorithm uses a set of seven rewrite rules for the transformation. The idea is to replace the partial applications of a function by a kind of closure. A closure contains a function identifier and the values of the free variables in the partial application.

Some of the seven rules deal with restricting polymorphism and become obsolete in our monomorphic setting. Our modified HOE algorithm [11] uses the following set of four rules:

1. **η-expansion.** This rule expands function definitions which return functions as *result* with as many additional formal arguments as the function returned expects. If the result was polymorphic before monomorphization, the number of additional arguments may depend on the call. Applications of the expanded function then include the application of the function returned and deliver a non-function result.

2. **Encode.** This rule encodes functional arguments using constructors and introduces apply functions which decode them.

3. **ApplVar.** If in a function application the function is represented by a variable which is marked to carry a closure value, a temporary type inconsistency occurs during the transformation because a closure cannot be applied. This rule wraps the closure in a call to an additional apply function which takes the closure as an argument.

4. **RemoveHOTypes.** To clean things up, all function types appearing in data type definitions are replaced by an algebraic data type, which is parametrized with an identifier of the encoded type and encompasses all closures.

The algorithm starts with a phase of applications of rule 1, followed by a phase in which rules 2 and 3 are applied repeatedly in any order, and terminates with a phase of applications of rule 4.

4 Experimental Results

Of paramount interest is the impact of the HOE algorithm on the target code. Our first example, Karatsuba, is an optimized multiplication of two polynomials, represented by a list of their coefficients [1, 4]. Our second example is the frequent set problem [12], a data mining application.

In Tab. 1, we have recorded some static characteristics of the code (row by row, as the compilation proceeds) and the effect of our optimizations.

The Karatsuba example is expressed with a skeleton whose parallelism is completely static, except for some parameters, e.g., the problem size. Thus, the compiler optimizations can only affect the local structure inside the customizing functions. The frequent set example is much more dynamic: optimizations can affect the structure of the entire implementation. Therefore, it pays to analyze the properties of the program after different phases of the compilation. We have built an $\mathcal{HDC}$ interpreter for this purpose. Tab. 2 shows the results of an interpretation of the abstract code with two different samples A and B. The improvements after optimization demonstrate the important role inlining plays after the HOE.

The large amount of work is due partly to the nature of the problem and partly to the lack of sophistication of our source program, derived from Alg. 3.7 of [12] (there are cleverer ways [7]). Regardless of that, note that the optimizations

number of	Karatsuba		frequent set	
1. source functions	7		21	
2. source lines	30		86	
3. functions before HOE	75		104	
4. functions after HOE	37		103	
5. tree nodes	416		968	
number of	no opt.	opt.	no opt.	opt.
6. DAG functions	31	11	86	25
7. total DAG nodes	202	269	492	455
8. total abscode nodes	212	343	534	563

Table 1. Effect of compilation and optimizations on the program

reduce the number of operations by up to 30% and that there is a high potential of parallelism.

input		no. of operations			no. of par. steps			average par.		
sample	threshold	no opt.	opt.	ratio	no opt.	opt.	ratio	no opt.	opt.	ratio
A	0.5	12075	8782	0.73	355	224	0.63	34.0	39.2	1.15
B	0.5	55935	39559	0.71	893	586	0.66	62.6	67.5	1.08
B	0.2	360963	252887	0.70	1854	1239	0.67	194.7	204.1	1.05

Table 2. Run-time characteristics of the frequent set example

We do not yet have data on the speedup through parallelism. But, compared to **GHC**-compiled code, the $\mathcal{HDC}$ Karatsuba example takes sequentially 20% longer, the frequent set problem roughly 2 to 2.5 times as long [6]. This is the price we pay for not having to deal with higher-orderness in the target code.

5   Conclusions

We purport that the elimination of higher-order functions is especially useful for a parallelization via the use of skeletons. We have succeeded in applying our compilation techniques without difficulty to two realistic, application-level functional programs.

The higher-orderness of skeletons permits the combination of static and dynamic techniques in program parallelization. E.g., the frequent set example requires many skeletons – some static, some dynamic.

Acknowledgements

This work has been funded by the DFG under project RecuR2 and by the DAAD under an exchange project in the ARC programme. Our former team member Robert Günz deserves special thanks for implementing the first two compiler phases. We are also grateful to Françoise Bellegarde, Christophe Darlot and John O'Donnell for fruitful discussions.

References

[1] Alfred V. Aho, John E. Hopcroft, and Jeffrey D. Ullman. *The Design and Analysis of Computer Algorithms*. Series in Computer Science and Information Processing. Addison-Wesley, 1974.

[2] Jeffrey M. Bell, Françoise Bellegarde, and James Hook. Type-driven defunctionalization. *ACM SIGPLAN Notices*, 32(8):25–37, 1997. *Proc. ACM SIGPLAN Int. Conf. on Functional Programming (ICFP'97)*.

[3] Richard Bird. *Introduction to Functional Programming using Haskell*. Series in Computer Science. Prentice Hall Europe, 2nd edition, 1998.

[4] Christoph A. Herrmann and Christian Lengauer. On the space-time mapping of a class of divide-and-conquer recursions. *Parallel Processing Letters*, 6(4):525–537, 1996.

[5] Christoph A. Herrmann and Christian Lengauer. Size inference of nested lists in functional programs. In Kevin Hammond, Tony Davie, and Chris Clack, editors, *Proc. 10th Int. Workshop on the Implementation of Functional Languages (IFL'98)*, pages 346–364. Department of Computer Science, University College London, 1998.

[6] Christoph A. Herrmann, Christian Lengauer, Robert Günz, Jan Laitenberger, and Christian Schaller. A compiler for HDC. Technical Report MIP-9907, Fakultät für Mathematik und Informatik, Universität Passau, May 1999.

[7] Zhenjiang Hu. Personal communication at the Dagstuhl Seminar on High-Level Parallel Programming, April 1999.

[8] Thomas Johnsson. Lambda lifting: Transforming programs to recursive equations. In Jean-Pierre Jouannaud, editor, *Proc. Conf. on Functional Programming Languages and Computer Architecture (FPCA'85)*, LNCS 201. Springer-Verlag, 1985.

[9] Lehrstuhl für Programmierung, Universität Passau. The HDC compiler project. http://www.fmi.uni-passau.de/cl/hdc/.

[10] Simon L. Peyton Jones. Compiling Haskell by program transformation: A report from the trenches. In Hanne Riis Nielson, editor, *Programming Languages and Systems (ESOP'96)*, LNCS 1058, pages 18–44. Springer-Verlag, 1996.

[11] Christian Schaller. Elimination von Funktionen höherer Ordnung in Haskell-Programmen. Diplomarbeit, Fakultät für Mathematik und Informatik, Universität Passau, September 1998. In German.

[12] Hannu Toivonen. *Discovery of Frequent Patterns in Large Data Collections*. PhD thesis, Department of Computer Science, University of Helsinki, 1996.

[13] Andrew Tolmach and Dino P. Oliva. From ML to Ada: Strongly-typed language interoperability via source translation. *J. Functional Programming*, 8(4):367–412, July 1998.

A Library to Implement Neural Networks on MIMD Machines[*]

Yann Boniface[1], Frédéric Alexandre[1], and Stéphane Vialle[2]

[1] LORIA, BP 239, 54506 Vandœuvre-lès-Nancy cedex, France
boniface@loria.fr, falex@loria.fr, Fax: +33 3 83 27 83 19
http://www.loria.fr/LORIA/EXT/equipes/CORTEX/
[2] Supelec, 2, rue Edouard Belin, 57078 Metz, France
Stephane.Vialle@supelec.fr, Fax: +33 387 76 47 00

1 Introduction

Executing a sequential implementation of an Neural Network (NN) generally induces a high cost in terms of computation time, including a large part for the learning phase. This cost together with the weak computation performance of a computer with regard to a human brain [3] make it difficult to test some complex large connectionist models, like some inspired from biological reality or some with a high dimensional input space or a large number of units. Moreover, NNs present a large amount of natural parallelism. Unfortunately, NN parallelism is very different from modern and general purpose parallel computer parallelism. NNs have a fine grain of parallelism and a natural message passing paradigm. On the opposite, modern parallel computers have a MIMD[1] architecture. Some recent hardware development led to use shared memory as an efficient parallel programming way with high number of processors. The main goals of this project are to speed up NN executions and to decrease development time of parallel NN implementations, in order to quickly implement various kinds of NN and to run on parallel computers more complex simulations than sequential computers allow. We offer to connectionists a tool to develop their models with fine grain parallelism and to execute them onto DSM[2] MIMD general purpose parallel computers.

2 Parallel Aspects of Neural Networks

Artificial neural models have been inspired from brain, often described as a parallel machine. As early as 1943, McCulloch and Pitts [4] have proposed a formal neuron as a simple calculus automaton with a set of weighted inputs and an output (cf figure 1). The automaton computes an internal activation as a weighted sum of its inputs. This activation is compared to a threshold to

---

[*] Y. Boniface is supported by a grant from the high performance computing centre Charles Hermite (CCH), which also furnished parallel computation facilities.
[1] Multiple Instructions Multiples Data
[2] Distributed Shared Memory

P. Amestoy et al. (Eds.): Euro-Par'99, LNCS 1685, pp. 935–938, 1999.
© Springer-Verlag Berlin Heidelberg 1999

decide whether the output is active or not. These simple distributed activities together with the principle of information widespreading lead to the characterisation of a fine grain parallelism and a message passing communication paradigm.

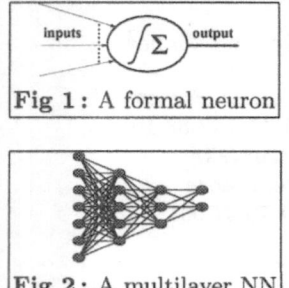

Fig 1 : A formal neuron

Fig 2 : A multilayer NN

From that date, a variety of artificial NNs has been built, as a set of interconnected formal neurons (cf figure 2) and many algorithms for the automatic tuning of the weights (learning algorithms) have been proposed[2]. Even if many researchers underline its inherent parallelism, most of them used to implement artificial NN on sequential machines. Nowadays, parallel machines are more widespread and easier to use and one can seriously think to implement artificial NNs to benefit from their natural parallelism [6].

3   Method

Numerous attempts have been already done to develop NNs onto parallel machines [5], most of them being concerned by a specific model. Beyond dedicated implementation, our approach consists of a library for adapting NN parallelism onto a general purpose computer parallelism [1].

We decided to design our library for Distributed Shared Memory (DSM) computers as this parallel architecture seems scalable and supports all of the currently used parallel paradigms. Our developments and benchmarks were made on an SGI-Origin2000. We chose to design and implement a library of functions that : (1) hides real parallel hardware, and respects natural NN parallel semantics, in order to make NN implementation easier, and (2) uses parallel algorithms well adapted to real parallel hardware in order to obtain efficient parallel execution. Several communication and synchronisation paradigms were experimented and compared : message passing with PVM and MPI, shared memory with processes and threads, and implicit parallelisation with parallel compiler directives. After comparison, explicit thread programming with shared memory communication method revealed to be the most efficient and the most usable on not always regular codes.

With our library, a connectionist developer thinks his network like a set of autonomous entities, working in parallel. Each entity has inputs and a unique output. These inputs and output build the network topology. Data consistency is managed by our tools. The library role is to ensure that each unit has been visited for each cycle, to drive messages through the connections and to ensure that all connections are consistent (during a cycle, a unit must read inputs valid for this cycle). Finally, the building by unit declaration makes the code clearer. A program based on our library is written in standard "C" language. This language was chosen because it is the usual language handled by the connectionist community, because it is portable and has a variety of libraries (like graphic tools). And, compared to sequential "C" programming, our tool makes it easier the building of network connectivity.

4   Parallel Design Description

Our work was implemented on a SGI-Origin2000 parallel computer based on cc-NUMA[3] architecture. With this architecture, each processor has a quick access to its local memory. Accessing remote memory adds an extra increment of time [7]. The library was implemented in "C" language. On the Origin system, the parallel implementation uses light processes (threads), and the barrier of synchronisation is based on semaphores which suspend waiting processes. Our library creates light processes and a unique light process would be run per processor.

In a DSM model, each processor has a direct access to the memory of every other processors in the system. A processor can directly load or store any shared address. When it needs data, it copies the cache line[4] in its own memory. When a processor modifies data in a cache line, each copy of this line is invalidated. So, when another processor wants to read data (modified data or other data in the line), it needs to update this line [7]. These characteristics reveal two difficulties which affect performance:

- **Cache contention** When a processor frequently modifies data in a cache line, and if other processors often read this data, they have to update after each modification. What will increase the data time access.
- **False sharing** Several processors frequently modify different data in the same cache line. They need to update the line before modifying their own data, if another processor changes data in the line.

To manage these problems, data are distributed in each processors in order to:

- help the Operating System to place data handled by a unique processor in its own memory or in a close memory.
- ensure no false sharing: if two data are modified by two different processors, they will not be in the same cache line.
- limit the cache contention. We try to avoid each period when a processor can modify data while another processor tries to read the same data.

Theoretically, every data are shared but they are not handled by all processes. So, two types of data are managed by a process: private data (the own of the process) and global data. If data are private, the process which handles them puts them in a data structure that contains all its private data, and which is aligned on cache lines and has a size multiple of the cache line size. As this structure is not accessed by other processes, there is no contention neither false sharing on private data. So, the cost of communication is weak to modify data (local memory) and null to export this cache line (no other process needs the cache line).

Global data are shared by all processes. But, in our model, there is no user datas updated by more than one process. They are located in the memory of this process, in a cache line containing only global data modified by the process.

---

[3] cache coherent Non-Uniform Memory Access
[4] In ccNUMA, memory is organised by cache lines

Moreover, stage when processes read global data and stages when they update the global data they managed are separated.

5   Results

The library supplies some simple functions to easily create neural units, to connect or disconnect them, and to start the neural computation. The same program developed with our library can be executed on different number of processors. The user just has to modify a processor number constant in the program and to recompile the code, no other modifications being necessary. So, our library is available on shared memory parallel computers, and on sequential ones. In terms of performance, training a Kohonen Map of 100×100 neurons, with 100,000 iterations by using our library with 8 processors was 7 times faster than a sequential execution of the same program and 3.6 times faster than the execution of a classical sequential "C" program. Beyond height processors, efficiency decrease, due to load balancing and the frequency of synchronisation instructions.

Description of use and performance of our library are described in [1].

6   Conclusion

The role of our library is to act as a transparent interface between both kinds of parallelism. We have searched for a semantic adapted to neural computation for a quick and natural implementation on DSM MIMD computers that avoid contention and false sharing problems. Experiments have shown that programs written with the first version of our library are easy to understand and scale well. Ongoing work is twofold. First, we are going to optimise the implementation of sequential parts, in order to limit the sequential cost of our library. Second, we will pursue applications to other connectionist models including complex biologically inspired models, to show the genericity of our library.

References

[1] Y. Boniface, F. Alexandre, and S. Vialle. A bridge between two paradigms for parallelism : Neural networks and general purpose mimd computers. IJCNN, 1999.

[2] J. Hertz, A. Krogh, and R. G. Palmer. *Introduction to the theory of neural computation*. Addison-Wesley, 1992.

[3] Y. Lallement. Intégration neuro-symbolique et intelligence artificielle, applications et implantation parallèle. *PhD. Thesis*, 1996.

[4] W.S McCulloch and W.P Pitts. A logical calculus in the ideas immanent in nerveous activity. *Bulletin of Matematical Biophysics*, 1943.

[5] M. Misra. Parallel environments for implementing neural network. *Neural Computing Survey*, 1:48–60, 1996.

[6] H. Paugam-Moisy. *The Handbook of Brain Theory and Neural Network.*, chapter Multiprocessor Simulation of Neural Networks. The MIT Press., 1995.

[7] Silicon Graphics, Inc. *Performance Tuning Optimization for Origin2000 and Onyx2.* http://techpubs.sgi.com/library/manuals/3000/007-3511-001/html/O2000Tuning.0.html.

Topic 12
Architectures and Algorithms for Vision and Other Senses

Alain Ayache, Virginio Cantoni, Concettina Guerra, and Pieter Jonker

Co-chairmen

The following are the proceedings of one of the topics of Euro-Par'99, held in Toulouse, France, on August 31 - September 3, 1999. A wide spectrum of topics are covered in this conference with 23 workshops dedicated to the promotion and advancement of all aspects of parallel computing and ranging from hardware, to software, algorithms, and applications. The theme of this workshop, the 12th of the series, is: Architectures and Algorithms for Vision and other Senses. In fact, due to its extremely high performance requirements, signal and image processing has always been one of the major application areas and one of the driving forces behind the design of special-purpose processors and massively parallel architectures. Presently, new challenges arise from areas like multi-media applications and visual communication in particular. The workshop is intended to provide a forum to present current work and new ideas in this challenging field.

The technical program that come out after the reviewing process consists of 12 regular papers and 1 short paper. The resulting papers deal with various issues that are crucial to meet the computational demands of vision and signal processing, and cover the mentioned topics. In more details, the papers center on the design and evaluation of parallel architectures and special-purpose processors and on the methodologies for efficiently designing algorithms for these architectures. The vision applications considered deal with both 2D and 3D data. They include among the others: stereo matching, identification of geometric primitives, detection of vanishing points, object recognition and volume rendering. The signal processing applications include signal and video compression, wavelet transforms. Two other papers deal with the design of architectures that implement in an integrated fashion basic image and signal processing tasks.

The topic committee would like to express its appreciation to the reviewers committee of the workshop for their qualified work. Special thanks should go to the organisers CERFACS and ENSEEIHT - IRIT and to the people for the local organisation for their precious help in managing electronically information, data, papers, and reviews.

P. Amestoy et al. (Eds.): Euro-Par'99, LNCS 1685, pp. 939–939, 1999.
© Springer-Verlag Berlin Heidelberg 1999

LUX: An Heterogeneous Function Composition Parallel Computer for Graphics

Stéphane Mancini and Renaud Pacalet

Ecole Nationale Supérieure des Télécommunications, Paris, France

Abstract. In this paper, we present an heterogeneous parallel computer dedicated to high realism computer graphics. A small network, with a reduced chip set, allows us to reduce rendering time by a very attractive factor. The low level mechanisms of the network are designed to manage the wide variety of data and algorithms used in computer graphics. Some nodes of the network may be specialized in the most time consuming parts of the algorithm and have specific data paths. Thanks to the function composition scheme, we unify both the management of specialization and of parallelism. Those mechanisms allow flexibility and easy design of programs.

1 Introduction

The graphic community is in need of very high computational power to achieve high realism pictures at interactive rates. We observe that single general purpose computers can't provide the needed performance, and general parallel computers hardly reach them, and for a high financial cost. In this paper, we present an heterogeneous parallel computer which may reach expected performances, with a reduced chip-set and this for a low financial cost.

We are particularly interested in the speed up of Ray-Tracing algorithms. They simulate the propagation of light in an environment, the scene, made of geometric primitives. The most time consuming part of such algorithms is the computation of the intersection between the rays and the scene. The computation of surface lighting characteristics can also be quite expensive. The graphic community has developed many algorithms to reduce the rendering time, essentially by reducing the number of ray-object intersections, but it's still quite high.

We propose to design specialized data paths to speed-up the most time consuming parts of the algorithm. Those specialized units are interconnected through a dedicated network to form an heterogeneous parallel computer.

In the second part of this paper, we briefly recall the Ray-Tracing algorithm and the way we want to accelerate it. In part 3, we show how we manage specialization and scheduling with the function composition scheme. In the two next parts (4 and 5), we precise the architecture and the load balancing strategy. At last, before the conclusion, we show the tool developed to help us to make architectural choices.

P. Amestoy et al. (Eds.): Euro-Par'99, LNCS 1685, pp. 940–949, 1999.
© Springer-Verlag Berlin Heidelberg 1999

2 Motivation

2.1 Ray Tracing

The Ray-Tracing algorithm simulates the propagation of light in a scene. More exactly, we follow the inverse path of light to determine which object is seen by the observer at a given pixel of the picture. A primary ray is sent, originating at the observer and passing through the considered pixel, and we compute its intersection with the scene to determine which object is visible. Once the intersection is found, we determine the amount of light at this point by sending secondary rays. We send rays to each light source to compute the intensity of direct lighting and we send refracted and reflected rays, depending on the local illumination law of the surface, to determine the amount of indirect lighting. The Ray-Tracing algorithm is recursive, and we stop the recursion when we reach a given criteria (recursion depth, contribution threshold, ...).

This basic algorithm is improved in several ways to deal with effects like indirect lighting through glasses and other realistic effects but the main point is that they're built over the computation of the intersection between a ray and the scene.

The Ray-Tracing algorithm needs very high computational power. Indeed, we need to compute the intersection of several millions of rays with millions of geometric primitives. That's why the graphic community has developed many algorithms to reduce the number of intersections to compute but the rendering times are still quite high. To accelerate it again, we need to increase computational power.

2.2 Parallelization

The Ray-Tracing algorithm can easily be parallelized. For example, the computation of two pixels is independent, and the intersection of a ray with two objects can be done in parallel. In the literature, we find two main strategies, implemented on general purpose parallel computers:

- "Control parallelism": part of the picture are mapped to nodes which exchange objects
- "Data parallelism": the scene is split over nodes which exchange rays

The choice between such solutions is often empirical but some ([2]) proposed a mix strategy where rays and objects move at the same time. The main advantage of general parallel computers is that programmers can reuse existing graphic software, especially in the first strategy. But the speed-up is at best linear with the number of nodes and mainly depends on the amount of memory of each node and on the buses bandwidth.

2.3 Specialization

Mainly all previous specialized solutions ([4], [3], [1]) failed because they didn't offer enough flexibility, which is essential in computer graphics. For example, ([3]) hard-wired Constructive Solid Geometry (CSG) but didn't provide algorithmic acceleration. [1] tried to design very specialized processors but that time technology didn't allow their realization. Also, the bottleneck would have been the host computer because their processors didn't accelerate all the parts of the algorithm at the same rate.

To be attractive enough, custom hardware must provide a computation power difficult to reach with a general purpose processor and, this, for a low financial cost. A flattening and unification of the Ray-Tracing algorithm shows us the possibility to design different specialized data paths for various parts of the algorithm. For example, we are able to design a chip computing 25 millions ray-triangle intersections per second, to be compared to 1 million/s on a general purpose processor at 400 MHz.

So, we propose the design of a MIMD parallel computer built on a network of specialized processors. The network provides low level mechanisms to manage parallelism, specialization of the nodes and flexibility. The load balancing strategy should allow to increase the computation power dedicated to a given part of the rendering process by multiplying the number of processors specialized in it. We also want it to take into account the possibility to process a task at various computing powers because we may update the nodes to follow technological improvements.

3   The Function Composition Scheme

3.1   Grain of Parallelism

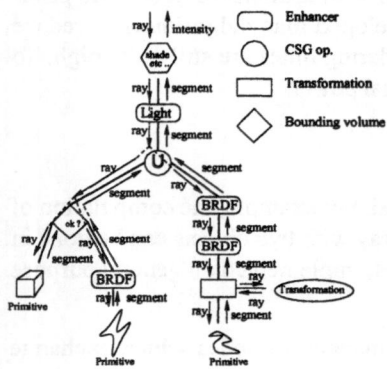

Fig. 1. Scene tree

The main problem of the design of parallel programs is to find an adapted level of parallelism because it has a direct impact on the ratio between computation time and data transfers. We have seen that the Ray-Tracing algorithm provides many levels of parallelism, so we have to choose one suitable for specialization.

The Ray-Tracing algorithm has something specific in the way that it is data driven: the sequence of computations depends on the scene description. The scene is represented by a graph, like shown on figure 1. This graph contains geometric primitives, colorimetric informations and algorithmic accelerations. The computation of the intersection of a ray with the scene is performed by a run on the tree. On figure 1, the rays propagate from top to bottom and the results of intersections propagate from bottom to top.

At this point, we note that the same function may have different behaviors, depending on the types of data which describe leaves or nodes (CSG operation, triangle, quadric, etc ...). So, we type data and we distinguish a function and its instantiation. Once we know the function to compute and its arguments, we select the corresponding instantiation depending on the type of the arguments and available hardware computational resources.

So, we choose to set the level of parallelism at the level of the nodes and leaves of the scene graph. This allows us to have an uniform grain of parallelism at each step of the Ray-Tracing algorithm, from rendering to shading. To increase the grain of parallelism and make a better use of Ray-Tracing properties, clusters of rays propagate in the scene graph instead of single rays.

```
I_Intersect_Ray_Triangle(ray,triangle,destination)
segment=intersect_triangle(ray,triangle)
Set_Argument(destination,segment)

I_Intersect_Ray_CSG_Add(ray,node,destination)
T_add=Create_Task(Add_segment,nil,nil,destination)
T_left=Create_Task(Intersect,ray,node-left,(T_add,0))
T_right=Create_Task(Intersect,ray,node-right,(T_add,1))
```

Fig. 2. Sample pseudo code

3.2 Tasks

Our system provides mechanisms to manage the stack created by the recursive run on the scene tree. As we don't want to centralize the stack, we give a task informations on the task to compose with.

To call a function, we create a task which is a data structure composed of two elements: an identifier of the function it uses and an array of arguments. We will note:

$$F = (f, (\text{arg}_0, \text{arg}_1, \ldots, \text{arg}_n))$$

f is the function associated with the task F. Arguments are pointers on the data needed by the function. A task is said sleeping until all the arguments are set, otherwise the task is said activated and, then, may be processed. Task composition is achieved by indicating a function the argument of the task to compose with. When a task produces a result, it sets the corresponding consumer task argument to the result's address. The consumer task may then be activated and move over the network to be processed somewhere.

Figure 2 illustrates this mechanism with a program sample. It shows the instantiation's pseudo-code of the ray/primitive or ray/CSG object intersection. The function Create_Task returns a pointer on the created task and the notation (*Task*,n) points on the nth argument of the task. The function Set_Argument creates a low level task which sets the argument of a task to a pointer on the specified list. We note that the two tasks *T_left* and *T_right* can be processed in parallel because they are both activated. The two segments "left" and "right" are destroyed by the task *T_add*.

3.3 Task Migration

The management of specialization is distributed over the nodes. When a node receives a tasks, it decides or to send it to a distant node or to process it locally. The diversity of load balancing strategies and migration law is coded in the function's genus. We have the types: "Sedentary", "Universal", "Specialized" and "Follower". A "Sedentary" task can only be executed on a specific node. At opposite, an "Universal" task can be processed everywhere. The management of specialized hardware is done through "Specialized" tasks which migrate according to the type of their arguments. And, at last, a "Follower" task migrates until it reach its first argument. The type of a task can have a great impact on the performances of the system.

3.4   Data Structures

We type data with three fields: (Class, Species, Type), called a genus. The data structures are unified in lists which are coded linearly, like we would write them by hand. Tasks are also coded within lists. A task is coded in a list of three elements: a function, the list of arguments and a data list. An argument can be a pointer on a list or a reference to a sub list of the data list. This late one allows us to move task and data together. We distinguish system tasks from applicative tasks with the function's genus.

4   Architecture

4.1   General

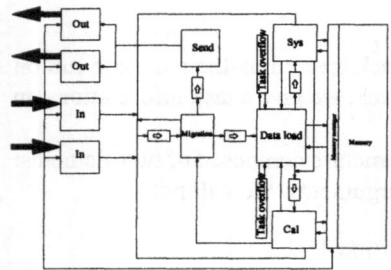

Fig. 3. Node architecture

LUX is an heterogeneous MIMD parallel computer and nodes of the network are specialized in parts of the rendering process. Nodes are interconnected through a service network and a data network. The service network interconnects all the processors through a ring and is dedicated to system communications like load balancing or specialization. The data network is a hierarchical network and is dedicated to local communications. In order to unify communications, processors only exchange tasks; even simple data are enclosed in tasks. The topology of the data network is free and we can build sub networks which can be considered by the rest of the network as single nodes.

Memory is distributed over the nodes of the network and addressing is unified through pointers; A pointer refers to a processor and to the local address of the data. Consequently, we have to transfer data and store them during computation of tasks. To do so, the private memory of each node is split between local data and cached data. This cache is made of copies of lists originally located on other processors. We note that the function composition scheme doesn't need the management of cache coherency. A data propagates on the tree and is destroyed when it has run everywhere it's needed. So, a list can be created or destroyed but not modified, except tasks. This policy may increase data transfers and memory resources but allows simple and fast mechanisms. Memory resources of processors is free and the larger they are the more we reduce communications. Nevertheless, high memory quantities, like mass memory, can be managed by specific processors to deal with large databases.

4.2   Node Architecture

The main functionality of a node is the task pipeline execution which steps are "Migration" step, "Load" step and "Compute" step.

At the "Migration" step, the node determines the target node depending on the migration law of the task and on the load balancing strategy. A task may continue the pipeline or be sent to another node.

Next, at the "Load" step of the pipeline, the node loads distant data. Cached lists are stored in cache cells. When there's a cache miss, the task is removed from the task pipeline and the node creates tasks to load the data. When the cell is loaded, the tasks which use it continue the pipeline. If there aren't enough available cells, the task is put on the bottom of a stack for a further try. We note that cache cells are locked until the end of the computation of tasks which use them.

The task function is computed at the third step. To simplify the design of processors, we suggest to standardize the units which manage all the system mechanisms and adapt the applicative unit, like shown on figure 3.

We introduce FIFOs between each stage of the task pipeline to smooth access on the different units. To prevent interlocking, we use FIFOs of virtual infinite capacity. Indeed, if the stacks were of finite size, when they would be full, the node couldn't receive neither create any active task which may solve the situation. So, to prevent interlocking, when one stack is full, a part of its content is saved in external memory. The saved content of the stack is restored when the stack is above a threshold. So, stacks are managed both by a local unit and by a distant manager. This way, we can consider that the stacks are of potential very large size and we prevent interlocking efficiently.

5   Load Balancing Strategy

5.1   The Algorithm

The resolution of the function migration law of a task may give us several target processors where to compute the task. We have to make a local choice in order to minimize the global computing time. The difficulty is that the different processors may have various computing powers. We choosed a policy based on the dynamic evaluation of the computing time of tasks depending on specialization and target processor.

For each possible processor, we estimate the time it has to run and the target processor is the one with minimum total time. The estimated run time of each processor is given by the following equation:

$$T = \sum_i n_{I_i} * T_{I_i}$$

Where n_{I_i} is the number of instantiation i of the tasks in the pipeline and T_{I_i} the evaluation of the computing time of the instantiation. In practice, we only take into account the time of the applicative tasks and the run time of a node is incrementally estimated when a task passes the migration step and when the computing of a task ends.

So, each processor needs to have knowledge about the state of others. Those informations are propagated through the service network at low rate. To reduce communications, we only send the estimated run time when its difference with the previous one is superior to a given threshold.

To take into account transfers on the buses, we attribute a factor of correction to the evaluation of the run time of each distant node. This policy allows us both to avoid ping-pong effects while a node discharge its tasks to a node with same specialization and to maintain locality of computations on subnetworks.

5.2 Qualitative Study

We have to ensure us that this algorithm is stable and allows a good load balancing of the nodes. Here, we don't show a mathematical proof but give some qualitative results. Instability would come from the delay between the moment we take a decision and the moment we have knowledge of its results. We send tasks to the wrong node until we get an evaluation of the estimated run time. One of the determinant factor is the time of the travel of the task from its originating node to the target node. The more the buses are loaded, the more we make wrong decisions. The delay due to the communication of the time after its estimation, at the end of the processing of the task, also has influence until the estimated time of the instantiation is stable. To reduce the delay, we can try to predict the computing time of distant nodes. To do so, we modify the evaluation of the computing time at the moment we make the decision. When we set the target node of a task, we add to the distant node run time the task's distant time.

The presence of many nodes with same specialization on the network has a drawback: we increase the communications on the buses. Indeed, a data used by many tasks may be loaded on each node cache cell and travel through the bus each time. On the other hand, the Ray-Tracing algorithm produces many tasks which use the same rays in a short time. The wrong choice of the target node, previously disputed, may send all those tasks to the same node and save some bandwidth.

At this point, we understand that it's quite difficult to give an a priori estimation of the performances of our system. That's why we provide a simulation tool to help us to make architectural decisions.

6 Simulation Framework

6.1 Simulator Engine

We've developed a simulation framework of the system to help us to make choices about network architecture, specialization, computational power, load balancing strategies and various parameters value. The simulator engine of the framework is written in C and the configuration tool in Tcl/Tk. We choosed C rather than VHDL, which would allow a fine description from top level to gates, because VHDL would have been too slow. The C code is divided in three parts:

- The core simulation engine
- Configuration and instantiation of the network
- The rendering program, collection of function's instantiations

The simulator is an event driven simulator and we attribute instantiations a time model. This model allows to simulate the different computational powers of each processor, depending on its supposed specialization. Each action is timed and the simulator freezes the different units until they finished their job. So, with probes set at various places, we can have a fine analysis of task migration and data flows.

Table 1. Architecture and scene influence

Architecture	Bus width	Scene	Picture size	Time	Chip load	Bus load
(diagram)	32	scene$_1$	50*50	27 ms	$T = 1$	$B = 0.23$
			100*100	118 ms	$T = 1$	$B = 0.1$
		scene$_2$	100*100	41 ms	$T = 0.9$	$B = 0.7$
	64	scene$_2$	100*100	41 ms	$T = 0.9$	$B = 0.4$
(diagram)	32	scene$_2$	100*100	41 ms	$T = 0.9$	$B_r = 0.5$ $B_l = 0.3$
(diagram)	32	Id.	Id.	41 ms	$T = 0.9$	$B_r = 0.5$ $B_l = 0.1$ $B_u = 0.2$
(diagram)	32	Id.	Id.	44 ms	$T = (0.4, 0.5)$	$B = 0.9$
	64	Id.	Id.	30 ms	$T = 0.6$	$B = 0.6$
(diagram)	32	Id.	Id.	30 ms	$T = 0.6$	$B_r = 0.6$ $B_l = 0.4$
(diagram)	32	Id.	Id.	30 ms	$T = 0.6$	$B_r = 0.6$ $B_l = 0.2$ $B_u = 0.3$

6.2   Program Design

The first step to write a program is to list the different functions involved and attribute them an identifier. The second step is the design of each function's instantiations. Some instantiations may create sub tasks to be executed and that for, they create a task with the function field set to the corresponding identifier. The system will automatically manage task's activation, like shown with the sample code on figure 2. The third step is the compilation and installation of the different instantiations on their processors. Finally, we establish a local link between each function and its instantiation, depending on the migration law. We note that, as we need to have an overall view of the different kind of processors to identify the functions, the steps declaration, compilation and establishment of link are done by a supervisor.

Finally, each processor communicates its specialization to others through the service network. An important point is that the design of the program is independent from the data network. Task migration, data access and load balancing is automatically managed by low-level mechanisms.

7   Results

7.1   Simulation Parameters

In this section, we give some results of simulation to show the influence of various parameters like scene description, network architecture and program design.

The simulations are done with a scene which contains a ground plane, a tower and a space station and is lighted with a punctual light source. The tower is made of boxes CSG added and subtracted. The space station is a mesh of about 10000 triangles split in an octree. In $scene_1$, the octree has a maximum subdivision depth of 6, with a maximum of 20 triangles per node. In $scene_2$ the maximum depth is 8 with 10 triangles per nodes. The nodes of the network are:

T Ray ∩ triangle	P Ray ∩ polyhedra	C CSG
Tr Transformation	F Filter	R Rendering

The T node computes the intersection between a ray and a triangle in 4 clock cycles at a 100 MHz frequency, which is an extreme computational power. To simplify the paper, all the system is synchronous, at 100 MHz, and all the buses have the same bandwidth, specified on the table of results. The computing power of all other processors is not critical but they have side effects. The load balancing is the one in its basic version. On all the simulations, for the clarity of the paper, we only give measures for the triangle processor and buses (the buses subscripts are: l=left, r=right, u=up and b = bottom).

7.2 Analysis

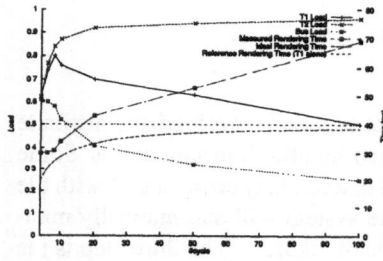

Fig. 4. Load balance

The table 1 shows us the influence of the relation between computation power and bus bandwidth. The results obtained with the first architecture shows us that the bottleneck is the T node when we do the computation with the scene $scene_1$ which contains a few complex primitives. In $scene_2$, we have many small primitives and the bottleneck becomes the bus which as to transfer the primitives and the results of intersections.

The three first architectures have approximately the same performances because the bottleneck is the triangle processor. Their behavior is different when we add a second triangle processor because the bottleneck becomes the bus bandwidth. To reduce the rendering time, we have to reduce the bus load. To do that, we can increase the bus bandwidth or split the bus.

We observe that the balance between the two triangle processors is quite good: we have less that 1% of difference of efficiency when they have the same computational power. The figure 4 shows the evolution of the balance between two processors of same specialization but with different computing power. We set the T_1 processor to the highest computing power (one intersection in 4 clock cycles) and we decrease the computational power of T_2. We observe that when the two processors have similar computational power, they behave the same. When the difference is too important, the slowest processor becomes the bottleneck. This behavior comes from bad evaluations of the computing time of tasks which leads to wrong choices. If the wrong decision of the migration step leads to a migration, the choice can be corrected by the target node. But, if it leads to local execution of the task, once it's in the pipeline of the slowest processor, it stays in it and becomes the bottleneck.

Simulations show us that we don't significantly increase the network performances by modifying simulation parameters. So, we have to improve the network mechanisms. A solution would be to have a better model of the computing time which would take into account the complexity of the arguments of the functions. We could insert a "refining" step in the pipeline, after the migration step and before the load of operands, which would give a better approximation of the computing time of the task. We are currently investigating another solution which performs regular correction of the load balance. In case of known unbalance, we remove tasks from the pipeline and send them back to the migration unit which would make better choices for a better balance. This late approach seems to be very promising.

8 Conclusion

We have presented a knew kind of parallel computer dedicated to a class of application. Rather than to parallelize Ray-Tracing in screen or object space, we choosed to parallelize on nodes and leaves of the scene tree. This allows us to have specialized processors dedicated on nodes or leaves and to preserve flexibility. The load balancing strategies are automatically managed by the system and the user can have a fine control on them by tuning the data flows and program design. The user may also add its own strategies with specific system tasks. However, our strategy allows us to reduce the time to ray trace pictures by a very attractive factor, like shown by simulations. We note that although initially designed for Ray-Tracing, our architecture could be used for a wide variety of algorithms.

References

[1] Kadi Bouatouch, Yannick Saouter, and Jean Charles Candela. A VLSI chip for ray tracing bicubic patches. In *Proc. of Eurographics'89*, pages 107–124. W. Hansmann and F. R. A. Hopgood and W. Strasser, 1989.

[2] Alan Heirich and James Arvo. A competitive analysis of load balancing strategies for parallel ray tracing. *Journal of Supercomputing*, 12(1 and 2):57–68, 1998.

[3] Gherson Kedem and Jonh L. Ellis. The raycasting machine. In *Proc. of the IEEE International Conference on Computer Design: VLSI in Computers ICCD '84*, pages 533–538. IEEE Computer Society Press, 1984.

[4] Tadashi Naruse, Masaharu Yoshida, Tokiichiro Takahashi, and Seiichiro Naito. Sight — a dedicated computer graphics machine. *Computer Graphic Forum*, 6(4):327–334, December 1987.

A Parallel Accelerator Architecture for Multimedia Video Compression

Bertil Schmidt and Manfred Schimmler

Institut für Datenverarbeitungsanlagen, TU Braunschweig,
Hans-Sommer-Str. 66, 38106 Braunschweig, Germany
{bschmidt, masch}@ida.ing.tu-bs.de

Abstract. This paper describes a parallel architecture for a variety of algorithms for video compression. It has been designed to meet the requirements of encoding and decoding according to the ITU-T standard H.263. The architecture is an implementation of the instruction systolic array (ISA) model which combines the simplicity of systolic arrays with the flexibility of a programmable parallel computer. Although the parallel accelerator unit is implemented on no more than 9 mm^2 of silicon it suffices to meet the compression rate necessary to send a compressed video stream through a standard ISDN terminal interface.

1 Introduction

Video compression for multimedia systems is a computationally intensive task. The desired compression factor determines the performance requirements of the underlying hardware. Several authors have suggested dedicated VLSI implementations [7] of video compression algorithms. Because of their regularity, high throughput rate, small silicon area, and low power consumption, systolic arrays have been proven as a good candidate structure for these dedicated solutions [2], [3]. Their disadvantage is the lack of flexibility with respect to the implementation of different algorithms and different problem sizes, i.e. each algorithm and each problem size requires an individual hardware solution. ISAs have been developed in order to combine the speed and simplicity of systolic arrays with flexible programmability [5]. Originally, the main application field of ISAs was supposed to be scientific computing. However, in the mid 90s the suitability of the ISA architecture for other applications was recognized, e.g. [6], [8]-[10]. In this paper we illustrate how an ISA architecture can solve all computationally intensive tasks of a multimedia video compression application efficiently. This ISA has been developed in order to compress a source data rate of 15 frames per second (fps) in CIF format according to the ITU-T standard H.263 [1] for a transmission with the given data rate of two ISDN channels.

This paper is organized as follows. The concept of the ISA is explained in Section 2. Section 3 gives an introduction to video compression with H.263. Section 4 presents the new video accelerator architecture. It is documented how the ISA is integrated on an accelerator for videophone applications. The parallel ISA algorithms

P. Amestoy et al. (Eds.): Euro-Par'99, LNCS 1685, pp. 950-960, 1999.
© Springer-Verlag Berlin Heidelberg 1999

for video compression are explained in Section 5 and their performance is evaluated in Section 6. The outlook to further research topics concludes the paper in Section 7.

2   Principle of the ISA

The ISA is a quadratic array of identical processors, each connected to its four direct neighbours by data wires. The array is synchronized by a global clock. The processors are controlled by instructions, row selectors and column selectors. The instructions are input in the upper left corner of the processor array, and from there they move step by step in horizontal and vertical direction through the array. This guarantees that within each diagonal of the array the same instruction is active during each clock cycle. In clock cycle $k+1$ processor $(i+1,j)$ and $(i,j+1)$ execute the instruction that has been executed by processor (i,j) in clock cycle k. The selectors also move systolically through the array: the row selectors horizontally from left to right, column selectors vertically from top to bottom. Selectors mask the execution of the instructions within the processors, i.e. an instruction is executed if and only if both selector bits, currently in that processor, are equal to one. Otherwise, a no-operation is executed.

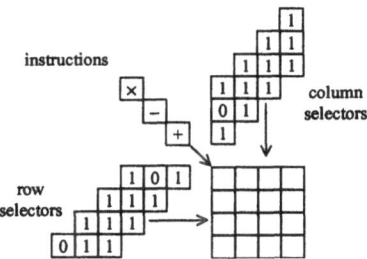

Fig. 1: Control flow in an ISA

Every processor has read and write access to its own memory. Besides that, it has a designated *communication register* (*C-register*) that can also be read by the four neighbour processors. Within each clock phase reading access is always performed before writing access. Thus, two adjacent processors can exchange data within a single clock cycle in which both processors overwrite the contents of their own communication register with the contents of the communication register of its neighbour. This convention avoids read/write conflicts and also creates the possibility to broadcast information across a whole row or column with one single instruction. This property can be exploited for an efficient calculation of row broadcasts, row ringshifts, and row sums which are the key-operations in many algorithms described in Section 5.

3   Multimedia Video Compression

The high amount of visual data associated with typical multimedia services establishes the need for efficient data compression schemes in order to facilitate transmission and storage applications. Several international standards have been introduced for video compression targeting different application fields. Communication applications (e.g. videophone, teleteaching) are covered by the ITU-T standard H.263. The H.263 codec codes video frames using a discrete cosine transform (DCT) on blocks of size 8 × 8 pixels. An initial frame is coded and transmitted as an independent frame. Subsequent frames, which are modelled as changing slowly due to small motions of objects in a scene, are coded efficiently in the inter mode using motion compensation (MC) in which the displacement of groups of pixels from their position in the previous frame are transmitted together with the DCT-coded difference between the predicted and original images. Fig. 2 shows a block diagram of the codec.

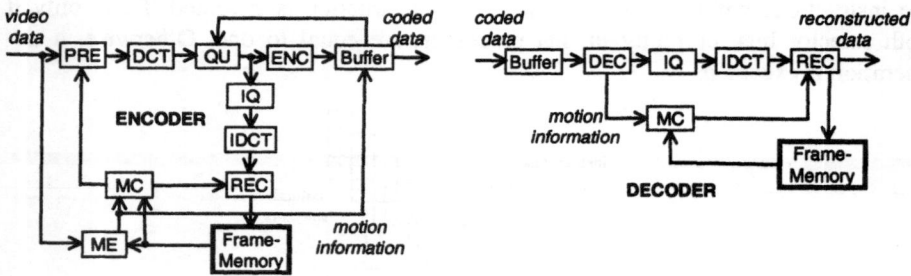

Fig. 2: Tasks and data flow of the H.263 encoder and decoder

The first step in the interframe coder is to calculate a motion vector for the current 16 × 16 pixels macroblock (**MB**) in ME (motion estimation). The motion vector is obtained by minimizing a cost function measuring the mismatch between a candidate MB in the previous frame and the current MB. Although several cost measures have been introduced [7], the most widely used one is the sum-of-absolute-differences (*SAD*) defined by $SAD = \Sigma_{k=0\ldots15}\Sigma_{l=0\ldots15} \, |c_{i,j}(k,l) - r_{i-u,j-v}(k,l)|$, where $c_{i,j}(k,l)$ represents the pixel (k,l) of a 16 × 16 MB from the current picture at the spatial location (i,j), and $r_{i-u,j-v}(k,l)$ represents the pixel (k,l) of a candidate MB from a reference picture at the spatial location (i,j) displaced by the vector (u,v). To find the MB producing the minimum mismatch error, we need to calculate SAD at several locations within a search window. The simplest, but the most compute-intensive search method, known as the full search or exhaustive search method, evaluates SAD at every possible pixel location in the search area. To lower the computational complexity, several algorithms that restrict the search area to a few have been proposed [3]. In baseline H.263, one motion vector per MB is allowed for motion compensation. Both, horizontal and vertical components of the motion vectors may be of half pixel accuracy, but their values may lie only in the ±15 range, limiting the search window used in ME.

The predicted MB represented by the calculated motion vector is loaded from the frame memory into MC (motion compensation). If the motion vector is of half pixel accuracy this operation requires an interpolation. The motion compensated prediction error is computed by the difference between the predicted and original MB in PRE (prediction). The resulting difference MB is transformed using a DCT of each 8×8 block, quantization by an adaptive quantizer (QU), entropy encoded using a variable-length coder (ENC), and buffered for a transmission over a fixed rate channel. The quantizer step size is calculated by evaluation of the buffer occupancy. After the quantization process, the original MB is reconstructed by the corresponding inverse operations (IQ, IDCT, REC) and stored in the frame memory. Of course, the reconstructed and original MB are not equal because of the performed lossy quantization. Since the reconstructed MB is available to encoder and decoder, it is used for prediction of the next frame.

The intra/inter mode selection is made at the MB level. If a MB does not change significantly with respect to the reference picture, an encoder can also choose not to encode it, and the decoder will simply repeat the MB located at the subject MB's spatial location. Only the processing instructions and capabilities of the decoder are standardized. The only necessary demand on the encoder is to produce a syntactically correct bitstream. The result is that the quality of H.263 video depends on the encoder implementation. In addition to the discussed baseline encoder algorithms, several negotiable options are offered by H.263 and H.263+. These optional modes allow developers to trade off between compression performance and computational complexity.

4   Accelerator Architecture for Video Compression

The analysis of the different video coding tasks and their processing efforts leads to a fixpoint processor architecture. The processor needs a small local memory for the storage and fast supply of local image data. The wordlength of the data items should be chosen to 8 bits because of the 8 bit input pixels and must also be able to process longer operands, e.g. DCT requires intermediate operands of length up to 24 bit. In order to achieve the real time requirements of the H.263 video codec several of these processors have to work in parallel. Thus, the particular processor has to be optimized with respect to its chip area and power consumption. Figure 3 depicts the processor architecture for the implementation of the H.263 video codec.

Due to the limited chip area the processor has to be very compact. In particular, it implies the choice of a bit-serial data organization. The size of the local memory is 32 internal registers plus 2 communication registers (C-registers). The word length of data items is 8 bits. Furthermore, each processor has two special constant registers 0 and −1. In addition to the registers, there are a zero flag and a negative flag that control the processing units depending on the state of the processor.

Two operand-multiplexers choose two registers of the own local memory or of the C-registers of the neighbours. The operands are propagated to the units required for execution of the current instruction. This units can be the multiplier and the arithmetic, logical, and conditional unit (ALCU). The corresponding instruction set consists of 24 instructions. The result of the execution is given to the result bus and from there they are written into the local registers or the C-registers or both.

During the execution of one instruction the processor receives the next instruction together with a column selector bit from the upper neighbour and a row selector bit from the left neighbour. The selectors are interpreted. If both are 1 then the execution of the instruction is prepared. If one of the selectors is 0 then the execution of a no-operation is initialized. Instruction and column selector are propagated bit serially to the lower neighbour and the row selector to the right neighbour during the execution of the instruction.

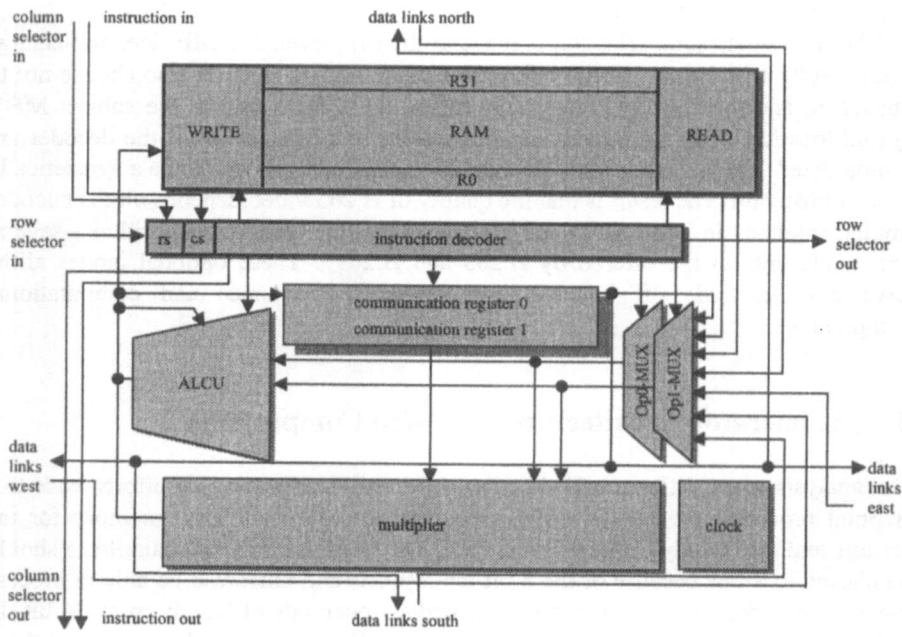

Fig. 3: Architecture of the array processor

The fine grained pipelined execution units are laid out for bitwise 300 MHz true single phase clock. Table 1 shows the areas of the different units of the processor. The full-custom design has been made with a 0.25μ digital CMOS process. An 12 × 8 ISA of the described processors has been structured in order to provide real-time processing for an H.263 codec at CIF resolution (352 × 288 pixels) and 15 fps. The resulting silicon area of the processor array is 3.7 mm². At a bitwise clock frequency of 300 MHz and a word format of 8 bits the theoretical peak performance of the array is 3.6 GIPS. The corresponding estimated power consumption is 120 mW. In order to exploit the computation capabilities of this unit, it is necessary to provide data and control information at an extremely high speed. Therefore, a cascaded memory concept is implemented on-chip that forms a fast input and output environment for the parallel processing unit (see Figure 4).

For the fast exchange of data with the processor array each processor has two memory banks. Each memory bank contains 8 interface registers. One of these banks is always assigned to the corresponding processor, the other to the internal RAM by means of a fast data channel. The exchange of data between ISA and the internal RAM is done by bank switching. Both memory banks can be active at the same time, i.e. data transfer between ISA and internal RAM can be done concurrently to the execution of an ISA program. The internal RAM can also communicate bidirectionally with the external SDRAM.

The data transfer is controlled by an on-chip controller. The controller is started by the sequential standard RISC core (e.g. Hitachi SH4) and operates autonomously afterwards. It receives instructions from an instruction queue. The controller supplies the processor array with instructions and selectors that are stored in the ISA program memory. The instruction queue and the ISA program memory are located in the external SRAM, separated from the data. This part of the SDRAM is exclusively available to the controller by means of a fast channel. The internal RAM is organized as single ported static RAM. Its size of 8 KByte is mainly determined by the H.263 implementation (see Section 5). The layout of the complete video accelerator with a 0.25µ digital CMOS process requires an area of 8.1 mm^2 as depicted in Table 2.

Table 1: Area of the individual processor.

RAM	12.144 µm^2
Multiplier	6.398 µm^2
Read-write-logic	2.874 µm^2
ALCU	5.125 µm^2
Instruction decoder	4.011 µm^2
Addressing	1.190 µm^2
Routing	1.674 µm^2
Operand multiplexers	2.133 µm^2
Clock	2.613 µm^2
Sum	**38.135 µ^2**

Table 2: Area of the accelerator.

ISA	3.7 mm^2
Controller	1.1 mm^2
Internal RAM	3.3 mm^2
Sum	**8.1 mm^2**

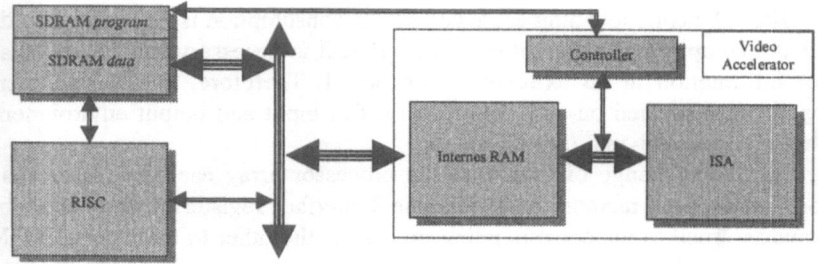

Fig. 4: Video accelerator architecture and data paths to the environment.

5   Parallelization of Video Compression Algorithms

The tasks of the hybrid coding scheme can be subdivided into low level, medium level, and high level tasks. Low level tasks typically operate on pixel data. They are highly regular and offer a large potential of data parallelism. Examples for low level tasks are full search motion estimation, DCT, and prediction. These tasks require a main percentage of the overall processing effort. Medium level tasks like quantization or entropy coding operate on the results of the low level tasks. Computation requirements are significantly lower than for low level tasks. The high level tasks comprise the bitstream handling and several control tasks, primary on encoder side, like quantizer control and decisions for coding strategy. High level tasks show an irregular control flow in conjunction with low computational rates. For control tasks a certain amount of flexibility with respect to modifications of the algorithms is advantageous. The proposed coprocessor architecture for multimedia comprise a flexible general-purpose RISC core with a less flexible accelerator adapted towards a specific type of processing. Thus, the RISC core performs the high level tasks of lower computational requirements, whereas the video accelerator executes the computation intensive but regular low level tasks and a part of the medium level tasks. The parallelization of the H.263 coding tasks on the proposed accelerator architecture is described in the following.

ME: In the 2D array structure of a 4×8 ISA the calculation of one SAD can be efficiently parallelized as follows: Assume, the current MB data $c(k,l)$ and the reference frame candidate MB data $r(k,l)$, $k,l = 0,\ldots,15$, are stored in the processor array such that processor (i,j), $i = 0,\ldots,3$, $j = 0,\ldots,7$, holds $c(4 \cdot i,2 \cdot j),\ldots,c(4 \cdot i+3,2 \cdot j+1)$ and $r(4 \cdot i,2 \cdot j),\ldots,r(4 \cdot i+3,2 \cdot j+1)$ in 16 internal registers. Now each node (i,j) computes the partial sum $\Sigma_{m=0,\ldots,3,n=0,\ldots,1}|c(4 \cdot i+m,2 \cdot j+n) - r(4 \cdot i+m,2 \cdot j+n)|$ within 30 instructions. The result is written into the two C-registers $C1$ (most significant byte) and $C0$ (least significant byte). The partial sums are added along each processor row within 2 instructions using two efficient row sum operations (see Section 2). The first operation adds up the values of $C0$ and the second operation adds up the values of $C1$ and the carry-bit of the previous addition. The total sum (SAD) for the candidate MB is completed by the corresponding two column sum operations in the rightmost processor column and the result is written into *C0* and *C1* of the lower right processor. This SAD is compared in the lower right processor to the stored SAD resulting from the previous comparison. The smaller one as well as the corresponding motion vector is kept for future comparisons.

The preceding computation procedure is repeated until all possible candidate MBs are compared and the final motion vector is obtained. To speed up the computation the calculated SAD value is compared to a preselected threshold value. The search process terminates if the calculated SAD value is smaller than the threshold. Because the processor array of the video accelerator is of size 12×8 three different SAD values for the current MB can be calculated in parallel by a simple replication of the program for the 4×8 array.

Shifting of the candidate MB within the search area depends on the chosen search strategy. In order to avoid delay times in the systolic architecture it is necessary to know these locations in advance, and thus the next candidate MB data can be preloaded from the internal RAM into the ISA during the processing of the current MB. The ±15 pixels 2-step search with subsequent half pixel refinement is suitable and leads to a very good coding quality [3]. In the first step, it evaluates the motion vector candidates with distance two between two nearby search points within the ±15 pixels search area. Thus, 225 SAD values of previously known locations have to be calculated in the worst case. In the second step, the SAD values of the 8 neighbours of the calculated minimum point are evaluated. A subsequent half pixel search around the full pixel accurate displacement vector evaluates the motion vector with half-pixel accuracy. This involves the additional evaluation of 8 candidate blocks per motion vector. The locations of candidate points in the second iteration step and the half pixel search are data dependent. Thus, the data locations to be used at the next step are transferred to the controller when the current step is completed. At the beginning of the new step, the controller loads the corresponding reference data into the ISA. MBs of adjacent candidate vectors overlap quite significantly. In order to reduce memory accesses (bandwidth) only the non-overlapping data have to be loaded into the processor array. Since SAD values for 3 horizontally neighboured motion vectors are computed in the processor array in parallel only 2×20 pixels have to be loaded when the search area moves vertically up in the first step of the 2-step search. Search areas of adjacent MBs also overlap. Thus, only the non-overlapped data of size 48×16 pixels is loaded from the external SDRAM to the internal RAM, which can be performed concurrently to the motion vector computation of the current MB.

MC: The complete reference frame MB corresponding to the calculated motion vector has to be loaded into the ISA. In contrast to the other processing tasks ME is only performed on the luminance data of the MB. For MC and the preceding tasks another mapping of MB data onto the ISA is used. Each processor holds a 2×2 subarray. Thus, each of the six 8×8 blocks of the MB is mapped onto a 4×4 processor subarray of the 12×8 ISA. If the motion vector is of half pixel accuracy linear interpolation is processed in the processor array.

PRE/REC: The motion compensated MB and the original MB are simply subtracted (prediction) resp. added (reconstructed). Only 4 subtraction resp. additions have to be performed within each processor in parallel.

DCT/IDCT: The DCT of an 8×8 block $x_{i,j}$, $i,j = 0,\ldots,7$, can be expressed as

$$y_{k,l} = \sum_{i=0}^{7} c_{i,k} \left[\sum_{j=0}^{7} x_{i,j} \cdot c_{j,l} \right], \text{ where } c_{j,l} = \cos\left((2j+1)l\pi / 16\right) \tag{1}$$

The core operation in DCT computation is the multiply accumulation (MAC).

$$y_{2l} = \sum_{j=0}^{3} \left(x_j + x_{7-j}\right) \cdot c_{j,2l}; \; y_{2l+1} = \sum_{j=0}^{3} \left(x_j - x_{7-j}\right) \cdot c_{j,2l+1} \qquad \text{for } l = 0,\ldots,3 \tag{2}$$

Therefore, the multiplier of the introduced ISA processor has been designed to perform MAC efficiently, i.e. simultaneously with multiplication, a third operand can be added to the result without additional delay. As already indicated by the brackets in (1), the 2D transform can be decomposed into cascade of two subsequent 1D transforms applied horizontally and vertically on each row, resp. column. Additional efficiency can be gained, if we split the 1D transform of a sequence x_j, $j = 0,\ldots,7$, into even and odd numbered frequency samples:

This algorithm is applied for the 8×8 DCT within each 4×4 processor subarray. The input values for the 1D DCT are initially permuted within each processor row of length 4, such that each processor j, $0 \leq j \leq 3$, holds x_j and x_{7-j} and the addition $x_j + x_{7-j}$ and the subtraction $x_j - x_{7-j}$ can be performed in parallel. Now each processor j, $0 \leq j \leq 3$, computes y_{2j} and y_{2j+1} according to (2) in 4 stages. The computation in each stage requires two MAC operations per processor. Permutation of the values $x_j + x_{7-j}$ and $x_j - x_{7-j}$ between every two stages involves routing of data, which is performed efficiently within each processor row by ringshifting. The coefficient values are precomputed and loaded into the processor array when needed. This load operation takes only a small number of instruction cycles since the values can be broadcasted along the columns of the ISA. The 1D DCT along the processor columns is computed by the *reflected ISA program* for the rows. The IDCT is implemented in a similar way.

QU/IQ: The division operation for quantization is evaluated by multiplication with the reciprocal value in each processor in parallel. The quantizer step size is controlled by loading the corresponding coefficients into the processor array.

ENC/DEC: Due its irregularity the variable length en/decoding and entropy en/decoding is not suitable for an efficient parallelization on the ISA. Thus, the sequential RISC performs these operations. Because of their low computational requirements, their runtime on the RISC is dominated by the processing time of the next/previous MB on the video accelerator.

6   Performance Evaluation

For the runtime determination of the parallel programs described in Section 5 on the introduced accelerator architecture the number of required instructions is multiplied with the corresponding clock cycle time of 26.7 ns. Additionally, the data transfer between ISA and internal RAM is considered with a throughput of 150 MByte/s. The results are shown Table 3. Note that the times for ME are worst cases, i.e. calculated SAD values are never smaller than the preselected threshold.

Table 3: Worst case runtime on the video accelerator for encoding and decoding per MB. It includes computing time on the ISA and data transfer time between ISA and internal RAM. Data transfer is divided into concurrent transfer, i.e. the transfer time is dominated by the computing time on the ISA, and not concurrent transfer, i.e. during the transfer time no computations are performed on the ISA.

Task		Runtime video accelerator	Data transfer ISA ↔ internal RAM	
			Concurrent to ISA	Not concurrent to ISA
En-coder	ME step 1	103 µs	Load 4080 Bytes	Load 640 Bytes
	step 2	6 µs	Load 36 Bytes	Load 288 Bytes
	half pixel	8.5 µs	Load 18 Bytes	Load 17 Bytes
	MC	5 µs	Load 128 Bytes	Load 451 Bytes
	PRE	0.5 µs		
	DCT and QU	10 µs		
	IQ and IDCT	10 µs	Store 384 Bytes	
	REC	3 µs		Store 384 Bytes
De-coder	IQ and IDCT	10 µs	Load 451 Bytes	Load 384 Bytes
	MC	2 µs		
	REC	3 µs		Store 384 Bytes
Sum per MB		161 µs	5097 Bytes	2584 Bytes
Sum per CIF		64 ms	2.02 Mbytes	1.0 MBytes

The amount of data transfer between external SDRAM and internal RAM per MB is

Encoder: current MB (384 Bytes), overlapping search area (768 Bytes), chrominance blocks of motion compensated MB incl. border pixels (162 Bytes), coded MB (384 Bytes), reconstructed MB (384 Bytes).

Decoder: current MB (384 Bytes), motion compensated MB incl. border pixels (451 Bytes), reconstructed MB (384 Bytes).

The resulting total amount of data transfer of 3301 Bytes per MB, resp. 1.3 MBytes per CIF frame, can be transferred concurrently to the accelerator activities with a reasonable throughput of the SDRAM bus, e.g. 100 MByte/s.

7   Conclusions

In this paper we have presented an ISA architecture for algorithms required for video compression. The accelerator unit has been implemented on an area of 8.1 mm^2 of silicon using a 0.25µ digital CMOS process. It is capable of on-line encoding and decoding of a video stream of 15 fps in CIF format with the H.263. The global architecture of the accelerator unit has been discussed as well as the detailed implementation of the single processing element of the 12 × 8 array. It has been shown how the new architecture can be programmed for applications as motion estimation, motion compensation, DCT, and quantization. Apart from its performance figures, the most promising property of the accelerator unit is its flexibility. It would be interesting to study the performance for the new architecture for applications like [8], [9], [10].

References

1. Video Coding for low bitrate communication, ITU-T Draft Recommendation H.263, Geneva, May (1996)
2. Chang, Y.-T., Wang, C.-L.: New Systolic Array Implementation of the 2-D Discrete Cosine Transform and Its Inverse, IEEE Trans. Circ. Syst. Vid. Tech. 5 (2) (1995) 150-157
3. Cheng, S.-C., Hang, H.-M.: A Comparison of Block-Matching Algorithms Mapped to Systolic-Array Implementation, IEEE Trans. Circ. Syst. Video Tech. 7 (5) (1997) 741-757
4. Kunde, M., et al.: The Instruction Systolic Array and its Relation to Other Models of Parallel Computers, *Parallel Computing* 7 (1988) 25-39
5. Lang, H.-W.: The Instruction Systolic Array, a parallel architecture for VLSI, *Integration, the VLSI Journal* 4 (1986) 65-74
6. Lang, H.-W., Maaß, R., Schimmler, M.: The Instruction Systolic Array - Implementation of a Low-Cost Parallel Architecture as Add-On Board for Personal Computers, *Proc. HPCN 94*, LNCS 797, Springer Verlag (1994) 487-488.
7. Pirsch, P., Stolberg, H.-J.: VLSI Implementations of Image and Video Multimedia Processing Systems, *IEEE Trans. Circ. Syst. Video Tech.*, 8 (7) (1998) 878-891
8. Schimmler, M., Lang, H.-W.: The Instruction Systolic Array in Image Processing Applications, Proc. *Europto 96*, SPIE 2784 (1996) 136-144
9. Schmidt, B., Schimmler, M., Schröder, H.: Long Operand Arithmetic on Instruction Systolic Computer Architectures and Its Application to RSA cryptography, *Proc. Euro-Par'98*, LNCS 1470, Springer Verlag (1998) 916-922
10. Schmidt, B., Schimmler, M., Schröder, H.: The Instruction Systolic Array in Tomographic Image Reconstruction Applications, Proc. *PART'98*, Springer Verlag (1998) 343-354

A Parallel Architecture for Stereoscopic Processing

Milton Romero and Bruno Ciciani

QSW - Quadrics Supercomputer World Ltd. miltonr@roma.quadrics.com
Department of Computer and System Engineering
University of Rome "La Sapienza", Via Salaria 113, 00198, Rome, Italy
miltonr,ciciani@dis.uniroma1.it

Abstract. An embedded pipeline/parallel architecture to support an extended quad-tree algorithm suitable for real-time estimation of the dense disparity map (DDM) for stereoscopic image processing is proposed. The system performance has been analyzed by several simulations to qualify the results by both an objective measurement (Mean Square Error) and a subjective assessment (output images). The proposed extended quad-tree is based on the block-matching algorithm, then a fine-grain granularity analysis to estimate the DDM leads us to a systolic array design for the basic Processor Element. This basic design has been utilized to the next levels quad-tree's Processor Elements design.

1 Introduction

The main task in creating a stereoscopic image is to reproduce two projections of the solid object obtained from two centers of perspectives lying in a horizontal line normal to the direction of vision. The two pictures projected on to planes from these two centers of perspective constitute a normal stereogram whose images differ in the horizontal parallax of corresponding points (Fig. 1) [1]. The stereoscopic vision parallax of a generic real point P is reconstructed by taking into account the horizontal displacement (disparity) between the two projections (correspondent points) from the real point P on to the left and right images (Fig. 1 right). The association of a disparity value with each pixel in one image of the stereo-pair defines a dense disparity map (DDM), otherwise, is called sparse. The disparity takes its values depending on the position of the shot point relative to the focal plane of the shooting system providing a depth information.

The stereoscopic processing has been found of much interest in many areas of applications such as: robot vision, multimedia, codecs 3DTV, etc. [2], [3], [4], [5]. A fundamental problem, common to these applications when tele-presence and an accurate depth information are involved, is to find the DDM estimation.

An estimation of the best correspondent points candidates can be obtained by the well known full search block matching algorithm (FBMA) based on the optimization of a cost function [6]. Block matching correlates each image block of a target image to a neighboring block of a reference image. When this technique

P. Amestoy et al. (Eds.): Euro-Par'99, LNCS 1685, pp. 961–968, 1999.
© Springer-Verlag Berlin Heidelberg 1999

Fig. 1. Stereo-pair "Tunnel" (courtesy of C.C.E.T.T. - F): left image (left), right image (middle). Horizontal disparity: $d_h = y_1 - x_1$; vertical disparity: $d_v = y_2 - x_2$ (right)

is applied to one image at time t and one image at time t+1 in the video sequence, the output map is called motion map.

Full resolution images (CCIR Rec. 601) are let to take the dimension of 720 pels x 576 lines/pel at 25-Hz frame rate according to the European standard digital video scan format. In order to meet the requirements due to a stereoscopic parallax [2], the FBMA for real-time processing of the DDM estimation requires a sustained computational performance of hundreds of giga operations per sec at full resolution. Thereby, special purpose hardware is broadly accepted as the most suitable approach. Therefore, the algorithm is here designed such that its computational complexity will be invariant with respect to the image content. That is, to have a fixed and few steps of computations and a regular data flow to be attractive for systolic array implementation. This design will be performed by following the same methodology utilized in [7] and described in detail in [8].

Real-time DDM estimation hardware for stereoscopic applications is presented in [9]. In [10] a fault tolerant extended quad-tree architecture for the DDM estimation is analyzed. Real-time block matching systolic architectures for motion estimation are presented in [7], [11].

In this paper a pipeline/parallel embedded architecture to support an extended quad-tree for the DDM estimation algorithm for stereoscopic signal processing is proposed. The system performance has been analyzed by several simulations to qualify the results by both an objective measurement (Mean Square Error - MSE) and a subjective assessment (output images).

This paper is organized as follows: in Section 2 the extended quad-tree and the systolic array design are presented. In Section 3 the results discussion is addressed and finally, in Section 4, the conclusions are presented.

2   Quad-Tree Algorithm

As well known, a quad-tree approach to the DDM estimation of stereo image pairs acts at different image resolution levels [10], [12]. The basic structure of the algorithm is reported in Fig. 2 left, taking in input one image stereo-pair (left and right images) and producing its DDM by two basic functional blocks: a bank of low-pass separable anti-alias Finite Impulse Response filters (LPF_k), and a cross-correlation processor (PRE-PROC, CORR_i) array. The coefficients of the basic filtering function -31 taps symmetrical- impulse response h(n) are:

$h0=0.63528158$; $h1=-0.20802693$; $h2=0.12030548$; $h3 = -0.81213146E - 1$; $h4=0.58646997E - 1$; $h5=-0.43450143E - 1$; $h8 = 0.17655253E - 1$; $h9=-0.1270347E - 1$; $h10= 0.87930853E - 2$; $h11 = -0.58282715E - 2$; $h12 =0.37617773E - 2$; $h13 =-0.21157524E - 2$; $h14= 0.10766329E - 2$; $h15 = -0.49893671E - 3$.

The filters stages act iteratively by convolutions with h(n) and a sub-sampling device $(L/k^2, R/k^2)$, in such a way that a further band reduction and decimation by 2 is introduced at each stage. The pre-processor (PRE-PROC) module is fed by two iconic representations (45x36 pixels) of the original images, thus producing the most-coarse DM_0 estimation (also 45x36 pixels) for the benefit of the next level (CORR_0).

$$MAD(m,n) = \sum_i \sum_j \parallel (x(i,j) - y(i+m,j+n)) \parallel \qquad (1)$$

The pre-processor performs the FBMA by optimizing the Means of Absolute Differences -MAD- cost function (Eq. 1), in which the indexes are let to take all values in the following ranges: search vector indexes m=± S_h and n=± S_v that control the search order performed sequentially in a column first order. Pixel coordinates indexes i=± W and j=± W that make each pixel within each block available sequentially pixel by pixel in a column first order. The blocks to be matched by the FBMA (square of 2*W+1 pixels) are indexed by h=0,mc and v=0,mr and will be loaded column by column sequentially [7]. mc, mr denote the number of columns and rows, and y, x stand for the pixels in the target icon and source icon, respectively. According to experimentation outcomes, a reasonable trade-off between disparity vector accuracy and computational load has been obtained with W=7. Such a big support let us to increase the reliability of the first estimation. The search area range is: n=±1 and m=±4 pixels that corresponds to ±16 x ±64 pixels at the highest resolution level.

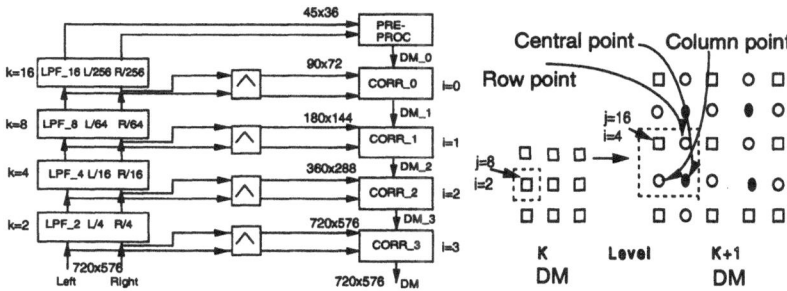

Fig. 2. Extended quad-tree algorithm (left). Quad-tree expansion (right)

In Fig. 2 right, the expansion of each pixel into a square block of 2x2 pixels is outlined; e.g., the disparity vector (DM) at position $j = 8$, $i = 2$, level=k will be put into the position $j = 16$, $i = 4$, level=**k+1** as twice of its value. These

pixels (represented as squares in Fig. 2 right) are further updated by applying the
FBMA with a search area range of $S_v = S_h = 1$ (Fig. 3 left - top). To estimate the
central pixel (represented as grey circles in Fig. 2 right), five points are selected
in the left image for the MAD operator, these are the four pixels pointed to
by the four disparity vectors originating at the four pixels surrounding the one
being considered, plus a fifth point which is evaluated as the barycenter of them
(Fig. 3 left - bottom). Analogously, to estimate the intermediate *rows/columns*
pixels (represented as light circles in Fig. 2 right), three points are selected on
the target image for the MAD operator, the two row/column neighbors and
the barycenter of them. After this five level update approach (PRE-PROC plus
$CORR\_i$) a full resolution DDM estimation is derived.

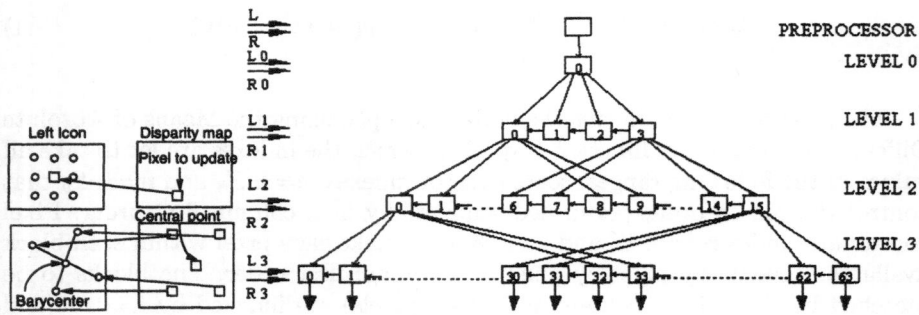

Fig. 3. Update of heired pixels -top- and the production of centered pixels
-bottom- (left). Pyramidal computing structure (right)

Due to the increasingly linear complexity with the level depth, the correlation processor is implemented by a regular replica in space of a Processing Element module -$PE_{i,j}$- charged with the basic functions identified above (where
i stands for the level and j denotes the number of the PE). Fig. 3 right shows
the pyramidal structure which builds on five levels: the first one represents the
pre-processor, which is appended to a four-level quad-tree, i.e., connecting its
root (level zero) to four nodes on level 1, also having sixteen nodes on level 2
and sixty-four nodes on level 3. The $PE_{i,j}$ has in charge the processing of all the
vectors located on columns h of the DM_i (see also Fig. 2 left) such that: h **mod**
$4^i = j$. Note that to produce new vectors each PE needs to know the vectors
updated by its right hand side neighbor only.

2.1 Preprocessor Systolic Array Design

In Section 2 the six-level FBMA based on the MAD cost function was outlined.
This algorithm is transformed to a three level algorithm by using b, l, k indexes,
where, in the horizontal dimension k we control the pixels belonging to each
block; in the l dimension the search area; in the vertical dimension b, the relationship between blocks and i_s, j_s denote the absolute coordinates (Eq. 2).

$$(2)$$

$$i_s = h + i; \quad j_s = v + j; \quad h = b \bmod mc; \quad b = v.mc + h, \quad 0 \leq b < N_{hv} = mc.mr$$
$$v = \lfloor \tfrac{b}{mc} \rfloor; \qquad l = (2S_v + 1)(m + S_h) + n + S_v, \quad 0 \leq l \leq (2S_v + 1)(2S_h) + 2S_v$$
$$m = \lfloor \tfrac{l}{(2S_v+1)} \rfloor - S_h; \qquad n = l \bmod (2S_v + 1) - S_v; \qquad i = \lfloor \tfrac{k}{N} \rfloor - \lfloor \tfrac{N}{2} \rfloor$$
$$j = k \bmod N - \lfloor \tfrac{N}{2} \rfloor; \qquad k = N(i + \lfloor \tfrac{N}{2} \rfloor) + (j + \lfloor \tfrac{N}{2} \rfloor), \qquad 0 \leq k < N^2$$
$$y(b, l, k) = y_s(l + k) = y(i_u, j_u); \qquad x(b, l, k) = x_s(k) = x(i_s, j_s)$$

$$(3)$$

$$i_s(b, l, k) = h + i = b \bmod mc + \lfloor \tfrac{k}{N} \rfloor - \lfloor \tfrac{N}{2} \rfloor; \quad j_s(b, l, k) = v + j = \lfloor \tfrac{b}{mc} \rfloor + k \bmod N - \lfloor \tfrac{N}{2} \rfloor$$
$$i_u(b, l, k) = i_s + m = h + i + m = b \bmod mc + \lfloor \tfrac{k}{N} \rfloor - \lfloor \tfrac{N}{2} \rfloor + \lfloor \tfrac{l}{(2S_v+1)} \rfloor - S_h$$
$$j_u(b, l, k) = j_s + n = v + j + n = \lfloor \tfrac{b}{mc} \rfloor + k \bmod N - \lfloor \tfrac{N}{2} \rfloor + l \bmod (2S_v + 1) - S_v$$

The partially localized FBMA in $[b, l, k]$ space, which means local communication and defines the dependence vectors, is shown next [8]. Note that by using Eq. 3, all these relations can be proved (not shown here).

```
for b = 0; b < N_hv /* N_hv = mc.mr */
  MV(b, 0, N²) = 0 /* disparity vector initialization */
  Dmin(b, 0, N²) = ∞ /* auxiliary initialization */
    for l = 0; l ≤ (2S_v + 1)(2S_h) + 2S_v /* search area l ε[0, 26] for W=7 */
    MAD(b,l,0) = 0 /* MAD initialization */
    for k = 0; k < N² /* window area N = 2 * W + 1, W = 7 */
      if 0 ≤ k < N(N − 1) && 0 ≤ l ≤ (2S_v + 1)2S_h + 2S_v && b mod mc ≠ 0
      x(b, l, k) = x(b − 1, l + (2S_v + 1), k + N)
      if 0 ≤ k < N(N − 1) && 0 < l ≤ (2S_v + 1)2S_h + 2S_v && b mod mc ≠ 0
      x(b, l, k) = x(b, l − 1, k)
      if 0 ≤ k mod N < N − 1 && l mod (2S_v + 1) ≠ 0
      y(b, l, k) = y(b, l − 1, k + 1)
      if 0 ≤ k < N(N − 1) && (2S_v + 1) ≤ l ≤ (2S_v + 1)2S_h + 2S_v
      y(b, l, k) = y(b, l − (2S_v + 1), k + N)
      if 0 ≤ k < N(N − 1) && 0 ≤ l ≤ (2S_v + 1)2S_h + 2S_v && b mod mc ≠ 0
      y(b, l, k) = y(b − 1, l + (2S_v + 1), k + N)
      MAD(b, l, k + 1) = MAD(b, l, k) + ‖ y(b, l, k) − x(b, k) ‖
    endk
    if Dmin(b, l, N²) > MAD(b, l, N²) /* MAD minimization */
    Dmin(b, l + 1, N²) = MAD(b, l, N²)
    MV(b, l, k) = l /* estimated disparity vector */
    endif
  endl
endb
```

It can be possible to represent the localized FBMA as a dependence graph (DG) in (b, l, k) space, not shown here, which must be further mapped (arcs, I/O, nodes) onto a few dimensions. Two steps are involved in mapping a DG to a Signal Flow Graph (SFG) array. The first step is processor assignment

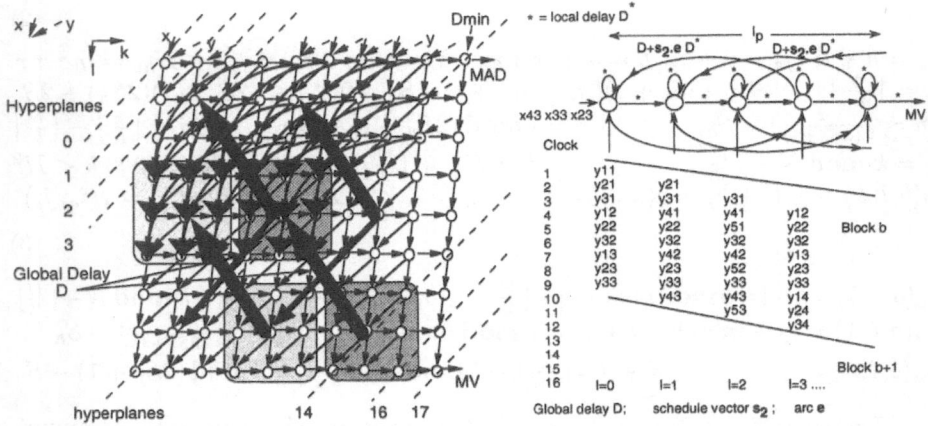

Fig. 4. 2-D SFG for the projection vector: $d_1^T = [1\ 0\ 0\ ]$ and the default scheduling vector: $s_1^T = [1\ 0\ 0\ ]$ and depiction of the 2-dimensional SFG projection in the k direction (left). One-dimensional DG in b,l,k space for $d_2^T = [0\ 1]$ and $s_2^T = [1\ 1]$ (right)

(definition of the number of processor) and the second is the scheduling in which nodes in parallel equitemporal hyperplanes in the DG, are scheduled at the same time step to be processed [8]. Therefore, the first projection vector is $d_1^T = [1\ 0\ 0]$ and the default scheduling vector is $s_1^T = [1\ 0\ 0]$ in $[b, l, k]$ space. This yields a 2-D SFG with the initial global delay D shown in Fig. 4 left. This delay acts on the four bigger arrows, that means that each node is connected with its homologous node. The second projection vector is $d_2^T = [0\ 1]$ (on the k axis) and the systolic scheduling vector is $s_2^T = [1\ 1]$ with a new local delay D^*, where $D = N^2 D^*$. In the Fig. 4 left, the new hyper-planes (0-17 as for example) are also shown. This second projection yields the 1-D SFG of Fig. 4 right. The l_p value is exactly equal to the search area, that is, from 0 to 26 (27 PEs); the asterisk denotes the local delays (D^*) and the arrows without any delay permit the propagation of the data. To overcome the cycles [8], the new delay is: $D + s_2.eD^*$ (Fig. 4 right). Note, that e stands for all (globally delayed) edges in the original SFG.

2.2   Quad-Tree Next Levels PEs Design

The next levels PEs design of the proposed architecture is a particular and simpler case of the design discussed in last subsection (Fig. 4 right). To proceed from one level to the next in the quad-tree, the resolution expansion is simply to multiply the disparity vector and its position by two, and the application of the FBMA with $S_v = 1$ and $S_h = 1$ to update the projected pixels. For the central points (row/column) estimation, the blocks of pixels on the target image do not have any common data among them (in the general case), thereby all $y(b, l, k)$ are independently treated. Therefore by using the same projection and schedule vectors as the pre-processing design, it will be found a similar SFG as Fig. 4

Fig. 5. Disparity map of Fig. 1 (left). Reconstructed right image (right)

right (with $l_p = 5$ for the central points, and $l_p = 3$ for the row/column points) in which the (Y) pixel for each $l=0,1 .., l_p$ would stand for different pixels.

As the reference and target icon are previously grouped in the scan order to be stored into two buffers, by determining the address of the x values:

$$A_X(b, l, k) = \lfloor \tfrac{b}{mc} \rfloor mc + (k \bmod N).mc - \lfloor \tfrac{N}{2} \rfloor mc + b \bmod mc + \lfloor \tfrac{k}{N} \rfloor - \lfloor \tfrac{N}{2} \rfloor$$

and applying a simple shift to this address, the y address is derived:

$$A_Y(b, l, k) = \lfloor \tfrac{b}{mc} \rfloor mc + (k \bmod N).mc - \lfloor \tfrac{N}{2} \rfloor mc + (l \bmod (2S_v + 1))mc$$
$$-S_v.mc + b \bmod mc + \lfloor \tfrac{k}{N} \rfloor - \lfloor \tfrac{N}{2} \rfloor + \lfloor \tfrac{l}{(2S_v+1)} \rfloor - S_h$$

In a similar way, the addresses for the level expansion can be generated.

3   Results and Discussion

The system performance has been analyzed by several simulations to qualify the results by both an objective measurement Mean Square Error (MSE) and a subjective assessment (output images). For illustration purposes, Fig. 5 represents the reconstructed right image by utilizing the DDM (right), as well as, the estimated DDM where objects near to the observer appear brighter (left). The subjective good quality of the reconstructed image is matched by the objective measurement of the MSE, evaluated with reference to the original right image. The MSE equal to 220 is still within an acceptable range, especially considering that most error energy is concentrated on the image left and right borders (due to the occlusions generated by the parallax of the pick-up geometry).

In Fig. 4 right, it can be seen that the pin count is equal to: 2 inputs (in the first PE), plus one input for each of the rest PEs, plus the output (disparity vector MV) with 8 bits each. This sum is equal to $(l_p+2)8 = 232$. In the same way, in the case of the projected pixel update, the pin count is 88. In the case of the central point estimation, the pin count is equal to 56; and for the other case, the pin count is equal to 40. It is possible to perform the update of central, row and column pixels in parallel and this implies a pin count of at most 136, depending on how the three basic systolic PEs array are connected. In the 2-D SFG, one can see that the latency is at most equal to $l_p + N^2$ clock cycles for each processed layer, as it is showed by the hyperplanes (Fig. 4 left). Each MAD value is provided after N*N clocks. Each PE is composed by one comparator to compute the maximal of Dmin and MAD; an adder to compute the difference between x and y values; and another adder to accumulate the partial sums along with the proper communication paths and the suitable control.

4   Conclusion

An embedded pipeline/parallel architecture to support an extended quad-tree algorithm for the estimation of the dense disparity map (DDM), that is attractive for real time stereoscopic image processing has been presented in this paper. The suitable quality results (objective measurement Mean Square Error -MSE- and a visual assessment) of the estimated map have shown a convenient depth information accuracy, adequate to applications when this characteristic is important, as vision, selective coding, region segmentation, among others.

High-speed hardware design guidelines to the computation of the DDM estimation in stereoscopic sequences based on a systolic array have been also presented. The feasibility of the design of the pre-processor module to cope with the critical step of the algorithm for real-time processing of the DDM estimation has been proved. It must be said that further analysis has to be made in order to optimize the design and to reduce hardware complexity and pin count. As in the parallel estimation of central, row and column points there are common data among neighbors, this characteristic can be exploited to improve the design.

References

[1] N.A.Valyus, "Stereoscopy", The Focal Press, London and New York, 1966.
[2] ISO/IEC/JTC1/SC29 Doc. MPEG 93/254, "Stereoscopic Video Transmission: A Future Application of Spatial Embedded Coding", March 1993.
[3] Horn B.K.P., "Robot Vision", Mc Graw-Hill, 1996.
[4] Chiari A., Ciciani B., Romero M., Rossi R. "Depth Controlled $3D - TV$ Image Coding", Proc. of the SPIE $IS\&T'98$, San Jose (CA), Jan. 1998.
[5] D. Tzovaras, N. Grammalidis and M. G. Strintzis "3-D Camera Motion Estimation and Foreground/Background Separation for Stereoscopic Image Sequences", Optical Engineering, Vol. 36, No. 2, Feb. 1997.
[6] Hans Georg Musmann, Peter Pirsch and Hans-Joachim Grallert, "Advances in Picture Coding", Proceeding of the IEEE, Vol. 73, No. 4, April 1985.
[7] Hang Yeo, Yu Hen Hu, "A Novel Modular Systolic Array Architecture for the Full-Search Block-Matching Motion Estimation", IEEE Transactions on Circuits and Systems for Video Technology, Vol. 5, pp. 407-416, Oct. 1995.
[8] S. Y. Kung, "VLSI Array Processors", Englewood Cliffs, Prentice Hall, Oct. 1988.
[9] http://www.tnt.uni-hannover.de. The ACTS project AC093 PANORAMA.
[10] Chiari A., Ciciani B., Romero M., "A Fault-Robust SPMD Architecture for 3D-TV Image Processing", 2nd IEEE Workshop on Fault Tolerance and Distributed Systems, Geneve, 1997.
[11] Yeu-Shen Jehng, Liang-Gee Chen, Tzi-Dar Chieu, "An Efficient and Simple VLSI Architecture for Motion Estimation Algorithms", Transactions on Signal Processing, Vol. 41, No. 2, Feb. 1993.
[12] P. J. Burt, "The Pyramid as a Structure for Efficient Computation", Multiresolution Image Processing and Analysis, NY Springer-Verlag, 1984.

A Robust Neural Network Based Object Recognition System and Its SIMD Implementation

Alfredo Petrosino[1] and Giuseppe Salvi[2]

[1] INFM - University of Salerno
Via S. Allende, 84081 Baronissi (Salerno), ITALY
alfredo@synapse.irsip.na.cnr.it
[2] Istituto per la Ricerca sui Sistemi Informatici Paralleli, IRSIP-CNR
Via P. Castellino 111, 80131 Naples, ITALY

Abstract. Recognition of objects is a particularly demanding problem, if one considers that each image must be interpreted in milliseconds (usually 30 or 40 frames/second). In this paper we propose a massively parallel object recognition system, which makes use of the multi polygonal approximation scheme for the extraction of rotation and translation invariant shape features, in connection with artificial neural networks for the parallel classification of the extracted features. The system has been successfully applied for recognizing aircraft shapes in different sizes, orientations, with the addition of noise distortion and occlusion. Timings on the Connection Machine 200 are also reported.

1 Introduction

Recognition of shapes is a fundamental task in computer vision [14]. In the last decades, successful attempts have been made for the recognition of isolated objects with complete and well-defined boundaries. The usual process consists in the boundary extraction by applying a segmentation technique to the original grey scale image and then applying edge detectors and thinning algorithms to the binary image to retain the boundary width as small as possible. Once the boundary has been established the feature extraction task begins with extracting relevant features from each object present in the scene. The features are chosen so as to be invariant with respect to the object position, size and orientation.

In addition, many applications, including automatic target recognition (ATR), characterization of biomedical images, identification of industrial parts for product assembly, etc. require the recognition to take place in real-time and also in strongly noisy and occlusion conditions. This problem is very interesting if we consider a sequence of images with a resolution of (512×512) pixels at colours (8-24 bits for pixel) which are to be handled at a rate of 30 frames/second. The processing and the recognition of the contents of such images can be compared to the processing of 23.6 million of bytes/second. Thus, the use of more and more sophisticated parallel architectures and procedures of computation is

P. Amestoy et al. (Eds.): Euro-Par'99, LNCS 1685, pp. 969–976, 1999.
© Springer-Verlag Berlin Heidelberg 1999

natural and strongly needed, capable to deal with different data structures at the different levels of processing and interpretation. However, the use of parallel architectures gives rise to new problems concerning the choice of the modality of parallel computation (SIMD, MIMD) and the most efficient programming language. Current literature on parallel object recongition essentially deal with graph matching approaches implemented in parallel [1] on tree search algorithms [5], geometric hashing [2, 16] and parallel hypothesis generation [15]. Here we show that a complete set of computer vision tasks implementing filtering, feature extraction and statistic classification can be efficientely implemented as a whole system on a massively parallel SIMD architecture.

There are many techniques available to describe an object based on their boundaries (see for instance [4]). Among others, the polygonal approximation (specially that of high order) [10] combined with the Circular AutoRegressive (CAR) model approach, formerly proposed by Kashyap and Chellappa in 1981 [8], represents a good solution to handle the problems of occlusion, distorsion and noise, while remaining a low computational cost method. We shall use a modified version of the CAR model in combination with Artificial Neural Networks (ANNs) as nonparametric classifiers. The modifications introduced in the CAR model are based on the consideration of one set of predictive parameters is not sufficient to describe a shape, the model has to take into account more than one set of parameters, each corresponding to a different polygonal approximation. We shall refer to it as the Multi-Polygonal AutoRegressive Model (MPARM). In addition, Artificial Neural Networks (ANNs) are adopted due to their ability to adjust when given new information, neuronal massive parallelism typical of the SIMD parallel mode, fault tolerance to missing, confusing and/or noisy data. The paper is organized as follows. Section 2 briefly decribes the proposed approach whereas the third section reports an ensemble of experimental recognition results and times taken on the Connection Machine 200.

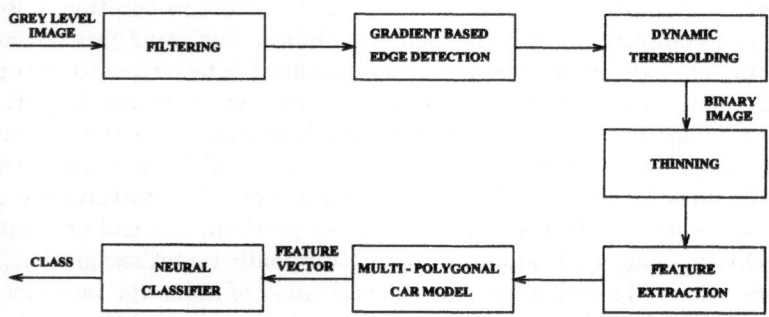

Fig. 1. The overall scheme of the object recognition system.

2   The Algorithm

The overall system is depicted in Fig. 1 and the operations can be summarized as follows. The first stage is dedicated to detect the object boundary, after cleaning out the noise from the image by applying a linear smoothing filtering and a strongly noise independent segmentation procedure, based on the image entropy optimization between foreground and background [7]. The boundary is then coded by using elementary but salient features. We adopt and compare two strategies for the feature extraction : the first approach uses the variational angle sequence [9] (see Fig. 2 (a))), while the second and more common approach uses the sequence of euclidean distances of points along the shape boundary from the centroid [3] (see Fig. 2 (b))).

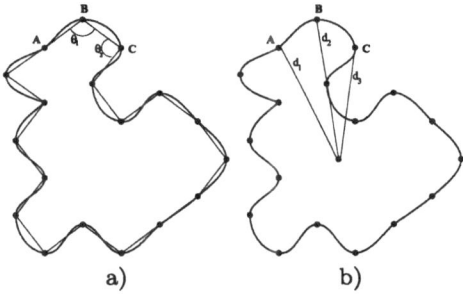

a)   b)

Fig. 2. The adopted shape description models: angle of variation a) and distance from centroid b)

The sequence of shape features is modelled as a Circular Auto-Regressive (CAR) process [8], which is a parametric equation that expresses each sample of an ordered set of data samples as a linear combination of a specified number of previous samples plus an error term. Since the sequence is assumed to be circular, then it is invariant to rotation and translation. The form of the model is:

$$y_i = \alpha_0 + \sum_{k=1}^{m} \alpha_k y_{i-k} \qquad \forall i = 0, 1, \ldots, n-1 \qquad (1)$$

where y_i is the current primary feature; y_{i-k} is the feature detected k times before the current features; $\alpha_0, \alpha_1, \alpha_2, \ldots, \alpha_m$ are the unknown CAR coefficients; m is the model order. Let us indicate with $\beta = [\alpha_0, \alpha_1, \ldots, \alpha_m]$ the least square error (LSE) estimate of the CAR model. To improve the representativity of this solution the following method is adopted. If we fix n as the number of line segments in the shape polygonal approximation, then $p = \lfloor (B/n) \rfloor$ pixels will lie on the contour between two end points, and $(p - 1)$ polygonal approximations of the shape through model (2) are possible, depending from the starting point.

The sequences of primary features generated for each of them may be slightly different. To overcome this problem we adopt an iterative scheme consisting of solving $p-1$ systems each having n equations and $m+1$ unknown parameters. It is based on the consideration that the $p-1$ sequences are obtained in some order (clockwise or counterclockwise); firstly, the CAR vector for the first sequence of primary features is computed, then for two sequences, three sequences and so on. The process is repeated $p-1$ times, according to the number of sequences or polygonal approximations. In particular, denoted with β_j the solution of the j-th system thus constructed, $j = 1, \ldots, p-1$, let us define

$$\varepsilon_j = \frac{\beta_j^T \beta_{j-1}}{\|\beta_j\|^2 \|\beta_{j-1}\|^2} \quad j = 2, \ldots p-1 \quad \text{and} \quad \varepsilon_1 = \frac{1}{\|\beta_1\|^2}$$

a measure of similarity. The final feature vector β^* is set to the solution of the s-th system, where $s = arg\min_{0 \le j \le p-1}\{\varepsilon_j\}$. We shall refer to this way of proceeding as the Multi–Polygonal Auto–Regressive Model (MPARM). For sake of clearness, we denote in the following, as *Scheme I* the MPARM applied to the variational angles, whereas *Scheme II* is the MPARM applied to euclidean distances for the centroid.

The sequences of CAR parameters are lastly fed into an Artificial Neural Network (ANN) trained to classify the planar objects with the Conjugate-Gradient (CG) algorithm [13]. The process is repeated for the overall set of reference images. After learning, an unknown object passing through all the previous stages is classified as correct or not from the ANN with frozen weights.

The overall system described above has been implemented on a reconfigurable SIMD machine. The implemented algorithms were: convolution on a $3D$ mesh, thresholding on a $2D$ mesh, thinning and feature description on a $2D$ mesh, conjugate gradient for neural learning on a $1D$ mesh. We report in the next section the recognition performance and the parallel excecution times of the overal system.

Fig. 3. Reference shapes.

3   Experimental Results

3.1   Recognition Performance

The task faced was the recognition of five differently shaped aircrafts depicted in Fig. 2. The data set was formed by changing each aircraft in size, orientation and position. Specifically, the objects were 5^0 rotated and translated in random positions within the image boundaries. The size of each aircraft was varied from 0.25 to 1.25 times the size of the original. By doing so, 5040 shapes were generated. For each shape the representative CAR vector was then extracted by using both feature angles of variation and euclidean distances from the centroid as elementary features.

The data set was consequently subdivided in 3600 shapes for the training and in 1440 shapes for the test of the neural classifiers.

We carried out various experiments with CAR models with different orders m. The value $m = 20$ turned out to be optimal for our purposes. Table 1 shows the classification performance on the training and test sets for selected classifiers with different number of hidden units and shape description schemes.

3.2   Recognition Results on Noisy and Occluded Objects

Noise and distortion effects were introduced by adding random noise to the boundary points of 360 object shapes. To each contour pixel wass assigned a probability p of retaining its original coordinates in the image plane and probability $q = (1 - p)$ of being randomly assigned to the coordinates of one of its neighboring pixels. The degree of noise was augmented with increased values of the noise level q. We choose q equal to 1/3, i.e. one pixel over three was selected to change its own position. The set of noisy shapes was subdivided in 270 shapes for the training and in 90 shapes for the test of the neural classifiers.

A moderate amount of occlusion is also accommodated in our experiments. The occluded shapes have been obtained by cutting off a portion of the generated shapes with a straight line. The restriction on occlusion is that all the major portions, or branches, should remain; therefore, no major geometric property is changed. The set of occluded shapes has been subdivided in 540 shapes for the training and 180 shapes for the test of the neural classifiers. We tested the trained system over the set of noisy and occluded shapes. As expected, the classification performance obtained was substantially degraded. o get better classification performance we re-trained the previously neural classifiers with the 270 noisy shapes and 540 occluded shapes and, after convergence, they were tested on the 90 test noisy patterns and the 180 test occluded patterns. The results obtained are reported in Table 2.

The best performance was achieved with 40 hidden neurons by using as elementary features both the angles of variation and the euclidean distances.

3.3   Timings

The parallel algorithms were implemented in C* language on the Connection Machine 200 (CM-200) [6], a fine-grain SIMD parallel computer configured with 8192 1-bit hypercube interconnected processors equipped with the Data Vault. We have configured the CM-200 to be used as a multi-dimensional mesh of processors.

As the computation progresses from one vision step to the next one, uniform partitioning is adopted. In the first stage of the shape recognition scheme (filtering, segmentation, edge extraction, thinning), the adopted re-mapping mode assures that performance gains are considerable, since at this stage there is no idle processor. This is not true for the second stage (feature extraction), where a contour following algorithm is involved, and the features are extracted only for the boundary pixels determined from the previous processing stage. There is of course evidence of a slowdown in performance, which is heavy in the context of the global performance reached. A MIMD solution should, in this case, give better performance. On the contrary, the expected performance gain due to the use of a neural classifier, where no idle processors are allowed, gives global efficient performance. Table 3 gives overall system performance from the first vision step, linear smoothing filtering, to the neural classification for the two schemes adopted for different image dimensions ($N = 64, 128, 256$).

A source of degradation in performance of the *Scheme I* against *Scheme II* is due to the need of calculating the trigonometric function involved in the computation of the angles of variation, which is slow on the Connection Machine-200. Commonly, a table lookup method is useful in computing an approximation to these functions. However, indexed addressing, which is required in lookup table, is as slow as indirect access on the Connection Machine-200, when each PE needs to access its table at a different address.

4   Summary and Conclusions

In this paper, a massively parallel system for the recognition of 2*D* shapes by translation, scale and rotation invariant boundary representations is reported. The noise and occlusion immunity of the system has also been tested. The aim of this paper is born from the consideration that in the last decades a lot of computational vision algorithms have emerged as solution for designing efficient computer vision systems. Many of them are inherently sequential and results as regards the efficiency and applicability on parallel platforms are not reported. Others, parallel in nature, need to be tested on existing parallel platforms in order to fix their advantages and/or disadvantages when considered in a more complex vision framework. Thus, selection of efficient vision steps and their validation and testing on existing parallel platforms are mandatory and profitable for architecture designers. Our aim was to propose efficient parallel versions on SIMD machines of a selected set of algorithms very efficient from a qualitative point of view. Specifically, two methods for extracting primary features are employed and compared: angles of variation and euclidean distances from central moments were used. According to our results the scheme based on centroidal profile representation is more efficient in terms of recognition performance and the time required from the initial image capture from Data Vault to the display of the recognized object is approximately 0.3 seconds on a 8k-processor Connection Machine running at 8Mhz, for images of dimension up to 128×128 (VP ratio = 2), strongly outperforming the timings reported in [15].

Table 1. The recognition performance by using *Scheme I* (a) and *Scheme II* (b).

NET	TRAIN	TEST
20-5-5	100.0	99.72
20-10-5	100.0	98.88
20-15-5	100.0	99.16
20-20-5	100.0	99.16
20-25-5	100.0	99.16
20-30-5	100.0	99.44
20-35-5	100.0	99.16
20-40-5	100.0	99.16

(a)

NET	TRAIN	TEST
20-5-5	100.0	99.44
20-10-5	100.0	98.88
20-15-5	100.0	98.88
20-20-5	100.0	98.88
20-25-5	100.0	99.44
20-30-5	100.0	99.16
20-35-5	100.0	99.16
20-40-5	100.0	99.16

(b)

Table 2. The recognition performance by using *Scheme I* (a) and *Scheme II* (b) for neural classifiers trained on occluded shapes.

NET	TRAIN	TEST
20-5-5	100.0	97.59
20-10-5	100.0	98.70
20-15-5	100.0	99.07
20-20-5	100.0	99.25
20-25-5	100.0	99.25
20-30-5	100.0	99.25
20-35-5	100.0	99.44
20-40-5	100.0	99.44

(a)

NET	TRAIN	TEST
20-5-5	97.43	96.11
20-10-5	100.0	98.14
20-15-5	100.0	96.85
20-20-5	100.0	98.33
20-25-5	100.0	97.59
20-30-5	100.0	97.22
20-35-5	100.0	97.40
20-40-5	100.0	98.33

(b)

Table 3. The experimental times (in seconds) for the whole process of computer vision as function of the size of the image $(N \times N)$.

Recognition Scheme	I	II
N = 64 (VP ratio = 1)	0.2595958	0.231238
N = 128 (VP ratio = 2)	0.384843	0.3357263
N = 256 (VP ratio = 8)	1.0183437	0.8930011

References

[1] R. Allen, L. Cinque, S. Tanimoto, L. Shapiro, D. Yasuda, 'A Parallel algorithm for graph matching and its MasPar implementation', *IEEE Trans. on Parallel and Dist. Systems*, vol. 8, n. 5, pp. 490-500, (1997).

[2] O. Bourdon, G. Medioni, 'Object Recognition Using Geometric Hashing on the Connection Machine', *Proc. of Intern. Conf. on Pattern Recognition*, pp. 596-600, 1988.

[3] C. C. Chang, S. M. Hwang, D. J. Buehrer, 'A shape recognition scheme based on the relative distances of feature points from the centroid', *Pattern Recognition*, **24**, No. 11, pp. 1053-1063, (1991).

[4] R. C. Gonzalez, P. Wintz, *Digital Image Processing*, Reading, MA, Addison-Wesley (1977).

[5] J. G. Harris, A. M. Flynn, 'Object recognition using the Connection Machine's router', *Proc. of Computer Vision and Pattern Recognition*, pp. 134-139, (1986).

[6] W. D. Hillis, *The Connection Machine*, MIT Press, Cambridge, MA, 1985.

[7] J. N. Kapur, P. K. Sahoo, 'A new method for grey level image thresholding using the entropy of the histogram', *Computer Vision, Graphics and Image Processing*, **29**, (1985).

[8] R. Kashyap, R. Chellapa, 'Stochastic models for closed boundary analysis: representation and reconstruction', *IEEE Trans. Inf. Theory*, **27**, 5, pp. 109-119 (1981).

[9] N. R. Pal, P. Pal, A. K. Basu, 'A new shape representation scheme and its application to shape discrimination using a neural network', *Pattern Recognition*, **26**, 4, pp. 543-551 (1993).

[10] T. Pavlidis, 'Algorithms for shape analysis of contour and waveforms', *IEEE Trans. Patt. Anal. Machine Intell.*, **2**, 4, pp. 301-312 (1980).

[11] A. Petrosino, G. Salvi, 'A two subcycle parallel thinning algorithm and its parallel implementation on SIMD machines', *IEEE Trans. on Image Processing*, to appear, (1999).

[12] A. Petrosino, G. Salvi, 'Recognizing planar objects with neural networks', *Proceedings of Intern. Conf. on Neural Networks ICANN '94*, M. Marinaro and P. Morasso eds., pp. 835-838, Springer Verlag (1994).

[13] M. J. D. Powell, 'Restart procedures for the Conjugate Gradient method', *Mathematical Programming*, **12**, pp. 241-254 (1977).

[14] P. Suetens, P. Fua, A. J. Hanson, 'Computational strategies for object recognition', *ACM Computing Surveys*, **24**, 1, (1992).

[15] L. W. Tucker, C. R. Feynman, D. M. Fritzsche, 'Object recognition using the Connection Machine', *Proc. IEEE Conf. Computer Vision and Pattern Recognition*, June 1988, pp. 871-877.

[16] C.-L. Wang, V. K. Prasanna, H. J. Kim, A. A. Khokhar, 'Scalable Data Parallel Implementations of Object Recognition Using Geometric Hashing', *Journal of Parallel and Distributed Computing*, 1994.

Multimedia Extensions and Sub-word Parallelism in Image Processing: Preliminary Results

Marco Ferretti and Davide Rizzo

Dip. Informatica e Sistemistica, University of Pavia,
Via Ferrata 1, 27100 Pavia, Italy
{ferretti,rizzo}@elzira.unipv.it

Abstract. General purpose microprocessors have long been considered a computing platform unsuited to image processing and vision tasks. The so-called Von-Neuman paradigm and the associated memory bottleneck have motivated the research into various forms of parallel processing and of special processors for vision. The outcome of this long standing effort is negligible, if one considers the computing platforms that became a true product. Recently, the micro-architecture of some general purpose microprocessors has been augmented with extensions to support multimedia processing. It is worthwhile considering how much speed-up can be actually obtained by the limited SIMD processing mode that is embedded in these extensions. This paper presents experimental results obtained on a very simple algorithm, the Haar transform, that has been coded for the HP and the Intel multimedia microengines. Preliminary results reported here show that the system environment (type and dimensions of first and second level caches, and compiler efficiency) affects considerably the theoretical speed-up due to the SIMD microengine.

1 Multimedia Extensions on General Purpose Processors

Modern processors are equipped with extended instruction sets (called *media instruction sets*) designed to cope with the increasing demand of processing power of multimedia applications. Sounds, images and video have become elementary data to many applications and their transmission and processing over Internet has changed the overall framework for evaluating the performance of processors. This process has led to changes in the architectures of the computing platform. We can distinguish two main directions: one approach relies on modifications to existing general purpose processors instruction set architectures such as HP MAX-2 [1], Intel MMX [2,3], UltraSparc VIS [4] and MIPS MDMX [5]; the other approach is represented by specific processors targeted to embed multimedia applications (Multimedia processors MMPs) like Philips TriMedia TM-1[6], Samsung [7]and Mpact [8].

In the past many efforts have been made to implement efficiently image processing and, to some extent, multimedia algorithms on specialised or general purpose parallel architectures. These efforts have brought to the design and construction of many prototypes, and to much fewer commercial systems. Most of these machines adopted a SIMD architecture; some followed the MIMD approach. Specialised processors exhibit systolic, wavefront and similar styles, down to the extremely narrow ASIC

P. Amestoy et al. (Eds.): Euro-Par'99, LNCS 1685, pp. 977-986, 1999.
© Springer-Verlag Berlin Heidelberg 1999

paradigm of an integrated- circuit-per-algorithm solution. General purpose parallel processors have not proved flexible enough, and above all have not found a market. Today multimedia application apparently justify the extensions to the micro-architecture of general purpose microprocessors; it is no surprise that they have recovered the SIMD paradigm, with a lower degree of parallelism, but with a much wider diffusion and specially much lower costs than in previous systems. Video and enhanced DSP processors are likely to fill the market share opened up by set-top-boxes, games and high quality audio. We concentrate here in the field of mainstream image and audio processing that is likely to be required by most applications.

Indeed, we can find some characteristics common to the various multimedia extensions, which emerge from the analysis of the of multimedia algorithms. When considering how to extend instruction sets, the starting point for all companies has been an analysis on operations common to simple tasks in image processing, communications and audio processing. The outcome of this analysis is somewhat different, since not all groups involved in the design of multimedia extensions have privileged the same algorithms. Nevertheless, most multimedia applications are characterised by small data types (in general 8, 16 bits width), large quantity of data to be processed in a continuous stream, operations with great data parallelism, real-time processing requirements and large I/O bandwidth requirements.

The paradigm adopted is *sub-word parallelism* (or *sub-word execution model*) [1,9] which exploits the wide paths available nowadays at the various levels of interconnection: at the frontside bus that connects main memory to the second level cache (when available), at the backside bus between the second level cache and primary caches, and at the processor core. At each of this interconnections, a minimum of 64-bit parallelism is available; this allows for parallel data transfer and parallel computations over lower-precision data (for example 8 bytes, 4 words, or 2 double words in the MMX implementation; MAX-2 only supports a 4 word parallelism).

Among the microarchitectures with a support for multimedia processing, we have worked on the Intel MMX and HP MAX-2. They provide two different sets of new instructions that can be subdivided in these classes: arithmetic, compare, shift, conversion/organisation, logic and multiply-accumulate (MMX only).

All the arithmetic operations are implemented in fixed point arithmetic, with support for either *wraparound* (or *modulo*), *signed* or *unsigned saturation*. Saturation arithmetic is one of the really new key features, since parallel fixed point operations need a way to manage overflow/underflow conditions [9] so common in image processing (consider, for example, the sum between two pixels: if the resulting value lies outside the representable range, using wraparound arithmetic, it would be a dark pixel in a linearly growing map of grey values, while using saturation, it is possible to clamp that value at the range's maximum, which makes much more sense).

Intel MMX instruction set is wider (57 new instructions) than MAX-2 (17 new instructions), which was released with the same goal, but rather different implementation costs. The MAX-2 extension is a minimal change to the existing PA-RISC micro-architecture, and it was designed in such a way to cause a minimal increase in the area requirements on the chip core. HP considered the 4-way parallelism the only really useful, and does not support the 8 byte data type available in MMX, nor the 32 bit one. The latter that can be handled with single precision floating point operations.

Another key feature of multimedia extension is the provision for parallel compare operations: data dependent comparison are very common in some image processing tasks. If handled with the usual RISC branch instructions and branch prediction logic, the resulting mispredictions add huge penalties to the execution within the superpipelined microarchitectures. Surprisingly, not all companies have considered the issue of data dependent computation with the same relevance. As an instance, HP MAX-2 has no parallel compare instruction, although the HP PA architecture does provide for a nullification mode that directly handles the negative outcome of a comparison. The discussion of this point is not central to this work, since we have considered initially an algorithm (a digital transform) that has no data dependent behaviour.

The microarchitecture and its support to SIMD processing is just one of factors for improving performance. At the opposite side of the processing chain, the compiler is the first actor, and often the determinant one. Effective, widespread use of an advanced microarchitecture is only possible if the software for application development does exploit the capabilities available. Currently no compiler is capable of automatically inferring and generating multimedia instructions in order to parallelise a high level fragment of C code, for example. At the present, compilers have facilities to embed assembly instructions inside the existing code, to parallelise the particular multimedia task/image processing operation chosen, but this technique relies heavily on the programmer's skills, rather on an optimising compiler technology back-end.

2   Algorithm: Haar Transform

For our preliminary evaluation of multimedia extensions we have chosen the Haar transform algorithm [10]. It's a local algorithm based on convolutions and subsampling. It is often used in image processing applications, when the original image is processed through a decomposition operator, to generate a multiresolution representation. A typical application is image compression based on transform coding, where the Haar transform is used to decorrelate the original image producing a different domain representation. The Haar belongs to the class of Wavelet Transforms [11,12], whose main goal is to obtain a new multiresoluton representation of a signal. Wavelet basis functions used to compute the transform can be chosen so that they have some important features and properties useful in image processing: a limited support and scale-space invariance.

There exist methods to compute a fast wavelet transform in a recursive way. However, the Haar transform has a very simple non separable implementation; it consists of four convolutions based on filters with a small support, that fit perfectly with the required subsampling by 2. There are four of them, to produce four different bands of the decomposition at every transform step. The 2x2 filters being considered are:

Band0	Band1	Band2	Band3
1 1 1 1	1 1 -1 -1	1 -1 1 -1	1 -1 -1 1

The decomposition results from iterating on Band0; it is composed of *logN* levels. At every step, the total number of coefficients to be considered is reduced by a factor of 4. In many practical cases, there is no need to go through the complete decomposition, which can be stopped after four or five levels. This is the case of image compression. To obtain the original signal, the Inverse Haar Transform uses the same filters (normalised with a scaling factor of ¼) to be applied to the corresponding bands of each level in a top-down process.

2.1   Factors Affecting Performance on the Reference Systems

The experiments carried out on the reference systems depend mainly on three factors: i) algorithm style; ii) compilers optimisation; iii) image sizes.

As to the algorithm description, we coded the most straightforward C program derivable from the algorithm definition (sequential version) and then we coded a corresponding parallel version, using the multimedia extension instructions available. We further decided to split the experiments in two cases: computations of first level only (algorithm Haar1), and computations of multiple levels, up to the fourth (algorithm HaarM).

Since there is no automatic generation of the new multimedia instructions directly from C through the available compilers, we had some alternatives for using these new instruction sets. With MAX-2, the options are macro definition or coding of assembly routines to be called within a C program. We followed this second option. With MMX, we used direct assembly inlining in C program to perform core computations.

Each C algorithm has been further optimised using the existing optimising compilers, in order to compare the best code obtainable from deep (and automated) compiler optimisation to what can be directly obtained adopting an assembly hand-coded core module embedded in a C program. We used the *cc* level 3 optimisation (+O3) on the PA-RISC system, which includes, among others, branch optimisation, faster register allocation, instruction scheduling, loop invariant code motion, software pipelining, register reassociation and function inlining. On the Pentium system, we selected Intel C compiler level 2 optimisation, which includes the following: optimised code selection, global register allocation, instruction scheduling, inline expansion, loop invariant code movement and copy propagation. Another option, available only on PA-RISC processors for MAX-2, is the dataprefetch facility, which could produce better results, in situations where the cache acts as the bottleneck of the chain; the available compiler adds dataprefetch instructions on writes only, while MAX-2 hand-coded algorithms use dataprefetch on reads as well.

The performances index is the time required to complete the kernel of the operations, once the image and all the other results have been allocated. We used two different ways to measure time, on the basis of the facilities that were provided on the particular reference system, and we repeated the experiment several times (more than 50), taking the minimum time in the end. For the MAX-2 measures, we used the *gettimeofday()* Unix system call, which provides a time resolution of microseconds.

For MMX, we used the Time Stamp Counter facility provided by Pentium II processors, to obtain the clock ticks elapsed since the start of the core algorithm: this allows to obtain more precise results than through the Unix system call.

Label	Image dim. (pixels)	Image size (Kb)	HP-PA L1 (1 Mb) Cache size/ image size	INTEL L1 (16Kb) Cache size/ image size	L2 (512 Kb) Cache size/ image size
T	128 x 128	16	n.a.	1	32
XS	256 x 256	64	n.a.	0.25	8
S	512 x 512	256	4	0.062	2
M	768 x 640	480	2.13	0.031	1.06
L	1280 x 1024	1280	0.8	0.012	0.4

Table 1. Ratio of cache size to image size. Labels T (Tiny) through L (Large) denote images of increasing dimensions. Cases T and XS apply only to INTEL MMX experiments.

As to images, we initially chose three 8-bit greyscale PGM images of different dimensions, denoted as Small (S), Medium (M) and Large (L) (see Table 1 for details), for the experiments on the HP-PA MAX-2. When working on MMX, we had to introduce two smaller images, eXtra Small (XS) and Tiny (T) because of the extremely different configurations and dimensions of the caches in the two reference systems.

3.2 Algorithm Coding

Independently of the language, in coding the algorithms we made some choices regarding memory allocation. For Haar1 algorithm, there is nothing particular, apart the need to reorganise the data read in order to fully exploit the parallelism (see MAX-2 experiments).

For algorithm HaarM, the problem regards the memory needed for intermediate computations. In order to exploit the inherent parallelism, the MAX-2 and MMX versions operate on a group of 4 adjacent result coefficients (on the same row) at a time, and this is true in each band, so that each iteration produces 16 coefficients. Each iteration reads in data from two consecutive input rows (initially 8 pixels in each row, then 8 coefficients values in each row), rearranges them and computes in parallel four coefficients in the same band. The parallel strategy can be realised in different ways. For example Intel [13] suggests a strategy based, among others, on the Parallel Multiply Accumulate instructions (PMAC), that handles a group of data and of coefficients stored in two parallel registers. Such a strategy cannot be used with MAX-2 extension, since no parallel multiply-add instruction is provided. PMAC is not needed really, since products factors are 1 or −1. It could also become a bottleneck, because its latency is 3 clock cycles, while all the others are 1-cycle latency instructions (assuming cache hits).

In the MAX-2 strategy the four data corresponding to the same band coefficient computation are placed in the same register word position and then accumulated according to the four Haar filters, using only additions, subtractions and logical operations. We implemented this scheme also in the MMX case. This approach can take advantage of the internal Pentium II microarchitecture. The MMX execution units are connected to two ports of the Reservation Station [14]: Port 0 has MMX ALU Unit and MMX multiplier Unit while Port 1 has MMX ALU Unit and MMX Shifter Unit. So the logical and arithmetical MMX instructions can be dispatched at the same time to two MMX ALU units, while multiplications can only be dispatched one at a time to a single port of the Reservation Station. Moreover these latter have a 3 cycle latency, while the others are one cycle.

A note on the precision of results. We have adopted a 16-bit precision for computation first because input data is 8-bit wide and we need a larger precision to accomplish exact computations; secondly, this allows a parallelism of four. A drawback is that we cannot compute the whole decomposition in all levels. The highest possible level, which still grants perfect reconstruction, is the third, because the growth in range of coefficients over higher levels, can cause a loss of precision that prevents an exact signal reconstruction. So when going to levels higher than third we may have to accept a potential loss of information and a potential reconstruction error (the loss happens in certain worst case images, that can represent however very extreme and not practical images). For MMX this situation could be avoided, reducing the parallelism to two double word (32-bit) computations at a time. Doing so, no loss of information is caused, but the speedup obtainable gets smaller. This however should not be a problem because the proportional relevance of higher level computations is very small with respect to total time needed to perform the first three levels. But with MAX-2 this is not feasible.

The output image is twice as large as the original one, since the transform of a NxN image is composed of NxN wavelet coefficients. As mentioned earlier, in order to perform correctly the filtering operation, the resulting coefficients have a doubled bit depth with respect to the input pixel values. The consequence is that we cannot superimpose the new values directly on the input image, because of data width mismatch. Some buffering is required for intermediate results. Once the first level transform is computed on a separate memory area (twice as large as that of the input image), in order to compute the second level one idea could be to avoid allocating further memory, trying to use the area allocated for Band0 to store both the approximation and the details of the second level. This is not possible without further data reorganization and temporary storage areas, since detail coefficients would replace approximation data not yet used in any computation. The strategy we have adopted is based on a *double buffering scheme*. During each level, the results are written to a different memory area and then, at the end of the transform, a final copy "merges" the results producing the usual wavelet structure. In this way no rearrangement operation, distinct from a copy, is needed, and the additional temporal buffer is equal to one half of the original input image (a quarter of the resulting transformed image).

4   Results

We have used two different hardware architectures to perform our evaluation on the chosen algorithm: a true RISC processor and an evolution from a CISC processor which adopts internal microarchitectural RISC features. The reference systems are an HP C180 series 700 workstation and an HP PC Vectra/233 Series DT.

The HP C180 is a Unix workstation running on HP/UX 10.20, with 128 MB RAM and 1 MB each of instruction and data L1 cache. The PA-RISC 2.0 processor runs at 180 MHz. The system has no L2 cache, a somewhat unique feature within RISC workstations. We used ANSI C with *cc* compiler v. A.10.32.

The HP Vectra system is a Windows NT workstation, equipped with a 233 MHz Pentium II, 96 MB RAM, 16 KB of first level (L1) data cache and the same value for instruction cache, and 512 KB of secondary (L2) cache. The compilers used are both Microsoft Visual C++ 5.0 and Intel C compiler plug-in v2.4. We also used VTune v.2.5, a tool provided by Intel to analyse Intel processor's performances.

While performing the experiments, the user had no particular priority privilege and the systems had all the services of a typical multi-user environment.

4.1 MAX-2 Experiments

At the beginning of the experiments we had to face the data re-organisation problem: since input images are in PGM format, with 8 bits per pixel, in order to fully exploit MAX-2 extensions, data must be rearranged to 16-bit width; this rearrangement operation can be performed either "externally" to the transform algorithm thus producing input data of the correct width or "internally", working directly on 8-bit input data. In the first case, the required memory is larger (*4mn* bytes, for *m* x *n* image against *3mn*); the algorithm has an additional step and its core still has to manage a partial data re-organisation, functional to the parallelism adopted. In the second case, the advantage is represented by the broader bandwidth in data memory transfer since every read operation can take 8 pixels at the same time, while in the other mode the bandwidth is halved, since only four data pixels are transferred at a time. This alternative regards only the first level computations, since after that results are 16-bit wide.

Experiments showed that the internal conversion is always better (see Table 2), so we have adopted this technique throughout all cases. The evaluations have been made also with a comparison of dataprefetch instructions, available in PA8000 processors.

Haar1MAX-2	S	M	L
External no dp	1.697	11.477	45.671
External dp	1.600	10.205	40.426
Internal no dp	1.404	5.672	36.087
Internal dp	1.291	4.965	33.155

Table 2. Times (msec) for 1 level Haar computation with external vs. internal input data width conversion, with (dp) and without (nodp) dataprefetching.

Table 3 shows the outcome of the execution of assembly coded algorithm for the direct and inverse 1-level Haar: the speed up is computed against the optimised C version.

We can observe that an acceptable speedup can be obtained only in the case of the small image. In fact this image can be completely contained in the L1 cache. With

larger images, the parallelism advantage is lost and the main cause is the great
number of cache misses. As an extreme case, no gain is obtained with large images.

Data prefetch instructions are useful with larger images, where the hints to data
load/store can reduce cache misses. Again cache misses with larger images prevent
full parallel ideal speedup.

Algorithm	S		M		L	
	T	Sp.up	T	Sp.up	T	Sp.up
Haar1MAX-2 no dp	1.404	2.84	5.672	1.97	36.087	1.17
Haar1MAX-2 dp	1.291	2.92	4.965	1.52	33.155	1.00
IHaar1MAX-2 no dp	2.638	3.03	8.157	2.10	32.795	1.68
IHaar1MAX-2 dp	1.828	3.69	5.836	2.23	27.285	1.34

Table 3. Execution times (msec) and speed up relative to C code for 1-level direct (Haar) and
inverse (Ihaar) transform with HP-PA MAX-2

Considering the multi-level Haar transform (HaarM), we obtained the results
shown in Table 4 (we have reported the computation times up to the fourth level, even
if these are related to Haar transform with reconstruction errors).

Algorithm	lev	S		M		L	
		T	Sp.up	T	Sp.up	T	Sp.up
HaarM MAX-2 no dp	2	1.844	3.22	10.518	1.74	52.246	1.29
	3	1.881	3.41	10.703	1.80	53.910	1.30
	4	1.910	3.46	10.727	1.83	54.158	1.32
HaarM MAX-2 dp	2	1.833	3.21	9.185	1.36	48.947	1.03
	3	1.930	3.15	9.374	1.39	50.279	1.03
	4	1.956	3.15	9.443	1.40	50.755	1.04

Table 4. Execution times (msec) and speed up relative to C code for multi-level (lev) direct
Haar transform with HP-PA MAX-2

We can see that the time increase to compute more than two levels is not really
relevant, since the number of new coefficients to be computed is relatively small
compared to first and second level.

4.2 MMX Experiments

For MMX experiments we used the internal data reorganisation mode, without
further checks. We report comparisons between optimised C codes and MMX
versions coded with inlining of parallel instructions. As previously mentioned, we
were forced to introduce smaller images (T and XS) in order to observe true
improvements from MMX parallel version of the algorithm with respect to the serial
C version. Considering Table 5 for one level computations, it's clear that the ideal
speedup can be obtained only when the images are completely contained in the cache

hierarchy. Indeed, when this condition does not hold, (see columns S, M and L), cache penalties become relevant and performance is greatly reduced. The forward and the inverse transform exhibit the same pattern.

Algorithm	T		XS		S		M		L	
	T	Sp.up	T	Sp.up	T	Sp.up	T	Sp.up	T	Sp.up
Haar1 MMX	0.190	3.41	0.737	3.56	10.75	1.45	22.37	1.45	61.51	1.30
IHaar1MMX	0.178	3.54	0.696	3.68	6.652	2.14	18.02	1.55	50.98	1.73

Table 5. Execution times (msec) and speed up relative to C code for 1-level direct (Haar) and inverse (Ihaar) transform with INTEL MMX

Considering multiple levels (algorithm HaarM), Table 6 reports execution times and speedups for computations up to the fourth level. Again cache penalties are heavy when working with large images.

Algorithm	lev	T		XS		S		M		L	
		T	Sp.up	T	Sp.up	T	Sp.up	T	Sp.up	T	Sp.up
HaarM	2	0.266	3.45	1.147	3.22	14.42	1.52	30.75	1.31	92.89	1.26
MMX	3	0.273	3.54	1.151	3.39	14.56	1.55	30.88	1.36	93.48	1.27
	4	0.276	3.58	1.182	3.36	14.87	1.52	30.95	1.39	94.61	1.28
IIaarM	2	0.279	2.94	1.114	2.98	10.47	1.73	29.79	1.27	86.16	1.47
MMX	3	0.283	3.02	1.174	2.97	10.95	1.72	29.16	1.34	88.64	1.49
	4	0.284	3.05	1.175	2.99	10.97	1.73	30.70	1.29	90.47	1.48

Table 6. Execution times (msec) and speed up relative to C code for multi-level direct (Haar) and inverse (Ihaar) transform with INTEL MMX

5   Conclusions

The Haar transform in the formulation used here has some distinct features: it does not require image transposition to compute the bi-dimensional convolution, as is usual with standard implementations of wavelets; the computations are not data dependent, and the output of each step are re-used partially in subsequent steps.

The effect of such characteristics on the performance on superpipelined, RISC microprocessors is distinctly different. On the positive side, the regular structure of the computations (the control structure of the kernel of the transform is a simple nested loop) minimises branch mispredictions.

As already shown, locality in memory references and cache usage is instead a critical point. Since the image need not be transposed in any of the phases of the algorithm, a major cause of cache miss is eliminated. The end effect is therefore the ratio between image and caches' dimensions. The pattern of the speed-up against the dimension of the image shows this effect clearly. At this stage of the work, it is uncertain whether tuning memory references can bring substantial benefits.

On the side of software development, the exploitation of the small effects of sub-word parallelism is still a matter of good assembly programming skill. Improvements in compiler technology could make the effects of the limited SIMD parallel processing visible only if cache usage is optimised first.

References

1. Lee, R.: Subword Parallelism with MAX-2, IEEE Micro, 16, n.4, (1996), 51-59
2. Peleg, A., Weiser, U., MMX Technology Extension to the Intel Architecture, IEEE Micro, July/Aug. (1996), 42-50
3. The complete Guide to MMX Technology, McGraw-Hill, (1997)
4. Tremblay et al.: VIS Speeds New Media Processing, IEEE Micro, July/Aug., (1996), 10-20
5. MIPS Digital Media Extension, Instruction Set Architecture Specification, http://www.mips.com/Documentation/isa5_tech_brf.pdf
6. Slavenburg, G.A., Rathnam, S., Dijkstra, H.: The Trimedia TM-1 PCI VLIW Media processor, Proc. Hot Chips VIII Symp., IEEE CS Press, Los Alamitos, Calif., (1996)
7. L.T. Nguyen et al: MSP: Multi-Media Signal Processor, Proc. Hot Chips VIII Symp., IEEE CS Press, Los Alamitos, Calif., (1996)
8. Kalapathy, P.: Hardware/Software Interactions on the Mpact, IEEE Micro, Mar./Apr. (1997), 20-26
9. Lee, R.: Accelerating Multimedia with Enhanced Microprocessors, IEEE Micro, 15, n. 2, (1995), 22-32
10. Haar, A.: Zur Theorie der Orthogonalen Functionensysteme, Math. Annal. 69, (1910), 331-371
11. Rioul, O., Vetterli, M.: Wavelets and Signal Processing, IEEE SP Magazine, October (1991), 14-37
12. Strang, G., Nguyen, T.: Wavelets and Filter Banks, Wellesley-Cambridge Press, (1996)
13. Using MMX Instructions to Compute the 2x2 Haar Transform, Intel Application Note AP-531, available at Intel Web site
14. Shanley, T.: Pentium Pro and Pentium II System Architecture, MindShare Inc., (1997)

Vanishing Point Detection in the Hough Transform Space

Andrea Matessi and Luca Lombardi

Dipartimento di Informatica e Sistemistica, University of Pavia,
Via Ferrata 1, I-27100 Pavia (Italy).
e-mail: matessi@stigma.it, llombardi@dis.unipv.it

Abstract. Depth estimation from monocular images can be retrieved from the perspective distortion. One major effect of this distortion is that a set of parallel lines in the real world converges into a single point in the image plane. The estimation of the co-ordinates of the vanishing point can be retrieved directly on the Hough Transformation space or polar plane. In fact the vanishing point in the image plane is mapped in the polar plane into a sine curve that can be estimated with a simple linear system.

1 Introduction

In monocular images depth estimation, if no particular prior knowledge of the scene is given, can be retrieved in many real images by the perspective distortion. One major effect is that a set of parallel lines in the three dimensional space in the image space converges to a single point called the *vanishing point*. This point in the image plane gives important information on the distance of the objects in the scene and of the three--dimensional structure. In Sedgwick [2] some features of the vanishing point are given. Since this point is the projection of a point at the infinite with respect to the point of observation, a finite motion of the point of observation produces a movement of the vanishing point in the image that is negligible. This is called *invariance to motion*. This feature must be taken into account when considering sequential images of a moving point of observation.

In this work an analytical method to determine the vanishing point in the Hough Parameters Space is presented. This method is computationally equivalent to others but has the advantage that in the Hough Plane the vanishing point has a precise shape that can be searched optimally. The algorithm to detect the vanishing point was also implemented and the results from real images are shown. This same algorithm can be used to detect the *focus of expansion* in moving scenes to determine the speed of objects in the three dimensional space.

The next section provides a brief background on the problem and a few previous solutions. Section 3 outlines the proposed approach. Section 4 shows our preliminary results. Finally, conclusions and directions for the future work are stated.

P. Amestoy et al. (Eds.): Euro-Par'99, LNCS 1685, pp. 987-994, 1999.
© Springer-Verlag Berlin Heidelberg 1999

2 Perspective Projection and Notations

Let us analyse the perspective projection and the transformation of a point of the real world into the image. The scene or the camera co-ordinates are set to define the Cartesian system XYZ where the Z-axis is the camera optical axis. Let the point $(0,0,-f)$ be the *viewpoint* or the centre of the lens of the camera. Thus the *image plane* (x, y) coincides with the XY-plane. The parameter f refers to the focal length.

In perspective projection a point in the real world (X, Y, Z) is mapped onto the intersection (x, y) of the XY-plane with the straight line that passes through the (X,Y,Z) point and the viewpoint. From Fig. 1 and from the similitude of triangles it is not difficult to show that the transformation (X, Y, Z) to (x, y) is given by:

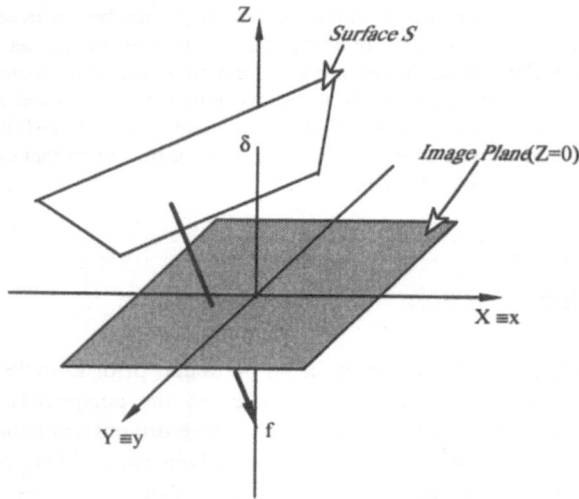

Fig. 1. The model of the perspective projection.

$$x = \frac{fX}{f+Z}, \qquad y = \frac{fY}{f+Z} \qquad (1)$$

Analysing the transformations in (1) and by experience it is clear that a set of parallel lines in the real world converges to a single point in the image plane. This point corresponds to the projection of the line in the set that passes through the point of observation. This point is called the *vanishing point*. In other words to the orientation of a set of parallel lines there corresponds a single vanishing point and, conversely, to a vanishing point there corresponds an orientation.

The detection of the co-ordinates of a vanishing point in the scene can provide relevant information on the depth of a single image, on the distance of objects in a plane, or also on the height of objects in the scene.

It is clear that to determine the vanishing point, at least two straight lines on the scene are necessary.

2.1 Hough Transformation

In this section we will describe how a straight line is detected in an image plane using the Hough Transform [6, 7].

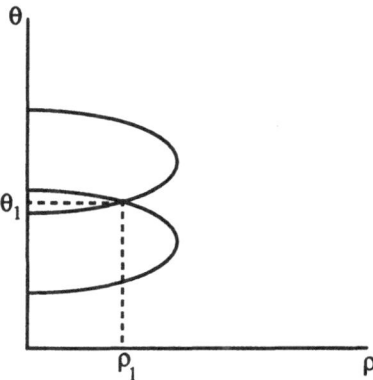

Fig. 2: Straight line detection with HT: the intersection of two sinusoidal curves determines the parameters of a line.

Hough Transform (HT) for straight lines detection consists in mapping the image plane into a parameters plane. A straight line can be therefore described through a parametric representation. The two most common representations are the following:
$$\rho = x\cos\theta + y\sin\theta, \qquad y = mx + q. \qquad (2)$$

The first formula in many cases is considered to be more convenient for the purposes of the present work, the two parameters ρ and θ are defined in limited ranges. In fact ρ is bound by the dimensions of the image, while θ takes values in the $[0, 2\pi[$ range.

Let us analyse the detection process of a straight line in the image plane through the parameter space that in the case of the first formula in (2) is called the *polar plane*. To each point (x_i, y_i) we can associate a sine curve $\rho(\theta, x_i, y_i)$ in the plane (ρ, θ). Two points in the image plane that belong to the same line determine a point in the polar plane (ρ_i, θ_i), given by the intersection of two sine curves. The straight line is therefore characterised by this point (Fig. 2).

This deterministic method is in real cases not useful because of noise in the image and because of the finite granularity of an image. In many cases, however, a rough probabilistic method is enough to determine a line with sufficient precision. A line then is determined by a threshold on the points in the polar plane that are mapped from the image plane. The threshold can be relative to the maximum number of points (x_i, y_i) that determine a single couple (ρ_i, θ_i).

2.2 Previous Work on the Detection of the Vanishing Point

Different algorithms have been implemented to determine the vanishing point in an image [7].

A first method implemented by Nakatani [1] uses the Hough Transformation to detect the straight lines in the polar plane as described above. The lines are then marked on a new image. The point where all the lines converge is the vanishing point. This is a deterministic method that is affected by many errors and in most cases there is not a single point where all the lines converge.

A second one is less deterministic. Here, the point that has the least distance forms all the lines in the image is considered to be the vanishing point. The minimum of the distance function is determined with a least square method.

What all these algorithms have in common is a double transformation. First the image is mapped in the polar plane with HT; then the polar plane is traced back to the image plane by means of some filtering.

In the new algorithm that is presented here, the vanishing point is detected directly on the polar plane. This has the advantage, as we will see later, that we search for a curve whose equation is clearly known and depends on the parameters that give us the position of the vanishing point. This fact helps us to do further approximations on the estimates. This algorithm is based on statistical assumptions that allows us to process noisy images. Further, acting directly on the polar plane, vanishing points that are outside the image do not need any special treatment. At present we are analysing how this algorithm must be modified so that multiple vanishing points can be recovered. Moreover, in the approach that detect the vanishing point from the original image, the choice of the hardness of filtering in the polar plane needed to detect straight lines has to be chosen in relation to the image. In the method that we present, filtering in the polar plane is not needed, thus the method is independent from the image.

3   Vanishing Point Estimation in the Polar Plane

A point given by the intersection of two straight lines in the image plane it is mapped into two points in the polar plane, each one of them representing one of the two straight lines in the image plane. These two points in the polar plane belong to the same arch of a sine curve. In fact, if we take the point (x_0, y_0), centre of a set of intersecting straight lines, on the polar plane the following function is traced:

$$\rho(\theta) = x_0 \cos\theta + y_0 \sin\theta = P(x_0, y_0, \Phi)\sin(\theta + \Phi(x_0, y_0)) \qquad (3)$$

This is a sine curve with amplitude $P = x_0 \cos(\Phi)(1 + y_0^2/x_0^2)$, and phase $\Phi = \arctan(y_0/x_0)$ (Fig. 3).

Thus the vanishing point is identified by the sine curve. In general in an image there are many straight lines that can not contribute to the vanishing point. Moreover there can be errors due mostly to the quantization of the image and to lines detection. Thus a deterministic approach to find the equation of the sine curve in not satisfactory. For this reason it is necessary to implement a statistical method. In this paper we adopted a least square method to find the pair of parameters (x_0, y_0) that gives us the sine curve.

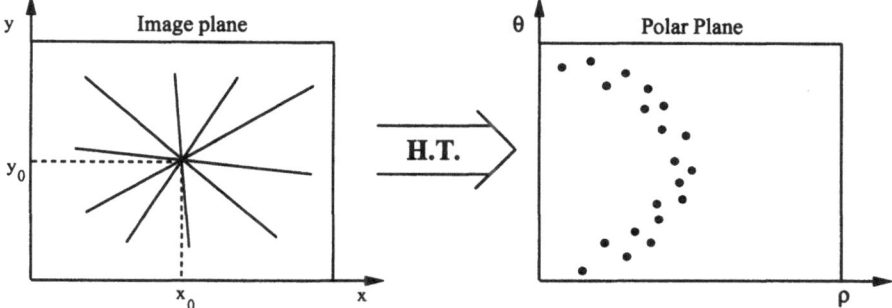

Fig. 3: Hough Transformation of a set of lines that intersect in a single point

The parameters that we have to estimate, x_0 and y_0, represent the co-ordinates of the vanishing point in the image. The estimation is made from a set of measures ρ_i and θ_i expected to lie on the curve defined by equation (3). The points with co-ordinates (ρ_i, θ_i) are weighted by W_i given by the ratio v_i/V, where v_i is the number of times that the pair (ρ_i, θ_i) is observed as a mapping point of a line in the image, and V is the sum of all these. The least square minimisation is thus:

$$\min_{x_0, y_0} \sum_{i=1}^{n} W_i \left(\rho_i - x_0 \cos \theta_i - y_0 \sin \theta_i \right)^2$$

where n is the total number of samples. If now we set $\cos\theta_i = a_i$ and $\sin\theta_i = b_i$ deriving with respect to x_0 and to y_0 and equating to zero, we get the following linear system:

$$\begin{cases} \sum W_i a_i (\rho_i - a_i x_0 - b_i y_0) = 0 \\ \sum W_i b_i (\rho_i - a_i x_0 - b_i y_0) = 0 \end{cases}$$

Summing the left hand side of the two equations and making the following substitutions:

$$A = \sum W_i a_i^2 \quad B = \sum W_i b_i^2 \quad C = \sum W_i a_i b_i$$

$$D = \sum W_i a_i \rho_i \quad E = \sum W_i b_i \rho_i$$

the following linear system is obtained:

$$\begin{cases} Ax_0 + Cy_0 = D \\ Cx_0 + By_0 = E \end{cases}$$

Thus the estimate is straightforward.

After the first estimate of the co-ordinates of the vanishing point is obtained, the previous process is repeated after elimination of the *outliers*. Outliers in the polar plane are identified in the following way. From the estimate of the vanishing point

(x_0, y_0) and for each θ_i the estimated value of ρ_i, $\overline{\rho}_i$, is calculated from (3); at this point the variance of the residuals $\varepsilon_i = \rho_i - \overline{\rho}_i$ is:

$$\sigma^2 = \sum_i W_i \varepsilon_i^2 .$$

Thus the outliers are all those points ρ_i of the polar plane that satisfy the following condition:

$$|\rho_i - \overline{\rho}_i| > k\sigma$$

where k is generally taken between 2 and 3 (the number of outlies increases as k decreases). Elimination of the outliers is done by setting to zero the points that are external to a band around the sine curve identified by the first estimate of the vanishing point.

Once the outliers are eliminated, a new estimate of the vanishing point is retrieved from, the result, obtaining the estimate (x_1, y_1). The loop ends when:

$$|x_{i+1} - x_i| + |y_{i+1} - y_i| \le \varepsilon$$

where ε can be chosen depending on the precision of the estimate desired.

If the polar plane is dimensioned dynamically the case of a vanishing point that is much outside the image can be resolved without any particular problem. The extension to the case in which the vanishing points are more than one can instead be quite difficult unless the exact number is known.

4 Results

In this section some of the results obtained with this algorithm are shown. Most of the images are taken form Internet. In figure 4 we have a first image. We display also the polar plane as it appears before the application of the algorithm (fig. 4-c) and after the elimination of the outliers (fig. 4-d).

In this example the vanishing point is marked with the white circle on the left side of the image in 4-b. It is clear from the polar plane that there are two vanishing points. One is more evident and is the one that has been estimated. The other one, which is on the extreme right of the image (outside the borders), is not very marked and then it does not influence the estimate.

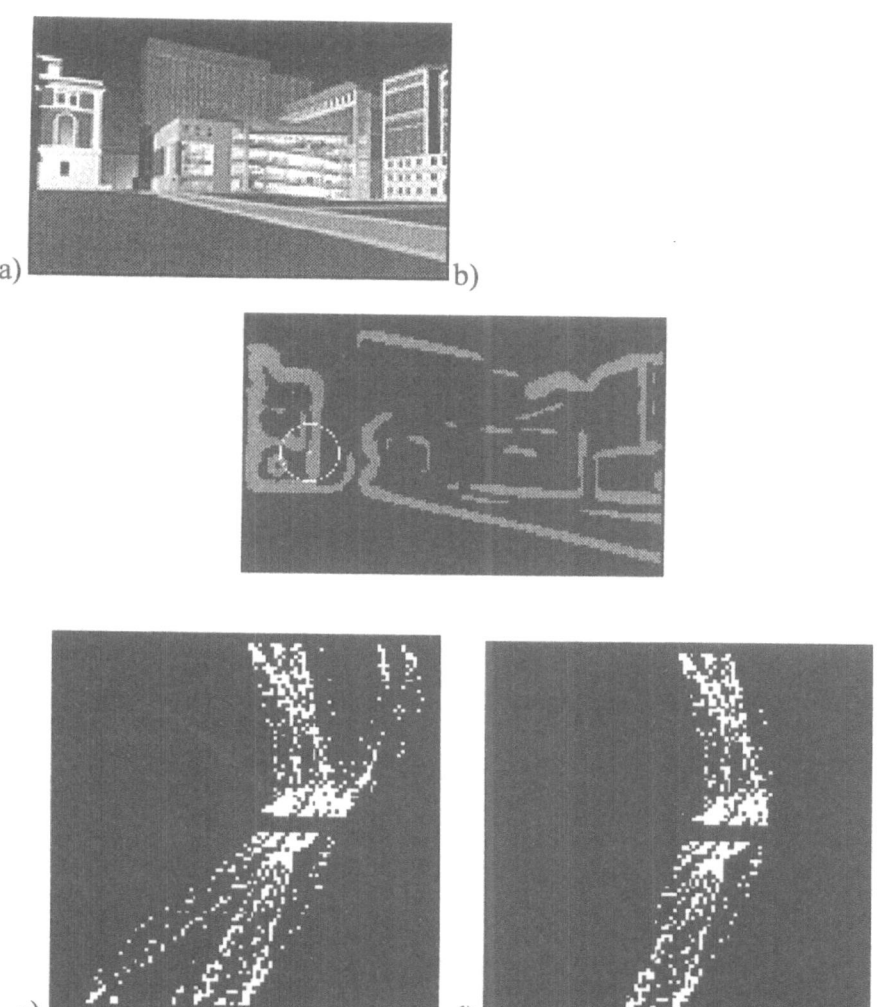

Fig. 4: In this artificial image we have a quite distinct vanishing point that is visible on image b). In the polar plane we can see in d) the sine curve that describes the vanishing point.

5 Conclusions

Through the estimation of the co-ordinates of the vanishing point many depth-related information's can be retrieved from a monocular image. The estimation of these co-ordinates can be done directly on the polar plane obtained with the Hough Transform. In fact in the polar plane the vanishing point is represented by a sine curve that can be detected with a statistical method. Filtering in the polar plane must not be done, hence the algorithm is independent from image structure and parameters.

References

1 H. Nakatani, S. Kimura, O. Saito, T. Kitahashi, Extraction of vanishing point and its application to scene analysis based on image sequence, *Proceedings of 5th International Conference on Pattern Recognition*, 1980, pp. 370-372.

2 H.A. Sedgwick, Environment-Centered Representation of Spatial Layout: Available Visual Information from Texture and Perspective, in: J. Beck, B. Hope and A. Rosenfeld (Eds.), *Human and Machine Vision*, Accademic Press, New York (1983), pp. 425-428.

3 M.J. Magee and J.K. Aggarwal, Determining vanishing points from perspective images, *Computer Vision, Graphics, and Image Processing*, vol. 26, pp.256-267, 1984.

4 B. Brillault and O'Mahony, New method for vanishing points detection, *CVGIP: Image Understanding*, vol. 54, no.2, pp. 289-300, 1991.

5 P. Parodi and G. Piccioli, 3D shape reconstruction by using vanishing points, *IEEE Transactions on Pattern Analysis and Machine Intelligence*, vol. 18, no. 2, 1996.

6 P. V. Hough, Methods and means to recognize complex pattern. U.S. Patent 3,069,654, 1962.

7 D. Ballard and C. Brown, *Computer Vision*. Prentice-Hall, Englewood Cliffs, NJ, 1982.

Parallel Structure in an Integrated Speech-Recognition Network

M. Fleury, A.C. Downton, and A.F. Clark

Department of Electronic Systems Engineering,
University of Essex, Wivenhoe Park,
Colchester, CO4 3SQ, U.K
tel: +44 - 1206 - 872795
fax: +44 - 1206 - 872900
fleum@essex.ac.uk

Abstract. Large-vocabulary continuous-speech recognition (LVCR) speaker-independent systems which integrate cross-word context dependent acoustic models and n-gram language models are difficult to parallelize because of their interwoven structure, large dynamic data structures, and complex object-oriented software design. This paper shows how retrospective decomposition can be achieved if a quantitative analysis is made of dynamic system behaviour. A design which accommodates unforeseen effects and future modifications is presented.

1 Introduction

Two varieties of LVCR system exist: a pipelined structure in which components of acoustic matching and language modelling are separated; and an approach which integrates cross-word context dependent acoustic models and n-gram language models into the search. The former has been thought to be more computationally tractable [1], while the latter has delivered a low mean error rate, 8.2% per word in ARPA evaluation, for a 65k vocabulary, tri-gram language model [2]. This paper examines whether an integrated system could also be parallelised as has been achieved [3] for the pipelined structure.

On a high-performance workstation, even after introducing efficient memory management of dynamic data structures, and optimising inner loops, timings on a 20k vocabulary application, perplexity[1] 145, indicate that a further fivefold increase in execution speed is needed to achieve real-time performance. Increasingly complex future applications are likely to maintain this requirement even as uniprocessor performance increases through Moore's law. This paper proposes a minimum cost redesign of such a sequential system aimed at achieving real-time performance for prototyping, rapid performance evaluation, and demonstrations in the development environment. The imminent ETSI standard for front-end processing enables such systems to act as servers to thin clients possibly on mobile stations. To this end a preliminary parallelisation of a 20k vocabulary application has been made.

---

[1] Perplexity is a measure of average recognition network branching.

P. Amestoy et al. (Eds.): Euro-Par'99, LNCS 1685, pp. 995–1004, 1999.
© Springer-Verlag Berlin Heidelberg 1999

2 Scale of the Problem

A standard stochastic modelling approach to speech recognition has both improved recognition accuracy and the speed of computation [4]. Mel-frequency cepstrum acoustic feature vectors, hidden Markov models (HMMs) [5] to capture temporal and acoustic variance, tri-phone sub-word representation, and Gaussian probability distribution mixture sub-word models [6] are amongst the algorithmic components that have led to the emergence of LVCR. Any parallelisation should seek to preserve this stable structure, onto which further algorithmic innovations have been conveniently hung. Tied states and modes within Hidden Markov models (HMM) for sub-word acoustic matching improve training accuracy for 'unseen' crossword triphones but imply shared data. Such common data also reduce computation during a recognition run on a uniprocessor or a multiprocessor with a shared address space but pose a problem for a distributed-memory parallel implementation.

Speaker-independent integrated systems are being contemplated for database enquiry by telephone services. Development of an LVCR system requires the considerable resources available to large organisations. It may be unrealistic to think that a parallel algorithm, for example [7], will now be newly applied to existing systems. British Telecom (BT) have developed a toolkit for constructing LVCR applications which employs a one-pass time-synchronous speech decoder where tokens carry scores (the sum of sub-word probabilities accrued), maintaining pointers to the n-best recognition network paths. The toolkit paradigm allows a variety of applications to be constructed from a core class library. One such application is considered in this paper. However, even when an application has been developed care must be taken that a subsequent unconsidered parallelisation prematurely fixes the algorithmic content. It is not always possible to predict the side-effects a change might have, or the restriction a change might have on future algorithmic development.

Achieving speaker-independent recognition in real time is significantly harder than speaker-dependent systems. Compare the IBM Tangora PC system [8] which uses an iterative search to reach real-time performance after the recognition network has been trained. Speaker-independent systems must model differences in speech intonation such as accent, dialect, age, and gender. Telephonic applications must also cope with a 10 kHz bandwidth restriction and noise reduction is needed on mobile stations. However, reduction of data-storage and algorithmic complexity is less of an obstacle than when squeezing a recognition interface onto a PC. A conversational interface, additionally requiring speech understanding, is possible on a larger system. The complete BT system is distributed, connected under the CORBA distributed object standard [9], with LVCR as one component. Portability, usually through standardized software, is also an important consideration if adding a parallel extension to the LVCR system. In this respect, the standard message-passing libraries, PVM and MPI, seem suitable.

Given the logistic difficulties of developing the BT system, there are also short-term benefits from parallelising a LVCR system. Though algorithmic de-

velopment is ongoing, there is an interim need to demonstrate real-time behaviour to potential clients. An additional benefit of speed up is a reduction in
performance-tuning run times. Test code in the BT system is included for all major modules to validate correct behaviour. However, in comparative performance
runs, the complete system must be reset even if a minor change is made. For this
purpose, speed-up need not be real time. In short, any improvement in speed
is welcome whether it results from code optimisations, algorithmic innovation
within the existing system, parallelisation or a combination of effects.

3   Effect of Object-Oriented Design

The BT LVCR system has been written in C++. In a run for a 20k word vocabulary, 1188 different functions were called. Object-orientation is a necessary
way of coping with this complexity. Even so, a class-browser, such as SNiFF
[10], is essential for retrospective analysis of the system. Object-orientation involves partitioning of data. Within an object, data are normally held privately or
in protected state (available only to derived classes). Partitioning of data deters
performance optimisations arising from merging functions.[2] Equally, careful consideration has to be given to how a system is decomposed as a prelude to parallelisation. Experimental systems to combine parallelism and the object-oriented
paradigm are documented in [11]. A class of synchronisation and communication primitives which do not extend the language is provided in [12], based
on Parmacs. However, [13] notes that simply adding these primitives does not
encapsulate parallelism suggesting adding path expressions to remove the inflex-
ibility. In the present design, we were constrained by pragmatic considerations
in introducing a parallel structure.

4   Processing Difficulties

Processing on workstations is an order of magnitude away from real time, assuming a 10 ms frame acquisition window, if an $n-$best single-pass search is made.
Formation of the initial feature vector is a task that is well understood and can
be delegated to Digital Signal Processors (DSP's). The Viterbi search algorithm
[14], based on a simple maximal optimality condition, has made the subsequent
network search at least feasible on uni-processors. The Viterbi search is breadth
first and synchronous, not asynchronous and depth first which might be more
suited to parallel computation. A beam search [15] is a further pruning option,
whereby available routes through the network are thresholded. Beam-pruning
with two-tiered score thresholding, signatures [16], and path merging [17] have
been added to the BT system to further reduce search complexity. However, it
is at the network decoding phase that more processing power still needs to be
deployed if no further radical pruning heuristics are forthcoming.

---

[2] In this context, the term 'function' seems more appropriate than the object-oriented
term 'method'.

BBN's HARC system took about twenty times real time for an n-best search on the 992-word vocabulary DARPA test with word pair perplexity of 60. Dual TI C30 DSP's were used to find the feature vectors, while a Sun4 workstation performed network parsing. To improve the speed [18] to double real-time, a two-pass iteration was made, though language parsing must be added on to this time. AT&T implemented a system for the same task on a symmetric multiprocessor (SMP) with four MIPS R3000 processors, but for a full search again recorded twenty times real time. A multicomputer was designed [3] with 128 DSP's in a proprietary interconnection topology, though the system employed to take a standard DARPA test was one node consisting of sixteen processors. Using localised memory and store-and-forward communication, on the AT&T system, results in a non-linear growth in communication overhead and hence in the number of processors needed for larger vocabularies.

The AT&T system has a pipeline architecture; possible because the component parts of the decoder system are separable as feature extraction, mixture scoring, phone scoring, word scoring, and phrase scoring. Other systems such as the Cambridge University HTK system, also with high accuracy scoring, are organised as a token-passing network [19], which is not easily broken into a pipeline. As mentioned in Section 1, the BT LVCR system is also of the token-passing network type.

5   The BT System

5.1   Problems to Be Overcome

The existing BT design, Fig. 1 [20], resists decomposition due to the close coupling of the network update procedures. 49-way acoustic feature vectors (frames) arriving every 10 ms, are applied to each active node of the recognition network. Real, noise, and null nodes embody models for respectively speech, noise, and word connections. The nodes are kept in global lists, necessary because a variety of update procedures are applied. In particular, dormant nodes are reused from application-maintained memory pools without variable delay due to system memory allocation. Large networks, for unconstrained speech or language models beyond bi-gram, are dynamically extended when a token reaches a network boundary. Network extension makes parallel decomposition by statically forming sub-networks problematic because of the need to load balance and hence repeatedly re-divide the network.

5.2   Execution Analysis of the LVCR System

The top-20 functions call graph, Fig. 2, for 97 utterances on the 20k Wall Street Journal test with bi-grams, showed 67% of total computation time including 3% load-time, was taken up by the 'feedforward' update. The branch of function calls, Fig. 3, resulting in the calculation of state output probabilities, bprobs, was uncharacteristically free of sub-function calls which otherwise can give rise

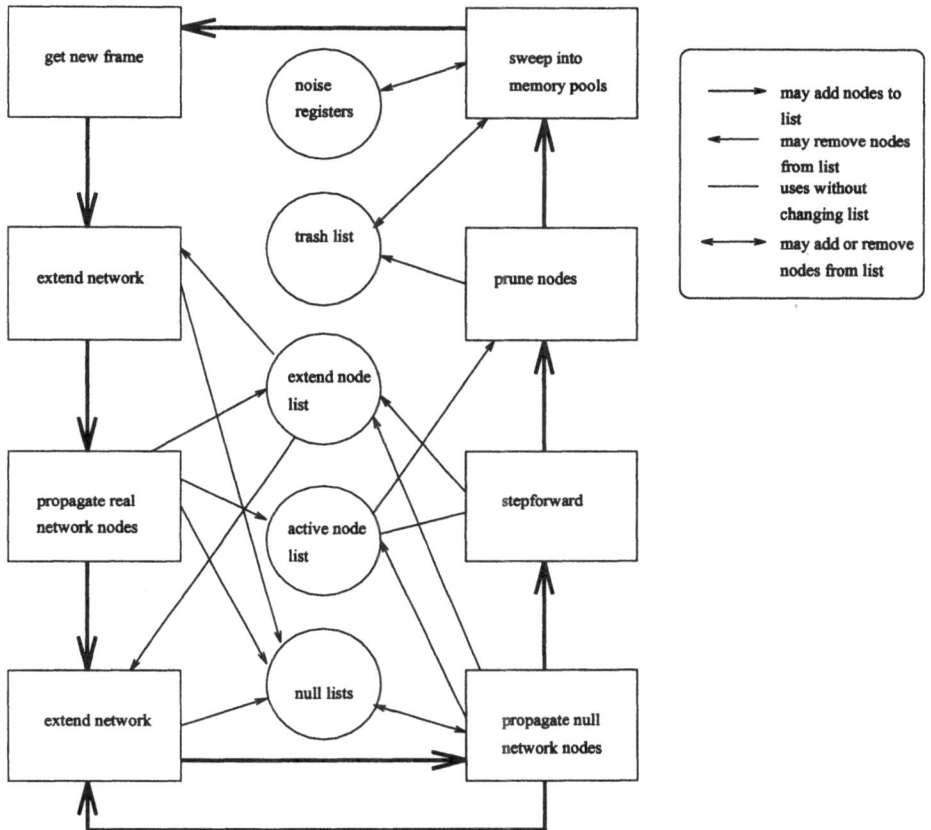

Fig. 1. Network processing cycle

to unforeseen data dependencies. Other parameters such as state transition probabilities, aprobs, remain fixed. The seemingly redundant level of indirection for mode-level checks enables future sharing of modes, which model variety in speech intonation. 44% of time is spent calculating a quadratic part of the sum forming the mixture of unimodal Gaussian densities which comprise the core of any state. 6,641 nodes were present in the mean for 395 frames representing 4s of speech.

6   Parallel Architecture

The parallel architecture that was arrived at can be considered to be a pipeline, Fig. 4, though no overlapped processing takes place across the pipeline stages because of the synchronous nature of the processing. The first of the two pipeline stages employs a data-farm. A data-manager farms out the computationally-intensive low-level probability calculations to a set of worker processes, with some work taking place local to the data-farmer while communication occurs.

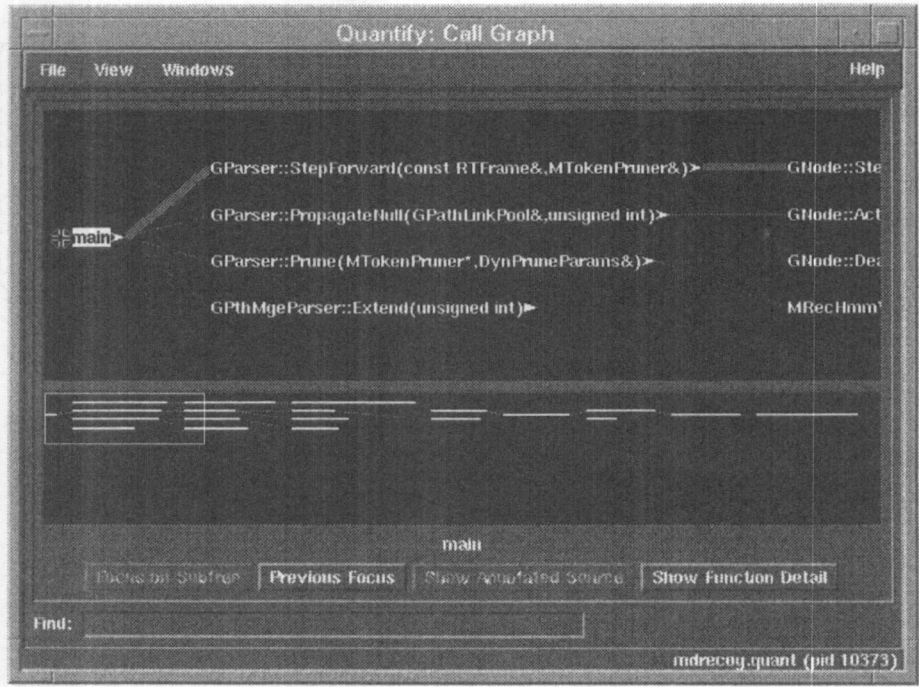

Fig. 2. Extract from high-level call graph, showing call intensity by link width

The standard PVM library of message-passing communication primitives was used in the prototype, run over a network of HP Apollo 700 series workstations. Worker processes each hold copies of a pool of 1,954 tri-state models (and 50 one-state noise models). State-level parallelism requires broadcast of the current model identity (4 bytes), the prototype system needing no global knowledge if started at the 55% point in Fig. 3. By a small breach of object demarcation, whereby at the node object level (with embedded HMM) the state object update history was inspected, the number of messages was sharply reduced, as only one in twelve state bprobs for any frame were newly calculated. A small overhead from manipulation of the node active list enables the parallel ratio to reach the 67% point, collection of thresholding levels then being centralised. Mode level checking when introduced would check for replications at the local level thus limiting the loss in efficiency.

POSIX-standard pthreads are proposed in the second pipeline stage whereby the residual system is parallelised. Propagation of null nodes and real nodes, sweeping-up nodes (thus avoiding over-use of free store), and recognition network pruning functions all have a similar structure. For example, the prune function first establishes pruning levels, which are then globally available for all spawned threads. Once spawned, the host thread of control, i.e. the prune

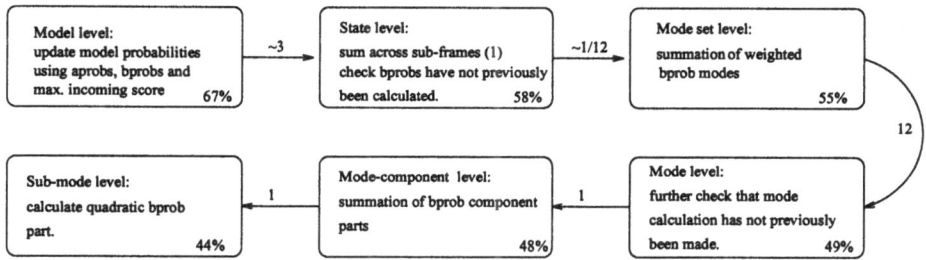

Fig. 3. Probability calculation function hierarchy with call ratios

function, is descheduled until it is reawoken by the completion of its worker threads. Worker threads proceed by taking a node(s) from the active list, deciding whether pruning should take place, and updating the trash list and active list if pruning takes place. The large number of active nodes allows granularity to be adapted to circumstances and the few points of potential serialization, requiring locks, increases the potential scale-up.

7   Future Implementation on an SMP

We considered whether a widely-available type of parallel machine would be sufficient to parallelize the complete system. On a symmetric multiprocessor (SMP), the thread manager would share one processor with the data manager. Efficient message-passing is available for SMPs [21] in addition to threads. Triphones, usual for continuous speech, restrict potential parallelism but with node level decomposition, Table 1, an eight processor machine would approach the required fivefold speed-up while a four processor machine would reduce turnaround during testing. The estimate assumes conservatively that half of the residual system is parallel, while scaling of the system to this level is irrespective of the frame processing workload distribution over time. Inlining of some functions is available as a further sequential optimisation.

parallelization	stage 1		stage 1 & 2	
level/processors:	4	8	4	8
state	1.58	1.92	1.88	1.94
node	2.01	2.42	2.68	3.71

Table 1. Speed-up estimate (Amdahl's law)

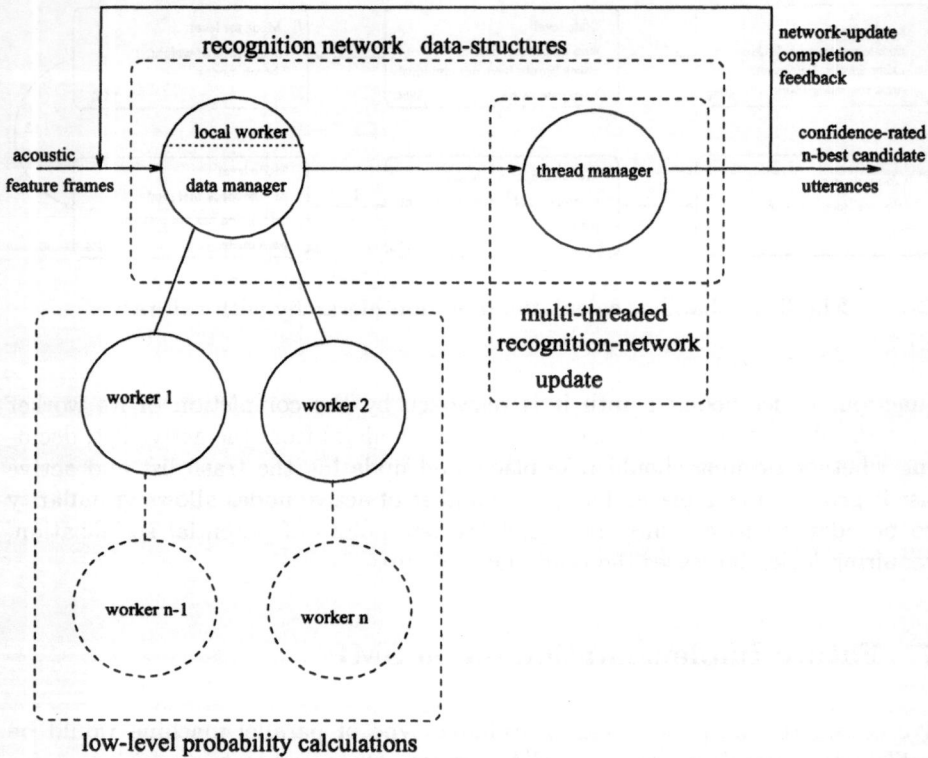

Fig. 4. Synchronous LVCR process pipeline

8   Conclusion

Speech recognition exhibits many features of present day systems that must be taken into account if parallelisation is used as a performance enhancing technique:

- The parallelisation should be conservative, i.e. preserve as much of the established structure as possible, given that the structure is well-proven and has been arrived at after considerable effort, under a sequential programming model.
- The system is complex resulting in considerable logistical difficulties at the testing phase, which a parallel structure may ameliorate.
- Object-oriented design is a way of managing complexity but can impose restrictions which should be considered. Given the advantages of object-oriented design in terms of managing large projects, a parallel design cannot ignore this issue.
- Software standardisation is important for this type of system, which means standards for parallel systems such as PVM and MPI and POSIX threads should be used and developed.

- Parallelisation is not the only way to improve performance and indeed algorithmic innovation in speech recognition, for example tied states, can result in greater improvements.
- The system exhibits algorithmic discontinuity which requires the combination of two different parallel programming paradigms.

Presented with a large-scale system, the prospect may seem daunting to the system analyst. It is important to examine the dynamic behaviour as well as the static structure of the system. Indeed doing this allowed a credible parallel architecture to be formed from the BT LVCR, combining two parallel programming paradigms: the data-farm and the multi-threaded shared-memory model. A high-level design with low-bandwidth message-passing has been developed and prototyped to cope with the major part of the processing. The complete parallel system, which is portable, hardly disturbs the existing integrated structure, but with modest hardware outlay is confidently estimated to bring significant performance gain.

Acknowledgements

This work was carried out under EPSRC research contract GR/K40277 'Parallel software tools for embedded signal processing applications' as part of the EPSRC Portable Software Tools for Parallel Architectures directed programme.

The software described in this paper was developed at BT Laboratories, Martlesham Heath, UK. The authors are indebted to BT Speech Technology Unit for permission to publish the experimental work on parallelising speech recognition. Opinions expressed in this paper are the authors' personal views, and should not be assumed to represent the views of BT.

References

[1] L. R. Rabiner, Juang B.-H., and C.-H. Lee. An overview of automatic speech recognition. In Lee C.-H., F. K. Soong, and K. K. Paliwal, editors, *Automatic Speech and Speaker Recognition Advanced Topics*. Kluwer, Boston, 1996.

[2] P. C. Woodland, C. J. Leggetter, J. J. Odell, V. Valtchev, and S. J. Young. The 1994 HTK large vocabulary speech recognition system. In *ICASSP'95*, volume I, pages 73–76, 1995.

[3] S. Glinski and D. Roe. Spoken language recognition on a DSP array processor. *IEEE Transactions on Parallel and Distributed Systems*, 5(7):697–703, July 1994.

[4] R. Moore. Recogition – the stochastic modelling approach. In C. Rowden, editor, *Speech Processing*, pages 223–255. McGraw-Hill, London, 1993.

[5] L. R. Rabiner. A tutorial on Hidden Markov Models and selected applications in speech recognition. *Proceedings of the IEEE*, 77:257–285, February 1989.

[6] L. A. Liporace. Maximum likelihood estimation for multivariate observations of Markov sources. *IEEE Transactions on Information Theory*, 28(5):729–734, September 1982.

[7] W. Turin. Unidirectional and parallel Baum-Welch algorithms. *IEEE Transactions on Speech and Audio Processing*, 6(6):516–523, November 1998.

[8] S. K. Das and M. A. Picheny. Issues in practical large vocabulary isolated word recognition: The IBM Tangora system. In C-H. Lee, F. K. Soong, and K. K. Paliwal, editors, *Automatic Speech and Speaker Recognition Advanced Topics*, pages 457–479. Kluwer, Boston, 1996.

[9] S. Baker. *CORBA Distributed Objects Using Orbix*. Addison-Wesley, Harlow, UK, 1997.

[10] TakeFive Software GmbH, Stichting Mathematisch Centrum, Amsterdam, the Netherlands. *SNiFF+ Release 2.2 User's Guide and Reference*, 1996.

[11] G. V. Wilson and P. Lu, editors. *Parallel Programming Using C++*. MIT, Cambridge, MA, 1996.

[12] B. Beck. Shared-memory parallel programming in C++. *IEEE Software*, 7(4):38–48, July 1990.

[13] Y. Wu and T. G. Lewis. Parallelism encapsulation in C++. In *International Conference on Parallel Processing*, volume II, pages 35–42. Pennsylvania State University, 1990.

[14] G. D. Forney. The Viterbi algorithm. *Proceedings of the IEEE*, 61(3):268–278, March 1973.

[15] R. Umbach and H. Ney. Improvements in beam search for 10,000−word continuous-speech recognition. *IEEE Transactions on Speech and Audio Processing*, 2(2):353–356, April 1994.

[16] S. P. A. Ringland. Application of grammar constraints to ASR using signature functions. In *Speech Recognition and Coding*, pages 260–263. Springer, Berlin, 1995. Volume 147 NATO ASI Series F.

[17] S. Hovell. The incorporation of path merging in a dynamic network parser. In *ESCA, EuroSpeech97*, volume 1, pages 155–158, 1997.

[18] S. Austin, R. Schwartz, and P. Placeway. The forward-backward search algorithm. In *International Conference on Acoustics, Speech, and Signal Processing*, volume 1, pages 697–700, 1991.

[19] S. Young. A review of large-vocabulary continuous-speech recognition. *IEEE Signal Processing Magazine*, pages 45–57, September 1996.

[20] D. Ollason, S. Hovell, and M. Wright. Requirements and design of the new continuous speech recognition parser – the Grid. Technical report, BT Laboratories, Martlesham Heath, Ipswich, IP5 3RE, UK, 1998.

[21] S. S. Lumetta and D. E. Culler. Managing concurrent access for shared memory active messages. In *IPPS/SPDP'98*, 1998. 7 pages from http://now.CS.berkeley.EDU/Papers2.

3D Optoelectronic Fix Point Unit and Its Advantages Processing 3D Data*

B. Kasche, D. Fey, T. Höhn, and W. Erhard

Friedrich Schiller University Jena
Faculty of Mathematics and Computer Science
Department for Computer Architecture and –communication
Ernst Abbe Platz 1-4
07743 Jena, Germany
Phone: ++49 3641 946373, Fax: ++49 3641 946372
kasche@informatik.uni-jena.de

Abstract. In this paper we show the design of a 3 dimensional optoelectronic hardware approach to realize a fix point processing unit. For that
we show the main ideas of the low level algorithm. We will introduce several concepts and evaluate them with regard to the highest throughput.
At the end we will focus on an application of our 3d approach, especially
on an algorithm for volume rendering of medical image sets.

1   Introduction

Optics is said to become one of the most important components for computing
hardware in the near future. This fact is motivated by the problems which are
generated by using pure electronics for data processing with a high demand on
communication.

There are some consequences of physics on pure electronic chip fabrication.
The so called MOORE's law [1] says that the transistor density is doubled every
12-18 monthes. The inference from this fact is both the steady increasing of the
number of transistors integrable into one chip and the decreasing of the transistor
switching time, thus the chip clock rate is increased and one can integrate more
logic within the same silicon.

RENT's rule [2] says that the number of pins necessary for in- and output increases in an exponential way with the number of transistors. The pin limitation
problem is known as the imbalance of a quadratic enlargement of the relative
chip area and the linear enlargement of the number of pins of a chip.

Finally it can be outlined that using only pure electronics the huge amount on
fast communication channels is hard to manage. Optics is said to help overcoming
those problems.

For this optics and electronics have to form a synergetic union. The chosen
algorithms have to be well adapted to the hardware, too. This is our way to
achieve high performance computing.

---

* This project is supported by a grant of DFG (Deutsche Forschungs Gemeinschaft)

P. Amestoy et al. (Eds.): Euro-Par'99, LNCS 1685, pp. 1005–1012, 1999.
© Springer-Verlag Berlin Heidelberg 1999

In our approach we have designed an arithmetic logical unit. For that we designed an integer and a fix point unit to be able to calculate standard functions for a data which is given in a floating point representation. By using optics we could employ the third dimension for data processing.

In this paper we emphasize the fix point unit. We have developed several approaches which are based on the so called *BICDIC* (bit completion digital computer) and *CORDIC* (coordinate rotating digital computer) algorithm belonging to the class of add and shift algorithm. These kind of low level algorithm were developed further and we could condense 8 of them into an unique structure.

Out of all developed concepts we want to determine the most efficient processing method. These concepts are characterized by the art of data processing. We used a bit serial, a bit parallel and a method using a redundant number representation. By using the redundancy it is possible to add any two numbers within a constant time.

The design process of a synergetic relationship between optics, electronics and last but not least between the low level algorithms is to be applied also to the application algorithms. These algorithms are called high level algorithms. As an example we will introduce algorithms which are necessary for 3d medical image processing. We will show that we are able to process 3 dimensional datasets by using our approach.

2   Synergy of Optics and Electronics

There are so called *Smart Pixel* elements which can guide to a solution of the communication problems using only pure electronics. It is a synergy of an optical and electronic signal processing, thus one Smart Pixel consists of an optical input and optical output to communicate and consists of an electronical processing unit like common VLSI chips. Figure 1 illustrates this fact.

But the electronical processing unit is not as complex as purchasable, common chips. Theoretical study shows [3] that there is a certain small size of such an Smart Pixel to ensure the highest efficiency. In general the following holds: the less sized a Smart Pixel is the more efficient is communication.

Thus the main task, designing an optoelectronic chip using Smart Pixel, is to find out the best ratio between chip area necessary for electronical processing and the chip area that is needed for the optical receiver and transmitter. In the following chapters hardware approaches with the best efficiency with respect to calculate standard functions are estimated.

3   Arithmetic Logical Unit

Our aim was an arithmetic logical unit (ALU) which is to realize using an optoelectronical approach to overcome problems in communication mentioned above.

In this paper we will focus on a fix point unit which is used in our ALU in conjunction with an integer unit. There will be a memory and an input output unit as well. All of them are partially controlled by the control unit.

All the units of each node consist of several processing elements. A certain number of processing elements form one pipeline which can fulfill the property of the dedicated unit. Thus parallel processing is realized not only by using several nodes but using several pipelines within each unit. In order to realize a fully synchronized processing of different data each pipeline must have the same time behaviour. Consequently, we will have a SIMD like structure, but the single instruction means in our case a floating point instruction. I.e. each pipeline could realize a different calculation of a standard function, but within the same time window. So we have a weaker SIMD structure.

4   Algorithm

In our arithmetic logical unit we have to calculate standard functions. There are several methods calculating standard functions. Since we want to get a Smart Pixel based approach we want to use only simple operation. Finally we are looking for an algorithm which is absolutely well geared with the hardware and vice versa.

Approaches based on table look up, power series or restoring algorithm are too space consuming and not as uniform as it would be necessary for a Smart Pixel based approach.

By looking for additional algorithm to realize standard functions we came across so-called *BICDIC and CORDIC algorithm*. Here we have to endeavour only simple operations.

All these algorithms are iterative procedures. We start with a triple (x_0, y_0, z_0). Each tuple is modified by applying a special transformation instruction to the successor triple.

If the CORDIC [4] algorithms are used, one will get a more unique structure from the beginning but it is necessary for some functions to execute additional procedures[5]. We were able to condense 8 different functions into one unique scheme, thus we can use always the same hardware with only slight modifications. We can calculate logarithm, exponential, square root, multiplication, division as well as sine, cosine and arctangent function.

All developed and adapted algorithm for the 8 standard functions can be found in literature[6].

Well adapted means that we use only simple operations SHIFT, ADD, SELECT. Thus, we have optimal starting conditions to design a Smart Pixel based approach. Here we have pursued two different ways.

First we investigated a so called multi chip version and evaluated the performance. This was followed by a single chip investigation.

5   Multi Chip Approach

We started to design a multi chip version in order to determine the estimated performance of a system which may require high technological equipment to be build up. That means we would have to design 4 different chips, each with a different functionality, and stack them together within a 3d setup. Figure 2 illustrates the composition.

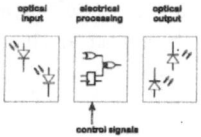

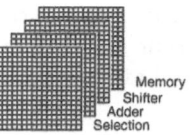

Fig. 1. Setup of a Smart Pixel

Fig. 2. Setup of the 3d multi chip approach

A pipeline is built up connecting all modules adjacent to each other. The largest module determines the over-all dilatation of a pipeline. For instance if one adder functionality covers the whole chip area of the add module only one pipeline could be realized.

We assumed, that the over-all dilatation of one pipeline is determined by the largest pipeline stage, i.e. by the adder functionality. Therefore we considered different approaches by using different methods to perform the adding.

To determine the best ratio between computing time and chip area we designed a *bit serial*, *bit parallel* and a *bit redundant* method. For each method we determined the required chip area and could determine the computing performance for the purpose of a maximized throughput.

The throughput Θ is determined by the number of parallel working pipelines (#Pipes.Chip), the number of steps (s) and the chip clock rate (Δt). The number of pipelines is determined by the whole chip area and the area occupied by one complete pipeline.

$$\Theta = \frac{\#\text{Pipes.Chip}}{\Delta t \cdot s} = \frac{\frac{A_{\text{Chip}}}{A_{\text{pipeline}}}}{\Delta t \cdot s} \tag{1}$$

The performance with respect to the given technological parameter was determined.

The throughput depends on the applied word length. If a word length of 32 bit is used, a maximum of about 35 giga operation per second could be performed. Giga operation per second means 10^9 finished calculations out of the 8 realized standard functions.

At the moment it is quite difficult to build just one optoelectronic chip, let alone four. It can hardly be justified to build up a large system just to show the principle technical feasibility. That's why we investigated a single chip as well. Here we can not expect as much performance as for the multi chip approach, but it might be possible to really get a realizable optoelectronic chip.

6   Single Chip Approaches

We have realized 3 different processing methods as we have done it for the multi chip approach.

In all approaches we have one chip to realize the functionality and another one to provide the table values. On the first chip there are all the pipelines each corresponding to one calculation. Due to different methods employed for the add functionality the pipelines require different chip area. Each iteration is calculated by the corresponding pipeline stage. Each time one iteration is finished, i.e. one calculation is done, the result is handed over to the north.

For the bit serial processing one processing element realizes the whole functionality of one iteration. That's why we expected to get the most pipelines next to each other onto the chip.

The behaviour of the 3 approaches mentioned above was described by a hardware description language (VHDL)[1], synthesized into a gate layout and finally into a transistor layout. The transistor layout is based on a $0.8\mu m$ CMOS process of the AMS company[7]. Thus we could determine the number of necessary transistors and the chip area as well as the critical path length in order to determine the maximum clock chip rate. All of this parameters are taken into consideration when we have evaluated the performance.

7   Performance Evaluation of the Single Chip Approaches

Our aim was maximizing the throughput when calculating function values of standard functions. So we determined the throughput using formula (1) again.

We have determined the bit serial approach as the most efficient one, with respect to our aim which was the maximized throughput. What does the supremacy means? We could see, that the gain of throughput by parallel processing is made up for the higher demand for the area size of the processing elements.

The bit serial approach was outlined as the best one, so we have determined the setup of the future hardware solution in more detail. The two necessary chips communicate optically. The left provides the necessary table values and the right one calculates the results of each iteration. At the top we get the results. Realizing 60 pipelines we get 60 results each clock cycle.

Knowing all the logical and technological parameters we were able to determine the real performance of the two chips working in a bit serial way. The performance is shown in Fig. 3. Here we can see, that our system outperforms existing signal processors, but not a super computer. But this was not our aim. We should mention as well, that the purpose of the digital signal processors used

---

[1] This was done in cooperation with the University of Erlangen

for comparison is not the maximized throughput. Such a chip have to realize a fast calculation of one single function value as well. But there is a problem if one wants to calculate function values more often. If we would have to design a chip calculating single function values as fast as possible we would have chosen the bit redundant approach. This is the fastest approach in terms of a single calculation.

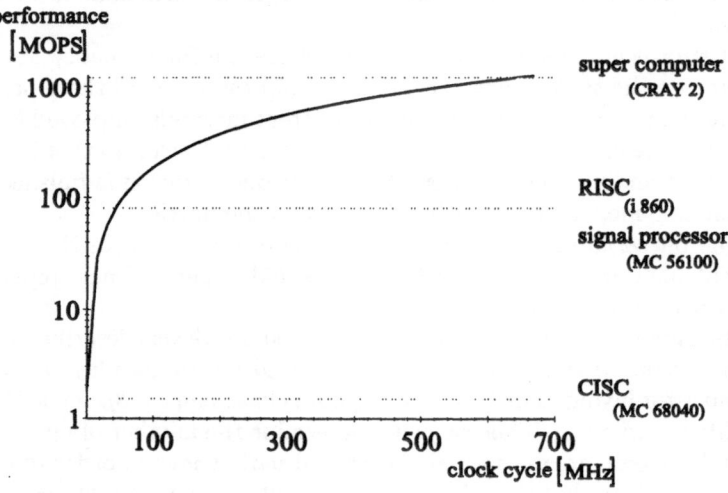

Fig. 3. Real performance of the bit serial chip approach in comparison to other existing fix point units

8 Application

As we have seen in the last section, we outperform existing system, even with the single chip approach, if we have a high demand on calculating function values. So we have pursued our investigations in the field of 3d imaging processing. Here we assign each voxel (pendant to pixel, but volume picture element) one calculation unit, i.e. one pipeline of the approaches described above. Thus we have the conformity between the computing task and our future hardware. Consequently we can more or less easily access the neighbours of each voxel within an one-step or a two step communication, see Fig. 4.

One application for our hardware could be a volume rendering or an artificial lighting. Here we have the volume data set and some light sources. There is a starting point, called x_0 and a given direction ω with the scaling $s \in [0, 1]$ (see Fig. 5).

The light intensity at one voxel within the 3d data set depends on the initial light intensity at the position zero in the direction ω, ($I(0, \omega)$) and the sum of

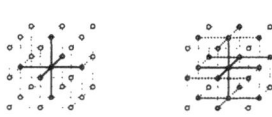

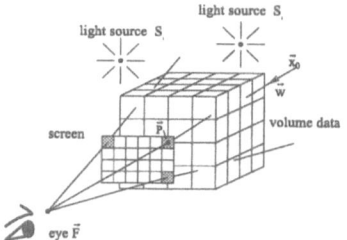

Fig. 4. The accessable neighbours using one or two steps to communicate

Fig. 5. General setup of a volume rendering scheme

all the light coming from the source $(J(s', \omega'))$ and all the points between. This light is determined by the extinction (κ), i.e. absorption, emission and scattering, the optical depth (τ) and the optical density (ϱ) [8] thus finally (2) holds.

$$I(s, \omega) = I(0, \omega) \cdot e^{-\tau(0,s)} + \int\limits_0^s J(s', \omega') \cdot \kappa \cdot \varrho(s') \cdot e^{-\tau(s',s)} ds \qquad (2)$$

One can see, that we integrate the radiance. This is modified by applying the exponential function. The scattering is determined by using sine and cosine function. At the end we need to realize a multiplication. The whole algorithm was described by an abstract description language and simulated on a MASPAR [7] multi processor system. Here we have determined that each Smart Pixel processing element have to have 8 register and uses six of the eight elementary functions mentioned in Chap. 4.

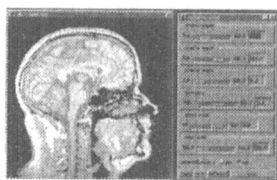

Fig. 6. Screenshot of the X11 application processing volume data

In order to prove the correctness of the algorithm manipulating 3d data sets we have designed a simulation tool. Fig. 6 shows a screen shot of the X11 application.

Another application is a 3d image rotating and an image correlation procedure. There the required amount of calculations is huge enough because common systems require still more calculation time than it would be necessary to realize a real time processing.

9   Summary and Outlook

The major problems in the current VLSI design are restrictions of both the number of available pins and the off–chip communication speed. The current approach of increasing integration density of VLSI chips keeps these problems alive and still increases the difficulties. Due to physical reasons the ability of a high speed off–chip communication in the same range of the on–chip communication is very difficult to achieve. Optoelectronic 3D circuits based on Smart Pixel technologies offer a principle solution for the problems mentioned above.

In our paper we presented a multi chip approach based on Smart Pixel technology as well as a single chip solution. We determined the necessary chip areas by describing the hardware using a hardware description language and synthesizing these into transistor layouts. Using this information we were able to determine the performance in terms of throughput. Our systems outperforms existing systems, but has still less computing power than super computers. But we don't have such a high expenditure of hardware as it is necessary for common super computers.

The most spectacular result was the supremacy of the bit serial approach over the bit parallel method using a conditional sum adder and even over a method adding two numbers by applying a redundant number representation.

We outlined the theory of BICDIC and CORDIC algorithm because an adapted kind of these algorithm was applied in our approaches.

We have finished our paper by presenting a volume rendering application in the field of medical image processing.

References

[1] G.E. Moore. Some Personal Perspectives on Research in the Semiconductor Industry. In A. Rosenbloom, S. Richard, and J.W. Spencer, editors, *Engines of Innovation*, pages 165–174. Harvard Business School Press, 1996.

[2] D. K. Ferry, L. A. Akers, and E. W. Greeneich. *Ultra Large Scale Integrated Microelectronics*. Prentice Hall, Englewood Cliffs, New Jersey, 1988.

[3] D. Fey and W. Erhard. Algorithms for High–Performance Computing with Smart Pixels. In G.A. Lampropoulos et al., editors, *Applications of Photonic Technology*, pages 97–100, New York, 1995. Plenum Press.

[4] J. Walther. A unified algorithm for elementary functions. In *Joint Computer Conference Proc.*, volume 38, 1971.

[5] Jean Duprat and Jean-Michel Muller. The CORDIC ALgorithm: New Results for Fast VLSI Implementation. *IEEE Transactions on Computers*, 42(2):168–177, February 1993.

[6] D. Fey, B. Kasche, C. Burkert, and O. Tschäche. A specification for a reconfigurable optoelectronic VLSI processor suitable for digital signal processing. *Applied Optics*, 37(2):284–295, January 1998.

[7] The mention of brand names in this paper is for information purposes only and does not constitute an endorsement of the product by the authors or their institutions.

[8] S. Chandrasekhar. *Radiative Transfer*. Oxford University Press, Dover, N.Y., 1960.

Parallel Wavelet Transforms on Multiprocessors*

Manfred Feil, Rade Kutil, and Andreas Uhl

RIST++ & Department of Scientific Computing
University of Salzburg, AUSTRIA
{mfeil,rkutil,uhl}@cosy.sbg.ac.at

Abstract. We discuss several issues relevant for parallel wavelet transforms and their possible implications on the choice of a proper programming paradigm for corresponding multiprocessor implementations.

1 Introduction

In this work we focus onto special problems associated with almost each parallel wavelet algorithm (here we compare pyramidal wavelet decomposition [3], wavelet packet decomposition [2], and the à trous algorithm [1]):

- Data decomposition strategies
- Handling of border data

Specifically we investigate the impact of these problems onto the choice of a proper programming paradigm for multiprocessors (i.e. shared memory programming vs. message passing). It is very interesting to see that very different results occur for different types of algorithms.

2 Wavelet Transform Algorithms

The fast wavelet transform (FWT) can be efficiently implemented by a pair of appropriately designed highpass and lowpass filters. A 1-D wavelet transform of a signal S is performed by convolving S with both filters and downsampling by 2. This operation decomposes the original signal into two frequency-bands (called subbands), which are often denoted coarse scale approximation and detail signal. Then the same procedure is applied recursively to the coarse scale approximations several times. Higher dimensional FWT is performed in separabel manner leading to $2^s - 1$ detail signals (and one coarse scale approximation) at decomposition level i in the s-dimensional case.

Wavelet packets (WP) represent a generalization of the FWT. Whereas in the wavelet case the decomposition is applied recursively to the coarse scale approximations only, in the wavelet packet decomposition the recursive procedure is applied to all the coarse scale approximations and detail signals, which leads

---

* This work has been partially supported by the Austrian Science Fund FWF, project no. P11045-ÖMA.

P. Amestoy et al. (Eds.): Euro-Par'99, LNCS 1685, pp. 1013–1017, 1999.
© Springer-Verlag Berlin Heidelberg 1999

to a complete wavelet packet tree (i.e. binary tree and quadtree in the 1-D and 2-D case, respectively) with 2^{si} frequency subbands at decomposition level i in the s-dimensional case.

The "à trous" algorithm represents a non-orthogonal discrete approach to the classical continuous wavelet transform. The basic idea behind the *à trous* algorithm is to design a discrete wavelet transform *without* a following decimation step. In the 1-D case a filtering operation is performed similar to the computation of the coarse scale approximation in the FWT case. Given the well-known *two-scale equation* found in classical wavelet theory, $\phi(k) = \sqrt{2} \sum_l h_l \phi(2k - l)$, the approximation coefficients are computed by $c_i(k) = \sum_l h_l c_{i-1}(k+l)$ (FWT case) and by $c_i(k) = \sum_l h_l c_{i-1}(k + 2^{i-1}l)$ (à trous case). Note that the expression "$(k + 2^{i-1}l)$" creates the "trous" (French for holes) in the computation which means that the distance between samples increases by a factor 2 from scale $i - 1$ ($i > 0$) to the next one. This fact has severe implications for possible parallelization approaches at the borders of the data.

In contrast to the FWT the detail signal is computed by $w_i(k) = c_{i-1}(k) - c_i(k)$. À trous decomposition in higher dimension is not performed in separabel manner but by direct convolution with a higher dimensional convolution kernel.

3 Programming Paradigms, Data Decomposition, and Border Treatment

We investigate two different programming paradigms on multiprocessors: shared memory programming and message passing. Shared memory programming can be performed very quickly by simply inserting parallel compiler directives into sequential programs. On the other hand, message passing requires an explicit programming of each communication event occurring among processors and is consequently very time demanding. However, message passing programs written in e.g. MPI or PVM may be used without changes on different architectures (no matter if multiprocessors or multicomputers) whereas shared memory programming mostly uses a native programming language.

The coarse grained programming style required for message passing demands a discussion of data decomposition strategies. Whereas there is nothing to discuss about data decomposition in the 1-D case, different possibilities exist for the 2-D and 3-D cases. The main distinction is between stripe partitioning and checkerboard partitioning (which are the 2-D cases, in the 3-D case simply a third dimension is added). Whereas checkerboard partitioning offers the obvious advantages of minimizing the block-border length at the cost of a larger number of neighbouring blocks, stripe partitioning requires only communication between two direct neighbours.

In the case of wavelet packet decompositions it has turned out that it is advisable to to perform a subband based data decomposition instead of the concepts mentioned before at a certain stage of the computation. This is explained briefly for the 2-D case. To do the decomposition in parallel, the data is redistributed according to the subband structure (after an initial stripe or checkerboard distribution - see Fig. 1 on the left side) at that specific decomposition level (denoted "distribution level") where the number of PE is lower or equal to the number of subbands. Fig.1 shows the data distribution onto 4 PE from level $j = 0$ to 2 (where the data redistribution takes place between level 0 and level 1 and f denotes the number of the subband).

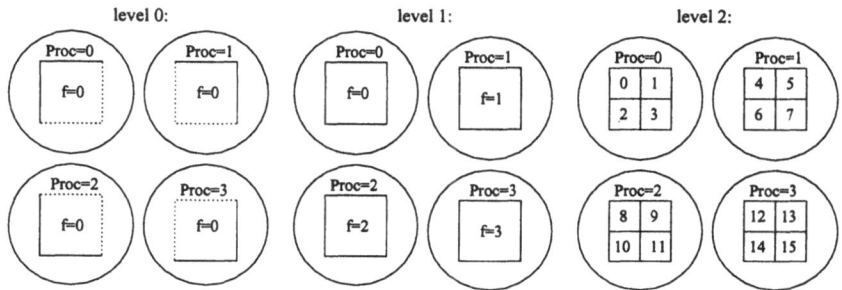

Fig. 1. Repartition of the wavelet packets onto 4 PE

Now we proceed with the discussion of border treatment for all types of wavelet transforms. Recall that the need for border data located on adjacent PE is caused by the nature of the filtering process which involves several neighbouring data points in order to compute a single transform coefficient. In order to provide the necessary border data to each PE we may distinguish between two approaches for border treatment (which trade off computation vs. communication demand):

- *Data swapping* method (also known as *non-redundant data calculation*): each PE computes only non-redundant data and exchanges these results with the appropriate neighbour PE in order to get the necessary data for the next calculation step (i.e. the next decomposition level).
- *Redundant data calculation* approach: in the initialisation step we do not only provide its share of the original signal to a PE but provide the entire data necessary to carry out the required decomposition steps on each PE locally without any communication demand (i.e. a highly redundant data distribution - overlapping blocks).

4   Experimental Results

For the 2-D case we employ as test image a 1024×1024 pixel version of the *Lena* image and perform a complete decomposition (i.e. 10 levels) with Daubechies $W20$ filters. In the 3-D case we use video data with 256×256 pixels per frame and 512 frames. All the computations have been performed on a SGI POWER-Challenge GR with 20 R10000 processors using either the native shared memory programming language PowerC or a native version of the PVM message passing library.

We start with the discussion concerning border treatment in the case of message passing. For 3-D wavelet decomposition (Fig. 2.a) and the 2-D á trous algorithm (Fig. 3.b) data swapping is clearly superior. This may be easily explained by the fact that the amount of redundant computations is too high especially related to the relatively inexpensive communication on the target architecture which is required by data swapping. However, almost no difference occurs in the case of 3-D wavelet packet decomposition (Fig. 2.b) – the obvious reason is that only 3 decomposition steps are performed with redundant data calculation or data swapping, the rest of the computation is performed using subband based data distribution. Therefore, the overall amount of the computations where border problems are involved is rather small and consequently it makes no difference which border-treatment approach to choose. This leads us directly to the question about data decompositions.

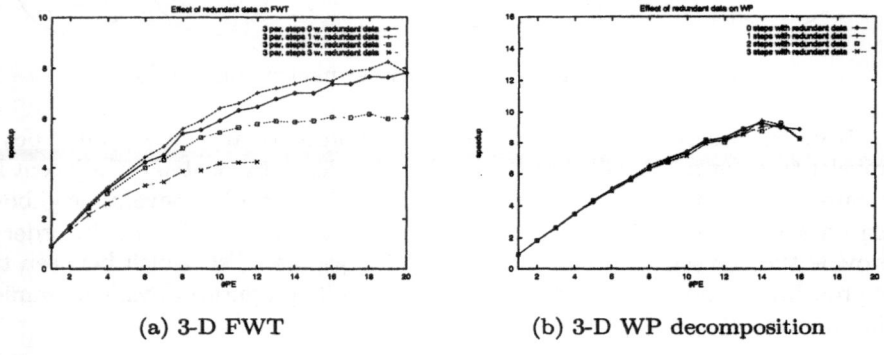

(a) 3-D FWT   (b) 3-D WP decomposition

Fig. 2. Data swapping vs. redundant data calculation

Concerning the question whether stripe or checkerboard partitioning is the better way to distribute data it turns out that both methods perform equally on the architecture considered. Fig. 3.a shows the advantage of subband based parallelization for wavelet packet decomposition in a drastic way – almost no speedup is achieved when "no distribution" (i.e. no subband based distribution) is performed, whereas monotonically increasing and significant speedup is achieved with subband based data distribution.

Now let us proceed to the question whether message passing or shared memory programming is the better way for parallel wavelet transforms on multiprocessors. Considering the results for 3-D wavelet packet decomposition (see Fig. 3.a) we clearly see that the algorithm implemented in PowerC exhibits worse scalability as compared to the PVM case. This trend is also observed for 3-D wavelet decomposition (and for both types of algorithms in lower dimension). The PowerC algorithm for 3-D wavelet packet decomposition where simply the loops corresponding to data rows or slices are distributed does not reach any speedup ("no distribution"). If for shared memory programming the subband

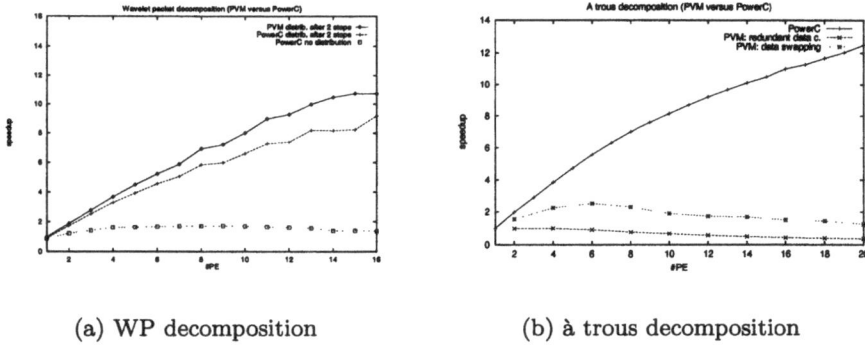

(a) WP decomposition   (b) à trous decomposition

Fig. 3. Comparison of programming paradigms: message passing (PVM) vs. shared memory (PowerC)

based data distribution concept is implemented as well (which is fairly complicated to do) we obtain a considerable speedup, but still lower as compared to the message passing case (e.g. for 3-D wavelet packet decomposition speedup 9 vs. 11 with 16 PE).

A very different situation may be observed for the 2-D à trous algorithm (Fig. 3.b). Whereas we do not reach speedup higher than 3 using message passing, speedup close to linear is achieved with shared memory programming. This effect is on the one hand due to the high amount of computation involved in the à trous algorithm (which allows the high speedup), on the other hand due to the expensive border treatment (see section 2 – this causes the bad performing message passing approach).

Whereas shared memory programming obviously is the paradigm of choice for the à trous algorithm and leads to acceptable (but clearly less scalable) results for the FWT, the performance of a straightforward shared memory programming of wavelet packet decomposition is extremely poor. Even if the subband based distribution concept is employed (which requires profound algorithmic knowledge and a significantly higher implementation effort) message passing still remains clearly superior.

References

[1] M. Feil and A. Uhl. Real-time image analysis using wavelets: the "à trous" algorithm on MIMD architectures. In D. Sinha, editor, *Real-Time Imaging IV*, volume 3645 of *SPIE Proceedings*, pages 56–65, 1999.

[2] A. Uhl. Wavelet packet best basis selection on moderate parallel MIMD architectures. *Parallel Computing*, 22(1):149–158, 1996.

[3] M-L. Woo. Parallel discrete wavelet transform on the Paragon MIMD machine. In R.S. Schreiber et al., editor, *Proceedings of the seventh SIAM conference on parallel processing for scientific computing*, pages 3–8, 1995.

Vector Quantization-Fractal Image Coding Algorithm Based on Delaunay Triangulation

Zahia Brahimi[1], Karima Ait Saadi[2], Noria Baraka[3]

128, Chemin Mohamed Gacem
B.P. 245 El Madania – Alger ALGERIE
☏ : (213) 02 27 68 68
Fax : (213) 02 27 59 37 / 27 93 93
[1]zbrahimi@yahoo.com,[2]ait_saadi@yahoo.com,[3]baraka2@yahoo.com

Abstract. In this paper, we present a flexible partitioning scheme for fractal image compression based on adaptive Delaunay triangulation. Such partition is computed on an initial set of points obtained with a split and merge algorithm in a grey level dependent way. The triangulation is fully flexible and returns a limited number of blocks allowing good compression Ratios Moreover, a vector quantization algorithm is implemented on pixel histograms directly generated from the triangulation. The aim is to reduce the number of comparisons between the two sets of blocks involved in fractal image compression by keeping only the best representative triangles in the domain blocks set. Quality coding results are achieved at rates between 0.06 b/pixel and 0.2 b/pixel for a PSNR between 22 and 25 dB depending on the nature of the original image and on the number of triangles refereed.

1 Introduction

This paper focuses on a fractal image coding technique based on the Delaunay triangulation, in association with a vector quantization of the triangles in order to reduce the complexity of the coding phase. The partitioning has the advantage of being fully flexible. It is computed on a set of points placed on images support in a grey level dependent way, with the help of a split and merge approach. The fractal method offers high compression ratio, good image quality, and resolution independence of the decoded image. Its disadvantage is the long encoding time due to the repeated search of the domain pool, much effort has been spent in devising ways to overcome this problem [1].

2 Fractal Image Coding

The principle of fractal image compression is to construct an eventually contractive operator W for which the fixed image A_t is a close approximation to a given image A to encode. The operator W is computed so that it minimizes a distances between W(A) and A, noted as d (W(A), A). In order to find W, Jacquin proposed [2] the use of a partition of the image A into N non-overlapping range squares R_i so that :

P. Amestoy et al. (Eds.): Euro-Par'99, LNCS 1685, pp. 1018-1021, 1999.
© Springer-Verlag Berlin Heidelberg 1999

$$A=\bigcup_{i=1}^{N}R_i \qquad with\ R_i\cap R_j=\varnothing\ \forall_{i,j}\ and\ i\neq j\ \cdot \qquad (1)$$

Each original square R_i is then approximated by a collage block produced by a linear combination of a constant block and another domain block noted $D_{\alpha(i)}$ larger than R_i and extracted from the original image itself. α is an application from [1, ...N] to [1, ...M]. The collage block is noted $W_i(D_{\alpha(i)})$. The compressed representation of the image A typically contains the partition R construction rule, the coefficients of the N local contractive transforms W_i associated to each block R_i and the coding of the shapes and locations in the image of the corresponding blocks $D_{\alpha(i)}$.

3 Partitioning

We propose a very flexible scheme based on the well-known Delaunay triangulation.
Principle : *Let S be a set of points in the plane. The Delaunay triangulation DT(S) associated to S is the unique triangulation with empty circles. More formally, we can write :*

$$DT(S)=\{(p,q,r)\in S^3,\ C(p,q,r)\cap\ S-(p,q,r)=\varnothing\}\ . \qquad (2)$$

Where $C(p, q, r)$ is the circle circumscribed by the three points p, q and r forming a Delaunay triangulation [3]. To compute the Delaunay triangulation (or the Voronoï diagram), the best known approaches are based on the "divide and conquer" [4] and the "incremental" algorithms [5]. For computing Delaunay triangulation we an incremental approach working by local modification of the diagram by insertion of new vertex. The method is based on split and merge approach initialized on a small number of regular triangles computed on regular grid of vertices [6]. The algorithm proceeds in the following three major steps :

Split and merge algorithm

```
1. Initialization
Construct a lattice S (triangle vertices) on the
image support.
2. Split
Repeat until convergence :
        a- Calculate the Delaunay triangulation of
        the set of vertices.
        b- For each triangle : if the triangle is not
        homogenous then insert a vertex on its
        barycenter.
3. Merge
Extract the useless vertices (merge the triangles
under the star polygon criteria).
```

A vertex $p_i \in S$ is said to be useless if all the triangles for which p_i is a vertex similar with respect to their grey level variance and mean. The aim of the third step is to decrease the number of triangles obtained after a new triangulation of all the star polygons [3]. Two parameters are used in this algorithm of split and merge : the grey level variance and the mean value. They are related to the concepts of homogeneity and similarity between adjacent triangles. Two partitions R and D will be extracted from the split and merge algorithm.

4 Implementation

4.1 Encoding Principle

Let N the number of blocks R_i of the partition R, obtained after an adaptive Delaunay triangulation which used the merge and split algorithm, and Q the number of block D_i of the partition D (a regular partition) (Fig. 1). The algorithm consists on finding for each block R_i, a block D_i, minimizing the Root Mean Square Error (RMS) d between grey scale values of R_i and those of the block image transformation $W_i(D_i)$. The triangles in the partition D have the same sizes and shapes, they are nonoverlappings. When the number of triangles is great, the transformed image W(A) can be very similar to the original one A. However, the time coding increase with respect to the number of inter-blocks comparisons (N*Q).

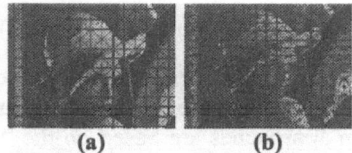

(a)   (b)

Fig. 1. Partition D (a) and partition R (b)

As the partition D is regular, and not depending on the image content, only the step of partitionement needs to be stocked. One octet is used to stock it. For the partition R and as we use the split and merge algorithm, it is not necessary to stock vertices of each triangle. The procedure consists on memorizing details of each step, starting with a partitionement known by coder and decoder. During the split step, adding a point to the barycentry of no homogeneous triangles, is coded on 1 bit. Extracting the useless vertex, during the merge step is also coded on 1 bit.

4.2 Improving Codec Speed

To speed up the matching process of the encoding algorithm, we propose to reduce the number of the triangles D_i resulting from the Delaunay triangulation. by using a classification algorithm that permits the construction of a codebook of triangles D_i. The goal of such an algorithm is to design a codebook containing the best representative triangles D_i. The classification algorithm that we use is the Linde, Buzo and Gray (LBG) algorithm [7]. A modified version of the Loyd algorithm was developed. The best representative element of a cluster is not the centroid but the training vector nearest to the centroid according to the criterion ρ (average MSE). The aim of the classification process is to keep the most significant grey level histograms.
Also in order to improve the speed of the decompression, we used an orthogonalization of the collage space proposed by Oien in [8].

5 Performances

The coder's performance is achieved by computing the PSNR between the original image f(x, y) and the decoded image F(x, y). The compression ratio for a $2^n x 2^n x 8$ b/pixel image is computed differently depending on whether we use the entiere triangulation D during the encoding process or not.

6 Experimental Results

Coding simulations have been performed on the image Lena of 256x256x8b/pixel. Decoded images are close to the original image and the obtained compression ratios are very interesting comparing to the classical fractal algorithm. It performs a reduced encoding complexity while preserving good decoding quality (PSNR-25db)at rates between 0.06 and 0.2 b/pixels depending on the nature of the image (Fig. 2).

Fig. 2. Left : the Original Lena Image, Middle Lena at the tenth iteration, and the Right the Reconstructed Image

7 Conclusion

A fractal coding algorithm. based on a Delaunay partitioning was proposed in this paper. Such a partition is very attractive because it provides a reduced number of blocks compared with square based partitions. Which minimizes the number of mappings. In a future work, we propose to merge neighboring stretched triangles into quadtrilaterals.

References

1. Behnam Bani-Eqbal, "Combining tree and feature classification in fractal encoding of images," Depart. of computer science - University of Manchester, U. K. (1995)
2. Arnaud E. Jacquin, "Fractal Image Coding," IEEE trans. review Proceeding, Vol. 81 (1993)
3. P. Volino, "Triangulation de Delaunay contrainte étude et implantation d'algorithmes," Rapport de stage de DEA (1992)
4. J. P. Prearation and M. I. S. Shamos, "Computational Geometry, an Introduction," S pringer verlay, New York (1988)
5. P. J. Green and R. Sibson "Computing Dirichlet tessellation in the plane," The computer J., Vol. 21, 173-1978
6. J. Vaisey and A. Gersho, "Image compression with variable block size segmentation," IEEE trans. Signal processing, vol. 40, (1992) 2040-2060
7. R. M. Gray, "Vector quantization," IEEEASSP Mag, (1986) 4-29
8. G. E. Oien, S. Lepsoy, "A new improved collage theorem with applications to mult-iresolution fractal image coding," Speech and signal processing, (1994)
9. Y. Fisher, "Fractal Image Compression," Springer-Verlag, 1995.

Topics 13 + 19
Numerical Algorithms for
Linear and Nonlinear Algebra

Robust, efficient and numerically reliable parallel algorithms for the solution of fundamental problems in numerical mathematics are essential components of most parallel software systems for scientific and engineering applications. This topic provides a forum for the presentation and discussion of new developments in the area of parallel numerical methods. All aspects of the design and implementation of parallel algorithms will be addressed, ranging from discussion of the ideas on which they are based to analyses of their complexities and performances on current parallel architectures. Methods for the solution of large linear systems are of particular interest because of their widespread occurrence in many fields, particularly in the numerical solution of partial differential equations. However, contributions dealing with new and improved parallel algorithms for the solution of other problems in numerical linear algebra, linear and nonlinear programming, error analysis, numerical quadrature, differential equations, fast transforms and non-linear systems have also been considered.

Among the 37 articles submitted, 11 have been selected as regular papers and 8 papers have been accepted as research notes. This topic is composed of five sessions mainly concerned with the following themes :

- iterative methods,
- sparse direct methods,
- dense linear algebra,
- nonlinear algebra and
- numerical reliability.

P. Amestoy et al. (Eds.): Euro-Par'99, LNCS 1685, pp. 1023–1023, 1999.
© Springer-Verlag Berlin Heidelberg 1999

mpC + ScaLAPACK = Efficient Solving Linear Algebra Problems on Heterogeneous Networks

Alexey Kalinov and Alexey Lastovetsky

Institute for System Programming, Russian Academy of Sciences
25, Bolshaya Kommunisticheskaya str., Moscow 109004, Russia
{ka,lastov}@ispras.ru

Abstract. The paper presents experience of using mpC for accelerating ScaLAPACK applications on heterogeneous networks of computers. The mpC is a language, specially designed for parallel programming for heterogeneous networks. It has facilities for distribution of participating processes over processors in accordance with performances of the latters. An mpC application carring out Cholesky factorization on a heterogeneous network of workstations is used to demonstrate that the heterogeneous process distribution has an essential advantage over the traditional homogeneous distribution. The application is implemented using calls to ScaLAPACK routines by means of the interface mpC - ScaLAPACK.

1 Introduction

ScaLAPACK [1] is the most famous library for solving linear algebra problems on distributed-memory, concurrent computers. The main target platforms for ScaLAPACK are distributed-memory supercomputers consisting of identical processors, because present high-performance scientific computations concentrate mostly on them.

On the other hand, progress in network technologies is making networks of computers (in particular, networks of PCs and workstations) more and more attractive for high-performance parallel computing. The main difference between supercomputers and networks is heterogeneity of the latters.

The heterogeneity is displayed at least in two forms. Firstly, in the form of heterogeneity of machine arithmetics of such parallel systems. Related challenges existing in writing reliable numerical library software for heterogeneous computing environments have been analyzed in [2].

Secondly, in the form of heterogeneity of hardware performance of individual processors. ScaLAPACK has been developed and tested with one process per processor running [3]. We will refer to that processes distribution as homogeneous. Let see what happens when a parallel linear algebra application, that provides good distribution of computations and communications when running one process per processor in homogeneous environments, runs with one process per processor on a heterogeneous network of computers. Since volumes of computations executed by different processors are approximately equal to each other,

P. Amestoy et al. (Eds.): Euro-Par'99, LNCS 1685, pp. 1024–1031, 1999.
© Springer-Verlag Berlin Heidelberg 1999

more powerful processors will wait for the slowest one at synchronization points. Therefore, the total time of computations will be determined by the time elapsed on the slowest processor.

We have done an experiment corroborating that statement. We considered two local subnetworks of our local network: homogeneous abcd consisting of four SUN workstations a, b, c, and d, and heterogeneous aEFG consisting of four SUN workstations a, E, F, and G. Workstation a belongs to the both networks and other workstations of the heterogeneous network are more powerful then workstation a. Performances of the workstations were estimated by means of Cholesky factorization of a matrix of the same dimension with a sequential LAPACK [4] Cholesky factorization routine dpotf2. The total power of the heterogeneous subnetwork is about 1.9 times greater then that of the homogeneous one. It could be expected that, for example, the parallel ScaLAPACK Cholesky solver [3] would be executed on the heterogeneous subnetwork about 1.9 times faster then on the homogeneous one. But the real situation, shown in figure 1, turned out quite different.

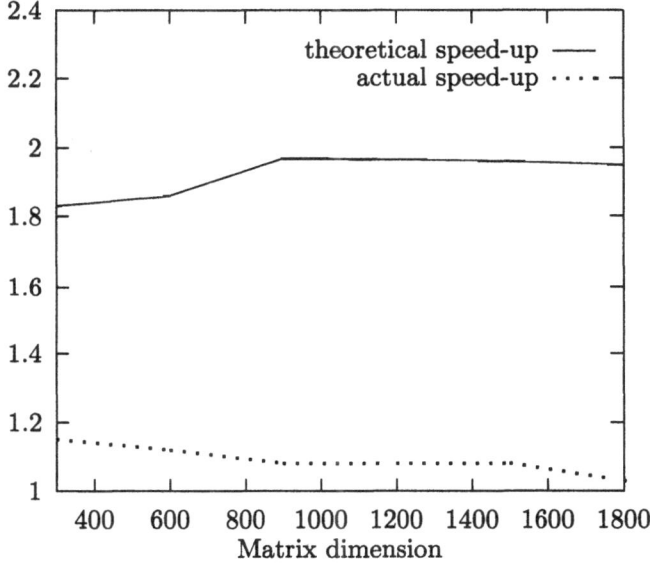

Fig. 1. Speed-up achieved by ScaLAPACK Cholesky solver on the heterogeneous network relative to the homogeneous one. The both networks consist of four workstations. One workstation of the heterogeneous network belongs to the homogeneous one. Other workstations of the heterogeneous network are more powerful. Drawn line represents increase in computing power of the heterogeneous network or theoretical speed-up. Dotted line represents actual speed-up.

A natural solution of this problem is heterogeneous distribution of processes of the parallel program over the processors, taking into account at least differences in performances of processors. The distribution may be done by means of configurational files. But it is a difficult task, and if the application topology is defined at run time (for example, process grid parameters depends on input data), this approach will not work.

An alternative approach is to write such applications that do not need a sophisticated process mapping to start up the applications efficiently. Designed specially to write efficient and portable parallel applications for heterogeneous networks of computers, the mpC language [5] allows to do that. This language is an ANSI C superset allowing to write applications adapting to differences in performances of both processors and communication links of any particular executing network. The basic idea is that an mpC application explicitly builds at run time an abstract heterogeneous computing network and distributes data, computations and communications over the network. The abstract network consists of virtual processors of different performances and different links. The mpC programming system uses this information at run time to map the abstract network to any real executing network of computers in such a way that ensures efficient running of the application on the real network. More about mpC as well as the mpC free software can be found at http://www.ispras.ru/~mpc.

In this paper, we consider only the heterogeneity of processor performances. We propose a heterogeneous distribution of processes over processors when the number of processes involved in computations on a separate processor depends on its performance. We investigate the distribution using a typical linear algebra problem - the Cholesky factorization of square dense matrices. The heterogeneous distribution of the involved processes is performed by an mpC program, while the latter calls a parallel ScaLAPACK solver to perform the parallel Cholesky factorization proper.

Section 2 shortly introduces the mpC language and describes an implementation of the heterogeneous distribution in mpC with calls to ScaLAPACK functions as well as interface mpC - ScaLAPACK. Section 3 gives experimental results of the Cholesky factorization on a network of heterogeneous workstations using the homogeneous and heterogeneous processes distribution.

2   Implementation of Heterogeneous Distribution of Processes over Processors in mpC

The language mpC [5] is a parallel language that allows an efficiently-portable modular programming heterogeneous networks of computers. It provides facilities for specification of requirements on resources, necessary for efficient execution of parallel application, and the mpC programming system tries to satisfy the requirements taking into account peculiarities of any particular heterogeneous network of computers.

The mpC language is an ANSI C superset that introduces a new kind of managed resource, the *computing space*, defined as a set of virtual processors of difference performances. At run time, the virtual processors are represented by actual processes of the particular running parallel application. The programmer manages the computing space by means of creating and discarding regions of the computing space, named *network objects*, just like he manages storage creating and discarding data objects (regions of storage). At any moment of program execution, just a set of defined network objects represents the abstract computing network.

The following mpC function HeHo implements the heterogeneous distribution of processes involved in computations and calls ScaLAPACK to perform parallel Cholesky factorization properly.

```
/*1 */ #define N 100
/*2 */ nettype Grid(nr,nc) {
/*3 */    coord I=nr, J=nc;
/*4 */ };
/*5 */ int [net Grid(nr,nc) v] mpC2Cblacs_gridinit(int *, char *);
/*6 */ void [*]HeHo(repl int P, repl int Q) {
/*7 */    {
/*8 */       int n=N,info;
/*9 */       double a[N][N];
/*10*/       init(a);
/*11*/       recon dpotf2_("U",&n,a,&n,&info);
/*12*/    }
/*13*/    {
/*14*/       net Grid(P,Q) w;
/*15*/       [w]: {
/*16*/          int ConTxt;
/*17*/          ([(P,Q)w])mpC2Cblacs_gridinit(&ConTxt,"R");
/*18*/          pdlltdriver1_(&ConTxt);
/*19*/          mpC2Cblacs_gridexit(ConTxt);
/*20*/       }
/*21*/    }
/*22*/ }
```

The heterogeneous strategy is implemented in three steps:

1. Performances of real processors for a relevant benchmark, Cholesky factorization by means of the LAPACK routine dpotf2, are determined in lines 7-12 with the help of statement **recon**. The statement (line 11) updates at run time the information about processor performances of the executing real network by means of execution of the corresponding computations as a benchmark (in our case, it is a call to function dpotf2). It is supposed that performances estimated when matrix dimension is 100 are not essentialy different from that estimated when matrix dimension is different. Our experiments confirm the supposition.

2. The network object w, executing the corresponding computations, is defined in line 14 (see also lines 2-4) as consisting of $P \cdot Q$ virtual processors of the

same performance (by default). Its parent, the virtual host-processor, has coordinates I=0, J=0 (by default). At run time, that definition of w leads to such a mapping of its virtual processors into processes of the running parallel program that the number of processes involved in computations on a separate real processor depends on its performance. The algorithm of the mapping is presented in [7].

3. A slightly modified version of the ScaLAPACK test driver for Cholesky factorization is called on the network object w (lines 15-20). This driver reads from a file problem parameters (matrix and block sizes), forms a test matrix and performs its Cholesky factorization. The only parameter of the driver is a context which is a ScaLAPACK analog of an mpC network object. The context is created by means of a call to function mpC2Cblacs_gridinit in line 17. This call creates the ScaLAPACK context ConTxt associated with the process grid in which network object w has been mapped. More details of the interface mpC - ScaLAPACK are described below in 2.1. A call to function mpC2Cblacs_gridexit in line 19 releases resources, allocated on creation of context ConTxt.

In mpC, there exist three kinds of functions: basic, network, and nodal.

Basic functions are called and executed on the entire computing space. Network objects can be defined only in basic functions. Function HeHo is a basic function, what is specified with construct [*] in line 6 placed just before the name of the function.

A network function is called and executed on a network object. Function mpC2Cblacs_gridinit, declared in line 5, is an example of a network function. It has 3 special formal parameters, v,nr,nc. Parameter v corresponds to a network object on which the function is executed. Parameters nr,nc is regarded as integer variables replicated over network object v. Network object v is of network type Grid(nr,nc). In line 17, this function is called with network object w and parameters P,Q of its type as arguments corresponding to the above formal parameters.

A nodal function can be executed on any separate virtual processor. In mpC, all C functions are considered nodal.

If declared without any special distribution specifier, a variable, declared in a basic function, is considered distributed over the entire computing space, while a variable, declared in a network function, is considered distributed over the corresponding network object. Distribution specifier [w] in line 15 specifies the network object executing the compound statement in lines 15-20.

2.1   Interface mpC - ScaLAPACK

The BLACS (Basic Linear Algebra Communication Subprograms) [6] provide a linear algebra oriented message passing interface that may be implemented efficiently and uniformly across a large range of distributed memory platforms. It is used as the communication layer of ScaLAPACK.

In the BLACS, there are two grid creation routines (Cblacs_gridinit and Cblacs_gridmap) which create a process grid and its enclosing context. These routines return context handles, which are simple integers. Subsequent BLACS routines will be passed these handles, which allow the BLACS to determine what context/grid a routine is being called from. Releasing contexts is done via the routine Cblacs_gridexit.

The mpC programming system allows to call parallel ScaLAPACK routines providing mpC analogs of the above BLACS routines. In particular, mpC provides the following network function

```
int [net Grid(nr,nc) v] mpC2Cblacs_gridinit(int *pConTxt, char
                          *order);
```

as an analog of the BLACS grid creation routine

```
int Cblacs_gridinit(int *pConTxt, char *order, int nr, int nc);
```

where pConTxt is a pointer to the context to be created, nr and nc are numbers of rows and columns in the process grid associated with the context, and order indicates how to map processes to the BLACS grid. The BLACS grid and corresponding context are created by network function mpC2Cblacs_gridinit from the processes representing virtual processors of the network object v. The created context can be used to call ScaLAPACK routines, for example, routine pdlltdriver1 in line 18.

The BLACS grid releasing routine Cblacs_gridexit has the mpC analog mpC2Cblacs_gridexit.

Currently, the mpC programming system uses MPI as a communication platform. So, the above interface works only for the BLACS implementation built on the top of MPI.

3   Experimental Results

We compared two processes distributions:

- The traditional homogeneous processes distribution (one process per processor) implemented in ScaLAPACK.
- The heterogeneous distribution of processes over processors implemented in mpC.

As a factor of the comparison, we used speed-up achieved by the heterogeneous process distribution relative to the homogeneous one when running with the same process grid parameters and block size.

The comparison was performed for the Cholesky factorization on a network of workstations. For our experiments, we used a part of a local network consisting of 8 uniprocessor Sun workstations of different performances interconnected via 10 Mbits Ethernet. MPICH 1.0.13 was used as a particular communication platform. All workstations executed the same copy of code. Performances of the workstations, obtained by means of execution of the LAPACK routine dpotf2 performing serial Cholesky factorization, is the following: a - 200, b - 200, c - 200, d - 200, e - 267, f - 267, G - 801, h̲ - 100. Only two workstations G (the fastest one) and h̲ (the slowest one) have performances essentialy different from others.

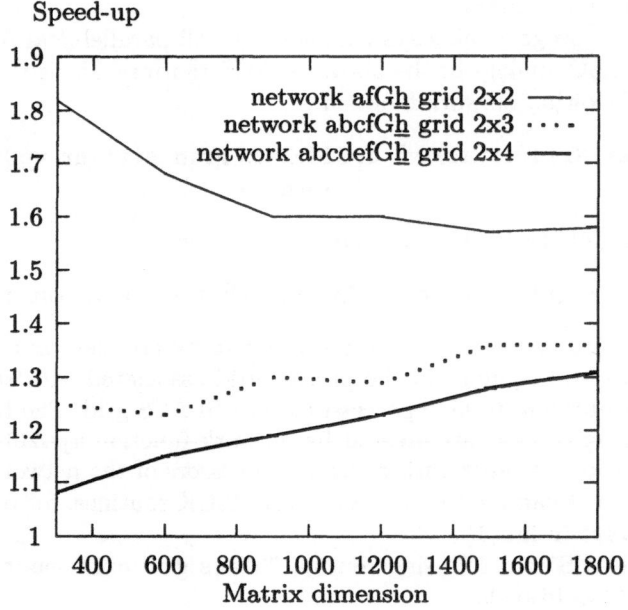

Fig. 2. Speed-up achieved by the heterogeneous processes distribution relative to the homogeneous one on networks afG̲h̲ consisting of 4 workstations a, f, G, and h̲, abcfG̲h̲ consisting of 6 workstations a, b, c, f, G, h̲, abcdefG̲h̲ consisting of 8 workstations a, b, c, d, e, f, G, h̲.

Figure 2 presents the speed-up achieved by the heterogeneous processes distribution when running on the following networks: afG̲h̲ consisting of 4 workstations a, f, G, and h̲) (process grid 2x2), abcfG̲h̲ consisting of 6 workstations a, b, c, f, G, and h̲ (process grid 2x3), and abcdefG̲h̲ consisting of 8 workstations a, b, c, d, e, f, G, and h̲ (process grid 2x4). It is interesting that in all cases the mpC run-time system distributes the involved processes with two processes running on the most powerful workstation G and no processes running on the slowest workstation h̲.

4 Conclusion

Numerical software developed for computations in homogeneous environments does not allow to utilize all performance potential of heterogeneous networks.

It has been demonstrated that in this case a heterogeneous network behaves as a homogeneous network obtained from the heterogeneous one by means of replacing all its processors with the slowest processor.

A natural way to answer this challenge is to develop dedicated numerical software aimed at heterogeneous environments. Such software should at least take into account the heterogeneity of processor performances. The paper presents a way to do that via adapting legacy numerical software .

The mpC parallel language is just aimed at portable and efficient programming for heterogeneous environments. It has facilities for heterogeneous distribution of the processes involved in computations over processors. ScaLAPACK is generally accepted numerical linear algebra package. Call to ScaLAPACK routines from mpC program gives a new facilities for efficiently solving linear algebra problems on heterogeneous environments. The mpC program can perform adaptation to pecularities of the particular network and ScaLAPACK routines can do calculations proper.

5   Acknowledgments

We would like to thank Jack Dongarra who gave us the idea of the research presented in this paper.

References

[1] L.S.Blackford, J.Choi, A.Cleary, E.D'Azevedo, J.Demmel, I.Dhillon, J.Dongarra, S. Hammarling, G. Henry, A. Petitet, K.Stanley, D.Walker, and R.C.Whaley "ScaLAPACK: A Linear Algebra Library for Message-Passing Computers" SIAM Conference on Parallel Processing, March 1997.

[2] L.S.Blackford, A.Cleary, J.Demmel, I.Dhillon, J.Dongarra, S.Hammarling, A.Petitet, H.Ren, K.Stanley, and R.C.Whaley, Practical Experience in the Dangers of Heterogeneous Computing UT, CS-96-330, July 1996.

[3] J.Choi, J.J.Dongarra, S.Ostrouchov, A.P.Petitet, D.W.Walker, and R.C.Whaley "The Design and Implementation of the ScaLAPACK LU, QR, and Cholesky Factorization Routines" UT, CS-94-246, September, 1994.

[4] E.Anderson, Z.Bai, C.Bischof, J.Demmel, J.Dongarra, J. Du Croz, A.Greenbaum, S.Hammarling, S.McKenney, S.Octrouchov, and D.Sorensen, "LAPACK Users' Guide, Second Edition", SIAM, Philadelphia, PA, 1995.

[5] A.Lastovetsky, The mpC Programming Language Specification. Technical Report, ISPRAS, Moscow, December 1994.

[6] R. Clint Whaley, "Basic Linear Algebra Communication Subprograms: Analysis and Implementation Across Multiple Parallel Architectures", Tech.Rep. LAPACK Working Note 73, University of Tennesee, TN, 1994.

[7] D.Arapov, A.Kalinov, and A.Lastovetsky, Resource management in the mpC Programming Environment, in "Proceedings of the 30th Hawaii International Conference on System Sciences (HICSS'30)", IEEE Computer Society, Maui, HI, January 1997.

Parallel Subdomain-Based Preconditioner for the Schur Complement

Luiz M. Carvalho[1*] and Luc Giraud[2]

[1] PESC-COPPE/UFRJ - Caixa Postal 68511
21945-970-Rio de Janeiro, RJ, Brasil
carvalho@cos.ufrj.br
[2] CERFACS - 42, av. Gaspard Coriolis
31057 - TOULOUSE - FRANCE
giraud@cerfacs.fr

Abstract. We present a new parallelizable preconditioner that is used as the local component of a two-level preconditioner similar to BPS. On 2D model problems that exhibit either high anisotropy or discontinuity, we demonstrate its attractive numerical behaviour and compare it with the regular BPS. To alleviate the construction cost of this new preconditioner, that requires the computation of the local Schur complements, we propose a cheap alternative based on Incomplete Cholesky factorization, that reduces the computational cost but retains the good numerical features of the preconditioner.

1 Introduction

The solution of elliptic problems is challenging on parallel distributed memory computers as their Green's functions are global. This problem is often tackled via domain decomposition techniques, using two-level preconditioners. In the framework of non-overlapping domain decomposition techniques, we refer for instance to BPS (Bramble, Pasciak and Schatz) [2], Vertex Space [7, 13], and to some extent Balancing Neumann-Neumann [10, 11], as well as FETI [8], for the presentation of major two-level preconditioners. We refer to [5] and [14] for a more exhaustive overview of domain decomposition techniques.

In this work, we consider non-overlapping domain decomposition techniques, and two-level preconditioners for the conjugate gradient method. These preconditioners can be written similarly to BPS [2], that is, as the sum of local and global components. We focus on a new local preconditioner that solves the assembled local Schur complement on the whole interface of each subdomain.

In Section 2, we briefly describe non-overlapping domain decomposition techniques and the class of two-level preconditioners we considered here. The main goal of that section is to formulate algebraically the sub-domain based preconditioner. In the next section, we introduce the 2D model problems used to

---

* The work of this author was partially supported by FAPERJ-Brazil under grant 130.117/98.

P. Amestoy et al. (Eds.): Euro-Par'99, LNCS 1685, pp. 1032–1039, 1999.
© Springer-Verlag Berlin Heidelberg 1999

benchmark the preconditioners, those model problems exhibit high anisotropy and high discontinuity. Numerical experiments are reported in Section 3 and, finally, some concluding remarks are reported.

2   Preconditioner Description

This section is two-fold. First, we formulate a two-dimensional elliptic model problem. Then, we introduce the preconditioners we studied.

We consider the following 2nd order self-adjoint elliptic problem on an open polygonal domain Ω included in $I\!\!R^2$:

$$\begin{cases} -\frac{\partial}{\partial x}(a(x,y)\frac{\partial v}{\partial x}) - \frac{\partial}{\partial y}(b(x,y)\frac{\partial v}{\partial y}) = F(x,y) \text{ in } \quad \Omega, \\ \qquad\qquad\qquad\qquad v = 0 \qquad\quad \text{on} \qquad \partial\Omega \end{cases} \tag{1}$$

where $a(x,y)$, $b(x,y) \in I\!\!R^2$ are positive functions on Ω. We assume that the domain Ω is partitioned into N non-overlapping subdomains $\Omega_1, \ldots, \Omega_N$ with boundaries $\partial\Omega_1, \ldots, \partial\Omega_N$; this defines a coarse mesh, τ^H, with mesh size H. We discretize (1) either by finite differences or finite elements resulting in a symmetric and positive definite linear system $Au = f$.

Let B be the set of all the indices of the discretized points which belong to the interfaces between the subdomains. Grouping the points corresponding to B in the vector u_B and the ones corresponding to the interior I of the subdomains in u_I, we get the reordered problem:

$$\begin{pmatrix} A_{II} & A_{IB} \\ A_{IB}^T & A_{BB} \end{pmatrix} \begin{pmatrix} u_I \\ u_B \end{pmatrix} = \begin{pmatrix} f_I \\ f_B \end{pmatrix}. \tag{2}$$

Eliminating u_I from the second block row of (2) leads to the following reduced equation for u_B:

$$S u_B = f_B - A_{IB}^T A_{II}^{-1} f_I, \quad \text{where} \quad S = A_{BB} - A_{IB}^T A_{II}^{-1} A_{IB} \tag{3}$$

is the Schur complement of the matrix A_{II} in A, and is usually referred to as the Schur complement matrix. For a stiffness matrix A arising from finite elements discretization the Schur complement matrix (3) can also be written as:

$$S = \sum_{i=1}^{N} S^{(i)}, \quad \text{where} \quad S^{(i)} = A_{BB}^{(i)} - (A_{IB}^{(i)})^T (A_{II}^{(i)})^{-1} A_{IB}^{(i)} \text{ only involves matrices}$$

computed locally on the finite elements in Ω_i. In Figure 1, we depicted a subdomain Ω_i with its edge interfaces E_m, E_j, E_k, E_ℓ; the local Schur complement matrix is dense and has the following block structure (for the sake of clarity, we do not consider the corner points):

$$S^{(i)} = \begin{pmatrix} S_{mm}^{(i)} & S_{mg} & S_{mk} & S_{m\ell} \\ S_{gm} & S_{gg}^{(i)} & S_{gk} & S_{g\ell} \\ S_{km} & S_{kg} & S_{kk}^{(i)} & S_{k\ell} \\ S_{\ell m} & S_{\ell g} & S_{\ell k} & S_{\ell\ell}^{(i)} \end{pmatrix} \tag{4}$$

The diagonal blocks represent the coupling between nodes on an edge interface. The off-diagonal blocks represent the coupling between each edge interface of Ω_i. Notice that the off-diagonals of $S^{(i)}$ are blocks actually existing in S, while the diagonal blocks are contributions to the diagonal blocks of the complete Schur complement matrix S. For instance, the diagonal block of the complete matrix S associated with the edge interface E_k is $S_{kk} = S_{kk}^{(i)} + S_{kk}^{(n)}$. We then obtain the local Schur complement assembled on the interface edges by:

$$\hat{S}^{(i)} = \begin{pmatrix} S_{mm} & S_{mg} & S_{mk} & S_{m\ell} \\ S_{gm} & S_{gg} & S_{gk} & S_{g\ell} \\ S_{km} & S_{kg} & S_{kk} & S_{k\ell} \\ S_{\ell m} & S_{\ell g} & S_{\ell k} & S_{\ell\ell} \end{pmatrix}. \tag{5}$$

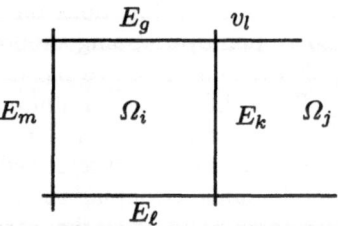

Fig. 1. Example of a regular 2D decomposition

We now describe the preconditioners, starting with BPS, followed by our new preconditioner. In this respect, we define a series of projection and interpolation operators. Specifically, for each E_i we define R_{E_i} as the standard pointwise restriction of nodal values on E_i. Its transpose extends grid functions in E_i by zero in the rest of the interface. Thus, $S_{ik} \equiv R_{E_i} S R_{E_k}^T$. Additionally, we define grid transfer operators between the interfaces and the coarse grid points in τ_H. R_0^T is an interpolation operator which corresponds to using interpolation between each set of edge endpoints (i.e. adjacent points in τ_H) to define values on the edge between the endpoints (i.e. edge interface E_i). R_0 is a projection operator and is the transpose of the interpolation operator. Finally, A_0 is a coarse grid approximation of the Schur complement operator on τ_H computed with to the Galerkin formula $A_0 = R_0 S R_0^T$. We refer to [3] and to the references therein for a more detailed description for the coarse grid component for this type of preconditioner.

With the above notation a close variant of the BPS preconditioner is given by $M_{BPS-E} = M_E + R_0^T A_0^{-1} R_0$ where $M_E = \sum_{E_i} R_{E_i}^T S_{ii}^{-1} R_{E_i}$ (in the original BPS A_0 is built from A and not from S as in our case). It can be interpreted as a generalized block Jacobi preconditioner for (3) augmented with a residual correction used on a coarse grid. The coarse grid correction term $R_0^T A_0^{-1} R_0$ allows global coupling to be incorporated between the interfaces. This global coupling is essential for scalability. In particular, it has been shown in [2] that, when applying the original BPS technique to a uniformly elliptic operator, the preconditioned system has a condition number

$$\kappa(M_{BPS}S) = \mathcal{O}(1 + \log^2(H/h)), \tag{6}$$

where h is the mesh size. This implies that the condition number depends only weakly on the mesh spacing and on the number of processors. Therefore, such a preconditioner is appropriate for large systems of equations on computers with a large number of processors.

The new preconditioner can be described in a similar way by defining another series of projection and interpolation operators. Specifically, for each subdomain Ω_i we define R_{Ω_i} as the standard pointwise restriction of nodal values on the interface of Ω_i. Its transpose extends grid functions on $\partial\Omega_i$ (the interface of Ω_i) by zero on the rest of the interface. Thus, $\hat{S}^{(i)} \equiv R_{\Omega_i} S R_{\Omega_i}^T$. With the above notation, we define the new preconditioner by $M_{BPS-S} = M_S + R_0^T A_0^{-1} R_0$ where $M_S = \sum_{\Omega_i} R_{\Omega_i}^T (\hat{S}^{(i)})^{-1} R_{\Omega_i}$.

In a distributed memory environment, the proposed preconditioner can be constructed almost at the same cost as regular BPS. More precisely, it requires the same amount of communication and a slight increase in the number of operations due to factorization of the dense assembled local Schur complement matrices rather than only their diagonal blocks. We address to [4] for details on the parallel implementation and time comparisons between regular BPS and our alternative.

Notice that the sub-domain based preconditioner M_S can be viewed as a Neumann-Neumann preconditioner [6] except that in our case the block diagonal coefficients of the local Schur complement matrices are assembled on each subdomain. Another difference with Neumann-Neumann is that the contribution of each subdomain is simply summed-up on each interface; for Neumann-Neumann a weighted sum is computed.

In fact, with the above notations, the Neumann-Neumann preconditioner, M_{NN}, can be written as: $M_{NN} = \sum_{\Omega_i} D_i(R_{\Omega_i}^T (S^{(i)})^{-1} R_{\Omega_i}) D_i$, where the matrices D_i are weight matrices defining a partition of unity (i.e. $\sum_{\Omega_i} D_i = I$). For internal domains, $S^{(i)}$ is singular and pseudo-inverses $(S^{(i)})^+$ should be used instead. Assembling the diagonal blocks of the local Schur complement matrices $S^{(i)}$ removes this singularity.

Finally, this new local preconditioner can also be viewed as an Algebraic Additive Schwarz preconditioner for the Schur complement, since it corresponds to a block diagonal preconditioner with an overlap between the blocks.

3   Numerical Experiments

With a first part dedicated to the description of the problems we are dealing with, this section presents the numerical behaviour of the local preconditioners introduced in Section 1. The central issue is to compare the number of iterations of a preconditioned conjugate gradient method when solving (3).

3.1 Model Problems

We mainly address the solution of Equation (1) discretized by linear finite elements. Convergence of the preconditioned conjugate gradient method is attained when the 2-norm of the residual of the current iteration, normalized by the 2-norm of the right hand side, is less than 10^{-5}. The grid is uniform.

The background of our study is the numerical solution of drift-diffusion equations for the numerical simulation of semi-conductor devices. In this respect, we intend to evaluate the sensitivity of the preconditioners to anisotropy and to discontinuity. With this in mind, we consider the following 2D model problems. In

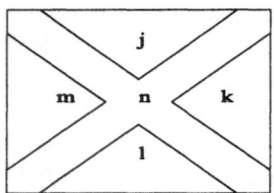

Fig. 2. A region where anisotropy and discontinuity are combined.

Figure 2, we represent the unit square divided into five regions where piecewise constant functions are defined. Let a and b be the functions of the elliptic problem as described in Equation (1). With these notations, we can define different problems with different degrees of difficulty:

- Poisson problem: $a = 1$ and $b = 1$,
- anisotropic and discontinuous problems: $a = 1$ and $b = j, k, l, m, n$.
  - Scot-flag1 (*SF1*): j=l= 10^{-2}, k=m= 10^{2} and n=1.
  - Scot-flag2 (*SF2*): j=l= 10^{-3}, k=m= 10^{3} and n=1.

In addition, we have considered another set of problems where we have only introduced anisotropy not necessarily aligned with the x or y axis but that makes an angle θ with the x-direction. For $\theta = 0$, this corresponds to the classical model anisotropic equation:

$$\varepsilon \frac{\partial^2 u}{\partial x^2} + \frac{\partial^2 u}{\partial y^2} = f \text{ with } \varepsilon \ll 1.$$

We have tested $\theta \in \{n\pi/8; n = 0, 1, 2\}$; because of the symmetry, the previous tests are actually for $n \in \{0, 1, 2, \dots, 15\}$.

3.2 Experimental Results

The proposed local preconditioners are computationally expensive to construct as we need to form explicitly the exact local Schur complement $S^{(i)}$. To alleviate the cost of this computation, cheap approximations can be obtained by replacing the exact solution of the local Dirichlet $(A_{II}^{(i)})^{-1}$ problem by some approximation based either on approximated inverse like AINV [1] (that preserves the symmetry of the preconditioner while SPAI [9] usually does not) or by an Incomplete Cholesky factorization ILL^T [12]. In both cases, an approximation of $S^{(i)}$ can be computed by sparse matrix-matrix computation at a lower cost than the one to pay even if an efficient sparse factorization were used. In order to study the effect of this approximation on the quality of the resulting preconditioner, we display in the next tables both the number of iterations using the exact local Schur and the approximation via an incomplete Cholesky factorization (denoted by $\tilde{M}_{BPS-E}$ for the approximated standard BPS preconditioner and $\tilde{M}_{BPS-S}$ for the approximated new preconditioner, respectively).

We first benchmark the preconditioner on the classical Poisson problem to establish that the new preconditioner satisfy the condition number estimate given by Equation (6). In Table 1, we can see that the number of iterations does not depends on the number of subdomains (that is $\frac{1}{H}$) when the size of the subdomain remains constant (that is $\frac{H}{h}$) as predicted by (6). For that problem, a simple ILL^T without fill-in is used for constructing the approximation of the local Schur complement matrices. Furthermore, we observe that the approximation of the local Schur complements do not affect the quality of the preconditioner and that the new preconditioner does not behave better than the regular block Jacobi used for standard BPS.

	Poisson		
# subdomains	4×4	8×8	16×16
M_{BPS-E}	9	11	11
M_{BPS-S}	10	10	11
$\tilde{M}_{BPS-E}$	12	13	13
$\tilde{M}_{BPS-S}$	12	13	13

Table 1. Number of iterations of the preconditioners for the Poisson problem. Each subdomain is discretized using a 16×16 grid.

For non-uniform elliptic operators, the theoretical bound does not work anymore. It can be seen in the next tables where we report the number of iterations for the anisotropic and anisotropic/discontinuous problems. In Table 2 and 3, we observe that the number of iterations increases when the number of sub-domains increases from 16 up-to 256. For both anisotropic and discontinuous problems, the new sub-domain based preconditioner ensures a convergence of the conjugate gradient method in less iterations than M_{BPS-E}. M_{BPS-S} converges in 25% less iterations on the anisotropic problem and reaches 40% less on the discontinuous problem.

Although the new preconditioner is still better when using approximations of the local Dirichlet problem, the figures show a deterioration in the convergence of both $\tilde{M}_{BPS-E}$ and $\tilde{M}_{BPS-S}$. We have used an ILL^T factorization with the fill-in controlled through a threshold strategy to construct cheap approximations of the local Schur complement matrices. This threshold induces some fill-in in the approximated factors; the extra storage is only a factor two or three more than the original matrices. However, this extra fill-in was necessary to avoid an undesirable deterioration in the numerical behaviour of the resulting preconditioner.

| # subdomains | Scot-flag | | | | | |
| | 4 × 4 | | 8 × 8 | | 16 × 16 | |
	$SF1$	$SF2$	$SF1$	$SF2$	$SF1$	$SF2$
M_{BPS-E}	25	42	34	69	42	105
M_{BPS-S}	19	30	22	44	27	64
$\tilde{M}_{BPS-E}$	26	47	38	76	48	110
$\tilde{M}_{BPS-S}$	24	40	29	56	36	77

Table 2. # iterations for problems combining high anisotropy and high discontinuity. Each subdomain is discretized using a 16×16 grid.

| # subdomains | Anisotropy ($\epsilon = 10^{-3}$) | | |
	4 × 4	8 × 8	16 × 16
M_{BPS-E}	20	28	35
M_{BPS-S}	17	21	26
$\tilde{M}_{BPS-E}$	21	29	35
$\tilde{M}_{BPS-S}$	17	21	25

Table 3. # iterations on the anisotropic problem with $\theta = \pi/4$. Each subdomain is a 16×16 grid.

4   Concluding Remarks

We have presented a new local parallelizable preconditioner that can be easily combined with the BPS coarse component to produce an efficient two-level preconditioner. We have shown for two difficult model problems that exhibit high anisotropy and/or discontinuity the effectiveness of this new preconditioner. To overcome its expensive construction, due to the exact explicit computation of the local Schur complements, we have proposed an alternative based on incomplete Cholesky factorization. This alternative enables a cheaper construction and only slightly deteriorates the efficiency of the resulting two-level preconditioners. Finally, the parallel implementation of the new preconditioner does not increase the complexity of the regular BPS, therefore it can be efficiently implemented on parallel distributed memory computers, we refer to [3] and [4] where parallel experiments are reported.

Acknowledgments

Part of this work was completed during a visit of the first author to CERFACS in January 1999. The hospitality and the support provided by CERFACS are greatly appreciated.

References

[1] M. Benzi, C. D. Meyer, and M. Tůma. A sparse approximate inverse preconditioner for the conjugate gradient method. *SIAM J. Sci. Comput.*, 17(5):1135–1149, 1996.

[2] J.H. Bramble, J.E. Pasciak, and A.H. Schatz. The construction of preconditioners for elliptic problems by substructuring I. *Math. Comp.*, 47(175):103–134, 1986.

[3] L. Carvalho, L. Giraud, and P. Le Tallec. Algebraic two-level preconditioners for the Schur complement method. Tech. Rep. TR/PA/98/18, CERFACS, Toulouse, France, 1998. Submitted to SIAM J. Scientific Computing.

[4] L.M. Carvalho and L. Giraud. Block diagonal preconditioners for the Schur complement method. In M. Papadrakakis and B.H.V. Topping, editors, *Innovative computational methods for structural mechanics*, pages 61–82, Edinburgh, UK, 1999. Saxe-Coburg publications.

[5] T.F. Chan and T.P. Mathew. *Domain Decomposition Algorithms*, volume 3 of *Acta Numerica*, pages 61–143. Cambridge University Press, Cambridge, 1994.

[6] Y.-H. De Roeck and P. Le Tallec. Analysis and test of a local domain decomposition preconditioner. In R. Glowinski, Y. Kuznetsov, G. Meurant, J. Périaux, and O. Widlund, editors, *Fourth International Symposium on Domain Decomposition Methods for Partial Differential Equations*, pages 112–128. SIAM, Philadelphia, PA, 1991.

[7] M. Dryja, B.F. Smith, and O.B. Widlund. Schwarz analysis of iterative substructuring algorithms for elliptic problems in three dimensions. *SIAM J. Numer. Anal.*, 31(6):1662–1694, 1993.

[8] C. Farhat and F.-X. Roux. A method of finite element tearing and interconnecting and its parallel solution algorithm. *Int. J. Numer. Meth. Engng.*, 32:1205–1227, 1991.

[9] M. Grote and T. Huckle. Parallel preconditioning with sparse approximate inverses. *SIAM J. Sci. Comput.*, 18:838–853, 1997.

[10] P. Le Tallec. *Domain decomposition methods in computational mechanics*, volume 1 of *Computational Mechanics Advances*, pages 121–220. North-Holland, 1994.

[11] J. Mandel. Balancing domain decomposition. *Communications in Numerical Methods in Engineering*, 9:233–241, 1993.

[12] T. A. Manteuffel. Shifted incomplete Cholesky factorization. In I. S. Duff and G. W. Stewart, editors, *Sparse Matrix Proceedings 1978*, Philadelphia, PA, 1979. SIAM Publications.

[13] B.F. Smith. *Domain Decomposition Algorithms for the Partial Differential Equations of Linear Elasticity*. PhD thesis, Courant Institute of Mathematical Sciences, September 1990. Tech. Rep. 517, Department of Computer Science, Courant Institute.

[14] B.F. Smith, P. Bjørstad, and W. Gropp. *Domain Decomposition, Parallel Multilevel Methods for Elliptic Partial Differential Equations*. Cambridge University Press, New York, 1st edition, 1996.

A Preconditioner for Improved Fermion Actions

Wolfgang Bietenholz[1], Norbert Eicker[1], Andreas Frommer[3], Thomas Lippert[2], Björn Medeke[3], and Klaus Schilling[1,2]

[1] NIC, c/o Research Center Jülich, 52425 Jülich, Germany
[2] Department of Physics, University of Wuppertal, 42097 Wuppertal, Germany
[3] Department of Mathematics, University of Wuppertal, 42097 Wuppertal, Germany

Abstract. SSOR preconditioning of fermion matrix inversions which is parallelized using a *locally lexicographic* lattice sub-division has been shown to be very efficient for standard Wilson fermions. We demonstrate here the power of this method for the Sheikholeslami-Wohlert improved fermion action.

1 Introduction

Recently, the SSOR preconditioner turned out to be parallelizable by means of the *locally lexicographic* (*ll*) ordering technique [1]. In this way, SSOR preconditioning has been made applicable to the acceleration of standard Wilson fermion inversions on high performance massively parallel systems and it outperforms odd-even preconditioning.

It appears intriguing to extend the range of *ll*-SSOR preconditioners such as to accelerate the inversion of improved fermionic actions, which became very popular in the recent years.

In Symanzik's on-shell improvement program [2], counter terms are added to both, lattice action and composite operators in order to reduce $\mathcal{O}(a)$ artifacts which spoil results in the instance of the Wilson fermion formulation. In the approach of Sheikholeslami and Wohlert (SWA) [3], the Wilson action is modified by adding a diagonal term, the so-called clover term with a new free parameter c_{SW}. The generic form of SWA is given by

$$M = D + A.$$

D represents diagonal blocks (containing 12×12 sub-blocks) and A is a nearest-neighbor hopping term. In the following, we will show that the *ll*-SSOR scheme applies not only to the couplings in A but also to the internal spin and color degrees of freedom of the block diagonal term D.

2 SWA

SWA is composed of A (Wilson hopping term) and D (SW diagonal),

$$D_{SW}(x,y) = \left(I + \frac{c_{SW}}{2}\kappa \sum_{\mu,\nu} \sigma_{\mu\nu} \otimes F_{\mu\nu}(x) \right) \delta_{x,y},$$

P. Amestoy et al. (Eds.): Euro-Par'99, LNCS 1685, pp. 1040–1043, 1999.
© Springer-Verlag Berlin Heidelberg 1999

$$A_{SW}(x,y) = -\kappa \left(\sum_{\mu} ((I - \gamma_\mu) \otimes U_\mu(x)) \, \delta_{x,y-e_\mu} \right.$$

$$\left. + \sum_{\mu} ((I + \gamma_\mu) \otimes U_\mu^H(x - e_\mu)) \, \delta_{x,y+e_\mu} \right),$$

where κ is the Wilson hopping parameter, c_{SW} couples the SW clover operator. This parameter is tuned to optimize $\mathcal{O}(a)$ cancellations. The local clover term $F_{\mu\nu}(x)$ consists of 12×12 diagonal blocks. Its explicit structure in Dirac space is given in Ref. [6].

3   Block SSOR Preconditioning

The preconditioned system is modified by two matrices V_1 and V_2,

$$V_1^{-1} M V_2^{-1} \tilde{\psi} = \tilde{\phi}, \; \tilde{\phi} = V_1^{-1}\phi, \; \tilde{\psi} = V_2\psi.$$

Let $M = D - L - U$ be the decomposition of M into its block diagonal part D, its (block) lower triangular part $-L$ and its (block) upper triangular part $-U$. Block SSOR preconditioning is defined through the choice

$$V_1 = \left(\frac{1}{\omega}D - L\right)\left(\frac{1}{\omega}D\right)^{-1}, \; V_2 = \frac{1}{\omega}D - U.$$

The *Eisenstat trick* [8] reduces the costs by a factor 2. It is based on the identity:

$$V_1^{-1}(D - L - U)V_2^{-1} =$$
$$(I - \omega L D^{-1})^{-1}\left[I + (\omega - 2)(I - \omega U D^{-1})^{-1}\right] + (I - \omega U D^{-1})^{-1}.$$

The preconditioned matrix-vector product, $z = V_1^{-1} M V_2^{-1} x$, is given by:

> solve $(I - \omega U D^{-1})y = x$
> compute $w = x + (\omega - 2)y$
> solve $(I - \omega L D^{-1})v = w$
> compute $z = v + y$

The "solve" is just a simple forward (backward) substitution process due to the triangular structure:

> for $i = 1$ to N
> $v_i = w_i + \sum_{j=1}^{i-1} L_{ij}s_j$
> $s_i = \omega D_{ii}^{-1}v_i$

Options for D of SWA take each block D_{ii} to be of dimension 12 ($D^{(12)}$), 6 ($D^{(6)}$), 3 ($D^{(3)}$) or 1 ($D^{(1)}$), as suggested by the structure of D. The blocks have to be pre-inverted, the cost depends on the block size [6].

Parallelism can be achieved by *locally lexicographic* ordering [1]. "Coloring" is the decomposition of all lattice points into mutually disjoint sets $C_1, \ldots, C_k$ (with respect to the matrix M), if for any $l \in \{1, \ldots, k\}$ the property $x \in C_l \Rightarrow y \notin C_l$ for all $y \in n(x)$ holds. $n(x)$ denotes the set of sites $\neq x$ coupled to x. A suitable ordering first numbers all x with color C_1, then all with C_2 etc. Thus, each lattice point couples with lattice points of different colors only. The computation of v_x for all x of a given color C_l can be done in parallel, since terms like $\sum_{y \in n(x),\, y \leq_o x}$ involve only lattice points of the preceding colors $C_1, \ldots, C_{l-1}$, with $x \leq_o y$ meaning that x has been numbered before y with respect to the ordering o.

Let the lattice blocks be of size $n^{loc} = n_1^{loc} \times n_2^{loc} \times n_3^{loc} \times n_4^{loc}$. A different color is associated with each of the sites of the n^{loc} groups. A *locally lexicographic* (*ll*) ordering is defined to be the color ordering, where all points of a given color are ordered after all points with colors, which correspond to lattice positions on the local grid that are lexicographically preceding the given color. The parallel forward substitution reads:

for all colors C_i, $i = 1, \ldots, \frac{n}{p}$, $\frac{n}{p} \in \mathbf{N}$
    for all processors $j = 1, \ldots, p$
        $x :=$ grid point of color C_i on processor j
        $v_x = w_x + \sum_{y \in n(x),\, y \leq_{ll} x} L_{xy} s_y$
        $s_x = \omega D_{xx}^{-1} v_x$

If the lattice point x is close to the boundary of the local lattice, then the set $n(x)$ will contain grid points y residing in neighboring processors. Therefore, some of the quantities s_y will have to be communicated from those neighboring processors. For SWA, up to 8 neighbors may be involved on the 4-d grid. The detailed communication scheme for this case was given in Ref. [1].

4   Improvement

The SWA has been implemented on an APE100 equipped with $p = 32$ processors. We use a de-correlated set of 10 quenched gauge configurations generated on a 16^4 lattice at $\beta = 6.0$ at 3 values of c_{SW}, 0, 1.0 and 1.769. We have applied BiCGStab as iterative solver. The stopping criterion has been chosen as $\|MX - \phi\| \leq 10^{-6} \|X\|$, with X being the solution. We used a local source ϕ.

We have determined the optimal over-relaxation parameter to be about $\omega = 1.4$ for all block sizes and c_{SW}. In Fig. 1, the results from three diagonal block sizes are overlaid, the 1×1, 3×3, and 6×6 blocks.

We plot the ratio of iteration numbers of the odd-even procedure vs. *ll*-SSOR as function of κ in Fig. 2. A gain factor up to 2.5 in iteration numbers can be found.

There is no dependence on c_{SW} or on the block size of D and only 10 % on the local lattice size. As to real CPU costs on APE100, the optimal block size of D is a 3×3 block whereas on a scalar system, the optimum is found for a 1×1 diagonal.

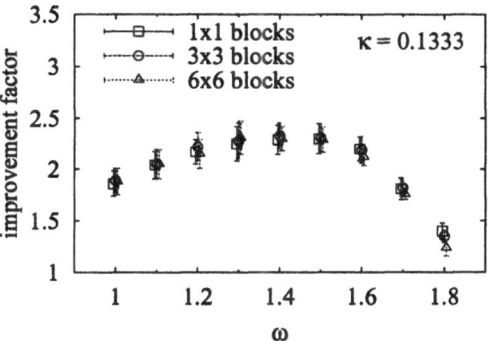

Fig. 1. Gain of ll-SSOR over odd-even preconditioning vs. ω for $c_{SW} = 1.769$.

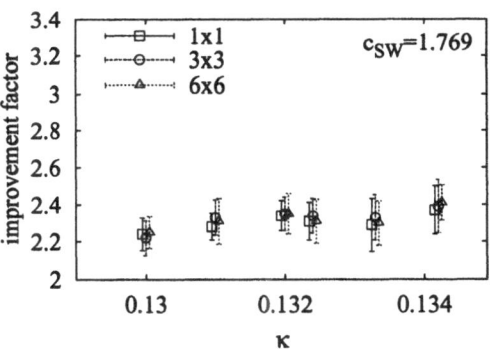

Fig. 2. Gain of ll-SSOR over odd-even preconditioning vs. κ.

References

[1] S. Fischer, A. Frommer, U. Glässner, Th. Lippert, K. Schilling: Comp. Phys. Comm. **98** (1996) 20

[2] K. Symanzik: Nucl. Phys. B **212** (1983) 1

[3] B. Sheikholeslami, R. Wohlert: Nucl. Phys. B **259** (1985) 572

[4] P. Hasenfratz, F. Niedermayer: Nucl. Phys. B **414** (1994) 785

[5] W. Bietenholz, R. Brower, S. Chandrasekharan, U.-J. Wiese: Nucl. Phys. B (Proc. Suppl.) **53** (1997) 921.

[6] W. Bietenholz, N. Eicker, A. Frommer, Th. Lippert, B. Medeke, K. Schilling: hep-lat 9807013, submitted to CPC

[7] W. Bietenholz, U.-J. Wiese: Nucl. Phys. B **464** (1996) 319

[8] S. Eisenstat: SIAM J. Sci. Stat. Comp. **2** (1981) 319

Application of a Class of Preconditioners to Large Scale Linear Programming Problems

Venansius Baryamureeba, Trond Steihaug, and Yin Zhang*

Department of Informatics,
University of Bergen,
5020 Bergen, Norway

Abstract. In most interior point methods for linear programming, a sequence of weighted linear least squares problems are solved, where the only changes from one iteration to the next are the weights and the right hand side. The weighted least squares problems are usually solved as weighted normal equations by the direct method of Cholesky factorization. In this paper, we consider solving the weighted normal equations by a preconditioned conjugate gradient method at every other iteration. We use a class of preconditioners based on a low rank correction to a Cholesky factorization obtained from the previous iteration. Numerical results show that when properly implemented, the approach of combining direct and iterative methods is promising.

Key Words. Weighted linear least squares, Parallel processing, Preconditioners, Linear programming, Primal-dual infeasible interior point algorithms.

1 Introduction

The class of preconditioners we will consider is a low rank correction of a Cholesky factorization of a weighted normal equation coefficient matrix. The combination of a direct method and an iterative method was reported by Karmarkar and Ramakrishnan [8]. A low rank correction to the matrix was used by Goldfarb and Liu [6] to prove a low complexity bound for convex quadratic programming. A frequently used iterative method [3, 8, 9] is the preconditioned conjugate gradient method on the normal equations.

Throughout this paper we use the following notation: $\min_i$ or $\max_i$ is for all i for which the argument is defined. For any matrix A, A_{ij} is the element in the i-th row and j-th column, A_j is the j-th column, and $A_{j\bullet}$ is the j-th row. The symbol I is used to denote the identity matrix; its size will always be apparent from the context. For any square matrix X, $\lambda_i(X)$ are the eigenvalues of X arranged in non decreasing order. The letters L and R represent lower triangular factors or lower Cholesky factors, and the type of factors will always be apparent from the context.

---

* Department of Computational and Applied Mathematics, Rice University, Houston, Texas 77005, USA.

P. Amestoy et al. (Eds.): Euro-Par'99, LNCS 1685, pp. 1044–1048, 1999.
© Springer-Verlag Berlin Heidelberg 1999

2   Problem

Consider a primal linear programming problem in standard form:

$$\begin{array}{l} \text{minimize} \quad c^T x \\ \text{subject to: } Ax = b, \ x \geq 0 \\ \qquad A \in \Re^{m \times n}, \ b \in \Re^m, \ c, x \in \Re^n. \end{array} \tag{1}$$

It can be shown (see [2]) that the normal equation form for a primal-dual interior point method corresponding to problem (1) is of the form

$$AGA^T y = AGh \tag{2}$$

where G is a diagonal and positive definite matrix. At every interior point iteration, (2) is solved.

In the following, the matrix A will be a full rank m-by-n matrix, $m \leq n$.

3   The Class of Preconditioners

Let $G, H \in \Re^{n \times n}$ be positive definite and diagonal matrices. For a given index set $\mathcal{Q} \subseteq \{1 \leq j \leq n | \ G_{jj} \neq H_{jj}\}$, let the $n \times n$ diagonal matrices $\bar{H}$ and $\bar{G}$ be given by

$$\bar{H}_{jj} = \begin{cases} H_{jj} & \text{if } j \in \mathcal{Q} \\ 0 & \text{otherwise,} \end{cases} \quad \text{and} \quad \bar{G}_{jj} = \begin{cases} G_{jj} & \text{if } j \in \mathcal{Q} \\ 0 & \text{otherwise.} \end{cases}$$

Then, the diagonal matrix K defined in (3) is positive definite.

$$K = H + \bar{G} - \bar{H}. \tag{3}$$

Note that $K_{jj} = G_{jj}$ if $j \in \mathcal{Q}$, otherwise $K_{jj} = H_{jj}$. In the mixed interior point method H is a weight matrix at a previous iteration where a direct method is used to solve (2), and G is the weight matrix at the current interior point iteration. We can thus assume that we have a factorization of $AHA^T = LL^T$.

We denote the number of elements in $\mathcal{Q}$ by $q = |\mathcal{Q}|$. Let $\bar{A} \in \Re^{m \times q}$ consist of all columns A_j such that $j \in \mathcal{Q}$. Further, let $\bar{D} \in \Re^{q \times q}$ be the nonzero diagonal submatrix of $\bar{G} - \bar{H}$ corresponding to the submatrix $\bar{A}$. Then

$$AKA^T = A(H + \bar{G} - \bar{H})A^T = AHA^T + \bar{A}\bar{D}\bar{A}^T. \tag{4}$$

An approximation to AGA^T is AKA^T. Observe that if $\mathcal{Q} = \{1 \leq j \leq n \ | \ G_{jj} \neq H_{jj}\}$ then $AKA^T = AGA^T$ so the approximation can be made arbitrary good.

Theorem 1. *[2]: Let $G, H \in \Re^{n \times n}$ be positive definite and diagonal. Let γ_j be the sorted elements of G_{jj}/H_{jj} in non-decreasing order*

$$\gamma_1 = \min_j \{G_{jj}/H_{jj}\} \leq \gamma_2 \leq \cdots \leq \gamma_n = \max_j \{G_{jj}/H_{jj}\}.$$

Let $\mathcal{Q}$ consist of the indices corresponding to the q_1 largest and the q_2 smallest diagonal elements of $H^{-1}G$. Then

$$\min\{\gamma_{q_2+1}, 1\} \leq \lambda_i((AKA^T)^{-1}AGA^T) \leq \max\{\gamma_{n-q_1}, 1\}.$$

Theorem 1 suggests choosing $\mathcal{Q}$ to consist of indices j corresponding to q_1 largest and q_2 smallest diagonal elements of $H^{-1}G$. Further, the condition number of the preconditioned matrix is bounded above

$$\kappa((AKA^T)^{-1}AGA^T) \leq \frac{\max\{\gamma_{n-q_1}, 1\}}{\min\{\gamma_{q_2+1}, 1\}}.$$

We can keep the condition number of $(AKA^T)^{-1}AGA^T$ bounded by a prescribed constant if we carefully choose q_1 and q_2 sufficiently large. The next result from [2] gives a lower bound on the condition number.

Theorem 2. Let $G, H \in \Re^{n \times n}$ be positive definite and diagonal. Let $|\mathcal{Q}| = q_1 + q_2 < m$, where q_1 is the number of indices in $\mathcal{Q}$ where $G_{jj}/H_{jj} > 1$. Let K be defined as in (3). Then

$$\kappa((AKA^T)^{-1}AGA^T) \geq \frac{\lambda_{m-q_1}((AHA^T)^{-1}AGA^T)}{\lambda_{q_2+1}((AHA^T)^{-1}AGA^T)}.$$

Wang and O'Leary [10] discuss a strategy of preconditioning the normal equation system based on formulation (4). In their implementation, they choose the index set $\mathcal{Q}$ to consist of the indices corresponding to the largest values of $|G_{jj} - H_{jj}|$.

3.1   Computing the Preconditioner

We consider two approaches of solving the linear system $AKA^T z = r$, namely the Sherman Morrison-Woodbury formula approach, and the approach based on updating the triangular factors.

Sherman Morrison-Woodbury Formula Let $AHA^T = LL^T$. Applying the Sherman Morrison-Woodbury formula ([7] pp. 50) to $AKA^T = LL^T + \bar{A}\bar{D}\bar{A}^T$ followed with some manipulation yields

$$(AKA^T)^{-1} = L^{-T}(I - VF^{-1}V^T)L^{-1}, \tag{5}$$

where $V = L^{-1}\bar{A}$ and $F = \bar{D}^{-1} + V^TV$. Solving $AKA^T z = r$ for z is equivalent to computing

$$z = (AKA^T)^{-1}r = L^{-T}[d - (V(F^{-1}(V^Td)))], \tag{6}$$

where $d = L^{-1}r$. To compute z in (6) we need to store L, V, d, and the factors of F. Note that F is a $q \times q$ symmetric matrix that may be indefinite.

Algorithm 3.2 Sparse case
Define $V = \bar{A}$; $C = \bar{D}$; $R = L$; $T = D$

Algorithm 3.1 Dense case
Define $V = \bar{A}$; $C = \bar{D}$

for $i = 1, \ldots, m$ do
$\quad p = (V_{i\bullet}C)^T$
$\quad T_{ii} = D_{ii} + V_{i\bullet}\,p$
$\quad u = (1/T_{ii})\,p$
$\quad C \leftarrow C - u\,p^T$
$\quad$ for $j = i+1, \ldots, m$ do
$\quad\quad V_{j\bullet} \leftarrow V_{j\bullet} - L_{ji}V_{i\bullet}$
$\quad\quad R_{ji} = L_{ji} + V_{j\bullet}\,u$

for $i = 1, \ldots, m$ do
$\quad$ if $V_{i\bullet} \neq 0$ then
$\quad\quad p = (V_{i\bullet}C)^T$
$\quad\quad T_{ii} = D_{ii} + V_{i\bullet}\,p$
$\quad\quad u = (1/T_{ii})\,p$
$\quad\quad C \leftarrow C - u\,p^T$
$\quad\quad$ for $j = i+1, \ldots, m$ do
$\quad\quad\quad$ if $\quad L_{ji} \neq 0$ then
$\quad\quad\quad\quad V_{j\bullet} \leftarrow V_{j\bullet} - L_{ji}V_{i\bullet}$
$\quad\quad\quad\quad R_{ji} = L_{ji} + V_{j\bullet}\,u$

Fig. 1. Updating the triangular factors

Updating the Triangular Factors   Suppose we have the factorization $AHA^T = LDL^T$. Now, we want to compute R and T so that $AKA^T \equiv LDL^T + \bar{A}\bar{D}\bar{A}^T = RTR^T$. Since the sparsity structure of AKA^T is the same as for AHA^T, sparsity structure of R is the same as for L. An algorithm for updating factors of dense and sparse matrices is given in Baryamureeba and Steihaug [1]. For algorithms updating factors of dense matrices see [1] and references therein. We state the algorithms in Figure 1. The sparse-case algorithm is very effective when there are either many zero rows in $\bar{A}$ or few nonzero elements in the unit lower triangular factor L.

The major computation in an interior point algorithm is in solving linear systems. Demmel et al. [4] discuss parallel algorithms for conjugate gradients and computing Cholesky factors. The construction of the preconditioner (updating of the triangular factors) can be done in parallel.

4   Numerical Results

Here, we give some numerical results using the same implementation details as in [2]. The test problems are from the Netlib set [5] of linear programs. The test code is implemented in MATLAB

The variable ϵ is a proximity measure of the interior point iterations to a solution of the linear programming problem and τ is the number of iterations (or corrections) allowed for the preconditioned conjugate gradient method. The percentages in Table 1 are based on that approximately half of the direct solves are replaced by an iterative solution. The gain is therefor approximately the ratio of the difference between the two methods and half of the total time. The results in Table 1 show that the mixed primal-dual Newton (mixed PDN) interior-point method, which alternatively uses a direct (Cholesky factorization) method and a preconditioned (with the preconditioner described in Section 3) conjugate gradient method to solve (2), competes favourably with the primal-dual Newton (PDN) interior-point method on large-scale problems.

Problem name	Rows (m)	Columns (n)	PDN		Mixed PDN		Gain %
			iter	cpu-time	iter	cpu-time	
czprob	737	3141	56	89.01	59	59.58	66
d2q06c	2171	5831	54	544.85	57	421.33	45
d6cube	404	6184	42	317.58	45	239.16	51
stocfor2	2157	3045	41	113.26	44	98.05	27
scsd8	397	2750	16	21.41	19	17.49	37

Table 1. Comparison of two implementations of primal-dual Newton interior point method. If $\epsilon \geq 0.1$ then $\tau = 7$, otherwise $\tau = 40$. Fixed value of $q = 6$, $q_1 = 3$.

The numerical results show that the mixed PDN method is promising and merits further study. In this paper the mixed PDN method is based on the odd-even alternation technique. Other dynamic alternation schemes [10] which could be based on the easily computable bounds on condition numbers should be investigated.

References

[1] V. Baryamureeba, and T. Steihaug, *Computational issues for a new class of pre-conditioners*, To appear in the Proceedings for the 2nd Workshop on Large-Scale Scientific Computations, 1999.

[2] V. Baryamureeba, T. Steihaug and Y. Zhang, *A class of preconditioners for weighted least squares problems*, Technical Report No. 170, Department of Informatics, University of Bergen, 5020 Bergen, Norway, April 30, 1999.

[3] T.J. Carpenter and D.F. Shanno, *An interior point method for quadratic programs based on conjugate projected gradients*, Computational Optimization and Applications, Vol. 2, pp. 5-28, 1993.

[4] J. W. Demmel, M.T. Heath, and H.A. Van der Vorst, *Parallel numerical linear algebra*, Acta Numerica, pp. 111-197, 1993.

[5] D.M. Gay, *Electronic mail distribution of linear programming test problems*, Mathematical Programming Society COAL Newsletter, No. 13, pp.10-12, 1985.

[6] D. Goldfarb, and S. Liu, *An $O(n^3L)$ primal interior-point algorithm for convex quadratic programming*, Mathematical Programming, Vol. 49, pp. 325–340, 1991.

[7] G. H. Golub, and Charles H. Van Loan, *Matrix computations*, Third Edition, 1996.

[8] N.K. Karmarkar and K.G. Ramakrishnan, *Computational results of an interior point algorithm for large-scale linear programming*, Mathematical Programming, Vol. 52, pp. 555-586, 1991.

[9] S. Mehrotra, *Implementations of affine scaling methods: Approximate solutions of systems of linear equations using preconditioned conjugate gradient methods*, ORSA J. Comput., 4(1992), pp. 103-118.

[10] W. Wang, and D.P. O'Leary, *Adaptive use of iterative methods in predictor-corrector interior point methods for linear programming* Technical Report No. CS-TR-4011, Computer Science Department, University of Maryland, April 1999. (Revised version of TR-3560, November 1995)

Estimating Computer Performance for Parallel Sparse QR Factorisation

David J. Miron* and Patrick M. Lenders

School of Mathematical and Computer Sciences, University of New England,
Armidale NSW 2350, Australia
Phone: 61 2 67732298 Fax: 61 2 67733312
david@turing.une.edu.au pat@turing.une.edu.au

Abstract. Performance estimates of a parallel computer during sparse matrix factorisation aid in the identification of overheads and the tuning of software. This paper proposes a technique which allows the computer parameters of computation speed, communication speed, latency and parallel efficiency to be estimated. The technique is based upon the use of mathematical models derived from a model problem in conjunction with experimental results. By combining the mathematical models with the experimental results, sets of simultaneous equations can be derived which can be solved for the above computer parameters. The technique is explained in the context of sparse QR factorisation.

1 Introduction

The primary focus of this work is the factorisation A=QR, where A is a sparse real $m \times n$ ($m \geq n$) matrix, Q is orthogonal, R is upper triangular and the number of nonzero entries in any row or column is less than 10%.

The sparse QR factorisation is important as it can be used to solve sparse linear least squares problems. These problems occur in areas such as geodesy [7], data mining [9], animal breeding [8] and finite element problems.

A sparse matrix is factorised by storing and operating only on the nonzeroes in the matrix, thereby realising significant savings in storage requirements and factorisation time. However, sparse matrices can be so large that factorisation on single processor computers is not practical, therefore requiring the use of parallel computers.

Pozo [13] provides a theoretical study of the distributed memory multifrontal LU factorisation. His analysis is based upon *completion time*, being the time taken by the *multifrontal method* [3] to perform the numeric factorisation of a sparse matrix. Using some example matrices from the Harwell Boeing Collection [2], Pozo was able to give estimates for the completion time and parallel speedup of various parallel distributed memory computers.

---

* This work was done as part of a PhD dissertation with the CRC for Advanced Computation in the Research School of Information Sciences and Engineering at the Australian National University.

P. Amestoy et al. (Eds.): Euro-Par'99, LNCS 1685, pp. 1049–1058, 1999.
© Springer-Verlag Berlin Heidelberg 1999

This work extends that of Pozo, applying a similar theoretical analysis to the sparse QR factorisation. A method is proposed for estimation of the computer parameters computation cost γ (seconds/operation), communication cost β (seconds/byte), latency α (seconds) and the parallel efficiency of the factorisation ϕ ($0 < \phi \leq 1$). In this context the computation cost γ is the average time for a computer to perform one operation, while the communication cost β is the time for the computer to transfer one byte of data between two processors.

The target architecture is a Fujitsu VPP300 computer at the Australian National University. This computer is a distributed memory vector parallel computer with 13 processors with each processor having a peak speed of 2.2 GFlops. Of the 13 processors 5 have 2 GB of RAM while the other 8 processors have 416MB of RAM with the processors connected by a crossbar switch that has a transfer rate of 570 MB/sec and a latency between processors of 570μ seconds.

To estimate the computer parameters computation cost γ, communication cost β, latency α and the parallel efficiency of the factorisation ϕ requires the construction of a model problem that is based upon a special case of a *relaxed supernodal elimination tree* [1, 3]. Using the model problem an equation can be derived that gives the theoretical completion time for parallel sparse QR factorisation using a multifrontal method. Actual completion times can then be found using an implementation of the parallel factorisation. This results in a set of simultaneous equations that can be solved for the above computer parameters.

In Section 2 a brief overview of multifrontal methods is given before proceeding to define the model problem in Section 3. Completion times are then presented in Section 4 with results given in Section 5. In Section 6 some conclusions are presented.

2   Multifrontal Methods

The multifrontal method of Duff and Reid [3] reduces the factorisation of a sparse matrix to the factorisation of a sequence of small submatrices called *frontal matrices*. A frontal matrix is associated with each column of the matrix A, with the factorisation of a frontal matrix corresponding to the elimination of a column in A. A factorised frontal matrix is dense and consists of a contribution to R and an *update matrix*, where the update matrix is made up of the rank-1 updates produced by the frontal matrix factorisation. The update matrix is stored on a stack to be later retrieved and summed into other frontal matrices.

The order in which frontal matrices are factorised and update matrices retrieved from the stack is governed by a graph structure called the *elimination tree*. An elimination tree is a tree of n nodes with each node corresponding to a column in the matrix A. The columns of the matrix A corresponding to the leaves of the elimination tree can be eliminated independently. The order in which columns are eliminated is determined by a postorder traversal of the tree.

Frontal matrices can be small, resulting in extensive overheads in stack operations and frontal matrix construction. These overheads can be reduced on vector computers by merging nodes of the elimination tree into "relaxed supernodes". Thus factorisation of a relaxed supernode involves elimination of all the columns associated with that node from the frontal matrix. Frontal matrices associated with relaxed supernodes explicitly allow zeroes into the factorised frontal matrix [1, 3]. In this work the term elimination tree shall refer to a relaxed supernodal elimination tree.

3 The Model Problem

The model problem is based upon a relaxed supernodal elimination tree with the following structure.

Firstly, the elimination tree T is a balanced binary tree consisting of $l = 1 + \log_2 p$ levels and p leaves. A node $t \in T$ at level i of the elimination tree has a *granularity* $\kappa_t = \frac{\kappa}{\nu^i}$ where κ is the number of columns to be eliminated at the root node and $\nu > \sqrt{2}$ is the rate at which the granularity of the nodes in the tree decreases between levels and is a constant.

Secondly, all the frontal and update matrices have magnitudes:

$$|\mathcal{F}^t| = \frac{c_1 \kappa^2}{\nu^{2i}} \tag{1}$$

and

$$|\mathcal{U}^t| = \begin{cases} 0 & \text{if } i = 0 \\ \frac{c_2 \kappa^2}{\nu^{2i}} & \text{if } i > 0 \end{cases} \tag{2}$$

respectively where the magnitude is the number of elements in the matrix M. The parameters c_1 and c_2 are constants and their ratio represents the overdeterminism of the frontal and update matrices respectively. An example of the elimination tree for the case $p = 4$ is given in Fig. 1.

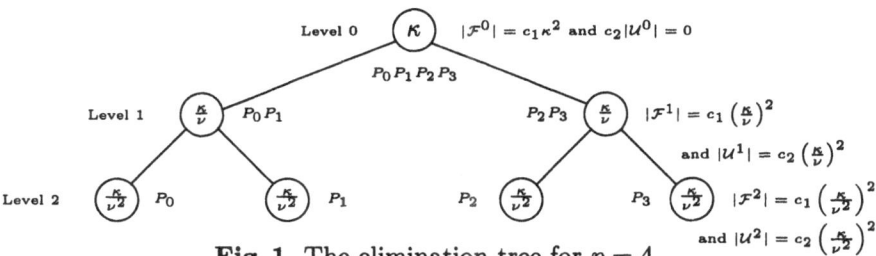

Fig. 1. The elimination tree for $p = 4$

The elimination tree in Fig. 1 corresponds to a *separator tree* generated by applying the Recursive Bisection [4] reordering to matrices arising from finite element problems. For finite element problems based on a 2 dimensional grid the granularity of the top level node κ would be proportional to $\sqrt{n}$ [5] while the rate at which the granularity changes between levels ν would be proportional to $\sqrt{2}$. For finite element problems arising from 3 dimensional grids it would be expected that κ would be proportional to $\sqrt[3]{n}$ and ν would be proportional to $\sqrt[3]{2}$.

Figure 1 also shows the processor to elimination tree mapping, the processors being labelled P_0, P_1, P_2 and P_3 respectively. In general terms the parallel factorisation of the elimination tree exploits the natural parallelism in the tree while at the same time distributing the factorisation of a frontal matrix $\mathcal{F}^t$ at level i over 2^{l-i} processors.

For each node $t \in T$ there is a *processor group* and for each processor group there is a *lead processor*. The lead processor is responsible for distributing a frontal matrix over the other processors within its group for parallel factorisation and the gathering of the resulting R contribution and the update matrix. On completion of the gathering, the lead processor then sends the update matrix to the lead processor of its parent node. It should be noted that R will not be distributed evenly over the processors, with processor P_0 containing the larger portion of R.

4   Completion Times

In this section equations for determining the completion times of the single processor and parallel factorisation are derived. Using these equations an equation for a two step factorisation is presented. A two step factorisation is used since in practice the number of leaves of the elimination tree is usually much greater than the number of processors available.

4.1   Single Processor Factorisation

On a single processor the multifrontal QR factorisation involves a postorder traversal of the elimination tree T. On visiting each node $t \in T$ a frontal matrix $\mathcal{F}^t$ is assembled and factorised.

The assembly involves the accumulation of update matrices $\mathcal{U}^s$ (for which s is a child node of t) into the frontal matrix $\mathcal{F}^t$. For each node $t \in T$ the assembly time is:

$$\mathcal{A}(t) = \gamma\lambda \left(\sum_{s \in \text{child}(t)} |\mathcal{U}^s| \right), \tag{3}$$

where $\lambda \geq 1$ relates to the efficiency of the indirect addressing associated with the assembly process, γ is the computation cost and $|\mathcal{U}^s|$ is the number of elements in the update matrix $\mathcal{U}^s$. In the ensuing discussion it is assumed that the cost of indirect addressing $\lambda = 1$.

After assembly the frontal matrix $\mathcal{F}^t$ is factorised. The factorisation involves the elimination of κ_t columns from the frontal matrix $\mathcal{F}^t$. This requires the computation of κ_t Householder vectors and rank 1 updates resulting in at least $4\kappa_t|\mathcal{F}^t|$ operations [6]. This gives an upperbound on the time to factorise a frontal matrix $\mathcal{F}^t$ as:

$$\mathcal{E}(t) = 4\gamma\kappa_t|\mathcal{F}^t|. \tag{4}$$

Thus an upperbound for the completion time for multifrontal QR factorisation on a single processor is the sum of the time required to factor each frontal matrix $\mathcal{F}^t$:

$$\mathcal{T}_{seq} = \sum_{t=1}^{\eta} (\mathcal{A}(t) + \mathcal{E}(t)), \tag{5}$$

where η is the number of nodes in the elimination tree. Substituting the frontal and update matrix sizes from (1) and (2) into (5) and simplifying gives a single processor completion time of:

$$\mathcal{T}_{seq} = 4\gamma\kappa^3 c_1 \left[\frac{1}{1 - \frac{2}{\nu^3}}\left(1 - \left(\frac{2}{\nu^3}\right)^{l+1}\right) + \frac{c_2}{2\kappa c_1\left(1 - \frac{2}{\nu^2}\right)}\left(1 - \left(\frac{2}{\nu^2}\right)^l\right)\right]. \tag{6}$$

4.2 Parallel Factorisation

The parallel factorisation involves sending update matrices from child nodes of the elimination tree to their parents. However this need only be done by one child. Further, the time to perform the parallel factorisation is governed by the height of the elimination tree $|h|$ where $|h| = \log_2 p + 1$ and h is a path from a leaf node to the root node of the elimination tree.

Let the time for sending k words of data from one processor to another be:

$$M(k) = \alpha + \omega\beta k, \tag{7}$$

where α is the startup cost, β is the per byte transmission time (communication cost) and ω the number of bytes per word. Therefore the cost of communicating an update matrix is:

$$C(t) = \alpha + \omega\beta|\mathcal{U}^t|. \tag{8}$$

If it is assumed that the frontal matrix $\mathcal{F}^t$ is on one processor before distribution and gathered back onto that same processor after factorisation, then an upperbound for the distributed factorisation time for a frontal matrix $\mathcal{F}^t$ is:

$$\mathcal{E}(t,p) = 2(p-1)\left(\alpha + \omega\beta\frac{|\mathcal{F}^t|}{p}\right) + \frac{4\gamma\kappa_t|\mathcal{F}^t|}{\phi p}. \tag{9}$$

The first term of (9) reflects the communication cost of distributing and gathering the frontal matrix $\mathcal{F}^t$, while the second term is an upperbound on the time required to factor the frontal matrix $\mathcal{F}^t$ over p processors with the distributed factorisation having a parallel efficiency of $0 < \phi \le 1$.

The parallel factorisation has a completion time of:

$$\mathcal{T}_{par} = \sum_{t\in h}\left[\mathcal{A}(t) + \mathcal{E}(t,p) + \mathcal{C}(t)\right]. \tag{10}$$

Making the appropriate substitutions in (10) and simplifying gives the following completion time for the parallel factorisation:

$$\mathcal{T}_{par} = 4\gamma\kappa^3 c_1\left[\frac{1}{2^l\phi\left(1-\frac{2}{\nu^3}\right)}\left(1-\left(\frac{2}{\nu^3}\right)^{l+1}\right) + \frac{\omega\beta}{2\gamma\kappa}\left(\frac{1}{1-\frac{1}{\nu^2}}\left(1-\left(\frac{1}{\nu^2}\right)^{l+1}\right) - \right.\right.$$
$$\left.\frac{1}{2^l\left(1-\frac{2}{\nu^2}\right)}\left(1-\left(\frac{2}{\nu^2}\right)^{l+1}\right)\right) + \frac{c_2}{2\kappa c_1\left(\nu^2-1\right)}\left(1-\left(\frac{1}{\nu^2}\right)^l\right)\left(\frac{\omega\beta}{2\gamma}+\nu^2\right) +$$
$$\left.\frac{\alpha}{4\gamma\kappa^3 c_1}\left(2^{l+2}-l-6\right)\right]. \tag{11}$$

4.3   Two Step Factorisation

Since the number of processors p is usually small compared to the number of leaves in the elimination tree T, the parallel algorithm uses two steps, a local step and a communication step [12]. The communication step is used for factorising that part of the matrix A which is associated with the nodes of the tree from level $2^i = p$ to level $i = 0$, while the local step is the sequential factorisation of the local subtrees that are on each processor.

Using (6) and (11) the completion time for the two step parallel factorisation can be derived. The completion time for the communication step $\mathcal{T}^l_{par}$ of the two step parallel factorisation is calculated by setting the number of levels in the elimination tree $l = \log_2 p$ in (11), while the completion time for the local step is given by:

$$\mathcal{T}^{l\to\infty}_{local} = 2\frac{4\gamma\kappa^3 c_1}{\nu^3\log_2 p}\left(\frac{1}{\nu^3-2} + \frac{c_2\nu^{\log_2 p}}{2\kappa c_1(\nu^2-2)}\right). \tag{12}$$

Equation (12) is derived from (6) by setting the rate at which the granularity decreases between levels of the elimination tree to $\frac{\kappa}{\nu^{\log_2 p+1}}$. The quantity $\frac{\kappa}{\nu^{\log_2 p+1}}$ is the granularity of the nodes of the separator tree at level $i = \log_2 p + 1$. The factor of 2 in (12) is required since there are two subtrees at level $i = \log_2 p + 1$ who share a parent node at level $i = \log_2 p$. The time to factorise a frontal matrix of a node at level $i = \log_2 p$ is not included in (12) as this time is accounted for in (11). Hence the completion time for the two step combined parallel factorisation is:

$$T_{2step} = T_{par}^{l=\log_2 p} + T_{local}^{l \to \infty}. \tag{13}$$

5 Results

The computer parameters computation speed γ^{-1}, communication speed β^{-1} and latency α, in addition to the parallel efficiency ϕ can now be estimated by performing a least squares fit of the model to the measured data.

The two step algorithm was implemented on the 13 processor Fujitsu VPP300 at the Australian National University using the Fujitsu Fortran 90 compiler with default options. The numerical factorisation was performed using the numerical linear algebra libraries of LAPACK [11] and ScaLAPACK [14]. LAPACK is used for the factorisation of frontal matrices associated with the nodes of the local subtrees while ScaLAPACK is used for the distributed factorisation of frontal matrices.

5.1 The Example Matrices

The number of rows m, columns n and nonzeroes nz of the example matrices used to gather completion times are given in Table 1.

Table 1. Statistics for test matrices

Identifier	m	n	nz
PIGS1	3140	1988	8510
PIGS2	9397	6119	25013
PIGS3	28254	17264	75018
RAND1	50000	10201	199818
RAND2	100000	22801	399883
RAND3	125000	40401	499101

The example matrices fall into two categories. The first category consists of the three overdetermined matrices prefixed with "PIGS". These matrices are from animal breeding statistics [8]. The second category of matrices are the three matrices prefixed with "RAND". The "RAND" matrices are randomly generated and are used in the study of additive models from "data mining" [9]. The normal equations matrices associated with the "RAND" matrices are block tridiagonal.

The Recursive Bisection used to generate the elimination tree was similar to that implemented by Karypis and Kumar [10] with the Recursive Bisection continuing until the frontal matrices had a maximum size of 50 columns.

5.2 The Numeric Phase

The times for the two step factorisation were found for an arrangement of the processors into a $p \times 1$ grid for the distributed factorisation. The frontal matrices were distributed by rows over the processors using a blocking factor of 200. On the VPP300 $n_{1/2} \approx 200$, where $n_{1/2}$ is the vector length that produces half the peak performance of the vector units.

Two runs of the numeric factorisation were made by seeding the random number generator of the Recursive Bisection implementation with two different values. For each run the best time for two trials of the factorisation was recorded for the processor configurations of $p = 1, 2, 4$ and 8. The results are given in Table 2.

Table 2. Timings for two step combined parallel factorisation.

	κ	c_1	p=1	p=2	p=4	p=8
PIGS1	122 *109*	10 *12*	0.32 *0.33*	0.27 *0.28*	0.23 *0.26*	0.22 *0.28*
				(1.19) (*1.19*)	(1.39) (*1.27*)	(1.45) (*1.18*)
PIGS2	186 *189*	19 *18*	1.49 *1.55*	0.91 *1.04*	0.63 *0.78*	0.54 *0.71*
				(1.64) (*1.49*)	(2.37) (*1.99*)	(2.76) (*2.18*)
PIGS3	172 *295*	65 *38*	5.30 *0.33*	3.32 *4.46*	2.44 *3.47*	2.05 *2.52*
				(1.60) (*1.52*)	(2.17) (*1.95*)	(2.59) (*2.69*)
RAND1	159 *136*	251 *293*	10.23 *8.99*	6.61 *5.86*	4.77 *4.12*	**** ****
				(1.55) (*1.53*)	(2.14) (*2.18*)	
RAND2	275 *382*	281 *203*	40.66 *45.25*	27.16 *30.99*	18.55 *22.40*	**** ****
				(1.50) (*1.46*)	(2.19) (*2.02*)	
RAND3	281 *268*	301 *316*	69.14 *56.76*	38.84 *34.62*	26.83 *23.39*	**** ****
				(1.78) (*1.64*)	(2.58) (*2.43*)	

The results for Run 2 in Table 2 are given in italics with the corresponding speedups given in parenthesis. Entries denoted by "****" indicate insufficient memory was available for the factorisation. The lack of memory was due to the fact that only 4 of the available processors had more than 416MB available. In addition to recording the completion times, the granularity of the top level node κ for each of the matrices was recorded for each run. With these values of κ, estimates for c_1 were computed using:

$$c_1 = \frac{|\mathcal{F}|}{\kappa^2} = \frac{(m - n + \kappa)}{\kappa}, \tag{14}$$

since $|\mathcal{F}| = (m - n + \kappa) * \kappa$ at the root of the elimination tree. The values for κ and c_1 are given in Table 2.

Using the values in Table 2 a set of four simultaneous equations can be constructed for each of the matrices for each run. For example, the equations (15), (16), (17) and (18) are those constructed for Run 1 of the matrix "Pigs1":

$$T_{seq}^{l \to \infty} = 0.32 \quad (p = 1), \tag{15}$$
$$T_{2step}^{l=1} = 0.27 \quad (p = 2), \tag{16}$$
$$T_{2step}^{l=2} = 0.23 \quad (p = 4), \tag{17}$$
$$T_{2step}^{l=3} = 0.22 \quad (p = 8). \tag{18}$$

In the case of the "RAND" matrices only three equations are available since the results for the case $p = 8$ ($l = 3$) are not available. Therefore an extrapolation is used to estimate the completion time for this equation. The extrapolation is based on the fact that the times are decreasing by approximately $\frac{1}{\sqrt{2}}$. Therefore to compute the time for $p = 8$ for each of the "RAND" matrices the times for $p = 4$ are multiplied by $\frac{1}{\sqrt{2}}$.

For each matrix the appropriate values for the top level node granularity κ and the frontal matrix constant c_1 from Table 2 were substituted into the four simultaneous equations. The update matrix constant c_2 had a range of $0 \le c_2 \le c_1$. Inspection of the size of the update matrices generated from both runs showed a wide variance in size, therefore a value of $c_2 = \frac{c_1}{2}$ was chosen. The rate at which the granularity decreased between levels of the separator tree ν was set to 1.5 (since $\nu > \sqrt{2}$). The sets of equations were solved using Mathematica [15] with the results for Run 2 given in Table 3.

Table 3. Estimated VPP300 parameters using Run 2 results.

	γ^{-1} (Mflops)	β^{-1} (MB/sec)	α (μ secs)	ϕ
PIGS1	466	27	3430	0.47
PIGS2	799	125	11130	0.49
PIGS3	1417	69	22589	0.55
RAND1	810	128	3209	0.53
RAND2	1328	95	< 0	0.97
RAND3	1056	190	107629	0.55

The value in Table 3 denoted by "< 0" indicates a value less than zero and consequently an infeasible solution. The values in Table 3 are in effect an average over all processor configurations for each matrix with the results for Run 1 omitted since infeasible solutions were generated for half the matrices indicating that the model is extremely sensitive to parameter values.

Half of the matrices exhibited a computation speed $\gamma^{-1} = 1100$ Mflops which is consistent with the setting of the vector length to $n_{1/2}$. The latency α is unexpectedly high, increasing with problem size. It is conjectured that the high latency α is associated with overheads in ScaLAPACK.

6   Conclusions

Using the mathematical models developed in Section 4 in conjunction with the experimental results in Section 5, estimates for the computer parameters computation speed γ^{-1}, communication speed β^{-1}, latency α and parallel efficiency

ϕ were made. The high latency was the only unexpected result. It was also found that the parameter values are problem dependent. This work has laid the foundation for estimating a parallel computer's performance during sparse QR factorisation and can also be adapted to Cholesky and LU factorisations. This method has also been applied in a forward manner to predict the completion time of parallel sparse QR factorisation [12].

References

[1] C. Ashcraft and R. Grimes. The influence of relaxed supernode partitions on the multifrontal method. *ACM Transactions on Mathematical Software*, 15(4):291–309, December 1989.

[2] I. Duff. Harwell Boeing sparse matrix collection release I, 1992.

[3] I. Duff and J. Reid. The multifrontal solution of indefinite sparse symmetric linear systems. *ACM Transactions on Mathematical Software*, 9:302–325, 1983.

[4] A. George. Nested disection of a regular finite element mesh. *SIAM J. Numer. Anal.*, pages 345–363, 1973.

[5] J. Gilbert and R. Tarjan. The analysis of a nested disection algorithm. *Numerische Mathematik*, 50:377–404, 1987.

[6] G. Golub and C. Van Loan. *Matrix Computations*. The Johns Hopkins University Press, 2nd edition, 1989.

[7] G. Golub and R. Plemmons. Large-scale geodetic least-squares adjustment by dissection and orthogonal decomposition. *Linear Algebra and its Applications*, 34:3–27, 1980.

[8] M. Hegland. On the computation of breeding values. In *Proceedings of the CONPAR-90-VAPPIV,Joint International Conference on Vector and Parallel Processing*, pages 232–242, Zurich, 1990.

[9] M. Hegland. Personal communication, February 1997.

[10] G. Karypis and V. Kumar. Multilevel k -way partitioning scheme for irregular graphs. Technical Report 95-064, University of Minnesota, University of Minnesota, Department of Computer Science, Minneapolis, MN 55455, August 1995.

[11] LAPACK users' guide. SIAM, May 1992.

[12] D. Miron. *The Parallel Solution of Sparse Linear Least Squares Problems*. PhD thesis, Computer Science Laboratory, Research School of Information Sciences and Engineering, Australian National University, 1998.

[13] R. Pozo. Performance modeling of sparse matrix methods for distributed memory architectures. In *Proceedings of Parallel Processing: CONPAR 92 - VAPP V*, number 634 in Lecture Notes in Computer Science, pages 677–688. Springer-Verlag, September 1992.

[14] ScaLAPACK user's guide. Preliminary Draft, May 1996.

[15] S. Wolfram. *Mathematica: A System for Doing Mathematics by Computer*. Addison-Wesley, 1988.

A Mapping and Scheduling Algorithm for Parallel Sparse Fan-In Numerical Factorization*

Pascal Hénon, Pierre Ramet, and Jean Roman

LaBRI, UMR CNRS 5800, Université Bordeaux 1 & ENSERB
351, cours de la Libération, F-33405 Talence, France
{henon|ramet|roman}@labri.u-bordeaux.fr

Abstract. We present and analyze a general algorithm which computes efficient static schedulings of block computations for parallel sparse linear factorization. Our solver, based on a supernodal fan-in approach, is fully driven by this scheduling. We give an overview of the algorithms and present performance results on a 16-node IBM-SP2 with 66 MHz Power2 thin nodes for a collection of grid and irregular problems.

1 Introduction

Solving large sparse symmetric positive definite systems $Ax = b$ of linear equations is a crucial and time-consuming step, arising in many scientific and engineering applications. Consequently, many parallel formulations for sparse matrix factorization have been studied and implemented; one can refer to [6] for a complete survey on high performance sparse factorization.

In this paper, we focus on the block partitioning and scheduling problem for sparse LDL^t factorization without pivoting on parallel MIMD architectures with distributed memory; we use LDL^T factorization in order to solve sparse systems with complex coefficients. More precisely, we present and analyze a general algorithm which computes an efficient static scheduling [7, 10] of the block computations for a parallel solver based on a supernodal fan-in approach [3, 11, 12, 13] such that the parallel solver is fully driven by this scheduling. This can be done by very precisely taking into account the computation costs of BLAS 3 primitives, the communication cost and the cost of local aggregations due to the fan-in strategy. Using this block partitioning and scheduling algorithm, combined with an ordering of the unknowns based on a tight coupling of Nested Dissection and Approximate Minimum Degree strategies [9], our parallel fan-in supernodal solver is almost as efficient on irregular problems as it is for regular 2D or 3D grid problems.

In order to achieve efficient parallel sparse factorization, three pre-processing phases are required:

• The *ordering* phase, which computes a symmetric permutation of the initial matrix A such that factorization will exhibit as much concurrency as possible

---

* This work is supported by the *Commissariat à l'Énergie Atomique* CEA/CESTA under contract No. 7V1555AC, and by the GDR ARP (iHPerf group) of the CNRS.

P. Amestoy et al. (Eds.): Euro-Par'99, LNCS 1685, pp. 1059–1067, 1999.
© Springer-Verlag Berlin Heidelberg 1999

while incurring low fill-in. In this work, we use a tight coupling of the Nested
Dissection and Approximate Minimum Degree algorithms [1, 8, 9]; the partition
of the original graph into supernodes is achieved by merging the partition of
separators computed by the Nested Dissection algorithm and the supernodes
amalgamated for each subgraph ordered by Halo Approximate Minimum Degree.

• The *block symbolic factorization* phase, which determines the block data
structure of the factored matrix L associated with the partition resulting from
the ordering phase. This structure consists of N column blocks, each of them
containing a dense symmetric diagonal block and a set of dense rectangular off-
diagonal blocks. One can efficiently perform such a block symbolic factorization
in quasi-linear space and time complexities [5]. From the block structure of
L, we can deduce the weighted elimination quotient graph that describes all
dependencies between blocks, as well as the supernodal elimination tree.

• The *block repartitioning and scheduling* phase, which refines the partition,
by splitting large supernodes in order to exploit concurrency within dense block
computations, and maps it onto the processors of the target architecture.

The rest of the paper is organized as follows. Section 2 presents our algorithmic framework for parallel sparse symmetric factorization, describes our
block repartitioning and scheduling algorithm, and outlines the supernodal fan-
in solver driven by this precomputed scheduling. Section 3 provides numerical
experiments on an IBM SP2 for a large class of sparse matrices, including performance results and analysis. Then, we conclude with remarks on the benefits
of this study and on our future work.

2   Block Mapping and Parallel Solver

Let us consider the block data structure of the factored matrix L computed by
the block symbolic factorization. Let us remind that a column block holds one
dense diagonal block and some dense off-diagonal blocks. From this block data
structure, we can introduce the boolean function $off\_diag(k, j)$, $1 \leq k < j \leq N$,
that returns *true* if and only if there exists an off-diagonal block in column block
k facing column block j. Then, we can define the two sets $BStruct(L_{k*}) := \{i <
k \mid off_diag(i, k) = true\}$ and $BStruct(L_{*k}) := \{j > k \mid off\_diag(k, j) = true\}$.
Thus, $BStruct(L_{k*})$ is the set of column blocks that update column block k,
and $BStruct(L_{*k})$ is the set of column blocks updated by column block k.

We consider now a parallel supernodal version of sparse LDL^t factorization with total local aggregation: all non-local column block contributions are
aggregated locally in block structures. This scheme is close to the Fan-In algorithm [4] as processors communicate using only aggregate update column
blocks. It is important to note that column block k will receive an aggregated update column block only from every processor in set $Procs(L_{k*}) :=
\{map(i) \mid i \in BStruct(L_{k*})\}$, where the $map()$ operator is the block mapping
function. These structures, denoted in the following $AUCB_k$, can be built from
the block symbolic factorization. The pseudo-code of LDL^t factorization can be
expressed in terms of dense block computations (see figure 1); these computa-

tions are performed, as much as possible, on compacted sets of blocks, for BLAS efficiency. Synchronizations are enforced thanks to r_k, which counts the number of contributions remaining for column block k. While $r_k \neq 0$, we need to receive contributions for column block k (lines 2–4). Then the diagonal block is factored and the off-diagonal blocks are updated (lines 6–8). Partial results (block F) are stored, to be re-used in the contribution computations. All of the contributions associated with column block k can now be computed; a contribution can be added locally (line 12), or in an aggregate update column block (line 14). An aggregate update column block is sent when all of its local contributions have been added (line 15). To limit memory consumption due to the storage of aggregates, these structures are allocated only when needed and are deallocated just after being sent. If memory is a critical issue, an aggregate update column block can be sent with partial aggregation to free memory space; this is close to the Fan-Both scheme [2].

Notations for a given k, $1 \leq k \leq N$:

- r_k : number of contributions yet to subtract to column block k,
- r_k^o : initial value of r_k,
- s_k : number of contributions yet to add to $AUCB_k$,
- s_k^o : initial value of s_k,
- symbol $*$ means $\forall j \in BStruct(L_{*k})$,
- let $j \geq k$; sequence $[j]$ means $\forall i \in BStruct(L_{*k}) \cup \{k\}$ with $i \geq j$.

On processor p :
1. **For** $k \in [1, N]$ such that $map(k) == p$ **Do**
2.   **While** $r_k \neq 0$ **Do** % receive_update step
3.   receive $AUCB_k$;
4. $A_{[k]k} = A_{[k]k} - AUCB_k$; $r_k \leftarrow r_k - s_k^o$;
5.   % compute_send step
6.   Factor A_{kk} into $L_{kk} D_k L_{kk}^t$;
7.   Solve $L_{kk} F^t = A_{*k}^t$;
8.   Solve $D_k L_{*k}^t = F^t$;
9.   **For** $j \in BStruct(L_{*k})$ **Do**
10.   compute $C = L_{[j]k} F_j^t$;
11.   **If** $map(j) == p$ **Then**
12. $A_{[j]j} = A_{[j]j} - C$; $r_j \leftarrow r_j - 1$;
13.   **Else** % aggregate step
14. $AUCB_j = AUCB_j + C$; $s_j \leftarrow s_j - 1$;
15.   **If** $s_j = 0$ **Then** send $AUCB_j$ to $map(j)$;

Fig. 1. Outline of the parallel factorization algorithm.

Before running the general parallel algorithm we presented above, we must perform a step consisting in the partitioning and mapping of the blocks of the symbolic matrix onto the set of processors. The partitioning and mapping phase aims at computing a static regulation that balances workload and enforces precedence constraints imposed by the factorization algorithm; the block elimination tree structure must be used there. Different levels of parallelism can be exhibited. The first level is induced by the sparsity of the matrix and corresponds to independent branches in the block elimination tree. The second level is induced by dense computation on large full blocks. The third level exploits instruction-level parallelism at the processor level by using BLAS subroutines. Algorithms like "subtree to subcube" or "proportional mapping" propose a partitioning and mapping strategy that balances workload based on a recursive traversal of the block elimination tree. Then, given the new partition and the mapping of blocks, the solver is responsible for scheduling block computations and communications.

This existing approach of a block partitioning and mapping strategy rises some problems. Those problems can be divided into two categories: on one hand problems due to the measure of workload and on the other hand those due to run-time scheduling of block computations in the solver. The measure of workloads is very rough because only numbers of operations are taken into account. Indeed, to be efficient, the solver algorithm is block-oriented, in order to use BLAS subroutines, whose efficiencies are very far from being linear in terms of numbers of operations. Moreover, the workload deals with block computations but does not take into account all the others phenomena in the parallel fan-in factorization: extra-workload generated by the fan-in approach and idle-waits due to the latency generated by message passing. The second kind of problem is run-time scheduling of concurrent block computations and communication in the solver is limited to rough criteria.

We tackle these problems by using an inspector-executor-like approach. In our scheme, the partitioning and mapping computations stand for the inspector phase, and the parallel solving computations for the executor phase. Thus, the partitioning and mapping step generates a fully ordered schedule used in parallel solving computations. This schedule aims at statically regulating all of the issues that are classically managed at run-time. These issues are:

- the order in which column blocks that have a contribution counter r_k equal to zero are processed can influence the efficiency of computations (line 1);
- the order in which to send aggregate update column blocks (line 3) and to process the received ones (line 15) determines what block will be ready next for local computation;
- the way to switch between receive_update and compute_send steps (line 2).

To make our scheme very reliable, we estimate the workload and message passing latency by using a BLAS and communication network time model. To be efficient, this model has to be calibrated on the target architecture. The proposed scheduling and mapping algorithm can yield 1D or 2D block distributions. At the time being, we only handle 1D distributions, and thus in all the following we will restrict our discussion to the 1D case. However, a 2D extension of our work is in progress, since it is well known that, for large numbers of processors, 2D distributions achieve better scalability.

Unlike usual algorithms, our partitioning and distributing strategy is divided in two distinct phases. The partition phase splits column blocks associated with large supernodes, and builds for each column block a set of candidate processors for its mapping. The scheduling phase optimally maps each column block onto one of these processors. The partitioning algorithm is based on a recursive top-down strategy over the block elimination tree. Pothen and Sun presented such a strategy in [10]. It starts by splitting the root and assigning it to the set of candidate processors Q. Then, recursively, each subtree is assigned to a subset of Q proportionally to its workload. It is important to notice that since a candidate processor is only a suggestion for the mapping and scheduling phase, we avoid any problem of rounding to integral numbers of processors by allowing a candidate processor to be in two sets of candidate processors for two subtrees having the same father in the block elimination tree.

Once the partitioning phase has built a new partition and the set of candidate processors for each column block, the task of electing an owner processor for each column block falls to the mapping phase. The idea that leads the mapping and scheduling strategy is to simulate parallel factorization as mapping comes along. Thus, we define for each processor a timer that will stand for the current time in computation of the processor and a ready block heap that contains at a time all blocks not yet mapped that have received their contribution and for which the processor is a candidate. Then we start mapping the leaves of the elimination tree (one candidate processor). When a block is mapped on a processor, three actions are processed:

- update the processor timer for local computation thanks to our BLAS model;
- compute the time a contribution from this block to another is computed;
- put a block in the ready block heaps of its candidate processors if all its contributions are computed.

After a block has been mapped we choose the next one to be mapped: we get the processor with lower timer and with a not empty ready block heap. Then we get the block in its ready block heap that is the sooner available. This one is the next block we are to map. Thus we are able to compute for each candidate processor the time it will have finished to treat this block if it is mapped on. This is possible thanks to:

- the processor timer;
- the date at which all contributions to this block have been computed;
- the communication cost modelization that gives the time needed to exchange the contributions.

The block is mapped on the candidate processor that will be able to compute it the sooner. Then we can loop on another block until the last.

The block computations are ordered in respect to the rank at which the column block have been mapped. Thus, for each processor p, we obtain a vector K_p of the N_p local column block numbers fully ordered by priority. Ordering column block computation is not sufficient to assure that the solver will respect all the predictions made in the mapping phase: emission and reception of aggregate update column blocks have to be ordered too.

The factorization algorithm is modified (see figure 2) to account for these new extensions. The emissions are performed on the set $AUCB_{ready}$ containing the ready aggregate update column blocks to send to $map(j)$ and whose priority is higher than the one of the first non ready aggregate update column block.

On processor p :
1. **For** $i = 1$ **To** N_p **Do**
$\quad k = K_p(i)$
$\quad [...]$
3-4. $\qquad$ subtract the $AUCB_k$ received
$\quad [...]$
15. $\qquad$ **If** $s_j = 0$ **Then** send $AUCB_{ready}$ to $map(j)$;

Fig. 2. Outline of the new parallel factorization algorithm.

3   Numerical Experiments

In this section, we describe experiments performed on a collection of sparse matrices; the values of the metrics in table 1 come from scalar column symbolic factorization.

These parallel experiments were run on an IBM SP2 with 66 MHz Power2 thin nodes (260 MFlops peak performance) having 128 MBytes of physical memory and 512 MBytes of virtual memory each, and switch interruptions are enabled with default delay to perform non-blocking communications efficiently.

Table 2 reports the efficiency of our parallel block factorization for our distribution strategy. Profiled executions confirm that BLAS level 3 block computations ($GEMM$ and $TRSM$) represent more than 75% of total BLAS computations. A multi-variable polynomial regression has been used to build an analytical model of these routines. This model and the experimental values obtained for communication startup and bandwidth are used by the partitioning and scheduling algorithm. It is important to note that the number of operations actually performed during factorization is greater than the OPC value because of amalgamation and block computations. A good scalability is achieved for all of the test problems, at least for moderately sized parallel machines, and even with 1D block mapping.

On both moderate and large size grid and irregular problems, the measured performance vary between 0.65 and 1.56 Gigaflops on 16 nodes, using double precision arithmetic, thereby obtaining about 35% of the processor peak theoretical performance. Our scheduling leads to efficient parallel execution, with very good communication locality, and minimizes idle-time (see figures 3, 4).

4   Conclusion and Perspectives

In this paper, we have presented an efficient static scheduling of the block computations of a parallel sparse direct solver. This work is still in progress; we are

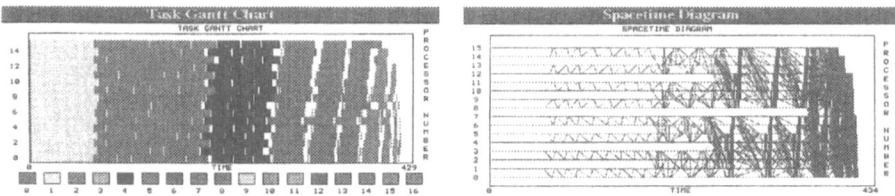

Fig. 3. Measured Gantt chart & Spacetime graph for CUBE39.

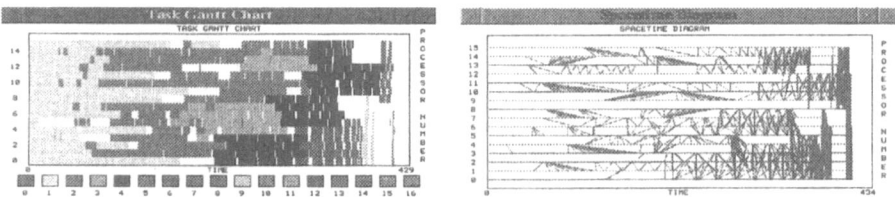

Fig. 4. Measured Gantt chart & Spacetime graph for CRANKSEG1.

currently developing a 2D mapping strategy for the uppermost column blocks of the block elimination tree, and studying an adaptive computation of optimal packet sizes to improve computation/communication overlap in asynchronous communications.

Table 1. Description of our test problems. NNZ_A is the number of off-diagonal terms in the triangular part of matrix A, NNZ_L is the number of off-diagonal terms in the factored matrix L and OPC is the number of operations required.

Name	Columns	NNZ_A	NNZ_L	OPC	Description
GRID511	261121	1041420	1.202166e+07	2.565341e+09	Regular 2D mesh
CUBE39	59319	730778	2.210534e+07	2.240674e+10	Regular 3D cube
BCSSTK30	28924	1007284	4.309003e+06	1.188915e+09	Boeing-Harwell
BCSSTK32	44609	985046	5.239146e+06	1.162900e+09	Boeing-Harwell
COL15	10428	701292	3.371109e+06	1.495460e+09	3D Cologne Challenge
COL30	20373	1394292	7.849464e+06	4.359803e+09	3D Cologne Challenge
COL75	50208	3473292	2.248613e+07	1.532035e+10	3D Cologne Challenge
CRANKSEG1	52804	5280703	3.142730e+07	3.007141e+10	PARASOL
OILPAN	73752	1761718	8.912337e+06	2.984944e+09	Boeing-Harwell
TOOTH	78136	452591	1.031143e+07	6.267094e+09	Boeing-Harwell
B5TUER	162610	3873534	2.541937e+07	1.530774e+10	PARASOL
BBMAT	38744	1274141	1.716094e+07	1.250040e+10	PARASOL
INVEXT	30412	906915	7.256566e+06	3.766788e+09	PARASOL
MIXTANK	29957	982542	9.280247e+06	7.316933e+09	PARASOL
MT1	97578	4827996	3.114873e+07	2.109265e+10	PARASOL
QUER	59122	1403689	9.118592e+06	3.280680e+09	PARASOL
SHIP001	34920	2304655	1.427916e+07	9.033767e+09	PARASOL
X104	108384	5029620	2.634047e+07	1.712902e+10	PARASOL

Table 2. Factorization performance results (time in seconds and **Gigaflops**) on the IBM SP2.

Name	Number of processors			
	2	4	8	16
GRID511	14.45 (0.18)	7.45 (0.34)	4.24 (0.61)	2.70 (0.95)
CUBE39	-	42.86 (0.52)	24.40 (0.92)	15.55 (1.44)
BCSSTK30	6.43 (0.18)	3.60 (0.33)	2.27 (0.52)	1.83 (0.65)
BCSSTK32	7.28 (0.16)	3.60 (0.33)	2.13 (0.55)	1.52 (0.77)
COL15	6.93 (0.22)	3.78 (0.40)	2.37 (0.63)	1.84 (0.81)
COL30	18.13 (0.24)	9.81 (0.44)	5.68 (0.77)	3.72 (1.17)
COL75	157.54 (0.10)	31.24 (0.49)	17.14 (0.89)	9.99 (1.53)
CRANKSEG1	-	58.14 (0.52)	32.11 (0.94)	19.29 (1.56)
OILPAN	14.03 (0.21)	7.61 (0.39)	4.41 (0.68)	3.12 (0.96)
TOOTH	32.97 (0.19)	18.05 (0.35)	10.17 (0.62)	6.49 (0.97)
B5TUER	-	31.79 (0.48)	17.58 (0.87)	10.77 (1.42)
BBMAT	61.54 (0.20)	30.21 (0.41)	17.07 (0.73)	11.17 (1.12)
INVEXT	20.37 (0.18)	10.99 (0.34)	5.88 (0.64)	4.10 (0.92)
MIXTANK	30.31 (0.24)	18.05 (0.41)	10.03 (0.73)	6.75 (1.08)
MT1	-	43.77 (0.48)	24.72 (0.85)	15.07 (1.40)
QUER	14.92 (0.22)	8.05 (0.41)	4.69 (0.70)	3.32 (0.99)
SHIP001	38.46 (0.23)	19.88 (0.45)	10.75 (0.84)	7.45 (1.21)
X104	-	38.21 (0.45)	21.24 (0.81)	13.97 (1.23)

References

[1] P. Amestoy, T. Davis, and I. Duff. An approximate minimum degree ordering algorithm. *SIAM J. Matrix Anal. and Appl.*, 17:886–905, 1996.

[2] C. Ashcraft. The fan-both family of column-based distributed Cholesky factorization algorithms. *Graph Theory and Sparse Matrix Computation, IMA, Springer-Verlag*, 56:159–190, 1993.

[3] C. Ashcraft, S. C. Eisenstat, and J. W.-H. Liu. A fan-in algorithm for distributed sparse numerical factorization. *SIAM J. Sci. Stat. Comput.*, 11(3):593–599, 1990.

[4] C. Ashcraft, S. C. Eisenstat, J. W.-H. Liu, and A. Sherman. A comparison of three column based distributed sparse factorization schemes. In *Proc. Fifth SIAM Conf. on Parallel Processing for Scientific Computing*, 1991.

[5] P. Charrier and J. Roman. Algorithmique et calculs de complexité pour un solveur de type dissections emboîtées. *Numerische Mathematik*, 55:463–476, 1989.

[6] I. S. Duff. Sparse numerical linear algebra: direct methods and preconditioning. Technical Report TR/PA/96/22, CERFACS, 1996.

[7] G. A. Geist and E. Ng. Task scheduling for parallel sparse Cholesky factorization. *Internat. J. Parallel Programming*, 18(4):291–314, 1989.

[8] F. Pellegrini and J. Roman. Sparse matrix ordering with SCOTCH. In *Proceedings of HPCN'97, Vienna, LNCS 1225*, pages 370–378, April 1997.

[9] F. Pellegrini, J. Roman, and P. Amestoy. Hybridizing nested dissection and halo approximate minimum degree for efficient sparse matrix ordering. In *Proceedings of IRREGULAR'99, Puerto Rico, LNCS 1586*, pages 986–995, April 1999.

[10] A. Pothen and C. Sun. A mapping algorithm for parallel sparse Cholesky factorization. *SIAM J. Sci. Comput.*, 14(5):1253–1257, September 1993.

[11] E. Rothberg. Performance of panel and block approaches to sparse Cholesky factorization on the iPSC/860 and Paragon multicomputers. *SIAM J. Sci. Comput.*, 17(3):699–713, May 1996.

[12] E. Rothberg and A. Gupta. An efficient block-oriented approach to parallel sparse Cholesky factorization. *SIAM J. Sci. Comput.*, 15(6):1413–1439, November 1994.

[13] E. Rothberg and R. Schreiber. Improved load distribution in parallel sparse Cholesky factorization. In *Proceedings of Supercomputing'94*, pages 783–792. IEEE, 1994.

Scheduling of Algorithms Based on Elimination Trees on NUMA Systems

María J. Martín[1], Inmaculada Pardines[2], and Francisco F. Rivera[2]

[1] Dept. Eléctrónica y Sistemas, Univ. A Coruña, SPAIN
maria@dec.usc.es
[2] Dept. Electrónica y Computación, Univ. Santiago de Compostela, SPAIN
{inma,fran}@dec.usc.es

Abstract. An important issue in the execution of programs on multi-processor systems with non-uniform memory access times is data locality. Most of the dynamic scheduling algorithms deal well with load balance, but fail to take locality into account and, therefore, behave poorly on NUMA systems. In this paper we present a scheduling algorithm which has as its objective to increase data locality, and therefore performance, in problems based on elimination trees. We applied this scheduling to the modified Cholesky factorization as a case study. Experimental results on the SGI O2000 are shown.

1 Introduction

The CC-NUMA (cache-coherent non-uniform memory architecture) multiprocessors are attractive systems due to their transparent access to local and remote memories. However, the remote access latency is 3 to 5 times the local access latency. Therefore, data locality is potentially one of the most important topics to obtain high performances. In this paper the scheduling problem is analyzed with the aim of improving data locality on a particular set of irregular problems, those based on elimination trees [5]. Several types of matrix factorization and solvers of triangular systems are examples of this kind of problems. The modified Cholesky factorization, as representative of this kind of problems, was considered [4]. Experimental results on the SGI O2000 are shown.

2 Modified Cholesky Factorization: The Right-Looking Algorithm

In the right-looking version of the standard Cholesky algorithm, once a column has been normalized, it is used to modify every posterior column that depend on it. This scheme is based on the successful parallel proposal developed in the Oak National Laboratory [3] for a distributed memory system. The Splash library [8] includes a supernodal version of this scheme.

We adapted this algorithm for the modified Cholesky factorization on shared memory systems. In the parallel program, each task consists of the computation

P. Amestoy et al. (Eds.): Euro-Par'99, LNCS 1685, pp. 1068–1072, 1999.
© Springer-Verlag Berlin Heidelberg 1999

of a column, that is to say, the diagonal calculation, its normalization and the updating of posterior columns. With each column, a counter that stores the number of remaining modifications is maintained. A shared global task queue stores the columns whose counters reach zero. These are the columns that have already received all their modifications, and therefore, are prepared to be normalized and to modify posterior columns. When a processor becomes free, it visits this queue to acquire a new task. Then, it calculates the diagonal, normalizes the column and carries out new modifications. The counters of the destination columns are decreased to reflect these modifications. If any of them reach zero, the column is placed in the global queue of available tasks.

This scheduling is dynamic, the task queue is created in run time and the processors have access to it as they become idle. This leads to a good load balance. The problem with this approach is that more than one processor can modify the same column and therefore will have to access and write on the same memory address. This produces a large number of memory conflicts, coherence operations and cache misses, and consequently an increase in the number of remote memory accesses which is the overwhelming factor affecting performance.

2.1 Experimental Results

The codes were implemented on the SGI 02000 using *parmacs* [6]. A set of symmetric matrices from the Harwell-Boeing library [2] was selected to analyze the behavior of the program. The minimum degree algorithm [1] was applied to all the matrices in order to reduce the fill. Table 1 shows the characteristics of the matrices and the execution times of the modified Cholesky factorization for each matrix in ms, where n is the number of rows and columns, n_z the number of nonzero entries and P is the number of processors. Note that these times increase from the sequential to the parallel execution. This overhead is due to the increase in the number of cache misses.

Table 1. Characteristics of the benchmark matrices and execution times

MATRIX	n	n_z in A	n_z in L	P=1	P=2	P=4	P=8
BCSSTK14	1806	36630	116071	330	490	410	390
BCSSTK15	3948	60882	707887	6000	9600	7300	6300
BCSSTK18	11948	80519	668217	4500	7500	5600	4800
BCSSTK23	3134	24156	450953	3800	6800	5200	4400
BCSSTK24	3562	81736	291151	1100	1700	1300	1200
BCSSTK30	28924	1036208	4677146	54500	65100	48000	40600
LSHP3466	3466	13681	93713	200	300	250	230

The external loop in the program is organized like a self scheduling loop. This leads to a good load balance as the distribution of tasks to the processors is made in a dynamic way. Figure 1 shows the load balance obtained for each matrix on 2, 4 and 8 processors. Load balance is defined as: $\bar{B} = I_t/(P * I_{max})$, where I_t is the total number of required instructions for the factorization and I_{max}

is the maximum number of instructions assigned to one processor. Therefore, $0 < B \leq 1$, and $B = 1$ represents a completely balanced load.

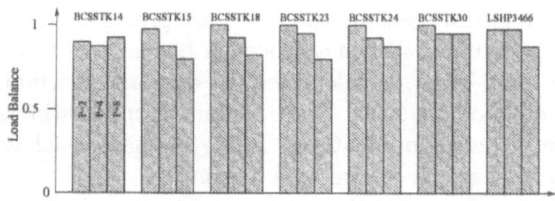

Fig. 1. Load balance in the right-looking program

3   New Scheduling for the Right-Looking Algorithm

In the previous strategy, the initial queue stores the leaf nodes of the elimination tree in ascending order from columns, and a processor accesses this queue following this order. Nevertheless, it would be more convenient that different processors work with nodes belonging to different subtrees. For this we introduce an ordering of the available tasks from the queue using a "distance function" between nodes of the elimination tree and selecting the nodes that have the major distance between them. To carry out this selection, the Prim's method [7] is applied taking the deepest tree node as the initial node until P nodes are obtained. The Prim algorithm was selected because is a fast and well known method.

The distance function must represent a measure of the nearness of the nodes in the elimination tree. For this, both the number of levels between nodes and the height of the first common predecessor should be taken into account. Let i and j be two nodes with their first common predecessor k. The distance between them is defined as:

$$d_{ij} = d_{ji} = \alpha \cdot min\{P_{ik}, P_{jk}\} + \beta \cdot (P_{ik} + P_{jk}) + \gamma \cdot L_k \qquad (1)$$

Where P_{ik} is the number of levels between nodes i and k and L_k is the height of node k, that is to say, the number of levels from the deepest node to node k. Therefore the first term of (1) reflects the work in parallel that can be carried out when processing the nodes i and j. This is what will have the greatest influence on the parallel program. The second term takes into account the number of levels existing between both nodes. The greater this number, the lesser the conflict between them will be. The third term reflects the height of the first common ancestor. The greater it is, the less nodes it will have to update and so the conflict will be lesser. α, β y γ are parameters which emphasize the relative influence of the three factors. We suggest: $\alpha = N$, $\beta = 1$ y $\gamma = 1$.

Once the initial assignation has been carried out, each processor should accede, wherever possible, to columns belonging to the same subtree. Then, if a processor is the last to modify a column, that column goes on to be processed by the same processor instead of being included in the global queue.

Several advantages can be obtained with the proposed scheduling. On one hand the reuse is greater. Through working with the parent node of the last processed column, the same column is used more often by the same processor, but principally, one column will modify columns belonging to the same subtree. In this way, the number of times that one processor modifies the same columns increases. On the other a lesser number of accesses to the global queue is needed, decreasing the overhead associated with synchronization.

3.1   Experimental Results

Table 2 shows the execution times with the new scheduling for different number of processors. The immediate observation is the drastic decrease in the execution time for more than one processor. The main problem of this approach is the load balance, which will limit the scalability of the parallel program. Figure 2 shows the results.

Table 2. Execution times

MATRIX	P=1	P=2	P=4	P=8
BCSSTK14	310	210	200	210
BCSSTK15	5600	4300	3900	3800
BCSSTK18	4200	3000	2800	2900
BCSSTK23	3700	3400	3300	3500
BCSSTK24	1000	790	620	610
BCSSTK30	41200	28700	24800	22600
LSHP3466	180	130	130	130

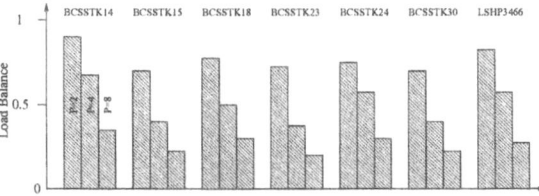

Fig. 2. Load balance in the right-looking program with the new scheduling

4   Conclusions and Future Work

In this work a new scheduling algorithm based on the elimination tree is proposed. Our proposal focus on the increase in the locality and thus the efficiency of the parallel code at a cost of producing load imbalance that limits its scalability. The results show that when a NUMA shared memory system is used, not only load balance and idle times have to be taken into account, but also that data locality plays a fundamental role in the performance of the program.

The strategies here described can be generalized to other sparse matrices problems like factorizations or the solvers of triangular systems. In general they are applicable to all those sparse codes whose dependencies can be represented by a tree.

Recently much more efficient parallel sparse algorithms to solve the standard Cholesky factorization based on blocking techniques have been proposed. We plan to apply the scheduling algorithm presented to a blocked right-looking sparse algorithm, as the found in the new Splash library [9], and analyze its impact on the memory locality, and to include strategies to reduce load imbalance.

Acknowledgements

The authors are grateful to the Universidad de Málaga (Spain) for providing access to their SGI O2000. This work was supported in part by CICYT under grant TIC96-1125-C03-02 and Xunta de Galicia under grant XUGA20605B96.

References

[1] T.A. Davis, P. Amestoy and I.S. Duff, *An approximate minimum degree ordering algorithm*, Technical report, Computer and Information Sciences Dept., University of Florida, 1995.

[2] I.S.Duff, R.G.Grimes, and J.G.Lewis, *User's guide for the harwell-boeing sparse matrix collection*, Technical Report TR-PA-92-96, CERFACS, October 1992.

[3] A. George, M. Heath, J. Liu, E. Ng, *Solution of sparse positive definite systems on a hypercube*, J. Comput. Appl. Math., 27, 1989, pp. 129-156.

[4] P.E. Gill, W. Murray and M.H. Wright, *Practical optimization*, Academic Press, London, 1981.

[5] J.W.H. Liu, *The role of elimination trees in sparse factorization*, SIAM Journal on Matrix Anal. Appl., 11, 1990, pp. 134-172.

[6] *Portable Programs for Parallel Processors*, Lusk et al, Holt Rinchart and Winston 1987.

[7] R.C. Prim, *Shortest connection networks and some generalisations*, Bell System Tech. J., 36, 1957, pp.1389-1401.

[8] Jaswinder Pal Singh, Wolf-Dietrich Weber and Anoop Gupta, *SPLASH: Stanford Parallel Applications for Shared-Memory*, Technical Report, Stanford University, Computer Systems Laboratory, Number CSL-TR-92-526, June 1992, p. 45.

[9] S. C. Woo, M. Ohara, E. Torrie, J. P. Singh and A. Gupta, *The SPLASH-2 Programs: Characterization and Methodological Considerations*, Proceedings of the 22nd Annual International Symposium on Computer Architecture, ACM Press, June 22-24 1995, pp. 24-37.

Block-Striped Partitioning and Neville Elimination

P. Alonso[1], R. Cortina[2], and J. Ranilla[2]

1 Edificio de Energía, Campus de Viesques, Universidad de Oviedo, E-33271 Gijón. Spain.
palonso@etsiig.uniovi.es
[2] Centro de Inteligencia Artificial, Universidad de Oviedo, E-33271 Gijón. Spain.
{raquel,ranilla}@aic.uniovi.es

Abstract. In this paper a message-passing parallel implementation of the solution of linear systems by means of the Neville elimination is described. This type of approach is especially suited to the case of totally positive linear systems, which appears in different application fields. Standard data partitioning techniques, such as block row and block column schemes, are considered on three different topologies: ring, mesh and hypercube. The theoretical performance of the proposed parallel algorithms, in terms of run time, speed-up and efficiency, is derived. Experimental results obtained on an IBM SP2 multicomputer confirm the high performance of the block row parallel algorithm.

1 Introduction

The Neville elimination has been used in some questions about the Theory of Approximation and in other fields (see [8]). In recent papers (see [5, 6, 7]) it has been confirmed that the Neville elimination has advantages over the Gaussian elimination when we work with certain kind of matrices; in particular, the totally positive matrices. A real matrix A is said to be totally positive if all its minors are nonnegative. Such matrices appear frequently in problems of Statistics, Theory of Approximation, and Computer Assisted Geometric Design, besides in other fields (see [4]).

The advantages we have mentioned refer to the fact that the Neville elimination provides algorithmic characteristics of these matrices and its subclasses. As far as the computational cost is concerned, it is in general the same for Neville elimination and for Gauss elimination. Though, in [7] it has been shown that the computational cost for Neville elimination can be lower than the computational cost for Gaussian elimination. To be exact, for certain special matrices, as for example the totally positive matrices that are a reverse of a band matrix.

In recent years, studies have been carried out about the analysis of the backward and forward error for Neville elimination (see [1, 2]). Finally, it is important to mention that in parallel computing the schemes of elimination similar to the Neville elimination have also proved very useful (see [3, 11]).

In this paper, we begin the study of the performance of Neville method in multiprocessor architectures. In section 2, we describe the Neville method and analyze the run time of its sequential algorithm. Then, in section 3 we undertake the study of this method taking into account that the data are distributed in blocks of

P. Amestoy et al. (Eds.): Euro-Par'99, LNCS 1685, pp. 1073-1077, 1999.
© Springer-Verlag Berlin Heidelberg 1999

consecutive rows or columns, storing these blocks in the k processors that take part in the elimination. In the last part, we will compare the theoretical results with the empirical ones, checking the advantages and inconveniences of the method.

2 The Neville Elimination

Let $Ax=b$ be a non-singular system of linear equations, with $A=(a_{ij})_{1\leq i,j\leq n}$ and $b=(b_i)_{1\leq i\leq n}$. The process of elimination of Neville applied to this system consists in n-1 successive stages, resulting in a sequence of matrices of the following form

$$A=\overline{A}_1 \rightarrow A_1 \rightarrow \overline{A}_2 \rightarrow A_2 \rightarrow ... \rightarrow \overline{A}_n = A_n = U , \qquad (1)$$

where U is an upper triangular matrix. For each k, the matrix $A_k = (a_{ij}^{(k)})_{1\leq i,j\leq n}$ has zeros in the k-1 first columns below its main diagonal. Therefore, the matrix A_k can be expressed in the following way

$$A_k = \begin{pmatrix} a_{11}^{(k)} & a_{12}^{(k)} & ... & ... & ... & ... & a_{1n}^{(k)} \\ 0 & a_{22}^{(k)} & ... & ... & ... & ... & a_{2n}^{(k)} \\ ... & 0 & ... & ... & ... & ... & ... \\ ... & ... & ... & a_{k-1,k-1}^{(k)} & a_{k-1,k}^{(k)} & ... & a_{k-1,n}^{(k)} \\ ... & ... & ... & 0 & a_{kk}^{(k)} & ... & a_{kn}^{(k)} \\ ... & ... & ... & ... & ... & ... & ... \\ 0 & 0 & ... & 0 & a_{nk}^{(k)} & ... & a_{nn}^{(k)} \end{pmatrix} . \qquad (2)$$

The matrix A_k is obtained from $\overline{A}_k$, placing in the last rows, if necessary, those rows that have one zero in the column k, according to

$$a_{ik}^{(k)} = 0, i \geq k \Rightarrow a_{hk}^{(k)} = 0, \forall h \geq i . \qquad (3)$$

To obtain $\overline{A}_{k+1}$ from A_k, we will make zeros in the column k below the main diagonal, subtracting a multiple of the row i-1 from the row i, for $i=n, n$-1, ..., k+1.

This elimination has been described in detail for finite matrices in [7], proving in [5] that when we work with totally positive matrices, this process can be carried out without row exchanges.

If we consider that a process of Neville elimination can be realized in the system $Ax=b$ without row exchanges, then $\overline{A}_k = A_k$, for k=1, 2, ..., n, and the elements of A_{k+1} are obtained according to the expression

$$
a_{ij}^{(k+1)} = \begin{cases} a_{ij}^{(k)} & \text{if } 1 \le i \le k, \\[2mm] a_{ij}^{(k)} - \dfrac{a_{ik}^{(k)}}{a_{i-1,k}^{(k)}} a_{i-1,j}^{(k)} & \text{if } k+1 \le i \le n \text{ and } a_{i-1,k}^{(k)} \ne 0, \\[3mm] a_{ij}^{(k)} & \text{if } k+1 \le i \le n \text{ and } a_{i-1,k}^{(k)} = 0 \ \big(\Leftrightarrow a_{ij}^{(k)} = 0\big). \end{cases} \tag{4}
$$

In the same way, the vector of independent terms is being modified stage by stage.

Taking into account this method, we can calculate the number of operations in float point that are carried out to compute the solution. The total cost is

$$
T(N;1) = \frac{4n^3 + 3n^2 - 7n}{6} t_c \approx \frac{2n^3}{3} t_c , \tag{5}
$$

where t_c is the time spent to realize one operation in float point. This cost coincides with the cost of Gaussian elimination in sequential (see [9]).

3 Striped Partitioning

Let A be a $n \times n$ matrix, b a vector of n components, and a multiprocessor system MIMD with k processors. We will consider that each processor has n/k consecutive rows of A and n/k consecutive elements of b. In this way, we are partitioning the information between the processors into blocks of consecutive rows.

In the next algorithm, we describe the stages to transform the system $Ax=b$ into an upper triangular system using the method of Neville.

```
Algorithm of Neville              for i=p*fi down j+1
for j=1 to n-1                        if a_ij<> 0 then
   fi=n/k                               m_ij=a_ij/a_{i-1,j}
   p=(j-1) div (fi+1)                   for r=j+1 to n
   STEP 1                                 a_ir=a_ir-m_ij*a_{i-1,r}
   STEP 2                               end for
end for                                 b_i=b_i-m_ij*b_{i-1};   a_ij=0
                                      end if
Step 1: In parallel do             end for
For l=p to k-1                     b) In P_s with s=p+1 to k
   h=l*fi                          for i=s*fi down (s-1)*fi+1
   for i=j to n                       if a_ij<> 0 then
      send a_hi from P_l to P_{l+1}      m_ij=a_ij/a_{i-1,j}
   end for                              for r=j+1 to n
   send b_h from P_l to P_{l+1}           a_ir=a_ir-m_ij*a_{i-1,r}
end for                                 end for
                                        b_i=b_i-m_ij*b_{i-1};   a_ij=0
Step 2: In parallel do                end if
a) In Pp                           end for
```

In this algorithm we distinguish two big steps, in the first, the information is transferred between the processors in each stage, sending the last rows from a processor to the next one. In the second step, the necessary computations appear to resolve the update.

Then, we will calculate the run time of the previous algorithm, regarding it as something synchronous. This time can be expressed in the following way $T(N;k) = T_{computation} + T_{communication}$ with

$$T_{computation} \approx \frac{n^3}{k} t_c , \quad T_{communication} \approx n t_s + \frac{1}{2} n^2 t_w , \qquad (6)$$

where t_s is the time required to handle a message at sending processor and t_w is the transfer time of a number in a float point.

As far as communication time is concerned, it has to be taken into account that in the type of communication between neighbour processors, a processor must communicate with its neighbour processor the rows that it needs in order to update the part of the matrix that it stores. The cost of this kind of communication is independent of the architecture that we use (see [10]).

Now, without going into details, we can sketch the case of column blocks. We consider that each processor has n/k consecutive columns of A and the last processor, processor k, has the vector of independent terms also. In this way, we are partitioning the information between the processors into blocks of consecutive columns. In this partitioning the total time of computation is

$$T_{computation} \approx \frac{n^3}{k} t_c . \qquad (7)$$

For communications time we need a broadcast model. The cost of this kind of communication depends on the architecture that we use, so, and taking into account chapter 3 of [10], the communication time is

- Ring: $\qquad\qquad T_{communication} \approx n \log k \, t_s + n^2 \log k \, t_w + nk t_h ,$

- Mesh: $\qquad\qquad T_{communication} \approx n \log k \, t_s + n^2 \log k \, t_w + n\sqrt{k} t_h ,$

- Hypercube: $\qquad T_{communication} \approx n \log k \, t_s + n^2 \log k \, t_w ,$

where t_h, per-hop time, is the time taken by the header of a message to travel between two directly connected processors in the network.

The speedup and efficiency of partitioning in rows and columns is

$$S = \frac{T(N;1)}{T(N;k)} \approx \frac{2k}{3} , \quad E = \frac{S}{k} \approx \frac{2}{3} . \qquad (8)$$

In order to check that the theoretic values are correct and to see the impact of the architectures in these parameters, we have programmed the algorithms in an IBM SP2 using, at most, 16 *thin160* processors and Parallel Virtual Machine (PVM) over native MPI (Message Passing Interface) for IBM.

The results obtained by rows are shown in the next graphics. We can observe the high degree of efficiency reached by our method when a partition of data in blocks is used.

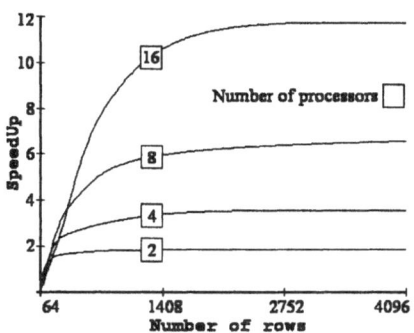
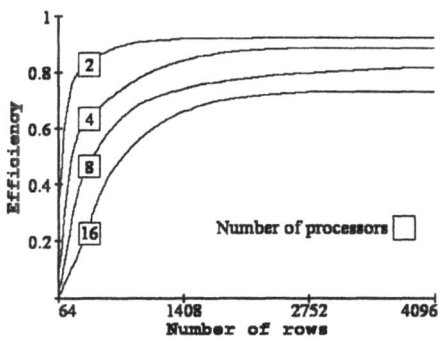

In general, the results obtained by row blocks are highly satisfactory, confirming the theoretical expectations. This allows us to affirm that this partitioning reaches a high level of efficiency when Neville elimination is applied.

As for column partitioning, the results obtained are similar to those found with rows. However, we observe a small cost increase due to the broadcast that appears in our method applications. This is not surprising taking into account the theoretical results.

Acknowledgement

This investigation was realized in part under grant NP-98-513-2 of the University of Oviedo (Spain), using the resources of CESCA (Computing and Communications Center of Catalonia).

References

1. Alonso, P., Gasca, M., Peña, J.M.: Backward error analysis of Neville elimination, Applied Numerical Mathematics 23, pp. 193-204 (1997).
2. Alonso, P., Gasca, M., Peña, J.M.: Estudio del error progresivo en la eliminación de Neville, Revista de la Real Academia de Ciencias Exactas, Físicas y Naturales, de Madrid, Vol. 92, num. 1, pp. 1-8 (1998).
3. Gallivan, K.A., Plemmons, R.J., Sameh, A.H.: Parallel algorithms for dense linear algebra computations, SIAM Rev. 32, pp. 54-135 (1990).
4. Gasca, M., Michelli, C.A.: Total positivity and its Applications, Kluwer Academic Publishers, Boston, 1996.
5. Gasca, M., Peña, J.M.: Total positivity and Neville elimination, Linear Algebra Appl. 165, pp. 25-44 (1992).
6. Gasca, M., Peña, J.M.: Total positivity, QR-factorization and Neville elimination, SIAM J. Matrix Anal. Appl. 14, pp. 1132-1140 (1993).
7. Gasca, M., Peña, J.M.: A matricial description of Neville elimination with applications to total positivity, Linear Algebra Appl. 202, pp. 33-45 (1994).
8. Gasca, M., Peña, J.M.: Neville elimination and approximation theory, En Approximation Theory Wavelets and Applications, ed. S.P. Singh, Kluwer Academic Publishers, pp. 131-151 (1995).
9. Golub, G.H., van Loan, C.F.: Matrix computations, Johns Hopkins University Press, 1989.
10. Kumar, V., Grama, A., Gupta, A., Karypis, G.: Introduction to Parallel Computing: Design and Analysis of Algorithms, The Benjamin/Cummings Publishing Company, 1994.
11. Trefethen, L.N., Schreiber, R.S.: Average-case stability of Gaussian elimination, SIAM J. Matrix Anal. Appl. 11, pp. 335-360 (1990).

A Comparison of Parallel Solvers for Diagonally Dominant and General Narrow-Banded Linear Systems II

Peter Arbenz[1], Andrew Cleary[2], Jack Dongarra[3], and Markus Hegland[4]

[1] Institute of Scientific Computing, ETH Zurich
arbenz@inf.ethz.ch
[2] Center for Applied Scientific Computing, Lawrence Livermore National Laboratory
acleary@llnl.gov
[3] Department of Computer Science, University of Tennessee, Knoxville
dongarra@cs.utk.edu
[4] Computer Sciences Laboratory, RSISE, Australian National University, Canberra
Markus.Hegland@anu.edu.au

Abstract. We continue the comparison of parallel algorithms for solving diagonally dominant and general narrow-banded linear systems of equations that we started in [2]. The solvers compared are the banded system solvers of ScaLAPACK [6] and those investigated by Arbenz and Hegland [1, 5]. We present the numerical experiments that we conducted on the IBM SP/2.

1 Introduction

In this note we continue the comparison of direct parallel solvers for narrow-banded systems of linear equations

$$A\mathbf{x} = \mathbf{b} \tag{1}$$

that we started in [2]. The n-by-n matrix A has a narrow band if its lower half-bandwidth k_l and upper half-bandwidth k_u are much smaller than the order of A, $k_l + k_u \ll n$.

We separately compare implementations of an algorithm for solving diagonally dominant and of an algorithm for solving arbitrary band systems. The algorithm for the diagonally dominant band system can be interpreted as a generalization of the well known tridiagonal *cyclic reduction* (CR), or more usefully, as Gaussian elimination applied to a symmetrically permuted system of equations $(PAP^T)P\mathbf{x} = P\mathbf{b}$. The latter interpretation has important consequences, such as it implies that the algorithm is backward stable [4]. The permutation enhances (coarse grain) parallelism. Unfortunately, it also causes Gaussian elimination to generate *fill-in* which in turn increases the computational complexity as well as the memory requirements of the algorithm [1, 6].

The algorithm for the arbitrary band system can be interpreted as a generalization of bidiagonal CR [10] which is equivalent to Gaussian elimination applied

P. Amestoy et al. (Eds.): Euro-Par'99, LNCS 1685, pp. 1078–1087, 1999.
© Springer-Verlag Berlin Heidelberg 1999

to a nonsymmetrically permuted system of equations $(PAQ^T)Q\mathbf{x} = P\mathbf{b}$. Here, the right permutation Q enhances parallelism, while the left permutation P enhances stability in that it incorporates the row-exchanges caused by pivoting.

Recently, the authors presented experiments with implementations of these algorithms using up to 128 processors of an Intel Paragon [2]. In this paper we complement those comparisons of the ScaLAPACK implementations with the experimental implementations by Arbenz and Hegland [1, 5] by timings on the IBM SP/2 at ETH Zurich. In section 2 we present our results for the diagonally dominant case. In section 3 the case of arbitrary band matrices is discussed. We draw our conclusions in section 4.

2 Experiments with the Band Solver for Diagonally Dominant Systems

An n-by-n diagonally dominant band matrix is split according to

$$
A = \begin{pmatrix}
A_1 & B_1^U & & & & \\
B_1^L & C_1 & D_2^U & & & \\
& D_2^L & A_2 & B_2^U & & \\
& & \ddots & \ddots & \ddots & \\
& & & B_{p-1}^L & C_{p-1} & D_p^U \\
& & & & D_p^L & A_p
\end{pmatrix}, \qquad (2)
$$

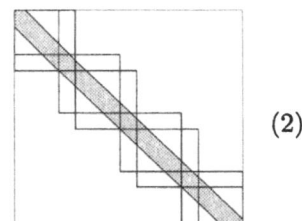

where $A_i \in \mathbb{R}^{n_i \times n_i}$, $C_i \in \mathbb{R}^{k \times k}$, $\mathbf{x}_i, \mathbf{b}_i \in \mathbb{R}^{n_i}$, $\boldsymbol{\xi}_i, \boldsymbol{\beta}_i \in \mathbb{R}^k$, and $\sum_{i=1}^p n_i + (p-1)k = n$, with $k := \max\{k_l, k_u\}$. The zero structure of A and its partition is depicted above to the right. This *block tridiagonal* partition is feasible only if $n_i > k$, a condition that restricts the degree of parallelism, i.e. the maximal number of processors p that can be exploited for parallel execution, where $p < (n+k)/(2k)$. The subscript of A_i, B_i^L, B_i^U, and C_i indicate on what processor the subblock is stored. As the orders n_i of the diagonal blocks A_i are in general much bigger than k, the order of the C_i, the first step of CR consumes most of the computational time. It is this first step of CR that has a degree of parallelism p. Therefore, if n is very big, a satisfactory speedup can be expected. After the first CR step the reduced systems are block tridiagonal with square blocks. The parallel complexity of this divide-and-conquer algorithm as implemented by Arbenz [3, 1] is

$$
\varphi_{n,p} \approx 2k_l(4k_u+1)\frac{n}{p} + \left(\frac{32}{3}k^3 + 4t_s + 4k^2 t_w\right)\lfloor \log_2(p-1)\rfloor. \qquad (3)
$$

Here, we assume that the time for the transmission of a message of n floating point numbers from one to another processor is independent of the processor distance. We represent its complexity relative to the floating point performance of the processor in the form $t_s + n t_w$ [11]. t_s denotes the startup time relative to the execution time of a floating point operation. t_w denotes the number of floating point operations that can be executed during the transmission of one word, here a 8-byte floating point number. Notice that t_s is much larger than t_w. On our target machine, the IBM SP/2 at ETH Zurich with 64 160 MHz P2SC processors, the startup time and bandwidth between applications are about 31 μs and 110 MB/s, respectively. Comparing with the 310 Mflop/s performance for the LINPACK-100 benchmark we get $t_s \approx 9600$ and $t_w \approx 22.5$. In the ScaLAPACK implementation, the factorization phase is separated from the forward substitution phase as in LAPACK. Therefore, there are more messages to be sent and the term $4t_s$ in (3) becomes $6t_s$ which can be relevant for small bandwidth. The complexity if the straightforward serial algorithm [9, §4.3] is

$$\varphi_n = (2k_u+1)k_l n + (2k_l+2k_u-1)rn + \mathcal{O}((k+r)k^2). \qquad (4)$$

The computational overhead is introduced as the off-diagonal blocks D_i^L and D_i^U are filled during Gaussian elimination.

Diagonally dominant case on the IBM SP/2								
(n, k_l, k_u)	(20000, 10, 10)			(100000, 10, 10)			(100000, 50, 50)		
p	t	S	ε	t	S	ε	t	S	ε
ScaLAPACK implementation									
1	128	1.0	4e-10	618	1.0	3e+8	2558	1.0	4e+8
2	120	1.1	4e-10	580	1.1	1e+8	5660	0.45	2e+8
4	69.8	1.8	3e-10	297	2.1	1e+8	2927	0.87	2e+8
8	38.5	3.3	2e-10	155	4.0	3e-9	2010	1.2	8e-9
12	30.6	4.2	2e-10	114	5.4	2e-9	1623	1.6	7e-9
16	25.1	5.1	2e-10	86.2	7.2	2e-9	1202	2.1	6e-9
24	27.1	4.7	2e-10	69.7	8.9	2e-9	855	3.0	5e-9
32	20.7	6.2	1e-10	53.1	12	1e-9	629	4.1	4e-9
48	20.4	6.3	1e-10	40.5	15	1e-9	479	5.3	4e-9
64	12.3	10	1e-10	28.4	22	1e-9	363	7.0	3e-9
Arbenz / Hegland implementation									
1	124	1.0	4e-10	609	1.0	5e-9	2788	1.0	1e-8
2	130	0.96	4e-10	666	0.91	4e-9	4160	0.67	1e-8
4	66.6	1.9	3e-10	326	1.9	4e-9	2175	1.3	1e-8
8	37.5	3.3	2e-10	164	3.7	3e-9	1007	2.8	8e-9
12	21.8	5.7	2e-10	109	5.6	2e-9	707	3.9	8e-9
16	17.5	7.1	2e-10	83.6	7.3	2e-9	509	5.5	6e-9
24	12.0	10	2e-10	60.5	10	2e-9	374	7.5	5e-9
32	9.41	13	1e-10	47.1	13	1e-9	271	10	4e-9
48	8.25	15	1e-10	30.2	20	1e-9	199	14	4e-9
64	9.57	13	1e-10	21.7	28	1e-9	180	16	3e-9

Table 1. Selected execution times t in milliseconds, speedups $S = S(p)$, and error for the two algorithms for the three problem sizes. ε denotes the 2-norm error of the computed solution.

We compare two implementations of the above algorithm, the ScaLAPACK implementation [7] and the one by Arbenz and Hegland (AH), by means of three test-problems of sizes $(n, k_l, k_u) = (100000, 10, 10)$, $(n, k_l, k_u) = (20000, 10, 10)$, and $(n, k_l, k_u) = (100000, 50, 50)$. The matrix A always has all ones within the band and the value $\alpha = 100$ on the diagonal. The condition numbers of A vary between 1 and 3, see [2]. The right-hand sides are chosen such that the solution gets $(1, \ldots, n)^T$ which enables us to determine the error in the computed solution. We compiled a program for each problem size, adjusting the arrays to just the size needed to solve the problem on one processor.

In Tab. 1 the execution times are listed for all problem sizes. For both implementations the one-processor times are quite close. The difference in this part of the code is that the AH implementation calls the level-2 BLAS based LAPACK routine dgbtf2 for the triangular factorization, whereas in the ScaLAPACK implementation the level-3 BLAS based routine dgbtrf is called. The latter is advantageous with the wider bandwidth $k = 50$, while dgbtf2 performs (slightly) better with the narrow band.

With the bandwidth $k = 10$ problems, the ScaLAPACK implementation performs slightly faster on two than on one processor. The AH implementation slows down by 5-10%. With the large problem there is a big jump from the one- to the two-processor execution times. From (3) and (4) one sees that the parallel algorithm has a redundancy of about 4. The additional work consists of computing the fill-in and the reduced system [2] which comprises a forward elimination, a backward substitution, each with k vectors, and a multiplication of a $k \times n_i$ with a $n_i \times k$ matrix. These operations are executed at very high speed such that there is almost no loss in performance with the small and the intermediate problem. In ScaLAPACK, for forward elimination and backward substitution the level-2 BLAS dtbtrs is called. In the AH implementation this routine is expanded in order to avoid unnecessary checks if rows have been exchanged in the factorization phase. This avoids the evaluation of if-statements. In the large problem size the matrices are considerably larger. There are more cache misses, in particular in the ScaLAPACK implementation where the matrices that suffer from fill-in are stored in $n_i \times k$ arrays. In the AH implementation these matrices are stored in 'lying' arrays which increases the performance on the RISC architecture of the underlying hardware considerably [1, 8].

The speedups of the AH implementation *relative to the 2-processor* performance is very close to ideal for the intermediate problem. With the small problem, the communication overhead begins to dominate the computation for large processor numbers. In the large problem this is effect in not yet so pronounced. The ScaLAPACK implementation does not scale as well. For large processor numbers the difference in execution times is about 2/3 which correlates with the ratio of messages sent in the two implementations.

Clearly, the speedups for the medium size problem with large n and small k are best. The $1/p$-term that containes the factorization of the A_i and the computations of the 'spikes' $D_i^U R_i^{-1}$ and $L_i^{-1} D_i^L$ consumes five times as much time as with the small problem size and scales very well. This portion is still increased with the large problem size. However, there the solution of the reduced system gets expensive also.

3   Experiments with a Pivoting Solver for Arbitrary Band Systems

The partition (2) is not suited for the parallel solution of (1) if partial pivoting is required in the Gaussian elimination to preserve stability. In order that pivoting can take place independently in block columns they must not have elements in the same row. Therefore, the separators have to be $k := k_l + k_u$ columns wide. As discussed in detail in [5, 2] we consider the matrix A as a *cyclic* band matrix by moving the last k_l rows to the top.

$$A = \begin{pmatrix} A_1 & & & & & & & D_1 \\ B_1 & & & & C_1 & & & \\ & A_2 & & & D_2 & & & \\ & B_2 & & & & C_2 & & \\ & & A_3 & & & D_3 & & \\ & & B_3 & & & & C_3 & \\ & & & A_4 & & & D_4 & \\ & & & B_4 & & & & C_4 \end{pmatrix} \qquad (5)$$

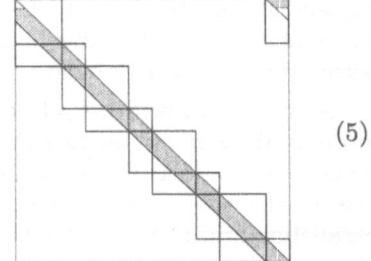

where $A_i \in \mathbb{R}^{m_i \times n_i}$, $C_i \in \mathbb{R}^{k \times k}$, $\mathbf{x}_i$, $\mathbf{b}_i \in \mathbb{R}^{n_i}$, $\boldsymbol{\xi}_i$, $\boldsymbol{\beta}_i \in \mathbb{R}^k$, $k := k_l + k_u$, and $\sum_{i=1}^p m_i = n$, $m_i = n_i + k$. If $n_i > 0$ for all i, then the degree of parallelism is p. Notice that the permutation that moves the last rows to the top is done for pedagogical reasons: it makes the diagonal blocks A_i and C_i square and the first elimination step gets formally equal with the successive ones. Also notice that A in (5) is block bidiagonal and that the diagonal blocks are lower triangular.

For solving $A\mathbf{x} = \mathbf{b}$ in parallel we apply a generalization of cyclic reduction that permits pivoting [10, 5, 2]. Its parallel complexity is

$$\varphi_{n,p}^{pp} \approx 4k^2 \frac{n}{p} + \left(\frac{23}{3} k^3 + 2t_s + 3k^2 t_w \right) \lfloor \log_2(p) \rfloor. \qquad (6)$$

The serial complexity of straightforward Gaussian elimination with partial pivoting is

$$\varphi_n^{pp} \approx (2k+1)k_l n, \qquad k := k_l + k_u, \qquad (7)$$

leading to a redundancy of about $2k/k_l$.

We again tested two versions of the algorithm, the ScaLAPACK implementation and the implementation by Arbenz and Hegland. We used the same test problems as above, however, we choose α, the value the diagonal elements of A, smaller. The condition number of the system matrix A, $\kappa(A)$, grows very large as α tends to one. For the problems with bandwidths 10, $\kappa(A) \approx 1$ for $\alpha = 10$ and $\kappa(A) \approx 3 \cdot 10^6$ for $\alpha = 1.01$. With the large bandwidth $k = 50$, we have

| \multicolumn{13}{c}{Non-diagonally dominant case on the IBM SP/2. Small problem size.} |||||||||||||
| | $\alpha = 10$ | | | $\alpha = 5$ | | | $\alpha = 2$ | | | $\alpha = 1.01$ | | |
p	t	S	ε	t	S	ε	t	S	ε	t	S	ε
\multicolumn{13}{c}{ScaLAPACK implementation}												
1	289	1.0	6e-10	294	1.0	3e-8	334	1.0	4e-7	333	1.0	7e-7
2	261	1.1	6e-10	274	1.1	4e-8	277	1.2	2e-6	276	1.4	7e-7
4	120	2.4	5e-10	154	1.9	4e-8	143	2.3	3e-7	136	2.4	1e-7
8	66.1	4.4	3e-10	86.3	3.4	3e-8	90.5	3.7	9e-7	90.3	3.7	2e-7
12	64.8	4.5	3e-10	62.4	4.7	4e-8	69.3	4.8	8e-7	66.1	5.0	4e-7
16	51.1	5.7	3e-10	51.5	5.7	3e-8	53.3	6.3	1e-6	52.3	6.4	7e-8
24	43.5	6.6	2e-10	41.3	7.1	9e-9	40.7	8.2	6e-7	41.0	8.1	3e-7
32	36.7	7.9	3e-10	33.9	8.7	9e-9	34.2	9.8	7e-7	34.1	9.8	1e-7
48	29.1	9.9	2e-10	29.5	10	1e-8	31.2	11	5e-7	36.8	9.1	7e-8
64	19.4	15	2e-10	19.1	15	1e-8	19.7	17	7e-7	19.9	17	5e-8
\multicolumn{13}{c}{Arbenz / Hegland implementation}												
1	193	1.0	7e-10	204	1.0	3e-8	242	1.0	6e-7	241	1.0	6e-7
2	171	1.1	6e-10	166	1.2	2e-8	183	1.3	1e-6	175	1.4	7e-7
4	87.6	2.2	5e-10	84.8	2.4	5e-8	95.0	2.5	1e-6	90.2	2.7	7e-7
8	48.8	3.9	4e-10	44.9	4.5	2e-8	49.4	4.9	1e-6	47.7	5.0	6e-7
12	37.1	5.2	3e-10	33.4	6.1	5e-8	35.0	6.9	1e-6	33.6	7.2	5e-8
16	30.2	6.4	3e-10	24.6	8.2	1e-8	29.7	8.1	7e-7	29.1	8.3	1e-7
24	18.7	10	3e-10	18.8	11	1e-8	21.1	11	1e-6	19.7	12	6e-7
32	15.2	13	3e-10	15.4	13	1e-8	16.3	15	5e-7	16.3	15	2e-7
48	13.8	14	3e-10	16.6	12	1e-8	13.3	18	7e-7	12.7	19	4e-8
64	11.1	17	3e-10	11.3	18	1e-8	11.7	21	3e-7	12.8	19	2e-7

Table 2. Selected execution times t in milliseconds, speedups S, and 2-norm errors ε of the two implementations for the small problem size $(n, k_l, k_u) = (20000, 10, 10)$ with varying α.

$\kappa(A) \approx 2 \cdot 10^5$ for $\alpha = 10$ and $\kappa(A) \approx 5 \cdot 10^8$ for $\alpha = 1.01$. Tables 2, 3, and 4 contain the respective numbers, execution time, speedup and 2-norm of the error, for the three problem sizes.

Relative to the AH implementation the execution times for ScaLAPACK comprise overhead proportional to the problem size, mainly zeroing elements of work arrays. This is done in the AH implementation during the building of the matrices. Therefore, the comparison in the non-diagonally dominant case should not be based primarily on execution times but on speedups. The execution times increase with the condition number $\kappa(A)$ of the problem which is of course hard or even impossible to predict as the pivoting procedure is unknown. At least the two problems with bandwidth $k = k_l + k_u = 20$ can be discussed along similar lines. (ScaLAPACK does *not* give correct results for processor numbers $p \leq 4$. This did not happen on the Intel Paragon [2]. Actually, the error occurs only on the last processor p. The execution times seem not to be affected.) The AH implementation scales better than ScaLAPACK. Its execution times

Non-diagonally dominant case on the IBM SP/2. Intermediate problem size.											
	$\alpha = 10$			$\alpha = 5$			$\alpha = 2$			$\alpha = 1.01$		
p	t	S	ε	t	S	ε	t	S	ε	t	S	ε
ScaLAPACK implementation											
1	1443	1.0	2e+9	1454	1.0	8e+11	1660	1.0	2e+12	1660	1.0	1e+10
2	1294	1.1	9e+8	1302	1.1	3e+11	1360	1.2	1e+12	1335	1.2	6e+9
4	799	1.8	9e+8	643	2.3	2e+11	685	2.4	7e+11	671	2.5	5e+9
8	412	3.5	4e-9	404	3.6	2e-6	416	4.0	1e-5	420	4.0	3e-6
12	279	5.2	4e-9	276	5.3	5e-7	290	5.7	6e-6	279	6.0	2e-6
16	210	6.9	3e-9	209	7.0	1e-6	219	7.6	6e-6	216	7.7	3e-6
24	152	9.5	3e-9	152	9.5	6e-7	151	11	5e-6	150	11	1e-6
32	115	13	2e-9	113	13	4e-7	123	14	3e-6	117	14	2e-6
48	84	17	2e-9	85.1	17	8e-7	85.9	19	2e-6	84.4	20	1e-6
64	63	23	2e-9	58.5	25	4e-7	61.8	27	2e-6	61.6	27	5e-7
Arbenz / Hegland implementation											
1	985	1.0	8e-9	978	1.0	2e-6	1252	1.0	2e-5	1173	1.0	7e-6
2	823	1.2	8e-9	826	1.2	2e-6	923	1.4	1e-5	878	1.3	5e-6
4	415	2.4	6e-9	421	2.3	5e-7	467	2.7	8e-6	439	2.7	5e-6
8	214	4.6	5e-9	213	4.6	3e-7	236	5.3	8e-6	225	5.2	2e-6
12	145	6.8	4e-9	145	6.8	9e-7	159	7.9	5e-6	152	7.7	1e-6
16	113	8.7	4e-9	109	8.9	6e-7	140	9.0	4e-6	115	10	9e-7
24	74.6	13	3e-9	75.1	13	6e-7	86.1	15	3e-6	80.3	15	1e-6
32	60.7	16	3e-9	57.3	17	1e-6	62.8	20	4e-6	62.1	19	9e-7
48	42.2	23	2e-9	41.0	24	4e-7	48.0	26	4e-6	43.3	27	6e-7
64	36.3	27	2e-9	32.4	30	6e-7	38.6	32	5e-6	34.5	34	7e-7

Table 3. Selected execution times t in milliseconds, speedups S, and 2-norm errors ε of the two implementations for the medium problem size $(n, k_l, k_u) = (100000, 10, 10)$ with varying α.

for large processor numbers is about half of that of the ScaLAPACK implementation except for $p = 64$. For a reason not yet clear to us, the ScaLAPACK implementation performs relatively fast for $p = 64$ when the execution time of the AH implementation is about 2/3 of ScaLAPACK, reflecting again the ratio of the messages sent. In contrast to the results obtained for the Paragon, the execution times for the pivoting algorithm for $\alpha = 10$ are clearly longer than for the 'simple' algorithm. Thus, the suggestion made in [4] to *always* use the pivoting algorithm can now definitively be rejected. The memory consumption of the pivoting algorithm is higher anyway. On the other hand, the overhead for pivoting in the solution of the reduced system by bidiagonal cyclic reduction is not so big that it justifies sacrificing stability.

With the large problem size, ScaLAPACK shows an extremely bad one-processor performance. In the ScaLAPACK implementation the auxiliary arrays mentioned above are accessed even in the one-processor run (when they are not needed) leading to an abundant memory consumption. The matrices do not fit

Non-diagonally dominant case on the IBM SP/2. Large problem size.											
$\alpha = 10$			$\alpha = 5$			$\alpha = 2$			$\alpha = 1.01$		
p	t	S^*	ε	t	S^*	ε	t	S^*	ε	t	S^*	ε
ScaLAPACK implementation											
1	92674	1.0	1e+11	78777	1.0	1e+12	69759	1.0	2e+12	54908	1.0	3e+11
2	7857	0.41	4e+10	7664	0.46	4e+11	11888	0.43	6e+11	9056	0.57	1e+11
4	3839	0.85	6e+10	3924	0.90	3e+11	7102	0.73	4e+11	4630	1.1	7e+10
8	3072	1.1	1e-6	3054	1.2	9e-6	4125	1.2	2e-4	3392	1.5	7e-4
12	1993	1.6	3e-6	1991	1.8	1e-5	2754	1.9	1e-4	2272	2.3	3e-5
16	2122	1.5	2e-6	2135	1.6	2e-5	2625	2.0	1e-4	2369	2.2	3e-4
24	1174	2.8	1e-6	1123	3.1	3e-5	1476	3.5	1e-4	1252	4.1	1e-4
32	1201	2.7	9e-7	1179	3.0	3e-6	1447	3.6	8e-5	1287	4.0	4e-4
48	710	4.6	2e-6	752	4.7	3e-6	896	5.8	2e-4	763	6.8	9e-5
64	636	5.1	8e-7	732	4.8	2e-5	786	6.6	2e-4	772	6.7	9e-5
Arbenz / Hegland implementation											
1	3257	1.0	6e-6	3514	1.0	2e-5	5155	1.0	2e-4	5170	1.0	1e-3
2	6102	0.53	4e-6	5893	0.60	9e-6	10140	0.51	1e-4	7305	0.71	7e-4
4	3074	1.1	2e-6	3079	1.1	6e-5	5182	1.0	1e-4	3786	1.4	1e-4
8	1671	1.9	2e-6	1669	2.1	1e-5	2723	1.9	2e-4	2023	2.6	9e-5
12	1261	2.6	2e-6	1254	2.8	1e-5	1944	2.7	8e-5	1496	3.5	2e-4
16	1008	3.2	1e-6	1015	3.5	1e-5	1514	3.4	1e-4	1176	4.4	2e-4
24	833	3.9	3e-6	836	4.2	6e-5	1162	4.4	1e-4	945	5.5	2e-4
32	724	4.5	1e-6	710	4.9	2e-5	952	5.4	1e-4	797	6.5	2e-4
48	668	4.9	2e-6	661	5.3	2e-5	1093	4.7	1e-4	717	7.2	5e-5
64	597	5.5	1e-6	598	5.9	2e-6	724	7.1	2e-4	652	7.9	5e-5

Table 4. Selected execution times t in milliseconds, speedups S, and 2-norm errors ε of the two implementations for the large problem size $(n, k_l, k_u) = (100000, 50, 50)$ with varying α. Speedups have been taken with respect to the one-processor times of the AH-implementation.

into local memory of 256 MB any more. The ScaLAPACK run times for processor numbers larger than 1 are comparable with the AH implementation. We therefore relate them to the one-processor time of the AH implementation (a call to LAPACK's routines dgbtrf and dgbtrs) to determine speedups. The AH implementation performs quite as expected by the complexity analysis. As the band is now relatively wide, factorization and redundant computation (fill-in, formation of reduced system) perform at about the same Mflop/s rate. The fourfold work distributed over two processors results in a 'speedup' of about 0.5. In this large example the volume of the interprocessor communication is big. A message consists of a small multiple of k^2 8-byte floating point numbers (20 kB). Thus, the startup time t_s constitutes only a small fraction at the interprocessor communication cost. The latter differ only little in the ScaLAPACK and AH implementation. In this large problem size, with regard to speedups ScaLAPACK performs slightly better than the AH implementation. The execution times are however longer by about 10-20%.

4   Conclusions

The execution times measured on the IBM SP/2 are shorter than on the Intel Paragon by a factor of about five for the small and 10 for the large problems on one processor. As the Paragon's communication network has a relatively much higher bandwidth we observed better speedups on this machine which narrowed the gap [2].

For systems with very narrow band, the implementations by Arbenz and Hegland which are designed to reduce the number of messages that are communicated are faster. The difference is however not too big. The flexibility and versatility of the ScaLAPACK justifies the loss in performance. We are convinced that a few little improvements in the ScaLAPACK implementation, in particular the treatment of auxiliary arrays, will further narrow the gap.

Nevertheless, it may be useful to have in ScaLAPACK a routine that combines the factorization and solution phase as in the AH implementation. Appropriate routines would be the 'drivers' pddbsv for the diagonally dominant case and pdgbsv for the non-diagonally dominant case. In the present version of ScaLAPACK, the former routine consecutively calls pddbtrf and pddbtrs, the latter calls pdgbtrf and pdgbtrs, respectively. The storage policy could stay the same. So, the flexibility in how to apply the routines remains.

On the IBM SP/2, in contrast to the Intel Paragon, the overhead for pivoting was always noticeable. Therefore, the suggestion made in [4] to *always* use the pivoting algorithm can now definitively be rejected. The memory consumption of the pivoting algorithm is about twice as high, anyway.

References

[1] P. ARBENZ, *On experiments with a parallel direct solver for diagonally dominant banded linear systems*, in Euro-Par '96, L. Bougé, P. Fraigniaud, A. Mignotte, and Y. Robert, eds., Springer, Berlin, 1996, pp. 11–21. (Lecture Notes in Computer Science, 1124).

[2] P. ARBENZ, A. CLEARY, J. DONGARRA, AND M. HEGLAND, *A comparison of parallel solvers for diagonally dominant and general narrow-banded linear systems*, Tech. Report 312, ETH Zürich, Computer Science Department, January 1999. (Available at URL http://www.inf.ethz.ch/publications/. Submitted to Parallel and Distributed Computing Practices (PCDP)).

[3] P. ARBENZ AND W. GANDER, *A survey of direct parallel algorithms for banded linear systems*, Tech. Report 221, ETH Zürich, Computer Science Department, October 1994. Available at URL http://www.inf.ethz.ch/publications/.

[4] P. ARBENZ AND M. HEGLAND, *Scalable stable solvers for non-symmetric narrow-banded linear systems*, in Seventh International Parallel Computing Workshop (PCW'97), P. Mackerras, ed., Australian National University, Canberra, Australia, 1997, pp. P2–U–1 – P2–U–6.

[5] ———, *On the stable parallel solution of general narrow banded linear systems*, in High Performance Algorithms for Structured Matrix Problems, P. Arbenz, M. Paprzycki, A. Sameh, and V. Sarin, eds., Nova Science Publishers, Commack, NY, 1998, pp. 47–73.

[6] A. CLEARY AND J. DONGARRA, *Implementation in ScaLAPACK of divide-and-conquer algorithms for banded and tridiagonal systems*, Tech. Report CS-97-358, University of Tennessee, Knoxville, TN, April 1997. (Available as LAPACK Working Note #125 from URL http://www.netlib.org/lapack/lawns/.

[7] ScaLAPACK is available precompiled for the SP/2 from the archive of prebuilt ScaLAPACK libraries at http://www.netlib.org/scalapack/.

[8] M. J. DAYDÉ AND I. S. DUFF, *The use of computational kernels in full and sparse linear solvers, efficient code design on high-performance RISC processors*, in Vector and Parallel Processing – VECPAR'96, J. M. L. M. Palma and J. Dongarra, eds., Springer, Berlin, 1997, pp. 108–139. (Lecture Notes in Computer Science, 1215).

[9] G. H. GOLUB AND C. F. VAN LOAN, *Matrix Computations*, The Johns Hopkins University Press, Baltimore, MD, 2nd ed., 1989.

[10] M. HEGLAND, *Divide and conquer for the solution of banded linear systems of equations*, in Proceedings of the Fourth Euromicro Workshop on Parallel and Distributed Processing, IEEE Computer Society Press, Los Alamitos, CA, 1996, pp. 394–401.

[11] V. KUMAR, A. GRAMA, A. GUPTA, AND G. KARYPIS, *Introduction to Parallel Computing*, Benjamin/Cummings, Redwood City CA, 1994.

Using Pentangular Factorizations for the Reduction to Banded Form

B. Großer[1] and B. Lang[2]

[1] Department of Mathematics, University of Wuppertal, D-42097 Wuppertal,
Germany
[2] Computing Center, Aachen University of Technology, D-52074 Aachen, Germany

Abstract. Most methods for computing the singular value decomposition (SVD) first bidiagonalize the matrix. The ScaLAPACK implementation of the blocked reduction of a general dense matrix to bidiagonal form performs about one half of the operations with BLAS3. If we subdivide the task into two stages *dense → banded* and *banded → bidiagonal*, we can increase the portion of matrix-matrix operations and expect higher performance. We give an overview of different techniques for the first stage.
This note summarizes the results of [9, 10].

Keywords: Linear algebra; Singular value decomposition; Bidiagonal reduction; Parallel BLAS

1 Introduction

Many algorithms for computing the singular value decomposition (SVD) of a general matrix $A \in \mathbb{R}^{m \times n}$ start with the reduction to bidiagonal form. That is, $A \to B = U^T A V$, where B is upper or lower bidiagonal, and $U \in \mathbb{R}^{m \times m}$ and $V \in \mathbb{R}^{n \times n}$ are orthogonal. We assume $m \geq n$ and consider only reduction to upper bidiagonal form. We can obtain B by alternately pre- and postmultiplying A with Householder transformations in order to introduce zeros in the columns and rows of the matrix. In [6], a block formulation of the bidiagonalization algorithm is given, which allows one half of the operations to be performed in matrix-matrix products (BLAS3 [7]) as well as a straightforward parallelization. We will call this reduction technique the *direct method*, because it reduces a dense matrix directly to bidiagonal form.

From Table 1 we see that subdividing the reduction to bidiagonal form into two stages *dense → banded* (cf. [10]) and *banded → bidiagonal* (cf. [11]) allows the design of *two-stage* bidiagonalization algorithms that do the vast majority of the calculations within matrix-matrix products and therefore can make full use of an optimized BLAS3 implementation. Due to better communication management this gain even increases if the algorithms are run on distributed memory parallel machines. The two-stage algorithms offer an attractive alternative, if only the singular values are required. If the orthogonal transformations must be accumulated explicitly (e.g., to compute all the singular vectors), then the *direct* method is superior.

P. Amestoy et al. (Eds.): Euro-Par'99, LNCS 1685, pp. 1088–1095, 1999.
© Springer-Verlag Berlin Heidelberg 1999

	reduction of A	update U	update V
direct			
overall flop	$4mn^2 - \frac{4}{3}n^3$	$2mn(2m-n)$	$2n^3$
BLAS3 portion	$2mn^2 - \frac{2}{3}n^3$	$2mn(2m-n)$	$2n^3$
two-stage			
first stage			
overall flop	$4mn^2 - \frac{4}{3}n^3 + \mathcal{O}(mnb)$	$2mn(2m-n)$	$2(n-b)^3$
BLAS3 portion	$4mn^2 - \frac{4}{3}n^3 + \mathcal{O}(mnb)$	$2mn(2m-n)$	$2(n-b)^3$
second stage			
overall flop	$8n^2b$	$2mn^2 + \mathcal{O}(n^2b)$	$2n^3 + \mathcal{O}(n^2b)$
BLAS3 portion	0	$2mn^2$	$2n^3$

Table 1. Approximate flop counts for the bidiagonalization methods.

In this note we briefly describe three implementations of the first stage, i.e., the reduction to banded form. Additional details can be found in [9] and [10].

2   Reduction to Banded Form

2.1   The Standard Algorithm

Our methods were designed to make extensive use of the ScaLAPACK [3], PBLAS [4], and BLACS [8] libraries. We assume that A is distributed over a process grid in a two-dimensional block cyclic data layout with block size $n_b \times n_b$. Algorithm 1 reduces A to b upper diagonals. A snapshot of the algorithm is given in Figure 1.

Algorithm 1 : standard (reduction to b upper diagonals)

$i = 0$
while $i < n$ **do**
   $s = i/b + 1$
   $\{\, A_s^{(QRfact)} \equiv A(i+1:m, i+1:i+b),\ A_s^{(QRupd)} \equiv A(i+1:m, i+b+1:n)\,\}$
   $A_s^{(QRfact)} = Q_s R_s$ $\{QR$ decomposition$\}$
   $A_s^{(QRupd)} \leftarrow Q_s^T A_s^{(QRupd)}$ $\{$update$\}$
   if $i + b < n$
      $\{\, A_s^{(LQfact)} \equiv A(i+1:i+b, i+b+1:n),\ A_s^{(LQupd)} \equiv A(i+b+1:m, i+b+1:n)\,\}$
      $A_s^{(LQfact)} = L_s P_s^T$ $\{LQ$ decomposition$\}$
      $A_s^{(LQupd)} \leftarrow A_s^{(QRupd)} P_s$ $\{$update$\}$
   endif
   $i = i + b$
enddo

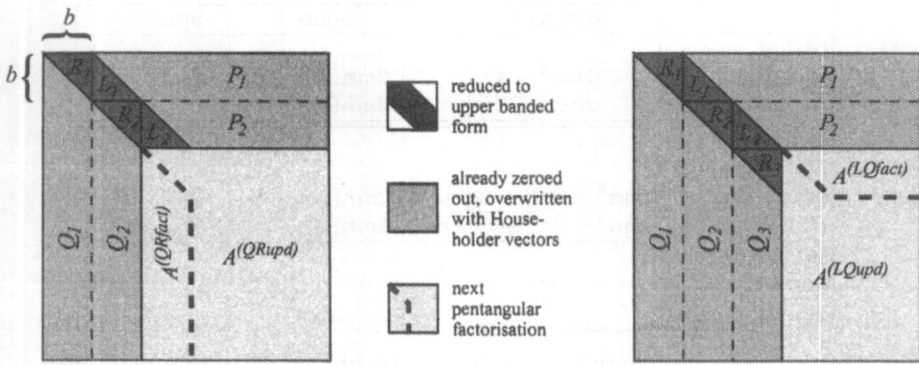

Fig. 1. Third step of Algorithm 1 (**standard**). The transformations Q_s and P_s are represented by the corresponding Householder pairs.

A typical subtask of the algorithm may be regarded as *pentangular factorization*: A given matrix $\tilde{A}$ is partitioned as $\tilde{A} = [A^{(QRfact)}, A^{(QRupd)}]$, where $A^{(QRfact)}$ contains the first b columns of $\tilde{A}$. A QR decomposition $A^{(QRfact)} = Q \cdot R \in \mathbb{R}^{\tilde{m} \times b}$ gives $\tilde{A} = Q \cdot [R, Q^T \cdot A^{(QRupd)}]$ where $[R, Q^T \cdot A^{(QRupd)}]$ has (upper) pentangular shape.

We inspect different methods to compute such factorizations (and their counterparts based on row-partitioning and LQ decompositions).

Our first approach is to modify the ScaLAPACK routine **PDGEQRF** [5], which is designed to compute a QR decomposition of a matrix, in a way that the complete upper pentangular factorization (i.e., not only the QR decomposition of $A^{(QRfact)}$) is computed. **PDGEQRF** partitions the matrix $A^{(QRfact)} \in \mathbb{R}^{\tilde{m} \times b}$ into panels of width $n_b \leq b$. For the k-th panel, the corresponding n_b Householder transformations are combined (cf. [12]):

$$H_{v_1}^{(k)} H_{v_2}^{(k)} \cdots H_{v_{n_b}}^{(k)} = I + V^{(k)} T^{(k)} V^{(k)T} , \tag{1}$$

and then applied to the trailing submatrix of $A^{(QRfact)}$. Our modification is based on two observations: First, note that these Householder transformations have also to be applied to $A^{(QRupd)}$. Second, $A^{(QRfact)}$ and $A^{(QRupd)}$ are stored consecutively. Thus we can slightly modify **PDGEQRF** such that $A^{(QRupd)}$ is updated with $I + V^{(k)} T^{(k)} V^{(k)T}$ immediately.

A naive implementation of Algorithm 1 would call the ScaLAPACK routines **PDGEQRF** for the blocked QR factorization of $A^{(QRfact)}$ and **PDORMQR** for the blocked update of $A^{(QRupd)}$. If we assume that both routines are called with identical block sizes n_b, the routine **PDORMQR** would compute the *block factors* $T^{(k)} \in \mathbb{R}^{n_b \times n_b}$ once again before they are applied to $A^{(QRupd)}$. It is exactly the computation of $T^{(k)}$ by **PDLARFT** which represents a major bottleneck of the whole algorithm. Thus the modification of **PDGEQRF** to compute the complete factorization results in significant savings.

2.2   Splitting the Factorizations

Another approach to compute the pentangular factorizations was inspired by [1], where reduction to block upper Hessenberg form was considered. In this approach, each QR and LQ decomposition is further subdivided into a "local" and a "global" phase.

For the QR decomposition, these two phases are best explained by taking a "row view" and "column view", resp., of the data distribution, see Figure 2.

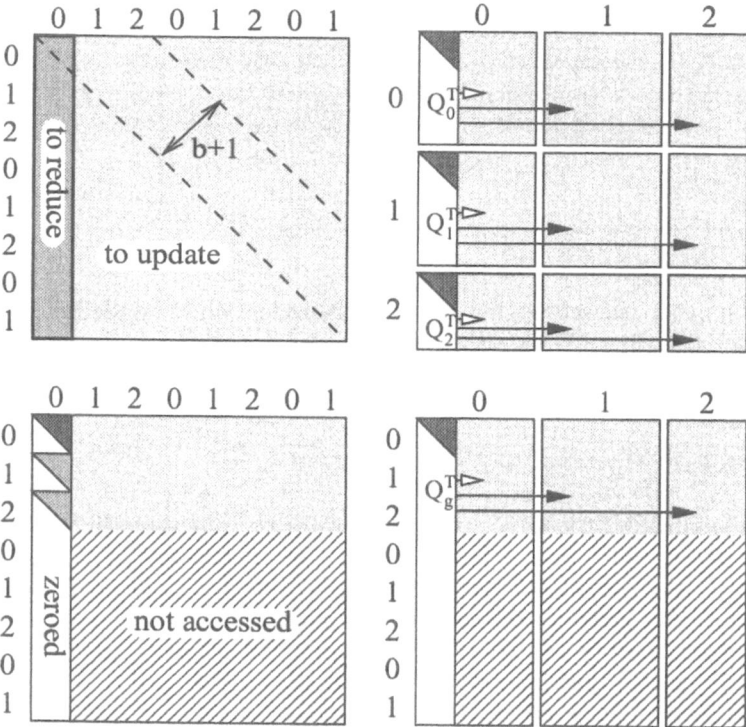

Fig. 2. Local (top) and global (bottom) phases in the "split" QR decomposition.

In the "local" phase, pentangular factorizations are computed for the matrix elements which belong to identical process rows. Here we assume $b = n_b$. This means that the QR decompositions can be carried out locally. Communication is only needed for the update of the trailing submatrices. The "global" phase computes a pentangular factorization for the first $nprows \cdot n_b$ rows of the matrix, where $nprows$ is the number of process rows. The "split factorizations" technique is summarized in Algorithm 2.

Algorithm 2 : splitfac (reduction to b upper diagonals)

$i = 0$
while $i < n$ **do**
   $s = i/n_b + 1$
   $\{\ A_s^{(QRfact)} \equiv A(i{+}1{:}m, i{+}1{:}i{+}n_b)\ \}$
   $\{\ A_s^{(QRupd)} \equiv A(i{+}1{:}m, i{+}n_b{+}1{:}n)\ \}$
   local QR decompositions for $A_s^{(QRfact)}$ and local updates for $A_s^{(QRupd)}$
   global QR decomposition for $A_s^{(QRupd)}$'s first $nprows$ blocks
     and global update of $A_s^{(QRupd)}$'s first $nprows$ block rows
   if $i + b < n$
     $\{\ A_s^{(LQfact)} \equiv A(i{+}1{:}i{+}n_b, i{+}b{+}1{:}n)\ \}$
     $\{\ A_s^{(LQupd)} \equiv A(i{+}n_b{+}1{:}m, i{+}b{+}1{:}n)\ \}$
     local LQ decompositions for $A_s^{(LQfact)}$ and local updates for $A_s^{(LQupd)}$
     global LQ decomposition for $A_s^{(LQfact)}$'s first $npcols$ blocks
       and global update of $A_s^{(LQupd)}$'s first $npcols$ block columns
   endif
   $i = i + n_b$
enddo

Besides other differences, the fact that the reduction to banded form involves equivalence transformations rather than similarity transformations greatly simplifies the communication patterns as compared to the algorithm described in [1].

2.3   Rank–$2b$ Updates

The update phases of the upper and lower pentangular factorizations incorporate rank-b updates, which are performed as BLAS3 operations. Typically, the performance of these operations increases with the blocking factor, i.e., the smallest dimension of the matrices involved in the matrix-matrix products. In the following we will describe one way to double this dimension by replacing the two rank-b updates of Algorithm 1 with one rank-$2b$ update.

Algorithm 3 reduces the matrix A to banded form with b upper *and* b lower diagonals. Here both updates with Q_s^T and P_s are carried out on the same submatrix $A_s^{(upd)} = A_s^{(QRupd)} = A_s^{(LQupd)} \in \mathbb{R}^{q \times p}$, where $q = m - i - b$ and $p = n - i - b$.

Using the original WY representations $Q_s = I + W_L Y_L^T$ and $P_s = I + W_R Y_R^T$ (cf. [2]), the two updates $A^{(upd)} \leftarrow Q_s^T A^{(upd)} P_s$ in Algorithm 3 may either be done separately with about $8qpb + 2qp$ flop. Alternatively they may be combined to a single rank-$2b$ update

$$
\begin{aligned}
A^{(upd)} &\leftarrow Q_s^T A^{(upd)} P_s \\
&= A^{(upd)} + Y_L W_L^T A^{(upd)} + A^{(upd)} W_R Y_R^T + Y_L W_L^T A^{(upd)} W_R Y_R^T \\
&= A^{(upd)} + [X, Y_L][Y_R, Z_L]^T ,
\end{aligned} \tag{2}
$$

where $Z_L = A^{(upd)^T} W_L \in \mathbb{R}^{p \times b}$, $Z_R = A^{(upd)} W_R \in \mathbb{R}^{q \times b}$, and $X = Z_R + Y_L(Z_L^T W_R) \in \mathbb{R}^{q \times b}$. The computation of Z_L, Z_R, and X requires $4qpb + 2qb^2 + 2pb^2 + qb$ flop, while the rank-$2b$ update in (2) requires $4qpb + qp$ flop.

Formula (2) is based on the same idea of remodeling two-sided matrix updates as presented in [6].

Algorithm 3 : rk2b (reduction to b upper and b lower diagonals)

$i = 0$
while $i + b < n$ **do**
$\quad s = i/b + 1$
$\quad A(i+b+1:m, i+1:i+b) \equiv A_s^{(QRfact)} = Q_s R_s$ $\qquad$ {QR decomposition}
$\quad A(i+1:i+b, i+b+1:n) \equiv A_s^{(LQfact)} = L_s P_s^T$ $\qquad$ {LQ decomposition}
$\quad \{\ A_s^{(upd)} \equiv A(i+b+1:m, i+b+1:n) \equiv A_s^{(QRupd)} \equiv A_s^{(LQupd)}\ \}$
$\quad A_s^{(upd)} \leftarrow Q_s^T A_s^{(upd)} P_s$ $\qquad$ {update}
$\quad i = i + b$
enddo
cleanup: reduce $A(i+1:m, i+1:n)$ to banded form.

---

3 Numerical Results

In this section we compare our two-stage reduction techniques to the direct bidiagonalization routine **PDGEBRD** from ScaLAPACK. We present results from the IBM SP/1 located at the High Performance Computing Research Facility, Mathematics and Computer Science Division, Argonne National Laboratory, USA. All programs are coded in Fortran77 and use the portable BLACS library for doing the communication. Calls to the BLAS were directed to the assembly-coded optimized essl-library. The experiments were performed in double precision on random matrices with $m = n$.

The overall execution time of the two-stage methods includes the times for the stages *dense* $\rightarrow$ *banded* and *banded* $\rightarrow$ *bidiagonal* and for a re-distribution of the data between the two stages (cf. [9]). A parallel implementation of the second stage *banded* $\rightarrow$ *bidiagonal* is described in [11]. Performance of the algorithms strongly depends on the block size n_b. Thus, the numerical experiments were preceded by a calibration phase to determine optimum parameters. Depending on the size of the process grid optimal block factors $n_b = 12$ or 24 were found. Optimal choices for the intermediate bandwidth in the two-stage approaches lay between $b = 24$ and 36 with b being a multiple of n_b.

Our first approach, using the modified **PDGEQRF** routine (cf. Section 2.1), performed best. Using this algorithm for the reduction to banded form, the two-stage bidiagonalization proved to be superior to the direct method for almost all matrix dimensions and for all grid sizes considered, with speedups up to 1.6 (see Figure 3, left). The improved performance is due to the fact that—roughly speaking—the matrix-vector products of the direct method have been replaced by matrix-matrix products without significantly increasing the overall complexity. Note that the *speedup increases with the number of processors* because the matrix-matrix products can save on communication startup, too.

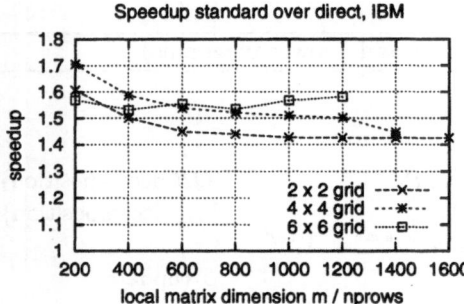

IBM, 4×4 grid				
m	t_{direct}	$t_{standard}$	$t_{splitfac}$	t_{rk2b}
800	12.7	6.4	9.6	16.5
1600	34.4	22.2	27.2	44.3
2400	80.7	53.0	67.0	95.2
3200	163.1	103.6	141.9	185.9
4000	273.7	179.8	244.5	312.0
4800	438.0	288.0	398.6	486.2
5600	653.8	452.1	608.0	712.6

Fig. 3. Left: Speedup of the two-stage reduction technique (**standard** variant) over the direct bidiagonalization algorithm **PDGEBRD**. Right: Overall execution times for the direct method vs. variants **standard**, **splitfac** and **rk2b**. The two-stage timings include the part *banded → bidiagonal*.

The two alternative approaches, **splitfac** and **rk2b**, do not reach the performance of the **standard** variant. The **splitfac** method can still outperform the routine **PDGEBRD** (see Figure 3, right). We will briefly discuss the most important reasons.

In the **splitfac** variant we can identify the global phases as a bottleneck. Here a large amount of communication is delaying a rather moderate number of floating-point operations. The global phases consume about 10 to 20 percent of the overall execution time. Asynchronous communication schemes may help to reduce this percentage, but at the cost of abandoning the BLACS library and therefore disabling fair timing comparisons with ScaLAPACK.

Our approach to maximize the dimension of the matrices involved in the update in order to squeeze some more speed out of the PBLAS3 routine **PDGEMM** as in Algorithm 3 (**rk2b**) was not successful for several reasons. First, detailed measurements show that the rank-$2b$ update according to (2) performs only marginally better than a rank-b update. But this small gain is more than neutralized by the costly extra computations, as setting up W_L or W_R requires substantially more time than building T via **PDLARFT**. In addition, the second stage must now bidiagonalize a banded matrix that has double bandwidth as compared to the one resulting from Algorithm 1.

We have already mentioned that the two-stage approach is not competitive if explicit updates of the transformation matrices U and V are needed. Tests with updates are still of interest to monitor rounding errors by computing deviation from orthogonality and residual. The errors from both the direct and the two-stage algorithms were always of the same magnitude.

4 Conclusions

Subdividing the bidiagonalization algorithm into two stages allows us to increase the portion of matrix-matrix operations. If only the reduction is needed, arithmetic costs are comparable to the direct method. Of the three techniques to reduce a dense matrix to banded form using pentangular factorizations, the method described in Section 2.1 is the easiest one to implement and yields the best performance. Combined with an effective algorithm for stage *banded* → *bidiagonal*, the two stage algorithm significantly outperforms the direct method.

References

[1] M. W. Berry, J. J. Dongarra, and Y. Kim. A parallel algorithm for the reduction of a nonsymmetric matrix to block upper-Hessenberg form. *Parallel Comput.*, 21(8):1184–1200, August 1995.

[2] C. Bischof and C. Van Loan. The WY representation for products of Householder matrices. *SIAM J. Sci. Stat. Comput.*, 8(1):s2–s13, January 1987.

[3] J. Choi, J. Demmel, I. Dhillon, J. Dongarra, S. Ostrouchov, A. Petitet, K. Stanley, D. Walker, and R. C. Whaley. ScaLAPACK: A portable linear algebra library for distributed memory computers—design issues and performance. *Computer Phys. Comm.*, 97:1–15, 1996.

[4] J. Choi, J. Dongarra, S. Ostrouchov, A. Petitet, D. Walker, and R. C. Whaley. A proposal for a set of parallel basic linear algebra subprograms. In J. Dongarra, K. Masden, and J. Waśniewski, editors, *Applied Parallel Computing*, pages 107–114. Springer Verlag, 1995.

[5] J. Choi, J. J. Dongarra, L. S. Ostrouchov, A. P. Petitet, D. W. Walker, and R. C. Whaley. The design and implementation of the ScaLAPACK LU, QR, and Cholesky factorization routines. *Scientific Programming*, 5:173–184, 1996.

[6] J. Choi, J. J. Dongarra, and D. W. Walker. The design of a parallel dense linear algebra software library: Reduction to Hessenberg, tridiagonal, and bidiagonal form. *Numer. Alg.*, 10:379–399, 1995.

[7] J. J. Dongarra, J. Du Croz, S. Hammarling, and I. Duff. A set of level 3 basic linear algebra subprograms. *ACM Trans. Math. Soft.*, 16(1):1–17, March 1990.

[8] J. J. Dongarra and R. C. Whaley. LAPACK Working Note 94: A user's guide to the BLACS v1.0. Technical Report CS-95-281, University of Tennessee at Knoxville, March 1995.

[9] B. Großer. Parallele zweistufige Verfahren zur Reduktion auf Bidiagonalgestalt. Diplomarbeit, Fachbereich Mathematik, Bergische Universität GH Wuppertal, 1997.

[10] B. Großer and B. Lang. Efficient parallel reduction to bidiagonal form. *submitted to Parallel Comput.*

[11] B. Lang. Parallel reduction of banded matrices to bidiagonal form. *Parallel Comput.*, 22:1–18, January 1996.

[12] R. Schreiber and C. Van Loan. A storage-efficient WY representation for products of Householder transformations. *SIAM J. Sci. Stat. Comput.*, 10(1):53–57, January 1989.

Experience with a Recursive Perturbation Based Algorithm for Symmetric Indefinite Linear Systems*

Anshul Gupta[1], Fred Gustavson[1], Alexander Karaivanov[2], Jerzy Wasniewski[2], and Plamen Yalamov[3]

[1] IBM Watson Research Center, P. O. Box 218, Yorktown Heights, NY 10598, USA,
fax: + 1 914 945 3434, anshul@watson.ibm.com, gustav@watson.ibm.com
[2] UNI•C, Building 304 - DTU, DK-2800 Lyngby, Denmark,
fax: + 45 3587 8990, alex@uni-c.dk, unijw@uni-c.dk
[3] University of Rousse, 7017 Rousse, Bulgaria,
fax: + 35 982 455 145, yalamov@ami.ru.acad.bg

Abstract. We consider recursive algorithms for symmetric indefinite linear systems. First, the difficulties with the recursive formulation of the LAPACK SYSV algorithm (which implements the Bunch-Kaufman pivoting strategy) are discussed. Next a recursive perturbation based algorithm is proposed and tested. Our experiments show that the new algorithm can be about two times faster although performing about the same number of flops as the LAPACK algorithm.

1 Introduction

Recursive algorithms for dense linear algebra problems are proposed and studied in [1, 2, 7]. It is shown that recursion leads to better performance on modern processors. Also, the codes using recursion are very simple, and easy to write in languages that support recursion (e. g. Fortran90).

In [1, 2, 5, 7] recursion is applied to three widely used algorithms, the LU and QR decompositions for general dense matrices, and the Cholesky decomposition for symmetric and positive definite matrices. In the present work we discuss the recursive approach to the decomposition of symmetric but indefinite matrices. It is well-known that such matrices require pivoting as the decomposition algorithms can be unstable, or can break down, even for well-conditioned matrices. Therefore, in LAPACK [3], the method Bunch-Kaufman pivoting is applied. As we will see in the next section the same type of pivoting is possible to apply in the recursive algorithm but doing so makes the algorithm more complicated and time consuming.

In practice there are different approaches to avoid the break down of accuracy in practice. In this paper we propose and test a perturbation approach; i. e., we

---

* This research is supported by the UNI•C collaboration with the IBM T.J. Watson Research Center at Yorktown Heights. The last author was partially supported by the Grant I-702/97 from the Bulgarian Ministry of Education and Science.

perturb pivot elements whenever they are small. In this way we move them away from zero, and improve the stability to some extent. Because of the perturbation the obtained decomposition is only accurate to a few digits. Nevertheless, it can be used for the solution of linear systems by adding 1-2 steps of iterative refinement. The cost of iterative refinement is tiny because it only uses triangular solves. Previously, the same approach has been applied to other types of matrices in [4] and [8], for example.

When the matrix of the problem is kept in full storage (like in _SYSV of LAPACK) our algorithm does not need additional memory. At the same time it can be up to three times faster than the corresponding LAPACK subroutine _SYSV.

The outline of the paper is as follows. In Section 2 we explain the difficulties with the recursive algorithm that uses Bunch-Kaufman pivoting. Then in Section 3 the algorithm with the perturbation approach is given. Numerical experience is presented and discussed in Section 4.

2   Recursive Factorization with Pivoting

It is well-known that the Cholesky factorization can fail for symmetric indefinite matrices. In this case some pivoting strategy can be applied (e. g. the Bunch-Kaufman pivoting [6, §4.4]). The algorithm can be given as follows.

LDL^T factorization
$L = I$ (identity matrix); $k = 1$;
while $(k < n)$
        Apply pivoting: choose a $1 \times 1 (s = 0)$, or $2 \times 2 (s = 1)$ pivot block,
        and exchange the corresponding rows and columns;
        $E = A_{k:k+s,k:k+s}$
        $C = A_{k+s+1:n,k:k+s}$
        $B = A_{k+s+1:n,k+s+1:n}$
        $L_{k+s+1:n,k:k+s} = CE^{-1}$
        $D_{k:k+s,k:k+s} = E$
        $A_{k+s+1:n,k+s+1:n} = B - CE^{-1}C^T$
        k = k+s+1
end

As a result we get
$$PAP^T = LDL^T,$$

where L is unit lower triangular, D is block diagonal with 1×1, or 2×2 blocks, and P is a permutation matrix.

Now let us look at the recursive formulation of this algorithm. This is given below. The recursion is done on the second dimension of matrix A; i. e., the algorithm works on full columns as in LU factorization.

Recursive Symmetric Indefinite Factorization (RSIF) of $A_{1:m,1:n}$
$k = 1$
if $(n = 1)$
    Define the pivot: 1×1, or 2×2.
    Apply interchanges if necessary
    $k = k + 1$, or $k = k + 2$
    If the pivot is 2×2: FLAG=1
else
    $n1 = n/2$
    $n2 = n - n1$
    RSIF of $A_{:,k:k+n1-1}$
    if (FLAG $= 1$)
        $n1 = n1 + 1$
        $n2 = n - n1$
    end
    update $A_{:,k:k+n2-1}$
    RSIF of $A_{:,k:k+n1-1}$
end

Since matrix A is square, we must set $m = n$ when we first call RSIF. The advantage of the recursive formulation is that the updating step is a matrix-matrix operation, and BLAS Level 3 subroutine can be used. Thus if the algorithm is properly implemented some speedup can be expected from this formulation. But this does not occur.

The fact that the recursive algorithm does not fully update the lower right part of A forces us to incorporate updating and downdating of the exchanged by the pivoting strategy columns. Such a step may bring more computation depending on the position of the columns. Additionally, the computation must be done as a Level 2 computation as we are updating and downdating single columns. To summarize: the application of the Bunch-Kaufman pivoting strategy forces one to do Level 2 computations (in some cases undoing a previously done computation at a Level 3 rate).

3   Perturbation Approach

An alternative to pivoting is our perturbation approach. It is applied in cases when there can be a large growth of elements (or breakdown), and pivoting is not desirable for some reason. The reason for avoiding the Bunch-Kaufman pivoting is that the performance suffers; this was illustrated in Section 2. This approach is applied in [4] to a parallel algorithm (where pivoting is not desirable because it adds more communication between the processors), and in [8] to an algorithm for inversion of Toeplitz matrices (where pivoting spoils the Toeplitz structure, and slows down the algorithm).

The idea of the perturbation approach is simple. Usually, growth of elements (or breakdown) happens when we pivot with a small number (or zero). Therefore, we add a small number δ to each pivot a,

$$a = a + \text{Sgn}(a)\delta, \quad \text{Sgn}(a) = \begin{cases} \text{sign}(a), & a \neq 0, \\ 1, & a = 0, \end{cases}$$

in case $|a|$ is small, more precisely, $|a| < \delta$.

With this approach the factorization for symmetric indefinite matrices looks as follows:

Perturbed Recursive Symmetric Indefinite Factorization (PRSIF)
of $A(1:n, 1:n)$
if $(n = 1)$
  if $(|A_{1,1}| < \delta)$
    $A_{1,1} = A_{1,1} + \text{Sgn}(A_{1,1})\delta$
  end
  $D_{1,1} = A_{1,1}$
else
  $p = n/2$
  PRSIF of $A_{1:p,1:p} = L_1 D_1 L_1^T$
  solve $X D_1 L_1^T = A_{p+1:n,1:p}$ for X
  update $A_{p+1:n,p+1:n} = A_{p+1:n,p+1:n} - X D_1 X^T$
  PRSIF of $A_{p+1:n,p+1:n} = L_2 D_2 L_2^T$
end

Let us note that the algorithm can be easily modified so that the case $n = 1$ is changed to $n \leq n_0$. In such a case we do not go to the deepest level of recursion, and decompose the block of size $\leq n_0$ with some appropriate algorithm. The numerical experiments at the end of the paper are done in this way (with $n_0 = 64$ which is the best choice for our architecture).

As a result of this algorithm we get $\tilde{A} = \tilde{L}\tilde{D}\tilde{L}^T$, where $\tilde{D}$ is diagonal, $\tilde{L}$ is unit lower triangular, and we put a tilde because of the perturbations.

Because of adding perturbations it is possible that we might change the input matrix A dramatically. So, the question is, how much does LDL^T change when we add perturbations. To answer this question we consider the product $LDL^T = A$. For any diagonal entry we have

$$A_{i,i} = D_{i,i} + \sum_{j=1}^{i-1} L_{i,j}^2. \tag{1}$$

If a perturbation is necessary then we add $\pm\delta$ to $D_{i,i}$, and we have

$$\tilde{A}_{i,i} = D_{i,i} \pm \delta + \sum_{j=1}^{i-1} L_{i,j}^2. \tag{2}$$

By comparing (1) and (2) we see that

$$\tilde{A}_{i,i} = A_{i,i} \pm \delta.$$

For the whole algorithm we operate not on the original matrix A but on $\tilde{A}$, where

$$\tilde{A} = A + \Delta A, \quad |\Delta A| \leq \delta I, \tag{3}$$

i. e. ΔA is diagonal. In the example problems we chose 1-2 perturbations seemed to be enough, so, only 1-2 entries of ΔA were nonzero and equal to $\pm\delta$. Thus if δ is small the changes in A are small, too. If A is well-conditioned this will lead to small changes in A^{-1} because we are only doing 1-2 small rank corrections. So, the perturbation approach is expected to work well for well-conditioned matrices.

Because of the changes in A the LDL^T factorization is no longer accurate but we can use this factorization for solution of linear systems $AX = B$. The idea is to apply iterative refinement [6, §3.5.3]:

solve $\tilde{A}X^{(0)} = B$;
for $k = 1, 2, \ldots$ until convergence do
$\qquad R^{(k-1)} = B - AX^{(k-1)}$;
$\qquad$ **solve** $(A + \Delta A)\tau^{(k)} = R^{(k-1)}$;
$\qquad X^{(k)} = X^{(k-1)} + \tau^{(k)}$;
end

The perturbed factorization is used when solving the linear systems above. Thus the iterative refinement needs $O(n^2)$ additional operations, and does not essentially increase the total operations count.

For the iterative refinement we need to keep the original matrix A. However, we do not need additional memory (except for one vector of size n where we store the diagonal of A) for this because we assume that matrix A is kept in full storage, and the original matrix A stays untouched by the algorithm in the upper triangle of the array. Thus, the algorithm essentially does not require more memory.

At present the perturbation δ is difficult to estimate theoretically. From our practical experience we suggest that the best value for δ is $\delta = \sqrt{\rho_0}$, where ρ_0 is the machine roundoff unit. In a Cholesky factorization $A = LL^T$ and $L_{ii} = \sqrt{\ldots}$. This analogy suggests why $\sqrt{\rho_0}$ works. With this value of δ 1-2 iterative refinement steps usually produce satisfactory results. In case δ is not chosen properly, we will have a large residual R. Since we compute the residual explicitly, we can notice such a situation, and produce a warning message.

4   Numerical Experiments

The experiments are produced on an IBM SMP node which has four CPUs (we use all of them in the experiments). The codes are written in Fortran90. We use double precision (roundoff unit $\approx$ 2.22E-16). Thus, our choice is $\delta =$ 1E-8. We compare our PRSIF algorithm with iterative refinement (denoted by RPSYSV) to the LAPACK SYSV algorithm. In order to see the advantage of the recursive algorithm we coded also a version of the SYSV algorithm in which the pivoting part was replaced by a perturbation part in the same way as for RPSYSV. This algorithm is denoted by PSYSV. This means that in PSYSV we also use iterative refinement. The number of iterative refinement steps in RPSYSV and PSYSV was fixed to be 1. Since IR doubles the number of digits and the precision we work with is approximately $\delta = 1e - 8$ the accuracy we obtain in our examples

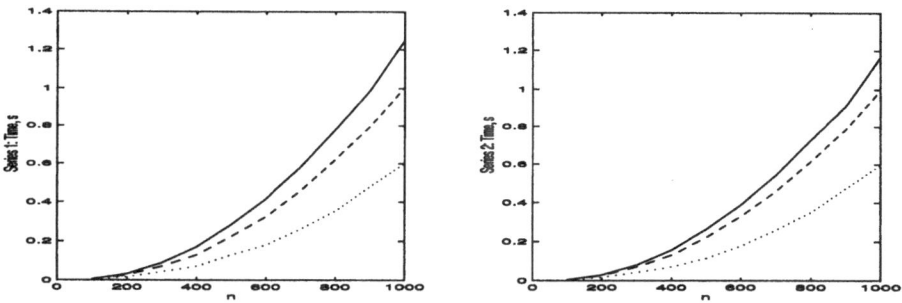

Fig. 1. Timing results: SYSV(-), PSYSV(--), RPSYSV($\cdots$)

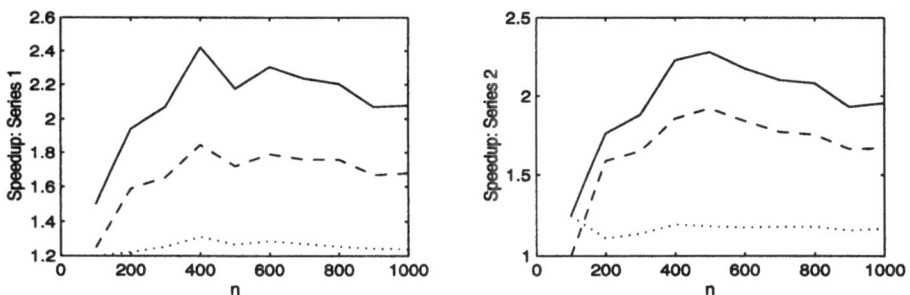

Fig. 2.   Speedup   results: SYSV/RPSYSV(-),   PSYSV/RPSYSV(--), SYSV/PSYSV($\cdots$)

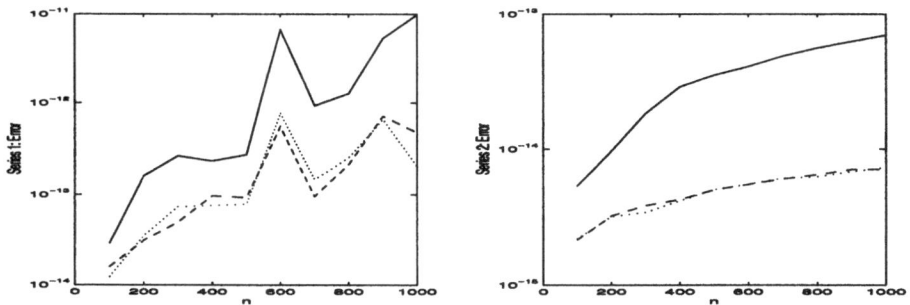

Fig. 3. Error results: SYSV(-), PSYSV(--), RPSYSV($\cdots$)

is close to $1e - 16$. We present results for the time, the speedup, and the forward relative error $\|\tilde{x} - x\|_\infty / \|x\|_\infty$ of the solution x for the three algorithms.

In the first series of experiments (denoted by Series 1) a random matrix is generated by the LAPACK subroutine DLAGGE.

In the second series of experiments (denoted by Series 2) matrix A is of the following type[1]):

$$A = \begin{pmatrix} \Delta & C \\ C^T & I \end{pmatrix},$$

where Δ is diagonal with small entries, C has random elements with large entries, and I is the identity. The entries of Δ are chosen to be less than δ. In this way the PRSIF algorithm is forced to make perturbations.

The results are presented in Fig. 1–3 for different sizes of matrix A.

The experiments show that 1) the recursive algorithm (in the best comparison) is about two times faster on average although performing almost the same number of flops as the LAPACK subroutine, and 2) the error in the recursive algorithm is similar to the error produced by the LAPACK subroutine.

As we mentioned the recursion is stopped when $n_0 = 64$. We tested also other choices of n_0. The performance degrades when decreasing or increasing n_0. The change is slight when we choose values of n_0 close to 64 (e. g. 56,60,68, or 72). When n_0 is significantly different (e. g. 8, or 200) the performance can be much worse. The corresponding blocking factor for LAPACK is chosen to be 32. This is the best choice (with highest performance) for most matrix sizes on our architecture.

Let us note that essentially we have two types of block operations in Algorithm PRSIF: 1) triangular solves with L_1, and 2) updating $A_{p+1:n,p+1:n}$. For the first operation we use the ESSL BLAS-3 subroutine which is compiled for the four CPUs. The second block operation is implemented by ourselves because there is no appropriate BLAS operation for symmetric matrices (this operation is included in the next version of BLAS). In the LAPACK subroutine the ESSL BLAS-3 routines are used wherever possible. The difference between the recursive and LAPACK algorithm is that the recursive algorithm works on larger and larger blocks while the LAPACK algorithm works on blocks with a fixed size. As a result the advantage of BLAS operations is better utilized by the recursive algorithm.

The performance of PSYSV is better than SYSV because perturbation is used instead if row and column exchanges in the Bunch-Kaufman pivoting which are slower. But the experiments show that RPSYSV has a significantly better performance than SYSV and PSYSV. The influence of recursion on the performance is especially illustrated by the difference between PSYSV and RPSYSV where the only difference is the recursion.

The timing behavior of RPSYSV is quite promising. At present we do not have theory about the application of the perturbation idea, and so we need to do more research to justify when its application will give accurate results.

---

[1] Suggested to us by John Reid.

References

[1] Andersen, B., Gustavson, F., Waśniewski, J.: A recursive formulation of the Cholesky factorization operating on a matrix in packed storage form, in. Parallel Processing for Scientific Computing, Proceedings of the Ninth SIAM Conference on Parallel Processing for Scientific Computing, San Antonio, TX , USA, March 24-27, 1999

[2] Andersen, B., Gustavson, F., Waśniewski, J., Yalamov, P.: Recursive formulation of some dense linear algebra algorithms, in. Parallel Processing for Scientific Computing, Proceedings of the Ninth SIAM Conference on Parallel Processing for Scientific Computing, San Antonio, TX , USA, March 24-27, 1999

[3] Anderson, E., Bai, Z., Bischof, C., Demmel, J., Dongarra, J., Du Croz, J., Greenbaum, A., Hammarling, S., McKenney, A., Ostrouchov, S., Sorensen, D.: LAPACK Users' Guide Release 2.0. SIAM, Philadelphia, 1995

[4] Balle, S., Hansen, P.: A Strassen-type matrix inversion algorithm for the Connection Machine. Report UNIC-93-11, October 1993

[5] Elmroth, E., Gustavson, F.: New serial and parallel recursive QR factorization algorithms for SMP systems. In: Applied Parallel Computing, (Eds. B. Kågstrm et. al.), Lecture Notes in Computers Science, v. 1541, Springer, 1998.

[6] Golub, G., Van Loan, C.: Matrix Computations, 3rd edition. John Hopkins University Press, Baltimore, 1996

[7] Gustavson, F.: Recursion leads to automatic variable blocking for dense linear-algebra algorithms. IBM J. Res. Develop. 41(1997), pp. 737–755

[8] Hansen, P., Yalamov, P.: Stabilization by perturbation of a $4n^2$ Toeplitz solver. Preprint N25, Technical University of Russe, January 1995

Parallel Cyclic Wavefront Algorithms for Solving Semidefinite Lyapunov Equations*

José M. Claver[1], Vicente Hernández[2], and Enrique S. Quintana-Ortí[1]

[1] Depto. de Informática, Universitat Jaume I, E-12080 Castellón, Spain.
[2] Depto. de Sistemas Informáticos y Computación, Universidad Politécnica de
Valencia, E-46071 Valencia, Spain.

Abstract. In this paper, we describe new parallel cyclic wavefront algorithms for solving the semidefinite discrete-time Lyapunov equation for the Cholesky factor using Hammarling's method by the message passing paradigm. These algorithms are based on previous cyclic and modified cyclic algorithms designed for the parallel solution of triangular linear systems. The experimental results obtained on an SGI Power Challenge show a high performance for large scale problems and better scalability than previous wavefront algorithms for solving these equations.

1 Introduction

Discrete-time Lyapunov equations arise in a great variety of problems of control theory and signal processing like, e. g., model reduction of linear control systems by means of the design of balanced realizations, Hankel-norm approximation problems, frequency domain approximation problems, solution of Riccati equations using Newton's method, etc. [9, 11].

Among the different solvers for these equations [1, 5], Hammarling's algorithm [6] is specially appropriate for model reduction via balanced realizations, as it directly computes the Cholesky factor of the solution. All these methods present a cubic computational cost and, already for medium-size problems, require the use of parallel computers. Thus, several wavefront algorithms based on Hammarling's algorithm have been implemented for shared memory multiprocessors in [2].

When dealing with large-scale problems, parallel distributed memory multiprocessors present the advantage of their scalability. In the last few years, parallel algorithms have been proposed for solving triangular linear systems on this type of architectures [4, 10]. These parallel algorithms can be classified as fan-in/fan-out, wavefront, and cyclic algorithms. Following these ideas, parallel distributed fan-in/fan-out and cyclic algorithms, based on the Schur method [1] or the Hessenberg-Schur method [5], have been developed for solving Sylvester equations [8]. More recently, parallel distributed wavefront Lyapunov solvers, based on Hammarling's method, have been presented in [3].

---

* This research was partially supported by the CICYT Project TIC96-1062-C03-01-03.

P. Amestoy et al. (Eds.): Euro-Par'99, LNCS 1685, pp. 1104–1111, 1999.
© Springer-Verlag Berlin Heidelberg 1999

The cyclic algorithms potentially offer the best performance, due to their minimal communication, on parallel distributed memory multiprocessors with a high latency. However, their performance deteriorates when the number of processors is increased. In this paper we describe several modifications of the cyclic algorithms based on the combination of cyclic and wavefront algorithms to overcome this deficiency.

In section 2 we review Hammarling's algorithm and its data dependency graph. In section 3 we present the parallel cyclic algorithms and, in section 4, a new sort of parallel cyclic wavefront algorithms. In section 5 we report experimental results on message passing based multiprocessors. Finally, the conclusions of the work are outlined in section 6.

2 Hammarling's Method

Consider the semidefinite discrete-time Lyapunov equation,

$$\bar{A}\bar{X}\bar{A}^T - \bar{X} + \bar{B}\bar{B}^T = 0, \tag{1}$$

where $\bar{A} \in \mathbb{R}^{n \times n}$ is the coefficient matrix, $\bar{B} \in \mathbb{R}^{n \times m}$ is part of the right-hand side matrix, $\bar{B}\bar{B}^T$, and $\bar{X} \in \mathbb{R}^{n \times n}$ is the matrix of unknowns. Hereafter, we assume that $n \leq m$. Note that in case $n > m$, it is possible to apply the same algorithm described in [6]. If the eigenvalues of matrix $\bar{A}$, denoted by $\{\lambda_1, ..., \lambda_n\}$, satisfy $|\lambda_i| < 1, i = 1, 2, \ldots, n$, then a unique, non-negative definite solution matrix $\bar{X}$ exists. In such case, it is possible to obtain the Cholesky decomposition of the solution $\bar{X} = \bar{L}\bar{L}^T$, where $\bar{L} \in \mathbb{R}^{n \times n}$ is a lower triangular matrix. However, using Hammarling's algorithm [6], equation (1) can also be solved directly for the Cholesky factor $\bar{L}$.

In the first step of Hammarling's algorithm, equation (1) is transformed into a (simpler) *reduced Lyapunov equation*. For this purpose, the real Schur decomposition of $\bar{A}$ is computed as $\bar{A} = QSQ^T$. Here, $Q \in \mathbb{R}^{n \times n}$ is an orthogonal matrix and $S \in \mathbb{R}^{n \times n}$ is a block lower triangular matrix with 1×1 and 2×2 diagonal blocks. Each 1×1 block contains a real eigenvalue of the coefficient matrix $\bar{A}$, and each 2×2 block is associated with a pair of complex conjugate eigenvalues. Algorithms for computing the real Schur decomposition on parallel computers are described in [7]. Applying the orthogonal similarity transformations defined by Q, we obtain the reduced Lyapunov equation

$$SXS^T - X = -BB^T,$$

where $X = Q^T\bar{X}Q$ and $B = Q^T\bar{B}$. Next, the product BB^T is reduced to a simpler form by computing an LQ factorization of B,

$$B = \begin{pmatrix} G\,0 \end{pmatrix} P,$$

where $G \in \mathbb{R}^{n \times n}$ is lower triangular and $P \in \mathbb{R}^{m \times m}$ is orthogonal. Finally, the solution L of the reduced Lyapunov equation

$$S\left(LL^T\right)S^T - \left(LL^T\right) = -GG^T, \tag{2}$$

provides the Cholesky factor of the original equation as $\bar{L} = QL$.

2.1   *The Serial Algorithm*

Following the method described in [6], the matrices S, L and G in (2) are initially partitioned as

$$S = \begin{pmatrix} s_{11} & 0 \\ \mathbf{s} & S_1 \end{pmatrix}, \quad L = \begin{pmatrix} l_{11} & 0 \\ \mathbf{l} & L_1 \end{pmatrix}, \quad \text{and} \quad G = \begin{pmatrix} g_{11} & 0 \\ \mathbf{g} & G_1 \end{pmatrix}, \quad (3)$$

where s_{11} is either a scalar or a 2×2 block. In the scalar case, l_{11} and g_{11} are also scalars, and $\mathbf{s}$, $\mathbf{l}$, and $\mathbf{g}$ are column vectors of $n - 1$ elements. Otherwise, s_{11}, l_{11}, and g_{11} are 2×2 blocks, and $\mathbf{s}$, $\mathbf{l}$, and $\mathbf{g}$ are $(n - 2) \times 2$ blocks.

For the sake of simplicity, hereafter we assume that all the eigenvalues of S are real. This problem will be denoted as the *real case* of the Lyapunov equation. Hence, the next three equations are obtained from (2) and (3)

$$\begin{aligned} l_{11} &= g_{11}/\sqrt{1 - s_{11}^2}, \\ (s_{11}S_1 - I_{n-1})\mathbf{l} &= -\alpha\mathbf{g} - \beta\mathbf{s}, \quad \text{and} \quad (4) \\ S_1\left(L_1 L_1^T\right) S_1^T - \left(L_1 L_1^T\right) &= -\tilde{G}\tilde{G}^T = -G_1 G_1^T - \mathbf{y}\mathbf{y}^T, \end{aligned}$$

where $\alpha = g_{11}/l_{11}$, $\beta = s_{11}l_{11}$, $\mathbf{y} = \alpha\mathbf{v} - s_{11}\mathbf{g}$, $\mathbf{v} = S_1\mathbf{l} + \mathbf{s}l_{11}$, and I_{n-1} stands for the identity matrix of order $n - 1$.

The diagonal element l_{11} is directly computed from the first equation in (4). The lower triangular linear system in the second equation is then solved for $\mathbf{l}$ by forward substitution. Finally, the last equation is a discrete-time Lyapunov equation of order $n - 1$, where $\tilde{G}$ has the following structure

$$\tilde{G} = (G_1, \mathbf{y}) ;$$

that is, $\tilde{G}$ is a block matrix composed of an $(n - 1) \times (n - 1)$ lower triangular matrix, G_1, and an $n - 1$ column vector $\mathbf{y}$. Therefore, it is possible to obtain the Cholesky decomposition of the product $\tilde{G}\tilde{G}^T$ using the LQ factorization,

$$\tilde{G} = (\bar{G}\, 0)\, \bar{P},$$

where $\bar{G} \in \mathbb{R}^{(n-1)\times(n-1)}$ is lower triangular and $\bar{P} \in \mathbb{R}^{n \times n}$ is orthogonal. This procedure can be repeated with the reduced Lyapunov equation,

$$S_1\left(L_1 L_1^T\right) S_1^T - \left(L_1 L_1^T\right) = -\bar{G}\bar{G}^T,$$

of order $n - 1$, until the problem is completely solved.

2.2   *Study of the Data Dependencies.*

Hammarling's algorithm is column-oriented; that is, when the j-th column of the solution is to be computed, it is necessary to obtain the elements $L(j : i - 1, j), j \leq i$, before computing the element $L(i, j)$. Consider now the computation of the $(j + 1)$-th column of L. The first element that must be computed is $L(j + 1, j + 1)$ but, according to step 1 of the serial algorithm, $G(j + 1, j + 1)$ is required in iteration j to nullify the $(j+1)$-th element of $\mathbf{y}$. The next element to be computed is $L(j+2, j+1)$, which requires $L(j+1, j+1)$ and the updated element $G(j+2, j+1)$.

It is important to outline from the analysis of the data dependencies, that the highest inherent parallelism is achieved when the elements on the same antidiagonal of L are computed simultaneously. This idea was previously introduced by O'Leary [12] in the context of the Cholesky decomposition problem and it was also used to design triangular linear system solvers on distributed memory multiprocessors in [4].

3   Parallel Cyclic Algorithms

The parallel cyclic algorithms described in this paper are based on previous work by Eisenstat *et al.* [4] for solving triangular linear systems on distributed memory multiprocessors. In these algorithms the matrices of the problem are partitioned and distributed among the processors by rows (or columns). This data layout presents good load balancing properties for matrix factorization procedures that generally precede the solution of triangular linear systems. In order to simplify the presentation of our algorithms we define a function **map**(j) that indicates the processor which stores the j-th row (column) of a matrix.

In the cyclic parallel triangular linear system solvers, all necessary information is stored in a segment of constant size $p-1$ that circulates among a ring of p processors. After receiving the segment, processor j computes an element of the solution. Next, the segment is updated and sent to the next processor $(j+1)$. The goal here is to overlap the circulation of the segment with the updating of several variables needed to compute the next element belonging to the same processor. Other kind of interconnection topologies like a bus can be appropriate for these algorithms since, as only one message is circulating at each time, no communication bottleneck exists.

Using the same ideas, Lyapunov equations can be solved by columns, with all the processors collaborating to compute each column. The solution of a column is divided into two stages: In the first stage, denoted as *substitution*, all the elements of a column are computed by solving a triangular linear system. In the second stage, denoted as *triangularization*, the right-hand side matrix $\tilde{G}$ is updated by computing an LQ factorization that nullifies vector $\mathbf{y}$. In this algorithm, the solution L is stored cyclically by rows, like matrices S and G.

The proposed algorithm, *PHCF*, is shown in Fig. 1. In the *substitution* stage, the segment consists of the last $p-1$ computed elements of a column of L, and during segment circulation the unknown elements not computed are updated. In the *triangularization* stage the segment stores $2(p-1)$ elements with the $p-1$ sines and $p-1$ cosines computed. The circulation of the segment must be overlapped with the updating of the right-hand side matrix G and the vector $\mathbf{y}$.

A different approach is to overlap the two solving stages, substitution and triangularization. In this algorithm (denoted as *PHCF2*), when a new element of the Cholesky factor is computed, the right-hand side matrix is immediately updated. Thus, two segments of size $(p - 1)$ and $2(p - 1)$ are simultaneously circulating among the processors and possible idle times are reduced.

Algorithm *PHCF.*
Memory: $S(n/p,n), G(n/p,n), L(n/p,n), seg(n), Cos(n), Sin(n)$
for $j = 0 : n - 1$
1. Substitution stage of $L(:, j)$
    1.1. Compute the diagonal element.
       broadcast$(L(j,j), S(j,j))$
    1.2. Compute the subdiagonal elements.
       for $i = j + 1 : n - 1$
       1.2.1. Receive the segment $seg(p - 1)$.
          receive$(\mathbf{map}(i - 1), seg(\max(i - p + 1, j + 1) : i - 1))$
       1.2.2. Compute $L(i, j)$.
       1.2.3. Send updated segment.
          send$(\mathbf{map}(i + 1), seg(\max(i - p + 2, j + 2) : i))$
       1.2.4. Update $L(:, j)$.
       end for
2. Triangularization stage.
    2.1. Compute vector **y**.
    2.2. Update matrix G.
       for $i = j + 1 : n - 1$
       2.2.1. Receive segments Sin and Cos.
          receive$(\mathbf{map}(i - 1), Cos(\min(i - p, j + 1) : i - 1))$
          receive$(\mathbf{map}(i - 1), Sin(\min(i - p, j + 1) : i - 1))$
       2.2.2 Apply the previous Givens rotations.
       2.2.3. Compute the Givens rotations $(\sin \theta_i, \cos \theta_i)$
       2.2.4. Send segments Sin and Cos.
          send$(\mathbf{map}(i + 1), Cos(\min(i - p + 1, j + 2) : i - 1))$
          send$(\mathbf{map}(i + 1), Sin(\min(i - p + 1, j + 2) : i - 1))$
       2.2.5. Apply the Givens rotations to the rest of elements.
       end for
end for
end *PHCF.*

Fig. 1 Algorithm *PHCF.*

We also propose a different approach to increase the granularity of the algorithm while maintaining the initial data distribution. In our algorithm *PHCR* each processor computes at each step q elements of the same row. Due to the data dependency of Hammarling's algorithm, we cannot separate the substitution and triangularization stages as was previously done in algorithm *PHCF*. In this algorithm, **y** is actually a matrix of size $n \times q$ and the circulating segment is of size $3q(p - 1)$.

4   Cyclic Wavefront Algorithms

Following the analysis by Eisensat *et al.*, we have developed similar pipelined and short-cut algorithms that are variants of those in [4]. The algorithms in [4] were designed for a type of multiprocessor systems where the ratio between the computational speed of the processors and the communication bandwidth of interconnection network was low. Current parallel architectures have experimented an increase of computational speed higher than that of the communication bandwith. On the other hand, many current bus-based multiprocessor systems have reduced the number of processors (4-12) since a high number leads to a dramatic reduction in the performance.

We propose a new sort of parallel algorithms, denoted as cyclic wavefront algorithms (*PHCWF*), where each processor solves simultaneously r columns, and r segments are circulating among the processors (with $r \leq p$, $rk = n$, and k a positive integer number). A row cyclic distribution of S and G is used, and the solution matrix is also stored cyclically by rows. Thus, the Lyapunov equation is solved by using an antidiagonal wavefront of size r. When $r > p$, there are only a maximum of p segments circulating among the processors. The segments will be sent either as a unique message of size $3(p-1)$ or as two segment messages of size $(p-1)$ and $2(p-1)$.

5   Experimental Results

These parallel algorithms have been implemented on an SGI Power Challenge (PCh). This computer reflects a current tendency in the construction of high performance computers. The PCh is a shared memory bus-based multiprocessor (the main memory has 1 GByte) with 12 superscalar R10000 processors at 200 MHz, and 64kBytes and 2MBytes per processor of primary and secondary cache memory, respectively.

The parallel algorithms have been implemented using C and the PVM message-passing library. Communications are tuned and implemented through the main memory. The parallel algorithms were compared with a serial block version of Hammarling's algorithm. We used double-precision arithmetic in our experiments and all the algorithms were compiled with the appropriate optimization flags.

The cyclic algorithms *PHCF*, *PHCF2* and *PHCR* report a low performance as in the case of cyclic triangular linear systems solvers [4, 10]. As we found out in the theoretical analysis of our algorithms, idle times are not reduced when the granularity is increased. The only way to reduce idle times is increasing the number of segments circulating among the processors while taking advantage simultaneously of data locality in the cache memory. The latter can be achieved by using cyclic wavefront algorithms.

Fig. 2 shows the efficiencies obtained for algorithm *PHCWF* on the PCh using 4 and 8 processors. In this implementation, for each column of the solution that is computed, two segments circulate among the processors. In this figure, r is the number of columns solved simultaneously (there are $2r$ segments circulating); when $r = 1$ the behavior of *PHCWF* is equal to that of algorithm *PHCF2* (performance in this case are lower than the performance obtained for *PHCWF* with $r = 2$, and therefore these results are not reported). We can observe a low performance for large problems, $r = 2$ and 4 processors. This low performance arises for larger problems and $r = 1$. We are performing further experiments to find out the reasons for this behavior.

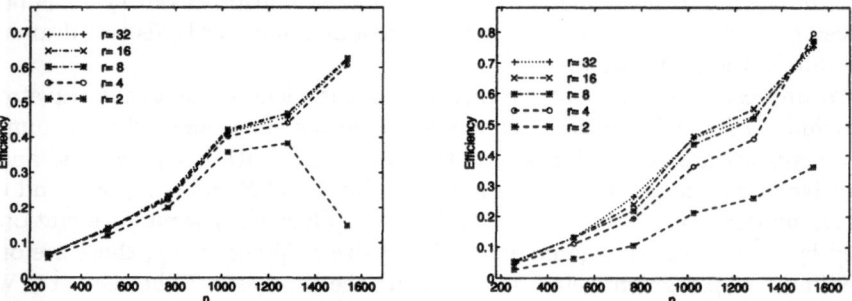

Fig. 2. Efficiency obtained of the algorithm *PHCWF* on the PCh using 4 (left) and 8 (right) processors, for different problem sizes and values of r.

Notice that a high performance is obtained only for large-scale problems (the reasons are the large size of the cache memories and the high cost of the communication start-up). It is also possible to observe that the efficiencies obtained are very similar when r is higher than a given value.

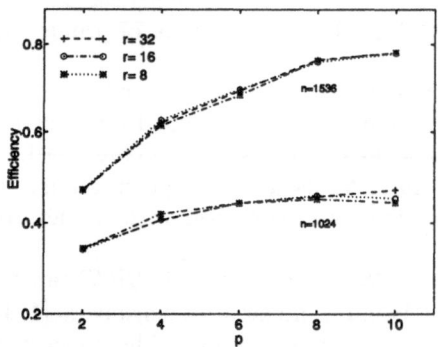

Fig. 3. Efficiency obtained of the algorithm *PHCWF* on the PCh for $n = 1024$ and $n = 1536$, using different number of processors and values of r.

Fig. 3 shows that, in some cases, a higher performance is obtained when the number of processors is increased. Hence, cyclic wavefront algorithms have better scalability than wavefront algorithms designed to solve Lyapunov equations [3], in which the efficiency decreases as the number of processors is increased.

6 Conclusions

New parallel cyclic wavefront algorithms have been presented for the solution of large and dense discrete-time Lyapunov equations using Hammarling's method on message passing multiprocessors. These algorithms can be easily adapted to the continuous-time version of the Lyapunov equation.

The algorithms combine two techniques previously used to solve triangular linear systems and introduce new ideas to solve problems arising in the algorithms. Cyclic wavefront algorithms show a good performance when the problem order and the number of processors is increased. Efficiencies near to 80% have been obtained for 10 processors and $n = 1536$. This behavior has been tested on a bus-based multiprocessor, the SGI Power Challenge, and excellent scalability has been reported for these algorithms.

References

[1] Bartels, R., Stewart, G.: Algorithm 432. Solution of the matrix equation AX+XB=C [F4]. Comm. ACM **15** (1972) 820–826

[2] Claver, J., Hernández, V., Quintana, E.: Solving discrete-time Lyapunov equations for the Cholesky factor on a shared memory multiprocessor. Parallel Processing Letters, **6** No 3 (1996) 365–376

[3] Claver, J., Hernández, V.: Parallel wavefront algorithms solving Lyapunov equations for the Cholesky factor on message passing multiprocessors. The Journal of Supercomputing **13** No. 2 (1999) 171–189

[4] Eisenstat, S., Heath, M., Henkel, C., Romine, C.: Modified cyclic algorithms for solving triangular systems on distributed-memory multiprocessors. SIAM J. Sci. Statist. Comput. **18** No. 3 (1988) 598–600

[5] Golub, G., Nash, S., Van Loan, C.: A Hessenberg-Schur method for the problem AX+XB=C. IEEE Trans. A. C. **24** (1979) 909–913

[6] Hammarling, S.: Numerical solution of the discrete-time, convergent, non negative definite Lyapunov equation. System & Control Letters **17** (1991) 137–139

[7] Henry, G., Watkins, D., Dongarra, J.: A parallel implementation of the nonsymmetric QR algorithm for distributed memory architectures. Lapack Working Note #121 (1997)

[8] Kågström, B., Poromaa, P.: Distributed block algorithms for the triangular Sylvester equation with condition estimator. Hypercube and Distributed Computers (1989) 233–248.

[9] Laub, A., Heat, M., Paige, G., Ward, R.: Computation of system Balancing transformations and other applications of simultaneous diagonalization algorithms. IEEE Trans. A. C. **32** (1987) 115–122

[10] Li, G., Coleman, T.: A new method for solving triangular systems on distributed memory message-passing multiprocessors. SIAM J. Scientific & Statistical Computing **10** (1989) 382–396

[11] Moore, B.: Principal component analysis in linear systems: Controlability, observ- ability, and model reduction. IEEE Trans. A. C. **26** (1981) 100–105

[12] O'Leary, D., Stewart, G.: Data-flow algorithms for parallel matrix computations. Comm. ACM **28** (1986) 840–853

Parallel Constrained Optimization
via Distribution of Variables

Claudia A. Sagastizábal[1] and Mikhail V. Solodov[2]

[1] INRIA-Rocquencourt, BP 105, 78153 Le Chesnay, France.
Also UFRJ/COPPE-PESC, CP 68501, Rio de Janeiro, RJ 21945-970, Brazil.
`sagastiz@ele.puc-rio.br`
[2] Instituto de Matemática Pura e Aplicada,
Estrada Dona Castorina 110, Rio de Janeiro, RJ 22460-320, Brazil.
`solodov@impa.br`

Abstract. In the parallel variable distribution (PVD) approach for solving optimization problems, the variables are distributed among parallel processors with each processor having the primary responsibility for updating its block of variables while allowing the remaining "secondary" variables to change in a restricted fashion along some easily computable directions. For constrained nonlinear programs, convergence in [4] was established in the special case of convex block-separable constraints. In [11], the PVD approach was extended to problems with general convex constraints by means of utilizing the projected gradient directions for the change of secondary variables. In this paper, we propose two new variants of PVD for the constrained case. For the case of block-separable constraints, we develop a parallel sequential quadratic programming algorithm. This is the first PVD-type method which does not assume convexity of the feasible set for convergence. For inseparable convex constraints, we propose a PVD method based on suitable approximate projected gradient directions. Using such approximate directions is especially important when the projection operation is computationally expensive.

1 Introduction and Motivation

We consider parallel algorithms for solving optimization problems of the form

$$\min_{x \in C} f(x), \tag{1}$$

where C is a nonempty closed set in $\Re^n$, and $f : \Re^n \to \Re$ is a continuously differentiable function which is bounded below on C. Our approach consists of partitioning the problem variables $x \in \Re^n$ into p blocks $x_1, \ldots, x_p$, such that $x_l \in \Re^{n_l}$ and $\sum_{l=1}^{p} n_l = n$, and distributing them among p parallel processors.

In parallel algorithms, an iteration usually consists of two steps: parallelization and synchronization [1]. Consider for the moment the unconstrained case, corresponding to $C = \Re^n$. Usually, the synchronization step aims at guaranteeing a sufficient decrease in the objective function, while the parallel step

P. Amestoy et al. (Eds.): Euro-Par'99, LNCS 1685, pp. 1112–1119, 1999.
© Springer-Verlag Berlin Heidelberg 1999

produces candidate points (or candidate directions) by solving subproblems (P_l) for $l = 1, \ldots, p$. Each (P_l) is defined on a subspace of dimension (significantly) smaller than n, in such a way that these p subspaces "span" the whole variable space $\Re^n$. Most methods, such as Block-Jacobi [1], updated conjugate subspaces [6], coordinate descent [12], and parallel gradient distribution [8], define (P_l) as a minimization problem on the l-th block of variables, i.e., in $\Re^{n_l}$. More recently, Parallel Variable Distribution (PVD) algorithms, introduced in [4] and further studied in [10, 11, 5], advocated subproblems (P_l) of slightly higher dimensions than n_l. In the l-th subproblem, in addition to the associated "primary" optimization variables x_l, there are $p - 1$ "secondary" variables representing in a *condensed form* all the remaining $n - n_l$ problem variables (see Algorithm 1 below). Since those remaining $n - n_l$ variables are allowed to change only in a restricted fashion (in a subspace of dimension $p-1$), the additional computational burden of solving such enlarged subproblems is not big. Moreover, in computational implementation, this approach results in better practical behaviour [4].

1.1 Parallel Variable Distribution Algorithms. General Scheme

To formalize the notion of primary and secondary variables, we need to introduce some notation. Let $\bar{l}$ denote the complement of l in the index set $\{1, \ldots, p\}$. Given some direction $d^i \in \Re^n$, that we shall call *PVD-direction*, the matrix $D^i_{\bar{l}}$ is an $n_{\bar{l}} \times (p - 1)$ block-diagonal matrix formed by placing the blocks $d^i_1, \ldots, d^i_{p-1}$ $(d^i_t \in \Re^{n_t}, t = 1, \ldots, p - 1)$ of the chosen direction d^i along its block diagonal.
  We describe next the basic PVD algorithm.

Algorithm 1 (PVD)
  Start with any $x^0 \in C$. Choose a PVD-direction $d^0 \in \Re^n$. Set $i := 0$.
  Having x^i, check a stopping criterion. If it is not satisfied, compute x^{i+1}:
Parallelization: *For each processor l compute a solution $(y^i_l, \mu^i_{\bar{l}}) \in \Re^{n_l} \times \Re^{p-1}$ of the subproblem*

$$(P_l) \qquad \begin{cases} \min_{x_l, \mu_{\bar{l}}} \psi^i_l(x_l, \mu_{\bar{l}}) := f(x_l, x^i_{\bar{l}} + D^i_{\bar{l}} \mu_{\bar{l}}) \\ (x_l, x^i_{\bar{l}} + D^i_{\bar{l}} \mu_{\bar{l}}) \in C. \end{cases}$$

Synchronization: *Compute x^{i+1} such that $f(x^{i+1}) \leq \min_{l \in \{1, \ldots, p\}} \psi^i_l(y^i_l, \mu^i_{\bar{l}})$. Set $i := i + 1$; choose the new PVD-direction d^i; and repeat.* $\square$

An important feature of this algorithm is the presence of the "forget-me-not" term $x^i_{\bar{l}} + D^i_{\bar{l}} \mu_{\bar{l}}$ in the parallel subproblems (P_l), which allows for a change in *all* variables, not only the primary variables x_l. The idea is to balance the reduction in the number of variables for each (P_l) with allowing just enough freedom for the change of other (secondary) variables. Due to this, the parallel subproblems better approximate the original problem that has to be resolved. Note that because the secondary variables can change only along chosen fixed directions, their inclusion does not significantly increase the dimensionality of the parallel subproblems (P_l). Indeed, the number of variables is $n_l + p - 1$, only $p-1$ more than if the secondary variables were excluded (in parallel computation, it is reasonable to assume that p is small relative to n_l).

When C is convex, the synchronization step in Algorithm 1 may also consist of minimizing the objective function in the affine hull, subject to feasibility, of all the points computed in parallel by the p processors. In principle, any point with the objective function value at least as good as the smallest computed by all the processors is acceptable. As for PVD-directions, they are typically some easily computable feasible descent directions for the objective function f at the current x^i, e.g., quasi-Newton or steepest descent directions in the unconstrained case.

Algorithm 1 is a rather general framework, which has to be further refined and specialized to obtain implementable/practical versions. In this respect, the two principal questions are:

– how to solve each parallel subproblem, including criteria for inexact resolution,
– how to choose PVD-directions for secondary variables.

When (1) is unconstrained, some of these issues have been addressed in [10] and [5]. The former work contains convergence results stronger than those in [4], including linear rate of convergence. These results, as well as useful generalizations, such as algorithms with inexact subproblem solution and a certain degree of asynchronization, were obtained by imposing natural restrictions (of sufficient descent type) on the PVD-directions. An even more general framework was developed later in [5].

When (1) is a constrained optimization problem, many questions remain unresolved, especially for a nonconvex feasible set C. For convex C with block-separable structure (i.e., C is a Cartesian product of closed convex sets), convergence was shown in [4]. It was further stated that in the case of inseparable convex constraints, the PVD approach may fail. While this is true for an arbitrary choice of PVD-directions in D_I^i, it seems clear that one of the keys to success of this approach should be precisely a reasonable choice of PVD-directions. In this respect, some progress have been made in [11], where it was proposed to use the projected gradient direction $d(x) := x - P_C[x - \nabla f(x)]$, where $P_C[\cdot]$ stands for the orthogonal projection onto C. Specifically, it was established that setting $d^i := d(x^i)$ would ensure convergence of PVD for problems with general (inseparable) convex constraints. Some criteria for inexact subproblem solution were also given in [11]. When C is a polyhedral set, computing the projected gradient direction requires solving at every synchronization step of Algorithm 1 a single quadratic programming problem. For this, a wealth of fast and reliable algorithms is available [3]. However, in the case of nonlinear convex constraints, the task of computing the projected gradient directions is considerably more computationally expensive. Actually, even in the affine case, computing those directions exactly (or very accurately) may turn to be rather wasteful, especially when far from the solution of the original problem.

In this paper, we propose two new versions of PVD for the nonlinearly constrained case. The first one applies to nonlinear programs with block-separable *nonconvex* feasible sets. It is based on the use of *sequential quadratic programming* techniques (SQP). Our second proposal is for *inseparable* convex feasible sets. We introduce a *computable approximation* criterion which allows us to employ *inexact* projected gradient directions. We emphasize that this criterion is readily implementable, and it preserves global convergence of PVD based on exact directions.

Our notation is fairly standard. The usual inner product of two vectors $x, y \in \Re^n$, is denoted by $\langle x, y \rangle$, and the associated norm is given by $\|x\|^2 = \langle x, x \rangle$. Analogous notation will be used for the reduced subspaces $\Re^{n_i}$. For a differentiable function $f : \Re^n \to \Re$, ∇f will denote the n-dimensional vector of partial derivatives with respect to x. For a differentiable vector function $c : \Re^n \to \Re^m$, Jc will denote the $m \times n$ Jacobian matrix, whose rows are the gradients of the components of c.

2   Using Sequential Quadratic Programming

Suppose the feasible set C in (1) is described as a system of inequality constraints

$$C := \{x \in \Re^n \mid c(x) \le 0\}, \tag{2}$$

where $c : \Re^n \to \Re^m$. In this section, we assume that C is block-separable, i.e.,

$$c_l : \Re^{n_l} \to \Re^{m_l} \quad \text{with } \sum_{l=1}^{p} m_l = m \quad \text{for } l = 1, \ldots, p. \tag{3}$$

SQP techniques are essentially constrained (quasi)-Newton methods. For a nonlinear program (1), the idea is to iteratively solve *quadratic tangential problems* (QTP) of the form

$$\begin{cases} \min_{\delta \in \Re^n} \langle \nabla f(x^i), \delta \rangle + \frac{1}{2} \langle \delta, M^i \delta \rangle \\ c(x^i) + Jc(x^i)\delta \le 0, \end{cases}$$

where M^i is a positive definite matrix. Locally, SQP methods are superlinearly convergent under appropriate assumptions. They can be globalized, for instance, by making a line search along $x^i + t\delta^i$ using some merit function. Local convergence requires the matrices M^i to asymptotically approach the Hessian of the Lagrangian of (1) at a Karush-Kuhn-Tucker (KKT) point. Global convergence only requires boundedness of M^i and $M^{i^{-1}}$, see [2, Ch. 13].

To produce parallel versions, it seems natural to define a (QTP_l) for each processor and then perform a line-search in the synchronization step. Note that the inclusion of forget-me-not terms does not seem to be crucial in the block-separable case (for example, such terms were not specified in the analysis of [4]). Thus, we drop here secondary variables in (P_l), by taking null PVD-directions. However, the use of secondary variables in SQP deserves further study for feasible sets with inseparable nonconvex constraints.

For our line-search, we shall use the exact penalty function [7]

$$\theta_\sigma(x) := f(x) + \sigma \sum_{l=1}^{p} \|c_l^+(x_l)\|, \tag{4}$$

where σ is a positive parameter, and $y^+ := \max(0, y)$ with maximum taken component-wise. Throughout, we assume that the QTP has a nonempty feasible set for every iteration (this is guaranteed, for example, if for each l the Jacobian Jc_l maps $\Re^{n_l}$ onto $\Re^{m_l}$, or if c_l is convex).

Let I_l be the $n_l \times n_l$ identity matrix. Then given x^i, the l-th block of $\nabla f(x^i)$ is denoted by $g_l^i := (I_l \quad 0_{n_l \times n_l}) \nabla f(x^i)$.

Algorithm 2 (Parallel SQP)

Start with any $x^0 \in C$. Choose $\underline{\sigma} > 0$, $\tau \in (0, 1/2)$, and p positive definite $n_l \times n_l$ matrices M_l^0. Set $i := 0$.

Having x^i, check a stopping criterion. If it is not satisfied, compute x^{i+1}:
Parallelization: *For each $l \leq p$ compute $(\delta_l^i, \lambda_l^i) \in \Re^{n_l} \times \Re_+^{m_l}$ a KKT point of the subproblem*

$$(QTP_l) \qquad \begin{cases} \min_{\delta_l \in \Re^{n_l}} \langle g_l^i, \delta_l \rangle + \frac{1}{2} \langle \delta_l, M_l^i \delta_l \rangle \\ c_l(x_l^i) + Jc_l(x_l^i) \delta_l \leq 0. \end{cases}$$

Synchronization: *Define $\delta^i := (\delta_1^i, \ldots, \delta_p^i) \in \Re^n$, $\lambda^i := (\lambda_1^i, \ldots, \lambda_p^i) \in \Re_+^m$.*
Line-search: *Take $\sigma_i \geq \|\lambda_l^i\| + \underline{\sigma}$ for $l = 1, \ldots, p$.*
Compute $\Delta^i := \langle g^i, \delta^i \rangle - \sigma_i \sum_{l=1}^p \|c_l^+(x_l^i)\|$.
Find $t_i > 0$ such that

$$\theta_{\sigma_i}(x^i + t_i \delta^i) \leq \theta_{\sigma_i}(x^i) + \tau t_i \Delta^i. \tag{5}$$

Set $x^{i+1} := x^i + t_i \delta^i$, $i := i + 1$; choose new matrices M_l^i, $l \leq p$ and repeat.
$\square$

It is sometimes argued that SQP is not a convenient technique for solving large-scale (with n and m large) nonlinear programs when there are inequality constraints present. Specifically, it is argued that the combinatorial aspect introduced by the complementarity condition in the KKT system associated to each QTP makes the solution of subproblems very costly. In relation to Algorithm 2, it is important to note that each subproblem (QTP_l) is a quadratic programming program of relatively small dimension (n_l and m_l are small). There exist a number of very fast and reliable algorithms for solving such problems (see [3]). Hence, our parallel version makes it possible to use SQP to solve large scale optimization problems with block-separable constraints. With respect to the original constrained PVD algorithm [4], two remarks are in order. First, our parallel SQP can be regarded as PVD with inexact solution of parallel subproblems. Indeed, "hard" nonlinear PVD subproblems of Algorithm 1 are approximated by "easy" quadratic programming subproblems in parallel SQP. And secondly, the algorithm is extended to the case of nonconvex constraints.

We state next that δ^i is a descent direction for the merit function θ_σ. This in turn implies that the line-search procedure is well-defined.

Lemma 1. *Let $\theta_\sigma(x; d)$ denote the directional derivative of the line-search function (4). With the definitions of Algorithm 2, the following holds:*
$$\theta_{\sigma_i}(x^i; \delta^i) \leq \Delta^i \leq -\sum_{l=1}^p \langle \delta_l^i, M_l^i \delta_l^i \rangle - \underline{\sigma} \sum_{l=1}^p \|c_l^+(x_l^i)\|.$$
$\square$

For the sake of brevity, we omit proofs in this presentation.

For the scheme above to be globally convergent, the line-search has to fulfill the following rules:

- The stepsize is found using an Armijo-like iterative process that keeps t_i bounded away from zero.
- The penalization parameter σ_i is kept bounded.

Our convergence result follows.

Theorem 1. *Let the feasible set C have block-separable structure as in (3), and let $\{(x^i, \lambda^i)\}$ be a sequence generated by Algorithm 2. Assume there exists an iteration index I such that $\sigma_i = \bar{\sigma}$ for all $i \geq I$. Suppose that for $l = 1, \ldots, p$ the matrices M_l^i and their inverses are kept bounded along iterations. Then*

$$\begin{cases} \nabla_x L(x^i, \lambda^i) \to 0 \\ c^+(x^i) \to 0 \\ \lambda^i \geq 0 \ and \langle \lambda^i, c(x^i) \rangle \to 0 , \end{cases}$$

and every accumulation point of the sequence $\{x^i\}$ satisfies the first-order necessary optimality conditions of problem (1). $\square$

The proof uses Lemma 1 and follows the technique in [2, Theorem 13.3], adapted to our parallel context.

3   Using Inexact Projected Gradient Directions

Suppose now that the feasible set C is defined by a system of convex inequalities, i.e., $c : \Re^n \to \Re^m$ in (2) has convex components. To compute the projected gradient directions which ensure convergence of the PVD algorithm (see [11]), one has to solve problems of the following structure:

$$\begin{cases} \min_{z \in \Re^n} \frac{1}{2} \|z - (x - \nabla f(x))\|^2 \\ c(z) \leq 0 . \end{cases} \tag{6}$$

Assuming some constraint qualification condition, it holds that $z = P_C[x - \nabla f(x)]$, if and only if, there exists some Lagrange multiplier $u \in \Re^m$ such that the pair z, u satisfies the KKT system

$$0 = \nabla_z L(z, u) = z - x + \nabla f(x) + Jc(z)^\top u, \\ u \geq 0, \ c(z) \leq 0, \ \langle u, c(z) \rangle = 0. \tag{7}$$

To solve (6), one has to apply an iterative algorithm. Suppose $z \in C$ and $u \in \Re_+^m$ are some current approximations to the primal and dual optimal solutions of (6), generated by this algorithm. The next lemma establishes an *error bound* for the distance from z to $P_C[x - \nabla f(x)]$ in terms of the violations of the KKT conditions (7) by the pair z, u. This result will be the key for devising our algorithm.

Lemma 2. *For any $z \in C$ and $u \in \Re^m_+$, it holds that*

$$\|z - P_C[x - \nabla f(x)]\| \leq \varepsilon(z, u),$$

where

$$\varepsilon(z, u) := \frac{1}{2}\|\nabla_z L(z, u)\| + \frac{1}{2}\sqrt{\|\nabla_z L(z, u)\|^2 - 4\langle u, c(z)\rangle}. \tag{8}$$

$\square$

Because of space limitations, we shall not include the proof here. A very nice feature of this estimate is that, unlike most error bound results in the literature [9], it does *not* involve any constants which may be unknown or any expressions which are not readily computable. Furthermore, this error bound holds *globally*, i.e., not just in some neighbourhood of the solution point.

We propose the following algorithm.

Algorithm 3 (PVD with inexact projected gradient directions)
*Choose $\sigma_1 \in (0, 1)$ and $\sigma_2 \in (0, (1 - \sigma_1)^2)$. Start with any $x^0 \in C$. Set $i := 0$.
Having x^i, check a stopping criterion. If it is not satisfied, proceed as follows:*
PVD-direction choice: *Compute $z^i \approx P_C[x^i - \nabla f(x^i)]$ such that $z^i \in C$ and the associated approximate Lagrange multiplier $u^i \in \Re^m_+$ satisfy*

$$\varepsilon(z^i, u^i) \leq \min\left\{\sigma_1, \frac{\sigma_2\|z^i - x^i\|}{\|\nabla f(x^i)\|}\right\} \|z^i - x^i\|, \tag{9}$$

where $\varepsilon(z^i, u^i)$ is given by (8). Set $d^i := x^i - z^i$. Compute x^{i+1} as follows:
Parallelization: *For each processor $l \leq p$ compute an approximate solution $(y^i_l, \mu^i_{\bar{l}}) \in \Re^{n_l} \times \Re^{p-1}$ of (P_l) as defined in Algorithm 1.*
Synchronization: *Compute x^{i+1} such that $f(x^{i+1}) \leq \min_{l \in \{1,\ldots,p\}} \psi^i_l(y^i_l, \mu^i_{\bar{l}})$.
Set $i := i + 1$; and repeat.*

$\square$

Regarding the tolerance criterion (9), we note that if z^i is the exact projection point then $\varepsilon(z^i, u^i) = 0$, while $z^i - x^i \neq 0$ (if $z^i = x^i$ then x^i satisfies the first-order necessary optimality condition $x^i = P_C[x^i - \nabla f(x^i)]$). From this considerations, it is easy to see that the projection problem (6) always has inexact solutions satisfying (9). Hence, the method is well-defined.

Suppose ∇f is Lipschitz continuous, with constant $L > 0$. Then the inexact approximate solution of the parallel subproblems can be further specified as the following : compute $(y^i_l, \mu^i_{\bar{l}}) \in \Re^{n_l \times (p-1)}$ such that

$$(y^i_l, x^i_{\bar{l}} + D^i_{\bar{l}}\mu_{\bar{l}}) \in C \quad \text{and} \quad \psi^i_l(y^i_l, \mu^i_{\bar{l}}) \leq \psi^i_l(x^i_l - \eta_i d^i_l, -\eta_i e_{\bar{l}}),$$

where $e_{\bar{l}}$ is the vector of ones of appropriate dimension, and $\eta_i \in (0, 2/L)$. We note that this sufficient descent condition is easily satisfiable by a single step of a variety of feasible descent methods applied to the given parallel subproblem, if we take $(x^i_l, 0)$ as a starting point when minimizing $\psi^i_l(\cdot, \cdot)$. This relaxation of the exact subproblem solution requirement, compared to the original Algorithm 1, is significant for several reasons. First of all, the global solution requirement is impractical if the objective function is not convex. And in any case, insisting on exact subproblem solution can be undesirable because it is likely to result in considerable idle times for processors that have already completed their work. It can be especially wasteful on the initial stages of minimization process. The sufficient descent criterion above shows that we are allowed a lot of flexibility in devising PVD algorithms. In particular, we can allow each of the p parallel processors to take as many steps as desired. Synchronization can be performed at any time provided every processor has achieved the sufficient descent condition. This certainly makes load balancing much easier.

As for the convergence properties of Algorithm 3, our result is the following.

Theorem 2. *Let f in (1) have a Lipschitz continuous gradient and let the feasible set C be convex. Then every accumulation point of the sequence $\{x^i\}$ generated by Algorithm 3 satisfies the first-order necessary optimality conditions.*   □

The proof uses Lemma 2 and the technique of [11, Theorem 2.1], albeit it is technically more involved, because the directions are computed only approximately.

References

[1] D.P. Bertsekas and J.N. Tsitsiklis. *Parallel and Distributed Computation.* Prentice–Hall, Inc, Englewood Cliffs, New Jersey, 1989.

[2] J.F. Bonnans, J.Ch. Gilbert, C. Lemaréchal, and C. Sagastizábal. *Optimisation Numérique: aspects théoriques et pratiques.* Springer Verlag, 1997.

[3] R. W. Cottle, J.-S. Pang, and R. E. Stone. *The Linear Complementarity Problem.* Academic Press, New York, 1992.

[4] M.C. Ferris and O.L. Mangasarian. Parallel variable distribution. *SIAM Journal on Optimization*, 4:815–832, 1994.

[5] M. Fukushima. Parallel variable transformation in unconstrained optimization. *SIAM Journal on Optimization*, pages 658–672, 1998.

[6] S.-P. Han. Optimization by updated conjugate subspaces. In D.F. Griffiths and G.A. Watson, editors, *Numerical Analysis*, number 140 in Pitman Research Notes in Mathematics, pages 82–97. Longman Scientific & Technical, Burnt Mill, England, 1986.

[7] S.-P. Han and O.L. Mangasarian. Exact penalty functions in nonlinear programming. *Mathematical Programming*, 17:251–269, 1979.

[8] O.L. Mangasarian. Parallel gradient distribution in unconstrained optimization. *SIAM Journal on Control and Optimization*, 33:1916–1925, 1995.

[9] J.-S. Pang. Error bounds in mathematical programming. *Mathematical Programming*, 79:299–332, 1997.

[10] M.V. Solodov. New inexact parallel variable distribution algorithms. *Computational Optimization and Applications*, 7:165–182, 1997.

[11] M.V. Solodov. On the convergence of constrained parallel variable distribution algorithms. *SIAM Journal on Optimization*, 8:187–196, 1998.

[12] P. Tseng. Dual coordinate ascent methods for non-strictly convex minimization. *Mathematical Programming*, 59:231–248, 1993.

Solving Stable Stein Equations on Distributed Memory Computers*

Peter Benner[1], Enrique S. Quintana-Ortí[2], and Gregorio Quintana-Ortí[2]

[1] Zentrum für Technomathematik, Fachbereich 3 – Mathematik und Informatik, Universität Bremen, D–28334 Bremen, Germany; benner@math.uni-bremen.de
[2] Depto. de Informática, Universidad Jaume I, 12080 Castellón, Spain; {quintana,gquintan}@inf.uji.es

Abstract. We investigate the parallel performance of numerical algorithms for solving stable Stein equations arising in discrete-time control problems. Our methods are based on the Smith iteration and the matrix sign function. We report experimental results of these algorithms on a cluster of Intel processors.

1 Introduction

We study the numerical solution of the *Stein equation*,

$$AXA^T - X + C = 0, \tag{1}$$

(also referred to as *discrete Lyapunov equation*), where $A, C \in \mathbb{R}^{n \times n}$, $C = C^T$, and $X \in \mathbb{R}^{n \times n}$ is the sought-after solution. It easily follows that if there exists a unique solution to (1), then this solution has to be symmetric.

Stein equations play a fundamental role in linear control and filtering theory for discrete-time systems (see references in [3]). Throughout this paper we assume (1) to be *Schur stable*, that is, if $\rho(A)$ denotes the spectral radius of A, then $\rho(A) < 1$. It is well known that under this assumption, the Stein equation (1) has a unique solution; see, e.g. [7]. Schur stable Stein equations appear in many computational problems for linear control systems [8].

The need for parallel computing in this area can be seen from the fact that already for a system with state-space dimension $n = 1000$, the corresponding Stein equations represent a set of linear equations with one million unknowns.

The standard methods for solving (1) are based on the Bartels-Stewart method, see [1]. In a first stage, the matrix A is reduced to real Schur form. Hence, in order to use these methods on parallel computers, it is necessary to have an efficient parallelization of the QR algorithm. However, several experimental studies report the difficulties in parallelizing the double implicit shifted QR algorithm on parallel distributed multiprocessors (see, e.g., [5, 6]). The parallelism and scalability of these algorithms are far from those of matrix multiplications, matrix factorizations, triangular linear systems solvers, etc.; see, e.g., [4] and the references given therein.

---

* Supported by the DAAD programme *Acciones Integradas Hispano-Alemanas*.

P. Amestoy et al. (Eds.): Euro-Par'99, LNCS 1685, pp. 1120–1123, 1999.
© Springer-Verlag Berlin Heidelberg 1999

Our methods are purely based on easy to parallelize computational kernels. These are the squared Smith iteration and the sign function method applied to the Lyapunov equation resulting from a Cayley transformation of (1). The Smith iteration will be reviewed in Section 2. For a description of the sign function method for the Lyapunov equation, see [3]. The algorithms considered here are implemented using the ScaLAPACK library [4] to ensure their portability. The computational performance and scalability of the implemented algorithms will be reported in Section 3. Some final remarks are given in Section 4.

2   The Smith Iteration

We can rewrite equation (1) as $X = AXA^T + C$, and form the fixed point iteration
$$X_0 := C, \qquad X_{k+1} = C + AX_kA^T, \quad k = 0, 1, 2, \ldots.$$
Then this iteration converges to X iff $\rho(A) < 1$, i.e., convergence is guaranteed under the given assumptions. The convergence rate of this iteration is linear. A quadratically convergent version of this fixed point iteration is suggested, among others, in [9]. Setting $X_0 := C$, $A_0 := A$, this iteration can be written as

$$X_{k+1} := X_k + A_kX_kA_k^T, \qquad A_{k+1} := A_k^2, \qquad k = 0, 1, 2, \ldots. \qquad (2)$$

The above iteration is referred to as the *squared Smith iteration*.

The most appealing feature of the squared Smith iteration regarding its parallel implementation is that all the computational cost comes from matrix products. These are known to be highly parallelizable; see, e.g., [4].

Remark 1. The convergence theory of the Smith iteration derived in [9] yields that for $\rho(A) < 1$ there exist real constants $0 < M$ and $0 < r < 1$ such that

$$\|X - X_k\|_2 \leq M\|C\|_2(1 - r)^{-1}r^{2^k}.$$

This shows that the method converges for all equations with a Schur stable coefficient matrix A. Nevertheless, if the coefficient matrix A is highly non-normal such that $\|A\|_2 > 1$, then overflow may occur in the early stages of the iteration due to increasing $\|A_k\|_2$ although eventually, $\lim_{k\to\infty} A_k = 0$. Hence we apply the squared Smith iteration only if $\|A\|_F < 1$ as this ensures a sequence of A_k with decreasing 2-norm.

3   Performance Results

In this section we evaluate the performance of several solvers for Stein equations of the form (1). A comparison of the numerical accuracy will be reported elsewhere. All the experiments were performed on a cluster of 16 nodes, connected with a *myrinet* cross-bar switch. Each node consists of an Intel Pentium-II processor at 300MHz and 128MBytes of RAM. The algorithms were coded in

Fortran 77, using IEEE double-precision arithmetic ($\varepsilon \approx 2.2 \times 10^{-16}$), a tuned BLAS library for Intel Pentium-II processors, and the LAPACK 2.0, BLACS 1.1, PBLAS 2.0α, and ScaLAPACK 1.6 libraries [4].

We compare the performance of the following Stein equation solvers: SB03PD from the Subroutine Library in Control Theory (SLICOT) [2] which is an implementation of the Bartels-Stewart method, our implementation of the squared Smith iteration denoted by DGEDLSM, and DGEDLSG which implements the Cayley transformation applied to (1), followed by solving the resulting Lyapunov equation with the sign function-based solver described in [3]. Following the naming convention in ScaLAPACK we use the prefix "P-" for the parallel versions of these routines.

The matrices in the experiments were generated as follows. A stable matrix A was generated by dividing the entries, uniformly distributed in $[0, 1]$, by the 1-norm of that matrix. The solution matrix was then generated to be symmetric and positive semidefinite as $X = G^T G$ with a random uniform matrix G. Finally, C was set to $C := X - AXA^T$. Although the execution time of the iterative solvers (routines DGEDLSM and DGEDLSG) depends on the number of iterations necessary for convergence, we always perform a fixed amount of 10 iterations. Our experiments showed that for these data sets, 8–10 iterations are enough for convergence.

In our first experiment, we evaluate the performance of the serial solvers SB03PD, DGEDLSM, and DGEDLSG. The left plot in Figure 1 reports the execution time of these solvers for Stein equations of size (n) varying from 100 to 1000. Here, we consider the execution time of SB03PD as the unit time and we report how much "faster" are the iterative solvers. The figure shows that routine DGEDLSM requires only 60–65% of the execution time of SB03PD. Routine DGEDLSG requires (except for the smaller problem sizes) around 90–95% of the execution time of SB03PD.

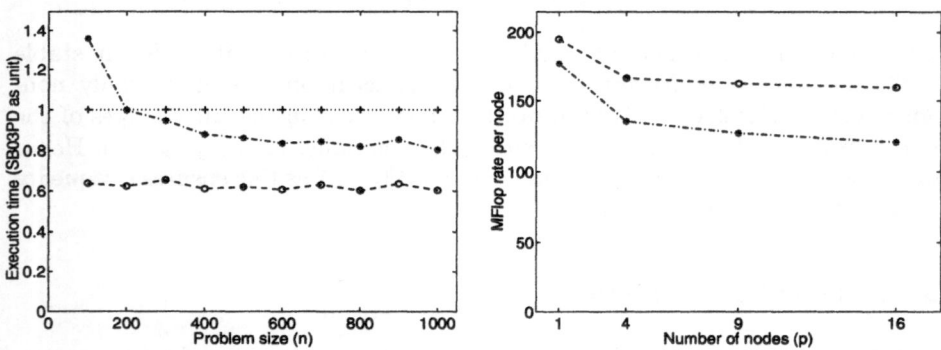

Fig. 1. Execution times of the serial Stein equation solvers (left plot) and performance of the parallel Stein equation solvers with $n/\sqrt{p} = 1000$. Legend: "$\cdots + \cdots$" = SB03PD, "$- \cdot - * - \cdot -$" = (P)DGEDLSG, and "$- - \circ - -$" = (P)DGEDLSM.

Our next experiment is designed to analyze the scalability and performance of the parallel solvers PDGEDLSM and PDGEDLSG. No parallel implementation of solver SB03PD is included as that requires a parallel kernel for solving Stein equations with A reduced to real Schur form which is not available in the current version of ScaLAPACK (version 1.6).

The right plot in Figure 1 reports the MFlop rate (millions of floating-point arithmetic operation per second) of the serial algorithm and the MFlop rate per node of the parallel implementations. We evaluate the parallel algorithm on $p=1$, 4, 9, and 16 nodes and we set n so that $n/\sqrt{p}$ is constant and equal to 1000. The figure reports a high scalability of the parallel routines as the performance remains almost constant as p is increased.

4   Concluding Remarks

We have compared the performance of iterative algorithms for solving Stein equations on a cluster of Intel processors. The Smith iteration basically consists of matrix products. The experimental results show the scalability and efficiency of this algorithm. The solver based on a Cayley transformation of the Stein equation and a sign function-based solution of the resulting Lyapunov equation requires only scalable parallel kernels. As these are readily available from ScaLA-PACK, the performance of this algorithm is only slightly worse than that of the Smith iteration.

References

[1] A. Y. Barraud. A numerical algorithm to solve $A^T X A - X = Q$. *IEEE Trans. Autom. Contr.*, AC-22:883–885, 1977.

[2] P. Benner, V. Mehrmann, V. Sima, S. Van Huffel, and A. Varga. SLICOT - a subroutine library in systems and control theory. *Applied and Computational Control, Signals, and Circuits*, 1:505–546, 1999.

[3] P. Benner and E.S. Quintana-Ortí. Solving stable generalized Lyapunov equations with the matrix sign function. *Numer. Algorithms*, 20(1):75–100, 1999.

[4] L.S. Blackford, J. Choi, A. Cleary, E. D'Azevedo, J. Demmel, I. Dhillon, J. Dongarra, S. Hammarling, G. Henry, A. Petitet, K. Stanley, D. Walker, and R.C. Whaley. *ScaLAPACK Users' Guide*. SIAM, Philadelphia, PA, 1997.

[5] G. Henry and R. van de Geijn. Parallelizing the QR algorithm for the unsymmetric algebraic eigenvalue problem: myths and reality. *SIAM J. Sci. Comput.*, 17:870–883, 1997.

[6] G. Henry, D.S. Watkins, and J.J. Dongarra. A parallel implementation of the nonsymmetric QR algorithm for distributed memory architectures. Technical Report LAPACK Working Note 121, University of Tennessee at Knoxville, 1997.

[7] P. Lancaster and M. Tismenetsky. *The Theory of Matrices*. Academic Press, Orlando, 2nd ed., 1985.

[8] V. Sima. *Algorithms for Linear-Quadratic Optimization*, volume 200 of *Pure and Applied Mathematics*. Marcel Dekker, Inc., New York, NY, 1996.

[9] R. Smith. Matrix equation $XA + BX = C$. *SIAM J. Appl. Math.*, 16(1):198–201, 1968.

Convergence Acceleration for the Euler Equations Using a Parallel Semi-Toeplitz Preconditioner

Andreas Kähäri and Samuel Sundberg

Information Technology, Department of Scientific Computing,
Uppsala University,
Box 120, SE-751 04 Uppsala, Sweden,
Andreas.Kahari@tdb.uu.se, Samuel.Sundberg@tdb.uu.se
http://www.tdb.uu.se/

Abstract. We have studied a preconditioning technique for Krylov subspace methods on a fluid dynamics problem in 2-D. By discretizing the time-dependent Euler equations with a finite volume method in space and using the trapezoidal rule in time, we get a nonlinear system which is solved using a Newton–Krylov method. We precondition the linear iterates using a parallel semi-Toeplitz preconditioner to reduce the number of iterations. The experiments show a substantial reduction in the number of iterations required for convergence.

1 Introduction

We have studied a semi-Toeplitz preconditioner that has been extended by Domain Decomposition for use in parallel computations [3]. The use of semi-Toeplitz preconditioners for Krylov subspace methods has been thoroughly investigated by Hemmingsson [1, 2, 4] and have been found very efficient. Therefore it is of interest to parallelize them in an efficient way in order to use them in large scale computations.

This paper is focused on the convergence behavior of a Newton–Krylov method for a 2-D flowproblem. We first show how the global preconditioner system is solved, then continue by sketching the application and the discretization, and finally we present results.

2 The Preconditioner

By using Domain Decomposition we obtain a partitioned preconditioner system where we distinguish between interior unknowns x^I, boundary unknowns $x^{B,k}$ and $x^{B,j}$, and corner unknowns x^C, see Fig. 1. With these definitions the global preconditioner system $Mx = y$ reads

P. Amestoy et al. (Eds.): Euro-Par'99, LNCS 1685, pp. 1124–1127, 1999.
© Springer-Verlag Berlin Heidelberg 1999

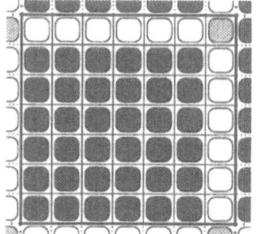

Fig. 1. The partitioning of the computational domain is shown in the picture to the left, and the right picture shows one of these blocks. Corner cells are grey, boundary cells are white and interior cells are black.

$$\begin{bmatrix} M^I & M^{IB,\bar{k}} & M^{IB,\bar{\jmath}} & 0 \\ M^{BI,\bar{k}} & M^{B,\bar{k}} & 0 & M^{BC,\bar{k}} \\ M^{BI,\bar{\jmath}} & 0 & M^{B,\bar{\jmath}} & M^{BC,\bar{\jmath}} \\ 0 & M^{CB,\bar{k}} & M^{CB,\bar{\jmath}} & M^C \end{bmatrix} \begin{bmatrix} x^I \\ x^{B,\bar{k}} \\ x^{B,\bar{\jmath}} \\ x^C \end{bmatrix} = \begin{bmatrix} y^I \\ y^{B,\bar{k}} \\ y^{B,\bar{\jmath}} \\ y^C \end{bmatrix}. \tag{2.1}$$

2.1 Solution of the Preconditioner System

We use the following solution strategy: (Step 1) solve for x^C, (Step 2) insert x^C in (2.1) and solve for $x^{B,\bar{k}}$ and $x^{B,\bar{\jmath}}$ and (Step 3) insert $x^{B,\bar{k}}$ and $x^{B,\bar{\jmath}}$ in (2.1) and solve for x^I.

Interior unknowns If $x^{B,\bar{k}}$ and $x^{B,\bar{\jmath}}$ are known, the solution for the interior unknowns decouples into the solution of as many independent systems of equations as there are subdomains in our computational domain. For a block ℓ we obtain the following system of equations

$$M^I_\ell x^I_\ell = y^I_\ell - M^{IB,\bar{k}}_\ell x^{B,\bar{k}}_\ell - M^{IB,\bar{\jmath}}_\ell x^{B,\bar{\jmath}}_\ell \tag{2.2}$$

Since M^I_ℓ is defined such that it has a semi-Toeplitz structure, (2.2) is solved by means of modified sine transforms [2].

Block boundary unknowns For the boundary unknowns we start by eliminating the corner unknowns in (2.1). The resulting system of equations decouples into two independent systems of equations by, first assuming that $M^{IB,\bar{\jmath}}$ is small and next assuming that $M^{IB,\bar{k}}$ is small, yielding the corresponding system for $x^{B,\bar{k}}$ and $x^{B,\bar{\jmath}}$. By block Gaussian elimination we get

$$C^{B,\bullet} x^{B,\bullet} = y^{B,\bullet} - M^{BC,\bullet} x^C - M^{BI,\bullet} \left(M^I \right)^{-1} y^I, \tag{2.3}$$

where $C^{B,\bullet} \equiv M^{B,\bullet} - M^{BI,\bullet} \left(M^I \right)^{-1} M^{IB,\bullet}$.

The systems defined by (2.3), the so-called block boundary Schur complement systems, decouple into as many independent systems of equations as there are rows and columns of subdomains respectively. These are solved using modified sine transforms and the solution of narrow-banded systems.

Corner unknowns Finally, we study the solution for the corner unknowns. The system for these unknowns is difficult to solve, and we have to make several approximations to obtain a system of equations that is easy to solve.

We start by making the approximation $M^{BI,\bullet} \approx 0$, which gives a system of equations that is decoupled from the interior unknowns. Using block Gaussian elimination we obtain

$$C^C x^C = y^C - M^{CB,\bar{k}} \left(M^{B,\bar{k}} \right)^{-1} y^{B,\bar{k}} - M^{CB,\bar{\jmath}} \left(M^{B,\bar{\jmath}} \right)^{-1} y^{B,\bar{\jmath}},$$

where $C^C \equiv M^C - M^{CB,\bar{k}} \left(M^{B,\bar{k}} \right)^{-1} M^{BC,\bar{k}} - M^{CB,\bar{\jmath}} \left(M^{B,\bar{\jmath}} \right)^{-1} M^{BC,\bar{\jmath}}$. We solve this system using a direct method.

3   Model Problem

We have studied the Euler equations on a backwards-facing step, where we use Crank–Nicholson for the time discretization. Spatial discretization is done by means of a finite volume method on a uniform grid. We let (k, j) denote the center of a cell. In each cell (k, j) with area $S_{k,j}$ our method is defined by

$$S_{k,j} \frac{q^{n+1} - q^n}{\Delta t} +$$
$$\sum_{\ell=1}^{4} \left[\theta \left(f n_x + g n_y \right)^{n+1} + (1 - \theta) \left(f n_x + g n_y \right)^n \right] \Delta S_\ell = 0, \quad (3.1)$$

where the ΔS_ℓ's are the walls of the cell. The fluxes are computed using third order upwind approximations.

3.1   Newton–Krylov Method

The discretization in (3.1) above gives us a nonlinear system, $F(q^{n+1}) = 0$, to solve in each time-step. We utilize Newton's method to solve this system, thus solving

$$\nabla F(q^{n+1,\mu}) \Delta q^{n+1,\mu} = -F(q^{n+1,\mu}), \quad (3.2)$$

and then adding

$$q^{n+1,\mu+1} = q^{n+1,\mu} + \Delta q^{n+1,\mu}. \quad (3.3)$$

We choose $q^{n+1,0} = 2q^n - q^{n-1}$ as initial guess, and let $q^{n+1} = q^{n+1,\mu+1}$ at convergence.

The linear system in 3.2 is solved using GMRES(6) where we use left preconditioning,

$$M^{-1} \nabla F \Delta q = -M^{-1} F, \quad (3.4)$$

to achieve faster convergence [5]. Here M is the semi-Toeplitz preconditioner described above.

4   Results

To study the performance of the preconditioner a series of experiments was run on a SMP cluster with three Digital Alpha servers, each hosting four Alpha EV5 processors. By taking 20 time-steps for different grid sizes, we found that the

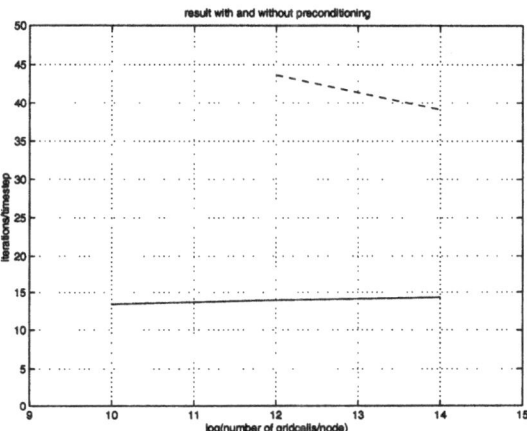

Fig. 2. Number of iterations/time-step with preconditioned (solid) and unpreconditioned (dashed) GMRES(6).

number of iterations per time-step was substantially reduced for the preconditioned method compared to the unpreconditioned method. Furthermore we found that the number of iterations was constant as the grid was refined, as shown in Fig. 2.

References

[1] L. HEMMINGSSON, *A fast modified sine transform for solving block-tridiagonal systems with Toeplitz blocks*, Numer. Algorithms, 7 (1994), pp. 375–389.

[2] ———, *Toeplitz preconditioners with block structure for first-order PDEs*, Numer. Linear Algebra Appl., 3 (1996), pp. 21–44.

[3] L. HEMMINGSSON AND A. KÄHÄRI, *A new parallel preconditioner for the Euler equations*, in Applied Parallel Computing. Large Scale Scientific and Industrial Problems, B. Kågström, J. Dongarra, E. Elmroth, and J. Waśniewski, eds., vol. 1541 of Lecture Notes in Comput. Sci., Springer-Verlag, Berlin, 1998, pp. 230–238.

[4] L. HEMMINGSSON AND K. OTTO, *Analysis of semi-Toeplitz preconditioners for first-order PDEs*, SIAM J. Sci. Comput., 17 (1996), pp. 47–64.

[5] Y. SAAD AND M. H. SCHULTZ, *GMRES: A generalized minimal residual algorithm for solving nonsymmetric linear systems*, SIAM J. Sci. Statist. Comput., 7 (1986), pp. 856–869.

A Stable and Efficient Parallel Block Gram-Schmidt Algorithm

Denis Vanderstraeten

Katholieke Universiteit Leuven, Computer Science Dept.,
Celestijnenlaan, 200A, B-3001 Heverlee (Belgium)
Denis.Vanderstraeten@@cs.kuleuven.ac.be

Abstract. The Modified Gram-Schmidt (MGS) orthogonalization process — used for example in the Arnoldi algorithm — constitutes often the bottleneck that limits parallel efficiencies. Indeed, a number of communications, proportional to the square of the problem size, is required to compute the dot-products. A block formulation is attractive but it suffers from potential numerical instability.

In this paper, we address this issue and propose a simple procedure that allows the use of a Block Gram-Schmidt algorithm while guaranteeing a numerical accuracy similar to MGS. The main idea is to dynamically determine the size of the blocks. The main advantage of this dynamic procedure are two-folds: first, high performance matrix-vector multiplications can be used to decrease the execution time. Next, in a parallel environment, the number of communications is reduced. Performance comparisons with the alternative Iterated CGS also show an improvement for moderate number of processors.

1 Introduction

The numerical simulation of physical problems often requires the construction of an orthogonal basis of a matrix or a set of vectors. Among the existing techniques, the Gram-Schmidt algorithms have proven to be successful candidates. In particular, the Modified Gram-Schmidt algorithm is often used in GMRES to built an orthogonal basis of the Krylov subspace. Unfortunately, in a parallel environment, the number of communications grows like the square of the number of vectors.

Various attempts have been proposed to overcome this problem. The Classical Gram-Schmidt algorithm or a block formulation of MGS provide a substantial reduction of the number of communications but a reorthogonalization procedure must be introduced to ensure the numerical accuracy [5, 8, 9]. In the particular context of Krylov basis, much attention has also been devoted and we refer the reader to [10, 11, 12] for further readings.

Our work departs from other works by the assumptions that are made: (a) we do not assume that GMRES (or any Krylov method) is our target application, and (b) we wish to avoid any reorthogonalization step to limit the additional cost compared to MGS.

P. Amestoy et al. (Eds.): Euro-Par'99, LNCS 1685, pp. 1128–1135, 1999.
© Springer-Verlag Berlin Heidelberg 1999

2 Gram-Schmidt Algorithms

The Modified Gram-Schmidt algorithm produces a QR factorization of a matrix $A = [a_1 \ldots a_n] \in \mathbb{R}^{m \times n}$ with $m \geq n$ [3, 4]. The pseudo-code for MGS is given in Alg. 1 where q_i represents the ith column of Q.

for $i = 1$ to n	for $k = 1$ to p
1. $q_i = a_i$	1. $W_k = A_k$
for $j = 1$ to $i - 1$	for $j = 1$ to $k - 1$
2. $\delta = q_j^T q_i$	2. $\Delta = Q_j^T W_k$
3. $q_i = q_i - q_j \delta$	3. $W_k = W_k - Q_j \Delta$
end	end
4. $q_i = q_i / \|q_i\|_2$	4. $Q_k = MGS(W_k)$
end	end

Alg. 1. Modified Gram-Schmidt   **Alg. 2.** Block Gram-Schmidt

On a parallel computer where the components of a_i and q_i are distributed over the processors, a synchronization, followed by the global accumulation of the local dot-products must be performed at step 2 of the algorithm. The application of MGS thus requires $n^2/2$ communications.

To reduce the number of communications, one can use the Classical Gram-Schmidt algorithm but its numerical accuracy is not guaranteed. To regain the accuracy, the Iterated Classical Gram-Schmidt algorithm (ICGS) performs two steps of Classical Gram-Schmidt if a loss of orthogonality between the vectors is detected [8]. The number of communications of ICGS is in $\mathcal{O}(n)$ but the computational work may be doubled compared to MGS.

Another approach towards the reduction of communication cost is to focus on a block formulation [9]. To this end, the matrix $A = [A_1 \ldots A_p]$ is partitioned in p blocks $A_k \in \mathbb{R}^{m \times s}$ of s consecutive columns (we assume for simplicity that n is a multiple of s). We partition Q accordingly. The Block Gram-Schmidt algorithm (BGS) is given in Alg. 2. Although BGS achieves better parallelism than MGS, it is not as stable. Indeed, Jalby and Philippe [9] show that the matrix $\hat{Q}$, computed by BGS, satisfies

$$\|\hat{Q}^T \hat{Q} - I\|_2 \leq \rho u \max_{k=1 \ldots p} \kappa_2(W_k) \, \kappa_2(A), \tag{1}$$

where W_k is the matrix used in step 4 of Alg. 2 and $\kappa_2(\cdot)$ denotes the condition number. Hence, if any block W_k has a condition number close to $\kappa_2(A)$, the bound (1) is of the order $\kappa_2^2(A)$ and BGS may produce an inaccurate result. A simple remedy to this problem consist in applying MGS (or CGS) twice to every block W_k. This is referred as B2GS in [9].

In the next section, we present a simple modification of BGS to ensure numerical stability still benefiting from the desirable parallel properties of BGS but without introducing a reorthogonalization step.

3   Dynamic Block Gram-Schmidt (DGS)

We have implicitly assumed that all blocks have the same size. This is not a requirement and allowing different block sizes gives room for improvement of BGS. Indeed, if we are able to find a decomposition of A such that every block W_k satisfies

$$\kappa_2(W_k) \leq \tau \qquad\qquad k = 1 \ldots p, \qquad\qquad (2)$$

where the parameter τ is a small threshold, then, we will have an algorithm with essentially the same bound as MGS.

We proceed iteratively to determine the block sizes. Assume that the size of the first $p - 1$ blocks have been determined and that the current block $W_p \in \mathbb{R}^{m \times (s-1)}$ satisfies (2). The next vector w_s is first orthogonalized with respect to Q_k, ($k = 1 \ldots p - 1$). Then, if the extended matrix $W_p^* = [W_p \ w_s]$ still satisfies (2), we can safely add w_s in the current block. Otherwise, a new block W_{p+1} is created with w_s as unique column.

Usually, the block Q_p is created along with the iterations and we should avoid keeping a copy of W_p for memory reasons. However, step 4 of Alg. 2 produces a factorization $W_p = Q_p R_p$. Since,

$$\kappa_2(W_p) = \kappa_2(R_p), \qquad\qquad (3)$$

if suffices then to keep the matrix R_p. Strictly speaking, in finite precision arithmetic, (3) is not exact but $\kappa_2(R_p)$ provides a sufficient approximation.

Computing the condition number is an expensive operation, even for small matrices. Instead of computing it exactly, we use the algorithm proposed by Bischof [2] that estimates the condition number of a triangular matrix. The cost of this algorithm is in $\mathcal{O}(s^2)$ for a matrix of order s. The main idea is to exploit the relation

$$\|R^{-1}\|_2 = \max_{R^T x = d, d \neq 0} \frac{\|d\|_2}{\|x\|_2}, \qquad\qquad (4)$$

together with the fact that R is constructed column by column. Assume that we have evaluated a vector $x \in \mathbb{R}^{s-1}$ with small (preferably the smallest) norm and a unit-norm vector d such that $R^T x = d$. The orthogonalization of w_s leads to the creation of an extended matrix

$$R^* = \begin{pmatrix} R & r \\ 0 & \rho \end{pmatrix}. \qquad\qquad (5)$$

We need to determine x^* and d^* solution of $R^{*T} x^* = d^*$ and such that the norm of x^* is small. If we choose $d^* = [cd, s]^T$ with $c = \cos\theta$ and $s = \sin\theta$, then the solution x^* is easily expressed in terms of x by

$$x^* = \begin{pmatrix} cx \\ (s - cr^T x)/\rho \end{pmatrix}. \qquad\qquad (6)$$

The minimization of $\|x^*\|_2$ reduces to the one-dimensional problem of finding the optimal θ. It requires only a few dot-products and the computational cost is in $\mathcal{O}(s)$.

The pseudo-code for the Dynamic Block Gram-Schmidt algorithm is given in Alg. 3. The parameter s_{max} is introduced to limit the sizes of the blocks. Note also that the function MGS needs only to orthogonalize w_s with respect to Q_p and returns $Q_p^* = [Q_p \ q_s]$.

$p = 1, \ s = 1, \ Q_1 = [\ ]$
for $i = 1$ to n
    $w_s = a_i$
    for $k = 1$ to $p - 1$ *Orthogonalize wrt. the previous blocks*
        $d \ = Q_k^T w_s$
        $w_s = w_s - Q_k d$
    end
    $[Q_p^*, \ R_p] = MGS([Q_p \ w_s])$ *Update the MGS of the current block*
    if $(\kappa_2(R_p) > \tau$ or $s \geq s_{max})$ then
        $Q_{p+1} = [q_s]$ *Create a new block*
        $s \ \ \ \ = 2$
        $p \ \ \ \ = p + 1$
    else
        $Q_p = Q_p^*$ *Increase the current block size*
        $s \ \ = s + 1$
    end
end

Alg. 3. Dynamic Block Gram-Schmidt

4 Complexity Analysis

On a parallel computer, every vector v is decomposed into P sub-vectors $v^{(k)}$ of size m/P that are distributed to the P processors. A parallel dot-product $v^T w$ requires the accumulation of the P local dot-products $v^{(k)^T} w^{(k)}$. As usual, a point-to-point communication of size s between two processors is modelled by

$$T_{sr}(s) = \alpha + \beta s.$$

where α is the latency and β^{-1} denotes the bandwidth. The communication graph for the accumulation of the local dot-products is represented by a binary tree resulting in a total accumulation time, for s concurrent dot-products, of

$$T_{acc}(s) = 2 \log_2 P(\alpha + \beta s).$$

Let s denote the average block size of DGS. A count of the communications gives a total communication cost of

$$T^{(comm)} = \frac{n}{2} \log_2 P \left((s + \frac{n}{s})\alpha + (n+1)\beta \right). \tag{7}$$

The algorithm DGS behaves like MGS but with a reduced startup time. If $s = 1$ or $s = n$, we obtain the communication cost of MGS. In addition, it is easy to show that the communication cost is minimized for $s = \sqrt{n}$.

Using DGS also gives a gain in terms of computational cost. Indeed, matrix-vector products with a high flop rate can be used instead of the vector-vector operations of MGS. If we model the cost of evaluating the condition number by δs^2, the computational cost of DGS writes

$$T^{(comp)} = (\gamma_1(s+1) + \gamma_s(n-s)) \, mn + \delta ns, \qquad (8)$$

where γ_s^{-1} denotes the flop rate to perform a matrix-vector product with a matrix of size $m \times s$, and γ_1 is the cost of a dot-product (this corresponds to the BLAS2 dgemv and the BLAS1 ddot operations). The first term corresponds to the orthogonalization and the second represents the overhead due to the evaluation of the condition number. It can be neglected.

The above analysis highlights the two potential improvements of DGS compared to MGS: a higher flop rate associated with the reduction of the number of communications. On a specific platform, an estimate of the parameters α, β and γ_s enables a detailed analysis of the execution time and the determination of the optimal block size. Remember however, that the size of the blocks is primarily set by the problem itself and stiff problems may require blocks of size 1.

5 Numerical Experiments

5.1 Loss of Orthogonality

Table 4 presents accuracy results obtained with MGS, BGS and DGS. The parameters were set as $\tau = 10$ and $s_{max} = 8$. For comparison purposes, we also show results produced by B2GS and ICGS using reorthogonalization.

Three matrices are considered: (a) a random matrix of size 1024×512 with entries uniformly distributed in $[-1, 1]$, (b) a Lauchli matrix of size 65×64 with $\epsilon = 10^{-4}$, and (c) a Hilbert matrix of size 20×10. The orthogonality is measured by $\|Q^T Q - I\|_2$ and compared with a "reference" orthogonal matrix Q_{ref} obtained by applying MGS twice.

	Random		Lauchli		Hilbert	
	$\|Q^T Q - I\|_2$	$\|Q - Q_{ref}\|_2$	$\|Q^T Q - I\|_2$	$\|Q - Q_{ref}\|_2$	$\|Q^T Q - I\|_2$	$\|Q - Q_{ref}\|_2$
MGS	8.7e-15	6.8e-15	3.8e-13	3.8e-13	2.4e-6	2.4e-6
BGS	8.9e-15	7.0e-15	6.5e-9	6.5e-9	6.0e-4	6.0e-4
DGS	1.1e-14	7.8e-15	3.8e-13	3.8e-13	3.5e-6	3.9e-6
B2GS	9.6e-15	6.6e-15	3.8e-13	3.8e-13	2.4e-6	3.3e-6
ICGS	1.6e-14	8.2e-15	2.9e-16	3.8e-16	1.3e-14	3.7e-6

Table 4. Loss of orthogonality of Q computed by the various Gram-Schmidt algorithms

As expected, B2GS gives results of the same order as MGS while ICGS produces an output that is within the machine accuracy.

The Random matrix is fairly well conditioned and all three algorithms produce accurate results. This is no longer the case for the ill-conditioned problems where BGS fails to construct an orthonormal basis. For the Lauchli matrix, the average block size is 7.11. Indeed, all the vectors are nearly aligned with the first vector, but their projections, parallel to this vector, turn out to be (almost) orthogonal. As a consequence, DGS begins by creating a block of size 1 while the subsequent blocks have size s_{max}. With the Hilbert matrix, with the exception of the first block of size 2, all blocks of DGS have size 1.

5.2 Parallel Performance

Parallel performance were measured on an IBM/SP2 with 8 processors using the vendor version of MPI. Results are presented for three different matrices of size 100 000 × 60, 100 000 × 120, and 200 000 × 60. These sizes are typical for the orthogonalization of Krylov bases arising in the solution of large linear systems with GMRES.

	s_{max}	100000 × 60				100000 × 120				200000 × 60			
		$P=1$	2	4	8	$P=1$	2	4	8	$P=1$	2	4	8
MGS		1.00	0.91	0.74	0.44	1.00	0.92	0.73	0.46	1.00	0.92	0.83	0.51
DGS	1	1.00	0.91	0.73	0.44	0.99	0.91	0.74	0.46	1.00	0.94	0.83	0.59
DGS	2	1.11	1.02	0.85	0.64	1.10	1.03	0.61	0.60	1.12	1.06	0.85	0.66
DGS	4	1.70	1.53	1.31	0.80	1.77	1.59	1.37	1.00	1.72	1.58	1.43	1.12
DGS	8	1.94	1.75	1.49	1.08	2.15	1.88	1.67	1.15	1.97	1.77	1.58	1.14
DGS	16	1.85	1.63	1.33	0.97	2.22	1.91	1.49	1.14	1.87	1.69	1.30	1.15
B2GS		1.91	1.70	1.46	1.05	2.11	1.86	1.69	1.18	1.90	1.72	1.48	1.18
CGS		2.86	2.68	2.28	1.72	3.10	2.88	2.44	2.08	2.92	2.70	2.36	1.90

Table 5. Parallel efficiencies compared to the sequential MGS algorithm

Table 5 presents the efficiencies computed using the sequential time obtained with MGS. Efficiencies above 100% result from the reduction in communication time *combined* with the gain due to the more efficient BLAS2 operations.

Random matrices are usually well conditioned. No *single* reorthogonalization step is required for ICGS and this algorithm become thus identical to CGS. Depending on the matrices, the efficiency of ICGS can drop to half of the efficiency of CGS.

5.3 Application: GMRES

The dynamic block Gram-Schmidt procedure has been tested for the solution of an advection-diffusion problem defined on a square, discretized by a finite-element method. The resulting linear system has 160 000 unknowns. It is solved iteratively with GMRES, preconditioned by a perfectly parallel block-ILU(0). The restart parameter is set to 30 vectors and convergence is attained when the *true* residual satisfies $\|b - Ax\|_2 \leq 10^{-4}\|b\|_2$.

Table 6 presents performance results obtained on an a 8-processors IBM/SP2 parallel computer, using the PETSc framework [1]. Both the total solution time of the linear system and the orthogonalization time are measured. The parallel efficiency is also measured.

	$P = 1$	$P = 2$		$P = 4$		$P = 8$	
	$T^{(tot)}$	$T^{(tot)}$	Eff.	$T^{(tot)}$	Eff.	$T^{(tot)}$	Eff.
Matvec	65	37	0.87	19	0.85	10	0.81
Precond	62	31	1.00	15	1.03	8.5	0.91
MGS	83	47	0.88	30	0.69	22	0.47
DGS	57	32	0.89	20	0.71	14	0.50
ICGS	95	50	0.95	28	0.84	20	0.59
Total (MGS)	288	153	0.94	92	0.78	63	0.57
Total (DGS)	263	140	0.93	83	0.79	55	0.60
Total (ICGS)	299	159	0.94	91	0.82	60	0.62

Table 6. Performance of the Dynamic Gram-Schmidt algorithm with GMRES

For 8 processors, DGS is 1.5 times faster than MGS, resulting in a reduction of 17% of the total computation time for the solution of the linear system. Of course, the global improvement depends on the relative cost of the orthogonalization compared to the other operations.

For this problem, about 80% of the Krylov vectors are reorthogonalized with ICGS. Following [8], modifying some parameters would decrease this percentage and produce a more competitive algorithm.

With GMRES, the value of τ has little influence as long as the matrix is reasonably well conditioned. Indeed, even for moderately large τ, the same number of iterations was needed to obtain convergence. First, restarting the iterations gives a smoothing effect on the loss of orthogonality. Next the linear iterations reach convergence to the desired value long before the effect of τ becomes apparent on the residual norm. This observation is in accordance with the results in [6, 7]. Indeed, these authors note that the important feature is the linear independence of the Krylov vectors and not the orthogonality. For this matrix, with the restart parameter and termination criterion, all three algorithms converge in the same number of iterations.

6   Concluding Remarks

We have presented an improved version of the Block Gram-Schmidt algorithm that enables its use on parallel computers while guaranteeing a numerical stability. The main idea is to dynamically determine the size of the blocks during the computation according to their condition number. An efficient algorithm for estimating the condition number of small matrices makes the overall procedure very attractive.

In sequential mode, DGS already outperforms the widely used Modified Gram-Schmidt algorithm. Furthermore, in a parallel environment, an additional gain is achieved by the reduction in the number of communications.

Acknowledgements

This research is partially funded by the UIAP P4/02 Interuniversity Poles of Attraction, initiated by the Belgian State, Prime Minister's Office for Science Technology and Culture. The author is a post-doctoral researcher funded by the *Fonds voor Wetenschappelijk Onderzoek*, Flanders, Belgium. The scientific responsibility rests with its author.

References

[1] S. Balay, W. D. Gropp, L. Curfman McInnes, and B. F. Smith. PETSc home page. http://www.mcs.anl.gov/petsc, 1998.

[2] C. H. Bischof. Incremental Condition Estimation. *SIAM J. Mat. Anal. Appl.*, No. 2, pp. 312–322, 1990.

[3] Å. Björck. Numerics of Gram-Schmidt Orthogonalization. *Linear Alg. Appl.*, pp. 297–316, 1994.

[4] Å. Björck and C. C. Paige. Loss and Recapture of Orthogonality in the Modified Gram-Schmidt Algorithm. *SIAM J. Mat. Anal. Appl.*, No. 1, pp. 176–190, 1992.

[5] J. W. Daniel, W. B. Gragg, L. Kaufman, and G. W. Steward. Reorthogonalization and Stable Algorithms for Updating the Gram-Schmidt QR Factorization. *Mathematics of Computations*, Vol. 30, No. 136, pp. 772–795, 1976.

[6] J. Drkošová, M. Rozložník, Z. Strakoš, and A. Greenbaum. Numerical Stability of GMRES. *BIT*, Vol. 35, pp. 309–330, 1995.

[7] A. Greenbaum, M. Rozložník, and Z. Strakoš. Numerical Behaviour of the Modified Gram-Schmidt GMRES Implementation. *BIT*, Vol. 37, No. 3, pp. 706–719, 1997.

[8] W. Hoffmann. Iterative Algorithms for Gram-Schmidt Orthogonalization. *Computing*, Vol. 41, pp. 335–348, 1989.

[9] W. Jalby and B. Philippe. Stability Analysis and Improvement of the Block Gram-Schmidt Algorithm. *SIAM J. Sci. Stat. Comput.*, No. 5, pp. 1058–1073, 1991.

[10] R. B. Sidje. Alternatives for Parallel Krylov Subspace Basis Computations. *Num. Lin. Alg. with Appl.*, Vol. 4, No. 4, pp. 305–331, 1997.

[11] E. De Sturler and H. A. van der Vorst. Reducing the Effect of Global Communication in GMRES(m) and CG on Parallel Distributed Memory Computers. *Applied Numerical Mathematics*, Vol. 18, pp. 441–459, 1995.

[12] H. F. Walker. Implementation of the GMRES Method Using Householder Transformation. *SIAM J. Sci. Stat. Comput.*, Vol. 9, No. 1, pp. 152–163, 1988.

On the Extension of the Code GAM for Parallel Computing[*]

Felice Iavernaro and Francesca Mazzia

Dipartimento di Matematica, Università di Bari.
Via Orabona, 4. I-70125 Bari (Italy)
mazzia@dm.uniba.it

Abstract. The code GAM numerically solves initial value ordinary differential equations by means of a family of variable-step variable-order block Boundary Value Methods. Here we consider the possibility of performing the code on parallel machines. Some numerical tests and comparisons are presented.

Key words. linear multistep formulas, Runge-Kutta methods, stiff initial value problems
AMS(MOS) subject classifications. 65L05, 65L20

1 Introduction

The code GAM [8] implements the Generalized Adams Methods (GAMs) of orders 3,5,7 and 9 [3, 6, 7] to solve initial value problems of the form

$$\begin{cases} \mathbf{y}'(t) = \mathbf{f}(t,\mathbf{y}), & \mathbf{f} : [a, a+T] \times H \to H, \\ \mathbf{y}(a) = \mathbf{y}_0, \end{cases} \tag{1}$$

where $H = \mathbb{R}^r$. Its effectiveness has been established on the basis of its good behaviour on a wide variety of test problems as compared to the performance of some well-known codes such as RADAU5 and MBDFDAE (see [7]). At the moment there exist two different versions of the code, both conceived to run on sequential machines. In this context our interest is in deriving some techniques that may allow an efficient implementation of the code on a parallel computer. It is our experience that most of the time (about 85 %) spent to advance the solution is devoted to solving the nonlinear systems underlying the integration procedure: the whole performance of the code will substantially benefit from an efficient parallelization of this part and our efforts will be concentrated to face this problem. In particular the result presented here will concern one of the above mentioned versions of the code as specified in the next section. The organization of the paper is as follows. In the next section the GAMs are briefly introduced

---

[*] Work supported by MURST (40% project).

P. Amestoy et al. (Eds.): Euro-Par'99, LNCS 1685, pp. 1136–1143, 1999.
© Springer-Verlag Berlin Heidelberg 1999

together with an outline on how these implicit formulae have been solved, with a closer inspection of those steps that will be subject to parallelization, as described in section 3. Here a possible approach is reported and analyzed for achieving parallelization at different levels: this matter forms the background for a future gradual development of the parallel code. In section 4 we present some numerical results related to the very first experiments concerning the parallel version of GAM. In the sequel, when needed, an $N_1 \times N_2$ matrix T, will be viewed as a linear operator $H^{N_2} \longrightarrow H^{N_1}$, that is for example, given a block vector $Y \in H^{N_2}$, AY will stand for $(A \otimes I_r)Y \in H^{N_1}$ with I_r the r-dimensional identity matrix.

2 Inside the Code GAM

Hereafter we give a short account of those sections of the code that, via a suitable modification, will be performed in parallel by a number of processes. Details about definitions, properties, implementation techniques and the overall functionality of the code may be found in [3, 6, 7]. During the integration process the continuous problem (1) is solved over adjacent time intervals

$$W_s = [t^{(s-1)}, t^{(s)}], \qquad s = 1, \ldots, M, \ t^{(0)} = a, \ t^{(M)} = a + T.$$

In each W_s the solution of (1), say $\hat{y}(t)$, is approximated by a vector $\widetilde{Y}^{(s)} = [y_0^{(s)}, \ldots, y_{N_s}^{(s)}]^T \in H^{N_s+1}$, that is $y_i^{(s)} \simeq \hat{y}(t_i^{(s)})$, where $\{t_i^{(s)}\}_{i=0}^{N_s}$ is a uniform mesh over W_s such that $t_0^{(s)} = t^{(s-1)}$ and $t_{N_s}^{(s)} = t^{(s)}$; the positive number $h_s = t_i^{(s)} - t_{i-1}^{(s)}$ is the stepsize of integration. It must be observed that since $W_{s-1} \cap W_s = \{t^{(s-1)}\}$, we also have $y_0^{(s)} = y_{N_{s-1}}^{(s-1)}$ that is the first component of the block vector $\widetilde{Y}^{(s)}$ should not be treated as an unknown because it takes information from the preceding step. The other N_s components of $\widetilde{Y}^{(s)}$ are the solution of the following algebraic system of dimension N_s:

$$\widetilde{A}_s \widetilde{Y}^{(s)} - h_s \widetilde{B}_s \widetilde{F}(\widetilde{Y}^{(s)}) = 0, \qquad s = 1, \ldots, M, \tag{2}$$

where $\widetilde{A}_s = \{\alpha_{ij}\}$ and $\widetilde{B}_s = \{\beta_{ij}\}$, $i = 1, \ldots, N_s$, $j = 0, \ldots, N_s$, are $(N_s+1) \times N_s$ real matrices and $\widetilde{F}(\widetilde{Y}^{(s)}) = [f(t_0^{(s)}, y_0^{(s)}), \ldots, f(t_{N_s}^{(s)}, y_{N_s}^{(s)})]^T$. Formula (2) may be viewed as a set of linear combinations of $y_i^{(s)}$ and $f(t_i^{(s)}, y_i^{(s)})$; it defines a block-GAM of odd order p and dimension N_s if the following conditions are fulfilled (to simplify the notation we omit in the sequel the superscripts (s)):

(i) each component of (2) assumes the form

$$y_i - y_{i-1} = h \sum_{j=-k_1^{(i)}}^{k_2^{(i)}} \beta_{ij} f_{i+j}, \qquad i = 1, \ldots, N_s, \tag{3}$$

with $k_1^{(i)}$ and $k_2^{(i)}$ nonnegative integers such that $k_1^{(i)} + k_2^{(i)} = p - 1$ and

$$k_1^{(i)} = \begin{cases} i & \text{for } i = 1, \ldots, (p-3)/2, \\ (p-1)/2 & \text{for } i = (p-1)/2, \ldots, N - (p-1)/2 \,, \\ i - N + p - 1 & \text{for } i = N - (p-3)/2, \ldots, N; \end{cases}$$

(ii) assuming that $y_0 = \hat{y}(t_0)$, then for each $i = 1, \ldots, N$, $y_i = \hat{y}(t_i) + O(h^{p+1})$ that is the coefficients β_{ij} are (uniquely) determined in order to provide a $(p+1) - st$ order approximation to the true solution at each time t_i.

From the $(p-1)$-step linear multistep formulae (3) it is deduced that $\widetilde{A}$ is bidiagonal and Toeplitz with $\alpha_{ii} = 1$ and $\alpha_{i+1,i} = -1$ as diagonal and lower diagonal entries. We remark that the dimensions N_s of each system (2) could be in principle arbitrarily large so as to cover (under the same h_s) wider or smaller intervals W_s. All the same, inside the code, once the order has been selected, the dimension of the corresponding formula remains fixed and is equal to 4, 6, 8, 10 for the orders 3, 5, 7 and 9 respectively. These dimensions allow the estimation of the error and the order changing routines to operate (for an explicit list of the coefficients β_{ij} see [6]).

Performing the partitions

$$\widetilde{Y} = [y_0, Y^T]^T, \qquad \widetilde{F}(\widetilde{Y}) = [f(t_0, y_0), F(Y)^T]^T,$$
$$\widetilde{A} = [a_0, A], \qquad \widetilde{B} = [b_0, B],$$

with $Y = [y_1, \ldots, y_N]^T$, $F(Y) = [f(t_1, y_1), \ldots, f(t_N, y_N)]^T$ and a_0, b_0 the first column of $\widetilde{A}$ and $\widetilde{B}$ respectively, we can move in (2) all the known terms to the right hand side, thus obtaining

$$AY - hBF(Y) = \mathbf{b}, \tag{4}$$

with $\mathbf{b} = -a_0 y_0 + h b_0 f(t_0, y_0)$ (a_0 and b_0 are used as linear operators $H \to H^N$). We now consider one of the two approaches adopted to solve equation (4).

3   Simplified Newton Iteration

The system (4) may be recast as $Y - hCF(Y) = \delta$, with $C = A^{-1}B$ and $\delta = A^{-1}\mathbf{b}$, thus obtaining an expression analogous to that used to derive the internal stages of a Runge-Kutta formula. It follows that methods for handling the nonlinear systems arising from the application of a R-K method may be as well applied in this context. Indeed we followed the approach used in RADAU5 (see [4] pages 118-122), although we preferred to maintain the expression (4) because it allows an easy estimation of the error. The modified Newton method is used to linearize (4) whose solution is consequently obtained as the limit of the sequence $\{Y^k\}$ defined as

$$\begin{cases} Y^0 & \text{given}, \\ (A \otimes I_r - h(B \otimes I_r)D_J)\,\Delta Y^k = \mathbf{b} - AY^k + hBF(Y^k), \\ Y^{k+1} = Y^k + \Delta Y^k, \end{cases} \tag{5}$$

where $D_J = diag[\frac{\partial f}{\partial y}(t_1, \mathbf{y}_1^0), \ldots, \frac{\partial f}{\partial y}(t_N, \mathbf{y}_N^0)]$. The starting value Y^0 is obtained by extrapolation considering the solution computed at the previous step (see [7]). To avoid more than one Jacobian evaluation per step we make the approximation

$$\frac{\partial f}{\partial y}(t_i, \mathbf{y}_i^0) \simeq J \equiv \frac{\partial f}{\partial y}(t_0, \mathbf{y}_0);$$

consequently the linear systems to be solved become

$$(A \otimes I_r - hB \otimes J)\, \Delta Y^k = G(Y^k), \tag{6}$$

with $G(Y^k) = \mathbf{b} - AY^k + hBF(Y^k)$. It is possible to further reduce the computational cost recasting (6) into block-diagonal form. This is done by first considering the block diagonal form of $A^{-1}B$ (its existence and well conditioning has been verified for all the considered GAMs)

$$T^{-1}A^{-1}BT = \Lambda, \qquad \Lambda = \begin{pmatrix} \alpha_1 & -\beta_1 & & & \\ \beta_1 & \alpha_1 & & & \\ & & \ddots & & \\ & & \ddots & & \\ & & & \alpha_{N/2} & -\beta_{N/2} \\ & & & \beta_{N/2} & \alpha_{N/2} \end{pmatrix}$$

and then performing the transformations of variables $Z^k = T^{-1}Y^k$. Setting $S = AT$, multiplying both sides of (6) by S^{-1} and exploiting the relation $(\Lambda T^{-1}) \otimes J = (\Lambda \otimes J)(T^{-1} \otimes I_r)$, the iteration scheme becomes:

$$(I_N \otimes I_r - h\Lambda \otimes J)\, \Delta Z^k = S^{-1}G(TZ^k), \tag{7}$$

which consists of $N/2$ decoupled systems having the form

$$\begin{pmatrix} I_r - h\alpha_n J & -\beta_n J \\ \beta_n J & I_r - h\alpha_n J \end{pmatrix} \begin{pmatrix} \mathbf{u}_n \\ \mathbf{v}_n \end{pmatrix} = \begin{pmatrix} \mathbf{c}_n \\ \mathbf{d}_n \end{pmatrix}, \qquad n = 1, \ldots, N/2, \tag{8}$$

where $\mathbf{u}_n = \mathbf{z}_{2n-1}^{k+1} - \mathbf{z}_{2n-1}^k$, $\mathbf{v}_n = \mathbf{z}_{2n}^{k+1} - \mathbf{z}_{2n}^k$ and analogously $\mathbf{c}_n$, $\mathbf{d}_n$ are the $(2n-1)$-st and $2n$-th block components of the vector $S^{-1}G(Y^k)$. Further amount of computation is gained transforming these $2r$-dimensional real subsystems into the r-dimensional complex systems

$$((I_r - h\alpha_n J) + i\beta_n J)(\mathbf{u}_n + i\mathbf{v}_n) = \mathbf{c}_n + i\mathbf{d}_n, \qquad n = 1, \ldots, N/2. \tag{9}$$

The total work to obtain the solution of (4) is then shared as follows: one Jacobian evaluation; $N/2$ complex LU factorizations; for each iteration (9) the computation of $Y^k = TZ^k$, $F(Y^k)$, $S^{-1}G(Y^k)$ and the solutions of N triangular complex systems of dimension r.

4   Parallel Simplified Newton Iteration

A basic level parallelization is easily achieved solving the $N/2$ decoupled systems
(9) in parallel. The number of processors required for the generic step will therefore depend on the order of the GAM chosen to advance the solution: 2, 3, 4,
5 processors are needed for the orders 3, 5, 7 and 9 respectively. Obviously, if
the changing order rule is allowed to select the method between the minimum
and the maximum order, five processes must be initialized at the start up of
the program but some of them will remain idle unless the order 9 formula is directly involved in the integration step. When evaluating the expected speed-up
one should therefore get information about the global work carried out by each
process. Suppose a problem has been solved and the orders 3, 5, 7, 9 formulae
contributed to the solution by n_3, n_5, n_7 and n_9 steps respectively (the rejected
steps must also be included). Then for that particular execution an estimation
of the expected speed-up S_e is

$$S_e = \frac{2n_3 + 3n_5 + 4n_7 + 5n_9}{n_3 + n_5 + n_7 + n_9} \in [2, 5]. \tag{10}$$

More precisely formula (10) represents an upper bound of the theoretical
speed-up since it does not take into account the sequential nature of some parts
of the code such as the Jacobian evaluation, the stepsize selection and the computation of the initial guess of the Newton iteration.

We see that, besides being problem dependent, once a problem has been fixed,
the value of S_e is also related to the input tolerances and other possible input
values. However comparisons between the real speed-ups S_r and expected speedups S_e are still possible because the numbers n_i together with the execution
times are available as output variables of the code. In detail the iteration scheme
(7) executed in parallel proceeds as follows. Initially each of the $N/2$ processes
involved at the current step performs the LU factorization of the coefficient
matrix of the corresponding system (9). Later, the elements of the sequence
$\{Z^k\}$ are generated until convergence is attained. Assume that the process n
has evaluated its own piece $[z_{2n-1}^k, z_{2n}^k]^T$ of the solution Z^k; let us see how the
construction of $[z_{2n-1}^{k+1}, z_{2n}^{k+1}]^T$ is carried out. For $n = 1, \ldots, N/2$, we set

$$Z_n^k = [0, z_{2n-1}^k, z_{2n}^k, 0]^T \in H^N,$$

$$Y_n^k = [0, y_{2n-1}^k, y_{2n}^k, 0]^T, F_n^k = [0, \mathbf{f}(t_{2n-1}, y_{2n-1}^k), \mathbf{f}(t_{2n}, y_{2n}^k), 0]^T \in H^N,$$

$$\mathbf{b}_n = [0, b_{2n-1}, b_{2n}, 0]^T \in \mathbb{R}^N,$$

$$G_n^k = \mathbf{b}_n - AY_n^k + hBF_n^k,$$

and observe that

$$G(Y^k) = \sum_{n=1}^{N/2} G_n^k. \tag{11}$$

The program now performs the following steps (points 1),3) and 5) are performed for $n = 1, \ldots, N/2$:

1) the process n computes $W_n = TZ_n^k$;
2) each process takes part in the computation of $Y^k = \sum_{n=1}^{N/2} W_n$ and receives the corresponding block $[\mathbf{y}_{2n-1}^k, \mathbf{y}_{2n}^k]^T$ (and therefore Y_n^k);
3) the process n computes $[\mathbf{f}(t_{2n-1}, \mathbf{y}_{2n-1}^k), \mathbf{f}(t_{2n}, \mathbf{y}_{2n}^k)]^T$ and then $G_n(Y_n^k)$;
4) exploiting the relation (11), all processes contribute to the computation of $S^{-1}G(Y^k)$ and receive the corresponding blocks of the known term $[\mathbf{c}_n, \mathbf{d}_n]$ (see formula (9));
5) the process n can finally solve the system (9) and get the solution $[\mathbf{u}_n, \mathbf{v}_n]$ and hence $[\mathbf{z}_{2n-1}^{k+1}, \mathbf{z}_{2n}^{k+1}]^T$.

The implementation described above represents a parallelization across the method similar to that used for Runge-Kutta methods with associated matrix having real and distinct eigenvalues (see for example [2]). There are a number of starting points to take into consideration for subsequent developments of the parallel code. As an instance, to avoid that some processors remain idle during the integration, one may chose $N = 10$ as the dimension of all considered GAMs. As a generalization, more than five processors could be activated considering $N > 10$, even though, in this case, the dependence of the convergence properties of the simplified Newton iteration on the dimension N, should be carefully studied.

5   Numerical Tests

In this section we present some numerical results related to the parallelization of the code GAM as described in section 3. The experiments were performed on a Cray T3E machine with distributed memory, using at most five processors. In the following the parallel code GAM will be referred to as P-GAM or P-GAM(n), where n is the number of processors used. The communications are performed by the MPI routines [10]. For numerical comparisons we choose the codes RADAU in the version of April 1998 [5] and GAM in the version of September 1997 [8].

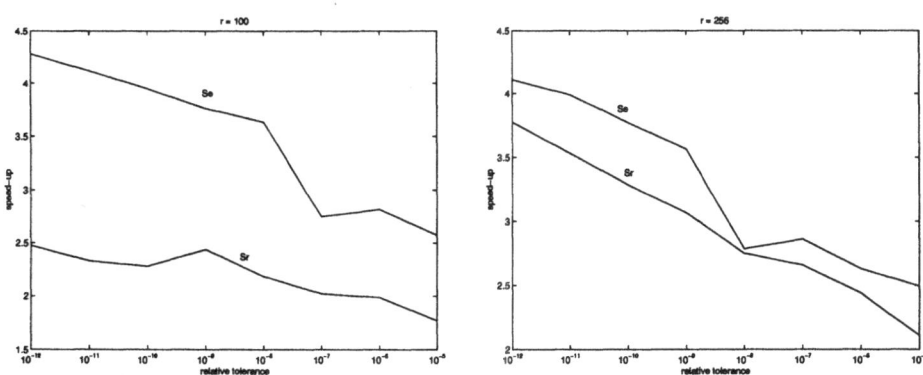

Fig. 1. SALESMAN problem

The first test problem is the travelling salesman problem described in [1]. We chose this problem in order to compare the expected speed-up (S_e) to the numerical one (S_r) by changing both the input parameters ($rtol = atol = h_0$) and the dimension r of the problem. In Figure 5 we show the results obtained for the dimensions $r = 100$ and $r = 256$ (the problem has a full Jacobian). The end is to quantify the dependence of the performance of the code on the communication times. We see that, how one should expect, overloading the processors, the obtained speed-ups are very close to the expected ones; this means that the communication times, which are $O(r)$, are almost negligible compared to the working times per step of each processor (which depends on $O(r^3)$). Better speed-ups for lower dimensions should be expected on shared memory parallel computers. We finally observe that the rise in speed-up (either S_e and S_r) when the accuracy of the solution is increased, is due to the involvement in the integration process of high order formulae which keep at work a greater number of processors.

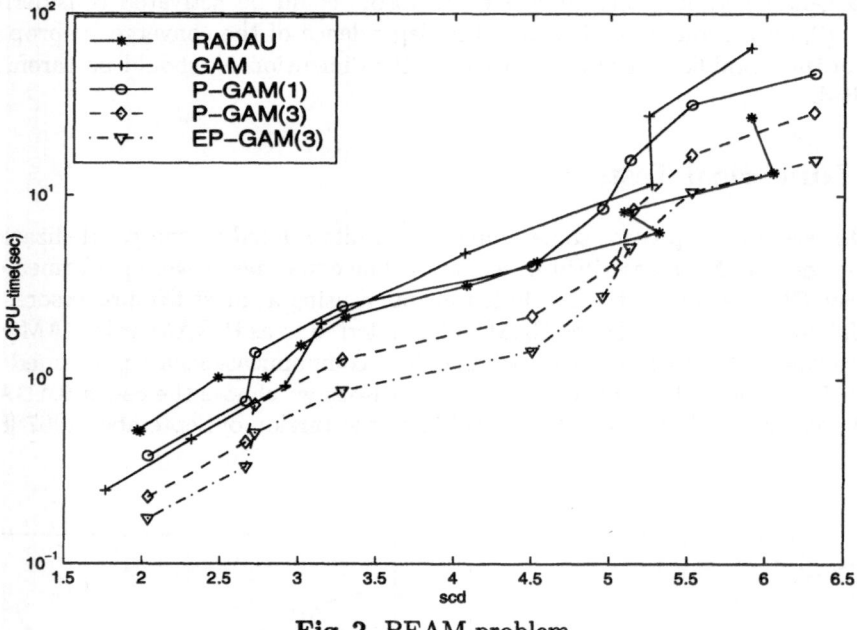

Fig. 2. BEAM problem

The second test is the BEAM problem described in [9], an ODE of dimension 80. In Figure 5 we report the work precision diagram, that is a range of input tolerances and a range of initial stepsizes were used to produce a plot of the resulting minimum number of significant digits in the components of the numerical solution at the endpoint (scd) against the CPU time in seconds needed for the run (observe that a logarithmic scale is used for the y-axis). The format of this diagram is as in [4]; naturally it strongly depends on the input tolerances and on the default parameters of each code; the variation of a parameter may considerably change the diagrams. For all problems we fixed $atol = rtol = h_0$ while the values of $rtol$ were chosen as follows: $10^{-(2+m)}$, $m = 0, \ldots, 10$, for RADAU and $10^{-(2+m)}$, $m = 0, \ldots, 8$ for GAM and P-GAM. A lower bound for the expected execution times is obtained dividing the CPU times of P-GAM(1) by S_e and is drawn as EP-GAM(3). The expected speed-up is for all tolerances less than 3 since the maximum order used is 5, the numerical speed-up is about 2.

References

[1] Bellen, A.: Pade test - A set of real-life test differential equations for parallel computing, Technical Report 103, Università di Trieste (1992)

[2] Bendtsen, C.: A parallel stiff ODE solver based on MIRKs, Adv. Comput. Math. **7** 1-2 (1997) 27-36

[3] Brugnano L., Trigiante D.: Solving Differential Problems by Multistep Initial and Boundary Value Methods, Gordon & Breach, Amsterdam, (1998)

[4] Hairer, E., Wanner, G.: Solving Ordinary Differential Equations II. Stiff and Differential–Algebraic Equations, Springer Series in Computational Mathematics, **14**, Springer-Verlag, Berlin, (1996)

[5] Hairer, E., Wanner, G.: RADAU, April 1998. Available via WWW at URL ftp://ftp.unige.ch/pub/doc/math/stiff/radau.f

[6] Iavernaro, F., Mazzia, F.: Block-Boundary Value Methods for the solution of Ordinary Differential Equations, SIAM J. Sci. Comput. (to appear)

[7] Iavernaro, F., Mazzia, F.: Solving ordinary differential equations by Generalized Adams Methods: properties and implementation techniques , Appl. Num. Math. **28** 2-4 (1998) 107–126

[8] Iavernaro, F., Mazzia, F.: F GAM August 1997. Available via WWW at URL http://www.dm.uniba.it/~mazzia/ode/readme.html.

[9] Lioen, W. M., de Swart, J.J.B., van der Veen, W. A.: Test Set for IVP Solvers, CWI, Department of Mathematics, Amsterdam, Report NM-R9615, (1996)

[10] Message Passing Interface Forum. MPI: A Message-Passing Interface Standard, (1995)

PAMIHR. A Parallel FORTRAN Program
for Multidimensional Quadrature
on Distributed Memory Architectures

G. Laccetti and M. Lapegna

Center for Research on Parallel Computing and Supercomputers - CNR
University of Naples "Federico II"
via Cintia - Monte S. Angelo. 80126 Napoli (Italy)
(laccetti,lapegna)@matna2.dma.unina.it

Abstract. PAMIHR: a parallel adaptive routine for the approximate computation of a multidimensional integral over a hyperrectangular region is described. The software is designed to efficiently run on a MIMD distributed memory environment, and it's based on the widely diffused communication system **BLACS**. PAMIHR, further, gives special attention to the problems of scalability and of load balancing among the processes.

1 Introduction and Software Description

PAMIHR (Parallel Adaptive Multidimensional Integration on Hyper-Rectangles) is a parallel FORTRAN package for the computation of:

$$I[f] = \int_{\Omega} f(t_1, ..., t_{IDIM}) \, dt_1 \cdots dt_{IDIM} \qquad (1.1)$$

on a MIMD distributed memory environment with P processes, where
$$\Omega = [A(1), B(1)] \times \cdots \times [A(IDIM), B(IDIM)]$$
is a IDIM-dimensional hyperrectangular region with $2 \leq IDIM \leq 10$.

The software we are describing is designed to return an approximation RESULT of $I[f]$ and an estimate ERROR of the absolute error $|I[f] - RESULT|$ such that:
$$|I[f] - RESULT| \leq ERROR < \max\{ABSACC, RELACC \times |RESULT|\}$$
where ABSACC and RELACC are absolute and relative input provided tolerances, respectively. In such a way the algorithm ends if the "fairest" accuracy is reached. PAMIHR is based on a typical global adaptive algorithm. For the computation of the integral and of the error estimates in each subdomain, PAMIHR uses an imbedded family of fully symmetric rules very similar to the well known integration rules developed by Genz and Malik in [7]. A common strategy to compute the error estimate in an automatic routine for the quadrature is based on the difference between two imbedded integration rules. However this procedure is very unreliable and PAMIHR uses the error estimate procedure developed in [1]. Is is well known that the main problem in the development of a parallel adaptive algorithm is the workload balancing, because, for this kind of algorithm, it is

P. Amestoy et al. (Eds.): Euro-Par'99, LNCS 1685, pp. 1144–1148, 1999.
© Springer-Verlag Berlin Heidelberg 1999

impossible to determine an *a priori* optimal decomposition of the integration region [3, 5, 10]. Therefore, in order to balance the workload, our algorithm configures the processes as a 2-dimensional periodic grid. In this topology each process p_i has 4 neighbours, 2 for each direction. In the generic direction dir, $(dir = 0, 1)$, we define $p_{i+}^{(dir)}$ and $p_{i-}^{(dir)}$, the next and the previous process of p_i, respectively. Therefore, at a generic iteration j, each process attempts to send to $p_{i+}^{(dir)}$ its subdomain $\hat{s}_i^{(j)}$ with largest error estimate, in order to share, with the other processes, its "hard" subdomains. More precisely, in a given direction $dir = mod(j, 2)$ let $\hat{e}_i^{(j)}$, $\hat{e}_{i+}^{(j)}$ and $\hat{e}_{i-}^{(j)}$ be respectively the largest error estimates in the current process p_i, in the process $p_{i+}^{(dir)}$ and in the process $p_{i-}^{(dir)}$; if $\hat{e}_i^{(j)} > \hat{e}_{i+}^{(j)}$ then the current process p_i sends the subdomain $\hat{s}_i^{(j)}$ with error estimate $\hat{e}_i^{(j)}$ to the process $p_{i+}^{(dir)}$, and $\hat{s}_i^{(j)}$ is removed from the memory of p_i. In the same way if $\hat{e}_{i-}^{(j)} > \hat{e}_i$ the current process receives the subdomain $\hat{s}_{i-}^{(j)}$ with error estimate $\hat{e}_{i-}^{(j)}$ from the process $p_{i-}^{(dir)}$. Then each process p_i can compute an approximation of the integral defined only over the subdomains located in its own local memory. Only one global sum among the processes is then required, at the end of the algorithm, for the computation of the final approximations RESULT and ERROR. Note that with this communication strategy, the scalability is improved by avoiding global communications and, at the same time, the workload is balanced by a distribution of the subdomains only among pairs of processes directly connected in a 2-dimensional mesh. For the communications among the processes, PAMIHR uses the BLACS (Basic Linear Algebra Communication Subprograms) [4] message-passing library. This library is the key to the portability of PAMIHR, because several versions of this library are available for different architectures (Intel Paragon, IBM SP2,...) as well for clusters of workstations and *pile of PC* (the so called Beowulf systems), so our software can run on all architectures where this communication system is available. However a prototype of the MPI version of PAMIHR is available too. The present version of the package is written in double precision ANSI Standard FORTRAN 77 except only for the calls to the BLACS routines written in non standard style. The BLACS communication system in not part of the package, because the efficiency depends strictly on the implementation of this library. So a vendor optimized version of BLACS should be used. However several versions of BLACS are available from *netlib*. The specification of PAMIHR is:

```
SUBROUTINE PAMIHR (ICNTXT, IDIM, F, A, B, RELACC, ABSACC, MAXFUN,
*        ITEST, NSPLIT, RESULT, ERROR, FCOUNT, WS, LD, IFLAG)
 INTEGER ICNTXT, IDIM, MAXFUN, ITEST, NSPLIT(IDIM), FCOUNT, LD, IFLAG
 DOUBLE PRECISION A(IDIM), B(IDIM), RELACC, ABSACC, RESULT, ERROR,WS(LD,*)
 EXTERNAL F
```

A detailed description of the input/output parameters can be found in the internal documentation of the routine[1]. In this contest we would like to emphasize

---

[1] refer to http://pixel.dma.unina.it/RESEARCH/pamihr.html

that, except for an initial explicit call to BLACS routines, the user calls PAMIHR in a simple and straightforward way, (in a sequential fashion, let say) without any explicit reference to the parallel environment and/or to communications among the processes.

2   Performance of PAMIHR

To test the performance of PAMIHR, we used the technique developed in [6] and a set of 8 functions families from the Genz package [6]. These families are characterized by some peculiarity (several kinds of peaks, oscillations, discontinuities, singularities,...). The families have been tested using several dimension up to 10 and different tolerances on the region $\Omega = [0,1]^{IDIM}$. Each family $f^{(j)}(\underline{x}), (j = 1,\ldots,8)$ is composed by 20 different functions where some minor parameters change. The parameters determine the sharpness and the location of the difficulty. A complete and detailed report of such intensive test activity may be found in [9]. Here, just some results related to the function with corner singularity:

$$f^{(8)}(\underline{x}) = (\textstyle\sum_{i=1}^{IDIM} \beta_i x_i)^{-IDIM/2.7}$$

are showed. All test are performed on the 512-node Intel Paragon XP/S operated by the Center for Advanced Computing Research. Access to this facility was provided by the California Institute of Technology.

The first set of experiments is aimed to measure reliability and accuracy of PAMIHR with just 1 process, by comparing the obtained results in 3 dimensions with the results from the sequential routine DCUHRE [2] that uses a similar computational kernel. Table 1 reports the required relative tolerance; then, for both PAMIHR and DCUHRE, there are: FCOUNT, the average, over the 20 functions of the family, of the number of function evaluations needed to reach the tolerance; OK, the number of cases where the tolerance is reached with FCOUNT≤MAXFUN = $100000 \times$ IDIM (where MAXFUN is the maximum number of function evaluations); NU, the number of cases where the actual error is bigger than the estimated one; E1, the average, over the 20 functions of the family, of the actual relative errors. From this table we can say that 1-process PAMIHR execution, produces very accurate results with a moderate number of function evaluations as well as DCUHRE. Further PAMIHR gives about the same results, in terms of accuracy and efficiency, of DCUHRE. However, from the complete results in [9], PAMIHR seems to privilege the accuracy, with respect to the efficiency, in the sense that, in general, PAMIHR computes more exact digits than DCUHRE with few more function evaluations. In Table 2 are reported the results of the second set of experiments performed to measure the efficiency of the algorithm when the number of processes P increases and the workload (the number of function evaluations) is constant in the processes (the so called *scalability*) [8]. We note good value of scalbility for the $f^{(8)}(\underline{x})$ functions family. Further, in Table 2, we can see which average error is possible to obtain with PAMIHR in a fixed time (the time required for the execution with 1 process and a given tolerance) when the number of processes P grows. For all problems we note a significant error reduction. That means that the "hard" subdomains are rapidly distributed across the processes so they do not waste time on unimportant subdomains.

	PAMIHR				DCUHRE			
RELACC	FCOUNT	OK	NU	E1	FCOUNT	OK	NU	E1
.10D+00	600	20	0	0.503D-02	277	20	0	0.111D-01
.10D-01	1771	20	0	0.336D-03	1840	20	0	0.319D-03
.10D-02	4335	20	0	0.208D-04	5782	20	0	0.131D-04
.10D-03	7053	20	0	0.246D-05	10826	20	0	0.124D-05

Table 1 - Accuracy test with 1 process for the $f^{(8)}$ functions family

RELACC	P=4	P=16	P=32	P=64
1.E-1	0.84	0.73	0.70	0.68
1.E-2	0.93	0.89	0.87	0.86
1.E-3	0.95	0.94	0.93	0.93
1.E-4	0.96	0.95	0.95	0.94

RELACC	P=4	P=16	P=32	P=64
1.E-1	2.E-3	1.E-3	8.E-4	7.E-4
1.E-2	1.E-4	7.E-5	5.E-5	4.E-5
1.E-3	3.E-7	1.E-7	8.E-7	7.E-7
1.E-4	1.E-8	1.E-9	5.E-10	2.E-10

Table 2 - Scaled efficiency and error reduction for the $f^{(8)}$ functions family

Finally some few words about a comparison with an early version of PAMIHR: the subroutine D01FAFP in the NAG Parallel Library. PAMIHR is a more robust software item than D01FAFP, in the sense that PAMIHR performs successfully the whole Genz test package, where D01FAFP shows some invalid result. For example in 2 dimensions with 1 process, for the *discontinuous function* of the Genz package:

$$f^{(6)}(\underline{x}) = \begin{cases} 0 & \text{if } x_1 > \beta_1 \text{ or } x_2 > \beta_2 \\ exp(\sum_{i=1}^{IDIM} \alpha_i x_i) & \text{otherwise} \end{cases} \quad \text{with} \quad \begin{array}{|c|c|c|} \hline \alpha_i & 5.8213 & 19.178 \\ \hline \beta_i & 0.85403 & 0.62471 \\ \hline \end{array} \quad (2.1)$$

(with such a values, the exact result, correct to 9 digits, is 204955.199) and RELACC = 0.10E-2 the two routines give the results reported in Table 3: the error estimate is correct for PAMIHR and not determined for D01FAFP. Such improvement is obtained in PAMIHR avoiding instructions leading to divisions by 0 and/or indeterminant forms in the error estimation procedure. Another interesting issue of PAMIHR is related to the maximum number of function evaluations MAXFUN. This number is an input parameter, and it refers to the "total" number of function evaluations. More precisely to each process is assigned the integer part of MAXFUN/P function evaluations. The node-algorithm is designed to avoid a next iteration if such (possible) iteration exceeds MAXFUN/P; but in general P does not divide MAXFUN, so each process still has available some function evaluations, Let M_i be the number of unused function evaluations in each process p_i. If $\sum_i M_i \geq 2 \times VALFUN$ (where $VALFUN$ is the number of function evaluation to compute the integration rule in a subdomain), some of the P processes can perform a further iteration. So, before returning to the calling program, PAMIHR distributes the function evaluations still available, assigning them to one or more processes, so that some of the processes can continue to work. Let us focus the attention, for example, on the *product peak* function of the Genz package in 10 dimensions with 8 processes:

$$f^{(2)}(\underline{x}) = \prod_{i=1}^{IDIM}(\alpha_i^{-2} + (x_i - \beta_i)^2)^{-1}$$

with RELACC = 0.10E-2, MAXVAL=200000 and

α_i	.401	.408	.832	.339	1.33	1.21	$3.16E-3$	1.35	$3.38E-2$	$7.89E-2$	
β_i	.910	.510	.150	.942	.503	.490	.275	.903	$4.71E-2$	.902	(2.2)

The routines give the results reported in Table 4, where we observe that IFLAG=
100 (maximum number of function evaluations exceeded) for DO1FAFP with
FCOUNT much smaller than MAXFUN, whereas IFLAG=0 (successful exit) for PAMIHR.

	DO1FAFP	PAMIHR
RESULT	2070E+02	2060E+02
ERROR	NaN	.4357E-03
FCOUNT	957	4257
IFLAG	0	0

Table 3: PAMIHR and DO1FAFP
results for $f^{(6)}(\underline{x})$ functions
family. α_i and β_i as in (2.1)

	DO1FAFP	PAMIHR
RESULT	.3567E-12	.3567E-12
ERROR	.6058E-04	.6292E-04
FCOUNT	104200	145880
IFLAG	100	0

Table 4: PAMIHR and DO1FAFP
results for $f^{(2)}(\underline{x})$ functions
family. α_i and β_i as in (2.2)

References

[1] Berntsen J. - *Practical error estimation in adaptive multidimensional quadra-ture routines* - J. Comput. Appl. Math., vol. 25 (1989), pp. 327-340.

[2] Berntsen J., T.O. Espelid, A.C. Genz - *Algorithm 698: DCUHRE-An adaptive multidimensional integration routine for a vector of integrals* - ACM Trans. on Math. Software, vol. 17 (1991), pp. 452-456,

[3] de Doncker E., J. Kapenga - *Parallel Cubature on Loosely Coupled Systems* - in *Numerical Integration: Recent developments, software and Applications* (T. Espelid and A. Genz eds.), Kluwer, 1992, pp. 317-327.

[4] Dongarra J., R.C. Whaley - *A user's guide to the BLACS v1.0* - Tech. Rep. CS-95-281, LAPACK Working Note no. 94, Univ. of Tennessee, 1995.

[5] Genz A.C. - *The numerical evaluation of multiple integrals on parallel comput-ers* - in *Numerical Integration* (P. Keast and G. Fairweather eds.), D. Reidel Publishing Co., 1987, pp. 219-230.

[6] Genz A.C. - *A Package for Testing Multiple Integration Subroutines* - in Nu-merical Integration, (P.Keast, G.Fairweather, eds.), D. Reidel Publishing Co., 1987, pp. 337-340.

[7] Genz A.C., A.A. Malik - *An imbedded family of fully symmetric numerical integration rules* - SIAM J. Num. Anal., vol. 20 (1983), pp. 580-588.

[8] Gustafson J., G. Montry and R. Benner - *Development of parallel methods for a 1024 processor hypercube* - SIAM J. on Scientific and Statistic Computing, Vol. 9 (1988), pp. 580-588.

[9] Laccetti G., M. Lapegna, A. Murli - *DSMINT. A Scalable Double Precision FORTRAN Program to Compute Multidimensional Integrals* - Tech. Rep. CPS-96-9 Center for Research on Parallel Computing and Supercomputers, 1996.

[10] Lapegna M. - *Global adaptive quadrature for the approximate computation of multidimensional integrals on a distributed memory multiprocessor* - Concur-rency: Practice and Experiences, vol. 4 (1992), pp. 413-426

Stability Issues of the Wang's Partitioning Algorithm for Banded and Tridiagonal Linear Systems*

Velisar Pavlov and Plamen Yalamov

Center of Applied Mathematics and Informatics,
University of Rousse, 7017 Rousse, Bulgaria
velisar@ami.ru.acad.bg, yalamov@ami.ru.acad.bg

Abstract. The main results of a componentwise error analysis for a parallel partitioning algorithm [4] in the cases of banded and tridiagonal linear systems are presented. It is shown that for some special classes of matrices, i.e. diagonally dominant (row or column), symmetric positive definite, and M-matrices, the algorithm is numericaly stable.

1 Introduction

A well-known algorithm for solving tridiagonal systems in parallel is the method of Wang [4]. Full roundoff error analysis of this algotithm can be found in [5]. Generalized versions of the partitioning algorithm of Wang for banded linear systems are presented in [1, 3], and full roundoff error analysis in this case can be found in [6].

In this paper we make a review of the main results on the stability of Wang's algorithm for banded and tridiagonal linear systems.

First, we present a brief description of the algorithm (for banded systems only). Let the linear system under consideration be denoted by

$$Ax = d, \tag{1}$$

where $A \in \mathcal{R}^{n \times n}$, which bandwith is $2j + 1$. For simplicity we assume that $n = ks - j$ for some integer k, if s is the number of the parallel processors we want to use. We partition matrix A and the right hand side d of system (1) as follows:

$$
\begin{pmatrix}
B_1 & \bar{c}_1 \\
a_k & b_k & c_k \\
& \bar{a}_2 & B_2 & \bar{c}_2 \\
& & a_{2k} & b_{2k} & c_{2k} \\
& & & \ddots & \ddots & \ddots \\
& & & & \bar{a}_{s-1} & B_{s-1} & \bar{c}_{s-1} \\
& & & & & a_{(s-1)k} & b_{(s-1)k} & c_{(s-1)k} \\
& & & & & & \bar{a}_s & B_s
\end{pmatrix}
\begin{pmatrix}
X_1 \\ x_k \\ X_2 \\ x_{2k} \\ \vdots \\ X_{s-1} \\ x_{(s-1)k} \\ X_s
\end{pmatrix}
=
\begin{pmatrix}
D_1 \\ d_k \\ D_2 \\ d_{2k} \\ \vdots \\ D_{s-1} \\ d_{(s-1)k} \\ D_s
\end{pmatrix},
$$

---

* This work was supported by Grants MM-707/97 and I-702/97 from the National Scientific Research Fund of the Bulgarian Ministry of Education and Science.

P. Amestoy et al. (Eds.): Euro-Par'99, LNCS 1685, pp. 1149–1152, 1999.
© Springer-Verlag Berlin Heidelberg 1999

where $B_i \in \mathcal{R}^{(k-j) \times (k-j)}$ are band matrices with the same bandwith as matrix A, $\bar{a}_i, \bar{c}_i \in \mathcal{R}^{(k-j) \times j}$, $a_{ik}, b_{ik}, c_{ik} \in \mathcal{R}^{j \times j}$, $X_i, D_i \in \mathcal{R}^{(k-j) \times 1}$, $x_{ik}, d_{ik} \in \mathcal{R}^{j \times 1}$.

After suitable permutation of the rows and columns of matrix A we obtain the system

$$\mathcal{A}\mathcal{P}x = \mathcal{P}d, \quad \mathcal{A} = \mathcal{P}A\mathcal{P}^T = \begin{pmatrix} A_{11} & A_{12} \\ A_{21} & A_{22} \end{pmatrix}, \tag{2}$$

where P is a permutation matrix, $A_{11} = \text{diag}\{B_1, B_2, \ldots, B_s\} \in \mathcal{R}^{s(k-j) \times s(k-j)}$, $A_{22} = \text{diag}(b_k, b_{2k}, \ldots, b_{(s-1)k}) \in \mathcal{R}^{j(s-1) \times j(s-1)}$, and $A_{12} \in \mathcal{R}^{s(k-j) \times j(s-1)}$, $A_{21} \in \mathcal{R}^{j(s-1) \times s(k-j)}$ are sparse matrices.

We will distinguish between the two matrices A (original) and $\mathcal{A}$ (permuted). Evidently, the permutation does not influence the roundoff error analysis.

The algorithm can be presented as follows.

Stage 1. Obtain the block LU-factorization

$$\mathcal{A} = \begin{pmatrix} A_{11} & A_{12} \\ A_{21} & A_{22} \end{pmatrix} = LU = \begin{pmatrix} A_{11} & 0 \\ A_{21} & I_{j(s-1)} \end{pmatrix} \begin{pmatrix} I_{s(k-j)} & R \\ 0 & S \end{pmatrix} \tag{3}$$

by the following steps:

1. Obtain the LU-factorization of $A_{11} = \mathcal{P}_1 L_1 U_1$ with partial pivoting, if necessary. Here $\mathcal{P}_1$ is a permutation matrix, L_1 is unit lower triangular, and U_1 is upper triangular.
2. Solve $A_{11}R = A_{12}$ using the LU-factorization from the previous item, and compute $S = A_{22} - A_{21}R$, which is the Schur complement of A_{11} in $\mathcal{A}$.

Stage 2. Solve $Ly = d$ by using the LU-factorization of A_{11} (Stage 1).
Stage 3. Solve $Ux = y$ by applying Gaussian elimination to the block S.

2   Main Stability Results

In the following by a hat we denote the computed quantities. By ΔT we denote an equivalent perturbation in matrix T, and by ρ_0 we denote the roundoff unit. The matrix inequalities are understood componentwise.

Proofs of the following theorems can be found in [6].

Theorem 1. *For the partitioning algorithm we have that* $(\mathcal{A} + \Delta\mathcal{A})\mathcal{P}\hat{x} = \mathcal{P}d$, *where*

$$|\Delta\mathcal{A}| \leq |\mathcal{A}|h_1(\rho_0) + |\mathcal{A}||N|h_2(\rho_0),$$

here

$$h_1(\rho_0) = K_1 f(\rho_0) + K_2 h(\rho_0) + K_1 K_2 f(\rho_0)h(\rho_0)$$
$$+ K_1 f(\rho_0)g(\rho_0) + K_2 h(\rho_0)g(\rho_0) + K_1 K_2 f(\rho_0)h(\rho_0)g(\rho_0),$$
$$h_2(\rho_0) = 3K_1 f(\rho_0) + 2K_2 h(\rho_0) + 2K_1 K_2 f(\rho_0)h(\rho_0)$$
$$+ 3K_1 f(\rho_0)g(\rho_0) + 3K_2 h(\rho_0)g(\rho_0) + 3K_1 K_2 f(\rho_0)h(\rho_0)g(\rho_0)$$
$$+ K_1 f(\rho_0)g^2(\rho_0) + K_2 h(\rho_0)g^2(\rho_0) + K_1 K_2 f(\rho_0)h(\rho_0)g^2(\rho_0),$$

and for the forward error it is true that

$$\frac{\|\delta x\|}{\|\hat{x}\|} = \frac{\|\hat{x} - x\|_\infty}{\|\hat{x}\|_\infty} \leq cond(A,\hat{x})h_1(\rho_0) + cond^*(A,x^*)rh_2(\rho_0).$$

In the above theorem $r = \max\{\|\hat{R}\|_\infty, 1\}$, $K_1 = \max\{k_1, 1\}$, $K_2 = \max\{k_2, 1\}$, where k_1 bounds the growth of elements when we obtain the LU factorization of A_{11} (Stage 1), k_2 bounds the growth of elements of the Gaussian elimination for the reduced system (Stage 3), $f(\rho_0), g(\rho_0)$ are as follows:

$$f(\rho_0) = \gamma_{j+1} = \gamma_{2j+1}, \qquad g(\rho_0) = \gamma_{j+1} + \rho_0,$$

where $\gamma_n = n\rho_0/(1 - n\rho_0)$, and $N = \begin{pmatrix} 0 & \hat{R} \\ 0 & I_{j(s-1)} \end{pmatrix}$. The condition number $cond^*(A, x^*)$ is defined below

$$cond^*(A,x^*) = \frac{\| |A^{-1}| |A| x^* \|_\infty}{\|\hat{x}\|_\infty},$$

where the vector x^* is constructed in the following way

$$x^* = (\|\hat{x}_k\|_\infty e, |\hat{x}_k^T|, \max\{\|\hat{x}_k\|_\infty, \|\hat{x}_{2k}\|_\infty\} e, \ldots,$$
$$|\hat{x}_{(s-1)k}^T|, \max\{\|\hat{x}_{(s-2)k}\|_\infty, \|\hat{x}_{(s-1)k}\|_\infty\} e)^T,$$

Here $e = (1, 1, \ldots, 1) \in \mathcal{R}^{1 \times (k-1)}$. The other condition number is known as the Skeel's conditioning number:

$$cond(A, \hat{x}) = \frac{\| |A^{-1}| |A| |\hat{x}| \|_\infty}{\|\hat{x}\|_\infty}.$$

The condition number $cond^*(A, x^*)$ is introduced to make the obtained bounds more realistic in some cases. As we shall see in the bounds of the forward error the condition number $cond^*(A, x^*)$ is multiplied by the factor r (which can be large sometimes) while the condition number $cond(A, \hat{x})$ is not. So, when $cond^*(A, x^*)$ is small the influence of r should be negligible.

A similar general result for tridiagonal matrices is obtained in [5].

Now we consider more precisely the case when the matrix A belongs to one of the following types: diagonally dominant, symmetric positive definite, or M-matrix.

For the following bounds of $\|\hat{R}\|_\infty$ and k_2 we need to analyze what is the type of the reduced matrix S if matrix A belongs to one of the above mentioned classes. We need this to bound $\|\hat{R}\|_\infty$. Then at the end of this section we comment on the growth of the constant k_2. An estimate on the constant k_1 come directly from the roundoff error analysis of the Gaussian elimination for banded linear systems [2, p.182].

Theorem 2. *Let $A \in \mathcal{R}^{n \times n}$. If the band nonsilngular matrix A is either symmetric positive definite, or row diagonally dominant, or a nonsingular M-matrix, then the reduced matrix S (the Schur complement) preserves the same property.*

As we saw in Theorem 1 the error bound depends not only on the growth factors K_1 and K_2, but also on the quantity r, which measures the growth in the matrix $\hat{R}$. Clearly, when some of the blocks B_i are ill conditioned (although the whole matrix A is well conditioned) the factor r can be large. This will lead to large errors even for well conditioned matrices. So, we need some bounds for r, or , equivalently $\|\hat{R}\|_\infty$.

Theorem 3. *Let $A \in \mathcal{R}^{n \times n}$ be nonsingular band M-matrix and $k_1 \, cond(A) f(\rho_0) < 1$. Then it is true that*

$$\|\hat{R}\|_\infty \leq \frac{cond(A)}{1 - k_1 cond(A_{11}) f(\rho_0)} \leq \frac{cond(A)}{1 - k_1 cond(A) f(\rho_0)}.$$

Theorem 4. *Let $A \in \mathcal{R}^{n \times n}$ be nonsingular, row diagonally dominant, band matrix and $k_1 \, cond(A) f(\rho_0) < 1$. Then we have*

$$\|\hat{R}\|_\infty \leq \frac{1}{1 - k_1 cond(A_{11}) f(\rho_0)} \leq \frac{1}{1 - 2k_1 cond(A) f(\rho_0)}.$$

Theorem 5. *Let $A \in \mathcal{R}^{n \times n}$ be a symmetric positive definite band matrix and $k_1(k-1) cond_2(A) f(\rho_0) < 1$, where $cond_2(A) = \|A^{-1}\|_2 \|A\|_2$. Then we have*

$$\|\hat{R}\|_\infty \leq \frac{\sqrt{j(s-1) cond_2(A)}}{1 - k_1 cond(A_{11}) f(\rho_0)} \leq \frac{\sqrt{j(s-1) cond_2(A)}}{1 - k_1(k-1) cond_2(A) f(\rho_0)}.$$

Theorems 3 - 5 show that $\|\hat{R}\|_\infty$ (and r, respectively) is not large for the three types of matrices when the original matrix A is well conditioned. In order to bound k_2 we can use Theorem 2 and the already obtained bounds for the Gaussian elimination in [2, p. 182].

So, the main conclusion is when A belongs to the one of the above mention classes the algorithm is numericaly stable.

References

[1] Conroy, J.: Parallel Algorithms for the solution of narrow banded systems. Appl. Numer. Math. **5** (1989) 409–421
[2] Higham, N.: Accuracy and Stability of Numerical Algorithms. SIAM, Philadelphia, 1996
[3] Meier, U.: A parallel partition method for solving banded linear systems. Parallel Comput. **2** (1985) 33–43
[4] Wang, H.: A parallel method for tridiagonal linear systems. ACM Transactions on Mathematical Software **7** (1981) 170–183
[5] Yalamov, P., Pavlov, V.: On the Stabilty of a Partitioning Algorithm for Tridiagonal Systems. SIAM J. Matrix Anal. Appl. **20** (1999) 159–181
[6] Yalamov, P., Pavlov, V.: Stability of a partitioning algorithm for special classes of banded linear systems. Preprint N 38, University of Rousse, March 1998 (submitted to LAA)

Topic 14
Emerging Topics in Advanced Computing in Europe

Renato Campo and Luc Giraud

Co-chairmen

The ESPRIT programme has now been active for over a decade. One of the areas to benefit substantially from the injection of ESPRIT research and development funding has been the area of parallel computing and high performance computing in the framework of industrial applications for large and small enterprises. It is therefore appropriate that at this conference there should be two sessions dedicated to outcome of ESPRIT projects. A wide range of papers were submitted, and those selected represent state-of-the-art projects in various areas of interest. The projects break down into those which deal with the software required for making high performance computers more efficient and simpler to use and applications which can benefit from the use of HPC. In the former category are papers like the one on the HPF+ project that succeeded to demonstrate that HPF, with a small set of language extensions and an appropriate compiler and tool infrastructure, has the potential to be efficient for advanced industrial applications, sometimes approaching the performance of explicit message passing code.

The second paper deals with solutions for fault tolerance in embedded automation. Those solutions are often based on strong customisation, have impacts on the whole life-cycle, and require highly specialised design teams, thus making dependable embedded systems costly and difficult to develop and maintain. The TIRAN project develops a framework which provides fault tolerance capabilities to automation systems, with the goal of allowing portable, reusable and cost-effective solutions. Application developers are allowed to select, configure and integrate in their own environment a variety of software-based functions for error detection, confinement and recovery provided by the framework.

The OCEANS project deals with optimising compilers for embedded applications. The objective of the OCEANS project is to investigate and develop advanced compiler infrastructure for embedded VLIW processors. This combines high and low level optimisation approaches within an iterative framework for compilation.

The fourth paper deals with the performances evaluation of the parallel ALM code developed by SMART and Prometeia Calcolo for the EU project PALMA (Parallel Asset and Liability MAnagement). The code implements a stochastic approach based on a dynamic ALM model specially tailored for the Italian financial market. Very good scalability and efficiency results are shown for execution on the Cray T3E present at CINECA, using a real world data set of Credito Italiano.

P. Amestoy et al. (Eds.): Euro-Par'99, LNCS 1685, pp. 1153–1154, 1999.
© Springer-Verlag Berlin Heidelberg 1999

The HYPERBANK project aims at providing the banking sector the requisite toolset for the increased understanding of existing and prospective customers. The approach exploits and integrates three areas: business knowledge modelling, data warehousing and data mining, together with parallel computing. Business knowledge modelling formally describes the enterprise in terms of roles, goals and rules.

The first of the last two papers dedicated to parallel multibody simulation software targeted for heterogeneous network of workstations presents an overview of the software tools developed in the framework of HIPERCOMBATS and MYSHANET projects. These are High Performance Computing and Networking (HPCN) projects funded by the European Commission in the frame of the CAPRI initiative and the Technology Transfer Nodes (TTN) initiative, respectively. These tools are aimed at helping the engineers in the complex task of designing new multi-body systems. The main objective of both projects has been the development of cost effective software tools that allow the designer performing parametric simulations on a non-dedicated network of heterogeneous computers.

The final paper is devoted to the ODESIM project, a High Performance Computing and Networking project funded by the European Commission within ES-PRIT programme. The project acronym stands for Optimum DESIgn of Multi-Body systems. The project main objective has been the development of a set of software tools for the kinematic and dynamic optimum design of multi-body systems. Several types of design procedures are targeted: interactive design, designer-driven optimisation and fully automatic optimisation. ODESIM has been implemented using high performance computing and networking techniques to perform parallel computations in the most CPU-intensive optimisation problems. Promising results on the solution of practical problems are presented.

On behalf of the programme committe we would like to thanks all the contributors and referees for the excellent work they performed that contribute to the organisation of this attractive session.

The HPF+ Project: Supporting HPF for Advanced Industrial Applications[*]

Siegfried Benkner[1], Guy Lonsdale[1], and Hans Zima[2]

[1] C&C Research Laboratories, NEC Europe Ltd.
Rathausallee 10, D-53757 St. Augustin, Germany
{benkner,lonsdale}@ccrl-nece.technopark.gmd.de
[2] Institute for Software Technology and Parallel Systems
University of Vienna, Liechtensteinstr. 22, A-1090 Vienna, Austria
zima@par.univie.ac.at

Abstract. High Performance Fortran (HPF) is a data-parallel language providing the user with a high-level interface for programming scientific applications, while delegating to the compiler the task of producing explicitly parallel code. In this paper, we give an overview of the motivation and the results of the ESPRIT project "HPF+". The project succeeded in demonstrating that HPF, with a small set of language extensions and an appropriate compiler and tool infrastructure, has the potential to be efficient for advanced industrial applications, sometimes approaching the performance of manually written message-passing code. We introduce the applications which were used to guide and evaluate the development work in the project, provide an overview of the HPF+ language and discuss the Vienna Fortran Compiler (VFC) as well as the performance obtained for the project benchmarks.

1 Introduction

The emergence of scalable parallel architectures brought two issues into focus: the necessity of controlling locality and the complexity of parallel programming. One way to control locality is to use an explicitly parallel approach, e.g., C or Fortran coupled with message passing. However, the resulting programs tend to be complex and rather low-level. It became clear that a higher level approach was also possible. High Performance Fortran (HPF) is such an approach, providing a high-level data-parallel programming paradigm with a single thread of control in a global address space, explicitly parallel constructs, and user-specified distribution and alignment directives for the control of locality.

However, the original definition of the language, HPF-1 [14], supported only applications with regular data distributions and access patterns efficiently. The motivation for the ESPRIT project "HPF+" came from the realization that (1) HPF would only be accepted by the user community if it proved to be able to

---

[*] The work described in this paper was partially supported by the ESPRIT IV Long Term Research Project 21033 "HPF+" of the European Commission.

P. Amestoy et al. (Eds.): Euro-Par'99, LNCS 1685, pp. 1155–1165, 1999.
© Springer-Verlag Berlin Heidelberg 1999

adequately handle advanced industrial codes as well, and (2) that early languages such as Vienna Fortran [8, 21] had already shown the feasibility of this goal by providing a proof of concept.

The HPF+ project was conducted in the timeframe January 1996 through April 1998, and involved the following partners: AVL List Gmbh (Austria), ECMWF (UK), ESI SA (France), NA Software Ltd. (UK), NEC Europe Ltd. (Germany), University of Pavia (Italy), and University of Vienna (Austria). The major results of the HPF+ project include

- the definition of a set of benchmark kernels representing typical features of advanced industrial applications;
- the definition of an extension of the HPF-1 language, HPF+, guided by the requirements of the benchmark kernels;
- the implementation of HPF+ in the *Vienna Fortran Compiler (VFC)*, supported by a range of software tools, in particular for performance analysis.

The HPF+ project was carried out in parallel to the development of the recent HPF standard, HPF-2 [15], which resulted in a convergence for some features. However, HPF+ went beyond HPF-2 by providing specialized control features, in particular for the manipulation of communication schedules, and also, on the other hand, simplified the language by eliminating some features with little applicability but troublesome semantics and implementation problems.

The paper is structured as follows. In the next section, we give an overview of the application benchmarks developed for the project. This is followed in Section 3 by a short overview of the HPF+ language. Section 4 provides a brief description of VFC, including a discussion of the performance results obtained for the benchmark kernels. The paper concludes with an outlook to potential future developments of HPF (Section 5).

2   The HPF+ Application Kernels

This section provides a short description of the application codes targeted within the HPF+ project. From these applications representative kernels were extracted to address the key challenges for the language and compiler development.

2.1   PAM-CRASH

PAM-CRASH [11], from ESI (France), is an explicit time-marching Finite Element program used for the numerical simulation of the highly non-linear, dynamic phenomena arising in short-duration contact-impact problems. It uses a central difference explicit time-marching scheme with unstructured meshes comprised of mechanical elements. The major part of the computational cost of the algorithm comes from the calculation of the internal forces at the nodal points. These force calculations can be broken down into two parts: stress-strain calculations and contact-impact calculations. Both parts display very different levels of data locality, when considering parallel implementations.

The stress-strain calculations are performed over the elements. The calculation on each element requires as input the latest co-ordinates and velocities from only those nodal points defining the element. Once calculated, the force on the element is distributed as individual forces at the nodal points. These calculations produce the largest contribution to the overall computational cost (between 60% and 80%, depending on the particular model). In contrast to the stress-strain calculations, the contact-impact algorithms used within the code have, in terms of data access, a pseudo-global nature. Two sets of HPF+ crash kernels have been developed. All kernels resemble the explicit time-marching scheme but the first set of kernels is restricted to stress-strain calculations only, whereas the second set includes also contact-impact algorithms.

2.2   FIRE

FIRE [1] from AVL (Austria) is a fully interactive general purpose computational fluid dynamics program. It was developed specially for computing compressible and incompressible turbulent fluid flows as encountered in engineering environments. Two- or three-dimensional unsteady or steady simulations of flow and heat transfer within arbitrary complex geometries with moving or fixed boundaries can be performed. For the discretization of the computational domain a finite volume approach is applied. The resulting system of strongly coupled non-linear equations has to be solved iteratively by an outer non-linear cycle and an inner linear cycle. The matrices in the linear cycle are extremely sparse and have a large and greatly varying bandwidth. In order to save memory, only non-zero matrix elements are stored in linear arrays and are accessed by indirect addressing.

The HPF+ FIRE kernel contains the complete time marching solution of a passive scalar transport equation, representing key requirements of the entire FIRE application.

2.3   IFS

The Integrated Forecasting System [2] (IFS) of the European Centre for Medium Range Weather Forecasts uses a spectral forecast model for the prediction of weather for a period of up to 10 days ahead. This model is highly parallel and has been implemented on both distributed and shared memory systems. Given a strong desire to protect the scientific code from details of the parallel implementation, a transposition strategy is used. With this approach, the complete data required is redistributed at various stages of a time step so that the arithmetic computations between two consecutive transpositions can be performed without any interprocess communication. Such an approach is feasible because data dependencies in the forecast model exist only in one coordinate direction, this direction being different for each algorithmic component. An overwhelming practical advantage of this technique is that the interprocessor communication is localized in a few routines. The transpositions are executed prior to the appropriate algorithmic stage, so that the computational routines (which constitute the vast bulk of the IFS source code) need have no knowledge of this activity.

In terms of HPF, the transpositions are no more than redistributions that would permit more efficient data distributions to be used during an algorithmic stage. It is the representation of the data spaces used in the IFS that present a technical difficulty for HPF as they are all irregular. Grid point space, Fourier space and spectral space cannot be represented by simple BLOCK or CYCLIC distributions without the severe overhead of additional interprocessor communication during the algorithmic stages.

Three HPF+ kernels representing the key challenges of the IFS application have been developed. The LG kernel to address the IFS program issues of gridpoint space, the SL kernel to represent the semi-Lagrangian (SL) calculations, and the TS kernel to represent the Fourier and Legendre transforms to transform data between grid-point space and spectral space.

2.4   HPF Language Requirements

All applications outlined above can be classified as highly irregular. They are based on unstructured grids and extensively use loops with indirect array accesses. For an efficient implementation of these codes in HPF it is essential to distribute data at runtime in an irregular manner reflecting the structure of the underlying grids and to dynamically balance the computational load of the processors.

For all applications considered in the HPF+ project, the requirement for generalized block distributions with run-time defined block-sizes is an absolute must. In addition indirect distributions were necessary for both the FIRE and IFS benchmarks, and all kernels utilized at some stage dynamic data re-distribution.

The need for efficient parallelization of independent DO loops with indirect addressing, conditional statements, and subroutine calls - and in some cases a nesting of such loops - has been the central subject of the language and compiler developments. It is not at all realistic to expect that complex, irregular codes will be able to be re-written in F90 array statements or simple FORALL constructs.

3   High Performance Fortran

The HPF Forum [16], a group of about 40 researchers from academia and industry, with the aim of producing a standardized proposal for a data parallel language based on Fortran. The Forum released Version 1.0 of HPF in May 1993; in November 1994, HPF Version 1.1, mainly incorporating corrections to the language, was produced. First commercial compilers for a language subset appeared on the market in 1995. A number of projects and studies demonstrated the usefulness of HPF 1.1 for regular codes. However, at the same time it became clear [9] that the language could not express advanced applications such as multiblock codes, unstructured meshes, adaptive grid codes, or sparse matrix computations, without incurring significant overheads with respect to memory or execution time [18].

As a consequence, the HPF Forum continued a third round of meetings in 1995 and 1996, resulting in the release of HPF 2.0 in January 1997. HPF 2.0 is based on the current Fortran standard, Fortran 95. It consists of three parts: a) the Base Language, b) the Approved Extensions, and c) Recognized Extrinsic Interfaces.

3.1   The HPF Base Language

The features of the HPF Base language are basically those of HPF-1. They include:

- *Data mapping directives* for the regular distribution of data across explicitly specified sets of abstract processors: block, cyclic, and block-cyclic. Furthermore, an extensive set of mechanisms allows the control of data alignment.
- *Data-parallel directives* extending the array statements and the *forall* constructs of Fortran 95. The INDEPENDENT directive can be used to assert that iterations of a loop do not have loop-carried dependences and thus can be executed in parallel. A REDUCTION clause can be used with this directive to identify variables which are updated by different iterations using associative and commutative operators.
- *New intrinsic and library functions* include system functions to inquire about the underlying hardware, mapping inquiry functions to inquire about the distribution of arrays and a set of computational intrinsic functions.
- *Extrinsic procedures* allow the accommodation of programming paradigms different from the HPF paradigm.

3.2   HPF+

HPF+ essentially includes the HPF-2 Base language, except that templates are not supported, alignments and mechanisms for passing distributed arrays to procedures are simplified, and additional features are provided that allow the explicit equivalencing of processor arrays.

The advanced features of HPF+ allow more complex applications to be expressed using HPF. In particular, they allow greater control of the mapping of data objects. Users can map pointers and components of derived types, and can map objects to subsets of processors directly. The GEN_BLOCK distribution generalizes the block distribution by allowing non-equal sized blocks and the INDIRECT distribution allows each element of an array to be mapped individually using a mapping array.

Another important feature is the support of dynamic remapping of data. If an object has been declared DYNAMIC then it can be remapped at runtime using the REDISTRIBUTE directive.

The ON directive allows users to map computation onto processors. The RESIDENT directive allows the specification of information about accesses to data objects within the scope of an associated ON block.

The above features are also included in the HPF-2 Approved Extensions. HPF+ provides a number of additional features, in particular for the explicit control of communication and locality.

The REUSE clause can be used to express the redundancy of a communication schedules associated with an independent loop. It asserts to the compiler that the schedules for all arrays are invariant for all loop executions and thus have to be computed only once, upon first execution of the loop. This clause can be guarded by a condition, implying that schedules should be only reused if the condition yields true. Furthermore, the language provides *schedule variables*, which may be explicitly bound to schedules by the user. This mechanism allows the reuse of schedules beyond a single loop [3].

The HALO directive of HPF+ extends the HPF-2 SHADOW directive. It allows the explicit specification of the set of all non-local elements of a distributed array that are accessed during program execution. In contrast to shadows, halos may be specified for any distribution and may be changed at runtime whenever the distribution of an array is changed [5].

Finally, the PUREST directive can be used to characterize a pure procedure with the additional property that its invocations do not require communication.

4   VFC

VFC [4] is a source-to-source parallelization system that translates HPF+ programs into parallel F90/MPI message-passing programs. VFC is currently available on several parallel platforms, including the QSW CS-2, the NEC Cenju-3 and Cenju-4, the NEC SX-4, the IBM SP2, PC clusters, and networks of workstations. VFC implements the HPF+ features as discussed in Section 3.2, including general block and indirect distributions, dynamic data distribution, the reuse clause, and the halo directive. In contrast to most commercial compilers, VFC provides powerful parallelization strategies for non-perfectly nested irregular independent loops that may contain conditional statements or procedure calls.

4.1   Parallelization Strategies

The parallelization strategy of VFC is based on the *Single-Program-Multiple-Data (SPMD)* programming model. VFC translates a source program into an SPMD message-passing target program, which is usually parameterized in such a way that it can be executed on an arbitrary number of processors. In order to provide support for dynamic memory allocation, dynamic distribution/redistribution, parameterization of programs by the number of processors, communication schedule reuse, separate compilation, and other features, a general, dynamic parallelization methodology has been realized.

Distributed arrays are transformed by VFC into allocatable arrays such that, according to the user-specified distribution directives, each processor only allocates memory for those parts of an array that are owned by it. Work distribution ensures that each processor executes only parts of the computations of the

HPF+ program and is performed either automatically based on the owner computes rule, or may be controlled by the user by means of an on-clause. Potential accesses to non-local data are handled by automatically allocating temporary variables and generating the required communication to transfer data from the owner processors into these temporary variables.

A major focus within the HPF+ project has been on the development of efficient parallelization techniques for loops with indirect (vector-subscripted) array accesses that cannot be analyzed at compile-time. This is briefly outlined below. Moreover, VFC provides optimized parallelization strategies based on regular section intersection for regular array assignments and loops.

4.2   Parallelization of Irregular Loops and Schedule Reuse

Irregular loop nests are transformed by VFC based on the Inspector/Executor paradigm [12, 20] into three phases: the *work distribution phase*, the *inspector phase* and the final *executor phase*.

In the work distribution phase, the iteration space of a loop nest is partitioned among the available processors based on the ON HOME clause or using heuristics.

In the inspector phase each processor analyzes all its accesses to distributed arrays, filters out all non-local accesses, and derives communication schedules required for gathering/scattering non-local data from/to the owner processors.

In the executor phase all non-local data read in the loop nest is gathered according to the computed schedules, followed by a local computation phase executing a transformed version of the original loop. Finally, if non-local data has been written, it is scattered back to their owner processors.

The inspector phase, in particular the computation of communication schedules, is usually very time-consuming and may by far exceed the sequential execution time of a loop. Therefore, the inspector/executor scheme may be applied with reasonable efficiency only if the communication schedules are invariant and can be reused over many subsequent executions of a loop, amortizing the overall preprocessing costs. Eliminating redundant inspectors and reusing communication schedules is performed by VFC based on the REUSE clause. For an independent loop with a REUSE clause the work distributor and inspector of a loop are guarded by means of conditional statements to enforce that they are executed only if the reuse-condition is false or if the loop is executed for the first time. Moreover, for distributed arrays for which the user has specified the HALO attribute the inspector phase can be drastically simplified since communication schedules can be computed directly from the halo at the time the distribution is evaluated.

As shown in Section 4.4, communication schedules reuse is a key feature to achieve acceptable performance for codes parallelized with runtime techniques.

4.3   Code Instrumentation and Performance Analysis

The instrumentation component [6] of the VFC compiler allows a selective analysis of parallelized codes by automatically inserting calls to a measurement run-

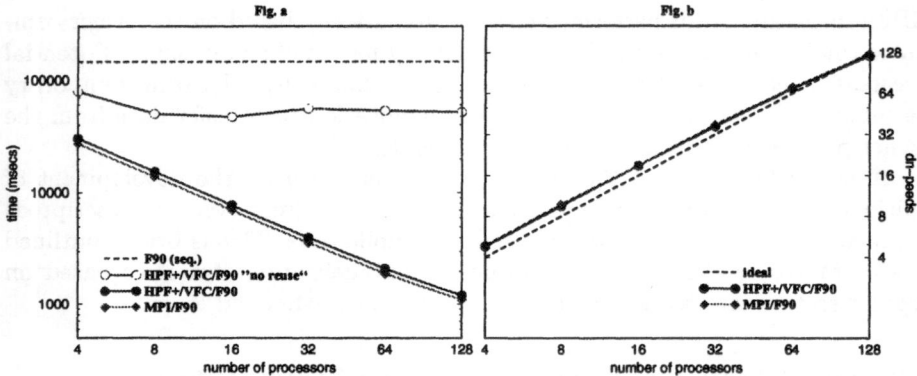

Fig. 1. HPF+ Weather Prediction Kernel (LG). Figure (a) shows the average time of 1 time-step of the serial kernel, of the HPF+ kernel without and with schedule reuse, and of the hand-written message-passing kernel. Figure (b) shows the speed-up of the MPI and HPF+ kernel with respect to the same serial version of the kernel.

time library. During execution of an instrumented program a tracefile is generated and can be analyzed by post-execution performance analysis tools such as, for example, MEDEA [7]. The code regions to be instrumented by VFC may be selected by the user by means of command-line options. Using environment variables, different tracefiles can be generated without recompiling the code.

One of the most important features of performance analysis tools is to relate performance indices back to the original source code. The instrumentation component addresses this issue by keeping track of the transformations performed at compile time and storing this information in a *measurement description file*. By combining the measurement description with the tracefile produced during runtime, MEDEA is capable of providing the user with detailed performance information at the HPF+ source code level.

4.4   Performance Results

We present performance results for two HPF+ benchmark programs and the equivalent hand-written MPI message-passing programs on the NEC Cenju-3 distributed memory parallel computer. In order to assess the importance of communication schedule reuse timings for a variant of the HPF+ kernels that did not reuse communication schedules are also given.

The first kernel extracted from the Integrated Weather Forecasting System (IFS) [2] addresses key issues arising in the representation of the IFS grid-point data spaces and the transpositions performed therein. The kernel has been coded using modules with allocatable arrays allowing to use the same executable for different resolutions, and resembles exactly the structure used in the IFS production code. The kernel used in our evaluation employed a grid with 134028

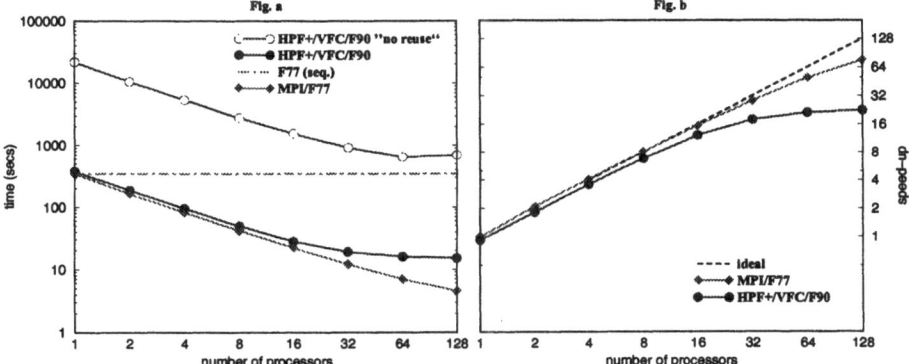

Fig. 2. HPF+ Crash Simulation Kernel. This figure shows the elapsed times
(a) and the speed-up (b) of the HPF+ crash kernel (with and without schedule
reuse) and an equivalent hand-coded MPI/F77 kernel on the NEC Cenju-3.

grid points and performed a total of 1000 time-steps. As shown in Figure 1 the
HPF+ kernel performs almost as well as the hand-coded message-passing kernel
exhibiting a speed-up of 120 on 128 processors. As can also be seen, without
reusing communication schedules no acceptable performance is achieved.

The second kernel represents the basic stress-strain calculation of a crash-
simulation code [11] based on 4-node shell elements. The kernel implements an
explicit time-marching scheme which is represented by an outer time-step loop
performing 1000 iterations. From within this loop, subroutines to calculate the
shell-element forces and to update the velocities and displacements according to
the computed forces are called. The communication schedules required within
the force calculation are invariant and thus only need to be computed once, in
the first time-step. In the code this is expressed by means of the REUSE clause.
In Figure 2 two variants of the HPF+ kernel, with and without schedule reuse,
are compared to a hand-written F77 message-passing code.

5   Conclusion

The data-parallel programming paradigm supported by HPF allows the efficient formulation of an important class of regular and, as demonstrated in the
HPF+ project, irregular applications. The HPF+ evaluations set the goal of
achieving absolute times approaching those possible with MPI implementations
in F77. The HPF+ language implementation within VFC has made major steps
towards reaching this goal. For the HPF+ LG kernel from ECMWF, absolute
performance essentially matched that of the MPI kernel. Otherwise the achieved
scaling with processor number is similar to that exhibited by the MPI codes.
However, for some of the kernels a major drawback remains the computational
overhead in the generated code. Another issue is that the pre-processing of com-

plex independent loops may lead to unacceptable performance penalties if communication schedules cannot be reused, as, for example, in the contact-impact version of the crash kernel.

In summary, there are two forces dominating the further development of HPF. First, the requirement to achieve performance comparable to the message-passing paradigm leads to enhancements of the HPF language within the constraints of the data-parallel paradigm. Secondly, the emergence of new architectures, such as clusters of SMPs including those with vector processors, as well as the growing importance of heterogeneous systems, will require a generalization of the HPF model[10] and the associated compilation and runtime technology.

References

[1] G. Bachler, R. Greimel, H. Schiffermueller, and M. Bernaschi. Benchmarking the parallel FIRE code on IBM SP1-2 scalable parallel platforms. *Lecture Notes in Computer Science*, Vol. 919, 1995, Springer Verlag.

[2] S. Barros, D. Dent, L. Isaksen, G. Robinson, G. Mozdzynski, and F. Wollenweber. The IFS model: A parallel production weather code. *Parallel Computing* 21, 1995.

[3] S. Benkner, P. Mehrotra, J. Van Rosendale, and H. Zima. High-Level Management of Communication Schedules in HPF-Like Languages. *Proceedings ACM International Conference on Supercomputing*, Melbourne, Australia, July 1998.

[4] S. Benkner. VFC: The Vienna Fortran Compiler, *Scientific Programming*, Vol.7, No.1, pp. 67-81, 1999.

[5] S. Benkner. Optimizing Irregular HPF Applications Using Halos. *Workshop Proceedings - International Parallel Processing Symposium (IPPS/SPDP)*, San Juan, Puerto Rico, April 1999, Springer Verlag.

[6] M. Calzarossa, L. Massari, A. Merlo, M. Pantano, D. Tessera. Integration of a Compilation System and a Performance Tool: The HPF+ Approach. *Proceedings International Conference on High Performance Computing and Networking*, Amsterdam, April 1998, Springer Verlag.

[7] M. Calzarossa, L. Massari, A. Merlo, M. Pantano, D. Tessera. Medea: A Tool for Workload Characterization of Parallel Systems. *IEEE Parallel and Distributed Technology*, 3(4):72-80, 1995.

[8] B. Chapman, P. Mehrotra, and H. Zima. Programming in Vienna Fortran. *Scientific Programming* 1/1,31-50, Fall 1992.

[9] B. Chapman, P. Mehrotra, and H. Zima. Extending HPF for Advanced Data Parallel Applications. *IEEE Parallel and Distributed Technology*, Fall 1994, pp. 59-70.

[10] B.Chapman,P.Mehrotra, and H.Zima. Enhancing OpenMP with Features for Locality Control. *Proc. ECMWF Workshop "Towards Teracomputing - The Use of Parallel Processors in Meteorology"*. Reading, England, November 1998.

[11] J. Clinckemaillie, B. Elsner, G. Lonsdale, S. Meliciani, S. Vlachoutsis, F. de Bruyne, M. Holzner. Performance Issues of the Parallel PAM-CRASH Code. *The International Journal of Supercomputing Applications and High-Performance Computing*, Vol.11, No.1, pp.3-11, Spring 1997.

[12] R. Das, J. Saltz, A manual for PARTI runtime primitives - Revision 2. Internal Research Report, University of Maryland, Dec. 1992

[13] G. Fox, S. Hiranandani, K. Kennedy, C. Koelbel, U. Kremer, C. Tseng, and M. Wu. Fortran D language specification. Department of Computer Science Rice COMP TR90079, Rice University, March 1991.

[14] High Performance Fortran Forum. *High Performance Fortran Language Specification. Version 1.1* Technical Report, Rice University, November 10, 1994.

[15] High Performance Fortran Forum. *High Performance Fortran Language Specification. Version 2.0* Technical Report, Rice University, January 31, 1997.

[16] High Performance Fortran Forum: http://www.crpc.rice.edu/HPFF/home.html

[17] ISO. Fortran 90 Standard, May 1991, ISO/IEC 1539 :1991 (E)

[18] P.Mehrotra,J.Van Rosendale and H.Zima. High Performance Fortran: History, Status and Future. In: E.Zapata and D.Padua(Eds.): *Parallel Computing*, Special Issue on Languages and Compilers for Parallel Computers, Vol.24, No.3-4,pp.325–354 (1998).

[19] Message Passing Interface Forum. MPI: A Message-Passing Interface Standard Vers. 1.1, June 1995. MPI-2: Extensions to the Message-Passing Interface, 1997.

[20] J. Saltz and K. Crowley and R. Mirchandaney and H. Berryman. Run-Time Scheduling and Execution of Loops on Message Passing Machines. *Journal of Parallel and Distributed Computing* 8(4), pp. 303-312, April 1990.

[21] H. Zima, P. Brezany, B. Chapman, P. Mehrotra, and A. Schwald. Vienna Fortran – a language specification. Internal Report 21, ICASE, Hampton, VA, March 1992.

TIRAN: Flexible and Portable Fault Tolerance Solutions for Cost Effective Dependable Applications

O. Botti[1], V. De Florio[2], G. Deconinck[2], F. Cassinari[3], S. Donatelli[4],
A. Bobbio[4], A. Klein[5], H. Kufner[5], R. Lauwereins[2], E. Thurner[5], E. Verhulst[6]

[1] ENEL S.p.A., R&D, Via Volta 1, I-20093 Cologno Monzese (Milano), Italy Tel:
+39-02-7224-5553, Fax: +39-02-7224-5465, botti@pea.enel.it
[2] K.U.Leuven, Dept. Elektrotechniek, Kard. Mercierln 94, B-3001 Heverlee, Belgium
[3] TXT Ingegneria Informatica, Via Socrate 41, I-20128 Milano, Italy
[4] Università di Torino, Dip. di Informatica, C.so Svizzera 185, I-10149 Torino, Italy
[5] Siemens AG, Dept. ZT SE 2, Otto-Hahn-Ring 6, D-81730 München, Germany
[6] Eonic Systems, Nieuwlandln 9, B-3200 Aarschot, Belgium

Abstract. Available solutions for fault tolerance in embedded automation are often based on strong customisation, have impacts on the whole life-cycle, and require highly specialised design teams, thus making dependable embedded systems costly and difficult to develop and maintain. The TIRAN project[1] develops a framework which provides fault tolerance capabilities to automation systems, with the goal of allowing portable, reusable and cost-effective solutions. Application developers are allowed to select, configure and integrate in their own environment a variety of software-based functions for error detection, confinement and recovery provided by the framework.

1 Industrial Motivations and Answers to the Market Needs

Market investigations with users and producers of automation systems have recognised the benefits offered by dependable systems, not only in the classical area of safety-critical tasks, but also in mission-critical ones, due to the high economic impact of failures. Moreover, dependable embedded systems were judged costly and difficult to develop and maintain, as they generally need strong customisation, impacting on the whole life-cycle, and requiring highly specialised design teams.

New proposals for fault tolerance (FT) should be based on pre-built and easily reusable solutions, customisable for different applications and portable on commercial and proprietary real-time operating systems (RTOS). The TIRAN project addresses these urgent needs raised from the market of real-time embedded automation systems.

---

[1] ESPRIT Project 28620 TIRAN — "Tailorable fault tolerance frameworks for embedded applications".

P. Amestoy et al. (Eds.): Euro-Par'99, LNCS 1685, pp. 1166–1170, 1999.
© Springer-Verlag Berlin Heidelberg 1999

TIRAN is a 24-month ESPRIT project. It provides a FT solution that will be developed on different platforms to allow an estimation of the reusability, and will be used on three different pilot applications to study its efficacy in making the applications more dependable. This paper describes the TIRAN solution and positions it with respect to the emerging standards and market trends.

The TIRAN solution is built around a software framework which provides fault tolerance capabilities to automation systems. Application developers are allowed to select, configure and integrate in their own environment a variety of software-based FT functions for fault masking, error detection, isolation and recovery, available in the framework. The framework includes techniques, tools and documentation to support the developers when combining the selected functions into their own fault tolerance strategies to obtain the required level of dependability.

The use of formal techniques to support requirement specification and predictive evaluation, together with the intensive testing on pilot applications, aims at guaranteeing the correctness of the framework, and to quantify the fulfilment of real-time, dependability and cost requirements, providing also valuable guidelines to the configuration process for the different users.

2 The Industrial Environment: Pilot Applications, Exploitation, and Market Access

The TIRAN Consortium includes large European suppliers and end-users of mission critical systems in their application fields. ENEL (the main Italian electricity supplier, the 3rd largest world-wide) and SIEMENS (the leading German company in the field of electrical engineering and electronics, one of the largest worldwide), EONIC Systems (B) (supplier of very high performance and application specific small RTOS for DSP processors and embedded ASIC cores), TXT Informatica (I) (provider of systems and services in many fields of real-time automation), the University of Leuven (B) and the University of Turin (I), particularly active on the R&D edges of fault tolerance and performance evaluation.

The project results, driven by industrial users' requirements and market demand, will be integrated with an existing off-the-shelf tool supplied by its producer (Virtuoso of Eonic Systems), will be ported onto commercial and proprietary RTOS (Windows CE, VxWorks, TEX), that will represent the basis for services offered by a system integrator (TXT), and are to be tested and adopted by two large end-user companies (ENEL, SIEMENS) within their application fields.

Pilot applications from Energy Distribution (ENEL, primary substation automation system) and Industrial Automation (SIEMENS, airfield lightcontrol system), bring a wide variety of requirements to the framework development and allow a deep validation of project results. Broad exploitation of results and market access are guaranteed in different sectors by the involved users, the technology providers and by third party association with market leaders—e.g., Wind River.

Functional	Error processing / Fault treatment	Error detection
		Fault containment / Error isolation
		Fault masking
		Error recovery
	Monitoring	Target monitoring
		Fault injection
Non-Functional	Portability	Platform and RTOS independence
	Flexibility	Configurability and scalability
	Performance	Hard real-time
		Soft real-time

Table 1. Classification of user requirements.

3   The TIRAN Approach to Fault Tolerance

TIRAN bases its fault tolerance strategy on the concept of "framework", which translates into the conjoint use of a layered system of fault tolerance mechanisms arranged into a library and of a sort of configuration tool by means of which the user expresses a number of recovery strategies. This is combined with a formal assessment of the performance of the framework approach. Industrial requirements have been summarized in Table 1. This section briefly describes the role and structure of the FT library, the configuration tool, and the techniques for framework validation and evaluation.

The FT Library. This is the central component of the TIRAN strategy. The system is made of three "layers": a *basic layer*, with entities related to fault containment, fault masking, error detection, isolation, and recovery mechanisms, a *control layer*, constituting a sort of "backbone" component for a coherent coordination of the entities in the basic layer, and a *monitoring and fault injection layer*, introduced for debugging purposes on embedded targets and substantially given by a hypermedia distributed tool. In particular the TIRAN control layer collects all relevant information concerning the current structure and status of the user application and of the basic TIRAN tools. This layer also coordinates recovery and interprets user-defined recovery strategies. The monitor describes the structure of the system (user application plus TIRAN framework components) and reports in a hypertextual structure all the events taking place at the basic layer and at the control layer. This component is also responsible for managing software fault injection services.

Configuration. A special component of the framework is a tool which offers a secondary application level that comes into action the moment an error is detected in the system. This level executes recovery scripts, to be coded by the user in a special "Recovery Language". With that, the user can configure strategies for recovery and reconfiguration. Special support towards design diversification (e.g., by means of recovery blocks) is under consideration.

Framework Validation and Evaluation. To prove the efficacy of the proposed framework it is important to assess the performance of the generic library and of its portings running on the different supported platforms. To reach these goals two techniques will be used in TIRAN: a probabilistic approach centred around modelling and measurements associated with fault injection.

Probabilistic modelling will be carried out using stochastic Petri nets [1]. Measurements of the system using software and physical fault injection on the pilot target environments will complement predictions providing a deep characterisation of the framework behaviour.

4   Comparison with Other Works

This section positions TIRAN with respect to other emerging fault tolerance approaches. In order to compare different approaches we selected a number of attributes, i.e., positive aspects and qualities of each approach, and assessed each approach with respect to these attributes. Doing this we can position each approach and research project into a unique spectrum (see Fig. 1). Five attributes have been selected: *efficiency* of the approach, measured by response times; *transparency*; *portability*; *cost* of adoption; *flexibility*, i.e. the effectiveness of the approach. Four orthogonal approaches have been considered (see below). Five ranking levels have been defined to measure the different approaches, ranging from low to high. The attributes and approaches constitute, in a sense, a *base* for discussing in more formal terms current fault tolerance trends.

System Approach. Embedding fault tolerance in the hardware and/or in the operating system is typically characterised by: high efficiency; high transparency; medium portability (good for the functional part of the user-application, bad portability of the features offered by the approach); no development or maintenance costs (for the customer), typically high costs for acquiring the target system with respect to general-purpose systems (ranking: low); bad flexibility. Project GUARDS [5] is partly based on this approach.

Library Approach. This approach consists in embedding a number of fault tolerance mechanisms and tools in the form of a library. The approach is characterised by: medium efficiency; medium transparency; high portability (in principle); low-to-medium development and maintenance costs; medium flexibility. An Example of this approach can be found in Project Isis.

Metaobject Protocols and Reflection. The idea of MOP's [4] is to open the implementation of an object-oriented language so that the developer can adopt and program different, custom semantics, adjusting the language to the needs of the user and to the requirements of the environment. It is characterised by medium efficiency; high transparency and full separation of functional concerns from dependability concerns; medium-to-high portability; medium development and maintenance costs; medium flexibility. An architecture including this approach is Friends [3].

The Recovery Language Approach. This approach is based on a language that constitutes a sort of secondary application layer, devoted to the execution of system-wide recovery strategies, which comes into action whenever an error is detected by one of the basic fault tolerance elements of a FT framework or of the underlying runtime systems. This approach is characterised by: medium efficiency; medium-to-high transparency; high portability; medium development costs and low maintenance costs; medium flexibility. Figure 1 summarizes and compares the above orthogonal approaches.

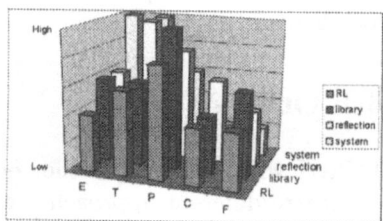

Fig. 1. A ranking of the orthogonal approaches according to five attributes.

TIRAN Positioning. TIRAN adopts a mixture of the library and of the recovery language approach for managing confinement/masking-level, detection-level and isolation-level tasks, with specific hooks to the underlying run-time system. The approach does not require object orientation and can be effectively used to add fault tolerance features to existing distributed or parallel applications.

5   Current Status and Outlook

Previous experiences within the EFTOS project [2] showed the feasibility of the TIRAN framework approach and pointed out the major required improvements. TIRAN started with a revision of the industrial requirements to allow a wide usability of results and to promote the development of an industrial product. A suitable general architecture was identified to allow the satisfaction of a wide range of performance requirements, from soft to hard real-time. Relevant novelties concern the flexibility and portability issues, the adoption of market standards, the integration of the framework within commercially available solutions supplied by their producers, as well as the adoption of formal validation techniques. The development of the FT framework has started: a concept demonstrator is available, the first framework prototype will be ready in October 1999, portings on VxWorks and Windows CE are scheduled for March 2000, while the final release of the product and its portings is planned for October 2000.

References

[1] M. Ajmone Marsan et al. *Modelling with GSPNs.* J. Wiley, NY, 1994.
[2] G. Deconinck et al. Industrial embedded HPC applications. *Supercomp.*, 69:23–44.
[3] J.-C. Fabre and T. Pérennou. FRIENDS: A flexible architecture for implementing fault tolerant and secure applications. *Proc. of EDCC-2*, Taormina, Oct. 1996.
[4] G. Kiczales et al. *The Art of the Metaobject Protocol.* MIT Press, MA, 1991.
[5] D. Powell. Preliminary definition of the GUARDS architecture. Technical Report 96277, LAAS-CNRS, January 1997.

OCEANS – Optimising Compilers for Embedded Applications*

Michel Barreteau[1], François Bodin[2], Zbigniew Chamski[4],
Henri-Pierre Charles[1], Christine Eisenbeis[5], John Gurd[6], Jan Hoogerbrugge[4],
Ping Hu[5], William Jalby[1], Toru Kisuki[3], Peter M.W. Knijnenburg[3],
Paul van der Mark[2,3], Andy Nisbet[6], Michael F.P. O'Boyle[7], Erven Rohou[2],
André Seznec[2], Elena A. Stöhr[6], Menno Treffers[4], and Harry A.G. Wijshoff[3]

[1] Laboratoire PRiSM, Université de Versailles, 78035 Versailles, France.
[2] IRISA, Campus Universitaire de Beaulieu, 35042 Rennes, France.
[3] LIACS, Leiden University, Niels Bohrweg 1, 2333 CA Leiden, The Netherlands.
[4] Philips Research, Information and Software Technology, Prof. Holstlaan 4,
5656 AA Eindhoven, The Netherlands.
[5] INRIA, Rocquencourt, BP 105, 78153 Le Chesnay Cedex, France.
[6] Department of Computer Science, The University, Manchester M13 9PL, U.K.
[7] Division of Informatics, The University, Edinburgh EH9 3JZ, U.K.

Abstract. This paper presents an overview of the activities carried out within the second year of the ESPRIT project OCEANS whose objective is to combine high and low-level optimisation approaches within an iterative framework for compilation. In this paper we discuss our approach to iterative compilation.

1 Introduction

Within the OCEANS project, the consortium intends to design and implement an optimising compiler that utilises aggressive analysis techniques and integrates source-level transformations with low-level, machine dependent optimisations [4, 5]. A major objective is to provide a prototype framework for iterative compilation, where feedback from the low-level is used to guide the selection of a suitable sequence of source-level transformations.

In general, compiler optimizations rely on static analysis, simplified processor and cache models and sometimes profiling information. Static analysis and models are necessarily pessimistic approximations. Profile based analysis produces averages of the observed behaviour of the system for a limited number of benchmarks/input sets. Compiler analysis determines the best parameters for each compiler optimisation separately (e.g., tile size). However, optimizations are not independent in their effect. We observe that the traditional approach to optimization only gives suboptimal results [6]. An alternative approach to optimisation, namely, iterative compilation, has been proposed in the OCEANS

---

* This research is supported by the ESPRIT IV reactive LTR project OCEANS, under contract No. 22729.

P. Amestoy et al. (Eds.): Euro-Par'99, LNCS 1685, pp. 1171–1175, 1999.
© Springer-Verlag Berlin Heidelberg 1999

project. It consists of searching for a good transformation sequence. This approach has also been adopted by some authors. Whaley and Dongarra [8], and Bilmes et al. [1] describe a system for generation highly optimised versions of BLAS routines by probing the underlying hardware to find optimal values for blocking factors, unroll factors etc. These systems are capable of producing code that is more efficient than the vendor supplied, hand optimised library BLAS routines. Wolf, Maydan and Chen [9] have described a compiler that searches for the best optimisation. This compiler uses a static cost model to evaluate the different optimisations in contrast to the present approach. Bodin et al. [2] search for the best optimisation on the assembly level, taking into consideration both execution times and code size.

In this paper, we present an overview of the work that has been carried out during the second year of the project. An overall description of the system is given in Section 2. In section 3 we discuss approaches to iterative compilation. Finally, in section 4 we present some conclusions and directions of future work.

2   An Overview of the OCEANS Compiler System

The OCEANS [4, 5] compiler is centered around two major components: a high-level restructuring system, MT1, and a low-level system for supporting assembly language transformations and optimisations, SALTO. SALTO is coupled with SEA, a set of classes that provides an abstract view of the assembly code, and tools for software pipelining (PiLo) and register allocation (LoRa). Their interaction is illustrated in figure 1. The OCEANS compilation process is driven by a global driver which select optimisations at the source-level and the low-level iteratively until a certain level of performance is reached.

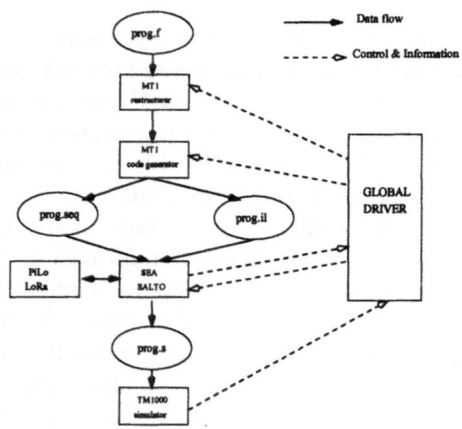

Fig. 1. The Compilation Process.

3   Iterative Compilation

We selected three important kernels and examined their behaviour across seven separate commodity processors and different data sizes: Matrix-Matrix Multiplication, Matrix-Vector Multiplication, and Successive Over Relaxation. We restricted attention to loop unrolling (with unroll factors from 1 to 20) and loop tiling (with tile sizes from 1 to 100). From the results reported in [3, 6] we can conclude that if static techniques are to find the local minima, they need to model program/processor interaction extremely closely. Given the difficulty of statically finding the minima, the next section considers the use of iterative compilation to search through the transformation space.

3.1   High Level Searching

Our compiler searches for the best transformation, by sampling the transformation space and measuring execution times. The algorithm used is grid based: we define an initial grid over the transformation space and refine around good candidate points, that is, points with a low execution time. See [6] for details. In figure 2, we have given the average percentage of how close to the absolute minimum the search algorithm comes across all platforms, benchmarks and data sizes. The x-axis shows the number of evaluations and the y-axis shows the distance to the minimum. The figure shows a monotonic decreasing graph that reaches high levels of optimization rapidly. Hence, we conclude that by visiting only a small fraction of the entire transformation space we can obtain good optimisations.

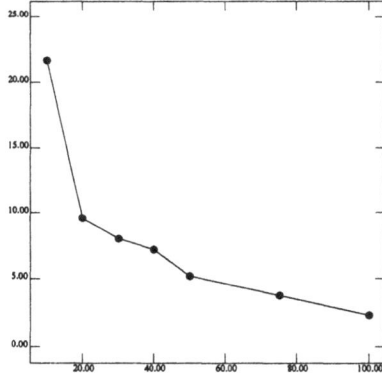

Fig. 2. Average percentage difference minimum

3.2   Alternative Search Techniques

Code Size–Performance Trade-Off An additional constraint of compilers for embedded applications compared to traditional compilers is that code size is important. We therefore not only search for the best optimisation concerning code performance but try to find a trade-off between these two aspects, using a cost model which takes dynamic and static feedback into account.

Genetic Algorithms We are also investigating the application of genetic algorithms (GA) as a means of determining the best transformation sequence. This has the potential benefit of investigating transformation spaces which cannot easily be described as a cartesian domain and is extremely robust in the presence of local minima. The OCEANS GA search is implemented as part of the GAPS compiler framework described in [7].

4   Conclusions and Open Problems

In this paper we have described the activities within the second year of the Esprit project OCEANS. We have addressed the problem of finding the best optimisation for a given processor, program and data size. We have shown that an iterative compilation approach based on a simple search algorithm may be able to find a good optimisation by visiting a relatively small fraction of the entire optimisation space. However, for real applications, the search spaces that need to be considered are extremely large. Hence aggressive pruning strategies need to be developed. The optimisation approach described in this paper is intended to be used for (kernels of) embedded applications. In such cases, long compilation times can be afforded since highly efficient code is required for these systems and compilation times can be amortised over a large number of shipped products.

References

[1] J. Bilmes, K. Asanović, C.W. Chin, and J. Demmel. Optimizing matrix multiply using PHiPAC: A portable, high-performance, ANSI C coding methodology. In *Proc. ICS'97*, pages 340–347, 1997.

[2] F. Bodin, Z. Chamski, C. Eisenbeis, E. Rohou, and A. Seznec. GCDS: A compiler strategy for trading code size against performance in embedded applications. Technical Report 1153, IRISA, Rennes, 1997.

[3] F. Bodin, T. Kisuki, P.M.W. Knijnenburg, M.F.P. O'Boyle, and E. Rohou. Iterative compilation in a non-linear optimisation space. In *Proc. Workshop on Profile and Feedback Directed Compilation*, 1998. Organised in conjuction with PACT'98.

[4] B. Aarts et al. OCEANS: Optimizing compilers for embedded applications. In *Proc. Euro-Par 97, LNCS 1300*, pages 1351–1356, 1997.

[5] M. Barreteau et al. OCEANS: Optimizing compilers for embedded applications. In *Proc. Euro-Par 98, LNCS 1470*, pages 1123–1130, 1998.

[6] T. Kisuki, P.M.W. Knijnenburg, M.F.P. O'Boyle, F. Bodin, and H.A.G. Wijshoff. A feasibility study in iterative compilation. In *Proc. ISHPC'99*, 1999.

[7] A. Nisbet. GAPS: Genetic algorithm optimised parallelization. In *Proc. Workshop on Profile and Feedback Directed Compilation*, 1998. Workshop organised in conjunction with PACT'98.

[8] R. C. Whaley and J. J. Dongarra.   Automatically tuned linear algebra software. In *Proceedings of Alliance 98*, Illinois, US, April 1998. Available through `http://www.netlib.org/atlas/`.

[9] M.E. Wolf, D.E. Maydan, and D.-K. Chen. Combining loop transformations considering caches and scheduling. *Int'l. J. of Parallel Programming*, 26(4):479–503, 1998.

Cray T3E Performances of a Parallel Code for a Stochastic Dynamic Assets and Liabilities Management Model*

G. Zanghirati[1], F. Cocco[2], F. Taddei[3], and G. Paruolo[3]

[1] Mathematics Department, University of Ferrara, via Machiavelli 35,
I-44100 Ferrara, Italy, phone (+39) 0532291004, fax (+39) 0532247292
and SMART s.r.l., via Don Minzoni 25, I-40057 Bologna, Italy,
g.zanghirati@dns.unife.it
[2] Prometeia Calcolo s.r.l., via G. Marconi 43, I-40122 Bologna, Italy,
phone (+39) 0516480912, fax (+39) 0516480911,
flavioc@prometeia.it
[3] CINECA, via Magnanelli 6/3, I-40033 Casalecchio di Reno (Bologna), Italy,
phone (+39) 0516171411, fax (+39) 0516132198,
taddei@notsomad.org, paruolo@cineca.it

Abstract. A parallel Asset & Liability Management (ALM) code was developed by SMART and Prometeia Calcolo for the EC project PALMA (Parallel Asset and Liability MAnagement). The code implements a stochastic approach based on a dynamic ALM model specially tailored for the Italian financial market. This paper reports the performances obtained on the Cray T3E at CINECA, running the code on real data provided by Credito Italiano. Very good scalability and efficiency have been achieved. Anyway the code is easily portable on other, possibly heterogeneous, high performance computing platforms.

1   Introduction

Asset and Liability Models represent an important tool for banks and financial firms, to measure the volatility of expected revenues. These mathematical and statistical models are traditionally static and deterministic: this limits the computational requirements and fits the resources of conventional computers. A dynamic and stochastic approach in forecasting the future expected revenues and measuring the related volatility represents a relevant improvement: this makes it possible to increase the precision of risk estimation, optimize capital allocation and enhance pricing and portability of financial products. This helps in reaching the main goal of an ALM system, that is to support the management in improving the performance of the investor by optimizing capital allocation among different business opportunities, taking the risk level into account (see just for example [2, 3, 4, 15]). ALM systems are build up on three main blocks: data

---

* This work is supported by the Esprit - PST European Community Project PALMA, n. 26001, http://www.smart.it/Palma.

P. Amestoy et al. (Eds.): Euro-Par'99, LNCS 1685, pp. 1176–1186, 1999.
© Springer-Verlag Berlin Heidelberg 1999

warehouse, analysis tools and reporting tools. In this paper we consider only the analysis phase and report the performances results of the parallel code developed for the EC – PST project PALMA (Parallel Asset and Liability MAnagement). The topic of the PALMA project is the core of the analysis tool: Different risk kinds are considered (market risk, credit risk, liquidity risk, etc.), together with multiple products and services embedding risk and multiple business units where risk is taken.

The model defines the stochastic process driving the dynamics of the environmental variables (the *scenario*) in order to compare strategies not only on the basis of their expected outcomes but also for the uncertainty (i.e. the risk) linked up to them. The parallel implementation outperforms the traditional ALM systems by increasing the precision by which the risk/return profile is simulated and by enhancing the pricing of financial products and hence their profitability.

The issues related to parallel implementation are discussed in section 2, section 3 reports the performance results and section 4 summarize conclusions and further developments.

2   Implementation Issues

The dynamic ALM model developed by Prometeia is based on a system of delay differential equations and was first implemented on a traditional scalar system by D. Borelli, G. Gervasio and G. Zuccaro: it gives a deterministic forecast for each fixed evolution of the scenario during a given time period called *temporal horizon*. The stochastic model considers the scenario as a random vector variable, thus a stochastic delay differential equations system is obtained. The discretization of this system on a finite set of time periods produces the discrete evolution laws of the interest variables. The simulations compute their values at each time step of the temporal horizon with respect to a set of scenarios, then a statistical analysis extracts the relevant information.

Given the independence of the computations related to different scenarios, the common intrinsic parallelism of the approach could be logicallydescribed as of SPMD type. Anyway, to produce a scalable and portable code we implemented the model in standard object oriented C++ programming language and we based parallel communications on MPI [16, 17] (therefore the real execution is of MIMD type). The parallel code utilizes a porting of the scalar Delphi version and it is structured into three main tasks:

- read firm's data. These data have been extracted from the firm's database and qualify the patrimonial status as well as the situation of the financial products at the starting date. The data are organized into several files, all to be read at the beginning of the execution;
- generation of new scenarios and elaboration of the data through the temporal horizon;
- collection of the results for the post-processing.

Table 1. Parallel pseudo-code.

```
for i = 1, ..., N_pes do in parallel
    1. compute environmental variables
    2. for each input data file do
        2.1 alloc suitable MPI local buffer
        2.2 if PE = root then
            | 2.2.1 read and decode data from disk in the MPI loc. buf.
        end
        2.3 broadcast data in root MPI buffer to all PEs
            /*** implicit synchronization barrier ***/
        2.4 store data in classes and free MPI local buffer
    end
    3. compute N_scn^(i) = number of scenarios to be elaborated on i-th PE
    4. compute N_rnd^(i) = amount of random numebrs needed by i-th PE
    5. initialize random numbers generator and alloc local buffer for output
    6. for j = 1, ..., N_scn^(i) do
        /*** main local loop ***/
        6.1 generate a new scenario and flush model structures
        6.2 elaborate current scenario trough temporal horizon
        6.3 store computed data in local memory output buffer
        6.4 if memory buffer is full then
            | 6.4.1 write buffer to disk and clean it
        end
    end
    7. if memory buffer not empty then write buffer to disk
end
```

The time step length can be freely chosen: in the case presented each time step represents one month. The skeleton of the algorithm is shown in Tab. 1, where N_{pes} is the number of processors used in the simulation.

Given the generally large amount of data to be elaborated, to achieve an efficient parallel code the optimization of I/O requests and disk access is mandatory. We paid particular attention to this tasks, exploiting both software and hardware[1] issues, but mainly designing code branches in order to reduce the concurrent I/O requests to the minimum [18, 1]. The version we present here is tailored to the Cray T3E at CINECA, a 3D torus MIMD architecture with 128 processors [9, 10]. However, other versions differ only in details.

Specifically, we allow only one processor (the *root*) to read the firm's input data from compressed disk files and then duplicate these data on other processors. Each processor maintains in the local memory copies of the data needed to start every simulation, so that only one broadcast is needed for all the execution. Then, each PE performs the assigned simulations autonomously way, with no communication and/or synchronization with the others. During this phase

---

[1] To maximize portability, we made a very limited use of hardware issues, allowing a simple tuning of this parts for each particular machine

it resets data structures (*lists flushing*), performs the computations and stores the results in a local memory output buffer, avoiding any disk access. When the computation ends, it stores the local output buffer on the physical device with an unformatted disk access. Two issues have to be pointed out. First, point 6.4 allows the elaboration of an unlimited number of scenarios and ensure portability: it may slow down the execution, but only for a large number of scenarios, or for an extremely large set of firm's data, or for a low number of concurrent processors in the system. Second, the random number generator initialization is a non trivial matter to be considered in order to obtain meaningful results: among several potential good choices (see e.g. [11]), we used the PRNGlib MPI Fortran library [12] to accomplish this crucial point.

Analysis results are based on two data sets of main financial indices of interest, respectively called endogenous and exogenous variables: the former have to be computed, while the latter are supposed to be known along the temporal horizon from previous estimations.

In the implementation of the scenario generator we followed Prometeia's analysis on both the Italian financial market and firm's model, as described in [5]. As Tab. 1 shows, this task is really "local" to each PE during the execution of the main loop, the only parallel issue being the generation of random numbers. The PRNGlib library ensure that they do not overlap nor are correlated with the ones generated on other PEs (see also [8, 13, 19]).

3   Performance Results

In order to evaluate the effectiveness of the code we performed an extensive simulation based on a subset of real data provided by Credito Italiano. The code, aggressively optimized by the compiler, was ran from a directory enabling a fast disk access. However, this affects only the write-to-disk phase, so that very little differences should be expected on the overall behaviour when output buffers fully fits in local PE memories. When this is not the case, only the loop time of the iteration involved in the disk access is affected.

We performed two types of simulations: in the first one we used 2048 scenarios per run and we repeated each run 10 times to obtain confidence statistics about the time sampling; in the second one we used 10240 scenarios per run with only one repetition. This two-stage approach was motivated by the aim of estimating system-dependent deviations on measures, due to other processes active on the system at the same time. The basic statistics we consider would lead to a sort of "system immunization". Note that jobs involving more than 64 PEs ran during the night or the week-ends.

In the following, with "averages of standard deviations" we mean different things for each monitored time sample: in the case of 2048 scenarios and 10 repetitions we computed the standard deviation with respect to the current number of PEs and then averaged the 10 resulting values, while in the case of 10240 scenarios simply the standard deviation average across the processors was taken.

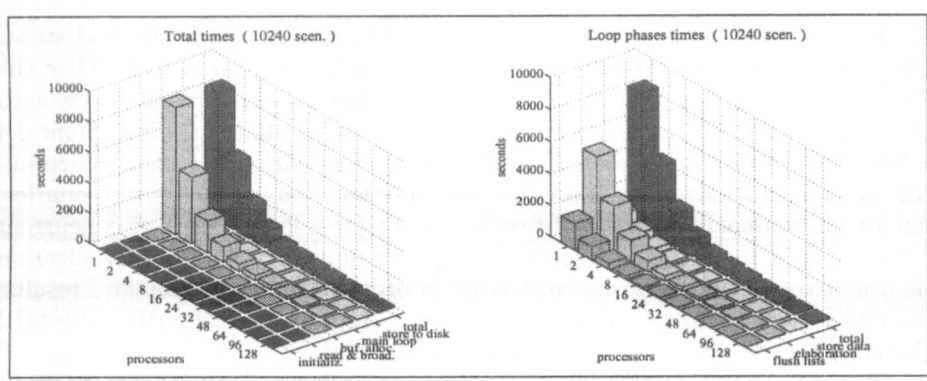

Fig. 1. Mean of total times and main loop times for 10240 scenarios.

Consider the second type of simulations: the number of scenarios elaborated by each processor ranges from 5120 for the 2 PEs configuration to 80 with 128 PEs. Figure 1 reports the overall behaviour of the whole program together with the details of the main loop local to each PE, both averaged with respect to N_{pes}. They clearly show how well the code scales, as expected given the "embarrassing" parallelism of the main body of the program. The first plot in Fig. 1 shows the success of the optimization strategies: communication times and the parallelization overhead remain almost constant and negligible with respect to the computational time, at least until the latter remains of a considerable amount. Things may change for a larger number of processors, since the time spent in computation becomes smaller: however, the good qualitative behaviour is preserved with very low level of standard deviation. We noticed that measure distortions are concentrated in the store-to-disk phase, where conflicts with other system jobs appear. Using the first kind of simulations, with 2048 scenarios, we consider that the mean values of code branches time are well representative since small values of standard deviations were observed. Furthermore, the time values remained almost constant across all PEs, thus the average of the local main loop time samples over all the simulations gives a quite good estimate of the time needed by *each* scenario, shown in Tab. 2. Different situations appear for store-to-disk statistics: higher variability is shown due to the disk access

Table 2. Mean values of local loop times.

local main loop phases	seconds	std. dev.
scenario generation and lists flushing	0.16	0.0052
data elaboration	0.58	0.0331
store output data in memory buffer (no disk access)	0.11	0.0006
total	0.85	0.0358

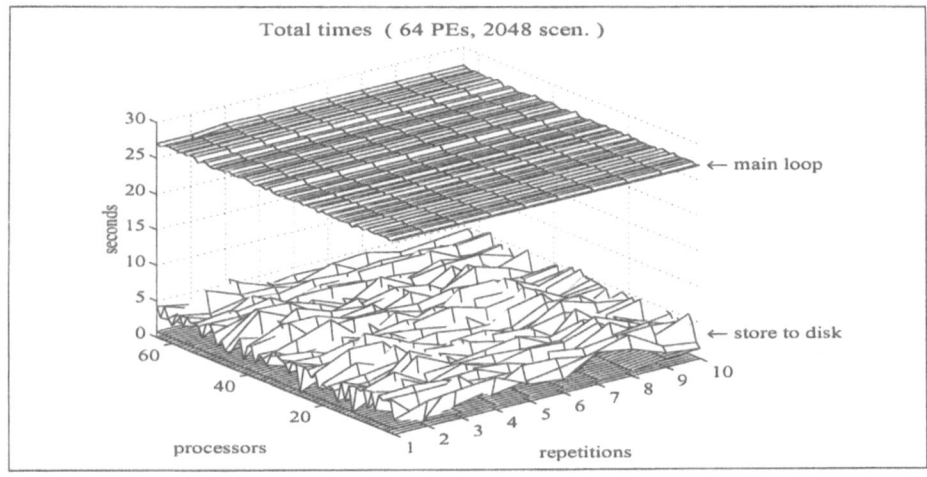

Fig. 2. Comparison of global execution times for 64 PEs, 2048 scenarios.

conflicts, compensated only in part by the decreasing amount of data to output. The maximum is reached for 128 PE: once again this is not only due to the system interaction, but also to the fact that the computational time could be of the same magnitude of store-to-disk time. Given that this is the most critical part of the code, it is interesting to see how it changes. Considering the 2048 scenarios simulations, Fig. 2 is representative of the influence of this phase with up to 64 PEs and Fig. 3 clearly shows the very critical last situation. The general conclusion is that the uncertainty in run times is due to the store-to-disk phase only, while read-from-disk and broadcast phases give a negligible contribution. The (averaged) performances of the code for this case are reported in Fig. 4. As it was expected from the type of code parallelism, the main loop exhibits very good scalability and efficiency. The small super-linear speed-up should be considered as a side effect of statistics. The other interesting contribution comes from the store-to-disk phase: not surprisingly, this is the worst performance case. In fact the low level mechanism to write binary data to disk is as faster as smaller is the number of processors requiring it, but physical location of PEs as well as available free output channel are also important. For all these sources of uncertainty, speed-up and efficiency analysis of the store-to-disk phase have a very low confidence level. Both the initialization and the read-and-broadcast phases efficiency is decreasing for N_{pes} increasing, as obvious: anyway, their contribution to the global behaviour is of no importance. Hence, globally, the code exhibits very good performances (solid line). The efficiency is quite good too, but for 128 PEs we have a minimum which can be clearly attributed to the critical store-to-disk phase.

While both the 2048 and the 10240 scenarios simulations present similar overall behaviour with respect to time, shown in Fig. 1, in the second case the local buffer needed by each PE to store output data does not fit in memory when

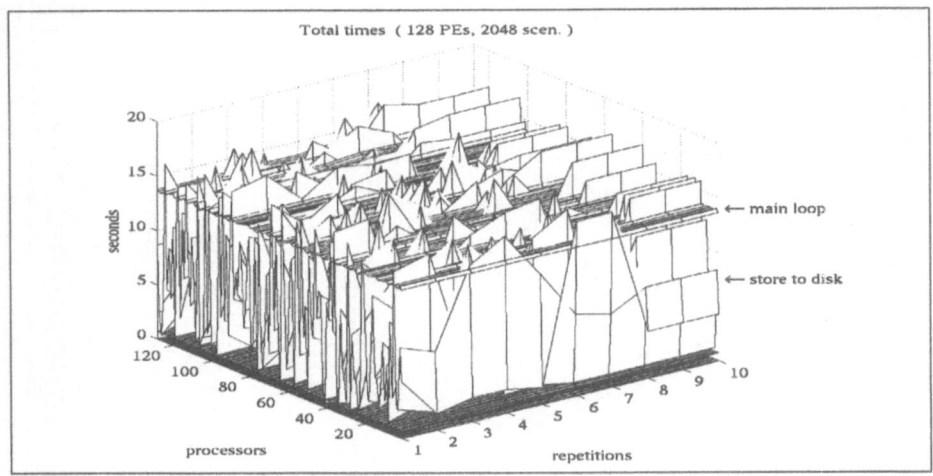

Fig. 3. Comparison of global execution times for 128 PEs, 2048 scenarios.

configurations with 1, 2 or 4 processors are used. This means that intermediate disk accesses take place to write the data when the buffer is full. While worser performances have to be expected, the results show an acceptable degradation with respect to the global times: this is shown by the very low super-linear speed-up of the main loop line in Fig. 6, from 8 to 128 PEs. Clearly, intermediate store-to-disk can be costly, but with the method adopted for disk output, only when 16 or more processors ask *simultaneously* for disk access then I/O conflicts become heavy. On the other hand, if 16 PEs (or more) need intermediate disk access, this means that a very large number of scenarios and/or a huge amount of firm's data have to be elaborated, so the computational time may hide the problem. Thus, up to such extreme situations, the code is expected to scale very well. Using MPI I/O cannot improve the performances, even if non-blocking calls from the new standard are used: this is because MPI calls rely on the underlying I/O subsystem and hence on hidden buffering strategies, mainly on large computers. Furthermore, to force MPI I/O to behave in non-standard way is not simpler than directly working with programming options for system resources, while the results can be far more unstable. Once again we point out that disk access conflicts at the end of the simulation become noticeable only for a large number of PEs: see Fig. 5 for an example. Figures 6 and 7 show scaled speed-ups and efficiency for the 10240 scenarios simulations. There are very limited differences with respect to the 2048 scenarios case, but two things should be noticed: first, the computational effort in the main loop is much greater than in the previous case, and so the degradation for the final store-to-disk phase is much less important; second, the super-linear speed-up of the main loop is not due to statistical side effects, but it appears, as already mentioned, because the code executed is not exactly the same in scalar and in parallel when more than 4 processors are used, given the intermediate disk swap of the memory buffer.

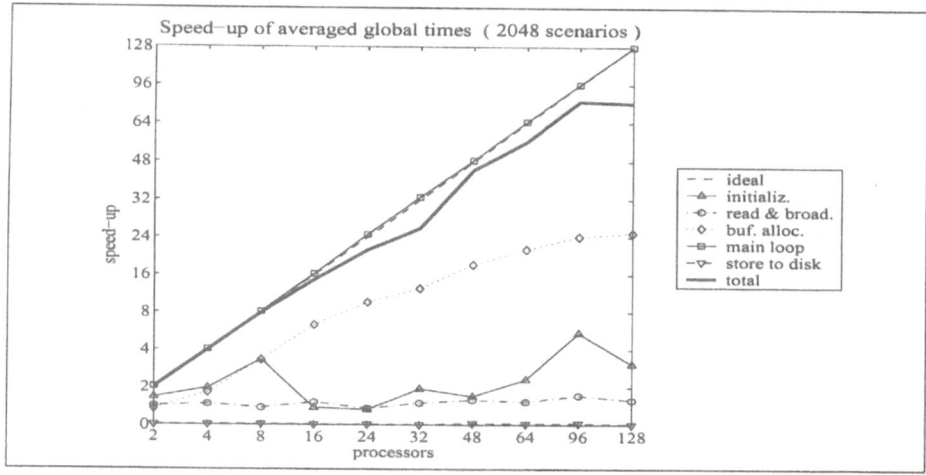

Fig. 4. Scaled speed-up statistics of averaged total times for 2048 scenarios.

Finally we mention, without reporting results, that we tried a simulation near the maximum allowed to fit the output buffer in memory: over 158000 scenarios were elaborated on 128 PEs (1250 per PE) in less than 1000 seconds. At the end, each processor had to store a buffer of more than 200 Mbytes to the disk and we experienced disk full problems.

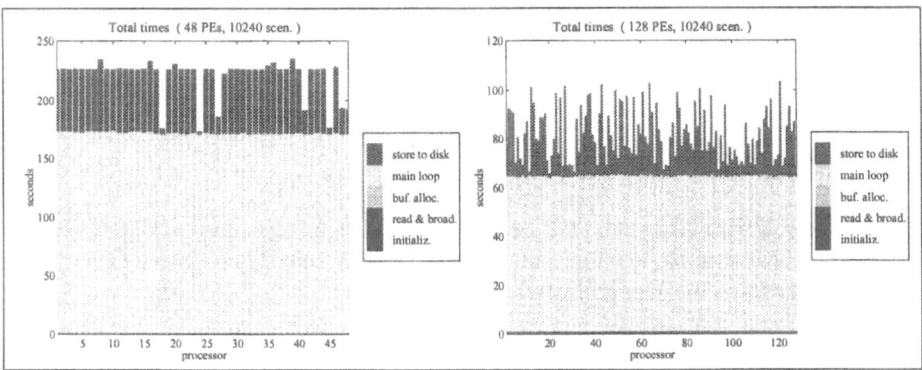

Fig. 5. Comparison of *per* PE total times for two simulations with 10240 scenarios.

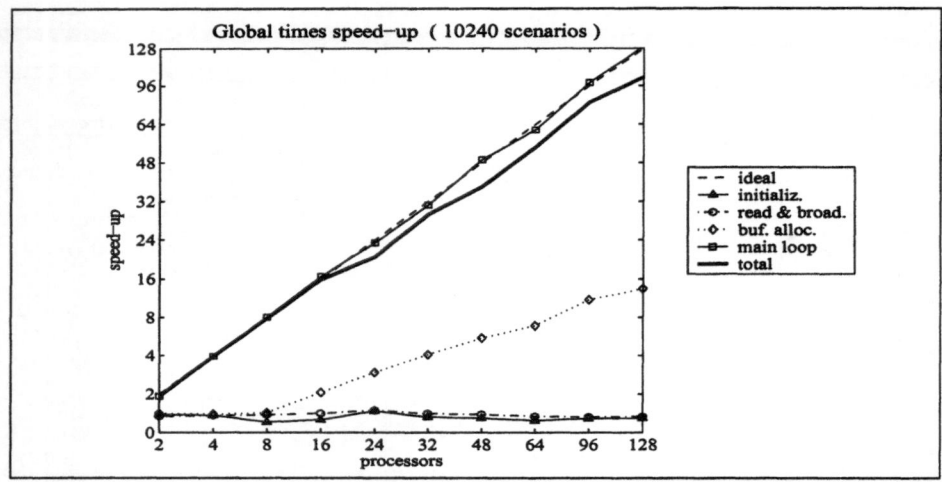

Fig. 6. Scaled speed-up for the total times, in the case of 10240 scenarios.

4   Conclusions and Developments

In this work we presented the performances of the parallel implementation of a stochastic dynamic ALM model on the Cray T3E at CINECA. The experiments performed clearly showed that good speed-up and efficiency results have been achieved. Further developments are planned in the near future, mainly concerning additional improvements of the computational core and of the web interface currently developed by SMART, to allow the remote utilization of the program trough the internet. Finally, the addition of a "strategies optimization" module is under study, to make the code a complete tool for financial decision making support.

Acknowledgments

The authors acknowledge the help received by various members of SMART (particularly M. Matteuzzi, developeder of the web interface), Prometeia Calcolo and CINECA staff during the completion of this work.

References

[1] S. Balay, W.D. Gropp, L.C. McInnes, B.F. Smith, *Efficient management of parallelism in object-oriented numerical software libraries*, in *Modern Software Tools in Scientific Computing*, E. Arge, A.M. Bruaset, H.P. Langtangen (eds.), Birkhäuser Press, 1997.

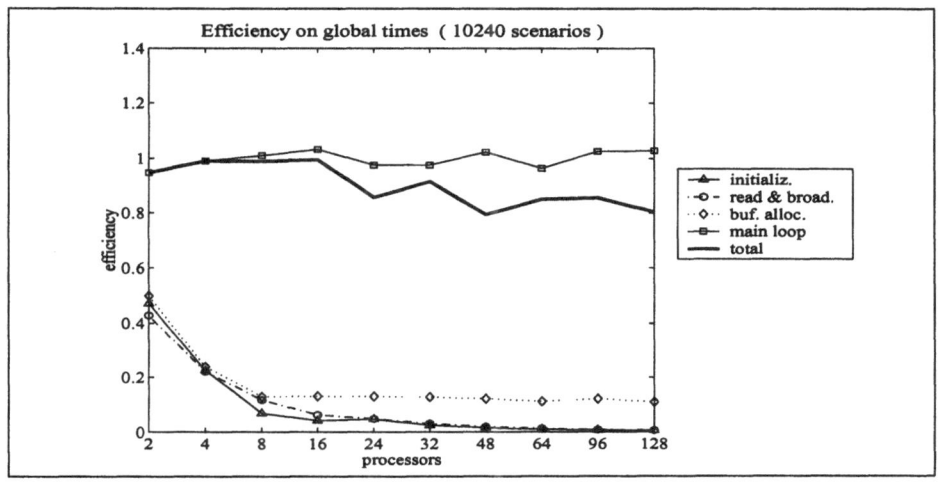

Fig. 7. Efficiency for the total times, in the case of 10240 scenarios.

[2] G.C.E. Boender, *A hybrid simulation/optimization scenario model for as-set/liability management*, Tech. Rep., ERASMUS University, Econometric Institute, Rotterdam, 1996.

[3] R. Camillo, *Modellizazione ad oggetti di attività finanziarie*, in *Atti del Ventesimo Convegno Annuale A.M.A.S.E.S.*, Urbino, 5–7 settembre 1996, (1997) 113–118.

[4] Y. Censor, S.A. Zenios, *Parallel optimization. Theory, algorithms and applications*, Oxford University Press, Oxford, 1997.

[5] F. Cocco, G. Zanghirati, P. Coriazzi, D. Borelli, G. Zuccaro, *D3.1 - Firm's model code completion* and *D3.2 - Scenario generator code completion*, deliverables of the EC - PST project PALMA, 1998.

[6] A. Consiglio, S.A. Zenios, *High-performance computing for computer aided design of financial products*, in J.S. Kowalik, L. Grandinetti and M. Vajtersic, eds., *High Performance Computing: Technology and Applications*, vol. **30** NATO ASI series, Kluwer Academic Publisher, 1997.

[7] G. Consigli, M.A.H. Dempster, *Dynamic stochastic programming for asset–liability management*, Annals of Operations Research **81** (1998) 131–161.

[8] A. De Matteis, S. Pagnutti, *Controlling correlation in parallel Monte Carlo*, Parallel Computing **21** (1995) 73–84.

[9] G. Erbacci, *Proceedings of the 7th Summer School on Parallel Processing*, CINECA, Bologna, Italy, July 1998.

[10] J. Haataja, V. Savolainen (eds.) *Cray T3E User's Guide*, 2nd Edition, Center for Scientific Computing, Finland, 1998.

[11] M. Mascagni, A. Srinivasan, D. Ceperley, *Scalable Parallel Random Number Generators Library for Parallel Monte Carlo Computations. V. 1.0*, National Center for Supercomputing Applications, University of Illinois at Urbana-Champaign, 1998.

[12] N. Masuda, F. Zimmermann, *PRNGlib: a Parallel Random Number Generator Library*, Tech. Rep. TR-96-08, Swiss Center for Scientific Computing, 1996.

[13] H. Niederreiter, *Random number generation and quasi-Monte Carlo methods*, SIAM, Philadelphia, 1992.

[14] G. Zanghirati, F. Cocco, G. Paruolo, F. Taddei, *D2.2 - Successful run on CREDIT's portfolio data–set, D3.3 - Demonstrator application code completion, D4.2 - Optimization and performance report*, deliverables of the EC project PALMA, 1998.

[15] S.A. Zenios, *Financial Optimization*, Cambridge Univ. Press, Cambridge, 1993.

[16] *MPI: A Message-Passing Interface Standard*. Message Passing Interface Forum, version 1.1 June 12, 1995.

[17] *MPI-2: Extension to the Message-Passing Interface*, Message Passing Interface Forum, version 2.0 July 18, 1997.

[18] *Application Programmer's I/O Guide*, Cray Res. Inc., pub. SG-2168 3.0.

[19] *RNG on the Net*, http://random.mat.sbg.ac.at/~charly/server.

Parametric Simulation of Multi-body Systems on Networks of Heterogeneous Computers

Javier G. Izaguirre[1], and José M. Jiménez[1], Unai Martín[2], Bruno Thomas[2],
Alberto Larzábal[3], Luis M. Matey[3]

[1]CMP, S.L.
Plaza de Pinares, 1, San Sebastián, SPAIN
jmjimenez@sportstt.com
[2]CERFACS
42, Av. G. Coriolis, Toulouse, FRANCE
{unai,thomas}@cerfacs.fr
[3]CEIT
P. Manuel de Lardizábal, 15, San Sebastián, SPAIN
Phone +34943212800 Fax +34943213076
{alarzabal, lmatey}@ceit.es

Abstract. This paper presents an overview of the software tools developed in the
framework of HIPERCOMBATS and MYSHANET projects. These are High
Performance Computing and Networking (HPCN) projects funded by the
European Commission in the frame of the CAPRI initiative and the Technology
Transfer Nodes (TTN) initiative, respectively. These tools are aimed at helping
the engineers in the complex task of designing new multi-body systems. The
main objective of both projects has been the development of software tools that
allow the designer performing parametric simulations on a non-dedicated
network of heterogeneous computers. The simulations are performed
concurrently by the parallel task manager module. Finally, an interactive post-
processor allows the users to have an easy access to the computed results.

1. Introduction

This paper presents recent developments in the field of computer-aided multi-body
analysis and design using high performance computing techniques. In particular, the
most relevant results achieved in the development of the projects HIPERCOMBATS
and MYSHANET are presented. Both projects have been developed by international
consortia funded by the European Commission's ESPRIT program. The resulting
software tools are aimed at helping the engineers in the complex and cumbersome
task of designing multi-body systems (MBS). To achieve this goal, complementary
techniques and methodologies have been integrated: "virtual prototyping" of MBS,
development of parametric mechanical models, parallel and distributed computation
techniques, non-dedicated networks of Unix workstations in HIPERCOMBATS
project and personal NT computers in MYSHANET project, object-oriented
programming approach, and user friendly interfaces for pre- and post-processing
tasks.

P. Amestoy et al. (Eds.): Euro-Par'99, LNCS 1685, pp. 1187-1194, 1999.
© Springer-Verlag Berlin Heidelberg 1999

The design of MBS is a fairly difficult task due to the non-linear characteristics of the equations that govern the kinematic and dynamic behaviour of such a system. Due to the complexity of the equations that govern the motion of a MBS, simplified models and model-oriented programs are often used in the early stages of the design process to understand the behaviour of the MBS under different working conditions. The advantage of this approach is that those simple models and programs can provide very quickly the response of the mechanical system under study. Intensive parametric studies can be carried out on a single PC in a working day. In the design process, engineers are interested in visualising a whole set of successive responses of the mechanical system; they want to know how changes on design parameters affect the system behaviour and they need the answer in a short time. In fact, when a new MBS is nothing more than a blueprint, it is more important to have the possibility of doing many quick simplified investigations rather than a single detailed analysis.

In this context, the use of general-purpose MBS computer programs might be very profitable (see 1 to 4). These programs can provide a huge amount of useful and accurate data about the behaviour of the MBS. At present time, several programs like ADAMS, DADS or Pro/Mechanica are available in the market. The main drawback of these programs is that they are not design tools but analysis tools. They simulate the behaviour of a MBS once all of its geometric and dynamic characteristics have been defined. However, if an intensive parametric study were to be carried out using one of these codes, it would take some days of continuous work for one engineer, who should define the model for each different set of design parameters, run the simulation and review the results.

The aim of HIPERCOMBATS and MYSHANET projects has been to develop a new software tool that permits the use of detailed MBS models since the first design steps. The resulting code allows carrying out in a working day a complete parametric study of any MBS with the accuracy of a detailed multi-body analysis. The use of such a software tool at any stage of the design process provides an important amount of data to study the influence of the different design variables.

2. Kinematic and Dynamic Simulation of MBS

Roughly speaking, a MBS can be defined as a complex mechanical system composed of rigid or flexible bodies linked together by means of kinematic joints. These joints allow certain degrees of freedom between the bodies it links.

The motion of a MBS can be described in terms of a set of independent coordinates called degrees of freedom. In the most general case, those coordinates cannot define unequivocally the position of each body of the mechanical system. To overcome this difficulty, an extended set of coordinates is often used. These coordinates are not independent but they are related through a set of non-linear algebraic equations, called kinematic constraint equations. This set of algebraic equations can be written as:

$$\ddot{O}(q, t) = 0 \qquad (1)$$

The vector q represents a set of n unknown dependent coordinates, m is the total number of kinematic constraint equations and therefore f = n-m is the number of degrees of freedom of the MBS. The kinematic analysis of MBS requires the solution of Equation (1) in order to know the position of all the bodies of the mechanism, once the position of the input elements is known. Analogously, the solution of the velocity and acceleration analyses requires the solution of the time derivatives of Equation (1). These derivatives can be written as:

$$\ddot{O}_q (q, t) \dot{q} = -\ddot{O}_t \tag{2}$$

$$\ddot{O}_q (q, t) \ddot{q} = -\tilde{O}_t - \tilde{O}_q \dot{q} \tag{3}$$

The dynamic analysis of MBS takes account of the kinematic constraints as well as the forces acting on the system. Using the Lagrange's equations, it is possible to derive the following expression for the equations of motion:

$$M\ddot{q} + \ddot{O}_q^T \ddot{e} = Q \tag{4}$$

where Q contains the external forces and λ is the vector of reaction forces associated with the kinematic constraints.

Expression (4) has n equations and (n+m) unknowns: the n components of the vector of dependent accelerations and the m components of the vector of reaction forces. In order to have a sufficient number of equations it is necessary to supply m more equations. One choice is to use Equation (3), which along with (4) define a set of non-linear differential algebraic equations (DAEs) of index one, that is written as:

$$\begin{bmatrix} M & \ddot{O}_q^T \\ \ddot{O}_q & 0 \end{bmatrix} \begin{Bmatrix} \ddot{q} \\ \ddot{e} \end{Bmatrix} = \begin{Bmatrix} Q \\ -\tilde{O}_t - \tilde{O}_q \dot{q} \end{Bmatrix} \tag{5}$$

A second possibility of formulating the equations of motion is based on the use of a velocity transformation. It may be demonstrated that, for a given position, a basis of the subspace of possible motions may be computed as the columns of a matrix R. The columns of this matrix determine a basis of the subspace of the jacobian matrix of the kinematic constraints. Thus, the following relationships are satisfied:

$$\ddot{O}_q R = 0 \tag{6}$$

$$\dot{q} = R\dot{z} \tag{7}$$

Equation (7) represents the linear relationship that exists between the dependent velocities and the velocities of the degrees of freedom. Taking the time derivative of (7), substituting in (4) and pre-multiplying by the matrix R transpose, the equations of motion can be expressed as a set of non-linear ordinary differential equations (ODEs), in the form:

$$R^T M R \ddot{z} = R^T \left(Q - M \dot{R} \dot{z} \right) \tag{8}$$

Once the differential equations of motion have been formulated either using Equation (5) or (8), they have to be integrated in order to obtain the time response of the mechanical system. The numerical integration routine will demand the computation of the function derivative, which requires the computation of the different terms involved in Equation (5) or (8) as well as the solution of a set of linear equations. All these computations have to be performed thousands of times until the end of the simulation time interval is reached. It can be easily understood that this is a very time consuming task, which may largely benefit from the use of parallel computing techniques.

3. Parallel Implementation in an Industrial Context

From the user point of view the main requirement for the parallel code is the easiness of use and in this respect the parallel aspects have to be hidden as much as possible. In an industrial context, the following points have to be addressed due to their relevance:

- *Openness and maintainability*. The code must be easily used, maintained and upgraded, especially by those who are not parallel experts
- *Similar to run a sequential code*. This is particularly important when considering that end-users are not used to run codes on parallel platforms.
- *Portability*. The resulting parallel program must as little "machine dependent" as possible. In this particular case, the code has been implemented using a portable parallel strategy and portable message-passing library: PVM (5).
- *Scalability*. A coarse grain parallelism is implemented to run the simulation on independent sets of design parameters
- *Reverse porting should be possible*. In order to maintain the parallel version, it must be possible to derive easily a serial version for non-parallel environments on which more convenient development and debugging tools are available.
- *Load balancing and fault tolerances*. The parallel task manager module takes account of load balancing issues as the variability either in the load or in the computational power of the available processors, and detects the breakdowns and recover them.
- *Independent development*.

In order to exploit efficiently the computational power of the computers that compose the network, the proposed parallel strategy implements a dynamic load balancing scheme based on the so called pool of tasks paradigm. It is typically implemented in a master/slave program where the master program holds the pools and farms out tasks to slave programs as soon as they fall idle. The pool is usually implemented as a queue. With this method all the slave process are kept busy as long as there are tasks left in the pool. This strategy is particularly suitable for the problem addressed in the considered problem, where the interval of interest for each design parameter is decomposed in subintervals that define the pool of tasks to be performed. A detailed description of the parallel task manager can be found in (5).

4. Software Components

The final objective of both HIPERCOMBATS (for Unix) and MYSHANET (for NT) projects has been to develop an efficient parallel software tool for the design of MBS by performing parametric studies economically on a network of non-dedicated heterogeneous computers. As a result, heterogeneous networks of both Unix and Windows NT platforms can be used to carry out the parallel simulations.

In order to be scalable, portable and re-usable the software has been developed using an object oriented approach (C++ language), a message passing paradigm (namely PVM (6)) to manage the parallelism, and standard user interface libraries.

The software is composed of four different modules, which are depicted as boxes in Figure 1. A detailed description of these software models can be found in (7).

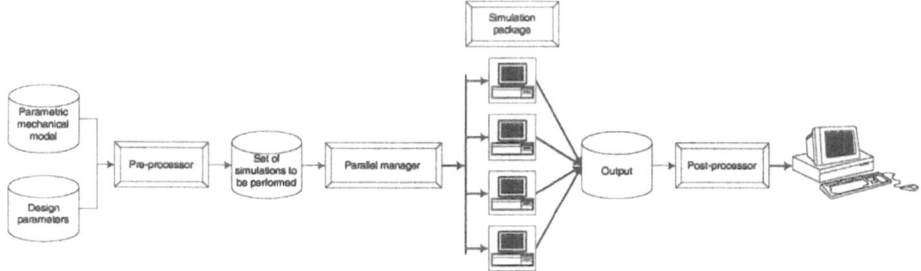

Figure 1. Software modules.

4.1 Pre-processor Tool

The pre-processor is an interactive software module that allows the end-user to define the parametric study session and to configure the network of computers for the parallel execution.

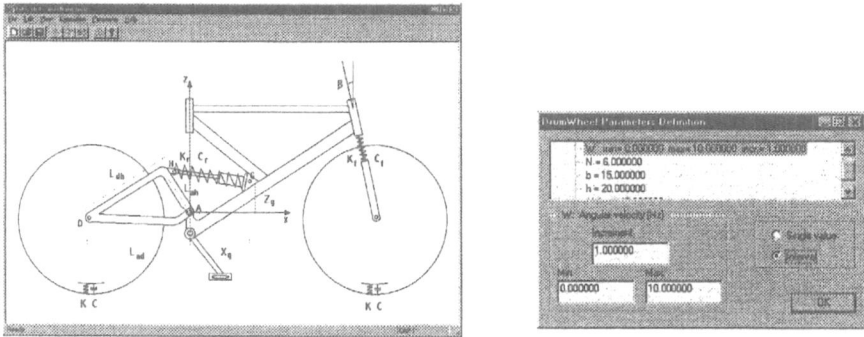

Figure 2. Windows user interface for the pre-processor.

Two different versions of this module have been implemented, one for Unix machines and the second for Windows NT (see Figure 2).

4.2 Parallel Task Manager

The parallel task manager module is in charge of the distribution of the different simulation tasks among all the computing nodes or processors available on the network. It is implemented using the message-passing paradigm.

During the computation of the parametric study, the user may interactively select new computers or release some of the processors allowing a large flexibility in the use of the software on a non-dedicated network of computers. This operation is as simple as clicking on a button on the pre-processor user interface.

4.3 MBS Simulation Package

The simulation package performs the requested analysis of the MBS model for each different combination of the design parameters and the specified working conditions. The simulation modules used within this software follow a fully numerical approach based on the use of the *natural co-ordinates* (3). The use of a numerical approach allows deriving automatically, without any intervention of the analyst, the kinematic constraint equations and the dynamic differential equations that govern the behaviour of the MBS system under consideration.

4.4 Post-processor Tool

The post-processor module manages the results obtained in the parametric study. This is usually a large amount of data corresponding to the different magnitudes computed in the simulation. The post-processing tool allows the user to interactively select the desired data and display them either as 2D plots or in tabular form.

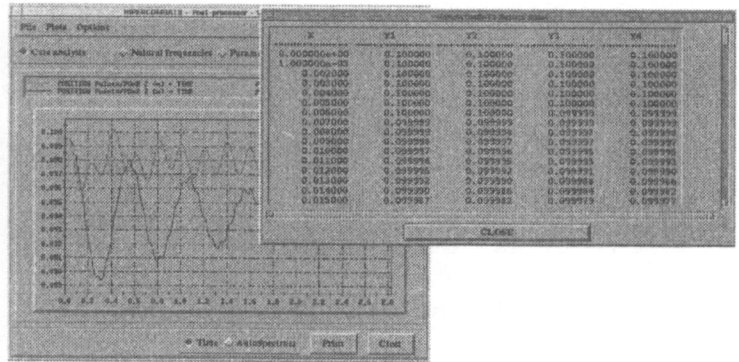

Figure 3. Motif based user interface for the post-processor.

5. Results

In this Section results obtained in practical examples are presented. The results presented have been measured using non-dedicated use of the net and CPUs.

The first test case corresponds to a parametric study performed on a network of Unix workstations. This case generates 60 tasks that are simulated using a heterogeneous network of six SGI MIPS 5000 workstations and a four MIPS 4400 multi-processor machine. The reference sequential time has been observed on a MIPS 5000 workstation. The speed-up scales linearly up to six MIPS 5000 workstations. Using more than six processors some of the MIPS 4400 processors of the multi-processor machine, which are less powerful, are used; this may partially explain the lost of the linear scalability. Figure 4 displays the speed-up factors obtained running the parametric study with different number of Unix processors connected via Ethernet in two cases, using NFS or "rcp" to write the output files on the master's hard disk.

A similar experiment has been run using the Windows NT version of the software. In this case, a parametric study of 60 tasks has been solved using a heterogeneous network of six PCs running under Windows NT 4.0. The processors available in the network are Intel processors of different power, from Pentium running at 267 MHz to Pentium II 400 MHz. The reference sequential time for ideal curve has been observed on a Pentium II 400 MHz. Two experimental curves are shown, one using the master computer to distribute tasks only, and the second one using the master computer to distribute and compute. In the last case, very high execution times are obtained, leading to super-linear speed-ups due to the poor performance of the time sharing management on Windows.

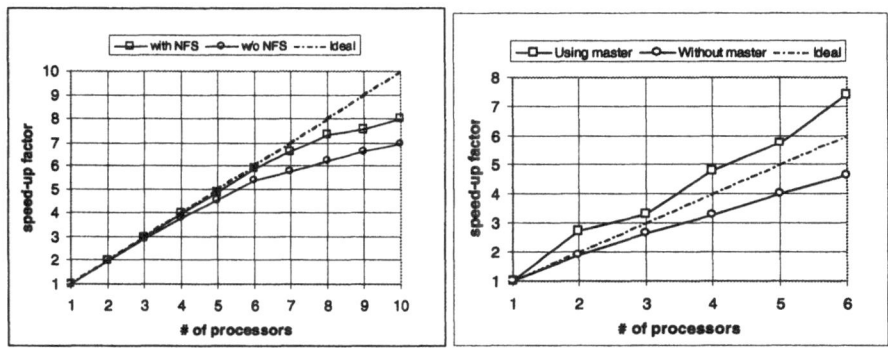

Figure 4. Speed-up on a network of Unix machines and on a network on PCs.

6. Conclusions

One of the main results of these projects has been to demonstrate that high performance computing techniques can sometimes be easily introduced in industrial companies without requiring any equipment investments.

The exploited parallelism can be considered as straightforward from a research point of view but it is nevertheless a very powerful tool for industry. The code has been designed to run efficiently on a non-dedicated network of heterogeneous computers. This is a cost-effective solution that could be used in any enterprise performing parametric studies ranging from SME to large companies.

Another important objective achieved in these projects has been the solution of the problems arisen in the use of general-purpose simulation programs in the design of new multi-body systems. The use of *virtual parametric models* and parallel computing allow carrying out in a working day a complete parametric study of the new mechanical system, with the accuracy of a detailed multi-body analysis.

7. Acknowledgements

The authors wish to thank the following people and their organizations for taking an active part during the development of these projects. In HIPERCOMBATS project, the authors would like to thank to M. Nuti, S. Baldini and P. Casalini (PIAGGIO V.E., S.p.A.); J. P. Halleux (ISPRA); S. Pagnutti (ENEA); and L. Giraud and C. Puglisi (CERFACS). In MYSHANET project, the authors would like to thank to P. Cassagnade and P. de Frenne (DONERRE); M. Nardi and L. Vignocchi (MARZOCCHI); and M. Bussuoli (ENEA). Special thanks are given to J. Celigüeta (CEIT) for his support and M. Valentini (ENEA) for her encouragement and permanent attention paid to the management of those projects.

8. References

1. Avello, A.L., Avello, A.N., Celigüeta, J.T., García de Jalón, J., García-Alonso, A. and Jiménez, J.M., COMPAMM - COMPuter Analysis of Machines and Mechanisms. User's Manual, 1995, publisher CEIT, ISBN 84-605-4064-2.
2. Jiménez, J.M., Avello, A., García-Alonso, A. and García de Jalón, J., "COMPAMM – A Simple and Efficient Code for Kinematic and Dynamic Numerical Simulation of 3-D Multi-Body Systems with Realistic Graphics", W. Schiehlen Ed., *Multibody Systems Handbook*, pp. 285-304, 1990, Springer-Verlag.
3. García de Jalón, J. and Bayo. E., *Kinematic and Dynamic Simulation of Multibody Systems-The Real-Time challenge*, 1994, Springer-Verlag.
4. Haug, E.J., *Computer-Aided Kinematics and Dynamics of Mechanical Systems, Volume I: Basic Methods*, 1989, Allyn and Bacon.
5. Baldini, S., Giraud, L., Izaguirre, J.G., Jiménez, J.M. and Matey, L., "High Performance Computing in Two-wheelers Suspension Design", accepted for publishing in *International Journal of Supercomputing Applications*.
6. Beguelin, A., Dongarra, J., Geist, A., Jiang, W., Manchek, R., and Sunderam, V., *PVM 3 user's guide and Reference Manual*, 1994, Tech. Rep. ORNL/TM-12187, Oak Ridge Nat. Lab., Tennessee 37831.
7. Jiménez, J.M., Matey, L., Izaguirre, J.G., Avello, A.N., Basldini, S., Giraud, L. and Hamel, L., "On the Use of High Performance Computing in Multi-Body Analysis for Two-Wheeled Suspension Design", H. Rajnehat and R. Whalley Eds., *Multi-Body Dynamics: Monitoring and Simulation Techniques*, pp. 295-304, 1997, Mechanical Engineering Publications Ltd.

Parallel Data Mining in the HYPERBANK Project[*]

S. Fotis, J.A. Keane, and R.I. Scott

Department of Computation UMIST, Manchester M60 1QD, UK.
{jak,rscott}@co.umist.ac.uk

Abstract. The aim of the High Performance Banking (HYPERBANK) project is to provide the banking sector with the requisite toolset for the increased understanding of existing and prospective customers. The approach exploits and integrates three areas: business knowledge modelling, data warehousing and data mining, together with parallel computing. Business knowledge modelling formally describes the enterprise in terms of roles, goals and rules. A generic customer-profiling model has been produced and has been instrumental in informing and guiding data mining experiments performed on the banks' data. Parallel computing is required to manipulate and analyse to maximum effect the vast amounts of data collected by banks. A parallel data warehousing tool has been produced and work is ongoing to integrate the customer profiling model with this tool. In this paper, we present work done in the development and implementation of a variety of parallel data mining techniques.

1 The HYPERBANK Project

The aim of the HYPERBANK project [11] is to provide the banking sector with the requisite toolset for increased understanding of customers and for better tailoring of products and services to those customers. The amount of data collected by modern financial institutions is vast. Knowledge hidden within this data can provide important understanding of customer behaviour and can be used to improve the performance and profitability of the bank. It is recognised that, to compete in todays global marketplace, banks must better understand and profile their customers.

Customer data is often generated by a number of diverse sources and spread across a number of databases. This data can be generated from bank account transactions, loan applications, loan repayments, credit card repayments, etc. The approach of the project towards profiling bank customers is to integrate the technologies of business knowledge modelling, data warehousing and data mining. Data warehousing and data mining are complex activities and so the large quantities of data necessitate the use of parallel computing.

Business Knowledge Modelling (BKM) is the process of describing an enterprise in a formal way in terms of its agents, work roles, goals, responsibilities

---

[*] This work was partially supported by ESPRIT HPCN project no 22693.

P. Amestoy et al. (Eds.): Euro-Par'99, LNCS 1685, pp. 1195–1198, 1999.
© Springer-Verlag Berlin Heidelberg 1999

and business rules together with the technological infrastructure that supports the enterprise. A generic model for customer profiling has been developed within the Hyperbank project, using the EKD approach to business modelling. In addition an approach to integrate business modelling with data mining has been developed [3].

A Data Warehouse has been defined as a single integrated data store which provides the infrastructural basis for information in an enterprise [8]. Data Warehousing is the application of various techniques and technologies to bring data from a variety of sources into one or more collections that are designed to improve information access. Information on the parallel data warehousing tool produced within the project can be found at [5].

Data Mining is a step in the KDD (Knowledge Discovery in Databases) process. The KDD process is defined as: the non-trivial process of identifying valid, novel, potentially useful, and ultimately understandable patterns in data [10]. The data mining step consists of building a model based on the data set. Data mining experiments have been performed on some of the banks' data [4], giving encouraging results.

In this paper, we will present the parallel data mining work being done within HYPERBANK. The techniques used were chosen because of their relevance to the banking sector, following analysis of available data. The paper is structured as follows: in §2 we outline the parallel data mining techniques considered and give some experimental results; conclusions are presented in §3.

2   Parallel Data Mining Techniques

The data mining techniques, outlined below, were selected because of their applicatbility to the banking institutions in the HYPERBANK consortium. The aim of the project is to show the utility of High Performance Computing in its broadest sense and to impact the operational activities of the banks. Not all of the banks had parallel machines available and so it was decided that one of the target architectures for the data mining algorithms would be a network of NT workstations. As portability was a prime concern, well-established association rule and classification methods were implemented using Java and JPVM [7]. Work on the more innovative link analysis method is ongoing and an IBM SP2 machine is being used.

2.1   Parallel Link Analysis

Link Analysis is applicable to many banking areas as it models relationships between entities (e.g. bank accounts) in a way other techniques do not. For example, the Financial Crimes Enforcement Network (FinCEN) in the US has used link analysis to control money laundering by examining wire transfers [9]. Link Analysis is not well supported by existing data mining tools because of its performance overheads, traversing links in relational databases requires SQL join operations that are computationally expensive on large tables.

The approach here to link analysis is to abstract the data into the form of a graph. The nodes of the graph represent the entities of interest and the edges represent the relationships. This graph can then be searched to find interesting groups of related objects or to derive new attributes. The source data sets in which these relationships reside are potentially huge and therefore the corresponding graph will consume a lot of memory. In addition, the number of interesting patterns in the graph is potentially enormous. The added memory available to parallel machines facilitates the construction of large graphs and the use of parallel processing has improved the response time of graph search operations [6].

2.2   Parallel Association Rule Discovery

An association rule is an expression that implies relationships among a set of items in a database. The well known parallel Count Distribution algorithm [2] for association rule discovery has been implemented on a network of 200 MHz Pentium PCs running Windows NT using Java and JPVM. Experiments were performed on data sets of varying size. Each data set contained 200 items, had an average transaction size of 15 and contained 2 maximal itemsets of length 10. The number of transactions ranged from 100000 to 400000.

The results obtained show good speed-up. The execution time on 8 processors is approximately $\frac{1}{7}$ of that on 1 processor. This near linear performanece is as expected because the amount of inter process communication to computation performed by each processor is small. Therefore, the results are expected to improve further for larger data sets.

2.3   Parallel Classification

The SPRINT [1] parallel decision tree construction algorithm has been implemented using Java and JPVM. Experiments were performed on an 8-processor IBM SP2 parallel machine. The data consisted of 3 attributes: a categorical attribute with 6 values; a continuous attribute with range from 0 to 10000 and a class variable with 2 values.

The results obtained on data sets of up to 200000 records showed reasonable speed-up. The program runs approximately 4 times faster on 8 processors when compared to 1 processor for a dataset of the same size. This is because of the large amount of communication required when splitting categorical attributes as transaction identifiers need to be communicated across all processors.

3   Conclusions and Further Work

In this paper we have given a brief overview of the HYPERBANK project which aims to provide integrated customer-profiling support for banking institutions. The project integrates the technologies of business knowledge modelling, data mining and data warehousing, and parallel computing. Here, the focus is on the parallel data mining algorithms being implemented and developed. Data Mining is the automated discovery of useful patterns in large amounts of data. It aims to find those important nuggets of knowledge hidden within large amounts of data that existing analysis techniques cannot find. Parallel processing is seen as vital in enabling the handling of large volumes of data. This paper has outlined three parallel data mining techniques. As a dedicated parallel machine was not available in all of the banks involved in the project, the association rule and classification data mining algorithms have been implemented using Java and JPVM so as to be highly portable. Experiments performed on data sets of varying size and have shown reasonable speed-up and size-up.

A parallel link analysis tool is currently under development. This tool models relationships inherent in some data sets using graphs. The graph abstracted from the data can then be used to identify interesting relationships in the data or to derive new attributes. These new attributes can then be used by other data mining methods. Preliminary results on an IBM SP2 are encouraging.

References

[1] J. Shafer, R. Agrawal and M. Mehta. SPRINT: A scalable Parallel Classifier for Data Mining. In *Proceedings of the 22nd VLDB Conference*, Mumbai (Bombay), India, 1996.

[2] R. Agrawal and J. Shafer. Parallel Mining of Association Rules: Design, Implementation and Experience. IBM Research Report RJ10004, 1996.

[3] R.Clarke, D. Filippidou, P.Kardasis, P.Loucopoulos and R. Scott. "Integrating the Management of Business Domain and Discovered Knowledge". In *Proc. of Panhellenic Conference on New Information Technology*, NIT'98, Athens, October 1998.

[4] D. Filippidou, J. A. Keane, S. Svinterikou and J. Murray. Data Mining for Business Process Management: Applying the HyperBank Approach. In *Proceedings of the 2nd International Conference on The Practical Application of Data Discovery and Data Mining*, The Practical Application Company Ltd., 1998.

[5] See www.metasuite.com, 1999

[6] J. A. Keane and R. Scott. Parallel Link Analysis. Department of Computation, UMIST, Manchester, UK, 1999. *in preparation*.

[7] A. J. Ferrari. JPVM: Network Parallel Computing in Java. Technical Report CS-97-29, Department of Computer Science, University of Virginia, USA. 1997.

[8] S. Anahory and D. Murray. Data Warehousing in the Real World: A Practical Guide for Building Decision Support Systems. Addison-Wesley, 1997.

[9] U.S. Congress, Office of Technology Assessment. Information Technologies for the Control of Money Laundering, OTA-ITC-630, Washington DC, US Government Printing Office, September, 1995.

[10] U. M. Fayyad, G. Piatetsky-Shapiro and P. Smyth. From Data Mining to Knowledge Discovery: An Overview. In *Advances in Knowledge Discovery and Data Mining*, AAAI Press, 1996.

[11] J. A. Keane. High Performance Banking. In *Proc. of RIDE '97*, IEEE Press, 1997.

High Performance Computing for Optimum Design of Multi-body Systems

José M. Jiménez[1], Nassouh A. Chehayeb[1], Javier G. Izaguirre[1], Beidi Hamma[2], and Yan Thiaudière[2]

[1] Computational Mechanics Applications, S.L., Miracruz, 10B, 1,
20001 San Sebastian, Spain
{jmjimenez, nchehayeb, jgizaguirre}@cmech.com
[2] CERFACS, 42, Av. G. Coriolis, Toulouse, France
hamma@cerfacs.fr
http://www.cerfacs.fr/algor/ODESIM/home.html

Abstract. This paper presents the most relevant results of ODESIM project, a High Performance Computing and Networking project funded by the European Commission within ESPRIT programme. The project acronym stands for *Optimum DESIgn of Multi-body systems*. The project main objective has been the development of a set of software tools for the kinematic and dynamic optimum design of multi-body systems. Several types of design procedures are targeted: interactive design, designer-driven optimization and fully automatic optimization. ODESIM has been implemented using high performance computing and networking techniques (HPCN) to perform parallel computations in the most CPU-intensive optimization problems. Results on the solution of practical problems are presented.

1 Introduction

Current multi-body system (MBS) simulation programs have powerful analysis capabilities that allow them to simulate complex mechanical systems with rigid and flexible bodies subject to different kinds of external forces, control laws, non-linear effects, etc. However, in the design process, the designer must perform kinematic and/or dynamic analysis in each design step, assess the performance of the system, and take a decision on what dimensions should be modified to improve the behavior of the MBS. This iterative process has two main drawbacks. First, it relies heavily on the designer's experience and thus can only be done by experts; and second, it is very time consuming since the designer has to repeat the assessment of performance, the decision making and the dimensional modification in every iteration step.

In order to reduce the time required for a new design and consequently shorten the time to market, a different approach should be followed. In this new approach, the MBS design problem must be addressed from the CAD definition of parts (bodies and joints). In the CAD program, the designer creates a parametric model of each part, from which ODESIM software generates a parametric MBS model. This parametric model will later be optimized according to a goal function and the design constraints.

P. Amestoy et al. (Eds.): Euro-Par'99, LNCS 1685, pp. 1199-1202, 1999.
© Springer-Verlag Berlin Heidelberg 1999

2   Optimum Design of MBS: The ODESIM Approach

Traditionally, MBS analysis codes consist of a separate pre-processor, a batch-oriented analysis engine and a post-processor. To make a change in the MBS design, the whole sequence of pre-processing, submitting the analysis job to the simulation module and post-processing the results must be repeated. In order to improve the convergence rate of this slow process, the MBS tools should be integrated within a CAD system, so as to generate and modify easily an MBS parametric model. In addition, modules for sensitivity calculation must be developed and integrated in the same tool, thus allowing the designer performing "*what if?*" studies interactively. Those studies consist in knowing how changes in the design variables affect the behavior of the MBS model. The use of optimization functions may also be very profitable, because may automatically lead to a design that satisfies the requirements.

Mathematical formulation of MBS kinematic and dynamic problems usually lead to small or medium size problems with just a few tens of degrees of freedom and a few hundreds of dependent co-ordinates. In spite of the small size, MBS problems present considerable specific difficulties. A description of the mathematical basis for MBS simulation and optimization can be found in [1]. In dynamic optimization of MBS, the goal function is computed as a result of a dynamic simulation that requires the numerical integration of the differential equations of motion. Very often, those equations are *stiff* and the numerical integration needs large CPU times. For this reason, and the fact that the gradient can be computed in parallel, MBS optimization is a problem especially suitable to be implemented in a HPCN environment.

The study of the user needs lead to consider three different levels in the use of the software, which are called *interactive loop* (or interactive simulation), *semi-automatic loop* (or manual optimization) and *fully automatic loop* (or automatic optimization). A more detailed description of ODESIM objectives can be found in [1].

3   Parallel Implementation of the Optimization Modules

In MBS optimum design, the computation of sensitivities is one of the most critical and time-consuming parts of the problem. The most suitable strategy to compute in parallel the gradients applying the finite differentiation method consists in the so call "pool of task" paradigm. This idea is based on the definition of a set of tasks that are stored in a queue managed by a software module called "the master" or parallel manager. The master farms out the tasks among the "the slaves", which are the processors available in the network. The finite differentiation method requires the solution of as many simulations as the number of function evaluations involved in the selected finite difference formula. Each simulation of the MBS model can be taken as a task and can be run independently from the others in a processor.

From the user point of view, the main requirement for the parallel manager is the easiness of use. The parallel aspects have to be hidden as much as possible to the final user. The main characteristics of ODESIM parallel manager are:

- Openness and maintainability.
- Similar to run a sequential code.
- Portability

- Availability of reverse porting to sequential code.
- Load balancing, fault tolerances and breakdown recovering.
- Scalability.

The current version of the ODESIM parallel manager was implemented according to the SPMD paradigm (Single Programme-Multiple Data). It ensures a unique interface between the parallel code and the end-user. However, by nature the parallel code is MIMD (Multiple Instruction-Multiple Data) implementing a master/slave scheme. PVM message-passing library [2] has been used in the current implementation of the parallel manager. In addition, the use of standards (UNIX, C++, X-Windows, Motif, etc.) and object oriented programming (OOP) techniques guarantee the openness, portability and scalability of the software.

4   Results

The example shown in this Section is a test case of ODESIM project defined and solved by Centro Ricerche Fiat. The test case consists in a multi-link rear suspension system that must be optimized in order to reach the desired target curves, being the design parameters the body point positions as well as the body bushing stiffness.

The traditional design procedure for this type of suspension is performed in two steps. First, a kinematic synthesis problem is solved. The target toe-in/toe-out and camber curves are defined as functions of the wheel vertical displacement. The body point positions have to be modified until the target curves are fulfilled. The second step is an elasto-kinematic problem in which the goal function is defined by the toe and wheel base curves, which are given as a function of the lateral and longitudinal forces on the tire, respectively. In this step, the design variables are the body bushing stiffness. The result of the second step affects the results of the first and an iterative procedure is followed repeating both steps until convergence is reached. This iterative process depends strongly on the designer experience.

Table 1. Speed-up factor using a network of computers.

Network	Time per iteration	Speed-up factor
1 MIPS R10000 (195 MHz)	24 min.	
3 MIPS R10000 (195 MHz) + 2 MIPS R8000 (75 MHz) + 4 MIPS R5000 (180 MHz)	6 min.	4

In this example, ODESIM software was used, in a first stage as a tool for helping the designer in the iterative processes described in the previous paragraph. In a second experiment, the two steps were solved in a single step. In this case, a single goal function was defined as a linear combination of the four target curves and the vector of design variables is composed of the body point positions and the body bushing stiffness, leading to 27 design variables. Table 1 shows the speed-up factor obtained in the solution of this problem using several processors. The reason for the relatively small speed-up factor can be found in the fact that the CPU-times were measured in normal working conditions in a non-dedicated network in the industrial site. In addition, the CPU time required to solve a single simulation is relatively small and there is a high overhead that must be considered.

Two conclusions derived from this test case must be highlighted. On one hand, it demonstrates that the duration of each step of the traditional design procedure can be shortened. On the other hand, the unified approach allows collapsing the two steps in a single optimization process that is run automatically. The intervention of the analyst is only sparely required to monitor the evolution of the optimization process, requiring less dedication than the traditional approach. Table 2 shows the design and designer times needed to perform the design of a new suspension using the traditional procedure with traditional tools or using the results of ODESIM project.

Table 2. Time required for the suspension design.

	Traditional prod. Design time = designer time	ODESIM Design time	ODESIM Designer time
Joint position definition	2 days	< 4 hours	3 hours
Bushing stiffness	2 days	< 8 hours	2 hours
Complete procedure (using ODESIM in each step)	6 days	3 days	1 day
Complete single step Optimization		< 2 days	5 hours

5   Conclusions

The most relevant result of the project is that the use of this software can reduce the time for the design process by a factor bigger than 200%. In addition, the use of HPCN capabilities as well as optimization functions may introduce more significant time reductions. The design process is conducted almost automatically and requires nearly no intervention from the analyst. Consequently, the designer can optimize the use of his own time, which in the end is the most expensive part of the design process.

Finally, a third result obtained from ODESIM experience is that high performance computing techniques can be easily introduced to solve practical industrial problems without requiring any equipment investments.

References

1.   Jiménez, J.M., Izaguirre, J.G., Suescun, A. and Avello, A.N, "Optimum Design of Multi-Body Systems - The ODESIM Experience", Multi-Body Dynamics: New Techniques and Applications, pp. 291-300, 1998, Professional Engineering Publishing Ltd.
2.   Beguelin, A., Dongarra, J., Geist, A., Jiang, W., Manchek, R., and Sunderam, V., *PVM 3 user's guide and Reference Manual*, 1994, Tech. Rep. ORNL/TM-12187, Oak Ridge Nat. Lab., Tennessee 37831.

Topic 15
Routing and Communication in Interconnection Networks

This topic is devoted to communication in parallel computers. All aspects of communication, including routing and communication algorithms, the design and packaging of interconnection networks, and the communication costs of parallel algorithms, will be examined. Contributed papers are sought that present significant, original advances in the theory and/or practice of communication in parallel computers.

Topics of interest include:

- Routing Algorithms
- Parallel algorithms with communication costs
- Interconnection networks
- Fault-tolerant communication
- Algorithms for collective communication
- Deadlock-free routing
- Call Admission
- Trade-offs in routing
- Synchronization in parallel computers
- Graph Embedding

P. Amestoy et al. (Eds.): Euro-Par'99, LNCS 1685, pp. 1203–1203, 1999.
© Springer-Verlag Berlin Heidelberg 1999

Optimizing Message Delivery in Asynchronous Distributed Applications*

Girindra D. Sharma, Nael B. Abu-Ghazaleh, Umesh Kumar V. Rajasekaran, and Philip A. Wilsey

Computer Architecture Design Laboratory,
Dept. of ECECS, PO Box 210030, Cincinnati, OH 45221–0030
Ph:(513) 556-4779, Fax:(513) 556-7326, phil.wilsey@uc.edu

Abstract. Since the message delivery time of asynchronous applications is unpredictable, each asynchronous process must probe (or poll) the network for new messages regularly. If polling is carried out too aggressively, a majority of the probes will be unsuccessful. However, the cost for probing the network is similar to the cost of communication operations. Thus, optimizing the message reception behavior has a significant impact on the performance of the application. This paper studies this problem, and develops a cost model to optimize the polling frequency. We also develop strategies to optimize the message delivery behavior, and study their impact on the performance of a parallel discrete event simulator.

1 Introduction

In this paper, we study the problem of asynchronous message delivery in distributed applications. We build a stochastic model of the operation of such algorithms, develop the criteria for optimal polling frequency, and compare it to the ideal case when no polling is necessary. In addition, we suggest some strategies (static, as well as adaptive) for optimizing the message delivery to minimize the polling overhead, and study their effect on the performance of a Time Warp simulator. The empirical studies show that, for the studied models, optimizing the message delivery yields significant improvement in the performance of the simulation. An ideal implementation for eliminating the polling overhead is to use *upcalls* — the underlying network signals the application when a message arrives [5]. For various reasons, upcalls are not supported in many communication libraries. The remainder of this paper is organized as follows. Section 2 develops a stochastic model of a process in an asynchronous distributed computation, compares it to the ideal (no polling cost) case, and uses it to develop the criteria for optimal polling frequency. Since the optimal polling frequency criteria could not be solved analytically, This section also presents heuristic algorithms for infrequent polling and empirically studies the success of the suggested algorithms in improving the performance of a Time Warp [2] discrete event simulator. Finally, Section 3 presents some concluding remarks.

* Support for this work was provided in part by the Advanced Research Projects Agency under contracts J–FBI–93–116 and DABT63–96–C–0055.

P. Amestoy et al. (Eds.): Euro-Par'99, LNCS 1685, pp. 1204–1208, 1999.
© Springer-Verlag Berlin Heidelberg 1999

2   The Analytical and Heuristic Models

In this section, a mathematical model of a process in a distributed application that communicates asynchronously is constructed. An execution cycle for such a process consists of polling the communication layer for messages, receiving the messages if any are available, executing some computation (event in the Time Warp case), and sending some messages. If the message probe operation successfully discovers a message, the message is received, and the network is polled again for messages until no more messages are detected. The time to execute one cycle can be expressed as

$$T_{base} = T_{process} + T_{poll} + \frac{T_{rcv}T_{base}}{\mu} = \frac{T_{process} + T_{poll}}{1 - \frac{T_{rcv}}{\mu}}, \qquad (1)$$

Equation 1 states that the time taken to execute one cycle (T_{base}) is the sum of the execution time ($T_{process}$), unsuccessful poll time (T_{poll}), and the cost of the receive of a message ($\frac{T_{rcv}T_{base}}{\mu}$), where μ is the average message arrival period. The reasoning by which this equation is obtained is as follows: under steady state operation, in every μ time units, an average of one message is received and $\frac{\mu}{T_{base}}$ cycles are executed. The message receive cost is assessed uniformly to these cycles. Note that successful poll time is amortized in T_{rcv} since it initiates some of the operations necessary for receiving the message. This is why the message receive time is significantly smaller than the message send time. Also, the equation is subject to the condition $0 \leq \frac{T_{rcv}}{\mu} < 1$; otherwise, messages arrive faster than they can be received. Define the granularity λ to be $\frac{T_{process}}{T_{poll}}$, and define the normalized message arrival period ρ to be $\frac{\mu}{T_{rcv}}$. Hence, the condition becomes: $1 < \rho \leq \infty$. Substituting these values in Equation 1, we obtain

$$T_{base} = \frac{(\lambda + 1)T_{poll}}{1 - \frac{1}{\rho}} \qquad (2)$$

It is interesting to compare this case to the case when no polling is necessary (upcalls) to understand the upper limit on optimizations that target minimizing the polling time. The execution time per cycle in an upcall environment (T_{upcall}) is the processing time per event, in addition to the receive cost if a message is received. More precisely

$$T_{upcall} = T_{process} + \frac{T_{rcv}T_{upcall}}{\mu} = \frac{T_{process}}{1 - \frac{T_{rcv}}{\mu}} = \frac{\lambda T_{poll}}{1 - \frac{1}{\rho}} \qquad (3)$$

Define the efficiency η, of the polling to be the ratio of the ideal cycle cost to the actual cycle cost. We obtain $\eta = \frac{\lambda}{\lambda+1}$. Surprisingly, η is not a function of ρ; ρ appears in the same multiplicative factor for each of the models and cancels out when their ratio is taken. For low granularity applications, polling costs dominate execution. Moreover, even for relatively high granularity applications, there is a significant drop in efficiency that can be reduced by optimizing the polling behavior. The optimization we are considering in this paper is to carry out the message polling periodically (instead of every cycle). The benefit of periodic polling is twofold: (i) the average cycle time is shorter, since polling is not done every cycle; and (ii) there is a lower chance that the polling is unsuccessful: because polling is carried out less frequently, there is a longer period of time available for messages to arrive between polls. We define the polling period f, as the number of cycles between consecutive polling operations. In addition, we assume that message arrival probability is uniform and memoryless. More precisely, the expected number of messages received during T time units is μT, where μ is the probability of arrival of a message in one time unit. This assumption is reasonable in an asynchronous application where the message arrival rate is independent of the state of the receiving processor. The time taken by f processing cycles becomes:

$$T_{f-cycles} = fT_{infreq} = fT_{process} + T_{poll} + T_{rcv}fT_{infreq}\mu. \tag{4}$$

where T_{infreq} is the time for one cycle of infrequent polling. In f cycles, f events are processed, a single probe operation is carried out, and a number of messages (expected $fT_{infreq}\mu$) are received. Solving equation (4) for T_{infreq} and letting it be $T_{infreq-base}$ we get:

$$T_{infreq-base} = \frac{T_{process} + \frac{T_{poll}}{f}}{1 - \frac{T_{rcv}}{\mu}} \tag{5}$$

Note that if f is set to infinity, the infrequent polling cost approaches the ideal case. The flaw in this model is that it assumes complete asynchrony and does not account for the effect of delaying the receipt of the messages. The exact effect of delaying message reception depends on the specific application. For analysis purposes, we assume a model where there is a critical message that the process will receive at a time governed by an exponential distribution with mean μ. More precisely, $P(t <= T) = \int_0^T \frac{e^{-\frac{t}{\mu}}}{\mu}dt$. Once the message is received, it will cause a drop in the efficiency proportional to a constant c that is dependent on the application. If c is 0, the message is not critical at all; if c is 1, all work after the arrival of the message is useless until it is received; if $c > 1$, not only is all work after the message arrives useless, but additional time has to be spent undoing the effect of not receiving the message immediately. The time to execute f cycles ($T_{f-cycles}$) becomes: $T_{f-cycles} = fT_{infreq} = fT_{process} + T_{poll} + \frac{T_{rcv}fT_{infreq}}{\mu} + \int_0^{fT_{infreq-base}} \frac{e^{-\frac{t}{\mu}}}{\mu}c(fT_{infreq-base} - t)dt$. In addition to the base execution time in Equation 5, this equation assesses a penalty cost if a critical message is delayed. The penalty is proportional to the delay time and the application specific factor c. Simplifying, $T_{infreq} = T_{infreq-base} + \frac{cT_{infreq-base} - \frac{\mu c}{f}(1 - e^{-\frac{fT_{infreq-base}}{\mu}})}{1 - \frac{T_{rcv}}{\mu}}$. Taking the derivative of T_{infreq} with respect to f and setting it to 0, we obtain the following optimality criterion for f: $2\mu c - 2T_{poll}(1 + \frac{c}{1 - \frac{T_{rcv}}{\mu}}) + ce^{-\frac{fT_{infreq-base}}{\mu}}(T_{poll} - fT_{infreq-base} - 2\mu) = 0$. This is a transcendental equation that cannot be solved analytically. In order to obtain $f_{optimal}$, the equation should be solved numerically, simultaneously with Equation 5. Clearly, numerical evaluation of the optimal period is not feasible as the basis for adapting the frequency at run-time. Therefore, we investigate heuristically converging on the optimal frequency in the next section. Developing the criteria for optimal polling frequency even under approximate conditions proved difficult. Even if it was possible to directly compute the optimal polling frequency, it is likely to vary in the lifetime of the application, and from process to process. Therefore, we present infrequent polling strategies that arrive at their polling frequency heuristically. Since these strategies are simple, they compute their polling frequency without adding a significant overhead to the computation. The following strategies were implemented: (a) *Static Polling Frequency*: In this strategy, an initial period for polling is selected at compile time, and enforced on all the processes. The advantage of this policy is that it adds no overhead for implementing an adaptive mechanism. However, if the polling frequency is not efficient, the performance of the application can be harmed; (b) *Fixed-Increment Adaptive Policy*: This strategy attempts to vary the polling frequency such that exactly one message is received in each period. If more than one message is received, then the period is too long, and it is shortened by a fixed number (determined at compile-time). If no messages are received, then it is likely that the process is polling too aggressively, and the period should be increased; and (c) *Variable-Increment Adaptive Policy*: This policy is different from the fixed-increment policy in that it increases (or decreases) the polling frequency by an increment (or a decrement) that is a function of the current polling frequency. More specifically, it is incremented (or decrement) by a fixed percentage of the current polling frequency. All three infrequent polling strategies were implemented in the communication module of WARPED [3], our Time Warp parallel discrete event simulator. The impact of infrequent polling on the performance of WARPED was studied using two simulation models. The simulations were conducted on a network of SUN workstations. Both applications communicate across the network by using MPICH [1]. The performance of the static polling frequency for the two applications are not reported (due to space constraints) but are available elsewhere [4]. The fixed frequency is enforced on all the processes; a given frequency may be suitable for some processes, but not for others and a global optimal frequency may not exist. For all the models, the adaptive strategies performed on par with the best static frequency. More details about these experiments are available in [4].

3   Concluding Remarks

A mathematical model of an asynchronous application was constructed. A model of the application using periodic infrequent polling that accounts for the effect of delaying the receipt of messages was also constructed. A criteria for the optimal polling frequency was obtained from the model. Unfortunately, the optimal polling frequency expression was transcendental and could not be solved analytically (however, it can be solved numerically). We suggested several strategies for heuristically converging on a near-optimal polling frequency. Three heuristic policies were suggested, and their performance effect on a Time Warp simulator was studied. The performance results showed that infrequent polling yields considerable improvement in performance for fine-grained models. In addition, it shows that an adaptive policy is necessary because the optimal polling frequency varies across processes in the same application.

References

[1] GROPP, W., LUSK, E., AND SKJELLUM, A. *Using MPI: Portable Parallel Programming with the Message-Passing Interface.* MIT Press, Cambridge, MA, 1994.

[2] JEFFERSON, D. Virtual time. *ACM Transactions on Programming Languages and Systems 7*, 3 (July 1985), 405–425.

[3] RADHAKRISHNAN, R., MARTIN, D. E., CHETLUR, M., RAO, D. M., AND WILSEY, P. A. An Object-Oriented Time Warp Simulation Kernel. In *Proceedings of the International Symposium on Computing in Object-Oriented Parallel Environments (ISCOPE'98)*, vol. LNCS 1505. Springer-Verlag, Dec. 1998, pp. 13–23.

[4] SHARMA, G. D. Time warp simulator designs for clusters of smps. Master's thesis, University of Cincinnati, May 1999.

[5] VON EICKEN, T., CULLER, D., GOLDSTIEN, S., AND SCHAUSER, K. Active messages: A mechanism for integrated communication and computation. Tech. Rep. CSD-92-675, Computer Science Division, University of California, Berkeley, CA 94720, March 1992.

Circuit-Switched Broadcasting in Multi-port Multi-dimensional Torus Networks*

San-Yuan Wang, Yu-Chee Tseng, Sze-Yao Ni, and Jang-Ping Sheu

Department of Computer Science and Information Engineering
National Central University, TAIWAN
{sywang, yctseng, nee, sheujp}@csie.ncu.edu.tw

Abstract. This paper studies the *one-to-all broadcast* problem in a *circuit-switched* torus with α-port capability, where a node can simultaneously send and receive α messages at one time. We show how to efficiently perform broadcast in 2-D and 3-D tori of any size, square or non-square, using near optimal numbers of steps. The main techniques used are: (i) a "span-by-dimension" approach, which makes our solution scalable to torus dimensions, and (ii) a "squeeze-then-expand" approach, which makes possible solving the difficult cases where tori are non-square. Existing results, as compared to ours, can only solve very restricted sizes or dimensions of tori, or use more numbers of steps.

1 Introduction

Efficient inter-processor communication is critical for a multicomputer network to deliver high performance. One primary communication is the *one-to-all broadcast*, where a source node needs to send a message to all other nodes in the network. In this paper, we study the scheduling of one-to-all broadcast in a circuit-switched torus. The network is assumed to use the α-*port* communication model, in which a node can send up to α messages and simultaneously receive up to α messages at a time, where $1 \leq \alpha \leq 2h$ and $h = 2, 3$ is the dimension of the torus. This is a generalization of the *one-port* model ($\alpha = 1$) and the *all-port* model ($\alpha = 2h$). Following the formulation in many works [1, 3, 4, 5, 8], this is achieved by constructing a sequence of *steps*, where a step consists of a set of congestion-free communication paths each indicating a message delivery. The goal is to minimize the total number of steps used.

One-to-all broadcast has been studied for meshes and tori based on different port models [2, 3, 4, 5]. Based on all-port circuit switching, the scheme in [5] uses optimal numbers of steps for any 2-D torus of size $5^p \times 5^p$ or $(2 \times 5^p) \times (2 \times 5^p)$, where p is any integer. Likewise, the schemes of [3, 4] remain optimal, but can be applied to any square k-D torus with $(2k+1)^p$ nodes on each side. Generalization to square 2-D/3-D tori/meshes supporting multi-port capability is shown in [2]. Drawbacks of these works include limitations on torus dimension, size, or that

* This work is supported by the National Science Council of the Republic of China under Grant # NSC88-2213-E-008-014 and #NSC88-2213-E-008-027.

P. Amestoy et al. (Eds.): Euro-Par'99, LNCS 1685, pp. 1209–1221, 1999.
© Springer-Verlag Berlin Heidelberg 1999

Table 1. Comparison of broadcast algorithms, assuming a network size of $n_1 \times n_2 \times \cdots \times n_k$. ($LB(k)_\alpha$ = the lower bound for α-port k-D tori in Lemma 1.)

Algorithm		Ours	Lee-Lee[2]	Park-Choi[3]	Peters[5]
Port Model		α-port	α-port	all-port	all-port
3-D	Size	$n_1 \times n_2 \times n_3$	$n_1 = n_2 = n_3$	$n_1 = n_2 = n_3 = 7^p$	N/A
	Steps	$\begin{cases} LB(3)_\alpha + 2 & \text{if } n_1 = n_2 = n_3, \\ LB(3)_\alpha + 6 & \text{otherwise} \end{cases}$	$LB(3)_\alpha + 2$	$LB(3)_6$	N/A
k-D	Size	$\Pi_{i=1}^k n$	N/A	$\Pi_{i=1}^k (2k+1)^p$	N/A
	Steps	$\begin{cases} LB(k)_\alpha & \text{if } n = (\alpha+1)^p, \\ LB(k)_\alpha + k - 1 & \text{otherwise} \end{cases}$	N/A	$LB(k)_{2k}$	N/A

the networks must be square. In Table 1, we compare these solutions against the yet-to-be-presented results in this paper. Our results can improve over existing results in either the number of communication steps required or the network dimension/size restrictions.

The difficulty of this problem lies in how we disseminate the broadcast message in a congestion-free manner. To achieve optimality, it typically relies on recursively partitioning the torus into $\alpha + 1$ smaller subnetworks. For instance, when the number of ports $\alpha = 1$ (resp., $\alpha = 3$), a recursively doubling (quadrupling) approach [6] can easily achieve optimality. Unfortunately, there is no known systematic solution to do so, especially when α is even (say, $\alpha = 2$ or 4 in a 3-D torus). One readily sees additional the difficulty when tori are non-square and of more dimensions.

We show how to efficiently perform broadcast in a torus of any dimension, any size, square or non-square, using near optimal numbers of steps. An earlier version of this paper has reported our solution to 2-D tori [9]. This paper extends the result to higher dimensional tori. The main techniques used here are: (i) a "span-by-dimension" approach, and (ii) a "squeeze-then-expand" approach similar to [7] (here we extend the technique in [7] from 2-D to 3-D). Technique (i) makes our results scalable to torus dimensions, while technique (ii) makes possible solving the difficult cases where tori are non-square.

Preliminaries are given in Section 2. Our solution to 3-D tori is in Section 3. Section 4 briefly summaries how to extend our results to higher-dimensional tori. Finally, conclusions are drawn in Section 5.

2   Preliminaries

A k-D torus of size $n_1 \times n_2 \times \cdots \times n_k$ is an undirected graph denoted as $T_{n_1 \times n_2 \times \cdots \times n_k}$. Each node is denoted as $p_{x_1, x_2, \cdots, x_k}$, $0 \leq x_i < n_i$, $1 \leq i \leq k$. Each node is of degree $2k$. Node $p_{x_1, x_2, \cdots, x_k}$ has an edge connecting to $p_{(x_1 \pm 1) \bmod n_1, x_2, \cdots, x_k}$ along dimension one, an edge to $p_{x_1, (x_2 \pm 1) \bmod n_2, \cdots, x_k}$ along dimension two, and so on. (Hereafter, we will omit using "mod" whenever the context is clear.)

In the *one-to-all broadcast* problem, a source node needs to send a message to the rest of the network. To achieve this, we will construct a sequence of *steps*, where a step consists of a number of *link-disjoint* paths each indicating one message delivery; paths of different lengths can co-exist in one step, but the corresponding communications are assumed to complete in about the same time due to the distance-insensitive characteristic of circuit switching. An α-*port* model will be assumed, in which a node can send up to α messages, and simultaneously receive up to α messages, along any α of its $2k$ outgoing and incoming channels, respectively.

Lemma 1 *In a k-D α-port torus $T_{n_1 \times n_2 \times \cdots \times n_k}$, a lower bound on the number of steps to perform one-to-all broadcast is $\lceil \log_{\alpha+1}(n_1 n_2 \cdots n_k) \rceil$.*

We will map a square $T_{n \times n \times \cdots \times n}$ into a modulo Euclidean integer space $\mathbb{Z}^k$, where $\mathbb{Z} = \{0, 1, \ldots, n-1\}$. We may interchangeably represent node $p_{x_1, x_2, \cdots, x_k}$ in $T_{n \times n \times \cdots \times n}$ as a point $(x_1, x_2, \ldots, x_k)$ in $\mathbb{Z}^k$. A vector in $\mathbb{Z}^k$ is a k-tuple $v = (v_1, v_2, \ldots, v_k)$. The i-th positive (resp., negative) elementary vector e_i (resp., e_{-i}) of $\mathbb{Z}^k$, $i = 1..k$, is the vector with all entries being 0, except the i-th entry being 1 (resp., -1). We may write $e_{i_1} + e_{i_2}$ as e_{i_1, i_2}, $e_{i_1} + e_{-i_2}$ as $e_{i_1, -i_2}$, and similarly $e_{i_1} + \cdots + e_{i_m}$ as $e_{i_1, \ldots, i_m}$. For instance, $e_{1,3} = e_1 + e_3$ and $e_{1,-3} = e_1 - e_3$. The *linear combination* of vectors (say $a_1 v_1 + a_2 v_2$, where a_1 and a_2 are integers) follows the typical definitions in linear algebra, except that a "mod n" is implicitly applied.

Definition 1 *In $\mathbb{Z}^k$, given a node x, an m-tuple of vectors $B = (b_1, b_2, \ldots, b_m)$, and an m-tuple of integers $N = (n_1, n_2, \ldots, n_m)$, we define the span of x by vectors B and distances N as a set of nodes denoted as $SPAN(x, B, N) = \{x + \sum_{i=1}^{m} a_i b_i | 0 \le a_i < n_i\}$.*

For example, the main diagonals of $T_{n \times n}$ and $T_{n \times n \times n}$ can be written as $SPAN(p_{0,0}, (e_{1,2}), (n))$ and $SPAN(p_{0,0,0}, (e_{1,2,3}), (n))$, respectively. Also, $T_{n \times n \times n}$ can be written as $SPAN(p_{0,0,0}, (e_{1,3}, e_2, e_1), (n, n, n))$ or $SPAN(p_{0,0,0}, (e_{1,3}, e_{1,2}, e_3), (n, n, n))$.

3   Broadcasting in 3-D Tori

The cases of $\alpha = 1$ and 2 can be solved by a simple recursive doubling/tripling on rows and then columns. So the following discussion will focus on the $\alpha \ge 3$ cases.

3.1   Case 1: The 3-D Torus Is Square

When $\alpha = 6$. Consider a 3-D $T_{n \times n \times n}$ with any n. Without loss of generality, let $p_{0,0,0}$ be the source node. The basic idea is to distribute the broadcast message M in three stages: (i) from $p_{0,0,0}$ to the line $SPAN(p_{0,0,0}, (e_{1,3}), (n))$, (ii) from the above line to the plane $SPAN(p_{0,0,0}, (e_{1,3}, e_{1,2}), (n, n))$, and then (iii) from

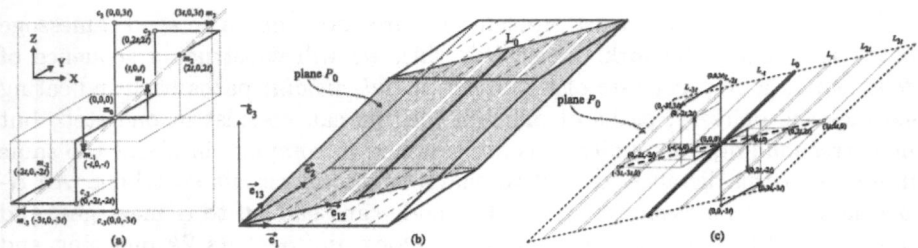

Fig. 1. Broadcasting in a square 3-D torus: (a) First step of stage 1. (b) Stage 2: viewing the torus from the perspective P_0 = $SPAN(p_{0,0,0}, (e_{1,3}, e_2, e_1), (n, n, n))$ which is partitioned into 7 strips. (c) Communication pattern in stage 2.

the above plane to the whole torus. For simplicity, we may use X-, Y-, and Z-axes to refer to the first, second, and third dimensions, respectively.

Stage 1: From the Source Node to a Line. To send M to $L_0 = SPAN(p_{0,0,0}, (e_{1,3}), (n))$, we use the following recursive structure. For simplicity, let n be a multiple of 7, $n = 7t$. We view L_0 as consisting of n nodes $p_{i,0,i}$, $i = -\lfloor\frac{n-1}{2}\rfloor$, $-\lfloor\frac{n-1}{2}\rfloor+1, \ldots, \lceil\frac{n-1}{2}\rceil$. We then partition L_0 horizontally into 7 segments $S_j, j = -3..3$, such that the first segment consists of the first t nodes, the second segment the next t nodes, etc. Let's identify the center node of S_j as m_j. In one step, node m_0 can forward M to $m_{\pm1}, m_{\pm2}, m_{\pm3}$, as illustrated in Fig. 1(a).

Clearly, we can recursively distribute M to nodes of S_j from $m_j, j = -3..3$. This stage will take $\lceil\log_7 n\rceil$ steps to complete.

Stage 2: From a Line to a Plane. In this stage, M will be distributed from the L_0 to the plane $P_0 = SPAN(p_{0,0,0}, (e_{1,3}, e_2), (n, n))$. However, we view the plane as consisting of n lines (see Fig. 1(b)):

$$L_i = SPAN(p_{i,i,0}, (e_{1,3}), (n)), \quad i = -\lfloor\frac{n-1}{2}\rfloor, -\lfloor\frac{n-1}{2}\rfloor+1, \ldots, \lceil\frac{n-1}{2}\rceil. \quad (1)$$

It is easy to send messages from a line to another parallel line in one communication step. For instance, to deliver messages from line $SPAN(p_{0,0,0}, (e_{1,3}), (n))$ to line $SPAN(p_{2,3,4}, (e_{1,3}), (n))$, we simply let each $p_{i,0,i}$ (of the former line) send to $p_{i+2,3,i+4}$ (of the latter line). One can easily generalize this to a line sending to six other parallel lines in one step.

This stage is based on a recursive structure as follows. For simplicity, let $n = 7t$. We partition the plane P_0 into 7 strips $S_j, j = -3..3$, such that S_{-3} consists of the first t lines in Eq. (1), S_{-2} the next t lines, etc. (refer to Fig. 1(c)). By having each $p_{i,0,i} \in L_0$ send M to the following six nodes: $p_{i+t,t,i}, p_{i,2t,i-2t}, p_{i,3t,i-3t}, p_{i-t,-t,i}, p_{i,-2t,i+2t}, p_{i,-3t,i+3t}$, we can distribute M to the following six lines in one step:

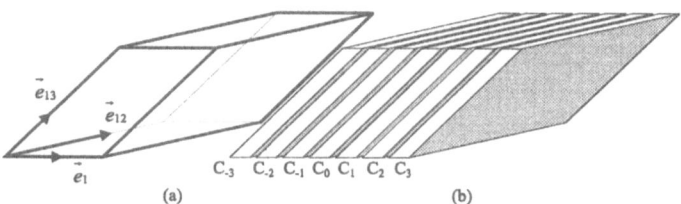

Fig. 2. (a) Viewing the torus as $SPAN(p_{0,0,0}, (e_{1,3}, e_{1,2}, e_1), (n, n, n))$, and (b) partitioning the torus into 7 cubes $C_j, j = -3..3$.

$$L_{\pm t} = SPAN(p_{\pm t, \pm t, 0}, (e_{1,3}), (n))$$
$$L_{\pm 2t} = SPAN(p_{0, \pm 2t, \mp 2t}, (e_{1,3}), (n))$$
$$L_{\pm 3t} = SPAN(p_{0, \pm 3t, \mp 3t}, (e_{1,3}), (n)).$$

This communication step, as illustrated in Fig. 1(c), is congestion-free. The resulting line L_{jt} is on the central line of S_j for all $j = -3..3$. To see this, let's prove the case $L_{2t} = SPAN(p_{0,2t,-2t}, (e_{1,3}), (n)) = SPAN(p_{0,2t,-2t} + 2te_{1,3}, (e_{1,3}), (n)) = SPAN(p_{2t,2t,0}, (e_{1,3}), (n)) \in P_0$, which is indeed the central plane of S_2. The other cases can be proved similarly.

Next, we can recursively perform the similar line-to-line distribution in each S_j using L_{jt} as the source. The recursion is repeated until each S_j is reduced to one or zero line. This stage takes $\lceil \log_7 n \rceil$ steps to complete.

Stage 3: From a Plane to More Planes. In this stage, we view the torus from another perspective: $SPAN(p_{0,0,0}, (e_{1,3}, e_{1,2}, e_1), (n, n, n))$, which is illustrated in Fig. 2(a). With this view, we partition the torus along the direction e_1 into n planes:

$$P_i = SPAN(p_{i,0,0}, (e_{1,3}, e_{1,2}), (n, n)), i = -\lfloor \frac{n-1}{2} \rfloor, \ldots, \lceil \frac{n-1}{2} \rceil. \qquad (2)$$

For simplicity, let $n = 7t$. Following the same philosophy as before, we divide the torus into 7 cubes $C_j, j = -3..3$, such that the first cube consists of the first t planes, the second cube the next t planes, etc. This is shown in Fig. 2(b).

The central plane P_0 in Eq. (2) already owns message M. In this stage, plane-to-plane message distribution will be performed. For instance, if every node on plane $SPAN(p_{0,0,0}, (e_{1,3}, e_{1,2}), (n, n))$ sends M along the Y- and Z-axes to nodes that are +3 and +5 hops way, respectively, then two planes will receive M: (i) $SPAN(p_{0,3,0}, (e_{1,3}, e_{1,2}), (n, n)) = SPAN(p_{-3,0,0}, (e_{1,3}, e_{1,2}), (n, n)) = P_{-3}$, and (ii) $SPAN(p_{0,0,5}, (e_{1,3}, e_{1,2}), (n, n)) = SPAN(p_{-5,0,0}, (e_{1,3}, e_{1,2}), (n, n)) = P_{-5}$.

Thus, by having each node $p_{a,b,c} \in P_0$ send M to the six nodes, $p_{a+t,b,c}$, $p_{a,b-2t,c}$, $p_{a,b,c-3t}, p_{a-t,b,c}, p_{a,b+2t,c}, p_{a,b,c+3t}$, we can distribute M to six other planes in one step:

$$P_{jt} = SPAN(p_{jt,0,0}, (e_{1,3}, e_{1,2}), (n, n)), j = -3..3.$$

The resulting planes P_{jt} is the central plane of C_j for all $j = -3..3$. Next, we recursively perform the similar plane-to-plane distribution in each C_j using P_{jt} as the source. This stage takes totally $\lceil \log_7 n \rceil$ steps.

Theorem 1 *In a circuit-switched all-port $T_{n \times n \times n}$ torus, broadcast can be done in $3\lceil \log_7 n \rceil$ steps, which number of steps is at most 2 steps more than the lower bound in Lemma 1.*

When $3 \leq \alpha \leq 5$. For an α-port torus $T_{n \times n \times n}$, $3 \leq \alpha \leq 5$, we modify the scheme developed in Section 3.1 as follows. In stage 1, the line $L_0 = SPAN(p_{0,0,0}, (e_{12}), (n))$ is evenly partitioned into $\alpha + 1$ segments recursively. The message delivery should be straight-forward. Similarly, in stage 2, the plane $P_0 = SPAN(p_{0,0,0}, (e_{1,3}, e_{1,2}), (n, n))$ is evenly partitioned into $\alpha + 1$ strips recursively; in stage 3, the torus $SPAN(p_{0,0,0}, (e_{1,3}, e_{1,2}, e_1), (n, n, n))$ is evenly partitioned into $\alpha + 1$ cubes recursively.

Theorem 2 *In a circuit-switched α-port $T_{n \times n \times n}$ torus, broadcast can be done in $3\lceil \log_{\alpha+1} n \rceil$ steps, which number of steps is at most 2 steps more than the lower bound in Lemma 1.*

3.2   Case 2: The 3-D Torus Is Non-square

Consider a non-square and α-port torus $T_{n_1 \times n_2 \times n_3}$, $3 \leq \alpha \leq 6$. Without loss of generality, let $n_1 = \min\{n_1, n_2, n_3\}$. We assume that n_1 is even; otherwise, we can translate the torus into $T_{(n_1-1) \times n_2 \times n_3}$, which will require one more step to perform broadcast.

When α Is Even. Here, our approach is still based on recursively partitioning lines/planes/cubes into $\alpha + 1$ parts. Below we first discuss the case of $\alpha = 6$. The case of $\alpha = 4$ can be solved similarly and we will briefly comment on it at the end of this section.

Definition 2 The dilated torus induced by $T_{n_1 \times n_2 \times n_3}$, denoted as $\check{T}_{n_1 \times n_2 \times n_3}$, is an $n_1 \times n_1 \times n_1$ torus consisting of n_1^3 nodes each denoted as $\check{p}_{i,j,k} = p_{i, \lfloor jn_2/n_1 \rfloor, \lfloor kn_3/n_1 \rfloor}$, $0 \leq i, j, k < n_1$.

Intuitively, in $\check{T}_{n_1 \times n_2 \times n_3}$, adjacent nodes along the same y-axis (resp. z-axis) are dilated by $\lfloor \frac{n_2}{n_1} \rfloor$ or $\lceil \frac{n_2}{n_1} \rceil$ (resp. $\lfloor \frac{n_3}{n_1} \rfloor$ or $\lceil \frac{n_3}{n_1} \rceil$) links in the original torus $T_{n_1 \times n_2 \times n_3}$, but there is no dilation along the x-axis. On $\check{T}_{n_1 \times n_2 \times n_3}$, we further define two (dilated) subtori:

$$\check{T}_{0,0,0} = SPAN(\check{p}_{0,0,0}, (2e_1, 2e_2, 2e_3), (\frac{n_1}{2}, \frac{n_1}{2}, \frac{n_1}{2}))$$

$$\check{T}_{1,1,1} = SPAN(\check{p}_{1,1,1}, (2e_1, 2e_2, 2e_3), (\frac{n_1}{2}, \frac{n_1}{2}, \frac{n_1}{2}))$$

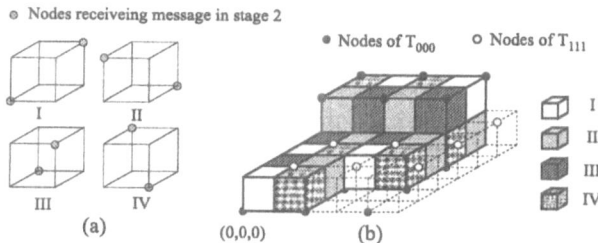

Fig. 3. (a) 4 types of unit cubes. (b) Their geometric relationship.

That is, $\check{T}_{0,0,0}$ and $\check{T}_{1,1,1}$ are dilated by two links along each dimension in $\check{T}_{n_1 \times n_2 \times n_3}$. Assuming $\check{p}_{0,0,0}$ as the source node, we have 5 stages.

Stage 1. The source $\check{p}_{0,0,0}$ sends M to $\check{p}_{1,1,1}$.

Stage 2. Node $\check{p}_{0,0,0}$ performs broadcast on $\check{T}_{0,0,0}$, and node $\check{p}_{1,1,1}$ performs broadcast on $\check{T}_{1,1,1}$. This can be done in parallel as $\check{T}_{0,0,0}$ and $\check{T}_{1,1,1}$ are independent of each other. We can apply the scheme in Section 3.1, which will take $3 \lceil \log_7 \frac{n_1}{2} \rceil$ steps to complete.

Stage 3. After stage 2, there are 1/4 nodes on $\check{T}_{n_1 \times n_2 \times n_3}$ already having M. If we look at each unit cube (size $1 \times 1 \times 1$) in $\check{T}_{n_1 \times n_2 \times n_3}$, two out of eight nodes in the unit cube already have M. According the distribution of the nodes having M, we can classify a unit cube into four types, as shown in Fig. 3(a). These cube's relationship is in Fig. 3(b).

Now we show how the recursion proceeds in one step. On each unit cube of type I, let its two corner nodes already owning M be: $p_{x,y,z}$ and $p_{x',y',z'}$ (refer to Fig. 4). Let the width of the unit cube on the original torus be $w = y' - y$. The recursion should proceed as long as $w \geq 7$. For ease of presentation, suppose w is a multiple of 7, $w = 7t$. Then we perform the following communications:

- $p_{x,y,z}$ sends M to three nodes $p_{x,y+2t,z}, p_{x,y+4t,z}$, and $p_{x,y+6t,z}$, and
- $p_{x',y',z'}$ sends M to three nodes $p_{x',y'-2t,z'}, p_{x',y'-4t,z'}, p_{x',y'-6t,z'}$.

The communication is illustrated in Fig. 4(a).

Now observe the type I unit cube in Fig. 4(a). More nodes have received M. If we consider the pattern of the nodes owning M, then after the above communication step the unit cube can be further partitioned into seven smaller unit cubes, four of which are of type I and three of which are of type III. Similarly, the type II, III, and IV unit cubes in Fig. 4(a) are now each partitioned into seven more smaller cubes of different types. The rest is shown in Fig. 4(b). An interesting property is that there are now totally 28 smaller unit cubes (which are evenly distributed to each type) obtained after the communication step. Also, the width of each cube along the Y axis is reduced by a factor of about $\frac{1}{7}$. As the initial value of w is no more than $\lceil \frac{n_2}{n_1} \rceil$, the total number of steps required in this stage is $\lceil \log_7 \frac{n_2}{n_1} \rceil - 1$.

Stage 4. In stage 3, we have "expanded" $\check{T}_{n_1 \times n_2 \times n_3}$ from one with n_1^3 unit cubes to one with more unit cubes along the Y axis. Furthermore, each unit

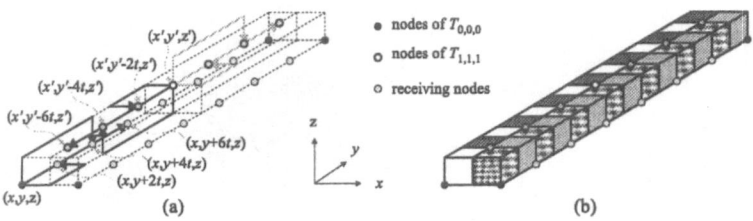

Fig. 4. (a) The communication pattern of one step in a type I unit cube. (b) Each unit cube is partitioned into 7 smaller unit cubes after performing the step in (a).

cube is dilated along the Y axis by no more than 6 links. In this stage, we will further "expand" the torus such that each unit cube is dilated along the Z axis by no more than 6 links, too.

Now we show how the recursion proceeds in one step. The geometric relationship of unit cubes is shown in Fig. 5(a). On each unit cube of type I, let its two corner nodes already owning M be: $p_{x,y,z}$ and $p_{x',y',z'}$ (refer to Fig. 5(b)). Let the height of the unit cube be $h = z' - z$. The recursion should proceed as long as $h \geq 7$. For ease of presentation, suppose h is a multiple of 7, $h = 7t$. Then we perform the following communications (see Fig. 5(b)):

– $p_{x,y,z}$ sends three messages to nodes $p_{x,y+2t,z}$, $p_{x,y+4t,z}$, and $p_{x,y+6t,z}$.
– $p_{x',y',z'}$ sends three messages to nodes $p_{x',y'-2t,z'}$, $p_{x',y'-4t,z'}$, and $p_{x',y'-6t,z'}$.

After the communication step, more nodes will own M. Using the pattern in Fig. 3, we can define new unit cubes. The result is shown in Fig. 5(c) and (d), where one can see that each unit cube is now partitioned into 7 smaller unit cubes.

The above recursion is repeated until $h < 7$. As the initial value of h is no more than $\lceil \frac{n_3}{n_1} \rceil$, the total number of steps required is $\lceil \log_7 \frac{n_3}{n_1} \rceil - 1$.

Stage 5. After stage 4, each unit cube of type I is dilated by at most 6 links along each of Y and Z axes (there is no dilation along the X axis). So there are 36 possible sizes of the cube. Due to space limitation, we only show how the largest case is solved in Fig. 6. This stage takes at most 3 steps.

Theorem 3 *In a circuit-switched 6-port non-square torus $T_{n_1 \times n_2 \times n_3}$, broadcast can be done within*

$$3 \left\lceil \log_7 \frac{n_1}{2} \right\rceil + \left\lceil \log_7 \frac{n_2}{n_1} \right\rceil + \left\lceil \log_7 \frac{n_3}{n_1} \right\rceil + c$$

steps, wher $c = 2$ (resp. 3) when n_1 is even (odd), which number of steps is at most 5 (6) steps more than the lower bound in Lemma 1.

Comment: When $\alpha = 4$, the communication patterns in stages 3 and 4 should be modified to ones as shown in Fig. 7. As can be seen, each unit cube is partitioned into 5, instead of 7, more smaller unit cubes after each communication

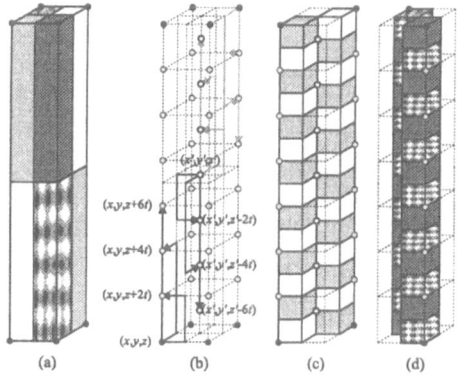

Fig. 5. (a) The geometric relationship of eight neighboring unit cubes. (b) The communication pattern in one step for a type I unit cube. (c) and (d) Each unit cube is partitioned into 7 smaller new cubes after performing the step in (a).

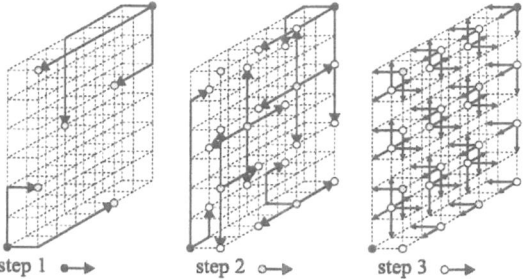

Fig. 6. The message distribution in a type I unit cube which is a $2 \times 7 \times 7$ mesh

step. Also, the recursion should be proceeded as long as the value of w or h is larger than 4.

Theorem 4 *In a circuit-switched 4-port torus $T_{n_1 \times n_2 \times n_3}$, broadcast can be done within*

$$3 \left\lceil \log_{\alpha+1} \frac{n_1}{2} \right\rceil + \left\lceil \log_{\alpha+1} \frac{n_2}{n_1} \right\rceil + \left\lceil \log_{\alpha+1} \frac{n_3}{n_1} \right\rceil + c$$

steps, wher $c = 2$ (resp. 3) when n_1 is even (resp. odd), which number of steps is at most 5 (resp. 6) steps more than the lower bound in Lemma 1.

When α Is Odd. Earlier, when α is even, we used the concept of unit cube in a recursive manner. After one recursive step, the number of nodes owning M in a unit cube remains as an *even* number. This is important to maintain the recursive structure. When α is odd, in order to maintain such structure, we

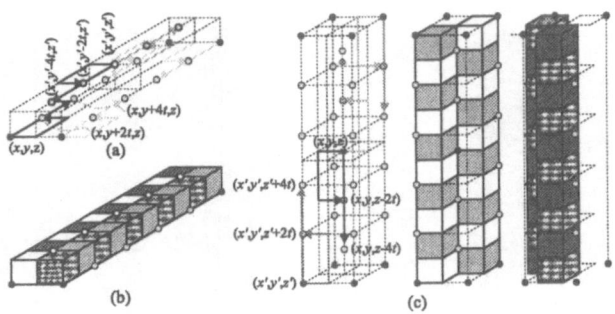

Fig. 7. Broadcasting in 4-port tori: (a) Stage 3. (b) Each unit cube is partitioned into 5 smaller unit cubes after stage 3. (c) Stage 4 and the result after each step.

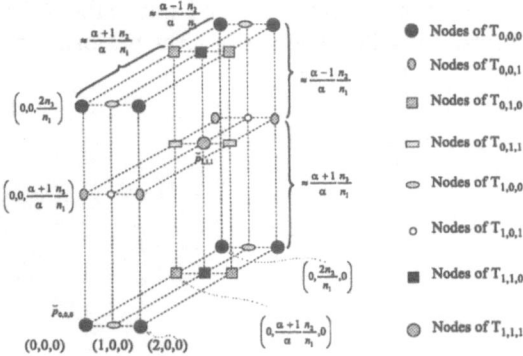

Fig. 8. Illustration of Definition 3.

have to redefine the dilated torus. To avoid the tedium of using floor and ceiling functions, we assume that n_2 and n_3 are each a multiple of n_1.

Definition 3 Given a non-square torus $T_{n_1 \times n_2 \times n_3}$, the dilated torus induced by $T_{n_1 \times n_2 \times n_3}$, denoted as $\check{T}_{n_1 \times n_2 \times n_3}$, is an torus consisting of nodes from the following eight $\frac{n_1}{2} \times \frac{n_1}{2} \times \frac{n_1}{2}$ tori:

$$T_{a,b,c} = SPAN(p_{a,b\frac{\alpha+1}{\alpha}\frac{n_2}{n_1},c\frac{\alpha+1}{\alpha}\frac{n_3}{n_1}}, B_3, N_3),$$

where $a, b, c = 0..1$, $B_3 = (2e_1, \frac{2n_2}{n_1}e_2, \frac{2n_3}{n_1}e_3)$ and $N_3 = (\frac{n_1}{2}, \frac{n_1}{2}, \frac{n_1}{2})$. $\check{T}_{n_1 \times n_2 \times n_3}$ has n_1^3 nodes, which are denoted by $\check{p}_{i,j,k}$ for $i, j, k = 0..n_1 - 1$.

Fig. 8 illustrates this definition. One important property is that the ratio of the dilations on the Y axis is $(\alpha + 1) : (\alpha - 1)$, and the same on the X.

Let $\check{p}_{0,0,0}$ be the source node. There are 5 stages to perform broadcast.
Stage 1. Node $\check{p}_{0,0,0}$ sends M to $\check{p}_{1,1,1}$.

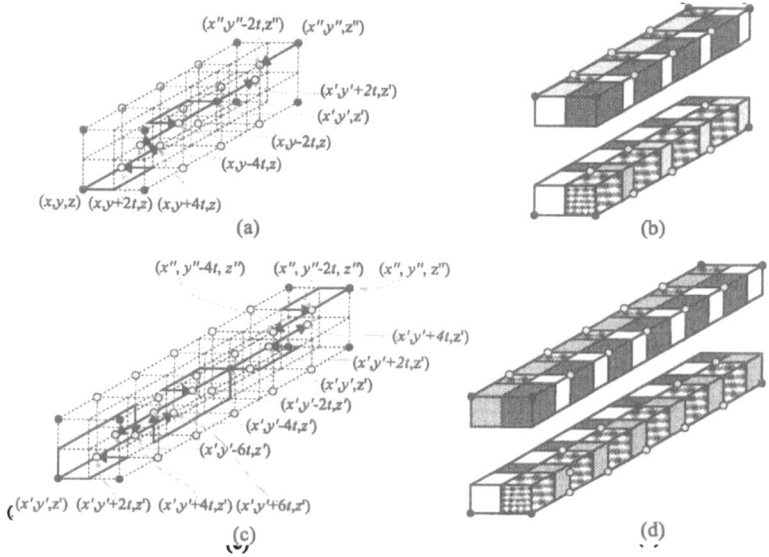

Fig. 9. One communication step of stage 3 in eight neighboring unit cubes: (a) the communication pattern of 5-port tori and (b) each unit cube is partitioned into six smaller cubes after one step; (c) the communication pattern of 3-port tori and (d) each unit cube is partitioned into four smaller cubes after one step.

Stage 2. Node $\breve{p}_{0,0,0}$ and node $\breve{p}_{1,1,1}$ concurrently perform broadcast on $T_{0,0,0}$ and $T_{1,1,1}$, respectively, by applying the result in Section 3.1. Hence this takes $3\lceil \log_{(\alpha+1)} \frac{n_1}{2} \rceil$ steps.

Stage 3. In this stage, we will "expand" nodes owning M along the Y axis. From nodes already owning M, we can define a number of unit cubes as in the previous section. Also, according to the distribution of the nodes owning M, we can classify the unit cubes into four types, I, II, III, and IV, acording to Fig. 3(a).

Now consider two consecutive unit cubes of type I as shown in Fig. 9(a). Let $p_{x,y,z}$, $p_{x',y',z'}$, $p_{x'',y'',z''}$ be the nodes already owning M. We can assume without loss of generality that $p_{x,y,z}$ is a node in $T_{0,0,0}$, and $p_{x',y',z'}$ one in $T_{1,1,1}$. By Definition 3, it must be that $(y'-y):(y''-y') \approx (\alpha+1):(\alpha-1)$. For easy of presentation, let $y''-y$ be a multiple of $2(\alpha+1)$. Then we perform the following communication: ($i = 1..\frac{\alpha-1}{2}$ and $j = 1..\frac{\alpha+1}{2}$).

- $p_{x,y,z}$ sends $\frac{\alpha+1}{2}$ messages to nodes $p_{x,y+2it,z}$,
- $p_{x',y',z'}$ sends $\frac{\alpha+1}{2}$ messages to nodes $p_{x',y'-2it,z'}$ and $\frac{\alpha-1}{2}$ messages to nodes $p_{x',y'+2jt,z'}$,
- $p_{x'',y'',z''}$ sends $\frac{\alpha-1}{2}$ messages to nodes $p_{x'',y''-2jt,z''}$.

For instance, assuming $\alpha = 3$, Fig. 9(a) shows the communication paths. Apparently, after this step, each unit cube will be partitioned into a number of smaller unit cubes. In this example, each unit cube will generate 3 or 5 smaller unit cubes. In Fig. 9(b), it is shown that totally 32 smaller cubes are generated from the 8 larger cubes.

As another example, Fig. 9(c) and (d) show the case of $\alpha = 5$. In general, the number of unit cubes will be multiplied by a factor of $\alpha + 1$ after each communication step. In order to regularly reduce the size of unit cubes, the recursion should maintain an important invariant:

I: For any two consecutive unit cubes of type I, the ratio of the width of the larger cube to that of the smaller one should be approximately $(\alpha + 1) : (\alpha - 1)$.

The recursion is repeated until $y'' - y < 2(\alpha+1)$. This stage takes $\lceil \log_{(\alpha+1)} \frac{2n_2}{n_1} \rceil - 1$ steps to complete.

Stage 4. This stage is similar to stage 3, except that we want to "expand" the nodes owning M along the Z axis. We omit the details.

Stage 5. Now each unit cube of type I will be of width at most $\alpha + 1$ and of height at most $\alpha + 1$. The approach is similar to the stage 5 of Section 3.2, so we omit the details. This stage takes 3 steps.

Theorem 5 *In a circuit-switched α-port 3-D torus $T_{n_1 \times n_2 \times n_3}$ such that α is odd, broadcast can be done in*

$$3 \left\lceil \log_{\alpha+1} \frac{n_1}{2} \right\rceil + \left\lceil \log_{\alpha+1} \frac{2n_2}{n_1} \right\rceil + \left\lceil \log_{\alpha+1} \frac{2n_3}{n_1} \right\rceil + c$$

steps, where $c = 2$ (resp. 3) when n_1 is even (resp. odd), which number of steps is at most 6 (resp. 7) steps more than the lower bound in Lemma 1.

4   Extensions to Higher-Dimensional Tori

We briefly show how to extend the result to a k-D torus $T_{n_1 \times n_2 \times \cdots \times n_k}$. If the torus is square, then a dimension-by-dimension approach can be used to distribute M. Otherwise, assuming the first dimension to be the one of smallest length, we can first "squeeze" the torus into a square, dilated one $T_{n_1 \times \cdots \times n_1}$. Then we try to reduce the dilation along the 2nd, 3rd, ..., k-th dimensions one by one. If the number of ports α is even, then we can simply partition the torus into sub-networks of even size. Otherwise, an invariant similar to **I** should be developed so that the recursion can proceed.

5   Conclusions

We have presented a systematical solution to solve the broadcasting problem in an α-port torus of any dimension and any size with circuit switching. The problem is challenging because a good solution should try to utilize as many of the available ports as possible. Further, when the torus is non-square, we should try to distribute the broadcast message to nodes in the network as evenly as possible to avoid congestion in the subsequent communications. The "dimension-by-dimension" and "squeeze-then-expand" approaches proposed in this paper have successfully conquered these difficulties and have delivered performance very close to the lower bound of this problem.

References

[1] C.-T. Ho and M.-Y. Kao. Optimal broadcasting in all-port wormhole-routed hypercubes. *IEEE Trans. on Paral. and Distrib. Sys.*, 6(2):200–204, Feb. 1995.

[2] S.-K. Lee and J.-Y. Lee. Optimal broadcast in α-port wormhole-routed mesh networks. In *Int'l Conf. on Paral. and Distrib. Sys.*, pages 109–114, 1997.

[3] J. L. Park and H.-A. Choi. Circuit-switched broadcasting in tori and meshes networks. *IEEE Trans. on Paral. and Distrib. Sys.*, 7(2):184–190, Feb. 1996.

[4] J. L. Park, S.-K. Lee, and H.-A. Choi. Circuit-switched broadcasting in d-dimensional torus and mesh networks. In *Int'l Parallel Processing Symp.*, pages 26–29, 1994.

[5] J. G. Peters and M. Syska. Circuit-switched broadcasting in torus networks. *IEEE Trans. on Paral. and Distrib. Sys.*, 7(3):246–255, March 1996.

[6] D. F. Robinson, P. K. Mckinley, and B. H. C. Cheng. Optimal multicast communication in wormhole-routed torus networks. *IEEE Trans. on Paral. and Distrib. Sys.*, 6(10):1029–1042, Oct. 1995.

[7] Y.-C. Tseng. A dilated-diagonal-based scheme for broadcast in a wormhole-routed 2d torus. *IEEE Trans. on Comput.*, 46:947–952, Aug. 1997.

[8] C.-M. Wang and C.-Y. Ku. A near-optimal broadcasting algorithm in all-port wormhole-routed hypercubes. In *ACM Int'l Conf. on Supercomputing*, pages 147–153, 1995.

[9] S.-Y. Wang, Y.-C. Tseng, S.-Y. Ni, and J.-P. Sheu. Circuit-switched broadcast in multi-port 2d tori. In *High-Performance Computing and Networking*, 1999.

Impact of the Head-of-Line Blocking on Parallel Computer Networks: Hardware to Applications*

V. Puente[1], J.A. Gregorio[1], C. Izu[2], and R. Beivide[1]

[1] Universidad de Cantabria
39005 Santander, Spain
{vpuente,jagm,mon}@atc.unican.es
[2] University of Adelaide
SA 5005, Australia
cruz@cs.adelaide.edu.au

Abstract. A fully adaptive router with hybrid buffers at the input and output channels was designed, which improves the throughput of its input buffer counterpart by up to 40% and has only 10% higher base latency. An in-depth analysis of different router buffer organization was carried out for a toroidal network, which uses either a deterministic (DOR) or a fully adaptive routing scheme. Each proposal is described in VHDL and evaluated with the *Synopsys* synthesis tool. Technological restrictions obtained were used to evaluate network performance under both synthetic loads and real applications.

1   Introduction

Most routers select a simple buffer organization as a strategy to limit their complexity. Buffer space is attached to each input channel, and a FIFO access policy is applied to route the incoming messages. Thus, it requires a memory with only one reading and one writing port. Notwithstanding, it is well known that this buffer organization can effectively reduce peak throughput due to the so called *head-of-line blocking*. Only the first packet of each FIFO competes for the output resources; thus, when one packet blocks due to network contention, all the remaining packets at its FIFO will also block, even when their possible outputs are available. In particular, a network switch with fixed length packets and a random distribution of their destinations could only achieve about 60% of its link capacity [4].

There are two possible solutions to this problem. The first one consists of applying flexible access policies to the input buffers. From the architectural point of view, this approach presents improved performance when compared with FIFO. However, any architectural gains are eliminated on its physical implementation because of the complexity of the buffer organization. A second solution is to allocate the router buffer space into a central queue or attach the buffers to each

---

* This work is supported in part by TIC98-1162-C02-01

P. Amestoy et al. (Eds.): Euro-Par'99, LNCS 1685, pp. 1222–1230, 1999.
© Springer-Verlag Berlin Heidelberg 1999

output. Both approaches require multiport structures which increment their silicon area. However, with current technology, the multiport option is not only a possible alternative, but, as this study will show, the small latency penalty of this solution is outweighed by its significant throughput gains.

This paper presents a comparative analysis of the different buffer organizations for either deterministic or adaptive toroidal routers. All logical combinations are explored in order to find their optimal buffer organization. The goal is to quantify the cost, in terms of latency and area, introduced by buffer proposals oriented to increase network throughput. Our evaluation ranges from the hardware design of each proposal up to the analysis of parallel application execution over cc-NUMA architectures with any of the proposed interconnection networks. Moreover, the network is evaluated under various types of synthetic traffic. This down-to-up methodology brings every factor of network design into consideration.

The rest of this paper is organized as follows. Section 2 presents the router architectures that have been considered as part of this analysis. Section 3 introduces the more significant details of their hardware implementation and Section 4 compares their performance at different levels. Finally, Section 5 collates the contributions of this paper.

2   Architectural Proposals for Torus Networks

This section introduces the deterministic and adaptive routing schemes. Then, it discusses how the different buffer organizations are applied to both routers. In all cases, the flow control is virtual cut-through (VCT). All routers use the Bubble mechanism [1] in order to avoid deadlock.

2.1   Routers with Head-of-Line Blocking

As our proposals focus on the reduction of head-of-line blocking (HLB), it is logical to introduce first our baseline routers, whose HLB motivated this study.

As mentioned before, this problem occurs in routers with the most basic buffer organization: input buffers and FIFO management policies. In fact, due to its simplicity, this type of architecture is widely used, including the network routers of commercial machines as described in [10].

The adaptive router for a *k-ary 2-cube* network, using the bubble mechanism to avoid deadlock. There are two virtual channels, attached to each physical input: a deterministic channel and an adaptive channel [3]. Messages progress in order of dimension through deterministic channels, and in any minimal route through adaptive channels. Buffering at both input channels applies a FIFO policy.

Each physical input channel has a synchronization unit. This is because the router design is synchronous but the communications between neighboring nodes are *self-timed*, in order to avoid any problems with the clock skew.

The basic architecture of the deterministic router is quite similar to the adaptive one. The main difference is that it only needs one (deterministic) channel per physical line. For a more exhaustive description of both routers, please refer to [8] and [2].

2.2   Routers without Head-of-Line Blocking

Deterministic Router The strategy followed in this design is simple: move the buffer space to the output ports. The output buffer has only one read port, but this won't cause unnecessary blocking because all packets stored in a given buffer are requesting the same output resource. However, multiple input ports may want to write simultaneously into the same output buffer. Thus, we should provide multiport memories. With DOR, we require 2 writing ports at the X dimension (x input port or injection port) and 4 writing ports at both the Y dimension (the y input port plus any x port or injection port) and consumption.

One of the problems with output buffering is the need to route the packet before applying the flow control to the selected output buffer. By providing a small buffer per input port, we can decouple the two processes, and reduce the internal node delay. This buffer cannot introduce HLB, because it has capacity for a single packet.

Adaptive Router Although we could apply the same approach to the adaptive router, both the addition of adaptive virtual channels and the higher flexibility to route packets from any input port to any output buffer, increase the number of writing ports to 5 for the output buffers in the X dimension and 7 writing ports for the output buffers on the Y dimension. The complexity of a 7-port buffer under current technology is considerable, so this is not a viable option. Besides, the access to the output port should be multiplexed between the deterministic and the adaptive output buffers, which may potentially add one more stage to the internal pipeline.

However, it is a known fact that in this type of adaptive routers, based on an adaptive virtual network plus a deterministic one used as a escape route [3], the deterministic buffer utilization is generally quite low. This fact led us to propose a hybrid buffering space as shown in figure 1. The buffering space for the deterministic channels is allocated at the input ports, and the buffering space for the adaptive channels is allocated at the output ports. Although this does not fully eliminate the HLB, it significantly reduces it because of the low occupancy of the deterministic input buffers. This scheme reduces the number of writing ports at the output buffer to 4 (3 adaptive inputs plus one deterministic) except for the consumption channel that requires 5 writing ports (any of the 4 adaptive input plus one deterministic), making their implementation technologically viable.

Finally, a crossbar allows for packet movement from the adaptive virtual network to the deterministic one and vice versa as shown in figure 1. The crossbar arbitrates between any virtual channel that wants to use a deterministic output channel, and any deterministic channel (or injection port) that wants to use

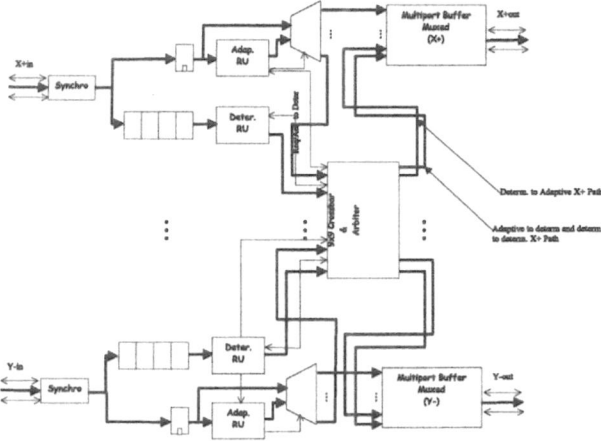

Fig. 1. Architecture of the adaptive Bubble router with hybrid buffering space.

an adaptive output channel. Although the crossbar is not contention free, the potential throughput lost is negligible, because of the low deterministic channel utilization.

3   Hardware Implementation

Once we have proposed the design alternatives to input buffering, we need to evaluate their performance, starting from their implementation cost. The implementation cost is measured by generating a VHDL description of each described router, which is then fed into *Synopsys*, a high-level synthesis tool. This design was mapped into 0.35 μm and 5-layer metal from MIETEC/ALCATEL foundry. The characteristics of each router have been obtained from the synthesis tool. Although we did not descend to the layout level, the results are very close to those of a physical implementation.

This section introduces the most relevant issues regarding the implementation of the routers with output and hybrid buffer space. Therefore, it focuses on the implementation of the multiport memory used for the output buffers. Other blocks present minor modifications from their input buffer router counterparts, whose detailed implementation can be found in [8], [9] and [2].

3.1   Implementation of the Pipeline Multiport FIFO

The design of our output buffers is based on [5] scheme using multiple input reading ports, but just one output port. Thus, messages can be sent out of the output buffer following a FIFO policy. So, the output buffers use a structure which we have called *Pipeline multiport FIFO* or PM-FIFO for short. Figure 2 shows the internal structure of a memory of this type, for a simplified case with 2 writing ports and packet length of 2 phits. The shadowed area (*VC Control*) shows the additional logic for the hybrid buffer scheme (see Fig 1), which multiplexes the adaptive channel (output of the PM-FIFO) with the deterministic channel.

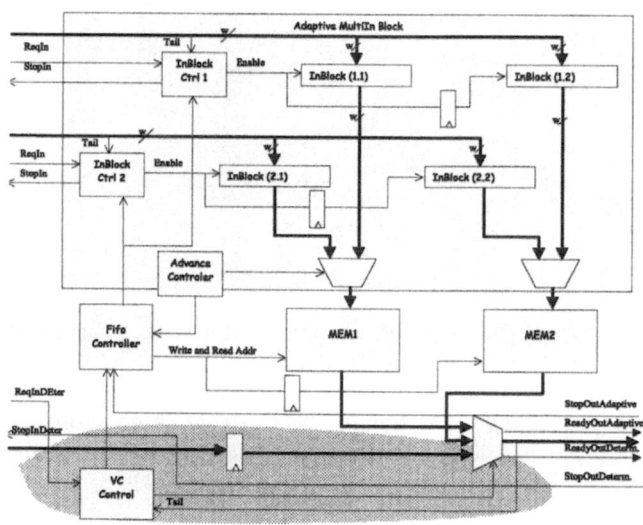

Fig. 2. Output buffer implementation as a Pipeline Multiport FIFO, plus a virtual channel multiplexer (shadowed area).

We can see from figure 2 that the number or stages of the pipeline memory matches the number of phits per packet. However, by setting a word size of 2 phits, we halved the pipeline length. This adds no penalty to the node delay, because the first phit (the header) requires at least one cycle at the RU to be routed to an output buffer.

3.2   Results of the Hardware Implementation

All router implementations share the following conditions: the phit width is 33 bits (4 bytes plus 1 bit tail), and the packet length is set to 10 phits (40 bytes/packet). The total buffer capacity was chosen by examining the gains due to increasing capacity of the adaptive router with hybrid buffers. Experimentally, we observed that little performance is gained by increasing the adaptive channel capacity above 4 packets. Thus, each router requires approx. 1.3 kilobytes. For input buffer routers the length of the pipeline is four cycles. For output buffer routers the length is 5 cycles.

Table 1 shows the characteristics of time and area for each router components for 0.35 μm technology. We should first note that the adaptive router with hybrid buffer improves the cycle time of its input buffer counterpart. The cycle time for the latter router is set by the 9 × 5 crossbar, which is slower than the 9 × 9 crossbar of the hybrid version because it arbitrates two virtual channels per physical output. In fact, the cycle time for the hybrid adaptive router is set by the multiport buffer.

Not only the multiport memory but also the demultiplexers, which provide the direct data paths from each input to any of the output buffers, are responsible for the higher area demands of the routers with output/hybrid buffering.

Module	Router							
	InDet.		**InAdapt.**		**OutDet.**		**OutAdapt.**	
	Critical Path (ns.)	Area (mm^2)	Critical Path (ns.)	Area (mm^2)	Critical Path (ns.)	Area (mm^2)	Critical Path (ns.)	Area (mm^2)
Synchr.	2.72	0.020(x5)	2.72	0.020(x5)	2.72	0.020(x5)	2.72	0.020(x5)
Fifo Iny.	3.40	0.462(x1)	3.67	0.444(x1)	3.35	1.405(x1)	3.49	0.504(x1)
Fifo InDet.	**3.40**	1.69(x4)	3.67	0.848(x4)	3.35	0.284(x4)	3.49	0.504(x4)
Fifo InAdapt.	–	–	3.67	0.848(x4)	–	–	3.49	0.282(x4)
RU Det.	2.88	0.016(x5)	3.66	0.117(x5)	3.07	0.043(x5)	3.49	0.118(x5)
RU Adapt.	–		3.66	0.117(x4)	–		3.49	0.145(x4)
Crossbar	3.17	0.192(x1)	**3.67**	0.597(x1)	–		3.49	0.937(x1)
Multiport	–	–	–	–	**3.35**	1.380(x2)x 1.760(x2)y	**3.49**	1.611(x 4)
Mulport Cons.	–	–	–	–	**3.35**	1.574(x1)	**3.49**	1.611(x1)
Total	3.40	7.594	3.67	9.305	3.35	10.710	3.49	13.910

Table 1. Time and area characteristics.

4   Performance Evaluation

The next step is to evaluate the performance of toroidal networks implemented with each of the routers, under the time constraints obtained above. We use a register-level transfer simulator called *SICOSYS* [7], which takes into account the key parameters of the low-level implementation and obtains results which are very close to those of VHDL simulator at a lower computational cost.

Each network has been evaluated under two different scenarios: synthetic loads which represent a variety of patterns present in real applications, and real loads generated by application benchmarks under cc-NUMA architectures.

4.1   Synthetic Loads

We have analyzed the behaviour of each network under random uniform patterns with two length distributions: fixed length (10 phits) and bimodal length(short and long messages). Our bimodal traffic combines 90% of short messages (10 flits) with 10% of long messages (50 phits). We have also considered three non-uniform permutations: *matrix transpose*, *bit-reversal* and *perfect-shuffle*.

Table 2 presents the base latency and peak throughput obtained for an 8×8 toroidal network, for each of the router alternatives. These results show the architectural differences (phits/cycles) of each router as well as their true performance when including their technological cost (phits/nanosecond).

Router	Random	M-Trans	Perfec-Shu	Bit- Rever	Bimodal
InBuffer Deter.	103.2	108.3	103.3	109.5	118.9
	11.50 (39.1)	3.92 (13.3)	5.27 (17.9)	3.58 (12.2)	7.33 (24.9)
InBuffer Adapt.	111.0	116.8	111.5	118.1	129.2
	10.87 (39.9)	7.60 (27.9)	10.14 (37.2)	8.82 (32.4)	7.82 (28.7)
OutBuff Deter.	115.7	124.0	116.2	123.0	130.7
	14.77 (49.5)	4.08 (14.5)	6.13 (22.3)	4.11 (13.7)	9.11 (30.54)
OutBuff Adap.	119.9	128.7	120.1	129.3	137.1
	16.30 (56.9)	9.7 (34.0)	13.13 (45.8)	12.35 (43.1)	11.21 (39.1)

Table 2. Base Latency in nanoseconds (for normalized load of 0.05% with respect to bisection) and maximum throughput accepted in phits/nanosecond (phits/cycle) for a 8×8 Torus.

Regarding the DOR router with output buffers, little gains are observed apart from uniform traffic, which does not justify the additional cost both in silicon area (30% and 15% more than the DOR and adaptive routers with input buffers, respectively) or its higher node delay.

On the other hand, the adaptive router with hybrid buffering achieves significant throughput gains under all traffic patterns in comparison with the other three alternatives. Not only does it outperform the DOR routers for non-uniform traffic, but it also achieves up to 40% higher throughput that its input buffer counterpart. This improvement indicates that the head-of-line blocking (HLB) and, to a lesser extent, the crossbar arbitration for the same output are limiting the performance of the adaptive routing algorithm. On the other hand, it increases base latency by less than 10% compared with the input buffer alternatives. In terms of cost, it requires 50% more area than the adaptive router with input buffers, and 90% (10%) more than the DOR router with input (output) buffering. This cost is easily justified for throughput-sensitive applications.

4.2   Real Loads

Finally, we have evaluated the impact of each network implementation on the performance of parallel applications running on a cc-NUMA architecture. This evaluation is carried out by using the tool ED-SYCOSYS [7] which is based on the RSIM simulator [6] replacing the original RSIM's network by our network simulator (SYCOSYS).

Due to space limitations we will only present the result for the FFT application which belongs to the SPLASH-2 suite [11]. The system is configured with the default values provided by RSIM except for the parameters discussed below. We set the cache line size to 32 bytes, and assume that network commands are 8 bytes long. Hence, data and command packets have a fixed length of 40 bytes (10 phits) and 8 bytes (2 phits) respectively. The network size is set to 64 nodes (8 × 8 torus). Three cases were simulated: two of them with separate data and control networks and processor speeds of 600 MHz and 1GHz respectively, and a third one in which data and control messages share a single network with 1GHz processor nodes. The latter network prevents fetch deadlock by providing sufficiently large consumption buffers at the network interface. The problem size is set to 64K complex doubles.

	600 MHz Processor Dual Network	1GHz Processor Dual Network	1GHz Processor Unified Network
InBuff. Deter.	621656	870482	961864
InBuff. Adapt.	589405	810486	870770
OutBuff. Deter.	629711	886967	956921
OutBuff. Adapt.	586535	799080	842776

Table 3. Execution time, in processor cycles, for FFT with 64K complex doubles over 64 nodes (Torus 8 × 8).

Table 3. shows the execution times for each experiment. The network traffic is only high for short periods of time, hence the small time variations observed in spite of the different network capabilities. As the volume of traffic increases with problem size, so do the time differences between the 4 network alternatives. Due to the complexity of the simulation environment, the study of larger problems was not feasible, but we could extract some conclusions from the above results.

5   Conclusions

An in-depth analysis of the router's buffer organization was carried out for a toroidal network which uses either a deterministic (DOR) or a fully adaptive routing scheme. Two alternatives to the simple input FIFO organization were proposed, one for each routing scheme. After a thorough evaluation, we can draw the following conclusions:

1. In both cases, the elimination of head-of line blocking (HLB) produces an improvement on network performance; but only the adaptive router presents significant gains under non-uniform traffic, which justify its cost in silicon area.
2. Output buffering increases node complexity, so throughput gains must outweigh the penalty on node delay. Such is the case for the adaptive router

which gains 40% throughput with less than 10% increment in base latency. The node delay penalty for the DOR router is similar, but the throughput gains are negligible.

3. The analysis of execution time for parallel applications shows, once more, the required balance between network latency and throughput. In a cc-NUMA environment, applications exhibit execution phases with either low or high network load. The former phases benefit from low latency and the latter ones from high throughput.

Finally, we should note the importance of the evaluation method which takes into account the architectural choices, the technological constraints and the application demands. None of them could be ignored when designing the interconnection network for a parallel system.

References

[1] C. Carrión, R. Beivide, J.A. Gregorio, F. Vallejo, "A flow control mechanism to avoid message deadlock in k-ary n-cube networks," *Fourth International Conference on High Performance Computing*, pp. 322-329, India, December,1997.

[2] C. Carrion, R. Beivide, J.A. Gregorio, "Performance Evaluation of Bubble Algorithm: Benefits for k-ary n-cubes. 7th Euromicro on Parallel and Distributed Processing, Madeira, Portugal 1999.

[3] J. Duato,"A necessary and sufficient condition for deadlock-free routing in cut-through and store-and-forward networks". *IEEE Trans. on Parallel and Distributed Systems*, vol.7, no.8, pp.841-854, August 1996.

[4] M. Karol, M. Hluchyj,S. Morgan "Input Versus Output Queuing on Space Division Packet Switch", IEEE Trans. On Communications, vol. COM-35, no. 12, Dec. 1987, pp. 1347-1356.

[5] M. Katevenis, P. Vatsolaki, A. Efthymiou "Memory Shared Buffer for VLSI Switches", ACM SIGCOMM, August 1995.

[6] V. S. Pai, P. Ranganathan, S. Adve "Rsim: An execution-Driven Simulator for ILP-Based Shared-Memory Multiprocessors and Uniprocessors", *IEEE TCCA Newsletter*, Oct. 1997.

[7] J.M. Prellezo, V. Puente, J.A. Gregorio, R. Beivide, "SICOSYS: a interconnection network simulator for parallel computers," available at http://www.atc.unican.es/REPORTS/TR-ATC2-UC98.pdf, June 1998.

[8] V. Puente, J.A. Gregorio,J. M. Prellezo, R. Beivide,J. Duato, C. Izu "Adaptive Bubble Router: a Design to Balance Latency and Throughput in Networks for Parallel Computers", *ICPP'99*, Sept. 1999.

[9] V. Puente, J.A. Gregorio, C. Izu, R. Beivide, F. Vallejo "Low-level Router Design and its Impact on Supercomputer System Performance", ICS'99, July 1999.

[10] S. L. Scott, G. Thorson, "The Cray T3E network: Adaptive routing in a high performance 3-D torus", *Hot Interconnects Symposium IV*, pp. 147-155, Aug. 1996.

[11] S. C. Woo, M. Ohara, E. Torrie, J.P. Singh, A. Gupta "The SPLASH-2 Programs: Characterization and Methodological Considerations". In Proceedings of the *22nd International Symposium on Computer Architecture*, pages 24-36. June 1995.

Interval Routing on Layered Cross Product of Trees and Cycles*

R. Královič[1], B. Rovan[1], and P. Ružička[2]

[1] Department of Computer Science, Faculty of Mathematics and Physics,
Comenius University, 84215 Bratislava, Slovakia.
{kralovic,rovan}@dcs.fmph.uniba.sk
[2] Institute of Informatics, Faculty of Mathematics and Physics,
Comenius University, 84215 Bratislava, Slovakia
ruzicka@dcs.fmph.uniba.sk

1 Introduction

Interval routing is an attractive space-efficient routing method for point-to-point networks (introduced in [13] and [15]) which has found industrial applications in the INMOS T9000 transputer design. Surveys of the principal theoretical results as well as recent trends in the area of interval routing can be found in [16, 5, 11].

Interval routing is based on compact routing tables, in which the set of nodes reachable via outgoing links is compactly represented in the form of intervals. The space efficiency can be measured by *compactness*, that is the maximum number of intervals per link.

Previous research mostly concentrated on *shortest path* interval routing schemes (IRS for short). Shortest path IRS of compactness 1 have been designed for a number of well-known interconnection networks including trees, rings, complete bipartite graphs, meshes, tori, and hypercubes. However, there are interconnection networks having provably large [8] compactness for shortest path IRS, for example shuffle-exchange, De Bruijn, cube-connected cycles, butterfly, pancake, and star graphs. Several generalizations of IRS were therefore proposed.

Multidimensional interval routing schemes (MIRS for short) were introduced in [3] and were used to represent the information on all shortest paths. MIRS with low memory requirements were proposed in [3] for hypercubes, grids, tori and certain types of chordal rings. Other efficient MIRS were designed in [12].

Certain graph operators have been found interesting in the design of communication networks. The impact of some graph operators on the compactness of interval routing has been previously studied in [6, 4, 10]. These results characterize the effect of the cartesian product, the composition, and the join of graphs on the minimum number of linear intervals needed for the optimal deterministic routing.

We present the study of another graph-theoretic operation, namely the layered cross product of graphs. The Layered Cross Product (LCP in short) was introduced in [2] as a technique for constructing some complex interconnection

---

* This research has been partially supported by VEGA 1/4315/97.

P. Amestoy et al. (Eds.): Euro-Par'99, LNCS 1685, pp. 1231–1239, 1999.
© Springer-Verlag Berlin Heidelberg 1999

networks on the basis of structurally simple multiplicands. Certain useful properties of networks decomposable as the layered cross product of simple graphs have been exploited. In [1], an efficient compact routing protocol was introduced for the LCP of trees. In [7], efficient deadlock-free packet and wormhole routing protocols have been proposed for interconnection networks constructed as the layered cross product of trees and series parallel graphs.

In this paper we consider the class of networks constructed as the layered cross product of trees and cycles. This class of networks is of interest, as it includes among others butterflies, mesh of trees, fat trees, and multiglobe graphs. We first prove that the classical shortest path interval routing schemes do not work efficiently on the class of interconnection networks, constructed as the LCP of regular complete trees or cycles. However, we show that there are efficient full information shortest path multidimensional interval routing schemes for this class of networks. By [12] there are other well-known interconnection networks (e.g., cube-connected cycles, stars) for which multidimensional schemes do not work as efficiently. These are not known to be composable by the LCP operation from trees and cycles. Our results thus indicate a possible explanation why improvement in the efficiency by using the multidimensional approach (instead of the deterministic unidimensional approach) is obtained for some well-known interconnection networks.

Due to space limitations the proofs are mostly omitted or sketched only.

2 Notions

A graph $G = \langle V, E \rangle$ is h-*layerable* if there exist h disjoint (non-empty) sets of vertices $V_1, ..., V_h$, where V_i is the set of vertices in the layer i, such that $V = V_1 \cup \cdots \cup V_h$ and every edge in E connects vertices of two adjacent layers. We shall call a graph with given $V_1, ..., V_h$ an h-*layered graph*. Let the layer with index 1 be the *top layer*, and the layer with index h the *bottom layer*.

The graph can be layered in many different ways. For example, each bipartite graph can be layered using just two layers. In what follows, we shall consider only "naturally" layered graphs in a sense to be apparent later.

Some well-known interconnection networks used in parallel computing can be viewed as layered graphs. Examples are grids, hypercubes, butterflies, Beneš graphs, mesh of trees or fat trees. On the other hand, examples of non-layerable networks are cycles of odd length or odd-dimensional cube-connected-cycles.

Let $G_1 = \langle V_1, E_1 \rangle$ and $G_2 = \langle V_2, E_2 \rangle$ be two h-layered graphs, where $V_1 = V_1^{(1)} \cup \cdots \cup V_h^{(1)}$ and $V_2 = V_1^{(2)} \cup \cdots \cup V_h^{(2)}$. The *Layered Cross Product* (LCP for short) of the two h-layered multiplicands G_1, G_2 (denoted by $G_1 \otimes G_2$) is an h-layered graph $G = \langle V, E \rangle$, where $V = V_1 \cup \cdots \cup V_h$ and V_i is the cartesian product of $V_i^{(1)}$ and $V_i^{(2)}$, $1 \leq i \leq h$, and an edge $((a_1, a_2), (b_1, b_2))$ belongs to E if and only if $(a_1, b_1) \in E_1$ and $(a_2, b_2) \in E_2$.

We are interested only in h-layered interconnection networks, in which every vertex is on a path of length h connecting a vertex of the top layer with a vertex of the bottom layer. Two layered graphs are considered to be equal if they are isomorphic, and the isomorphism preserves the layer to which a vertex belongs. Under this assumption, the LCP operation is commutative and associative. Thus, we may consider the LCP of more than two h-layered graphs (all with the same number h of layers) without regard to the order in which they are written, or the order in which the binary operations are applied. A simple h-layered path serves as the identity element of the LCP operation.

In what follows, we are interested in the class of networks constructed as the LCP of simple regular layered graphs. By the LCP of trees one can obtain some interesting interconnection networks, as butterflies, mesh of trees or fat trees. Globe graphs [6] can be constructed as the LCP of cycles. And multi-globe graphs [14] can be composed as the LCP of cycles and trees.

3   Path Systems and Routing

The main topic of our study is the routing in graphs. Hence, given a graph G we are interested in the all-to-all shortest paths system (i.e., the set of (all) shortest routing paths between all pairs of vertices in G).

The relationship between all-to-all shortest paths systems in the multiplicands G_1, G_2 and the one in the LCP product graph $G_1 \otimes G_2$ is given in the following two simple lemmas.

Lemma 1. [1] *The distance between* (a_1, a_2) *and* (b_1, b_2) *in* $G_1 \otimes G_2$ *is at least the maximum of the distance between* a_1 *and* b_1 *in* G_1 *and the distance between* a_2 *and* b_2 *in* G_2.

Lemma 2. *The distance between* (a_1, a_2) *and* (b_1, b_2) *in* $G_1 \otimes G_2$ *is at most the sum of the distances between* a_1 *and* b_1 *in* G_1 *and between* a_2 *and* b_2 *in* G_2.

The above lemmas do not give an exact estimate of the length of shortest paths. However, exact values can be achieved for some special topologies. We present results concerning graphs obtained as the LCP of trees and cycles.

We start by introducing some basic layered graphs. By an h-layered top-tree (bottom-tree) we mean a rooted tree for which the root is in the top (bottom) layer 1 (h) and each path from the root to a leaf passes through increasing (decreasing) layers. By an h-layered cycle we mean a cycle $C_{2h-2} = \langle V, E \rangle$, where $V = V_1 \cup ... \cup V_h$ with $|V_1| = |V_h| = 1$ and $|V_i| = 2$ for $i = 2, ..., h-1$.

The LCP of the $(h+1)$-layered complete d-ary top-tree and the $(h+1)$-layered complete d-ary bottom-tree can be viewed as a generalized butterfly graph.

A *generalized butterfly graph of degree h and alphabet size d* (denoted as $GBF(h, d)$) consists of $h + 1$ layers, each layer containing d^h vertices, each of them labeled by a unique d-ary string of length h. An edge connects two vertices in $GBF(h, d)$ if and only if they are in the consecutive p-th and $(p+1)$-st layer, respectively, and their labels are either equal or differ only in the p-th position.

Let $\alpha = a_h \ldots a_1$ be a d-ary string ($a_i \in \{0, 1, \cdots, d-1\}$) and let p, $1 \leq p \leq h + 1$, be an index of a layer. Operations $L^{(i)}, R^{(i)}$ are defined as $L^{(i)}((p, \alpha)) =$

$(p+1, a_h \ldots b_p^{(i)} \ldots a_1)$, and $R^{(i)}((p+1,\alpha)) = (p, a_h \ldots b_p^{(i)} \ldots a_1)$, respectively, where $b_p^{(i)} = (a_p + i) \bmod (d-1)$. An edge (u,v) in $GBF(h,d)$ is called an $L^{(i)}$–edge, $R^{(i)}$–edge, if $L^{(i)}(u) = v$, $R^{(i)}(u) = v$, respectively, $0 \le i < d$.

Formally, $GBF(h,d)$ is a graph $\langle V, E \rangle$, where $V = \{u \mid u \in \{1, \ldots, h+1\} \times \{0, 1, \ldots, d-1\}^h\}$ and $E = \{(u,v) \mid L^{(i)}(u) = v \text{ or } R^{(i)}(u) = v \text{ for } 0 \le i \le d-1\}$.

Lemma 3. *Let G_1 be the $GBF(h-1, d)$ and G_2 be the h-layered cycle. Consider $G = G_1 \otimes G_2$. Let $u = (l_1, p, a_h \ldots a_1)$ and $v = (l_2, q, b_h \ldots b_1)$, $p \ge q$, be vertices in G. The distance between u, v is*

1. $dist(u,v) = p - q$ if $l_1 = l_2$ and $a_i = b_i$ for $i = 1, 2, \ldots, h$
2. $dist(u,v) = r_{max} - r_{min} + |p - r_{max}| + |q - r_{min}| + c$ if $l_1 = l_2$ and there is i such that $a_i \ne b_i$
3. $dist(u,v) = p - q + 2 \cdot min\{r_{min}, h - r_{max}\} - 2$ if $l_1 \ne l_2$ and $a_i = b_i$ for $i = 1, 2, \ldots, h$
4. $dist(u,v) = r_{max} - r_{min} + |p - r_{max}| + |q - r_{min}| + c + 2 \cdot min\{r_{min}, (h - r_{max})\} - 2$ if $l_1 \ne l_2$ and there is i such that $a_i \ne b_i$

where $r_{max} = max_{1 \le i \le h}\{i \mid a_i \ne b_i\}$, $r_{min} = min_{1 \le i \le h}\{i \mid a_i \ne b_i\}$, and $c = 0$ if $r_{max} < p$, otherwise $c = 2$.

Note. Lemma 3 can be modified for the case where G_1 is the LCP of an arbirary h-layered top-tree with an arbitrary h-layered bottom-tree.

Interval Routing Schemes A *k-Interval Routing Scheme* (*k*-IRS) is a scheme of labeling each vertex in a graph $G = (V, E)$ by a unique integer from the set $\{1, 2, \ldots, n\}$, $n = |V|$ and each arc by at most k intervals $[a, b]$, allowing cyclic intervals $[a, b] = \{a, a+1, \ldots, n, 1, \ldots, b\}$ for $a > b$. The set of all intervals associated with the arcs incident to a vertex must form a partition of the set $\{1, 2, \ldots, n\}$. Messages to a destination vertex having a label l are routed via the arc labeled by the interval $[a, b]$ such that $l \in [a, b]$. In this paper we consider *k*-IRS specifying all-to-all (unique) shortest paths system only, called *optimal k*-IRS.

Multi-dimensional interval routing schemes (MIRS for short) are an extension of interval routing schemes. In $\langle k, d \rangle$-MIRS every vertex is labeled by a unique d-tuple $(l_1, \ldots, l_d)$ of integers and each arc by at most k d-tuples of (possibly cyclic) nonempty intervals (one interval per dimension). In any vertex a message with the destination $(l_1, \ldots, l_d)$ is routed along any outgoing arc containing a d-tuple of cyclic intervals $(I_1, \ldots, I_d)$ such that $l_i \in I_i$ for all i. Multiple paths between any two vertices u, v may be specified. In fact, we only consider *full information* MIRS, specifying a path system consisting of all all-to-all shortest paths only.

4   Lower Bound Results

Let $G = \langle V, E \rangle$ be a simple connected graph with maximum degree Δ. For a vertex $v \in V$ and an arc e outgoing from v, denote $S(v, e)$ the subset of vertices $w \in V$ which can be reached optimally from v over its outgoing arc e.

In the following lemma we present a lower bound on the number of intervals for an optimal interval routing scheme in G. The idea of the proof technique is based on the so called *wq-property*: Given a graph G, we choose two disjoint sets of vertices W and Q such that for any distinct vertices $w_i, w_j \in W$ there is a vertex $v \in Q$ such that in any optimal routing scheme the messages sent by v to w_i and w_j are routed along different outgoing arcs.

Lemma 4. [8] *Let G be a graph with maximum degree Δ and let us have an optimal k-IRS on G. Let Q and W be disjoint vertex subsets of G satisfying the wq-property, that means for $w_i, w_j \in W$, $w_i \neq w_j$, there is $v \in Q$ such that for each arc e outgoing from v it holds $w_i \notin S(v, e)$ or $w_j \notin S(v, e)$. Then it holds $k \geq |W|/(\Delta|Q|)$*

The previous lemma proved to be a quite powerful tool for certain interconnection networks. It can be effectivelly applied when there is a "large" set W and a relatively "small" set Q such that the system of all shortest paths between all pairs of vertices from $Q \times W$ satisfies the wq-property. In [8], this argument has been applied to some well-known constant degree interconnection networks (like shuffle-exchange, De Bruijn, cube-connected cycles, butterfly) to obtain superpolynomial lower bounds w.r.t. the diameter of the networks as well as to obtain near-optimal lower bounds for some non-constant degree interconnection networks (like star).

Now we first show that the argument is also suitable for the class of networks, constructed as the LCP of complete regular trees.

Theorem 1. *Let $G = \langle V, E \rangle$ be the LCP of an $(h+1)$-layered complete d_1-ary top-tree and an $(h+1)$-layered complete d_2-ary bottom-tree, where $d_1 \geq d_2 > 1$. Then every optimal k–IRS on G requires $k \geq d_1^{h-1} / \left(4 d_2^{\lfloor \frac{h}{2} \rfloor + 1} \right)$.*

Proof. Consider a d_1-ary alphabet $\mathcal{A} = \{0_{\mathcal{A}}, ..., (d_1 - 1)_{\mathcal{A}}\}$ and a d_2-ary alphabet $\mathcal{B} = \{0_{\mathcal{B}}, ..., (d_2 - 1)_{\mathcal{B}}\}$. Then every vertex in the i-th $(0 \leq i \leq h)$ layer of G can be labeled by a string $\langle x_1, ..., x_i, y_{h-i-1}, ..., y_1 \rangle$ where each $x_j \in \mathcal{A}$ and each $y_j \in \mathcal{B}$. The edges connect vertices of the form $\langle x_1, ..., x_i, y_{h-i-1}, ..., y_1 \rangle$ and $\langle x_1, ..., x_i, x_{i+1}, y_{h-i-2}, ..., y_1 \rangle$.

Let $h = p + q + 1$ where $p = \lfloor \frac{h}{2} \rfloor$. Consider the following sets Q and W.
$$Q = \bigcup_{i=1}^{p-1} \{\langle x_1, ..., x_i, 0_{\mathcal{B}}, ..., 0_{\mathcal{B}} \rangle\} \cup \bigcup_{j=1}^{q-1} \{\langle 0_{\mathcal{A}}, ..., 0_{\mathcal{A}}, y_j, ..., y_1 \rangle\},$$
$$W = \{\langle x_1, ..., x_p, 1_{\mathcal{A}}, y_{h-p-2}, ..., y_1 \rangle\}.$$
$|W| = d_1^p d_2^q$, $|Q| = \frac{d_1^p - d_1}{d_1 - 1} + \frac{d_2^q - d_2}{d_2 - 1}$, $\Delta = d_1 + d_2$, $|V| = \frac{d_1^{h+1} - d_2^{h+1}}{d_1 - d_2}$ if $d_1 > d_2$ and $|V| = (h+1)d^h$ if $d_1 = d_2 = d$.

We show that W and Q satisfy the wq-property expressed in Lemma 4. Let w_1 and w_2 be arbitrary vertices from W. W.l.o.g. suppose that w_1 and w_2 differ somewhere to the left of the middle 1. Then for some $|\alpha| \leq p - 1$ it holds $w_1 = \langle \alpha a_{\mathcal{A}} r_1 \rangle$ and $w_2 = \langle \alpha b_{\mathcal{A}} r_2 \rangle$. Choose $v \in Q$ as $v = \langle \alpha 0_{\mathcal{B}} ... 0_{\mathcal{B}} \rangle$.

Clearly every shortest path from v to w_1 must start with an edge that changes $0_{\mathcal{B}}$ to $a_{\mathcal{A}}$ and every shortest path from v to w_2 must start with an edge that changes $0_{\mathcal{B}}$ to $b_{\mathcal{A}}$.

As a consequence of Lemma 4 it holds $k \geq \frac{|W|}{\Delta \cdot |Q|} = \frac{d_1^p d_2^q}{(d_1+d_2)\left(\frac{d_1^p - d_1}{d_1 - 1} + \frac{d_2^q - d_2}{d_2 - 1}\right)}$.

If $d_1 = d_2 = d$ we get $k \geq \frac{1}{4} d^{q-1}$. W.l.o.g. suppose $d_1 > d_2$, we get $k \geq \frac{d^{p+q}}{4d_2^{p+1}}$.

The same argument as in Theorem 1 can be used to prove (reprove) lower bounds on k: butterfly [8] $\Omega(\sqrt{|V|/\log|V|})$, wrap-around butterfly $\Omega((|V|/\log|V|)^{1/4})$, fat tree [2] $\Omega(\sqrt{|V|})$, globe graph [8] $\Omega(\sqrt{|V|})$.

5   Upper Bound Results

In [1] an interval based routing scheme (not strictly MIRS) for a LCP of general trees was shown. We shall present a MIRS based solution for the d-ary trees and then generalize it for arbitrary trees.

At first we give an efficient multi-dimensional interval routing scheme for generalized butterfly networks. Consider the following GBF–machine. It has a work tape with h cells and a head which can be positioned between cells or at any of the ends of the tape. Each cell contains one d-ary digit (from $\{0, 1, ..., d-1\}$). In one step, the head moves to the left or to the right over a cell and writes a digit from $\{0, 1, ..., d-1\}$ to this cell. The state diagram of a GBF–machine with vertices corresponding to the states and arcs corresponding to the steps forms exactly the $GBF(h, d)$ graph. This allows us to consider the vertices of the $GBF(h, d)$ graph as being the states of the described machine.

Proposition 1. *Given the $GBF(h, d)$ graph, let w be a vertex of the form $(p, u\alpha v)$, where $p = |u\alpha|$, $\alpha \in \{0, ..., d-1\}$. There exists a shortest path from w to a vertex z starting with an arc e corresponding to moving the head to the left and writing zero iff the vertex z is of the form $(A) : (q, w_1 w_2)$, $w_1 \neq u\alpha$, $q \geq p = |w_1|$ or of the form $(B) : (q, w_3 0 v)$, $q \leq p = |w_3 0|$.*

If we want to design a full information shortest path routing scheme, it must route messages destined to these vertices precisely along the arc e. The characterization of vertices whose messages are to be routed along arcs of other types is similar.

Now we briefly describe a $\langle 2, 3 \rangle$-MIRS of the $GBF(h, d)$.

Lemma 5. *There exists a full information shortest path $\langle 2, 3 \rangle$-MIRS on the $GBF(h, d)$.*

Sketch of the proof.   Let us label the vertices in the individual dimensions as follows: The 1st dimension of the label represents the number written on the tape, the second dimension represents the number written on the tape read backwards and the third dimension represents the position of the head.

For any vertex w and any arc e from the previous Proposition 1 it is possible to select vertices of the forms (A) and (B) using two triples of intervals. The first triple selects the vertices not starting with $u\alpha$ (these form a cyclic interval in the 1st dimension) and not having the head to the left of w's head (these form a cyclic interval in the 3rd dimension). The second triple selects the vertices ending with $0v$ (these form a cyclic interval in the 2nd dimension) and having the head to the left of w's head. For other types of arcs the construction is similar. The bit length of the labels of the described routing scheme is $2h + \log h$ and therefore the space required per vertex in bits is $O(h)$.

Lemma 6. (Generalization of Lemma 5) *Let G be the LCP of an $(h+1)$-layered top-tree and an $(h + 1)$-layered bottom-tree. Then there exists a full information shortest path $\langle 2, 3 \rangle$-MIRS on G.*

Sketch of the proof. Let T_1 be an $(h + 1)$-layered top-tree and T_2 be an $(h + 1)$-layered bottom-tree. Let $G = T_1 \otimes T_2$.

Each vertex in the l-th layer of G can be represented by a pair $(l; x_1, ..., x_h)$, where $x_1, ..., x_{l-1}$ is the string representation of the path from the root of T_1 to the vertex at layer l and $x_h, ..., x_l$ is the string representation of the path from the root of T_2 to the vertex at layer l. Consider the T-machine (analogous to the GBF-machine above), for which each cell contains one symbol of the vertex representation. Thus the state diagram of this T-machine forms the graph G.

An analogue of the Proposition 1 also holds for the graph G. Let us label the vertices of G in the following dimensions: The first dimension forms the DFS-labeling of T_1 [13], the second dimension forms the DFS-labeling of T_2 and the third dimension contains the position of the head.

A $\langle 2, 3 \rangle$-MIRS on G can be described similarly as in the proof of Lemma 5.

Lemma 7. *Let G be a globe graph. Then there is a full information shortest path $\langle 2, 2 \rangle$-MIRS on G.*

We can now show that efficient MIRS exists for a class of graphs, which includes, e.g., the multiglobe graphs used in some lower bound proofs.

Theorem 2. *Let G be composed as the LCP of h-layered top-trees, h-layered bottom-trees, and h-layered globe graphs. Then there is a constant full information shortest path MIRS on G.*

Sketch of the proof. The LCP operator is commutative and associative. The LCP of two h-layered top-trees is an h-layered top-tree and the LCP of two h-layered bottom-trees is an h-layered bottom-tree. Clearly cycle is a globe graph. The LCP of two h-layered globe graphs is an h-layered globe graph. Hence, it is sufficient to consider the LCP of an h-layered top-tree, an h-layered bottom-tree and an h-layered globe graph.

By Lemma 6, a network composed as the LCP of an h-layered top-tree with an h-layered bottom-tree can be optimally routed using a full information shortest path $\langle 2, 3 \rangle$-MIRS. By Lemma 7, an h-layered globe graph can be optimally routed using a full information shortest path $\langle 2, 2 \rangle$-MIRS. Hence by combining the two above schemes using an analogous reasoning as for Lemma 3, an optimal routing on the resulting h-layered graph can be obtained by using a full information shortest path $\langle 4, 6 \rangle$-MIRS.

6 Conclusions

We have presented efficient multidimensional interval routing schemes for the class of networks constructed as the LCP of layered trees and cycles. It is a step forward in understanding the complexity of multidimensional routing. The question remains to identify other classes of networks with efficient MIRS and to characterize the exact border between the efficiency and inefficiency of MIRS.

We have also presented a superpolynomial lower bound on the compactness (w.r.t. the diameter of the network) for the class of networks composed as the LCP of complete regular trees and cycles. The question remains whether this lower bound can be improved.

References

[1] T. Calamoneri, M. Di Ianni: *Interval Routing and Layered Cross Product: Compact Routing Schemes for Butterflies, Mesh of Trees and Fat Trees.* In Euro-Par'98, Lecture Notes in Computer Science 1470, Springer-Verlag, 1998, pp. 1029–1039.

[2] S. Even, A. Litman: *Layered Cross Product – A Technique to Construct Interconnection Networks.* Networks, Vol. 29, 1997, pp. 219–223.

[3] M. Flammini, G. Gambosi, U. Nanni, R. Tan: *Multi-Dimensional Interval Routing Schemes.* Theoretical Computer Science, Issue 1-2, 1998, pp. 115–133.

[4] P. Fraigniaud, C. Gavoille: *Interval Routing Schemes.* Algorithmica 21, 1998, pp. 155–182.

[5] C. Gavoille: *A Survey on Interval Routing.* Research Report RR-1182-97, LaBRI, Université Bordeaux I, October 1997. To appear in Theoretical Computer Science, 1999.

[6] E. Kranakis, D. Krizanc, S.S. Ravi: *On Multi-Label Linear Interval Routing.* The Computer Journal, Vol. 39, No. 2, 1996, pp. 133–139.

[7] R. Kráľovič, P. Ružička: *Rank of Graphs: The Size of Acyclic Orientation Cover for Deadlock-Free Packet Routing.* Accepted to SIROCCO, 1999.

[8] R. Kráľovič, P. Ružička, D. Štefankovič: *The Complexity of Shortest Path and Dilation Bounded Interval Routing.* In Euro-Par'97, Lecture Notes in Computer Science 1300, Springer-Verlag, 1997, pp. 258–265. Full version of the paper will appear in Theoretical Computer Science.

[9] T. Leighton: *Introduction to Parallel Algorithms and Architectures: Arrays, Trees, and Hypercubes.* Morgan Kaufmann, San Mateo, CA, 1991.

[10] L. Narayanan, S. Shende: *Characterization of Networks Supporting Shortest-Path Interval Labeling Schemes.* In 3rd SIROCCO, Carleton Scientific, 1996, pp. 73–87.

[11] P. Ružička: *Efficient Communication Schemes.* To Appear in Theoretical Computer Science.

[12] P. Ružička, D. Štefankovič: *The Complexity of Multi-Dimensional Interval Routing.* To appear in Theoretical Computer Science, 1999.

[13] N. Santoro, R. Khatib: *Labelling and implicit routing in networks.* The Computer Journal 28, 1985, pp. 5–8.

[14] S.S.H. Tse, F.C.M. Lau: *A Lower Bound for Interval Routing in General Networks.* Networks, Vol. 29, No. 1, 1997, pp. 49–53.

[15] J. van Leeuwen, R.B. Tan: *Interval Routing.* The Computer Journal 30, 1987, pp. 298–307.

[16] J. van Leeuwen, R.B. Tan: *Compact routing methods: A survey.* In 1st International Colloquium on Structural Information and Communication Complexity (SIROCCO), Carleton Press, 1994, pp. 99–110.

Topic 16
Instruction-Level Parallelism and Uniprocessor Architecture

Pascal Sainrat and Mateo Valero

Co-chairmen

Research in Instruction-Level Parallelism (ILP) is concerned with architectural innovations in the processor to expose parallelism between the execution of instructions. Of course, the relationship with the research on the memory hierarchy and on compiler optimisation techniques is very strong. Another point is that such a research needs tools to simulate the mechanisms. Thus, researchers have to develop their tools. Such a tool is detailed in the paper on code cloning tracing by Lafage et al from IRISA.

Most of these topics are represented in this workshop although there are no papers on the lower levels of the memory hierarchy.

The memory hierarchy, and particularly, the first-level cache is highly related to ILP research since superscalar processors place higher demands on it for obtaining more instructions and more data per cycle. In addition to the requirement of higher bandwidths, latency is also an important issue. One way to reduce the latency is prefetching as proposed by Chi and Yuan. Another issue is the way the cache is managed. Software can afford hints for a better management, which might result in a good speedup as in the paper of Lebeck et al.

A big and old deal is what should be in the hardware and what should be left in the compiler. Returning to simpler processors while leaving part of the job to the compiler might arrive in the future. Thus, we should care of compiler studies. Moreover, compiler studies might have an impact on the architecture. The papers by Norris, Fenwick and Genius, Lelait concern compiler optimisations. The first one deals with register allocation that can have a great impact on the reordering of instructions while the second one apply techniques of register allocation to the data in memory in order to improve the use of the cache. The VLIW architecture highly depends on the quality of the compiler. The paper of Ebcioglu et al gives encouraging results for such an approach.

Increasing the ILP is, at last, limited by data dependencies. To overcome these dependences, we should predict the results of instructions. A sophisticated value predictor is proposed in the paper of Pinuel et al. However, predicting values need to recover efficiently a normal state when a misprediction occurs. Soti studies recovery in its paper. At a longer term, asynchronous processors might provide a solution to the problems of clock distribution on very large chips as expected in a dozen years. But, making processors asynchronous might need new architectural ideas as in Pessolano's paper.

Finally, the most important and most studied problem concerns control dependences which have a dramatic impact on performance. Predicated execution

P. Amestoy et al. (Eds.): Euro-Par'99, LNCS 1685, pp. 1241–1242, 1999.
© Springer-Verlag Berlin Heidelberg 1999

reduces the problem by removing control dependences. Embedded processors are, particularly in Europe, a hot topic. Using predication in embedded processors might be useful when performance is an issue as explained in the paper by Connors et al. A more classical approach is the prediction of the outcome of branches and many schemes have been proposed. A new approach where predictors are cascaded is proposed by Driesen and Hözle. The last paper of this topic is also related to the branch problem. It shows that executing a mispredicted path does not always result in a useful prefetching as suggested previously.

It was very difficult for the scientific committee to choose among the good papers that have been received. We hope you will share our enthusiasm for the papers that are presented here.

Design Considerations of High Performance Data Cache with Prefetching

Chi-Hung Chi, Jun-Li Yuan
School of Computing
National University of Singapore
Singapore 119260
Email: chich@comp.nus.edu.sg

Abstract. In this paper, we propose a set of four load-balancing techniques to address the memory latency problem of on-chip cache. The first two mechanisms, the **sequential unification** and the **aggressive lookahead** mechanisms, are mainly used to reduce the chance of partial hits and the abortion of accurate prefetch requests. The latter two mechanisms, the **default prefetching** and the **cache partitioning** mechanisms, are used to optimize the cache performance of the unpredictable references. The resulting cache, called the **LBD (Load-Balancing Data)** cache, is found to have superior performance over a wide range of applications. Simulation of the LBD cache with RPT prefetching (Reference Prediction Table - one of the most cited selective data prefetch schemes [2,3]) on SPEC95 showed that significant reduction in the data reference latency, ranging from about 20% to over 90% and with an average of 55.89%, can be obtained. This is compared against the performance of prefetch-on-miss and RPT, with an average latency reduction of only 17.37% and 26.05% respectively.

1.   Introduction

To improve the accuracy and coverage of data prefetching, current research emphasizes on the exploration of hybrid data address and value prediction. Data accesses in a program are partitioned into exclusive distinct reference classes, each of which is handled exclusively by one predictor/prefetch unit. Good examples of these predictors include linear memory reference predictor and pointer predictors.

With an accurate predictor and its supporting hardware/software, very good cache performance is expected. However, experiment showed that it might not be as simple as it appears. Even though the overall cache performance can be improved, there are still many data references who access patterns are predicted accurately but are missing from cache [5]. This percentage ranges from a few percents to over 95%, with an average of 54%. In other words, about half of the cache misses are actually due to data references that can be predicted accurately! The overall cache effectiveness is further bounded by the access behavior of unpredictable references in cache, which contributes to the other half of the cache misses.

In this paper, we propose a set of four load-balancing mechanisms to make up for the discrepancy between the ideal and the observable performance of accurate prefetching. They are (i) sequential unification of demand and prefetch requests, (ii)

P. Amestoy et al. (Eds.): Euro-Par'99, LNCS 1685, pp. 1243-1250, 1999.
© Springer-Verlag Berlin Heidelberg 1999

aggressive lookahead prefetching, (iii) default prefetching, and (iv) cache partitioning. The first two mechanisms are mainly used to reduce the chance of partial hits and the abortion of accurate prefetch requests by demand fetch requests while the latter two mechanisms are to optimize the cache performance of the unpredictable references. To ensure the performance gain, these mechanisms will only be triggered selectively, based on the level of confidence on the urgency, accuracy, and nature of the prefetched data. The resulting cache, called the **LBD (Load-Balancing Data)** cache, is found to have superior performance over a wide range of applications. Simulation of the LBD cache with RPT prefetching (Reference Prediction Table [2,3]) on SPEC95 showed that significant reduction in the data reference latency, ranging from about 20% to over 90% and with an average of 55.89%, can be obtained. This is compared against the performance of prefetch-on-miss and RPT, with an average latency reduction of only 17.37% and 26.05% respectively. To help the discussion in the rest of the paper, the reference prediction table (RPT) prefetching [2-4] will be chosen as the basic accurate data prefetch scheme, on which our mechanisms will be added on. RPT is chosen because it is one of the most cited schemes for accurate data prefetching.

Before we proceed, it will be helpful to give precise definition for the following terms used in the rest of the paper. A *partial hit* for a cache block I is said to occur if, while the block I is being prefetched from the next level of the memory hierarchy, a demand fetch for block I occurs. A *selective prefetch (SP) scheme* refers to a scheme that is based on one or more reference predictors for data prefetching. Furthermore, it will only handle one type of data sequences with some predefined characteristics such as the linear memory accesses. In the rest of this paper, the term SP and RPT might sometimes be used interchangeably because RPT is chosen as the SP scheme here.

1.1.   Basic Simulation Environment

Processor Specification:	
UltraSPARC ISA compatiable	
Superscalar with 3 integer units, 3 floating point units, 1 branch unit, 1 LOAD/STORE unit	
Out-of-order execution, register renaming, 2-bit branch predictor, reorder-buffer with 64 entries.	
1st Level Instruction Cache:	1st Level Data Cache
Cache Size: 32 Kbytes	Cache Size: 32 Kbytes
Block Size: 32 bytes	Block Size: 32 bytes
Associativity: direct-mapped	Associativity: direct-mapped
Replacement: LRU	Placement: write-back
Placement: write-back	Mem. Access Time to 2nd Level Cache: 6 cycles
Access Time to 2nd Level Cache: 6 cycles	Bandwidth to 2nd Level Cache: 8 bytes
Bandwidth to 2nd Level Cache: 8 bytes	Memory queue: 8 queue
Memory queue: 8 entries	Store buffer: 8 entries
Basic prefetch-on-miss scheme	Non-blocking cache
	Prefetching using Chen's RPT (i.e. linear memory access prefetching) with 512 entries & lookahead of 3 iterations
2nd Level Unified Cache:	
Cache Size: 256 Kbytes	Placement: write-back
Block Size: 64 bytes	Mem. Access Time: 60 cycles
Associativity: direct-mapped	Bandwidth to Main Memory: 8 bytes

Table 1: Simulation Parameters for the Baseline Architecture

In our study, SPEC95 was chosen as the benchmark suite for experimentation. Each of these programs was traced and simulated cycle-by-cycle for 100 million instructions. The baseline architecture together with its parameters used in the simulation study is given in Table 1. It is an UltraSPARC ISA compatible superscalar processor with rich details on its architectural features, including the pipeline, register renaming, reorder buffer, branch prediction, and multi-level memory hierarchy.

2.   Previous Research

Recent work in data prefetching concentrates on the predictability of references and the accuracy of prefetching. A direct consequence of this research direction is the introduction of hybrid address and value predictors [1,6,7,10,12]. Multiple predictors are implemented, each of which will be designed, optimized, and responsible for one type of references. Usually, they can achieve very high prefetch accuracy, but their coverages are quite restricted. Furthermore, there is always a non-negligible group of unpredictable references that they cannot handle. To overcome this problem, multiple prefetch schemes in a cache system are proposed [11]. In addition to the selective scheme(s), the prefetch-on-miss mechanism will be employed as the default prefetch scheme in case the references do not fall into the selective coverage set. Good results were reported. However, their interactions with other caching techniques such as cache partitioning/buffering still need to be studied.

Another technique in caching is the prefetch buffer. There are three interesting data buffer designs that are worth mentioning here. The first one is the HP's PA7200 multimedia processor [9]. Between the core processor and the off-chip cache, there is a data buffer, called the assist cache, on the processor chip. This assist cache is a 64 fully associative cache of size 2 Kbytes. All referenced data, both prefetched and demand-fetched, are moved into the assist cache. When data are replaced from the assist cache, there are supports (both in hardware and software) that allow the replaced data to bypass the off-chip cache. This is to reduce the interference between the prefetched spatial data and the temporal data in the off-chip cache.

The second one is the victim cache proposed by Jouppi [8]. A small buffer is inserted between the first and second level caches. When a cache block is replaced, it will be placed in the victim cache. This increases the "pseudo" associativity of the cache with much less hardware. Despite the performance improvement obtained, both designs do not address the problem of interference between spatial, prefetched data and temporal demand-fetched data.

Finally, in the same paper [8], a stream buffer is also described. The purpose of the stream buffer is to prefetch a number of linear memory reference sequences into a separate buffer instead of entering the cache. In this aspect, it functions like a prefetch buffer. However, the buffer design does not have any load-balancing mechanism to tradeoff space for bandwidth when partial hits or complete cache misses occur to the predictable reference group.

3.   Load-Balanced Data Cache

In the beginning of this paper, we pointed out that there are four aspects of performance loss due to memory accesses. Either the correctly predicted data cannot be prefetched in time for consumption or the accurate prefetch requests are aborted while they are being served. This observation is generally true, independent of the prefetch accuracy. For the unpredictable group of references, since they are ignored by the SP schemes, no performance improvement is expected. Even worse, their working set might be destroyed by the predictable group of references and this results in more cache misses.

To make up for this performance gap, four mechanisms are proposed to load balance the bus bandwidth and the cache space. They are: (i) sequential unification of prefetch and demand fetch requests, (ii) the aggressive lookahead prefetching, (iii) default prefetching, and (iv) cache partitioning with load-balancing. The resulting cache that supports all these mechanisms, is called the **Load-Balancing Data (LDB)** cache.

3.1.   Sequential Unification

In the current memory hierarchy system, the time required to transfer a block of data can roughly be divided into two components: (i) the memory startup latency between the sending out of a request and the receiving of the first byte of data, and (ii) the actual transfer time for the subsequent data. As the bus width between successive memory levels is getting larger, the overall fetch time for a block of data is mainly dominated by the memory startup latency. Thus, it will be significant to the system performance if the startup time of some memory requests can be minimized.

Given a memory access instruction (MAI), two possible types of data references can occur. They are the demand fetch and prefetch requests. Thus, one good way to eliminate the startup overhead of the memory requests is to combine these two types of requests for the same MAI aggressively. Whenever the memory blocks containing the demand fetch and prefetch data are consecutive to each other, they can be combined to form a single request with double block size. Since the combined request is still considered as a "demand" fetch, one side effect is the increase in the bus usage priority of the prefetch portion.

There is one side-effect of the sequential unification mechanism that one needs to be careful. It is about the bus usage priority. High priority is good to prefetch requests with high accuracy, but is bad to incorrect prefetch requests. The additional time required for the transfer of the prefetch data is likely to be seen by the processor. If the benefit for prefetching them is not high (again related to accuracy), it might actually degrade the performance instead. Consequently, this mechanism is triggered only when both requests are missed from cache. We choose this situation because the startup overhead of the unified request is the same as the unavoidable demand fetch, but that of the prefetch request is eliminated.

3.2.   Aggressive Lookahead

Current data prefetch schemes try to prefetch the next N^{th} iteration data while the current iteration datum is being referenced, where N is usually small. Under this situation, partial hits are found to occur quite often, especially in the situation of small stride linear memory accesses. This is not difficult to understand. The next block is prefetched only when the current referenced block is about to be used up. Thus, a memory latency of 1 to (Cache_Miss_Penalty - 1) cycles might potentially be visible to the program execution for each partial hit. The abortion of accurate, on-going prefetch requests by demand fetch requests also contributes to some performance loss.

It is observed that during the time period of accessing two elements in the same cache block, the chance to have some free bus cycles is quite high. Thus, it might be useful to perform more aggressive, yet accurate prefetching during this period. This can be achieved easily through dynamic adjustment of the prefetch block address. When the current reference and the prefetch data are mapped to the same cache block address i, the prefetch block address will be adjusted to the next sequential one. This kind of additional prefetching has similar accuracy as the original schemes do because they go through similar kind of confirmation test. Note that unlike the sequential unification, its triggering will be independent of the reference hit or miss.

3.3.   Default Prefetching

The coverage of a prefetch scheme refers to the portion of the data references in a program that the scheme tries to identify their reference patterns and to issue prefetch requests for them. Current SP schemes often trade off the prefetch coverage for accuracy and this puts an upper bound to their performance potentials.

To solve the coverage problem of the SP schemes, we propose a hybrid approach: highly accurate SP schemes should work co-operatively with less accurate (yet beneficial) prefetch schemes. Given a set of data references in a program, the cache prefetch unit divides them into two groups - a selected, predictable reference group and an unpredictable, non-spatial reference group. Data from the predictable reference group is handled by the SP scheme. The unpredictable reference group, instead of being ignored by the cache controller, is now handled by some default prefetch scheme such as POM. Additional performance improvement is possible because all data references in a program are now taken care of, although not to the same extent. Note that as far as each individual reference group is concerned, the prefetch accuracy obtained by its corresponding prefetch scheme is roughly unchanged.

3.4.   Cache Partitioning and Prefetch Buffer

Interference among data references in cache refers to the situation when there is no free space for the incoming data and some data in cache needs to be replaced out.

In practice, it is impossible to eliminate all the data interference in cache because of the finite cache size. However, due to the difference in localities of the predictable and the unpredictable references, certain portion of the data interference in cache can be minimized through cache partitioning. Generally speaking, predictable data from the highly accurate SP schemes have stronger spatial locality while the unpredictable, demand fetched data have relatively stronger temporal locality. To minimize the interference among them, it is better to separate the predictable, spatial data prefetched by the SP schemes from the unpredictable temporal data that are fetched by demand or by the default prefetch schemes. This suggests the use of a separate prefetch buffer.

Further study on the design of the data prefetch buffer reveals that two more issues need to be taken care of. The first issue is the insufficient bandwidth for the SP prefetching. This results in partial hits and the memory latency cannot be completely hidden from the processor execution even if the prefetching were 100% accurate. The second issue is about the predictable data references that are accessed during the transient period when the hardware reference prediction table tries to build up its content; these data should not be prefetched into the prefetch buffer. To enhance the cache performance further, some kind of load balancing for the cache space usage is found to be useful: when a miss or a partial hit occurs to a predictable reference, the corresponding cache block will be re-directed from the prefetch buffer back to the data cache. This is to reduce the bandwidth demand when the bus is expected to be overloaded.

4.   Experimental Result

To study the performance of our proposed mechanisms, cycle-by-cycle simulation on the baseline architecture was conducted for various cache configurations and control designs. In the analysis, the normal data cache without prefetching was used as the reference standard. And the memory latency reduction is defined by the following formula:

$$Mem\_latency\_reduction = \frac{Execution\_Time(Normal\_Cache\_No\_Prefetch) - Execution\_Time(New\_Design)}{Execution\_Time(Normal\_Cache\_No\_Prefetch) - Execution\_Time(Perfect\_Cache)} * 100\%$$

The overall result is summarized in Table 2. It shows that when the four load-balancing mechanisms are used collaboratively in the LBD cache, significant improvement in cache performance can be obtained for almost all benchmark programs simulated. Its reduction percentage ranges from about 20% to over 90%, with an average of 55.89%. This is compared to the performance of the POM and the RPT, with an average latency reduction of only 17.37% and 26.05% respectively. The huge performance improvement is actually within our expectation. In the POM case, its prefetch accuracy for the unpredictable reference group is the same as that of the LBD cache, but its accuracy for the predictable reference group is much lower than that of the LBD cache. In the RPT case, both the RPT and the LBD cache have similar performance for the predictable references, but the LBD cache takes care of the unpredictable reference group instead of ignoring them. In other words, the LBD cache can handle both predictable and unpredictable reference groups well. Furthermore, its cache space and bandwidth are well balanced and the interference between the predictable and unpredictable reference groups is greatly reduced.

| Overall Performance | POM | RPT | | | | | LB-Data Cache |
		Original	+ Unification	+ Aggressive	+ POM	+ Pref. Buffer	
Latency Improv.	17.37%	26.05%	30.18%	25.69%	29.59%	44.47%	55.89%
Prefetch Req. Sent (w.r.t. POM)	100.00%	67.03%	66.66%	68.06%	124.47%	106.15%	164.62%
Prefetch Accuracy	37.62%	67.18%	67.32%	65.99%	53.41%	75.03%	77.22%

Table 2: Overall Performance of Various Prefetch Schemes on SPEC95

One very interesting observation is that while the combined effect of the four mechanisms in the LBD cache is very impressive, their individual effects differ significantly. Among the four proposed mechanisms, the load-balancing prefetch buffer is found to have superior performance over the other three. An extra performance of 18.42% memory latency reduction is obtained by the prefetch buffer, as compared to the 5% additional performance gain obtained by the others. This suggests that *while improving prefetch accuracy and performing more aggressive prefetching are important, the interference among the predictable and unpredictable references cannot be ignored.* For the other three mechanisms, their individual effectiveness seems to be quite limited. The default prefetching and the sequential unification mechanisms can only provide an additional performance of 3.54% and 4.13% respectively while the aggressive prefetching mechanism even performs slightly worse than the original RPT scheme does. This is surprising because this means that the overall performance gain of the LBD cache is not accumulated from the individual results. Further investigation found that they actually enforce the effect of each other instead. A separate study on the sequential unification and aggressive lookahead mechanisms only [5] also observed similar enforcement effect as we have here.

5.   Conclusion

In this paper, we propose four mechanisms to improve the data cache performance. By integrating a prefetch request into an unavoidable demand fetch request, its startup memory latency can be eliminated and the bus priority of the prefetch request is increased. More aggressive prefetching is also carried out whenever the bus is expected to have relatively light workload. With default prefetching on top of the accurate SP prefetching, the unpredictable reference group can now be taken care of. Finally, the interference between the predictable and unpredictable reference groups in cache can be minimized with the help of a load-balanced prefetch buffer. Simulation shows that these four mechanisms enforce each other and provide significant performance improvement over current accurate prefetch schemes such as the RPT.

References

1. Black, B., Mueller, B., Postal, S., Rakvic, R., "Load Execution Latency Reduction," *Proceedings of the ACM International Conference on Supercomputing*, July 1998.
2. Chen, T.F., Baer, J.L., "Reducing Memory Latency via Non-Blocking and Prefetching Caches," *Proceedings of the Fifth International Conference on Architectural Support for Programming Languages and Operating Systems*, October 1992, pp. 51-61.
3. Chen, T.F., Baer, J.L., "Effective Hardware-Based Prefetching for High Performance Processors," *IEEE Transactions on Computers*, Vol. 44, No. 5, May 1995, pp. 609-623.
4. Chen, T.F., "Data Prefetching for High Performance Processors," *Ph.D Thesis, Department of Computer Science and Engineering, Washington University*, 1993.
5. Chi, C.H., Yuan, J.L., "Sequential Unification and Aggressive Lookahead Mechanisms for Data Memory Accesses," *Internal Report, School of Computing, National University of Singapore*, 1999.
6. Gonzalez, J., Gonzalez, A., "Speculative Execution via Address Prediction and Data Prefetching," *Proceedings of the ACM International Conference on Supercomputing*, July 1997, pp. 196-203.
7. Ibanez, P., Vinals, V., Briz, J.L., Garzaran, M.J., "Characterization and Improvement of LOAD/STORE Cache-Based Prefetching," *Proceedings of the ACM International Conference on Supercomputing*, July 1998.
8. Jouppi, N.P., "Improving Direct-Mapped Cache Performance by the Addition of a Small Fully-Associative Cache and Prefetch Buffers," *Proceedings of the 18th Annual Symposium on Computer Architecture*, May 1990, pp. 364-373.
9. Kurpanek, G., Chen, K., Zheng, J., DeLano, E., Bryg, W., "PA7200: A PA-RISC Processor with Integrated High Performancve MP Bus Interface," *IEEE Publications 1063-6390/94*, 1994, pp. 375-382.
10. Lipasti, M.H., Wilkerson, C.B., Shen, J.P., "Value Locality and Load Value Prediction," *Proceedings of the 7th International Conference on Architectural Support for Programming Languages and Operating Systems*, October 1996.
11. Manku, G.S., Prasad, M.R., Patterson, D.A., "A New Voting Based Hardware Data Prefetch Scheme," *Proceedings of 4th International Conference on High Performance Computing*, Dec. 1997, pp. 100-105.
12. Sazeides, Y., Smith, J.E., "The Predictability of Data Values," *Proceedings of the MICRO-30*, 1997, pp. 248-258.

Annotated Memory References: A Mechanism for Informed Cache Management

Alvin R. Lebeck, David R. Raymond, Chia-Lin Yang, and
Mithuna S. Thottethodi

Department of Computer Science
Duke University
Durham, NC 27708-0129 USA
alvy@cs.duke.edu

Abstract. As the importance of cache performance increases, allowing software to assist in cache management decisions becomes an attractive alternative. This paper focuses primarily on a mechanism for software to convey information to the memory hierarchy. We introduce a single instruction—called TAG—that can annotate subsequent memory references with a number of bits, thus avoiding major modifications to the instruction set. Simulation results show that annotating all memory reference instructions in the SPEC95 benchmarks increases execution time between 0% and 2% for both statically and dynamically scheduleded processors. We show that exposing cache management mechanisms to software can decrease the execution time of three media benchmarks (*epic*, *pegwit*, *ijpeg*) between 11% and 17% speedups on a 4-issue dynamically scheduled processor.

1 Introduction

One of the key challenges facing computer architects is the increasing discrepancy between processor cycle times and main memory access time. Conventional cache designs rely solely on hardware to manage the memory hierarchy, and the policies for cache management tend to be relatively naive. These limitations could be overcome by using information provided by software to help manage the memory hierarchy [1, 6, 3].

This paper presents a mechanism—*called annotated memory references*—for software to convey information to the memory hierarchy, and shows how it can be used to exploit information on replacement strategies and block size selection for improved performance. Our implementation adds a single instruction (TAG) that can annotate each of the following 6 memory references with a 4-bit annotation. The TAG instruction initializes an annotation register, and the appropriate annotation is extracted for each subsequent memory reference. We analyze our new instruction in a statically scheduled processor using a simplified model of the Digital Alpha 21164 [5] and using the SimpleScalar toolset to model a dynamically scheduled processor [2].

P. Amestoy et al. (Eds.): Euro-Par'99, LNCS 1685, pp. 1251–1254, 1999.
© Springer-Verlag Berlin Heidelberg 1999

Our simulation results show that using the TAG instruction to annotate all memory references increases the number of instructions executed by 5.5% to 16.2% for the programs we studied. The instruction count overhead does not necessarily translate directly into performance loss. Instead, the superscalar processor reduces execution time overheads to at most 2%. Using the TAG instruction to manage cache replacement and block size selection we can reduce execution times by 11% to 17% for three media benchmarks (epic, pegwit, ijpeg) executing on typical embedded system memory hierarchy and a 4-issue dynamically scheduled processor.

2   A Static Instruction Annotation Mechanism

Our goal in designing a mechanism is to minimize the impact on execution time and avoid gross modifications to the instruction set. We also want to provide a mechanism that can support a variety of annotations, and is not specific to any particular use of the annotations. In essence, we are defining an interface between the executing program and the memory hierarchy. For example, one application (or one phase of an application) may want greater control over Level-1 and Level-2 cache replacement decisions, while another application (or phase) may want several forms of sophisticated prefetching. Further discussion of these issues is beyond the scope of this paper. Instead, we focus on the interface for passing information from the executing program to the memory hierarchy.

The mechanism we propose does not require any modification to existing instructions and requires the addition of only a single instruction—*called TAG*. The TAG instruction initializes a new hardware register—*called the annotation register*—from which bits are extracted to annotate subsequently executed memory instructions.

Conceptually, the annotation register is a simple shift register. Every memory reference extracts the right most bits from the annotation register, which is then right shifted with zeros shifted into the left most bits. We assume a zero annotation indicates default cache operation, hence untagged memory references default to normal memory system operation.

We evaluate the overhead of TAG instructions using ATOM [5] to model a statically scheduled processor using the Digital Alpha 21164 issue rules and using SimpleScalar [2] to evaluate a 4-issue dynamically scheduled processor with 64 RUU entries. In both systems we assume a perfect memory system and perfect branch prediction. Our results show that TAG instructions increase execution time by at most 2% even when all memory references are annotated. For most applications, we expect to annotate a much smaller number of memory references, thus further reducing overhead.

3   Utilizing Annotated Memory References

The goal of this section is to provide an example of how software can use annotated memory references to help manage the memory hierarchy. We focus on multimedia applications typical of those that would execute on future embedded systems:

- **epic**: a lossy image compression program which is designed for extremely fast decoding on non-floating-point hardware at the expense of slower encoding,
- **ijpeg**: a lossy image compression program, and
- **pegwit**: a public-key encryption and authentication program

We use cache profiling [4] followed by source code inspection to obtain programmer specified annotations. Cache profiling provides key insights into an applications memory system behavior. For epic, we found that controlling cache replacement may improve performance. Similarly, cache profiling and source code inspection revealed that control of cache block size may improve the performance of ijpeg and pegwit.

To control cache replacment, we use annotations that allow an application to specify that the referenced cache block should be *retained* in a 4-way set-associative cache. This prevents the block from being replaced until the application uses a *release* annotation to unlock the cache block. Cache block size is controlled by using an annotation that specifies *wordmode*, causing only the accessed word to be fetched from main memory. For the blocksize studies, we use a direct-mapped cache, and by reusing the data portion of a 32-byte block we can dynamically increase the associativity so the given cache set can hold three 8-byte blocks.

3.1   Results

We use the SimpleScalar simulator in our experiments to evaluate the benefits of annotated memory references. We simulate a 4-way issue out-of-order processor with 64 RUU entries and 32 LSQ entries, perfect branch prediction, and a cache miss penalty of 28 cycles.

For epic, we compare the performance of systems with an 8 KB, 32-byte block, 4-way set-associative cache with and without *retain/release* annotations. Using the *retain/release* annotations provides a 17% reduction in execution time. The miss ratio drops from 26.5% to 20.8%.

For pegwit and ijpeg, we compare performances of systems with an 8 KB, 32 byte block, direct mapped cache with and without *wordmode* annotations. Using the *wordmode* annotations yields a 11.6% and a 12.4% reduction in execution times for pegwit and ijpeg, respectively. The miss ratio also decreases from 16.8% to 12.8% for pegwit and from 5.7% to 5% for ijpeg.

The results in this section show that annotated memory references can be used to improve memory hierarchy performance. We provided two examples of how software could assist in memory hierarchy management by conveying information about replacement strategies and block size requirements. However, the flexibility of annotated memory references permits a variety of annotations. We are currently investigating other sources of information and techniques for obtaining the appropriate annotations (including compile-time) and ways to exploit that information in the memory hierarchy.

4   Conclusion

Cache memories are an important component of the overall solution to the ever increasing discrepancy between processor clock cycle time and main memory access time. Unfortunately, conventional cache designs do not utilize any information that software may provide about expected reference patterns. This paper introduces a mechanism that enables software to provide hints about future memory references and assist in cache management. We propose adding a single instruction—called TAG—that can annotate six memory references with 4-bit annotations to be interpreted by the memory system. Our analysis showed that when annotating all memory references, execution time overhead ranges from zero to 2% on either a 4-issue dynamically scheduled or 4-issue statically scheduled processor. Using annotated memory references to exploit information on better cache replacement strategies and better block sizes, we can reduce execution time by 11% to 17%.

As the demands on cache performance increase, new techniques that complement, rather than replace, hardware cache management become an attractive alternative. The relatively low overhead of TAG instructions combined with the ability to support many different types of annotations can enable new approaches to cache management. In particular, new cache organizations and management techniques that utilize software assistance can easily be supported without requiring significant modifications to an instruction set.

References

[1] S. G. Abraham, R. A. Sugumar, D. Windheiser, B. R. Rau, and R. Gupta. Predictability of load/store instruction latencies. *Proceedings of the 26th Annual International Symposium on Microarchitecture*, pages 139–152, December 1993.

[2] D. C. Burger, T. M. Austin, and S. Bennett. Evaluating future microprocessors-the simplescalar tool set. Technical Report 1308, University of Wisconsin–Madison Computer Sciences Department, July 1996.

[3] A. Gonzalez, C. Aliagas, and M. Valero. A data cache with multiple caching strategies tuned to different types of locality. In *ACM 1995 International Conference on Supercomputing*, pages 338 – 347, 1995.

[4] A. R. Lebeck and D. A. Wood. Cache profiling and the spec benchmarks: A case study. *IEEE COMPUTER*, 27(10):15–26, October 1994.

[5] A. Srivastava and A. Eustace. Atom a system for building customized program analysis tools. In *Proceedings of the SIGPLAN '94 Conference on Programming Language Design and Implementation*, pages 196–205, June 1994.

[6] G. Tyson, M. Farrens, J. Matthews, and A. R. Pleszkun. A modified approach to data cache management. In *Proceedings of the 28th Annual International Symposium on Microarchitecture*, Dec. 1995.

Understanding and Improving Register Assignment*

Cindy Norris and James B. Fenwick, Jr.

Department of Computer Science
Appalachian State University
Boone, North Carolina, USA 28608
(828) 262-2359
{can,jbf}@cs.appstate.edu

Abstract. Register allocation can decrease instruction-level parallelism by prohibiting the scheduler from reordering instructions. The impact of register assignment strategies on a subsequent scheduling phase is explored. A new register assignment strategy and experimental results are presented.

1 Introduction

Register assignment is the phase of the register allocator that decides what values to put in each register. First-Fit register assignment [5] chooses the first available register in a sequential ordering of the registers. Round-Robin assignment [5] begins searching for an available register at the point where the last successful search ended. Scheduling increases run-time performance by rearranging the code to overlap the execution of independent instructions. A register assignment strategy cooperates better with a subsequent scheduling phase than another strategy if it (1) introduces less false dependences, or (2) introduces false dependences that don't prevent the scheduler from uncovering sufficient fine-grain parallelism.

This paper examines the impact of register assignment strategies within a *global* register allocator on a subsequent local instruction scheduling phase. The findings presented in section 3 indicate that register assignment of a global allocator *does* impact scheduling, particularly for functional units with higher latencies. Our Improved First-Fit strategy was found to cooperate better than First-Fit and better than the Round-Robin strategy when register pressure is high. The experimental study incorporates a postpass scheduler. However, in the absence of this scheduling phase, the Improved First-Fit strategy should still prove to be effective by reducing the number of hardware stalls caused by the false dependences added by register assignment.

As an illustration of this impact, figure 1(a) contains an unallocated section of code. References to $r2$, $r3$, $r4$, $r5$, $r6$, $r7$ are references to live ranges. It is the job of the register allocator to rewrite this code changing the references to

---

* This work was partially supported by NSF under grant CCR-9625219.

P. Amestoy et al. (Eds.): Euro-Par'99, LNCS 1685, pp. 1255–1259, 1999.
© Springer-Verlag Berlin Heidelberg 1999

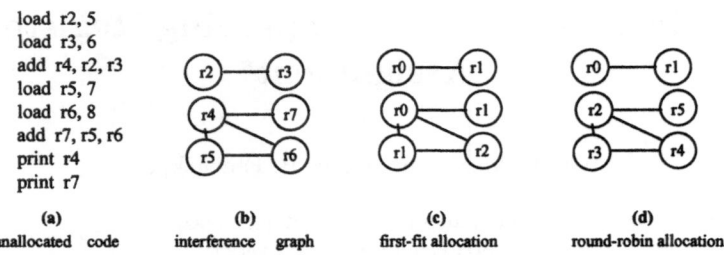

Fig. 1. Register Assignment Strategies

live ranges to references to physical registers. Graph coloring register allocators [4, 3] represent the register allocation problem as a graph coloring problem in which nodes in a graph represent live ranges and edges are added between two nodes (which are then called *neighbors*) if the corresponding live ranges overlap and must occupy different registers. The allocation is determined by coloring the nodes of the graph, called an *interference graph*, with K colors where K is the number of physical registers. Two nodes can not be colored the same color if they are connected by an edge in the interference graph. Figure 1(b) contains the interference graph corresponding to the code in figure 1(a).

Unlike local register allocation [7] in which the live ranges are assigned physical registers in order of proximity, nodes in an interference graph are assigned physical registers in a priority order based partially upon the number of uses and definitions in the code. In this example, each live range contains one definition and one use making the priority of all nodes the same, and thus they can be arbitrarily assigned physical registers in the order *r2*, *r3*, etc. Figures 1(c) and (d) demonstrate the register assignment decided by the First-Fit and Round-Robin strategies, respectively. First-Fit register assignment chooses the first available register in a sequential ordering of the registers. Round-Robin assignment begins searching for an available register at the point where the last successful search ended. This explains why *r4* in the interference graph is mapped to *r0* by First-Fit and *r2* by Round-Robin as shown in figure 1. Register assignment, in particular First-Fit register assignment, adds dependences known as *false dependences*. They are not present because of references to a single value, but because of accesses to a single register. False dependences can prevent a scheduler from reordering instructions in order for them to be executed in parallel.

When the register allocator is unable to assign a physical register to a live range, the allocator *spills* the live range to memory. Intuitively, it is expected that the two assignment strategies result in different amounts of spill code. Since First-Fit chooses the first free register available, it uses registers more sparingly. Assigning two live ranges the same physical register may save a physical register to be assigned later and prevent a spill.

2   Improved First-Fit Register Assignment

The Round-Robin strategy adds fewer false data dependences during register assignment than the First-Fit strategy. However, the false dependences that are not added are avoided only accidentally, if at all. In addition, Round-Robin may cause more spilling. This section discusses a register assignment strategy that *deliberately* attempts to avoid adding the false data dependences of First-Fit, avoid spill code, and *remain as simple* as the other register assignment strategies. More complicated strategies exist [8, 9, 1], however our goal is to improve register allocation without impacting the time required to do the allocation.

Our Improved First-Fit (IFF) register assignment uses an awareness of the sequential nature of register usage to make register assignment decisions. If two definitions are *near* each other in the sequential ordering of the statements, the IFF assignment strategy tries to avoid assigning those definitions to the same physical register. The IFF register assignment strategy takes as input the size of the *nearness window* which identifies how many definitions near a particular defining statement are to be examined. When assigning a physical register to a definition d, the definitions near to d are examined and if a physical register has been assigned to one of those near definitions, then that physical register is not assigned to d. For example in the code below, given a nearness window of size 3, the IFF strategy attempts to avoid assigning to $r13$ the same physical registers assigned to $r14$, $r15$, and $r16$.

```
ldi  r13 4
mul  r14 r13 r9
add  r15 r5 r14
ldi  r16 4
mul  r17 r16 r9
```

Why choose to examine nearby statements when making an assignment decision? First, techniques such as loop unrolling and global instruction scheduling precede register allocation and increase the number of statements within a basic block in order to increase the instruction-level parallelism. After increasing the available parallelism within a basic block, the IFF strategy avoids adding the false data dependences that prevent the simultaneous execution of sequential statements. Second, examining nearby statements when making the assignment decision makes the IFF strategy easy to implement and very efficient.

3   Experimental Study

An experimental study was performed to evaluate the performance of the three register assignment strategies. A C program is processed by the SUIF compiler system [6] to convert the program into an intermediate code format. This intermediate code is then translated into Iloc, which is a low-level intermediate code designed at Rice University for the development of optimizing compilers [2]. Our implementation of the optimistic allocator [3] performs register allocation on the Iloc code using any of the three register assignment strategies discussed previously. Allocated code is then fed to a local instruction scheduler, and the scheduled code is converted back to C, inserting instructions to simulate a fine-grain parallel machine.

Two different fine-grain parallel architectures were simulated: a pipelined machine with low latencies and a pipelined machine with high latencies. Twenty-five programs taken from the Livermore loops, SPEC, and Stanford benchmark suites were used as input to the experimental study of the three different register assignment strategies on each of the two architecture classes. Simulations were performed using three different register set sizes (8, 16, and 32). The register set sizes are not typical of current architectures, but were chosen to impact the register pressure of the benchmark programs. In other words, the register set of size 8 causes a great deal of register pressure for the selected benchmark programs while 32 registers generated little register pressure.

The performance of the assignment strategy was measured by calculating $totalcycles(FirstFit)/totalcycles(RoundRobin)$ and $totalcycles(FirstFit)/totalcycles(ImprovedFirstFit)$ where $totalcycles$ is the number of cycles required to execute the Iloc code as determined by the simulator. Due to page length restrictions, a table showing the performance speedups was omitted, but it may be viewed at http://www.cs.appstate.edu/~can/research/papers.html.

The study indicates that assigning registers with a Round-Robin strategy generally produces code that executes faster than code assigned registers using First-Fit. Two items in particular are noted. First, Round-Robin results in significantly better code for high-latency pipelined architectures with a register set large enough to keep spilling at a minimum. Second, for low-latency machines when register pressure is high, First-Fit is a better strategy. The reason First-Fit performs better under high register pressure is due to the increased amount of spill code inserted by Round-Robin. The simulations show that for 8 registers, Round-Robin resulted in 5% more spills than First-Fit.

In general, the Improved First-Fit strategy performs as well as or better than the Round-Robin strategy. Of particular note is that Improved First-Fit is significantly better than Round-Robin for high-latency machines under a high register pressure. In the case of low register pressure, the Round-Robin strategy has the potential to add fewer false dependences since it reuses a register only after all other registers have been used once. For the Improved First-Fit strategy, reuse of registers occurs more quickly when the nearness window is small. Even a window size of 7 can cause more false dependences than Round-Robin.

The Improved First-Fit strategy performs better as the size of the nearness window increases. For example on the high-latency machine with 32 registers, the average speedup is 1.20 when the nearness window size is 3 and 1.24 when the nearness window size is increased to 7. However, increases in the nearness window size have less impact on smaller register sets. With less registers overall, the algorithm is less successful in finding a register not used in the nearness window; thus, reverting to First-Fit.

It was expected that the Round-Robin strategy would result in the most spill code being inserted of the three assignment strategies, with First-Fit causing the least amount and Improved First-Fit falling in between. The experimental study supports this hypothesis. Using 8 registers 670, 684, and 703 spills occurred among the 25 benchmark programs for the First-Fit, Improved First-Fit with nearness window of 3, and Round-Robin assignment strategies, respectively. (The larger sized register sets cause dramatically fewer spills as more free registers were available.) As the nearness window size increased, the amount of spill code inserted by Improved First-Fit decreased on average although the results for individual benchmarks varied. This decrease occurs because the IFF strategy reverts more often to First-Fit when it has a larger set of instructions to examine when looking for an unused register.

4   Summary

The goal of this research was to understand better the effect of register assignment strategies within a global register allocator on instruction-level parallelism. A new register assignment strategy, Improved First-Fit, was presented that deliberately avoids the false dependences created by First-Fit, and inserts less spill code than Round-Robin. This new strategy is a simple, effective technique requiring only minor modifications to existing global register allocation algorithms.

References

[1] David A. Berson, Rajiv Gupta, and Mary Lou Soffa. Resource spackling: A framework for integrating register allocation in local and global schedulers. In *PACT '94: International Conference on Parallel Architectures and Compilation Techniques*, Montreal, Canada, August 1994.

[2] Preston Briggs, Keith D. Cooper, and Linda Torczon. R^n Programming Environment Newsletter #44. Dept. of Computer Science, Rice University, Sept. 1987.

[3] Preston Briggs, Keith D. Cooper, and Linda Torczon. Rematerialization. In *Proceedings of the SIGPLAN '92 Conference on Programming Language Design and Implementation*, 1992.

[4] G. J. Chaitin. Register allocation and spilling via graph coloring. In *SIGPLAN Symposium on Compiler Construction*, Boston, June 1982.

[5] James R. Goodman and Wei-Chung Hsu. Code scheduling and register allocation in large basic blocks. In *Supercomputing '88 Proceedings*, pages 442–452, Nov. 1988.

[6] Stanford SUIF Compiler Group. *The SUIF Parallelizing Compiler Guide*. Stanford University, 1994. Version 1.0.

[7] Wei Chung Hsu, Charles N. Fischer, and James R. Goodman. On the minimization of loads/stores in local register allocation. *IEEE Transactions on Software Engineering*, 15(10):1252–1260, 1989.

[8] Cindy Norris and Lori L. Pollock. A scheduler-sensitive global register allocator. In *Supercomputing '93 Proceedings*, Portland, OR, November 1993.

[9] S. S. Pinter. Register allocation with instruction scheduling: a new approach. In *Proceedings of the SIGPLAN '93 Conference on Programming Language Design and Implementation*, June 1993.

Compiler-Directed Reordering of Data by Cyclic Graph Coloring

Daniela Genius[1] and Sylvain Lelait[2]

[1] Institut für Programmstrukturen und Datenorganisation,
Universität Karlsruhe, D-76128 Karlsruhe,
genius@ipd.info.uni-karlsruhe.de
[2] Institut für Computersprachen, Technische Universität Wien,
A-1040 Wien, Austria,
sylvain@complang.tuwien.ac.at

Abstract. We show that cyclic graph coloring techniques from loop register allocation are successfully applicable to caches. Values of one color belong together, even if they stem from different data structures, resulting in a sytematic merging.

1 Introduction and Background

Compile-time cache optimizations for scientific computing applications exploit regular access patterns. For caches with limited associativity, it is often crucial to adjust the placement of data in memory. We show that compiler techniques for register allocation, namely graph coloring, support a systematic data placement to prevent the arrays from interfering with each other in the cache (*cross interference*). Innermost loops of perfect loop nests are considered. We concentrate on first-level data caches with limited associativity.

Merging by the programmer avoids interference of different data structures. *Padding* [1] inserts gaps into data structures. Characteristic of padding is the danger of paging due to increased memory usage. Graph coloring was first examined for cache analysis by Rawat [8]. He contributes the notion of *togetherness* in a cache line. In [5] we employed cyclic coloring for conflict reduction only.

We show in Section 2 how live ranges are derived and how graph coloring is applied. Section 3 contains measurement results, Section 4 outlines future work.

2 Coloring the Cyclic Interval Graph

A *(cache) live range* denotes the period of time in which a value is actually present in the cache [5]. The live range begins, in analogy to a register live range, when the memory location's content is first accessed (here: loaded into the cache). It ends when the value is no longer required to be in the cache.

Example 1: Livermore kernel 7 of Figure 1.a is rather well-behaved, except for cross interferece. Four arrays compete for the cache lines in only one loop. For u, seven subsequent elements are accessed in one iteration. Figure 1.b shows the

P. Amestoy et al. (Eds.): Euro-Par'99, LNCS 1685, pp. 1260–1264, 1999.
© Springer-Verlag Berlin Heidelberg 1999

intermediate representation of the innermost loop body. Accesses to x, y and z last one cycle. Live ranges for u last 7 cycles. The resulting cyclic interval family is shown in Figure 1.c, where a "dot" denotes an access. We have now created live ranges of the form required by cyclic interval graph coloring algorithms. A *variable or register live range* originally denotes the period during which a certain value is present in a register during the program execution. Finding a conflict layout for mapping possibly infinitely many live ranges onto a limited number of resources is also the aim of register allocation techniques.

Usual Cyclic Interval Graph Coloring Live ranges in innermost loops can be represented by a *cyclic interval family*. Cyclic interval graph q-coloring, also known as circular-arc graph q-coloring, is a polynomial problem [3]. But the algorithm is only practical for small values of q, therefore heuristics are used. One of the most efficient is the one introduced in [6]. The lower bound on the number of colors is the maximum number of live ranges overlapping a point in the interval family, often noted $MaxLive$ [7]. Unfortunately this method can not handle intervals overlapping themselves. Furthermore this method does not guarantee a coloring with $MaxLive$ colors.

Meeting Graph On the other hand, with the *meeting graph* method [2], intervals which overlap themselves can be handled. The meeting graph heuristic was designed to obtain a family of intervals that can be colored with $MaxLive$ colors by unrolling the loop if necessary. It only works for families having $MaxLive$ intervals at each cycle, therefore some fictitious intervals may be added. We have one node for each interval, and an arc between a node i and a node j if the interval corresponding to i ends at the point where the interval corresponding to j begins. A weight w corresponding to the length of the interval is added to each node. Similarly each circuit has a weight equals to the sum of the weights of its nodes divided by the number of cycles of the loop. To find a suitable unrolling degree, u, one has to take the lcm of the weights, W_i, of disjoint circuits which can be found in the graph. In all cases finding a decomposition is a polynomial problem [2] and the coloring with $MaxLive$ colors is ensured. For this step of coloring, W_i colors are cyclically allocated to the intervals belonging to the u copies of each circuit. In our context, the intervals depict the time that a value must stay in the cache. Minimizing the number of colors is equivalent to minimize the number of cache lines used. A decomposition of the meeting graph is computed, the lcm of the circuits weights gives the unrolling degree of the loop.

A Togetherness Criterion Cyclic interval graph coloring can be used to indicate togetherness. Togetherness for m live ranges is a result of cyclic graph coloring: if they bear the same color, they are allocated together to the same cache line. Assume that a loop nest was unrolled and k colors were used. Live ranges that have been identified as belonging to the same color stem from m different arrays. For one of these arrays, let x_n denote the offset in direction of the

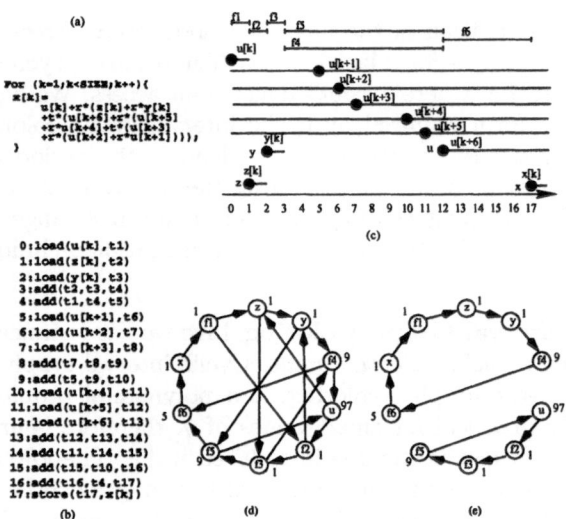

Fig. 1. Applying the meeting graph to Livermore loop 7

innermost loop index, s the element size. The position p of the array in the compound array determines which of the $m * s$ blocks is addressed, x_n mod $(m * s)$ yields the offset inside such a block. $merge : I\!N \rightarrow I\!N$ affects only the "innermost" dimension: $merge_{s,p}(x_n) = (m * s) * ((x_n * m)\ div\ s) + (p - 1) + (x_n$ mod $(m * s))$. Finally, we add starting address adjustments for the c_k compound arrays in order to avoid cross interference between the compound; for full detail see [4].

Example 2: Figure 1.d shows the meeting graph of Livermore Loop 7. Each live range is represented by a node labeled with its length in cycles. As *MaxLive* equals 7 only during the last cycle, we must add fictitious intervals, as shown in Figure 1.c, 1 during cycles 0-2 and 12-16, and 2 during cycles 3-11. As no interval begins or ends during cycles 4-10 and 13-15, the unit intervals were merged to build $f4, f5$ and $f6$. This does not change the results, but allows us to have less nodes in the graph. Figure 1.e shows one possible decomposition, yielding one circuit of weight 1 and one of weight 6. An unrolling factor of $lcm(1, 6) = 6$ is required to prevent interval u from overlapping itself. 6 colors are required for the 6 live ranges issued of u and one additional color for those belonging to the other arrays. The decomposition prescribes to merge z, y and x.

3   Experimental Results

Measurements were made in stand-alone mode on a 500 MHz 21164 DEC ALPHA AXP with 8K of direct mapped first-level data cache (32 Byte cache lines).[1] In Figure 2, we show measurements for LL7 and matrix multiply (MM); *merge* stands for our new method, *standard* for the usual layout method, *opt* for a

---

[1] In the long version [4], additional results on Intel Pentium, are shown.

successful optimization method for the benchmark, i.e. the coloring method of [5] for LL7 and blocking for MM, respectively. We also checked sizes of powers of two

program	size	miss rate %		
		none	opt	merge
LL7	100000	15.63	12.54	5.09
LL7	262144	20.3	17.8	5.83
MM	256*256	12.3	2.80	2.49
MM	300*300	7.57	1.11	2.75

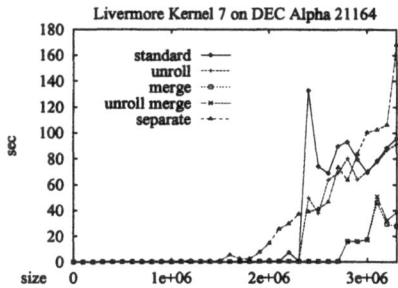

Fig. 2. Miss rates ans run times for LL7 on DEC Alpha

– known as pathological cases for caches; *merge* turns out to be quite insensitive to those, whereas blocking beats merging only for MM 300, while losing for size 256. The curves show how run times are improved for LL7; merging also reduces paging activity for large array sizes. In addition we compared our approach (*unrollmerge*) to unrolling without merging (*unroll*), merging without unrolling (*merge*) and the allocation of [5] (*separate*). Several other Livermore kernels and many image processing and compression algorithms can be classified along with LL7, making a more thorough examination clearly promising.

4   Conclusions and Future Work

By applying the meeting graph method, the compiler can compute an unrolling factor and determine the maximal number of colors, i.e. cache lines, required. Values with the same color are mapped to memory together. Since these values may stem from different data structures, our technique offers a natural way of dealing with conflicts between different arrays. For typical codes from the above areas, cache behavior is considerably improved. We are currently investigating the effects of reordering between different loop nests of a program and applying our method to larger scientific programs.

References

[1] D. F. Bacon, S. L. Graham, and O. J. Sharp. Compiler transformations for high-performance computing. *ACM Computing Surveys*, 26(4):345–420, December 1994.

[2] Christine Eisenbeis, Sylvain Lelait, and Bruno Marmol. The meeting graph: A new model for loop cyclic register allocation. In *Proc. PACT'95*, pages 264–267, Limassol, Cyprus, June 27–29, 1995. ACM Press.

[3] M.R. Garey, D.S. Johnson, G.L. Miller, and C. H. Papadimitriou. The complexity of coloring circular arcs and chords. *SIAM J. Alg. Disc. Meth.*, 1(2):216–227, 1980.

[4] D. Genius and S. Lelait. Improving Data Layout through Coloring-Directed Array Merging. Technical Report 1999-3, Universität Karlsruhe, Januar 1999.

[5] Daniela Genius. Handling Cross Interferences by Cyclic Cache Line Coloring. In *Proc. PACT'98*, pages 112–117, Paris, France, October 1998. IEEE.

[6] L. Hendren, G. Gao, E. Altman, and C. Mukerji. A register allocation framework based on hierarchical cyclic interval graphs. In *Proc. 4th CC*, pages 176–191, 1992.

[7] R.A. Huff. Lifetime-Sensitive Modulo Scheduling. *SIGPLAN Notices*, 28(6), 1993.

[8] Jai Rawat. Static analysis of cache performance for real-time programming. Technical Report IASTATECS//TR93-19, Iowa state university, Nov 1993.

Code Cloning Tracing: A "Pay per Trace" Approach

Thierry Lafage, André Seznec, Erven Rohou, and François Bodin

IRISA, campus de Beaulieu, 35042 Rennes cedex, France
{Thierry.Lafage, Andre.Seznec, Erven.Rohou, Francois.Bodin}@irisa.fr

Abstract. Code Cloning Tracing is a new software annotation method that makes it possible to collect traces from time consuming applications. To this end, Code Cloning Tracing provides instrumented programs with two execution modes: a low overhead "no-trace collection" mode which serves to position the application in an interesting state with regard to tracing, and a "trace collection" mode. This paper details the Code Cloning Tracing method and presents **calvin**, our first implementation. On the SPEC95 suite, **calvin** exhibits low execution slowdown factors in "no-trace collection" mode varying from 1.02 to 2.09.

1 Introduction and Motivations

Execution traces are needed for architecture simulations and validations in processor design, or for validations of new microarchitecture ideas. The process of collecting traces is time consuming (execution slowdowns in the 10-100 range [6]), and even more time consuming is microarchitecture simulation (slowdowns in the 1,000–10,000 range [1]). As a result, in practice, most microarchitecture studies are performed using the first instructions (maybe a few billion) of an application and skipping the first billion(s) to avoid the initialization phase (see [2]). This solution may still be very time consuming since the tested application is instrumented/simulated entirely: skipping the first instructions still induces a non-negligible execution overhead. Consequently, large workloads (running hours of CPU) are never used in microarchitecture studies.

In this paper, we introduce Code Cloning Tracing, a new software annotation method for collecting traces. Code Cloning Tracing is aimed at collecting large trace samples on large applications for microarchitecture studies and focuses on the execution slowdown of the traced applications when the trace is not used (collected). The Code Cloning Tracing's key feature is to provide intrumented programs with two execution modes: a low overhead "no-trace collection" mode and a "trace collection" mode through a static code duplication (cloning). The code duplication produces two clones which are part of the same executable program. During the execution, some *events* activate dynamic execution switches between both clones, enabling or disabling the trace collection.

In this paper, we only focus on the "no-trace collection" mode performance, knowing that current techniques (e.g. see [3]) may be applied to the "trace collection" mode.

P. Amestoy et al. (Eds.): Euro-Par'99, LNCS 1685, pp. 1265–1268, 1999.
© Springer-Verlag Berlin Heidelberg 1999

In the next section, we detail the Code Cloning Tracing method. Section 3 presents the first implementation called **calvin**. An evaluation of the execution overhead in "no-trace collection" mode is presented in Section 4. Section 5 summarizes this study and presents our directions for future development.

2   Principle of Code Cloning Tracing

The principle of Code Cloning Tracing is illustrated in Figure 1. The original code is duplicated. Both code copies (clones) are annotated in order to enable the execution to dynamically switch between each other.

One of the clones, called *P_inst*, is heavily instrumented by the trace user [1] and allows the trace to be generated.

The second clone, called *P_exec*, is kept nearly identical to the original code. *P_exec* provides the "no-trace collection" mode. This clone is expected to induce a very low execution overhead.

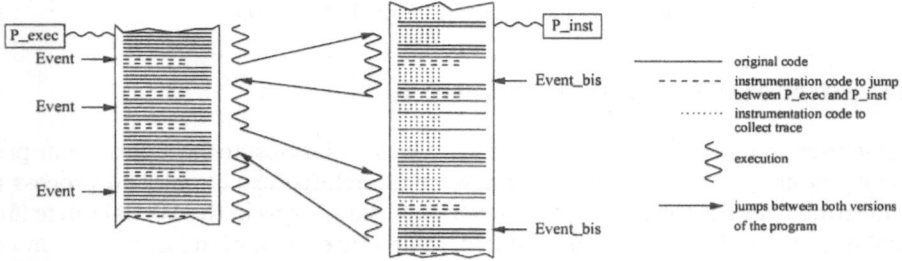

Fig. 1. Execution of cloned code to collect trace.

On a large application, *P_exec* will execute most of the original instructions ("no-trace collection" mode) to position the application in a state in which trace collection and simulation are interesting, at a very low execution overhead (compared to *P_inst*). This technique can be related to SimOS [5] with even less execution overhead when trace is not generated.

3   Code Cloning Tracing with Calvin

Code Cloning Tracing is a general technique and may be implemented at various levels: executable or assembly code. Our first prototype, **calvin** [2] uses SALTO, a

---

[1] For convenience, we call a *trace user* someone who collects program traces. He/she uses the framework we provide, but also needs to instrument the tested applications to get the traces he/she wants.

[2] cloning assembler and looking into veritable instrumentation needs.

retargetable framework for developing tools that manipulate programs expressed in assembly language [4].

3.1   Checkpoint Code

The annotation code added by **calvin** is composed of several *checkpoints*. Checkpoints are a few instructions (about 10) which check that the current clone running is the right one. When this is not the case (an activating *event* has occured), control is given to the other clone.

We inserted checkpoints at each procedure call and inside each loop. With a complete *ad-hoc* instrumentation of each SPEC95 benchmark program, we computed the number of executed checkpoints among the executed instructions. This computation shown a reasonable number of checkpoints among executed instructions (5.02 % on average for CINT95 programs, and 2.53 % on average for CFP95's) which allowed us to expect low execution slowdowns in "no-trace collection" mode, as seen in Section 4.

3.2   Activating Mode Switching

An *event* determines when to switch to the other clone of the tested application as shown in Figure 1. In the first implementation of **calvin**, the event used is a given number of executed checkpoints in each clone.

4   No-Trace Collection Mode Slowdown

In order to validate our approach, we cloned and instrumented the popular SPEC95 benchmark suite. These programs were compiled with **gcc** or **g77** with '-O3' optimization option [3]. Each run on this benchmark suite was tested with the *ref* input data set. **calvin** can annotate SPARC assembly code, and all the results were collected from a Sun Ultra 1 workstation with a 143 MHz processor and 256 MB of memory, running Linux.

Here, we also acted as trace users to obtain completely instrumented applications, like computer architects would do. We heavily instrumented the *P_inst* clone to produce instruction and data addresses. Since this study is not directed at the performance of the "trace collection" mode, we did not collect any traces, rather the programs entirely ran in "no-trace collection" mode.

Table 1 presents the execution times (user + system, in seconds) of each SPEC95 benchmark. The *base* time is the execution time of the original (i.e. non instrumented) program. The number in parentheses represents the slowdown of the concerned workload which is computed as follows: $slowdown = \frac{tested\_time}{base\_time}$. As expected, the execution overheads are quite low: from 1.02 to 2.09. Note that they are directly related to the number of checkpoints executed.

---

[3] *145.fpppp* only compiled with '-O1'.

CFP95	Base	P_exec only
101.tomcatv	625	696 (1.11)
102.swim	548	628 (1.15)
103.su2cor	619	709 (1.15)
104.hydro2d	772	1024 (1.33)
107.mgrid	1533	1630 (1.06)
110.applu	2592	3059 (1.18)
125.turb3d	3460	3899 (1.13)
141.apsi	1057	1213 (1.15)
145.fpppp	2048	2071 (1.02)
146.wave5	864	1164 (1.35)

CINT95	Base	P_exec only
099.go	276	438 (1.58)
124.m88ksim	554	1151 (2.09)
126.gcc	10	16 (1.57)
129.compress	413	784 (1.90)
130.li	575	1130 (1.96)
132.ijpeg	378	445 (1.18)
134.perl	267	351 (1.32)
147.vortex	728	998 (1.37)

Table 1. "No-trace collection" mode execution time (sec.) on the SPEC95 benchmarks.

5   Summary and Future Work

In this paper, we have presented the basic principles of Code Cloning Tracing. **calvin**, the prototype we built, has been currently tested on single process applications. It has shown to induce a very acceptable execution slowdown in "no-trace collection" mode (from 1.02 to 2.09). Such a low slowdown would be acceptable for computer architects to position large workloads (with their large data sets) in interesting tracing states.

Future developments will first include trace collection on multiprocess workloads using external events to switch between both execution modes, and dynamically linked libraries (libc, libg2c, ...). Finally, we plan to use Code Cloning Tracing to instrument all user applications and the operating system on a desktop computer.

References

[1] D. C. Burger and T. M. Austin. The SimpleScalar tool set, version 2.0. Technical Report CS-TR-97-1342, University of Wisconsin, Madison, June 1997.

[2] M. J. Charney and T. R. Puzak. Prefetching and memory system behavior of the SPEC95 benchmark suite. *IBM Journal of Research and Development*, 41(3), 1997.

[3] Alvin R. Lebeck and David A. Wood. Active memory: A new abstraction for memory system simulation. *ACM Transactions on Modeling and Computer Simulation*, 7(1):42–77, January 1997.

[4] E. Rohou, F. Bodin, and A. Seznec. SALTO: System for assembly-language transformation and optimization. In *Proceedings of the Sixth Workshop Compilers for Parallel Computers*, December 1996.

[5] M. Rosenblum, E. Bugnion, S. Devine, and S. A. Herrod. Using the SimOS machine simulator to study complex computer systems. *ACM Transactions on Modeling and Computer Simulation*, 7(1):78–103, January 1997.

[6] R. Uhlig and T. Mudge. Trace-driven memory simulation: a survey. *ACM Computing Surveys*, 1997.

Execution-Based Scheduling for VLIW Architectures

Kemal Ebcioğlu, Erik R. Altman, Sumedh Sathaye, and Michael Gschwind

IBM Thomas J. Watson Research Center
Yorktown Heights, NY 10598
{kemal,erik,sathaye,mikeg}@watson.ibm.com

Abstract. We describe a new dynamic software scheduling technique for VLIW architectures, which compiles into VLIW code the program paths that are actually executed. Unlike trace processors, or *DIF*, the technique executes operations speculatively on multiple paths through the code, is resilient to branch mispredictions, and can achieve very large dynamic window sizes necessary for high ILP. Aggressive optimizations are applied to frequently executed portions of the code. Encouraging performance results were obtained on **SPECint95** and **TPC-C**. The technique can be used for binary translation for achieving architectural compatibility with an existing processor, or as a VLIW scheduling technique in its own right.

Keywords: INSTRUCTION-LEVEL PARALLELISM, DYNAMIC COMPILATION, BINARY TRANSLATION, SUPERSCALAR

1 Background and Motivation

VLIW architectures are desirable because they offer a simple hardware design path toward achieving wide issue at high frequency. However, architectural incompatibility with existing architectures, and hence the requirement to make software changes when migrating to a new VLIW architecture, has been a problem. In prior papers on the **DAISY** (Dynamically Architected Instruction Set from Yorktown) project [1, 2], the authors have established techniques for achieving 100% architectural compatibility with an existing processor through software techniques applied to a wide issue VLIW. A number of difficulties were addressed, such as self modifying code, multi-processor consistency, memory mapped I/O, preserving precise exceptions while aggressively re-ordering VLIW code, and so on.

In the previous version of **DAISY**, the unit of translation was a page. Thus if execution reached a previously unseen page **P**, at address **X**, then all code on page **P** reachable from **X** — via paths entirely within page **P** — was translated to VLIW code. Any paths within page **P** that went offpage or that contained a register branch were terminated. At the termination point was placed a special type of branch that would (1) determine if a translation existed for the offpage/register location specified by the branch, and (2) branch to that translation if it existed, and otherwise branch to the translator. Once this translation

P. Amestoy et al. (Eds.): Euro-Par'99, LNCS 1685, pp. 1269–1280, 1999.
© Springer-Verlag Berlin Heidelberg 1999

was completed for address **X**, the newly translated code corresponding to the original code starting at **X** was executed.

In the previous **DAISY**, as well as in current version reported here, this translation/execution process begins at the bootstrap location for the emulated processor, for example, 0xFFF00100 for *PowerPC*. In this way, and as described in detail in [1, 2], the original base architecture can be properly emulated, with no need for any operating system or other changes.

We now define a few terms from our earlier work and make a few other observations in hopes of better illuminating what is new in our current work and what motivated it:

- As noted, the previous **DAISY** algorithm started at an entry point of a page, and scheduled along all paths reachable from that point. Because of the real-time constraints on the amount of time which may be spent scheduling operations, the scheduler did not build control-flow graphs and hence did not recognize join points — instead the code beyond join points was duplicated along however many paths pass through it. Because of this scheduling policy, the regions scheduled were trees which were in turn composed of tree VLIW instructions [3].
- We term these tree regions, **groups**. Likewise each leaf or exit of the tree, we term a **tip**. Since *groups* are trees, knowing by which *tip* the group exited, fully identifies the control path executed from the *group* entrance (or tree root).
- As befits **DAISY**'s real-time requirements, use of tree *groups* simplifies many areas of scheduling. For example, there is at most one reaching definition for each value. Trees also have drawbacks, in particular the duplicated code beyond join points can result in VLIW code that is many times larger than the code for the base architecture.
- The original **DAISY** had another big source of code explosion: since all paths from a given entry point were translated, *groups* could contain operations from paths which were rarely, if ever executed.

Thus, the previous **DAISY** technique was adequate for VLIW processors of modest width and large **ICaches**. However, for very wide machines, page crossings and indirect branches limited ILP. In this paper, we describe new techniques we have added to **DAISY** which overcome these ILP limitations and attack the code explosion problem. The present **DAISY**:

- Maintains the tree structure of groups, as well as tree instructions.
- Compiles only the executed portion of the code, by interpreting operations first. Thus **ICache** resources are spent more effectively.
- Crosses page boundaries and indirect branches, by making use of run time information to convert each indirect branch to a set of conditional branches. There is no limit on the number of pages a translated code fragment may cross. Only interrupts and code modification events are serializers.

— Conserves **ICache** and compile time resources by applying modest optimizations initially, and then scheduling aggressively with a large window size, only on the frequently executed portions of the code.

This **DAISY** approach can either be used as a binary translation system or as a VLIW scheduling technique in its own right. The rest of the paper is organized as follows. Section 2 describes the dynamic compilation algorithm. This algorithm includes not only the scheduling of operations, but rules for ending a scheduling region (Section 2.1), as well as a hardware/software mechanism for optimizing very frequently executed fragments of code (Section 2.2). Section 3 describes our performance evaluation experiments, Section 4 compares our approach to previous work, and Section 5 concludes.

2   The Dynamic Compilation Algorithm

In this section, we describe the execution based dynamic compilation algorithm. In what follows, the *"base architecture"* [4, 5] refers to the architecture with which we are trying to achieve compatibility, e.g., *PowerPC* or *S/390*. In this paper, our examples will be from *PowerPC*. To avoid confusion, we will refer to *PowerPC* instructions as *operations*, and reserve the term *instructions* for VLIW instructions (each potentially containing many *PowerPC operations*).

From the actually executed portions of the *base architecture* binary program, the dynamic compilation algorithm creates a VLIW program consisting of *tree regions*, which have a single entry (root of the tree) and one or more exits (terminal nodes of the tree).

The dynamic translation algorithm interprets code when a fragment of *base architecture* code is executed for the first time. As *base architecture* instructions are interpreted, the instructions are also converted to execution primitives (these are very simple RISC-style operations and conditional branches). These execution primitives are then scheduled and packed into VLIW tree regions which are saved in a memory area which is not visible to the *base architecture*. Any untaken branches, i.e., branches off the currently interpreted and translated trace, are translated into calls to the binary translator. Interpretation and translation stops when a stopping condition has been detected. (Stopping conditions are elaborated in section 2.1.) The last VLIW of an instruction group is ended by a branch to the next tree region.

Then, the next code fragment is interpreted and compiled into VLIWs, until a stopping condition is detected, and then next code fragment, and so on. If and when program decides to go back to the entry point of a code fragment for which VLIW code already exists, it branches to the already compiled VLIW code. Recompilation is not required in this case.

Looking at Figure 1(a), if the program originally took, path A through a given code fragment (where cr1.gt and cr0.eq are both false), and if the same path A through the code fragment (tree region) is followed during the second execution, the program executes at optimal speed within the code fragment — assuming a big enough VLIW and cache hits.

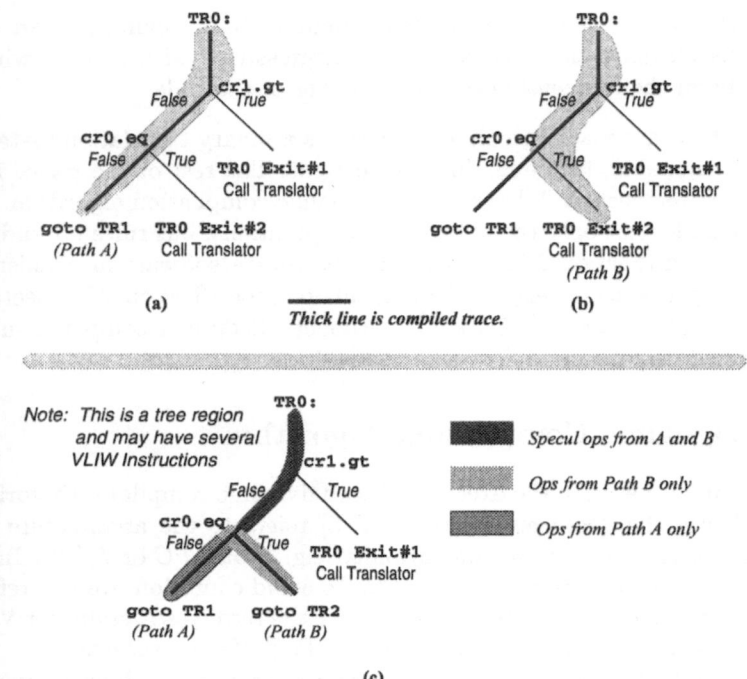

Fig. 1. Tree regions and where operations are scheduled from different paths.

If at a later time, when the same tree region labeled TR0 is executed again, the program takes a different path where cr1.gt is false, but cr0.eq is true (labeled path B), it branches to the translator, as seen in Figure 1(b). The translator may then start a new translation group at that point, or instead extend the existing tree region by interpreting *base architecture* operations along the second path B starting with the target of the conditional branch if cr0.eq. The *base architecture* operations are translated into primitives and scheduled into either the existing VLIWs of the region, or into newly created VLIWs appended to the region, as illustrated in Figure 1(c).

Assuming a VLIW with a sufficient number of functional units and cache hits, if the program takes path A or B, it will now execute at optimal speed within this tree region TR0, regardless of the path. This approach makes the executed code more resilient to performance degradation due to unpredictable branches.

The compilation of the tree region is necessarily never complete. It may have "loose ends" that may call the translator at any time. For instance, as seen in Figure 1(c), the first conditional branch if cr1.gt in tree region TR0 is such a branch whose off-trace target is not compiled. Thus, dynamic compilation is potentially a never-ending task.

In our previous work, indirect branches always ended a tree region. This serialization is a significant impediment to high ILP, as such branches can occur every 25 branches or even more frequently in some programs. To avoid this problem, we note the address being branched to when an indirect branch is scheduled. For example the *PowerPC* Link Reg may contain 0x1234 on a blr instruction. Then, as in the example below, the blr can be converted from an indirect branch to a direct branch.

```
cmplr   cr8=r33,0x1234      # r33 holds PowerPC link register
if      (cr8.eq) goto L1234 # If lr==0x1234, goto translated code
                            # for PowerPC addr 0x1234
else    call_interpreter    # Start interpreting ops at addr
                              in lr/r33
```

In this way, the operations found at 0x1234 can be scheduled into the current tree region. If other values of the Link Reg are encountered later in execution, explicit tests may be made for them as well.

2.1   Stopping Points for Paths in Tree Regions

Finding appropriate stopping points for a tree region is crucial for achieving high ILP, as well as for limiting the size of the generated VLIW code and translation time required for translation. Currently we consider ending a tree region at two types of operations:

- The **target** of a *backward branch*, typically a loop starting point, or
- A **subroutine entry** or **exit**, as detected heuristically through *PowerPC* **branch and link** or register-indirect branch operations.

Stopping (and hence starting) tree regions only at well-defined potential stopping points is useful, since if there was no constraint on where to stop, code fragments starting and ending at arbitrary *base architecture* operations could result, leading to unnecessary code duplication and increasing code expansion. Establishing well-defined starting points increases the probability of finding a group of compiled VLIW code when the translator completes translation of a tree region.

We emphasize that encountering one of the *stopping points* above does *not* automatically end a tree region. To actually end a tree region at a stopping point, at least one of the following *stopping conditions* must previously have been met:

- The desired ILP has been reached in scheduling operations, or
- The number of *PowerPC* operations on this path since the beginning of the tree region entry has exceeded a maximum *window size*.

The purpose of the ILP goal is to attain the maximum possible performance. The purpose of the *window size* limit is to limit code explosion — a high ILP goal may be attainable only by scheduling an excessive number of operations into a tree region.

2.2   Adaptive Scheduling Principles

In order to obtain the best performance, we do not make the ILP goal or maximum window size constants. Instead, a tree region is initially scheduled with modest ILP and window size parameters. If this region eventually executes only a few times, this represents a good choice for conserving code size and compile time.

If we later find that the time spent in a tree region tip is greater than a threshold fraction **thresh** of the total cycles spent in the program, then we optimize this area much more aggressively, e.g., using a much higher ILP goal and larger window size. Thus, if there are parts of the code which are executed more frequently than others (implying high re-use on these parts), they will be optimized very aggressively. If, on the other hand, the program profile is flat and many code fragments are executed with almost equal frequency, then no such optimizations occur, which could be good strategy for preserving **ICache** resources and translation time. Prior work on adaptive profiling-based optimizations includes the SUN *HotSpot* technology [6], for profile-guided optimization of frequently executed **JAVA** code fragments.

Determining whether a tree region tip is consuming a fraction greater than **thresh** of the total cycles can be done in a variety of ways. One possibility is to compile the profiling code into holes in the VLIW code. Another is to examine the current and previous program counter on timer interrupts enabled on transitions between tree regions.

In our results in Section 3, we employ a third way, namely an **8K** entry *8-way* set associative (hardware) array of cached counters indexed by the tip (exit point) of a tree region. These counters are automatically incremented upon exit from a tree region and can be inspected to see which tips are consuming the most time. They offer the additional advantages of not disrupting the **DCache** and being reasonably accurate.

3   Performance Evaluation

VLIW projects usually involve compilers and simulators to run the compiled code, they do not use traces as inputs. In this round of experiments, we have used a trace based evaluation methodology, which gives us access to kernel as well as application traces. Here, we report results for **SPECint95** and **TPC-C**. Our **SPECint95** traces are from **RS/6000** *PowerPC* machines. Each trace consists of **50**, *2 million* operation samples, uniformly sampled over a run of the benchmark. The **TPC-C** trace is slightly longer, but similarly obtained.

The performance evaluation tools implement the dynamic compilation strategy using a number of tools:

- A tree-region former reads a *PowerPC* operation trace and forms tree-regions according to the strategy described in this paper. However, to avoid translating short-lived groups (tree regions), groups are interpreted 30 times before translation. Also, a group is allowed to grow a new tree branch from an exit, only if that exit is executed frequently. The initial region formation parameters were: **ILP goal=3**, *window size limit=24* operations. When **5%** of the time is spent on a given tree region tip, the tip is aggressively extended with **ILP goal=10**, *window size limit=180*. An **8K** entry, **8-way** associative array of counters were simulated, to detect the frequently executed tree region tips, as described in Section 2.2.

- A VLIW scheduler schedules the *PowerPC* operations in each tree region and generates VLIW code according to the clustering, functional unit and register constraints, and determines the cycles taken by each tree region tip.
- A VLIW instruction memory layout tool lays out VLIWs in memory according to architecture requirements.
- A multi-level **ICache** simulator determines the **ICache** CPI penalty using a history-based prefetch mechanism.
- A multi-level **DCache** and **DTLB** simulator. The data references in the original trace are run through these simulators for hit/miss simulation. To account for the effects of speculation and joint cache effects on the off chip **L3**, we multiplied the **DTLB** and **DCache CPI** penalties by a factor of *1.7* when calculating the final **CPI**. We chose *1.7* based on speculation penalties we have previously observed in an execution-based model. To account for disruptions due to execution of translator code, we flush on-chip caches periodically based on a statistical model of translation events.

From the number of VLIWs on the path from the root of a tree region to a tip, and the number of times the tip is executed, we can calculate the total number of VLIW cycles. Empty VLIWs are inserted for long latency operations, so each VLIW takes one cycle. The total number of VLIWs executed, divided by the original number of *PowerPC* operations in the trace, yields the infinite cache, but finite resource **CPI**.

Stall cycles due to caches and TLBs, are tabulated using a simple *stall-on-miss* model for each cache or TLB miss. In the *stall-on-miss* model everything in the processor stops when a cache miss occurs, or data from a prior prefetch is not yet available.

To model **translation overhead**, we first define **re-use rate**:

$$\textbf{Re-use Rate} = \frac{\text{Number of Dynamic Ins in Trace}}{\text{Number of Unique Ins Addresses in Trace}}$$

Reuse rates are shown in the last column of Table 1, and are in millions. **SPECint95 rates** were measured through an interpreter based on the reference inputs although operations in library routines were not counted [1]. The **TPC-C** value was obtained from the number of code page faults in a benchmark run. **Re-use rates** may be used to estimate **translation overhead** (in terms of CPI) as follows:

- $\#P = \#$ of Times an Operation undergoes *Primary* Translation
- $\#S = \#$ of Times an Operation undergoes *Secondary* Translation

---

[1] We are indebted to Jay Leblanc for providing us this data.

- CP = Cycles per *Primary* Translation of an Operation
- CS = Cycles per *Secondary* Translation of an Operation

Then

$$\text{Overhead} = \frac{\#P \times CP + \#S \times CS}{\text{Re-use Rate}}$$

The translation (or *primary* translation) of a *PowerPC* operation occurs when it is being added to a tree region for the first time. A *secondary* translation of an operation occurs when it is already in a tree region while new operations are being added to the tree region. In this study we have used an estimate of 4000 cycles for a *primary* translation and 800 cycles for a secondary translation. Our **DAISY** experience yielded about 4000 PowerPC operations to translate one *PowerPC* operation [2]. Secondary translation merely requires disassembling VLIW code and reassembling it, something we estimate to take about 800 cycles.

Program	Inf Resrc CPI	Resrc CPI Adder	Inf Cache CPI	CPI Adders ICache	DCache	TLB	Xlate Overhd (CPI)	Final CPI	Avg Win-dow	Code Explo	Reuse Rate
li	0.37	*0.00*	0.37	*0.00*	0.01	*0.00*	0.00	**0.38**	0.8	21.7	*16.5*
m88k	0.19	*0.08*	0.28	*0.00*	0.00	*0.00*	0.00	**0.29**	0.7	36.7	*14.8*
ijpeg	0.18	*0.13*	0.31	*0.00*	0.01	*0.00*	0.00	**0.33**	1.2	50.2	*10.3*
vortex	0.22	*0.06*	0.28	*0.00*	0.13	*0.02*	0.00	**0.44**	1.0	41.1	*3.4*
perl	0.30	*0.08*	0.38	*0.00*	0.00	*0.00*	0.00	**0.38**	1.1	36.4	*6.8*
compr	0.39	-0.01	0.38	*0.00*	0.14	*0.01*	0.00	**0.52**	0.7	26.8	*69.2*
go	0.60	-0.04	0.56	*0.04*	0.06	*0.00*	0.00	**0.67**	7.2	16.0	*6.2*
gcc	0.38	*0.02*	0.41	*0.03*	0.01	*0.00*	0.01	**0.46**	2.7	20.2	*0.74*
GMean			0.36					**0.42**	1.8	30.8	*8.1*
TPC-C	0.29	*0.09*	0.39	*0.03*	0.20	*0.03*	0.00	**0.65**	0.8	27.7	*3.8*

Table 1. Performance on **SPECint95** and **TPC-C**.

Infinite cache CPI with the approach described here is roughly 30%-40% better than the old page-based **DAISY**. Table 1 details the performance of our current approach on a 16 issue machine configuration where the 16 total operations can include up to 8 **Load/Stores**. The machine is divided into 4 clusters of 4 functional units each, for high frequency operation. Within a cluster back to back dependent operations are allowed, but when a cluster is crossed an extra cycle is incurred. **L1 DCaches** are duplicated in each cluster, but stores are broadcast to all copies. Cache and TLB parameters are given in Table 2. The configurations used are quite aggressive, to tolerate speculation and the large code explosion that results from the present approach.

The *Infinite Resource* CPI column of Table 1 describes the CPI of a machine with infinite registers and resources, constrained only by serializations between

Cache	Size	Linesize	Assoc	Latency
L1–I	64K	1K	8	1
L2–I	1M	2K	8	3
L1–D	32K	256	4	2
L2–D	512K	256	8	4
L3	32M	256	8	42
Memory	–	–	–	150
DTLB1	128 Entries	–	2	2
DTLB2	1K Entries	–	8	4
DTLB3	8K Entries	–	8	10
Page Table	–	–	–	90

Table 2. Cache and TLB Parameters.

tree regions, and realistic operation latencies (including a load latency of 3 cycles for unsigned loads and 4 for algebraic loads). The *Finite Resource CPI Adder* describes the extra CPI due to finite registers and function units, as well as clustering effects, and possibly compiler immaturities. (This value can sometimes be negative, since the load latency for the finite resource ILP measurement is 2 cycles). *Infinite Cache CPI* is the sum of the first two columns. The **ICache**, **DCache** and **DTLB CPI** describe the additional **CPI** incurred due to **ICache**, **DCache**, and **TLB** misses, assuming the *stall-on-miss* machine model described above. *Translation Overhead* is determined using the formulas and values above. *Final CPI* is then the sum of the *Infinite Cache CPI*, **ICache**, **DCache**, **TLB**, and *Overhead* columns. The initial interpretation overhead is insignificant. Also, there are no branch stalls, due to our zero-cycle branching technique [3, 7].

Even though unlike previous infinite cache VLIW studies our model takes into account all the major CPI components, we have not modeled the VLIW machine at a very detailed (*e.g., RTL*) level. Hence performance could fall short of the numbers presented here. However, our model also omits some potential performance enhancers, such as software value prediction, software pipelining, tree-height reduction, and **DCache** latency tolerance techniques.

Note that the VLIW **ICache** consists of 8 independent mini-**ICaches** corresponding to $\frac{1}{8}$th of a VLIW supplying a pair of ALUs. Thus the **L2** mini-**ICache** size is logically 128K instead of 1M, and the mini-**ICache** linesize is 256 bytes instead of 2K bytes. But because each such mini-**ICache** has to have a redundant copy of the branch fields to reduce the wire delays, the physical size is larger than the logical size. The VLSI technology and packaging needed for this design will probably be realizable on a single chip within a few years.

The *Average Window Size* in Table 1 indicates the average dynamic number of *PowerPC* operations between tree region crossings. The *Code Explosion* indicates the ratio of translated VLIW code pages to *PowerPC* code pages. Our mean code explosion of 1.8 is more than 2× better than the old page-based **DAISY**. This improvement has come about largely because of our use of adaptive scheduling techniques and the fact that only executed code is translated.

Preliminary experiments on large multi-user systems indicate that a translation space of **2K-4K** *PowerPC* pages is sufficient to cover the working set for code. With a code explosion factor of 1.8×, such large multi-user systems would likely require **15 − 30 Mbytes** for VLIW code:

$(2K/4K)$ pages × $4K$ bytes per page × 1.8 Code Explosion ≈ $15M/30M$

15 − 30 Mbytes for translated code will probably be affordable on moderate size systems over the next few years.

4   Related Work

Previous work in inter-system binary translation has largely focused on easing migration between platforms. To this end, problem state executables were translated from a legacy instruction set architecture to a new architecture. By restricting the problem domain to a single process, a number of simplifying assumptions can be made about execution behavior and the memory map of a process. Dynamic binary translation of programs as a translation strategy is exemplified by caching emulators such as **FX!32** [8]. **FX!32** emulates only the user program space and depends on support from the OS (*Microsoft* **Windows NT**) to provide a native interface identical to that of the original migrant system.

The presented approach is more comparable to full system emulation, which has been used for performance analysis (e.g., **SimOS** [9]) and for migration from other legacy platforms as exemplified by **Virtual PC, SoftPC/SoftWindows** and to a lesser extent **WABI,** which intercepts **Windows** calls and executes them natively. Full system simulators execute as user processes on top of another operating system, using special device drivers for *virtualized* software devices. This is fundamentally different from our approach which uses dynamic binary translation to implement a processor architecture. Any operating system running on the emulated architecture can be booted using our approach.

The present approach is different from the **DIF** approach of Nair and Hopkins [10]. It schedules operations on multiple paths to avoid serializing due to mispredicted branches. Also, in the present approach, there is virtually no limit to the length of a path within a tree region or the ILP achieved. In **DIF**, the length of a (single-path) region is limited by machine design constraints (e.g., 4-8 VLIWs). Our approach follows an all software approach as opposed to **DIF** which uses a hardware translator. This all-software technique allows aggressive software optimizations hard to do by hardware alone. Also, the **DIF** approach involves almost three machines: the sequential engine, the translator, and the VLIW engine. In our approach there is only a relatively simple VLIW machine.

Trace processors [11] are similar to **DIF** except that the machine is out-of-order as opposed to a VLIW. This has the advantage that different trace fragments do not need to serialize between transitions between one trace cache entry and another. However, when the program takes a path other than what was recorded in the trace cache, a serialization can occur. The present approach solves this problem by incorporating an arbitrary number of paths in a software trace cache entry, and by very efficient zero overhead multiway branching hardware [7]. The dynamic window size (trace length) achieved by the present approach can be significantly larger than that of trace processors, which should allow better exploitation of ILP.

5   Conclusion

We have described the latest version of **DAISY**, which employs a dynamic software translation approach whereby operations from the actual execution path of a base architecture such as *PowerPC* are scheduled into VLIW instructions. The proposed technique allows operations from multiple code pages and can schedule operations through indirect branches. This technique can also schedule operations from multiple paths. The proposed technique is adaptive, and schedules more aggressively on frequently executed paths. This technique exposes significant ILP, with values reaching almost 2.5 instructions per cycle even after accounting for cache effects.

References

[1] K. Ebcioğlu and E. Altman. **DAISY**: Dynamic Compilation for 100% Architectural Compatibility. Research Report RC 20538, IBM T.J. Watson Research Center, Yorktown Heights, NY, 1996.

[2] K. Ebcioğlu and E. Altman. **DAISY**: Dynamic Compilation for 100% Architectural Compatibility. In *Proc. of the 24th Annual International Symposium on Computer Architecture*, pages 26–37, Denver, CO, June 1997. ACM.

[3] K. Ebcioğlu. Some Design Ideas for a VLIW Architecture for Sequential-Natured Software. In M. Cosnard et al., editor, *Parallel Processing*, pages 3–21. North-Holland, 1988. (Proceedings of IFIP WG 10.3 Working Conference on Parallel Processing).

[4] G. M. Silberman and K. Ebcioğlu. An Architectural Framework for Migration from CISC to Higher Performance Platforms. In *Proc of the 1992 International Conference on Supercomputing*, pages 198–215, Washington, DC, July 1992. ACM Press.

[5] G. M. Silberman and K. Ebcioğlu. An Architectural Framework for Supporting Heterogeneous Instruction-Set Architectures. *IEEE Computer*, 26(6):39–56, June 1993.

[6] Sun Microsystems. The Java Hotspot Performance Engine Architecture. http://java.sun.com/products/hotspot/whitepaper.html, April 1999.

[7] K. Ebcioğlu, J. Fritts, S. Kosonocky, M. Gschwind, E. Altman, K. Kailas, and T. Bright. An eight-issue tree-VLIW processor for dynamic binary translation. In *Proc. of the 1998 International Conference on Computer Design (ICCD '98) – VLSI in Computers and Processors*, pages 488–495, Austin, TX, October 1998. IEEE Computer Society.

[8] A. Chernoff, M. Herdeg, R. Hookway, C. Reeve, N. Rubin, T. Tye, S. B. Yadavalli, and J. Yates. FX!32–A Profile-Directed Binary Translator. *IEEE Micro*, 18(2):56–64, March 1998.

[9] M. Rosenblum, S. Herrod, E. Witchel, and A. Gupta. Complete Computer Simulation: The SimOS Approach. *IEEE Parallel and Distributed Technology*, 3(4):34–43, Winter 1995.

[10] R. Nair and M. Hopkins. Exploiting Instruction Level Parallelism in Processors by Caching Scheduled Groups. In *Proc of the 24th Annual International Symposium on Computer Architecture*, pages 13–25, Denver, CO, June 1997. ACM.

[11] E. Rotenberg, Q. Jacobson, Y. Sazeides, and J. Smith. Trace Processors. In *Proc. of the 30th Annual International Symposium on Microarchitecture*, pages 138–148, Research Triangle Park, NC, December 1997. IEEE Computer Society.

Decoupling Recovery Mechanism for Data Speculation from Dynamic Instruction Scheduling Structure

Toshinori Sato

Toshiba Microelectronics Engineering Laboratory
580-1, Horikawa-Cho, Saiwai-Ku, Kawasaki 210-8520, Japan
toshinori.sato@toshiba.co.jp

Abstract. In this paper, we propose to decouple the recovery mechanism for data speculation from dynamic instruction scheduling structure. Instruction reissue mechanism for data speculation has a serious impact on processor performance. The effective capacity of instruction window is reduced since instructions dependent upon a speculated instruction must remain in instruction window until they are committed. The decoupling of the recovery and scheduling mechanisms solves the problem. A small instruction window schedules instructions and its entry is released immediately when an instruction is dispatched. A large instruction buffer is active only when a misspeculation occurs and is used to reissue instructions dependent upon the misspeculated instruction. Using a cycle-by-cycle simulator, we evaluated the proposal and found that the decoupling is useful.

1 Introduction

Recently, there are many studies which try to speculate data dependences in order to extract more instruction level parallelism (ILP). An outcome of an instruction is predicted by value predictors and the instruction and its dependent instructions can be dispatched simultaneously, thereby ILP is exploited aggressively. If a speculation is mispredicted, it is necessary to recover processor state. A straightforward implementation of the recovery mechanism is instruction squashing which is already used for branch prediction. The instruction squashing is not adequate for the data speculation, because it throws away execution results of instructions which are independent of the mispredicted instruction and because these instructions should be fetched again from instruction memory. This wastes useful computations. Moreover, since penalty caused by data dependence misprediction is quite larger than that caused by control dependence misprediction, the overhead including instruction squashing and re-fetching is very serious. Instruction reissue is one of the promising solutions for this problem and several studies of the instruction reissue are done[7, 8, 10]. However, the instruction reissue mechanism for data speculation has a serious impact on processor performance. The effective capacity of instruction window is reduced since instructions dependent upon a speculated instruction must remain in instruction window until

P. Amestoy et al. (Eds.): Euro-Par'99, LNCS 1685, pp. 1281–1290, 1999.
© Springer-Verlag Berlin Heidelberg 1999

they are committed. In this paper, we propose to decouple the recovery mechanism for data speculation from dynamic instruction scheduling structure. The decoupling solves the problem. A small instruction window schedules instructions and its entry is released immediately after an instruction is dispatched. A large instruction buffer is active only when a misspeculation occurs and is used to reissue instructions dependent upon the misspeculated instruction.

The organization of the rest of this paper is as follows. In Section 2, previously proposed related works are surveyed. Section 3 explains the decoupling of the recovery and scheduling mechanisms. Section 4 presents the evaluation methodology and Section 5 evaluates the proposed mechanism. Finally, our conclusions are presented in Section 6.

2   Related Work

Data speculation[4, 5] is a technique which executes instructions speculatively using predicted data values. Data dependences are speculatively resolved and thus ILP is increased. When a misspeculation occurs, it is necessary to recover processor state.

There are two mechanisms for the recovery action. One is instruction squashing and the other is instruction reissue. The instruction squashing flushes all instructions following a mispredicted instruction. It is easy to implement the instruction squashing since the same mechanism is already implemented for branch prediction. However, it is found that data speculation relying upon the instruction squashing sometimes diminishes processor performance due to large misprediction penalties[8, 10]. On the other hand, the instruction reissue re-executes only instructions dependent upon a misspeculated instruction. Those instructions are detected selectively and reissued inside instruction window. In order to realize the instruction reissue, the dependent instructions are forced to retain in instruction window. When the predicted instruction produces an actual value, the predicted value must be compared with the actual one. If they match, the prediction is correct and the dependent instructions release instruction window. If the prediction fails, the dependent instructions are invalidated and reissued. Lipasti et al.[5] proposed the concept of the instruction reissue but did not mention its implementation.

Tyson et al.[10] evaluated the usefulness of the instruction reissue. They proposed a renaming-based load value predictor and found that the instruction reissue proposed in [5] can improve processor performance even for the applications whose performance is degraded when the instruction squashing is used. However, the practical implementation of the instruction reissue was not discussed. Rotenberg et al.[7] investigated an instruction reissue scheme on Trace Processor architecture and found it is useful for dataflow speculation. However, they did not evaluate it on superscalar processors which are currently in mainstream.

We proposed a practical implementation of the instruction reissue[8]. Register update unit (RUU)[9] is extended to realize the instruction reissue. The RUU is very suitable for the instruction reissue, since it forces each instruction to retain until the instruction has been committed. Our instruction reissue mechanism serially detects and reissues all instructions dependent upon a misspeculated instruction, and thus a lot of comparators working in parallel to detect the dependent instructions can be removed.

While the instruction reissue considerably reduces misspeculation penalty by selectively re-executing instructions, it has a negative impact on processor performance[3]. As explained above, in order to reissue only dependent instructions, those instructions must remain in instruction window until they are committed. This reduces utilization of instruction window and thus its effective capacity becomes small. The instruction window size affects processor performance significantly. Therefore, it is necessary to increase the instruction window size in order to maintain its effective capacity. However, the instruction window is one of the dominant of processor cycle time and its size severely affects its speed[6].

Recently, Akkary et al.[1] proposed two level instruction window structure. There are two instruction windows. One is small and its entry is released immediately when an instruction allocated to the entry is dispatched. The other is large and works as the backup of the small one when a misspeculation occurs. The backup process works just like cache refill process. Dependent instructions which should be reissued are injected to the small instruction window from the large one. Since most of the time the small instruction window is utilized, the negative impact on processor performance is reduced. They evaluated the two level instruction window only on a multithreading processor.

3   Decoupled Instruction Window

In this section, we propose a decoupled instruction window. As explained above, implementing the instruction reissue mechanism causes the following problem. Since every instruction must remain in instruction window until it is committed, the effective capacity of instruction window is reduced. In order to keep processor performance, it is necessary to increase the instruction window size. However, it is difficult to maintain processor cycle time for large instruction window. Thus, the wakeup and select logic of the large window should be pipelined, degrading processor performance[6].

For the purpose of solving the problem, we propose to decouple the recovery mechanism for data speculation from dynamic instruction scheduling structure. Fig.1 depicts a processor utilizing the decoupled instruction window. The decoupled window consists of a small instruction window for the scheduling and a large instruction buffer for the instruction reissue. After an instruction is fetched and decoded, it enters both the instruction scheduling window and the instruction buffer. When the instruction is dispatched to a functional unit, it leaves the instruction window and release its entry but remains in the instruction buffer. When it is committed, it leaves the instruction buffer and releases its entry. In the case that either the window or the buffer is full, issuing instruction into the decoupled window stalls.

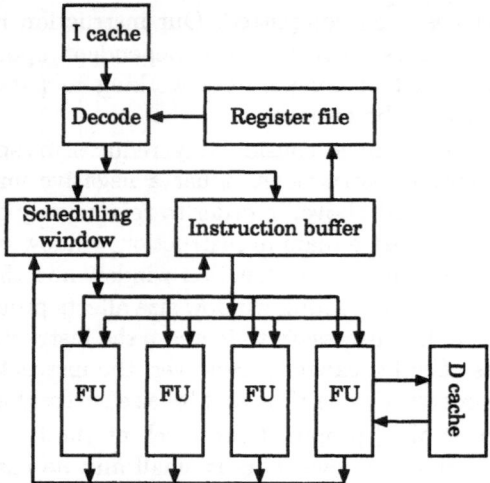

Fig. 1. Processor diagram

The small instruction window works for dynamic instruction scheduling. Thus, most of the times each instruction is dispatched from the small window. Since instructions are aggressively deallocated from the scheduling window when they are dispatched, the problem reducing its effective capacity is solved. In addition, its small size does not have serious impact on processor cycle time. However, it is impossible to reissue misspeculated instructions inside the scheduling window. The large instruction buffer works as the backup for the small window and performs the instruction reissue. Each instruction remains in the buffer until it is committed. When a misspeculation occurs, reissued instructions are obtained from the instruction buffer. In order to aggressively speculate instructions, the instruction buffer should be very large. Since it is difficult to access such a large buffer in one cycle, the wakeup and select logic of the buffer is pipelined to maintain high processor cycle time. It is expected that the pipelining does not degrade processor performance, since the instruction buffer is active only when misspeculations are detected.

The decision from which structures an instruction is dispatched to a functional unit works straightforwardly. When an instruction is misspeculated, its dependent instructions which have already dispatched and thus should be reissued are obtained from the instruction buffer only. They have already left the scheduling window. The dependent instructions which have not been dispatched remain in both the scheduling window and the buffer. Thus, both structures can dispatch the instructions. However, note that the instruction buffer is pipelined, and the instructions are obtained from the scheduling window earlier than from the buffer. And then, the same instruction provided by the buffer can be canceled to dispatch. Therefore, those instructions are scheduled by the window and release their entries.

From the explanations, it can be observed that the decoupled instruction window solves the problem caused by the instruction reissue and will maintain processor performance.

4   Evaluation Methodology

In this section, we describe the evaluation methodology by explaining a processor model and benchmark programs.

4.1   Experimental Model

An execution-driven simulator which models wrong path execution caused by misspeculations is used for this study. We implemented the simulator using the SimpleScalar tool set[2]. The SimpleScalar/PISA instruction set architecture (ISA) is based on the MIPS ISA.

The simulator models a realistic 8-way out-of-order execution superscalar processor based on RUU[9] which has 128 entries. We model two RUUs. One is a one cycle latency RUU and the other is a two cycle latency pipelined RUU. Each functional unit can execute all operations and has a latency of 1 cycle except for multiplication (4 cycles) and division (12 cycles). A 4-port, non-blocking, 128KB, 32B block, 2-way set-associative L1 data cache is used for data supply. It has a load latency of 1 cycle after the data address is calculated and a miss latency of 6 cycles. It has a backup of a 8MB, 64B block, direct-mapped L2 cache which has a miss latency of 18 cycles for the first word plus 2 cycles for each additional word. No memory operation can execute beyond a store whose data address is unknown. A 128KB, 32B block, 2-way set-associative L1 instruction cache is used for instruction supply and also has the backup of the L2 cache which is shared with data supply.

For control prediction, a 1K-entry 4-way set associative branch target buffer, a 4K-entry gshare-type branch predictor, and an 8-entry return address stack are used. The branch predictor is updated at instruction commit-time.

Value predictor used in this study is a 4096 entry direct-mapped stride predictor[4]. Fig.2 depicts the predictor. It is indexed by instruction address and each entry has a tag field (tag), a previous value field (prev_value), a stride field (stride), and a confidence field (conf). The tag field is used for distinguishing individual instructions from each other. The previous value field holds the last value generated by the instruction. The stride field keeps a difference of the last two values. The predicted value is predicted as the sum of the previous value and the stride. The confidence field is a 2-bit saturated counter and decides if the speculation using the predicted value should be initiated. When the count is larger than two, the speculation is initiated. In this paper, we call the processor model using the value predictor value prediction model.

The decoupled instruction window consists of a 64 entry centralized reservation station and a 128 entry RUU performing the instruction reissue[8]. The wakeup and select logic of the RUU is pipelined and have two cycle latency.

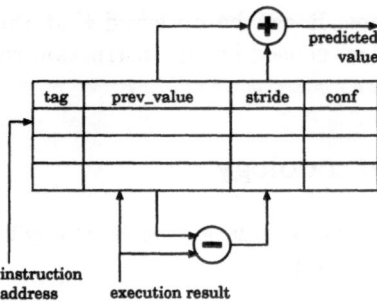

Fig. 2. Stride value predictor

4.2 Workloads

The SPEC95 CINT benchmark suite is used for this study. The **test** input files which are provided by SPEC are used. All programs are compiled by GNU GCC (version 2.6.3) with the optimization option, -03. Each program is executed to completion or for the first 100 million instructions. We count only committed instructions.

5   Simulation Results

In this section, we present simulation results. First, we evaluate the value predictor. Second, we show how the pipelined RUU affects processor performance. And last, the effect of the decoupling is presented.

5.1   Effect of Value Prediction

Fig.3(i) shows the characteristics of the value prediction. Each bar is divided into three parts. The bottom part (black) indicates the percentage of the instruction whose data value is correctly predicted. The middle part (white) indicates the percentage that is mispredicted. And the top part (gray) indicates the percentage that is not predicted. It is observed that 36.8% of dynamic instructions on average are correctly predicted.

Fig.3(ii) presents performance improvement when the value predictor is utilized. We use committed instructions per cycle (IPC) as a metric for evaluating processor performance. For each program, performance of the value prediction model is normalized by that of the baseline model. The processor performance is improved by 5.38% on average. This is a modest improvement but study of value predictors is beyond the scope of this paper.

Table 1 shows the utilization of the instruction window. The columns between 2 and 5 are for the baseline model and the columns between 6 and 9 are for the value prediction model. For each group of four columns, the first two columns indicate the numbers of instructions remaining in the instruction window. They

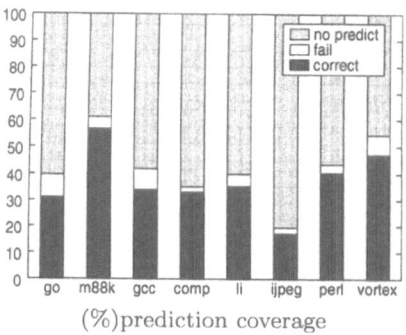

(%)prediction coverage

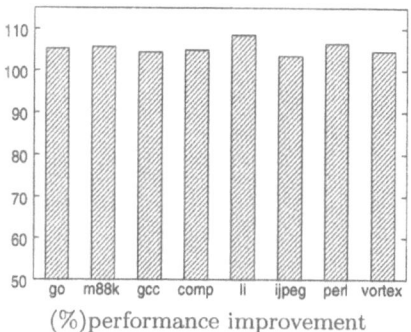

(%)performance improvement

Fig. 3. Characteristics of value prediction

Table 1. Instruction window utilization

program	w/o data prediction				with data prediction			
	total		waiting		total		waiting	
	avg	max	avg	max	avg	max	avg	max
099.go	21.6	128	11.0	126	21.5	128	9.6	126
124.m88ksim	18.3	128	8.3	121	16.2	128	5.6	127
126.gcc	21.7	128	11.9	127	21.6	128	11.1	127
129.compress	87.6	128	68.4	121	88.5	128	72.0	121
130.li	17.9	128	8.2	121	17.7	128	6.8	121
132.ijpeg	71.8	128	37.5	121	71.7	128	36.5	121
134.perl	23.8	128	10.9	121	22.0	128	9.1	121
147.vortex	27.8	128	15.0	121	24.8	128	12.8	121

include instructions dispatched to functional units. Note that the instruction window evaluated in this paper is the RUU, which forces each instruction to retain until the instruction has been committed. The remaining two columns show the numbers of instructions waiting for dispatch. They are candidates for dynamic scheduling. Furthermore, for each group of two columns, the left column presents the average number and the right one presents the maximum number. It is found that when data prediction is used, the average number of waiting instructions decreases except for 129.compress while that of total instructions changes little. This means that the effective capacity of the instruction window is reduced even if it is compared with the RUU which also holds dispatched instructions. Thus it is necessary to increase the window size to maintain the scheduling ability when data prediction is used. It is also found that the average utilization is significantly small compared with the total capacity. For most of the programs, it is less than 1/3.

5.2 Performance Impact of Pipelined Instruction Window

Fig.4 shows how pipelining of RUU affects processor performance. The wakeup and select logic is pipelined by two cycles. For each group of two bars, the left

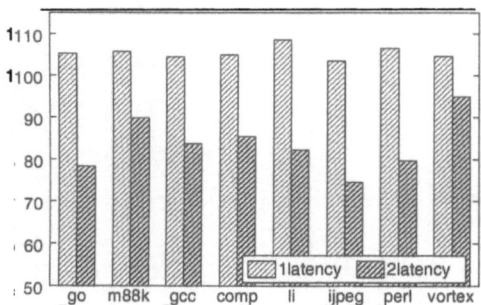

Fig. 4. (%)Performance degradation by pipelined wakeup and select logic

bar indicates the performance of the value prediction model and the right bar indicates that of the processor whose RUU is pipelined. Note that the processor model whose RUU is pipelined also utilizes data prediction. As can be easily seen, the performance degradation is very severe. The improvement gained by data prediction is lost, and furthermore the processor performance is lower than that of the baseline model. For example, comparing the baseline model, performance of 132.ijpeg is reduced by approximately 25%. In order to fill the performance gap, it is necessary to improve cycle time by 25%. But this is almost impossible to realize. On the other hand, it is an another answer to increase the instruction window size but it will cause further degradation of processor cycle time.

5.3   Effect of Instruction Window Decoupling

In this section, we investigate the effect of the decoupling. Since the average utilization of the RUU is less than 1/3 as we have seen in Section 5.1, we decide that the size of the scheduling window is half of the instruction buffer.

Fig.5 presents the performance contribution of decoupled instruction window. For each group of three bars, the first bar (see from left to right) indicates the performance of the value prediction model. The second bar indicates that of the processor whose RUU is pipelined. And the last bar indicates that of the processor which utilizes the decoupled window. It is observed that the performance degradation due to the pipelined instruction window is compensated by its decoupling. For all cases, processor performance is improved over the baseline model. Compared with the value prediction model, performance of the decoupled model is lower. However, it is expected that the clock cycle speed of the decoupling model is faster than that of the value prediction model. Therefore, the processor performance of the decoupling model will be comparable to that of the value prediction model.

Table 2 shows the utilization of the scheduling window and the instruction buffer. The columns 2 and 3 are for the scheduling window and the columns 4 and 5 are for the instruction buffer. For each group of two columns, layout and subject matter of Table 2 are the same as for Table 1. Since the scheduling window size of

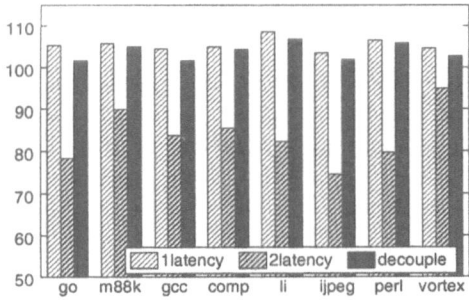

Fig. 5. (%)Performance contribution of decoupled instruction window

Table 2. Instruction buffer/window utilization

program	buffer		window	
	avg	max	avg	max
099.go	23.9	128	10.0	64
124.m88ksim	17.4	128	5.6	64
126.gcc	21.3	128	9.4	64
129.compress	57.8	128	40.7	64
130.li	18.7	128	7.0	64
132.ijpeg	68.1	128	32.4	64
134.perl	22.8	128	9.1	64
147.vortex	24.3	128	10.8	64

the decoupling model is smaller than the RUU size of the baseline and the value prediction models, it is natural that the window utilization of the decoupling model is lower than those of the other models. However, it is important to note that the utilization is improved over the value prediction model for 099.go and 130.li.

6   Concluding Remarks

In this paper, we have proposed to decouple instruction window. The decoupled window consists of a small instruction window for the dynamic instruction scheduling and a large instruction buffer for the instruction reissue. Since the large buffer is active only when misspeculations are detected, it is expected that the instruction buffer size is increased while the processor cycle time is maintained by pipelining the wakeup and select logic of the buffer. Using a cycle-by-cycle simulator, we have evaluated the decoupled instruction window. While processor performance is significantly diminished by pipelining instruction window, it is compensated by decoupling instruction window. In addition, since it is expected that the clock cycle speed of the decoupling model is fast, there is a possibility to increase processor performance.

Acknowledgment

The author is grateful to Dr. Mitsuo Saito, Dr. Haruyuki Tago and Dr. Shigeru Tanaka for their continuous encouragements. He also thanks anonymous reviewers whose comments and suggestions helped to improve the quality of this paper.

References

[1] Akkary,H., Driscoll,M.A.: A dynamic multithreading processor. 31st Int'l Symp. on Microarchitecture (1998)

[2] Burger,D., Austin,T.M.: The SimpleScalar tool set, version 2.0. ACM SIGARCH Computer Architecture News, 25(3) (1997)

[3] Chrysos,G.Z., Emer,J.S.: Memory dependence prediction using store sets. 25th Int'l Symp. on Computer Architecture (1998)

[4] Gabbay,F.: Speculative execution based on value prediction. Technical Report #1080, Dept. of Electrical Eng., Technion (1996)

[5] Lipasti,M.H., Wilkerson,C.B., Shen,J.P.: Value locality and load value prediction. Int'l Conf. on Architectural Support for Programming Languages and Operation Systems VII (1996)

[6] Palacharla,S., Jouppi,N.P., Smith,J.E: Complexity-effective superscalar processors. 24th Int'l Symp. on Computer Architecture (1997)

[7] Rotenberg,E., Jacobson,Q., Sazeidas,Y., Smith,J.: Trace processors. 30th Int'l Symp. on Microarchitecture (1997)

[8] Sato,T.: Data dependence speculation using data address prediction and its enhancement with instruction reissue. Euromicro'98 Conf., Workshop on Digital System Design (1998).

[9] Sohi,G.S.: Instruction issue logic for high-performance, interruptible, multiple functional unit, pipelined computers. IEEE Trans. Comput., 39(3) (1990)

[10] Tyson,G. Austin,T.M.: Improving the accuracy and performance of memory communication through renaming. 30th Int'l Symp. on Microarchitecture (1997)

Implementation of Hybrid Context Based Value Predictors Using Value Sequence Classification[†]

Luis Piñuel, Rafael A. Moreno, and Francisco Tirado

Departamento de Arquitectura de Computadores y Automática,
Universidad Complutense, Madrid 28040, Spain
{lpinuel,rmoreno,ptirado}@dacya.ucm.es

Abstract. Value prediction is as yet a very novel technique, whose efficiency has still to be proved. To take advantage of this emerging technique in the short term it is essential to design accurate and low cost value predictors. This work presents a new approach of implementing hybrid predictors that allows the maximum sharing of information between predictors. We show that the new hybrid predictor outperforms not only the accuracy of the others predictors, but also their hardware utilization.

1 Introduction

Over the last few years, many of the efforts in microarchitecture research have been focussed on attempting to counteract the program dependencies and thus improve the extractable instruction-level parallelism (ILP). Several techniques for eliminating control and data dependencies have been proposed, based mainly on branch and data prediction and speculative execution. While branch prediction [1], can be considered today as a widely accepted classical technique, value prediction, however, is as yet a very novel technique, whose efficacy has still to be proved.

The value prediction technique, like branch prediction, allows temporal violation of the program constraints without affecting its semantics. However, if the predictions are not accurate, the use of speculation is detrimental for processor performance. It is obvious that the more history the predictor captures the more accuracy it has. However, the use of large history tables is not realistic for the next generations of processors. Consequently, to take advantage of the value prediction in the short term it is essential to design accurate and low cost value predictors.

In this paper, we first study the relationship between accuracy and cost for different value predictors. Subsequently, we propose a new methodology for designing cost-effective hybrid data value predictors that allows us to combine the benefits of different predictors, without increasing the amount of hardware. The rest of the paper is organized as follows. Section 2 summarizes the previous work on value prediction. Section 3 describes the experimental framework. Section 4 presents a comparative analysis of actual value predictors. Section 5 introduces the low-cost hybrid predictor proposed by the authors. Finally, section 6 presents the conclusions.

[†]This work has been supported by the Spanish Ministry of Education under grant TIC96-1071.

P. Amestoy et al. (Eds.): Euro-Par'99, LNCS 1685, pp. 1291-1295, 1999.
© Springer-Verlag Berlin Heidelberg 1999

2 Previous Work

Recent work on value prediction has shown that data are highly predictable [2][3] and several models for value prediction have been proposed. Most of the value predictors proposed in the literature can be classified as one of the following types. *Last-value* predictors (LVP), which make a prediction based on the last outcome of the same static instruction, and can correctly predict constant sequences of data. [2][3]. *Stride predictors* (SP), which make a prediction based on the last outcome plus a constant stride, and can correctly predict arithmetic sequences of data [3][4]. *Context based predictors* (CBP), which learn the values that follow a particular context and make a prediction based on the last values generated by the same instruction. They can also correctly predict repetitive sequences of data [5]. *Hybrid predictors* (HP), which combine some of the previous predictors and include a selection mechanism, either hardware [6] or software [4]. Several different implementations of each kind of predictor have been proposed. If we compare the different proposals and their respective results, we can draw the following conclusion. The more complex the predictor, the higher the percentage of correct predictions is, but also the more expensive the hardware needed for its implementation.

3 Experimental Framework

The simulators used in this work are derived from the SimpleScalar 3.0 tool set [7]. To perform our experimental study, we have collected results for the integer Spec95 benchmarks‡. Table 1 shows a description of the benchmarks§.

Our architecture is derived from the architecture used by the SimpleScalar Out-of-Order simulator [7]. The modifications made to the baseline architecture are aimed at including the hardware mechanisms involved in value speculation: prediction, validation, and re-execution recovery. The architectural parameters employed in our simulations are the following: superscalar width of 8, instruction window of 32, a two level cache hierarchy (64KB/4MB) and perfect branch prediction, value predictor updating, and memory disambiguation. We must clarify at this point that only single precision instructions that write into general-purpose registers are eligible for value prediction.

Benchmark	Input Set	# Inst	# Skipped
compress95	30000 e 2231	95 M	0
cc1	gcc.i	203 M	50M
go	9 9	132 M	0
ijpeg	specmun.ppm	553 M	200M
m88ksim	train input	120 M	0
perl	train input	40 M	0
li	scrabbl.in	183 M	40M
vortex	train input	2520 M	1000M

Predictor	Cost Formulae
Last Value	$E * (N_{TAG} + N_{VALUE} + N_{CONFIDENCE})$
Stride	$E * (N_{TAG} + 2N_{VALUE} + N_{STATE})$
Context	$E_{VHT} * (3N_{VALUE}) + E_{VPT} (N_{VALUE} + N_{CONFIDENCE})$
Hybrid	$E_{LVT} * (N_{TAG} + N_{VALUE} + N_{STATE}) + E_{ST} * (N_{VALUE}) + E_{CVHT} * (2N_{VALUE}) + E_{CVPT} (N_{VALUE} + N_{CONFIDENCE})$

Table. 1. Description of benchmarks.   **Table. 2.** Predictor Cost (Kbits).

---

‡ The programs were compiled with the *gcc* compiler using the optimisation level -O3.
§ Due to time constraints, the simulations are limited to 100 million instructions.

4 Comparative Analysis of Value Predictors

The different value predictors under analysis are particular implementations of last value, stride and context predictors. The last value and stride predictors are implemented by means of direct mapped and tagged tables as in [3]. The context based predictor is derived from the work of Sazeides *et al.* [5] and it uses a 2-level table. The first level table called the *value history table* (VHT) is direct mapped and non-tagged and is responsible for storing the order-3 context of the instructions. The second level table, called the *value prediction table* (VPT) is indexed by a hash function, which uses context information from the VHT and provides the prediction that follows a particular context. The confidence mechanism employed in both the LVP and CBP is a 2-bit saturating counter whereas for the SP we have employed the mechanism presented in [3].

Previous work on data value prediction focuses on improving the number of predictor hits, but none of them analyze the relation between accuracy and cost. The hardware cost of the predictors depends mainly on the prediction tables. Consequently, in the following comparison, the size of the tables is considered an approximation of the predictor cost. Table 2 shows the formulae used for calculating the cost of the predictors: E represents the number of table entries and N represents the number of bits of an entry field. Figure 1 summarizes the experimental results obtained by last value, stride and the different configurations of the context predictor (it also includes the results of the hybrid predictor discussed later). From these results, we can conclude that the percentage of hits of the last value and stride predictors reach an asymptotic level at around 512 Kbits and only minor improvements are obtained above this point. In contrast, for the context-based predictor, the percentage of hits rises steeply between 512 Kbits and 1536 Kbits, and grows gradually beyond this cost. On the other hand, we observed that the context predictor misses much more than the last value and stride predictors, independently of the predictor cost.

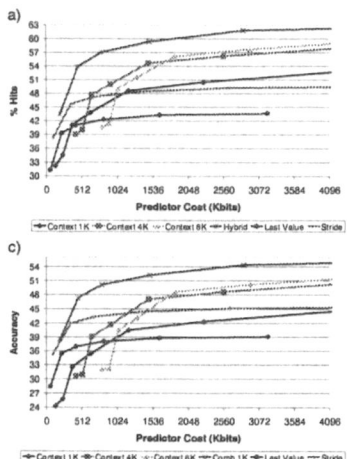
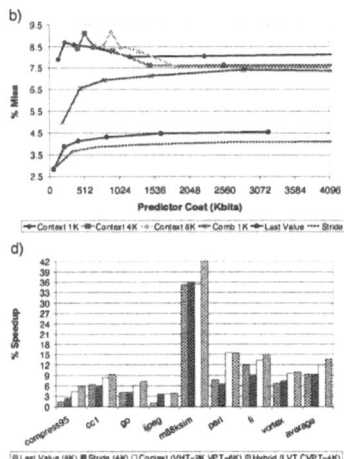

Fig. 1. Experimental results: % hits (a), % misses (b), accuracy value (c), and % speedup (d).

Predictor misses involve a re-execution of dependent instructions and thus entail a time penalty. Consequently, for comparing the efficiency of different prediction schemes we must give the same weight to their hits and misses. To undertake this, we propose using the difference between hits and misses as a measure of the predictor accuracy. From this figure, we can now conclude that the stride predictor is the most accurate predictor for any cost under 1024 Kbits, whereas above this cost, the context predictor with a VHT of 4K or 8K becomes more accurate. Our results confirm that context based predictors are necessary for achieving the highest prediction accuracy. However, a great portion of correct predictions is also captured by the last value stride predictor. Assuming that context based predictors are more expensive this suggests that a hybrid predictor might be useful for enabling high prediction accuracy at a lower cost.

5 Low-Cost Hybrid Predictor

Traditional hybrid prediction schemes make use of separated predictors and a selection mechanism [1]. Usually, the predictor with a higher confidence value is selected for producing the prediction. This mechanism is adequate for mixing non-overlapped sets of predictable instructions. However, when the sets of predictable instructions share a significant amount of elements, this approach leads to a waste of hardware. Why use duplicated hardware for predicting the same instructions? The selection mechanism that we suggest is based on the classification of value sequences. For each sequence type, it selects the predictor with lowest cost from all those that are able to predict this sequence type. In this way, the hybrid predictor should use the last value predictor for the constant sequences, the stride predictor for stride-only sequences (either repetitive or non-repetitive) and the context based predictor for the non-stride sequences. The finite state machine of the new hybrid predictor is described in figure 2 and its basic structure is shown in figure 3. The utilization of the hardware resources is optimal because the predictors share information instead of duplicating it, e.g., the last outcome of the instruction is needed by all the predictors but is only stored once. This property along with the selection mechanism constitute the principal differences in relation to other hybrid predictor [6].

Figure 1 (a-c) also shows the experimental results obtained by our hybrid predictor using the following configuration: LVT= CVPT= {1K, 4K, 8K, 16K , 32K} and ST= CVHT=1K. We can observe that the hybrid predictor not only improves on the percentage of hits of all the other predictors, but it also exhibits a lower percentage of misses than the context-based predictors. Furthermore, the hybrid predictor outperforms the accuracy of the other predictors; as a result, we can conclude that it is the most efficient one (highest accuracy for the same cost). Figure 1 (d) shows the speedup results obtained by realistic predictors ($\approx$ 512Kbits). Notice that the hybrid predictor is not only more efficient but it also produces a higher performance.

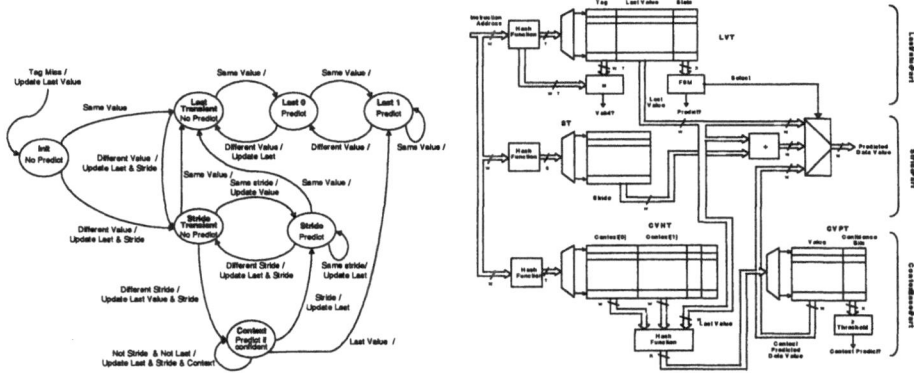

Fig. 2. Selection Mechanism.   **Fig.3.** Hybrid Predictor Block Diagram.

6 Conclusions

From the comparison of traditional predictors, we can conclude that, although the highest accuracy is obtained by the context-based predictor, in the case of small-scale prediction tables the stride predictor is the most accurate one. We have also shown that a hybrid predictor might be useful for enabling high prediction accuracy at a lower cost. The proposed hybrid predictor combines the hardware of the last value and stride predictors and the accuracy of the context-based predictor. It is clear from its description that its hardware utilization is optimal. Furthermore, as we have recently seen it also produces higher performance. In consequence, we can conclude that the new approach not only provides the best way of implementing a hybrid context-based predictor, but it also represents the best solution to implementing highly accurate and low cost data value predictors.

References

1. S. McFarling, "Combining Branch Predictors." Technical Report TN-36, Digital Equipment Corp., June 1993.
2. M.H. Lipasti and J.P. Shen, "Exceeding the Dataflow Limit via Value Prediction," Proc. of the 29th Int. Symp. on Microarchitecture, pp. 226-237, Dec. 1996.
3. K. Wang and M. Franklin, "Highly Accurate Data Value Prediction using Hybrid Predictors," Proc. of 30th Int. Symp. on Microarchitecture, pp. 281-290, Dec. 1997.
4. Y. Sazeides, J.E. Smith. "Implementations of Context Based Value Predictors". Technical Report #ECE-TR-97-8, University of Wisconsin-Madison, 1998.
5. F. Gabbay and A. Mendelson, "Can Program Profiling Support Value Prediction? ", Proc. of the 30th Int. Symp. on Microarchitecture, pp. 270-280, Dec. 1997.
6. B. Rychlik, J. Faisty, B. Krug, J.P. Shen, "Efficacy and Performance Impact of Value Prediction", Proc. of Int. Conf. on Parallel Architectures and Compilation Techinques, 1998.
7. D. Burger and T.M. Austin. "The SimpleScalar Tool Set, Version 2.0". Technical Report CS#1342, University of Wisconsin-Madison, 1997.

Heterogeneous Clustered Processors: Organisation and Design

Francesco Pessolano

South Bank University, 103 Borough Road, London SE1 0AA, UK
pessolf@sbu.ac.uk

Abstract. The rapid development of electronic technology and new trends in the software market are forcing micro-architects to explore new solutions to improve performance and reliability. In this paper, we describe the idea of *Heterogeneous Clustered Processors* as a viable alternative to well known proposals for future billion-transistor processors. The architectural model is described together with a test core and its preliminary performance evaluation.

Introduction

Recent multimedia applications, portable computing and sub-micron technologies are evolving in directions that may change the shape of computing. The computer architecture community faces the challenge of how to combine standard processor design with multimedia and mobile functionality within the same package [1,2], where delays are dominated by wires and no longer by gates [3]. The possibility of integrating one billion transistors makes it possible, but an efficient architectural approach for such systems on-a-chip has to be found.

In this paper, we propose a solution based on the idea of *unbalanced dependence chains*, called *Heterogeneous Clustered Processors (HCP)*. This solution is a possible way to improve performance and functionality for system-on-a-chip and standard general-purpose processor. An asynchronous general-purpose processor core, called *GRAVITY*, is being investigating in order to evaluate design complexity and effectiveness of the proposed approach. *GRAVITY* has also been used to evaluate both a synchronous and an asynchronous implementation [4]; preliminary results justify preference for the latter approach.

The paper is organized as follows: section 2 describes the ideas behind *Heterogeneous Clustered Processors*. The basic hardware properties are introduced in Section 3, some considerations about its instruction set features in Section 4. The *GRAVITY* core is analyzed in section 5, while preliminary performance results are analyzed in Section 6. Some conclusions are drawn in section 7.

P. Amestoy et al. (Eds.): Euro-Par'99, LNCS 1685, pp. 1296-1300, 1999.
© Springer-Verlag Berlin Heidelberg 1999

Heterogeneous Clustered Processors

Traditional approaches to processor design share the idea of code formed by a sequence of instruction blocks, where each instruction specifies one operation and its operands. The boundaries of these blocks are represented by control instructions (e.g. branches). Instructions are arranged to allow parallel execution with dependence as constraint; all instructions share global register files. Recently, code has started to change shape [5]: instructions are replaced by one or more operations sharing a predication field. This field is used to specify a condition that must be met if the operations are to be executed. Such approaches allow breaking of the classic block boundaries, but they introduce a new global resource: the conditional register file for predication. It represents a performance bottleneck just like the register file. Some attempts to split the register file have been made, but no attempt has been made to split the conditional register file among different clusters.

We propose *Heterogeneous Clustered Processors (HCP)*, which are based on the idea of *unbalanced dependence chains*. Dependence chains are defined as a sequence of predicated operations, which directly uses only a given subset (*cluster*) of the

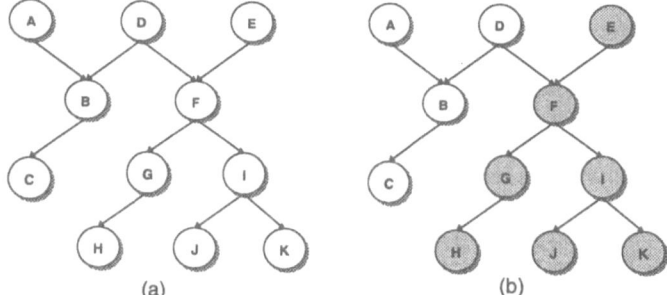

(a)   (b)

Fig.1: Data Dependence Graphs and manipulation into sub-graphs

available hardware resources and accesses the rest through external communication. If all chains may have different dimension, we define them *unbalanced dependence chains*. In order to clarify this concept, let us consider a simple example. In compiling our code, after optimizations for standard predicated machines, we obtain a set of *Data Dependence Graphs (DDGs)* like the one of Fig.1a [6]. Such graphs are used to represent data flow and dependencies. Each DDG is typically used in order to find independent instructions so to create horizontal microcode. In contrast, we statically handle the DDGs in order to generate dependence chains. The compiler partitions each DDG into sub-graphs, whose number must not exceed the number of clusters of local hardware resources. For example, we could get the two sub-graphs of Fig.1b: each sub-graph is then reduces into an unbalanced dependence chains. A single cluster executes each chain. Arcs going from one chain to another represent an inter-cluster communication. No knowledge of communication latency is needed either at scheduling time or at run-rime.

HCP systems also present another fundamental property: *heterogeneity*. Clusters have different functionality and compile-time cluster scheduling is needed so to allow instruction format reuse. The same binary opcode has different meaning depending on the receiving cluster; this is intrinsically incompatible with a run-time cluster scheduling strategy. Load balancing, which is needed in previous approaches, is no longer a constraint: in fact, we cannot impose load balancing without code dimension explosion when clusters have their own ISA-subset. This implies an efficient way of modifying the processor state so to allow unbalanced dependence chains to be correctly executed. Besides, clusters' decoupling state requires dedicated solutions such as decoupled predication and synchronization (i.e. predication and synchronization can involve down to a single cluster).

Hardware Model of Heterogeneous Clustered Processor

The minimum hardware model for HCP cores is described in Fig.2. Instructions are independently issued and executed by the clusters (*CI*). Register files, conditional register files, dispatch queues and functional units are distributed across multiple clusters. Each cluster is assigned a distinct architectural and conditional register file, which can be directly accessed. Whenever a cluster needs data from another one, an communication action takes place through explicit SEND/REC instructions or use of special registers. Each cluster independently accesses the *instruction pool*, in order to issue a new instruction. If no instruction is available, the unit is stalled. The instruction pool is the most critical part of the system: it has to fetch a block of fixed dimension from the cache and allow decoupled issue. We can think of it as a configurable routing network, whose configuration the software modifies in order to guarantee correct execution of unbalanced chains. Clusters also independently access data cache, which is therefore a shared resource. It may present two major problems: data coherency and indeterminate behavior (only in an asynchronous implementation). We have adopted a decoupled solution with each load/store operation split into two simpler ones (address calculation and data transmission). The external memory is also a shared resource: the Memory Management Unit may receive concurrent requests from both caches and this leads to indeterminate behavior of the MMU. In order to reduce penalties we adopted the major-vote strategy used in high-reliability system.

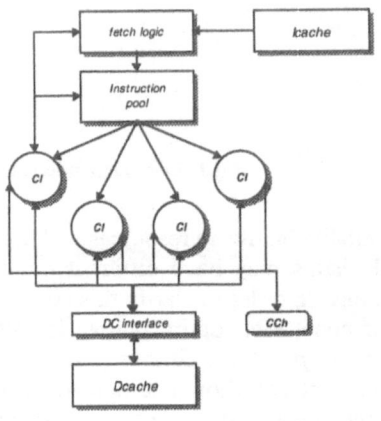

Fig.2: Basic hardware model of HCP cores

Considerations on HCP Instruction Set

The full instruction set of heterogeneous clustered processors depends on the targeted applications, since each cluster executes different instruction subsets sharing the same binary encoding. Nevertheless, these subsets must be provided with a basic common model and shared instructions. The former element is used for inter-cluster communication, predication and synchronization: each cluster includes a register subset used for fast and code efficient inter-cluster communication, which can be used as alternative to the standard bus-based communication scheme. At the same time, each instruction subset has the same predication tag that allows *local* (i.e. visible at cluster level) and *global predication/synchronization* (i.e. visible at system level). The latter element (i.e. shared instructions) is needed to correctly tune the system: these instructions are called *directives* and provide the hardware with proper run-time information to allow execution of unbalanced loads, decoupled branching and memory accesses, resource configuration and allocation.

Implementation of the GRAVITY Processor Core

In order to analyze design complexity and performance bottlenecks of HCP, we have designed an asynchronous test processor core, called *GRAVITY*. The organization of the processor core is shown in Fig.3.

The core is provided with two Arithmetic-Logic Clusters (*ALC*), one Memory and Branch Cluster (*MBC*) and one Floating-Point Cluster (*FPC*). It supports 32-bit integer arithmetic and 32/64-bit IEEE-754 floating-point arithmetic without traps. Clusters are provided with private and global predicative register files, local-global synchronization and communication mechanisms. We are currently performing detailed HDL simulations, where the HDL description has been refined to a level easily translated into circuits. All the delays used in the HDL description have been extracted from spice simulation based on 0.5um MOSIS CMOS process technology with over-estimated capacitance loads. Simulation results show that the core has a minimum IP-dependent cycle time varying from 3.6ns to 4ns executing up to 4 instructions per cycle.

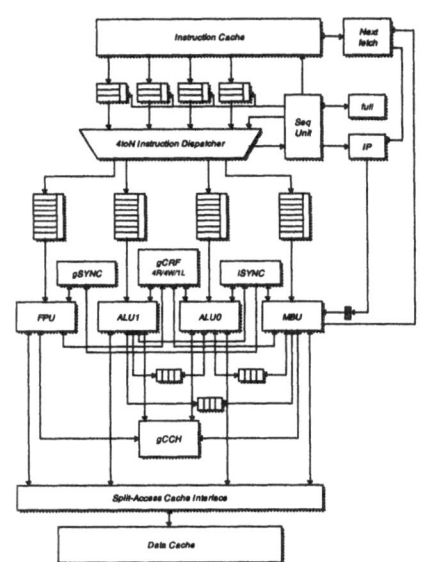

Fig.3: The GRAVITY HCP core structure

We have studied the effect on this value of an increase in the number of clusters, which is a measure of the instruction pool complexity. In case of 16 clusters, the peak throughput of each is reduced to 86%

whilst the peak processor one is increased up to 345% respect the *GRAVITY* values. Penalties associated with the directive execution (e.g. branching or run-time information passing) range from 1ns to 11ns, depending on the executed directive and efficiency of the decoupled branching mechanism (i.e. code structure). A comparison with a synchronous version has shown that the latter introduce larger penalties associated with directive execution and, therefore, is less efficient. We are currently evaluating performance through benchmark extracts from integer, floating-point and mixed applications. We are comparing the obtained results with the ones of other static scheduled machines, which have been designed with the same techniques and constrain. First results are in the domain of integer code density, relevant to embedded applications: the proposed solution shows better code density respect all other parallel models whilst having a similar parallelism. Full performance evaluation through benchmarks is still on progress.

Conclusions

The rapid development of electronics technology and new trends in the software market are forcing micro-architects to explore new solutions to improve performance and reliability. Reducing the transistor dimensions causes problems in clock distribution and synchronization between opposite ends of a chip. In this paper, we have described a solution based on the idea of *unbalanced dependence chains*, called *Heterogeneous Clustered Processors (HCP)*. The proposed solution is a viable alternative to well known proposals for future billion-transistor processors. We are also evaluating the use of dependence chains in a multithreading context, where a whole chain would replace the single instruction in building the execution sequences. Further work is needed to understand how effective this extension could be and to complete the performance evaluation.

References

[1] D Burger, D Goodman, "Billion-transistor architectures – Guest editors' introduction", IEEE Computer, Vol.30, No.9, Sept. 1997
[2] TM Conte, et al. "Challenges to combining general-purpose and multimedia processors into one package", *IEEE Computer*, pp.33-37, Dec. 1997
[3] D Matzke, "Will physical scalability sabotage performance gains?", *IEEE Computer*, vol.30, No.9, pp.37-39, 1997
[4] S Hauck, "Asynchronous design methodologies: an overview", *Proc. of the IEEE*, Vol.83, No.1, Jan. 1995
[5] WNW Hwo, RE Hank, et al., "Compiler technology for future microprocessors", *IEEE Proceedings*, Vol.83, No.12, pp.1625-1640, 1995
[6] M Wolfe, *High-performance compilers for parallel computing*. Addison-Wesley, 1996

An Architecture Framework for Introducing Predicated Execution into Embedded Microprocessors

Daniel A. Connors[1], Jean-Michel Puiatti[2], David I. August[1],
Kevin M. Crozier[1], and Wen-mei W. Hwu[1]

[1] Department of Electrical and Computer Engineering
The Coordinated Science Laboratory
University of Illinois, Urbana, Illinois (USA) 61801
{dconnors, august, crozier, hwu}@crhc.uiuc.edu
[2] Logic Systems Laboratory (DI-LSL)
Swiss Federal Institute of Technology Lausanne, CH-1015 Lausanne, Switzerland
puiatti@lslsun.epfl.ch

Abstract. Growing demand for high performance in embedded systems is creating new opportunities for Instruction-Level Parallelism (ILP) techniques that are traditionally used in high performance systems. Predicated execution, an important ILP technique, can be used to improve branch handling, reduce frequently mispredicted branches, and expose multiple execution paths to hardware resources. However, there is a major tradeoff in the design of the instruction set, the addition of a predicate operand for all instructions. We propose a new architecture framework for introducing predicated execution to embedded designs. Experimental results show a 10% performance improvement and a code reduction of 25% over a traditionally predicated architecture.

1 Introduction

Growing demand for high performance in embedded computing systems is creating new opportunities for Instruction-Level Parallelism (ILP) techniques that are traditionally used in high performance systems. In several ways, the needs of embedded computing differ from those of more traditional general purpose systems. Embedded systems have more stringent constraints on cost [6] that lead to the design of limited-sized instruction caches and physical memories. The limited nature of these instruction memory resources is more pronounced by current technological developments in embedded systems. In order to meet numerous requirements for embedded system features and functionality, compilers and high-level languages have been employed in ways to manage the size and complexity of system design. Unfortunately, this can increase program code size over previously used traditional methods of hand-coding programs. Thus, compiler technology not only has a large effect in enhancing the performance of these processors, but also in affecting the instruction memory utilization and

P. Amestoy et al. (Eds.): Euro-Par'99, LNCS 1685, pp. 1301–1311, 1999.
© Springer-Verlag Berlin Heidelberg 1999

code size. Although classic code optimizations decrease the number of executed instructions, superscalar optimization, inline expansion, loop unrolling, and superblock formation [5] often increase the execution performance at the cost of increasing the overall code size.

Current embedded processor research and development illustrate different strategies for dealing with memory size issues. The traditional strategy for reducing code size focuses on reducing the instruction encoding size in the design of the Instruction Set Architecture (ISA). For example, embedded processor such as M-Core, Thumb, Tiny RISC, and future ARM-10 designs [8] use this technique. A comparison of the cache performance in [1] shows that denser instruction sets have significantly lower miss rates for small caches, but that advantage disappears for larger caches.

Predicated execution is an emerging ILP technique that requires several changes to existing ISAs, which can affect program code size. Predicated execution is the conditional execution of an instruction based on the value of a Boolean source operand, referred to as the predicate of the instruction. Predicated execution, can be used to improve branch handling, reduce frequently mispredicted branches, and expose multiple execution paths to hardware resources. Although the performance benefits of full predicated execution are high, there is a major tradeoff in the design of the instruction set, namely the addition of a predicate source operand for all instructions.

We propose a new framework for introducing predication into embedded processors. The first contribution of this paper is to present the effect of predicated execution on program code size. This study concludes that the growth in binary size when adding predicate source operands to every instruction is wasteful since only 28% of instructions are predicated after applying aggressive predicate formation [2] and optimization techniques. The second, and the more important contribution of this paper is to propose a new instruction issue mechanism that supports predicated and non-predicated versions of instructions. Experimental results of the new predicated architecture model achieves an average 10% performance improvement over a traditionally predicated architecture and reduces the memory requirements of highly optimized code by 25%

2   Background and Motivation

2.1   Predication Background

Predicated execution allows conditional execution of instructions based upon a computed condition and may be supported by several different architectural models [3]. Each model must support a method of expressing the condition and a method for the condition to affect instruction execution. Full predication supports this using new instruction set and microarchitecture extensions.

The full predication model consists of four components: a predicate register file for holding 1-bit predicate values, an additional source operand for each instruction to specify a predicate for instruction execution, a conditional-execution stage to nullify instructions, and a set of predicate defining instructions for generating conditions. The values in the predicate register file are associated with each instruction through the use of an additional source operand, or predicate operand. This operand specifies which predicate register will determine whether the instruction should execute. A predicate register value of 1, or true, indicates the instruction is executed; a value of 0, or false, indicates the instruction is suppressed. An unconditional instruction is designated by a predicate register that is always true. The architectural support for predicated execution can be found in the HPL PlayDoh Architecture Specification [4].

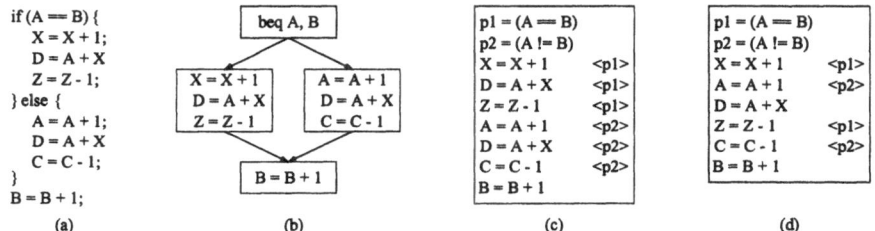

Fig. 1. A simple if-then-else C code construct (a), unpredicated code (b), predicated code (c), and optimized predicated code (d).

Predication support allows the compiler to use an *if-conversion* algorithm to convert conditional branches into predicate defining instructions, and instructions along alternative paths of each branch into predicated instructions [9]. Figure 1 demonstrates the limitation of the traditional control flow graph when applied to predicated code. A simple if-then-else construct is shown in Figure 1(a). The code generated for this segment without predication is shown in Figure 1(b). Here the control flow graph clearly shows that one and only one side of the if-statement may execute. The predicated code control flow graph is shown in Figure 1(c). In this case all the code falls into one basic block because there is no possibility of branching until the end of the set of instructions.

The most notable modification of predication to the instruction set encoding format is the addition of the predicate operand source for every instruction. The predicate operand increases the instruction size and has significant effects on overall program code size. One model [10] proposes a new set of predicate guarding instructions that would reduce the drawback of existing methods of specifying predicated execution through the use of predicate mask-setting instructions. Although the mechanism is useful in reducing the predicate operand overhead, the general mechanism constrains several aspects of predicated execution and dramatically alters the instruction issue logic of microprocessors.

2.2   Motivation

Predication Performance. There are two major benefits associated with applying if-conversion. First, a compiler can eliminate problematic branches from the program. In doing so, all the associated overhead with these branches is removed, including misprediction penalties, penalties for redirecting sequential instruction fetch, and branch resource contention. Second, predication facilitates increased ILP and speedup by allowing separate control flow paths to be simultaneously executed. Figure 2(a) shows the performance when predication support is provided by the architecture and a capable compiler is employed to take advantage of it. The 6-issue processor simulated utilizes profile-based static branch prediction, a 4-cycle misprediction penalty, and a perfect memory system. Across all benchmarks predication yields an average performance gain of 34%.

Predicate Utilization. Although the performance of predicated execution is significant, it is at the cost of adding a predicate source operand on every instruction. In full predication model, all instructions have a predicate source operand, even those which are not conditionally executed. Figure 2(b) illustrates the percentage of static instructions with conditional predicates relative to the overall number of instructions. The percentage of conditional instructions averages around 40% of the total instructions, meaning that a large portion of instructions do not require a predicate operand. Since the percentage of unconditional instructions is significant, the unnecessary increase in instruction format size can dramatically impact embedded system designs.

Predicated Instruction Cost. Figure 3 shows the code size expansion attributed to the predicate operand for three distinct models on the same predicated benchmarks. First, *Zero Size* shows the code size for predication when the predicate representation has zero cost. Next, *Predicate Only* shows the effect when the instruction size growth of the predicate operand is attributed to only the conditional instructions. Finally, *Full Size* shows the size of the operand added to every static instruction as designed in an architecture supporting full predication. All of the predicated code sizes are compared to a base architec-

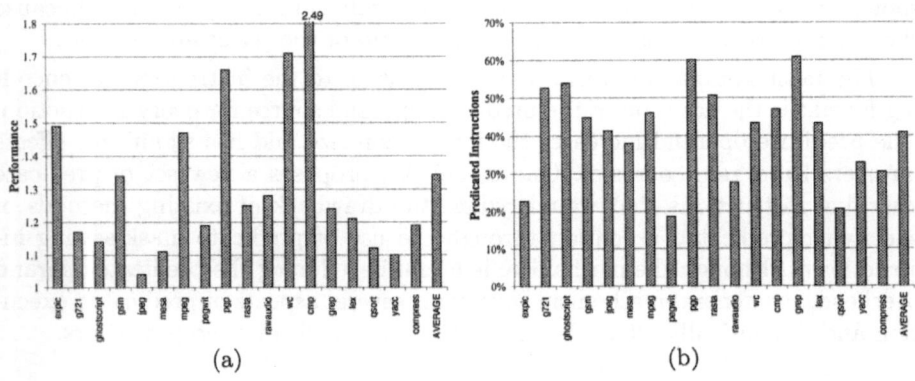

Fig. 2. Predicated execution performance (a) and utilization(b).

ture without predication support. Note that compilation for predication alone has some effect on code size. The size of the predicate operand was evaluated assuming a 24-bit base instruction format and a 5-bit predicate operand field.

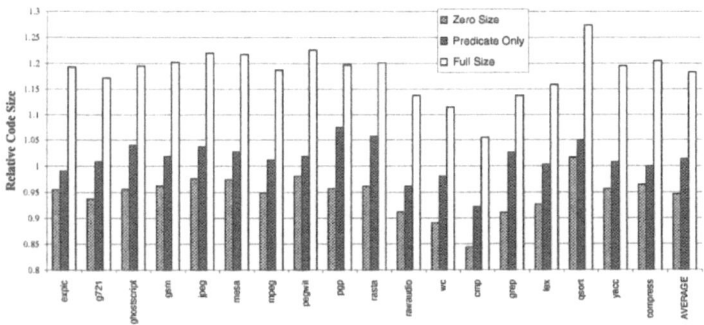

Fig. 3. Code expansion considering predication source operand.

Figure 3 indicates that predicated execution increases program code size by an average of 23%, and often as high as 30%. The results of the *Zero Size* model of code size evaluation indicate that for a large number of programs, predication effectively has fewer instructions and reduced code size. An interesting pattern is observed in Figure 3 for *Predicate Only* instructions. As a general rule, the code size for this model is significantly smaller than the *Full Size* code size, and averages near the base non-predicated code size. The difference between predicated and non-predicated results occur because predication has a fundamental ability to remove numerous control instructions and because compiler support of predicated execution can perform optimizations that allow the code to share instructions that are on different execution conditions. For example, in Figure 1(d) the instruction $D = A + X$ does not require a predicate operand since the compiler guarantees that it unconditionally executes in the block.

3   Prefix-Based Predication

There are several methods of adding predicated and non-predicated versions of instructions to an ISA design. This section details the addition of predication to a 24-bit instruction word for embedded processors. The proposed method is extendible to other instruction format constraints.

3.1   Architecture Model

Prefix-based predication uses opcode prefixing to add sufficient instruction bits to indicate a predicate operand exists for instructions which the compiler has designated to conditionally execute. As illustrated in the previous section, a significant amount of code size can be saved when only the predicated instructions incur the predicate operand overhead. Figure 4 illustrates the base 24-bit instruction format that includes an operation code, a destination register index, and two source operands (potentially register indexes or immediate data).

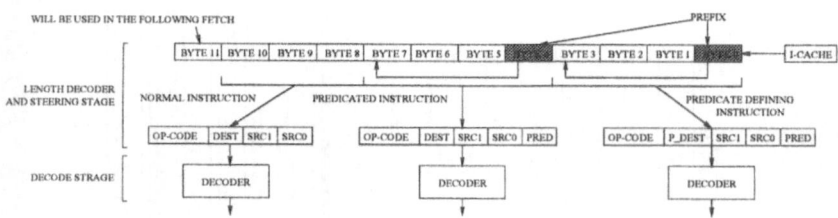

Fig. 4. Prefix-based predication decoding of normal and predicated instructions.

Figure 4 illustrates how a prefix opcode of the 24-bit instruction can designate that an additional 1-byte contains supplementary instruction information follows. The complete 32-bit instruction can then be decoded into a 26-bit instruction with a 6-bit operation code, a 5-bit predicate register index, a destination register index, and 2 source operands. The prefix opcode is then discarded. In this example architecture, the 5-bit predicate index can be used to access a 32-entry predicate register file. New predicate defining instructions for expressing predicate conditions are also added using the prefixng mechanism.

3.2   Microarchitecture Support

The primary microarchitecture component affecting prefix-based predication is the instruction decode methodology. Most prefix architecture designs integrate an additional instruction decode stage in the original pipeline design. In this model, the first stage is used to determine instruction lengths (prefix detection) and steer the instructions to the second stage where the actual instruction decoding is performed. Figure 4 illustrates this process. The multiple pipelined decode method is successful for several reasons. First, the design places the focus on resources other than instruction memory. A second reason for using an additional decode stage is that the number of branch instructions executed in a predicated architecture is significantly reduced, resulting in the number of mispredictions also being reduced. This limits the negative effect of adding more pipeline stages before branch resolution has on the misprediction penalty. The branch prediction accuracy for predicated architectures is about 7% higher than branch prediction for traditional architectures.

4   Techniques for Reducing Predicated Code Size

The compilation techniques utilized in this paper to exploit predicated execution are based on an abstract structure called a *hyperblock* [2]. Several additions were made to the existing hyperblock framework of our experimental compiler. First, infrequently executed blocks with instruction merging opportunities that might normally be excluded from hyperblocks are included. This includes new methods of forming hyperblocks using basic blocks with zero or low execution frequency. New predicate optimization routines were also developed.

The new optimization routines extend the techniques of predicate promotion and predicate merging. Predicate promotion refers to speculation performed by changing an instruction's predicate to a predicate whose expression subsumes that of the original predicate [2]. Promotion may result in the instruction being unconditionally executed, reducing the number of predicated instructions. Predicate merging allows identical instructions on intersecting predicate conditions to be combined. Merging thereby removes one instruction copy, and promotes the remaining instruction. These optimizations will be detailed in a future work, and are not presented here due to space limitations.

4.1   Predication Code Size and Execution Characteristics

The compiler's ability to affect program code size and the percentage of predicated instructions for several benchmarks is now examined. The selected benchmarks are the MediaBench suite [7] and UNIX utilities. Figure 5 indicates that predication reduces the total number of instructions for traditionally optimized code by 6.3%. A significant portion of the instructions eliminated were control instructions, which were reduced by 13%, where control instructions include predicate defining instructions and any traditional branch instructions. Other characteristics include a 7% reduction in the number of dynamically executed

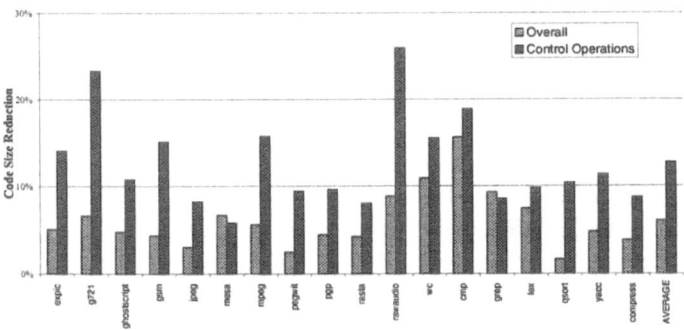

Fig. 5. Code reductions due to predicated execution.

Benchmark	Merging %	Code Merging Reduction %	Promotion %	Hyperblock Static Pred %	Pred-Optimization Static Pred %
expic	6.03	1.45	37.59	22.68	14.92
g721	1.60	0.85	43.31	52.64	29.77
ghostscript	0.26	1.01	32.68	41.31	24.79
gsm	3.21	1.80	51.44	44.78	23.28
jpeg	29.97	1.96	39.38	53.88	34.78
mesa	8.62	3.55	37.49	37.96	22.07
mpeg	5.05	2.40	34.52	46.03	26.13
pegwit	3.72	0.75	15.08	18.60	14.95
pgp	2.48	1.52	14.12	60.12	49.32
rasta	3.38	1.75	17.48	50.60	39.94
rawaudio	2.17	0.61	26.09	27.71	21.21
wc	16.92	7.91	10.77	43.33	40.29
cmp	22.12	11.57	16.81	46.89	37.04
grep	10.52	6.89	14.43	60.85	50.74
lex	11.87	5.44	14.97	43.15	32.75
qsort	8.00	1.61	48.00	20.49	11.90
yacc	7.30	2.48	26.26	32.83	21.11
compress	5.85	1.82	30.70	30.37	14.60
average	8.28	3.08	28.40	40.79	28.31

Table 1. Instruction merging and predicate promotion characteristics.

instructions in general code, and a 31% reduction in the number of dynamically executed instructions in code with superscalar optimization.

Table 1 summarizes the amount of predicate optimization that the compiler is able to perform on the hyperblocks. For the instruction merging category, the percentage of static predicated instructions averages 8% that can be merged. The additional code reduction attributed to merging is shown in the next column. The percentage of predicated instructions that are promoted to unconditionally executed instructions is shown in the next column. These numbers indicate that an average 28% of the originally predicated instructions may be promoted. The final two columns include the percentage of static predicated instructions relative to total program instructions for the original and predicate-optimized hyperblocks. The most important result of Table 1 is that only 28% of the static instructions remain predicated after predicate optimization.

5   Experimental Evaluation

5.1   Methodology

The IMPACT compiler and emulation-driven simulator were enhanced to support the proposed architecture framework. The base architecture modeled uses a 5 stage pipeline that can issue in-order 6 operations per cycle (up to the limit of the available functional units: four integer ALU's, two memory ports, two floating point ALU's, and one branch unit). The instruction latencies used match the HP PA-7100 microprocessor (integer operations have 1-cycle latency, and load operations have 2-cycle latency). The processor contains 32 integer and 32 floating point registers. To support prefix-based predication, 32 predicate registers and an additional decoding stage were modeled. The memory system simulated was either perfect or used a 2K, 4K, or 8K sized direct-mapped instruction caches and a 8K direct mapped, blocking data cache; both with 64-byte blocks and a miss penalty of 12 cycles. A static branch prediction strategy was employed.

5.2   Results and Analysis

Figure 6 shows the results of varying the instruction cache size for the non-predicated and prefix-based predicated architectures. Substantial performance improvement is established at small cache sizes; however, for larger increases in instruction cache size, the relative performance improvements of the base architecture are larger, and the relative performance saturates. This indicates that the base model is more dependent on instruction cache resources than the prefix-based predicated architecture. The results of cache simulations show that prefix-based predication has an average 7% higher hit rate for 2K instruction caches and 2.5% for 8K caches compared to the non-predicated model. Experiments also indicate that prefix-based predication has an average 10% higher speedup over traditional predicated architectures for small instruction cache models.

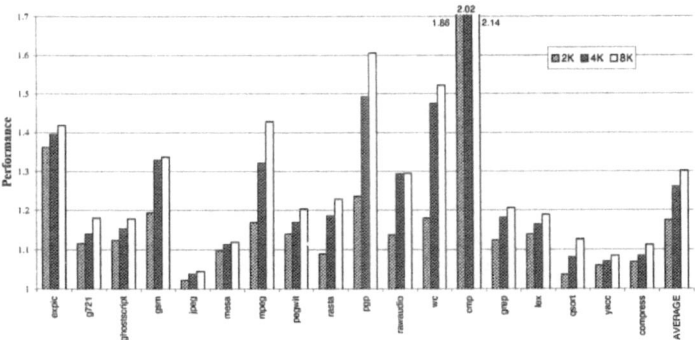

Fig. 6. Performance of varying instruction cache size for prefix-based predicated architecture relative to non-predicated architecture.

The relative performance of superscalar (superblock formation, loop unrolling) optimization for prefix-based predicated and non-predicated architectures is an average 63% better than general levels of optimization for the simulation of a perfect memory system. For superscalar optimization, the average speedup of the predicated architecture is only 12% more than the non-predicated architecture. The performance of the superscalar optimization indicates that the performance gains of predicated execution do not greatly exceed the non-predicated version. However, the corresponding code size of the predicated code for high performance code is significantly reduced. Figure 7 shows the code expansion of the superscalar optimization for the non-predicated, full-predicated,

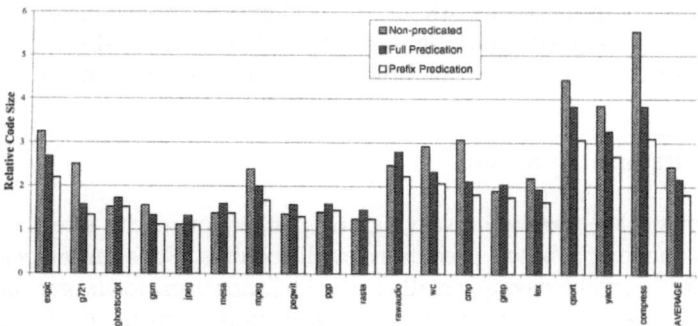

Fig. 7. Code expansion of superscalar relative to traditional optimization.

and prefix-based predicated architectures. Clearly the 12% performance improvement is substantial since the improvement requires a significantly smaller code size. The full predicated architecture has an average 11% smaller code size and the prefix-based predicated architecture has an average 25% smaller size.

6   Conclusions

The prefix-based predicated execution architecture framework proposed has the potential to significantly enhance the effectiveness of introducing predicated execution into embedded microprocessors. For regions of non-predicated code, the prefix-based method offers better code density characteristics than traditional models of predication support. For predicated regions, the prefix-based method offers performance improvement over an architecture without predication support. We illustrate that an optimizing compiler can enhance the prefix-based predication model by performing aggressive instruction merging and predicate promotion to reduce the number of predicated instructions by 30%. Overall, prefix-based predication achieves 12% performance improvement for code created with superscalar optimization and reduces code size by 25%.

References

[1] J. Davidson and R. Vaughan. The effect of instruction set complexity on program size and memory performance. In *Proceedings of the 2nd International Conference on Architectural Support for Programming Languages and Operating Systems*, pages 60–64, October 1987.

[2] S. A. Mahlke et al. Effective compiler support for predicated execution using the hyperblock. In *Proceedings of the 25th International Symposium on Microarchitecture*, pages 45–54, December 1992.

[3] S. A. Mahlke et al. A comparison of full and partial predicated execution support for ILP processors. In *Proceedings of the 22th International Symposium on Computer Architecture*, pages 138–150, June 1995.

[4] V. Kathail et al. HPL PlayDoh architecture specification: Version 1.0. Technical Report HPL-93-80, Hewlett-Packard Laboratories, Palo Alto, CA, February 1994.

[5] W. W. Hwu et al. The Superblock: An effective technique for VLIW and super-scalar compilation. *The Journal of Supercomputing*, 7(1):229–248, January 1993.

[6] R. Gonzalez and M. Horowitz. Energy dissipation in general purpose micropro-cessors. *IEEE Journal of Solid-State Circuits*, 31:1277–1284, 1996.

[7] C. Lee and W. Mangione-Smith. Mediabench. In *Proceedings of the 30th Annual International Symposium on Microarchitecture*, pages 330–335, December 1997.

[8] MicroDesign Resources. *Embedded Processor Forum*, San Jose, CA, October 1998.

[9] J. C. Park and M. S. Schlansker. On predicated execution. Technical Report HPL-91-58, Hewlett Packard Laboratories, Palo Alto, CA, May 1991.

[10] D. N. Pnevmatikatos and G. S. Sohi. Guarded execution and branch prediction in dynamic ILP processors. In *Proceedings of the 21st International Symposium on Computer Architecture*, pages 120–129, April 1994.

Multi-stage Cascaded Prediction

Karel Driesen and Urs Hölzle
Department of Computer Science
University of California
Santa Barbara, CA 93106
{karel,urs}@cs.ucsb.edu
http://www.cs.ucsb.edu/oocsb

Abstract. Two-level predictors deliver highly accurate conditional branch prediction, indirect branch target prediction and value prediction. Accurate prediction enables speculative execution of instructions, a technique that increases instruction level parallelism. Unfortunately, the accuracy of a two-level predictor is limited by the cost of the predictor table that stores associations between history patterns and target predictions. Two-stage cascaded prediction, a recently proposed hybrid prediction architecture, uses pattern filtering to reduce the cost of this table while preserving prediction accuracy. In this study we generalize two-stage prediction to multi-stage prediction. We first determine the limit of accuracy on an indirect branch trace using a multi-stage predictor with an unlimited hardware budget. We then investigate practical cascaded predictors with limited tables and a small number of stages. Compared to two-level prediction, multi-stage cascaded prediction delivers superior prediction accuracy for any given total table entry budget we considered. In particular, a 512-entry three-stage cascaded predictor reaches 92% accuracy, reducing table size by a factor of four compared to a two-level predictor. At 1.5K entries, a three-stage predictor reaches 94% accuracy, the hit rate of a hypothetical two-level predictor with an unlimited, fully associative predictor table. These results indicate that highly accurate indirect branch target prediction is now well within the capability of current hardware technology.

1   Introduction

Prediction of branch targets and load values side-steps control and data-flow dependencies, enabling speculative execution of instructions and increasing instruction level parallelism [HP95]. The importance of accurate prediction increases as the processor-memory gap grows, processor pipelines become deeper, and superscalar issue increases. Processor technology has followed these trends in the past and probably will do so in the foreseeable future [P+97].

Currently, highly accurate conditional branch prediction is achieved by variations of the two-level predictor architecture proposed by Yeh and Patt [YP91]. Two-level prediction increases prediction accuracy by correlating a history of taken/non-taken bits of recently executed branches with the direction of the current branch. Lipasti et. al. successfully applied two-level prediction to load value prediction [CHP97].

Two-level predictors also prove highly effective for the prediction of indirect branch targets [CHP97]. Indirect branches, which transfer control to an address recently loaded into a register, are hard to predict accurately. Unlike conditional branches, they can have more than two targets, so that prediction requires a full 32-bit or 64-bit address rather than just a "taken" or "not taken" bit. Furthermore, their behavior is often directly determined by data loaded from memory, such as in virtual function calls in C++ and Java. Since the popularity of these languages continues to grow, we expect that processors will execute indirect branches more frequently in the future. Even today,

P. Amestoy et al. (Eds.): Euro-Par'99, LNCS 1685, pp. 1312-1321, 1999.
© Springer-Verlag Berlin Heidelberg 1999

indirect branch misses can cause significant overhead. Without two-level prediction (using a simple branch target buffer or BTB), the overhead of virtual function calls in C++ programs is as high as 29% [DH96]. Similarly, Chang, Hao, and Patt show that for the SPECint95 programs perl and gcc the indirect branch overhead is approximately 15% and 8% [CHP97].

In this study we evaluate predictor architectures for indirect branches. We believe that our conclusions will also apply to conditional branch prediction and value prediction, for reasons discussed in section 6. The accuracy of two-level predictors depends on the size of the predictor table that stores associations between history patterns and predicted targets. Longer histories lead to higher prediction accuracy but also increase the number of different history patterns. This effect causes capacity misses, which deteriorate prediction accuracy even for large tables. In a recent study [DH98b], we reduced the required size of the two-level predictor by placing a small BTB in front of it. Many branches are perfectly predicted by this cheap first stage, so that their associated history patterns can be filtered out; only history patterns of branches that are hard to predict enter the second stage.

Here we investigate the accuracy of a natural generalization of this two-stage cascaded predictor by allowing any type of predictor in the first stage and any number of stages. First, we use the maximum number of stages, and unlimited, fully associative tables for each stage, to determine the limit of prediction accuracy reachable by this architecture. Secondly, we test two and three stage predictors for a wide range of table sizes, in order to study cost reduction for practical predictors.

This paper makes the following contributions:

- It demonstrates, for the first time, that idealized indirect branch predictors can exceed 95% prediction accuracy.
- It describes and evaluates a practical (4K) indirect branch predictor that achieves nearly 95% accuracy on average for our set of large C and C++ applications.
- It explains why cascaded predictors work so well, and quantifies the dramatic reduction in predictor table working set size achieved by cascading. This analysis suggests that conditional branch predictors or load value predictors could benefit from cascaded prediction as well.

The rest of this paper is organized as follows: in Section 2 we discuss the benchmark suite used, and Section 3 briefly reviews two-level and cascaded predictor architectures. Section 4 compares the accuracy of two-level and ideal cascaded predictors for various numbers of stages/path lengths, and Section 5 presents results for realistic predictors. Section 6 discusses related work, and we conclude in Section 7.

2   Benchmarks

We minimize misprediction rate using a reduced instruction trace consisting of indirect branch addresses and targets, and simulate only the indirect branch predictor. This allows us to explore two to three orders of magnitude more predictor configurations than a full cycle-level simulation would allow. Reductions in misprediction rate should lead to corresponding reductions in branch misprediction overhead (as demonstrated in [CHP97]).

Our main benchmark suite consists of large object-oriented C++ applications ranging from 8,000 to over 75,000 non-blank lines of C++ code each, and *beta*, a compiler for the Beta programming language [MMN93], written in Beta. We also measured the

Table 1. Benchmarks and commonly shown averages (arithmetic means)

Name	Description	Style	K lines of code	K # of indirect branches	instr. / indirect	virtual%	switch%	indirect%	1 target%	2 targets%	> 2 targets%	active branches 99%	active branches 100%
idl	IDL compiler[a]	OO	14	1,884	47	93.2	3.2	3.6	97.1	0.1	2.8	70	543
jhm	JHM[b] 6-12M	OO	15	6,000	47	93.6	1.2	5.2	58.7	1.4	39.9	34	155
self	Self-93 VM: 5-6M	OO	77	1,000	56	76.0	4.4	19.6	40.1	31.6	28.3	848	1855
xlisp	SPEC95	C	5	6,000	69	0.0	0.1	99.9	38.9	9.0	52.1	4	13
troff	GNU groff 1.09	OO	19	1,111	90	73.7	12.5	13.8	41.9	13.6	44.5	61	161
lcom	HDL[c] compiler	OO	14	1,738	97	63.2	36.8	0.0	33.5	54.0	12.5	87	328
AVG-100: instr/ind < 100			24	2,955	68	66.6	9.7	23.7	51.7	18.3	30.0	184	509
perl	SPEC95	C	21	300	113	0.0	31.7	68.3	41.2	0.0	58.8	7	24
porky	scalar optimizer[d]	OO	23	5,393	138	70.6	23.8	5.6	15.6	8.1	76.3	89	285
ixx	IDL parser[e]	OO	11	212	139	46.5	52.2	1.3	37.1	6.4	56.5	91	203
edg	C++ front end	C	114	549	149	0.0	62.4	37.6	7.9	29.6	62.5	186	350
eqn	equation typesetter	OO	8	296	159	33.8	66.2	0.0	4.2	37.8	58.0	58	114
gcc	SPEC95	C	131	865	176	0.0	31.5	68.5	0.8	1.7	97.5	95	166
beta	BETA compiler	OO	73	1,006	188	0.0	2.3	97.7	18.7	28.1	53.2	135	376
AVG-200: 100 < instr/ind < 200			55	1,232	152	21.6	38.6	39.9	17.9	16.0	66.1	94	217
AVG: instr/indirect < 200			40	2,027	113	42.4	25.3	32.4	33.5	17.0	49.5	136	352
AVG-OO: OO, instr/ind < 200			28	2,071	107	61.2	22.5	16.3	38.5	20.1	41.3	164	447
AVG-C: C, instr/ind < 200			68	1,928	127	0.0	31.4	68.6	22.2	10.1	67.7	73	138
m88ksim	SPEC95	C	12	300	1.8K	0.0	46.2	53.8	2.9	10.3	86.8	5	17
vortex	SPEC95	C	45	3,000	3.5K	0.0	30.7	69.3	23.1	16.9	60.0	10	37
ijpeg	SPEC95	C	17	33	5.8K	0.0	97.8	2.2	96.7	3.2	0.1	7	60
go	SPEC95	C	29	550	56K	0.0	99.0	1.0	0.2	0.0	99.8	5	14
AVG-infreq: instr/indirect > 200			26	971	17K	0.0	68.4	31.6	30.7	7.6	61.7	7	32

[a] SunSoft version 1.3
[b] Java High-level Class Modifier
[c] hardware description language compiler
[d] SUIF 1.0
[e] Fresco X11R6 library

SPECint95 benchmark suite with the exception of compress which executes only 590 branches during a complete run. Together, the benchmarks represent over 500,000 non-comment source lines[1].

For each benchmark, Table 1 lists the number of indirect branches executed, the number of instructions executed per indirect branch, and the source of the indirect branches (switch statements, virtual function calls, or indirect function calls). It also shows the percentage of indirect branches that during the entire run jump to one, two, and more targets, as well as the number of branch sites responsible for 99% and 100% of the branch executions. For example, only 5 different branch sites are responsible for 99% of the dynamic indirect branches in go. Four of the SPEC benchmarks execute more than 1,000 instructions per indirect branch. Since the impact of branch prediction will

---

[1] See technical report for compilation details [DH99].

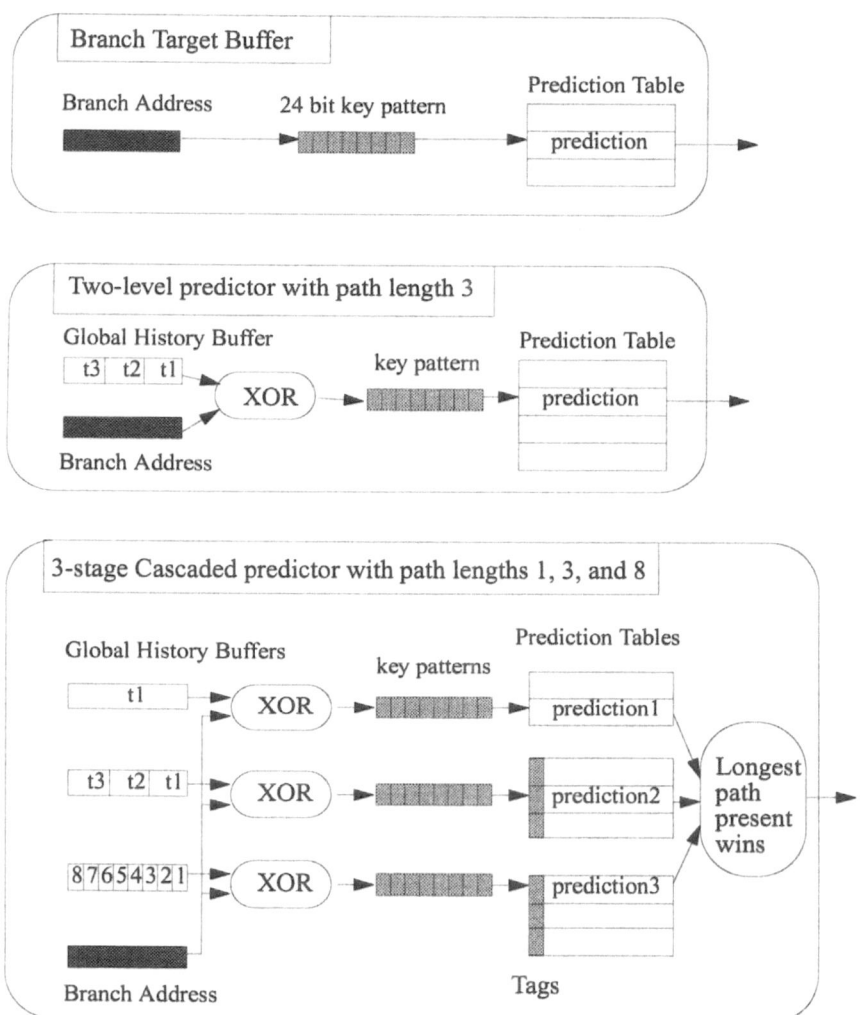

Figure 1. Representative examples of a branch target buffer, two-level predictor and cascaded predictor. A staged predictor looks the same as a cascaded predictor, but has a different update rule (every stage is updated, where a cascaded predictor prevents insertion of new patterns in later stages if an earlier stage predicts a branch correctly).

be very low for the latter four benchmarks, we exclude them when optimizing predictor accuracy (by minimizing the AVG misprediction rate).

3   Predictor architectures

Figure 1 shows representative examples of the predictor architectures tested in this study. The simplest architecture is a *branch target buffer* (BTB). A selection of bits from the branch address serves as a key pattern into a predictor table, which stores the last target observed for this branch. We use tagged tables to distinguish table misses (pattern is absent) from prediction misses (pattern is there, but the stored target is wrong). For the unlimited, fully associative tables in Section 4, the tag consists of the complete key pattern. For the 4-way associative, limited tables employed in Section 5, part of the 24-bit pattern is used as a table index, and the rest is stored as a tag, indicating which of the 4 entries in the associativity set has a prediction for the pattern, if any. A predictor table closely resembles a 4-way associative cache [HP95]. Note that only the second and later stages of the cascaded predictor need tags in order to function correctly. However, for easy comparison we use identical, tagged tables in all schemes.

A *two-level* predictor extends the BTB scheme by taking bits from the last p branch targets preceding the execution of the current branch and xor-ing these bits with the branch address. The parameter p is the *path length* of the two-level predictor. For longer path lengths, fewer bits are extracted from each target in order to fit into the 24-bit key pattern[1].

A *cascaded* predictor consists of several stages, each containing a two-level predictor with its own history buffer and predictor table. Successive stages use increasing path lengths (in [Dri99], we demonstrate the inferior accuracy of decreasing path lengths). The use of separate tables allows all stages to predict in parallel. In a final step, the predictor chooses the prediction from the last stage that did not encounter a table miss. This ensures that its target prediction is based on the longest available path history.

A cascaded predictor saves table space by using a *leaky filter* update rule: a new history pattern enters a long path length stage only if none of the shorter stages predicted the branch correctly. This rule prevents easily predicted branches from occupying table space in an expensive, long path length stage. For example, branches with only a single target are perfectly predicted by a BTB, after the initial compulsory miss, so they do not need a long history pattern (this is a substantial portion of all indirect branches, as shown in Table 1). As a result, the longer path length stage encounters fewer capacity misses, improving overall prediction accuracy.

We also measure the accuracy of a cascaded predictor without filtering. We call this a *staged* predictor. Staged predictor improve prediction accuracy compared to a two-level predictor because they reduce cold start misses. Longer path length two-level predictors are more accurate than short path length predictors, but they need a longer time to reach that potential since they store more patterns per branch. In a staged predictor, the early stages predict many branches accurately while the later stages are warming up.

In the next section we investigate predictor accuracy under ideal circumstances, in the absence of table interference (conflict misses) and capacity misses.

4   Ideal predictors

We use the term *ideal* for a predictor scheme with an unlimited, fully associative predictor table. The misprediction rate measured for such a predictor is thus free from the noise of conflict and capacity misses[2]. Ideal cascaded predictors have a full complement of stages. For example, an ideal cascaded predictor of path length 6 has 7

---

[1]   We use a two-bit counter update rule, reverse interleaving of target bits and 4-way associative tables with an LRU eviction policy, as in the two-level indirect branch predictor implementations in [DH98a].

[2]   In one respect, the predictors studied in this section are not ideal: they have limited 24-bit history buffers. A small buffer represents each target with few bits, and this can cause pattern interference, reducing prediction accuracy. However, a 24-bit history buffer suffices for near-ideal accuracy (see [DH98a]), as we found during a preliminary experiment with a 30-bit buffer and path lengths up to 15 (also see the technical report for more details [DH99]).

stages, consisting of ideal two-level predictors with path lengths 0 (a BTB) up to 6. Similarly, and ideal *staged* predictor also has the maximum number of stages, but does not employ pattern filtering.Table 2 shows the ideal configurations, and the path length

Table 2. Ideal predictors

Terminology	Description	Best P	Miss%
Ideal two-level	Two-level predictor with 24-bit history buffer, storing the xor of (24 div P) bits of the P most recent targets, with an unlimited, fully associative predictor table	6	6.0%
Ideal BTB	Ideal two-level predictor of path length 0	N/A	24.9%
Ideal staged	A non-filtering staged predictor, with P+1 stages, consisting of ideal two-level predictors of pathlength 0,1,..,P	8	4.5%
Ideal cascaded	An ideal staged predictor of pathlength P, with filtering of new patterns	8	5.2%

that minimizes the misprediction rate over the benchmark set (AVG). Two-level prediction reaches a minimum misprediction rate of 6.0% at path length 6. Longer path lengths show increasing misprediction rates, because cold start misses start to negate the advantage of capturing longer-term correlations. As a result, an ideal cascaded predictor is better than ideal two-level prediction at all path lengths. Staged prediction reaches lower misprediction rates than cascaded prediction at all path lengths. This is to be expected, since a cascaded predictor uses filtering and therefore stores strictly less information than a staged predictor. Cascaded prediction economizes on the number of table entries required, which does not increase accuracy for unlimited tables.

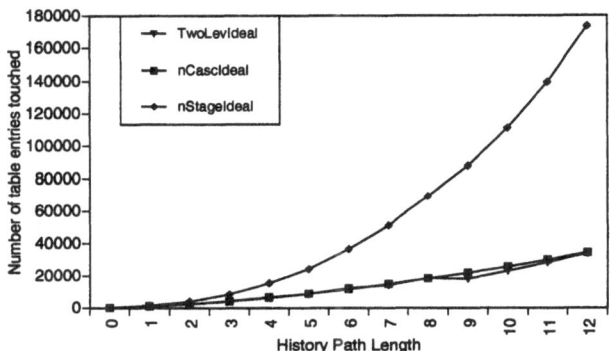

Figure 2. Total number of patterns stored by an ideal two-level, cascaded and staged predictor, for path lengths 0 to 12

However, the reduction of table entry cost is dramatic, as shown in Figure 2. The graph shows the total number of table entries occupied in a two-level, cascaded and staged predictor. As the path length grows, a staged predictor's size grows exponentially, while a cascaded and two-level predictor show nearly linear growth. In the next section we measure these benefits in the context of practical predictors.

5   Practical predictors

In this section we study practical cascaded predictor architectures with limited, 4-way associative tables and a small number of stages. Our aim is to reduce the number of table entries required to attain a given prediction accuracy. We also want to find out how close we can get to the prediction accuracy of the hypothetical ideal predictors of the previous section.

5.1   Practical predictor tables

The main cost of predictor architectures lies in the amount of on-chip memory required to store predictions. In the previous section we saw that the total number of patterns generated by an ideal predictor grows as its path length increases. For example, a two-level predictor reaches a minimal misprediction rate of 6.0% at path length 6, by storing 12324 pattern/target associations (averaged over the AVG benchmarks). Given a table entry size of about 60 bits (24-bit tag, 1-bit update counter, and 32-bit target address), the resulting data structure takes up about 800 K bits of memory, straining the capability of current processor technology. We want to reduce this memory cost while keeping misprediction rates as low as possible.

Reducing the size of a predictor table generates *capacity* misses. A capacity miss occurse when a pattern/target association, stored previously, was evicted from the table by a pattern/target of a more recently executed branch. For smaller tables, the path length must be shortened to prevent extensive capacity misses. However, shorter path length predictors are less accurate. For every given table size, there is some path length which forms the optimal compromise between these opposing effects. We use simulation to determine this optimal path length, and the resulting misprediction rate, for table sizes from 32 to 32K entries.

One further limitation is necessary for practical predictors: a limited associativity (see section 3). We use tables with associativity four, a common choice for memory caches.

5.2   Practical multi-stage predictors

A staged predictor must split up a given total table entry budget and allocate some part of it to each stage. Although it is conceivable to use one global predictor table for all stages, this would require an expensive multi-access table. Therefore each stage uses a separate single-access table. Given a limited number of stages, we cannot use each path length between 0 and 12. Instead, we need to choose the two or three path lengths that minimize misprediction rate for 2- and 3-stage predictors. We detemined the best path length combinations for all configurations shown in Table 3. We focus on predictors with a small number of large stages, since they outperform predictors with a large number of small stages for any given total number of entries (see the technical report [DH99] for path lengths and alternative configurations).

5.3   Results

Figure 3 shows misprediction rates that result from using the best path length combinations for each predictor configuration[1]. For comparison we also show the misprediction rate of a BTB and the best ideal two-level, cascaded and staged predictors

---

[1]   [DH99] contains the list of optimal path length combinations per table size, the precise data for all graphs shown in this paper, and per-benchmark misprediction rates for selected predictor schemes.

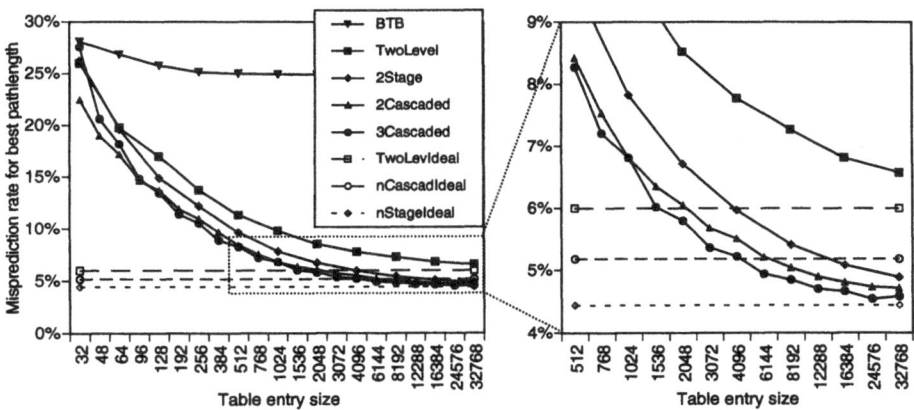

Figure 3. Misprediction rates for 2-stage, 2-stage cascaded, and 3-stage cascaded predictors

Table 3. Practical predictor configurations (with tuned path length combinations)

Terminology	Description
Two-level	Two-level predictor with a 4-way associative predictor table of size T
2-staged	A non-filtering staged predictor, using 2 two-level predictors of size T/2
2-staged cascaded	A 2-staged predictor of size T, with pattern filtering. If T is not a power of 2, the stages have size T/3 and 2T/3
3-staged cascaded	A 3-staged predictor with pattern filtering, using 3 two-level predictors of size T/3. If T is a power of 2, the stages have size T/4, T/4, and T/2

of the previous section (shown as dotted lines parallel to the x-axis since table size is not a factor).

Cascaded predictors perform better than two-level predictors at all table sizes in the explored range. In other words, any given misprediction rate is bought at a much lower cost. For instance, a 3-stage predictor of size 512 outperforms a two-level predictor of size 2K. At a fairly modest budget of 1.5K entries, a 3-staged cascaded predictor attains the same misprediction rate as an ideal two-level predictor with an unlimited table.

3-stage cascaded predictors consistently outperform 2-stage predictors, and 2-stage cascaded predictors outperform 2-stage (non-filtering) predictors. For limited budgets, cascaded prediction wins over staged prediction because pattern filtering reduces capacity misses, allowing a cascaded predictor to use longer path lengths for any given table size.

2-and 3-staged predictors seem to get asymptotically close to the accuracy of an ideal staged predictor for large, but still practical table budgets. A small number of stages clearly suffices to get almost arbitrarily close to ideal accuracy. A 3-stage cascaded predictor, at 12K table entries and higher, even improves upon the accuracy of an ideal cascaded predictor with a full complement of stages. Filtering seems to be responsible for the reduced accuracy, since a non-filtering ideal staged predictor still outperforms all other predictors.

6   Related work

Indirect branch prediction has been studied by Lee and Smith [LS84] (several forms of BTBs), Jacobson et al. [J+96] (path-based history schemes), Emer and Gloy [EG97] (single-level indirect branch predictors), and Chang et al. [CHP97] (two-level indirect branch prediction). In [CHP97], the resulting speedup of selected SPECint95 programs is measured by simulation for a superscalar processor. The misprediction rate of a BTB is reduced by half to 30.9% for gcc with a Pattern History Tagless Target Cache with configuration gshare(9), resulting in 14% speedup.

Kalamatianos and Kaeli [KK98] apply partial prefix matching (PPM) prediction to indirect branches, demonstrating excellent accuracy. A PPM predictor shortens a history pattern bit by bit, and looks it up in successively smaller stages. Each stage is half the size of its predecessor. The bits correspond to branch targets, so this scheme tests ever shorter path lengths. This resembles the prediction rule of a cascaded predictor. Cascaded prediction differs from PPM prediction because a cascaded predictor employs pattern filtering and uses a separate history buffer for each stage. The number of stages is also independent from pattern length, and each stage can use any table size. Although they demonstrate slightly better prediction performance on some of the benchmarks in this study, at least part of the improvement is due to dynamic classification of indirect branches into two classes: a class that correlates best with a history buffer which stores both conditional and indirect branch targets, and one that correlates best with only indirect branch targets. In this study we use a purely indirect branch target trace.

[EM98] proposed the YAGS architecture for conditional branch prediction. A YAGS predictor uses two kinds of predictor tables. A direct-mapped table (the *choice* table) stores the dominant direction of a branch using a 2-bit counter. Two tagged tables (*direction* tables) store a prediction for a pattern that represents history as taken/non taken bits. One of these is used for branches that are mostly taken, the other for branches that mostly not taken. Prediction in a direction table takes precedence over the prediction in the choice table, and patterns enter a direction table only if the choice table mispredicts. This scheme resembles a 2-stage cascaded predictor with a direct-mapped BTB as first stage. The main difference is that the second stage of a YAGS predictor consists of two separate tables. However, the authors agreed that this is not a requirement. A YAGS predictor shows better prediction accuracy than other conditional branch predictor schemes. We believe this is evidence that cascaded prediction is also likely to perform well on conditional branches.

7   Conclusions

We have studied the accuracy of a new hybrid predictor architecture, the multi-stage cascaded predictor, on a trace of purely indirect branches. Cascaded prediction delivers superior accuracy in the absence of resource constraints, by exceeding the accuracy reached by any other predictor scheme previously tested on these traces. In the context of limited transistor budgets, cascaded prediction also provides superior accuracy, this time by reducing the cost of two-level prediction by a factor of four or more.

More specifically:

- Ideal cascaded prediction with unlimited, fully associative tables reaches a hit rate of 94.8%. Ideal staged prediction, without pattern filtering, reaches 95.5%. We believe this accuracy is close to the limit of predictability, using a pure indirect branch history, of the indirect branches in our benchmark suite.

- Cascaded predictors with a small number of stages closely approach this limit when using large but practical table entry budgets. In particular, a 4K entry, 3-stage cascaded predictor attains 94.8% accuracy.

- At every table entry budget from 32 to 32K entries, multi-staged cascaded prediction delivers accuracy superior to two-level prediction. In particular, a 512-entry three-stage cascaded predictor reaches 92% accuracy, reducing table size by a factor of four compared to a two-level predictor. With only 1.5K entries, a 3-stage predictor reaches 94% accuracy, the maximum hit rate achievable by a hypothetical two-level predictor with an unlimited, fully associative predictor table.

We believe that cascaded prediction can also improve conditional branch prediction and load value prediction, because these applications suffer equally from cold start and capacity misses, and because recent related work [EM98] shows that a similar architecture delivers superior accuracy on conditional branches.It seems to be an idea whose time has come.

8   References

[CHP97] Po-Yung Chang, Eric Hao, Yale N. Patt. Target Prediction for Indirect Jumps. *ISCA '97 Proceedings*, July 1997.

[DH96] Karel Driesen and Urs Hölzle. The Direct Cost of Virtual Function Calls in C++. In *OOPSLA '96 Conference proceedings*, October 1996.

[DH98a] Karel Driesen and Urs Hölzle. Accurate Indirect Branch Prediction. *ISCA '98 Conference Proceedings*, pp. 167-178, Barcelona, July 1998.

[DH98b] Karel Driesen and Urs Hölzle. The Cascaded Predictor: Economical and Adaptive Branch Target Prediction. *Micro '98 Conference Proceedings*, Dallas, Texas, December 1998.

[DH99] Karel Driesen and Urs Hölzle. *Multi-stage Cascaded Prediction*. Technical Report TRCS99-05, Department of Computer Science, University of California, Santa-Barbara, February 12, 1999.

[Dri99] Karel Driesen. *Software and Hardware Techniques for Efficient Polymorphic Calls*. PhD dissertation, University of California, Santa Barbara (in preparation).

[EM98] A.N.Eden and T.Mudge. The YAGS Branch Prediction Scheme. *Micro '98 Conference Proceedings*, Dallas, Texas, December 1998.

[EG97] Joel Emer and Nikolas Gloy. A language for describing predictors and its application to automatic synthesis. *ISCA '97 Proceedings*, July 1997.

[HP95] Hennessy and Patterson. *Computer Architecture: A Quantitative Approach*. Morgan Kaufmann, 1995.

[J+96] Quinn Jacobson, Steve Bennet, Nikhil Sharma, and James E. Smith. Control flow speculation in multiscalar processors. *HPCA-3 proceedings*, February 1996.

[KK98] John Kalamatianos and David Kaeli. Predicting Indirect Branches via Data Compression. *Micro '98 Conference Proceedings*, Dallas, Texas, December 1998.

[LS84] J. Lee and A. Smith. Branch prediction strategies and branch target buffer design. *IEEE Computer 17(1)*, January 1984.

[MMN93] Ole Lehrmann Madsen, Birger Moller-Pedersen, Kristen Nygaard. *Object-Oriented Programming in the Beta Programming Language*. Addison-Wesley 1993.

[CHP97] Mikko H.Lipasti, Christopher B. Wilkerson, and John Paul Shen. Value Locality and Load Value Prediction. *Proceedings of the 7th International Conference on Architectural Support for Programming Languages and Operating Systems* (ASPLOS VII), October 1996, pp. 138-147.

[P+97] Yale N.Patt, Sanjay J. Patel, Marius Evers, Daniel H. Friendly, Jared Stark. One Billion Transistors, One Uniprocessor, One Chip. *IEEE Computer*, September 1997

[YP91] Tse-Yu Yeh and Yale N. Patt. Two-level adaptive branch prediction. *MICRO 24*, November 1991.

Mispredicted Path Cache Effects

Jonathan Combs, Candice Bechem Combs, and John Paul Shen

Carnegie Mellon Microarchitecture Research Team (CMuART)
Department of Electrical and Computer Engineering
Carnegie Mellon University
Pittsburgh, PA 15213 USA
{jcombs, cbechem, shen}@ece.cmu.edu

Abstract. As superscalar pipelines become wider and deeper, the percentage of dynamic instructions fetched into the machine from the mispredicted path significantly increases. This paper discusses how a new cycle-accurate performance simulator is used to accurately measure mispredicted path effects on the cache hierarchy. Previously published results based on less accurate tools indicated that mispredicted path instructions have the serendipitous positive effect of doing memory prefetching. Our results show that while such prefetching does occur for some benchmarks, it does not occur consistently for all benchmarks. Furthermore the IPC impact varies widely among the benchmarks. SPECint95 benchmarks show IPC changes ranging from -8% to +12%.

1 Introduction

The aim in building a wider and deeper pipelined microarchitecture is to enable the processor to execute more instructions in parallel, and at the same time achieve higher clock frequencies. One major design challenge is to keep issue rates high despite the fact that basic blocks are small and memory latencies are high. A design that contributes to a more uninterrupted instruction fetching and reduces the memory latency will help increase issue and completion rates with an overall gain in IPC (average instructions per cycle).

To achieve high instruction fetch bandwidth, modern microprocessors employ branch predictors to speculatively fetch instructions beyond conditional branches. If these speculative instructions are determined to be on a mispredicted path, they must be invalidated and removed from the machine.

Mispredicted instructions can affect many parts of the machine, particularly the functional units, branch predictors [5], and caches. This research focuses on the effects mispredicted path instructions have on the cache hierarchy, due to the increased number of instruction and data references. Previous research in this area showed that mispredicted path references have a prefetching effect, but the methods that measured these effects had serious limitations. Using our improved fMW [1,2] performance simulation tool we found that the magnitude of the effect varies with each benchmark and is not always positive.

Before we can accurately differentiate previous methodologies from our own, it is important to have a background on simulators. Functional simulators model the

P. Amestoy et al. (Eds.): Euro-Par'99, LNCS 1685, pp. 1322-1331, 1999.
© Springer-Verlag Berlin Heidelberg 1999

instruction set architecture (ISA). The primary concern of functional simulators is functional correctness. Performance simulators model the *micro*architecture. They model the machine organization and are concerned with machine performance. Sometimes these simulators are also referred to as cycle-accurate simulators to reflect their concern with timing issues. Traditionally, performance simulators are implemented as trace-driven tools; i.e. their inputs are traces of dynamic instructions, without full-function simulation capability.

Some of the new microarchitecture features, such as aggressive branch prediction, value prediction, and multi-path execution, cannot be accurately simulated in a trace-driven tool. To address these shortcomings, we use a new performance simulator with full-function capability called fMW (functional microarchitecture workbench) [1]. The new fMW builds on MW by incorporating a customized version of the PSIM [3] functional simulator and by extending the capabilities of the original MW [2].

2 Previous Work

A handful of studies [6,8,9] have examined the effects of mispredicted paths; however each of these efforts is hampered by inadequate modeling techniques. The simulator used in [8,9] is trace driven leading to several inaccuracies. Since trace-driven simulators can not execute mispredicted path instructions, Pierce and Mudge injected a fixed number of instructions to emulate the mispredicted path [8]. However, the number of cycles a given machine spends on each mispredicted path depends on the aggressiveness of the branch predictor and the branch resolution latency. Fixing the branch resolution latency at a constant number of instructions introduces significant error. Nevertheless, using this method, Pierce and Mudge found the mispredicted path instructions tend to prefetch the data cache.

A continuation work [9], using the same tool, focused on the instruction cache and found the prefetching effects of mispredicted path instructions far outweigh the pollution effects caused by them. Lee et al. studied instruction cache fetch policies using a cache simulator and found mispredicted path instructions did not cause any degradation in performance over fetching the correct path only [6].

Previous studies generally show mispredicted path execution to be a beneficial prefetching mechanism for the I-cache [6,8,9]. Similar benefit for the D-cache was also suggested [6]. However, the methods that measured these effects had serious limitations and are inherently inaccurate. Using fMW such inaccuracies are removed by directly simulating the mispredicted path instructions in the machine model. The following section summarizes our experimental results.

3 Experimental Methodology

All data reported in this paper are generated using the new fMW [1] tool which integrates a functional simulator and a cycle-accurate timing simulator that is built based on a validated PowerPC 604 model [2]. The machine model is based on published reports [4,7,10] and accurately models all aspects of the microarchitecture.

The machine model and the simulation framework used in this paper are discussed in the next two subsections.

Table 1. SPECint95 benchmarks

Name	Input Set	Instruction Count
compress	10000 e 2231	39,719,131
gcc	-f<all optimizations> -O regclass.I –s regclass.s	257,670,349
go	5 9	79,544,303
ijpeg	tinyrose.ppm	92,054,217
li	queen6.lsp	56,572,774
m88ksim	dhry.big.100iter, cache off	106,900,787
perl	trainscrabbl.in	50,039,056
vortex	tiny.in	153,084,257

3.1 The Machine Model

The SPECint95 benchmark suite is used for all experiments. The input sets and run lengths of each benchmark are summarized in Table 1. To focus the current study on the effects of speculative execution and to emphasize the effect of mispredicted path instructions in the pipeline, the PowerPC 604 microarchitecture is extended to remove resource constraints, and widened to allow a greater number of in-flight instructions. The instruction window is limited to 512 instructions with an unlimited number of functional units and rename registers. Instruction fetch and dispatch widths are increased to 16 instructions per cycle. There is no limit to the number of instructions that may complete in each cycle[1]. A 64-entry fully associative branch target address cache (BTAC) and a 512-entry branch history table (BHT) handle branch prediction.

The memory hierarchy includes a perfect main memory, a 32KB 4-way set associative Level-1 instruction cache (IL1), a 32KB 8-way set associative Level-1 data cache (DL1), and a 512KB 8-way set associative unified Level-2 cache (UL2). All caches use a write-back, write-allocate scheme. Access latencies are 1, 3, and 100 cycles for the L1, L2, and main memory respectively. An unlimited load-miss queue and an unlimited store queue handle all load and store execution. The store queue performs data forwarding, and load/store instructions execute out-of-order if no address aliasing is detected.

In terms of mispredicted branches and cache accesses made by mispredicted path instructions, there are three key stages that determine the effects of wrong path instructions. These stages are where different types of branch outcomes resolve in a PowerPC machine. Unconditional branches resolve in decode, branches to count or link registers resolve in dispatch, and conditional branches resolve in the branch execution unit within execution. The later the stage that a branch resolves, the greater number of mispredicted path instructions that are allowed into the machine. As soon as a mispredicted branch is resolved, however, the mispredicted path instructions are flushed from the machine. Since most branches are resolved before reaching the

---

[1] The authors are not proposing this as a realistic machine design, but a model that increases the effects of mispredicted path execution while enforcing register and memory data dependencies.

execution core, nearly all mispredicted path load accesses are flushed from the machine without ever affecting the caches. For those branches that do not resolve until reaching the execution unit, there is a limited window of opportunity for mispredicted path load accesses to execute and thus affect the DL1 and UL2.

The cache fetch policies process all cache misses. Given the limit study nature of this work, when mispredicted path loads are allowed to access the caches any complications due to memory mapped I/O are ignored. Also, the IL1 cache must become nonblocking. This is because correct path instruction fetching is allowed to continue immediately even if an I-cache miss is outstanding due to instruction fetches down the wrong path. This is referred to as the "Resume" policy as discussed in [6]. The Resume policy was determined to have the lowest miss ratio during speculative execution; thus one can infer this policy to have the most beneficial impact on IPC gain. In [6], however, only a single outstanding miss was allowed. The policy in this study is to allow unlimited outstanding misses.

3.2 Simulation Framework

To determine the exact effects mispredicted path instructions have on the cache hierarchy, there needs to be a record of what the accesses would have been if mispredicted path instructions were not simulated. To accomplish this, we duplicated the cache hierarchy in the simulation model. One set of caches is accessed by all instructions, including mispredicted path instructions, and is used for the timing simulation. The other set is accessed only by correct-path instructions. The delays of the correct-path-only copies are for accounting purposes only and do not affect the flow of instructions through the machine. When the delay of a cache access differs from the correct-path-only cache delay, that difference is caused by mispredicted path instructions. By recording the delay difference (in cycles) and the direction of the difference, we can accurately measure the exact prefetching and/or polluting count as well as the average number of cycles gained or lost due to the difference. Separate tallies are kept for the prefetch and pollution accesses for both the instruction cache hierarchy (fetch) and the data cache hierarchy (load). Given that individual accesses can have very different delay differences (i.e. one access can prefetch from the L2 while another prefetches from main memory), the average cycles gained or lost is a crucial metric.

The simulator can be instructed regarding whether or not the machine model should fetch down the mispredicted path. This allows the IPC difference caused by mispredicted path to be determined. Also, the simulator has the ability to stop mispredicted path loads from accessing the caches. This enables the effects on the data cache to be separated from those on the instruction cache. If the overall net effect of mispredicted path cache accesses is to prefetch data, then this should result in a lower average memory latency and a higher IPC.

Finally, to simulate the effects of the deeper pipelined machines, variable delay stages (VDS) are placed between pipe stages prior to the execute stage in the timing simulation model of the machine (Figure 1). This allows increases in the average branch latency and, proportionally, the number of mispredicted path instructions fetched into the machine. To view these effects, we allow the variable delay stages to be set to 0, 1, 2, or 3. Note that because there are three pipe stages prior to the execute stage, a setting of D=1 adds 3 additional cycles to the front end, a setting of D=2 adds

6 cycles, and so on. These modifications allow us to determine the general trend in IPC as the number of front-end stages increases. [8] did similar studies by noting the effects of injecting more mispredicted path instructions into the instruction trace.

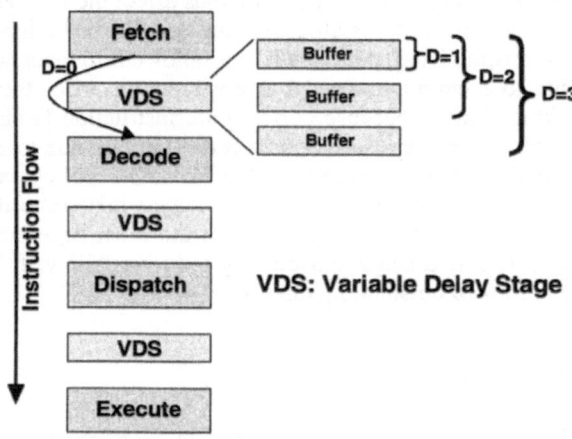

Figure 1. Variable delay stages for simulating deeper front end.

4 Experimental Results

Table 2 shows the cache access discrepancies on the instruction cache hierarchy for running the benchmarks to completion. For this part of the experiment we ignore the data cache accesses by load instructions in the mispredicted path. The amounts of prefetch and pollution accesses are presented along with the average number of cycles gained or lost. The "Net Change" column records the overall cycle change in cache delay cycles caused by the mispredicted path accesses. A positive number indicates a decrease in cycle count (good), whereas a negative number indicates an increase in the cycle count (bad).

Table 2. Cache access discrepancies caused by mispredicted path instructions.

Benchmark	Pollution Accesses	Avg. cycle Loss/access	Prefetch accesses	Avg. cycle gain/access	Net Change (cycles)	Projected % IPC Change
compress	207	1.00	110	24.44	2481	0.02%
gcc	2356036	14.92	543954	34.35	-16479381	-6.67%
go	349958	3.95	188388	50.51	8133593	16.23%
ijpeg	108252	2.88	42341	38.56	1320485	4.12%
li	659	7.96	203	34.36	1730	0.01%
m88ksim	857	6.28	361	34.25	6983	0.02%
perl	96656	29.60	18885	45.63	-1999238	-9.26%
vortex	278049	10.16	99737	34.84	650993	1.16%

Since the machine model used is designed to put pressure on the front-end, a delay caused by a missed fetch access should correspond directly to an increase in the total execution time of the benchmark. If, however, during benchmark execution the machine is not bottlenecked due to the latency of instruction fetching, the cycles gained or lost during this period will not impact the IPC. Assuming that the latter is not the usual case, a projected % IPC change is calculated and included in the table. Taking the execution time of the benchmarks with the mispredicted path disabled and changing it by the "Net Change" value allows us to compute a projected IPC change when the mispredicted paths are enabled. This column is based on the assumption that for each delay cycle difference, this cycle directly changes the total execution cycles of the benchmark. This projection is included to give some rough indication of the potential impact on IPC due to the net changes in total cycle count. Actual IPC changes based on cycle-accurate simulation are presented later.

The results in Table 2 show that most benchmarks have a positive net change due to mispredicted path cache accesses, however the extent varies greatly. Only *perl* and *gcc* show negative net changes. Examining the average cycles gained/lost for the two types of accesses, it can be seen that most gains on prefetch accesses are from main memory prefetches because the average cycles gained per prefetch access is in the range of 25-50 cycles. On the other hand most of the polluting accesses are hitting in the L2. This means the polluting wrong path accesses contaminate mainly the first level of the cache. *Perl* and *gcc* are the exceptions; both exhibit significant number of penalty cycles per pollution access, indicating significant number of misses to the main memory. For these benchmarks, polluting accesses have a tendency to not only remove data from the L1 cache, but from the L2 cache as well.

4.1 Mispredicted Path Impact on IPC

Figure 2 shows the actual percent change in IPC caused by mispredicted path instructions. The data in this figure are obtained by actually simulating the mispredicted path instructions using the fMW cycle-accurate simulator. The values range from the greatest increase in *go* of 12.0%, to the *perl* decrease of -7.93%. The average across all benchmarks is around 1.0%. These values correlate strongly with the projected IPC changes of Table 2 suggesting that, indeed, the machine is bottlenecked mainly in the front-end of the pipeline. It is important to notice the negligible change in IPC for *compress*, *li*, and *m88ksim*. This is because the data and instruction working sets for these benchmarks are small enough to be contained entirely within the caches.

Although the net cycles gained is positive for most benchmarks except for *perl* and *gcc*, the IPC changes vary significantly from benchmark to benchmark. Therefore, the findings of [9] seem to have been on the right track in terms of instruction prefetching, but the conclusion that the mispredicted path instructions always perform instruction prefetching is not substantiated by our results. [8,9] did not actually simulate any benchmark whose accesses caused more pollution than prefetching and thus concluded by saying that the cache effects of mispredicted path instructions would always be beneficial. It is also important to note that whether prefetching or polluting becomes the dominant factor in changing the total cycle count, depends not only on the number of accesses of each type but also on the cycle count impact by each access of each type.

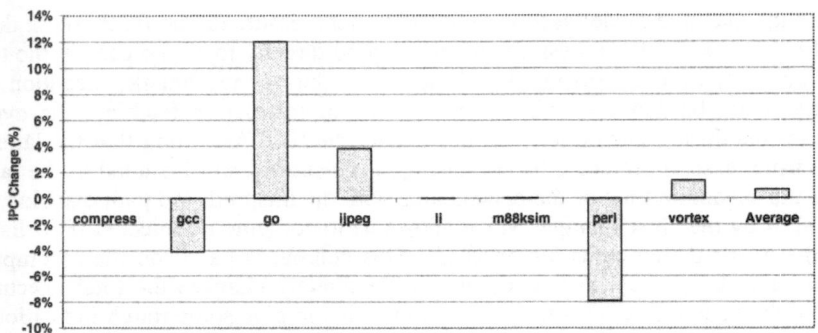

Figure 2. Percent IPC change due to mispredicted path instructions

Mispredicted path load accesses and their impact on the data cache hierarchy (studied in [8]) are also studied here by running the same benchmarks to completion, but allowing the mispredicted path load instructions to access the cache hierarchy if given the opportunity. The results show these loads to have no impact on the IPC of the machine. Of course, for a machine bottlenecked in the front-end, such results are expected. Studying data similar to Table 2 but for the load accesses only, it seems that if the mispredicted path loads could influence the machine performance, the overall trend tends to be polluting rather than prefetching as [8] suggested. Only *go* and *perl* have positive net cycle changes caused by the mispredicted path, and these numbers are negligible. However, it must be noted that such cache delay results are not as pronounced when dealing with the load accesses due to the nature of the out-of-order execution core coupled with an aggressive nonblocking cache hierarchy. Unlike the fetch unit, which must stall when the next predicted instruction misses in the cache, the execution core can be processing other instructions in parallel with outstanding load misses. Only accurate simulation of data dependency relationships for each load can determine the actual impact on IPC.

From our results, it can be seen that the amount and type of cache effects caused by mispredicted path instructions are strongly benchmark dependent. The actual IPC impact is not only a function of the relative frequency of prefetching vs. polluting accesses, but also the number of machine cycles gained or lost by each of such accesses. Applications should be analyzed individually to determine the cache effects of executing mispredicted path instructions. It is interesting to note that *go*, one of the most troublesome benchmarks for branch predictors, shows the best IPC gain due to such serendipitous prefetching. We surmise that this is due to the control flow structure of *go* such as the presence of many short "hammocks" in the code, the traversal of which is difficult to predict. With such control flow structure, a long mispredicted path is likely to converge with the correct path and/or the instructions fetched by the mispredicted path are soon after fetched as part of a correct path. It also appears that for *go,* significant number of such prefetches are prefetching from the main memory, while most of the polluting misses are between L1 and L2.

4.2 Effects Throughout Benchmark Execution

To get a better understanding of how the mispredicted path accesses affect performance, it is necessary to see how the cycle impact changes over the lifetime of the benchmark's execution. To do this, the benchmarks are run for incrementally increasing instruction counts beginning at 1 million and until the end of benchmark execution. We then study the percent IPC changes for runs with and without mispredicted path instruction effects. Temporal IPC changes can be seen in Figure 3. From this figure one can observe that the effects of mispredicted path cache accesses continue throughout the lifetime of the benchmark execution. Conventional wisdom says that these extra cache accesses would be the most beneficial during early part of execution because of compulsory misses. Such behavior is observed for most benchmarks, but its impact is smaller than expected.

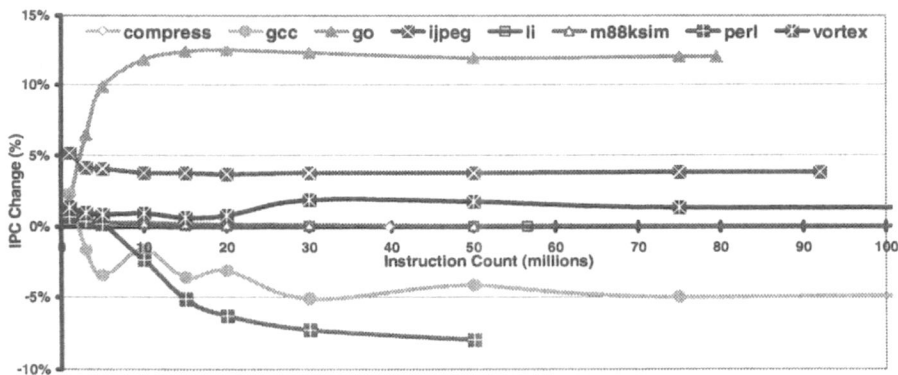

Figure 3. IPC change over benchmark lifetime

4.3 Sensitivity to Memory Latency and Pipeline Depth

Experiments on varying main memory latencies are done on all benchmarks running to an instruction count of 50 million. The results can be seen in Figure 4. As expected, *compress*, *li*, and *m88ksim* display no variation in IPC due to memory latency changes since these benchmarks basically only access main memory for compulsory misses (these three overlap on the 0% axis). The rest of the benchmarks show a linear progression of more dramatic changes in IPC as the latency to main memory is increased, and the slope of each progression is dependent on the specific benchmark. This slope variation is probably caused by the differing prefetch to pollution ratios of the mispredicted path accesses. Similar experiments are also conducted on variations of L2 latency. These results show negligible impact on the percent IPC change for delays varying from 3 cycles to 10 cycles.

Our other sensitivity analysis is done by adding stages in the front-end. This causes mispredicted branches to take longer to resolve, resulting in more fetch cycles being spent fetching instructions down the mispredicted path and increasing the number of mispredicted path instructions being brought into the machine. Figure 5 shows the resulting changes in IPC as the VDS parameter D is varied from 0 to 3. The experiments on varying front-end pipeline depths are done on all benchmarks and run to an instruction count of 30 million. In general, if the trend is for mispredicted path instructions to cause more prefetching than pollution, then increasing the number of mispredicted path instructions will simply cause more prefetching to occur. [8] showed similar results. If the accesses are more polluting (as it is for *perl* and *gcc*), then more instructions will cause more pollution. Yet this trend is not universal, as can be seen most dramatically in *vortex*. For some benchmarks, it seems that going down the mispredicted path to a certain extent causes prefetching effects. Going too far down the mispredicted path, however, starts polluting the caches more than prefetching.

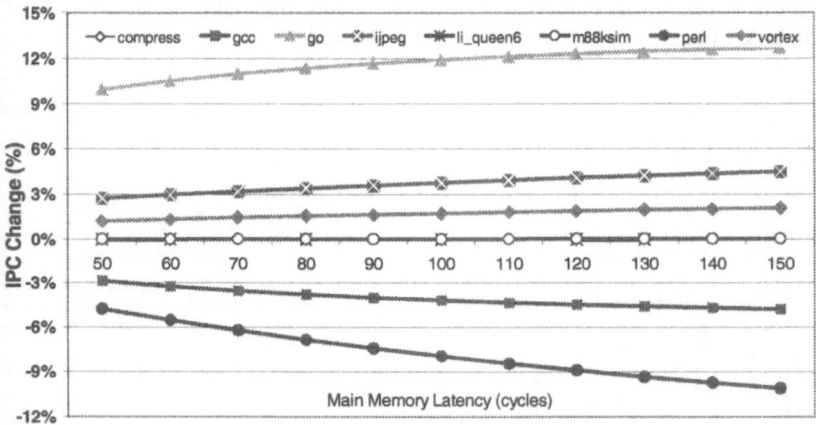

Figure 4. Memory latency effects.

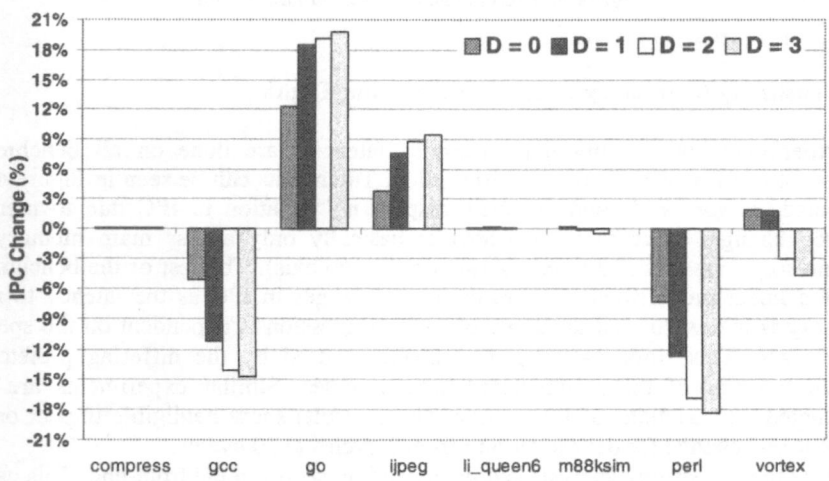

Figure 5. Variable delay stage effects.

5 Conclusion

Mispredicted path instruction fetches can serendipitously prefetch for subsequent correct path instructions. However this potential is highly benchmark dependent; for example *perl* and *gcc* show significant losses in IPC due to pollution. Furthermore, the overall performance impact is a function of the relative numbers of prefetching vs. polluting accesses as well as the actual cycle count impact by each of such accesses. The mispredicted path data accesses seem to display a more uniform trend of pollution over prefetching. Finally we note that the impact of mispredicted path instructions is likely to increase in the future, as shown in our sensitivity experiments. An interesting area of future research is to develop microarchitecture that can adapt the aggressiveness of processing mispredicted path instructions depending on whether the anticipated impact is prefetching or polluting.

Acknowledgement and Footnote

This work benefited from a cluster of simulation machines donated by Intel. We also benefited from discussions with Prof. Trevor Mudge at the University of Michigan. Jonathan Combs is now with Intel's Texas Development Center, and Candice Bechem Combs is now with Motorola in the Engineering Rotation Program.

References

1. C. Bechem, J. Combs, N. Utamaphethai, B. Black, R.D. Blanton, J.P. Shen. "An Integrated Functional Performance Simulator." IEEE Micro, May-June 1999, pp 2-11.
2. B. Black and J.P. Shen. "Calibration of Microprocessor Performance Models." COMPUTER, May 1998, pp. 59-65.
3. A. Cagney. PSIM User's Guide. ftp://cambridge.cygnus.com/pub/psim/index.html August 1996.
4. K. Diefendorf and E. Silha. "The PowerPC User Instruction Set Architecture." IEEE Micro, October 1994, pp. 30-41.
5. S. Jourdan, T. Hsing, J. Stark, and Y. Patt. "The Effects of Mispredicted-Path Execution on Branch Prediction Structures." Conference on Parallel Architectures and Compilation Techniques (PACT), October 1996.
6. D. Lee, J-L. Baer, B. Calder, and D. Grunwald. "Instruction Cache Fetch Policies for Speculative Execution." International Symposium on Computer Architecture, June 1995.
7. IBM Microelectronics Division, PowerPC 604 RISC Microprocessor User's Manual 1994.
8. J. Pierce and T. Mudge. "The Effect of Speculative Execution of Cache Performance." Proceedings of the International Parallel Processing Symposium, April 1994.
9. J. Pierce and T. Mudge. "Wrong Path Instruction Prefetching." Technical Report, University of Michigan, 1994.
10.S. Song, M. Denman, and J. Chang, "The PowerPC 604 RISC Microprocessor." IEEE Micro, October 1994, pp. 8-17.

Topic 17
Concurrent and Distributed Programming
with Objects

Patrick Sallé and Marc Pantel

Co-chairmen

As advocated by the widespread use of JAVA and CORBA based technology, the use of objects is recognized as a major improvement in the development of concurrent and distributed applications.

However, the recent successes of these technologies hide the fact that their use still requires many theoretical and practical studies in order to be fully applicable.

This workshop aims at improving the body of knowledge concerning their use. The papers selected for presentation (one distinguished and five regulars), which cover a wide spectrum of theoretical and practical aspects of concurrent programming including applications, are split into two sessions :

- The former, "Semantics", covering more theoretical aspects, contains two papers describing the verification of dynamic processes using static analyses based on types and on temporal logic. The third paper develops a denotational model for self-inflicted calls in the context of aliasing due to object migration.
- The latter, "Tools and applications", treating more practical subjects, is composed of three papers about UML, CORBA and object-oriented programming for numerical applications.

We would like to thank sincerely all the referees who contributed to the reviewing process.

P. Amestoy et al. (Eds.): Euro-Par'99, LNCS 1685, pp. 1333–1333, 1999.
© Springer-Verlag Berlin Heidelberg 1999

Non-regular Process Types

Franz Puntigam

Technische Universität Wien, Institut für Computersprachen
Argentinierstraße 8, A-1040 Vienna, Austria
franz@complang.tuwien.ac.at

Abstract. Process types specify sequences of acceptable messages. Even
if the set of acceptable messages changes dynamically, a type checker can
statically ensure that only acceptable messages are sent. As proposed so
far, all message sequence sets specified by types can be generated by
regular grammars. We propose to increase the expressiveness so that
non-regular message sequence sets can be specified. Type equivalence
and subtyping take possible type extensions into account.
Keywords: type systems, subtyping, active objects

1 Introduction

A process type [13, 14, 15] specifies not only a set of messages, but also constraints on acceptable sequences of these messages. Type safety is checked statically by ensuring that each object reference is associated with an appropriate
"type mark"; it specifies all message sequences that can be sent through the
reference.

An instance of a subtype can be used wherever an instance of a supertype is
expected [6]. In a process type system, a subtype can extend a supertype by specifying additional messages and less constraints on message sequences. Messages
accepted by objects of supertypes are also accepted by objects of subtypes.

It is decidable if two regular grammars generate the same language and if
the language generated by a regular grammar is contained in that generated
by another [5]. These relations are undecidable (or not known to be decidable)
for more expressive grammars like LR(1) and context-free grammars. Process
types shall be extended so that non-regular languages can be expressed. Since
equivalence and containment of message sequence sets are undecidable, stricter
notions of type equivalence and subtyping shall be used.

According to Liskov [6], a type is a partial specification of object behavior.
A subtype specifies the behavior in more detail. If a supertype allows a user
to rely on some property, a subtype has to allow the user to rely on the same
property. For example, if a supertype always allows a user to send a message
"put", a subtype must not allow the user to send "delete" if "delete" is the last
acceptable message. This restriction ensures that several users can safely send
messages to the object, although each user knows a different type.

Process types enforce this restriction in a different way: If a user can send
"delete", a type splitting rule ensures that no other user can send "put". The set

P. Amestoy et al. (Eds.): Euro-Par'99, LNCS 1685, pp. 1334–1343, 1999.
© Springer-Verlag Berlin Heidelberg 1999

of acceptable message sequences does not reflect the restriction. There is no loss in flexibility if the restriction is enforced by the type equivalence and subtyping relations. We shall show that such relations are sound and complete.

Syntax and informal semantics of non-regular process types are introduced in Section 2. The type equivalence and subtyping relations are defined in Section 3. A characterization of these relations concerning acceptable message sequences is given in Section 4. A discussion of related work follows in Section 5.

2   Specification of Non-regular Process Types

A process type (as presented in [14, 15]) consists of two parts – a set of message descriptors and a state. A message descriptor specifies the signature of a message, conditions for accepting such messages depending on the state, and updates of the state on message acceptance. States are represented by multi-sets of tokens, and conditions by the availability of tokens in states. The state of an object's type is updated when a message is accepted, that of the type associated with a reference (a type mark) when a message is sent through the reference. The tokens in the type marks' states are, essentially, contained in the state of the object's type. This concept works fine for regular process types, where all conditions correspond to the availability of a minimum number of tokens.

We must deal with aliasing: Let a server accept messages according to a type $\pi\|\rho\|\sigma\|\tau$; at least the sequences specified by $\pi, \ldots, \tau$ are accepted in arbitrary interleaving. Two clients have references to the server with the type marks $\pi\|\rho$ and $\sigma\|\tau$, respectively. If both clients send a message containing a reference to the server as argument to a further object, the type marks must be split:

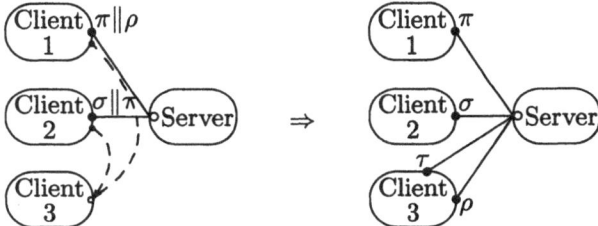

The type marks π and σ remain at the references in the first two clients, while Client 3 gets two new references with the type marks τ and ρ. Now, messages can be sent to the server through four independent references.

Static type checking is possible: The compiler has to ensure that (1) all message sequences specified by the objects' types are accepted by the objects, (2) only messages specified by the references' type marks are sent through the references, and (3) type marks are split when an object reference is used as argument. Parameter types in message signatures determine how to split types.

In non-regular process types, the acceptability of messages may depend on the exact number of available tokens of some kind. Such conditions can be used only if all tokens of a kind occur in a single type mark's state. States shall hold information about whether other type marks may contain further tokens.

Type representation. An infinite set of names (denoted by $u, v, w, \dots$) and a set of constant symbols ($x, y, z, \dots$) are considered given. Names are used as parameters, symbols as message selectors and tokens. Parameters are underlined when occurring at positions where types or integers shall be substituted for them. An underlined occurrence binds all following free occurrences. Some conventions simplify the notation: For each meta-symbol e, $\tilde{e}$ is an abbreviation of $e_1, \dots, e_n$ for arbitrary n. For example, $\tilde{u}$ denotes a list of names $u_1, \dots, u_n$. Likewise, $\{\tilde{e}\}$ denotes the set consisting of $e_1, \dots, e_n$, and $|\tilde{e}|$ the length of the sequence. For a binary operator $\circ$, $\tilde{e} \circ \tilde{g}$ stands for $e_1 \circ g_1, \dots, e_n \circ g_n$ ($|\tilde{e}| = |\tilde{g}|$). All names in an underlined parameter list $\underline{\tilde{u}}$ are regarded as pairwise different.

This is the syntax of non-regular process types (denoted by $\pi, \rho, \sigma, \tau, \dots$):

$$
\begin{aligned}
\tau &::= \{\tilde{\alpha}\}[\tilde{r}] \mid *\underline{u}\{\tilde{\alpha}\}[\tilde{r}] \mid u \mid \tau_1 \| \tau_2 & \mu &::= \text{type} \mid \text{int} \\
\alpha &::= a[\tilde{r}_1][\tilde{r}_2] & r &::= x^p \mid x^{p_1|p_2} \\
a &::= x\langle \underline{\tilde{u}}{:}\tilde{\mu}, \tilde{\tau} \rangle & p &::= n \mid \infty \mid u \mid p_1 + p_2
\end{aligned}
$$

A type $\{\tilde{\alpha}\}[\tilde{r}]$ consists of a set of message descriptors $\tilde{\alpha}$ and a state $[\tilde{r}]$. Each state descriptor $(r, s, t, \dots)$ is of the form x^p or $x^{p|q}$, where x is a symbol used as token, p a multiplication factor and q a splitting factor. $[\tilde{r}]$ is regarded as a multi-set containing p tokens x for each x^p or $x^{p|q}$ in $\tilde{r}$; x^1 is abbreviated by x. State descriptors $x^{p|q}$ specify that message sequences may depend on the exact number of tokens x. The state has been split q times. If a state contains $x^{p_1|q_1}, \dots, x^{p_n|q_n}$, the type has about $1/2^{q_1} + \dots + 1/2^{q_n}$ of the tokens x to be considered. The type has all available tokens of symbol x if $x^{p|0}$ is in the state.

Integers $(p, q, \dots)$ are nonnegative integer constants, ∞ denoting a very large (infinite) integer, parameters standing for integers, and integer addition.

Message descriptors $(\alpha, \beta, \gamma, \dots)$ are of the form $a[\tilde{s}][\tilde{t}]$, where a is a message signature, $[\tilde{s}]$ an in-set and $[\tilde{t}]$ an out-set. A message of signature a is acceptable if each state descriptor in the in-set is in the type's state $[\tilde{r}]$ and, for each state descriptor $x^{p|0}$ in the in-set, no further state descriptor for token x is in $[\tilde{r}]$; the type is updated by removing all state descriptors in the in-set from the state and adding those in the out-set. A state descriptor $x^{p|q}$ can occur in an in-set only with $q = 0$, and $x^{p|0}$ in an out-set only if an $x^{q|0}$ is in the in-set.

A message signature $(a, b, c, \dots)$ is of the form $x\langle \underline{\tilde{u}}{:}\tilde{\mu}, \tilde{\tau} \rangle$, where x is the message selector, the $\tilde{u}$ are type and integer parameters, the $\tilde{\mu}$ meta-types and the $\tilde{\tau}$ types of object parameters; the $\tilde{u}$ can occur in the $\tilde{\tau}$. Meta-types $(\mu, \nu, \dots)$ are "type", the type of all type expressions, and "int", the type of all integers.

A type $*\underline{u}\{\tilde{\alpha}\}[\tilde{r}]$ is recursive: The type parameter u can occur in the message signatures in $\tilde{\alpha}$. Type parameters can occur wherever types can occur. A type $\sigma \| \tau$ denotes the combination of two types; it can be split into σ and τ.

Examples show static parts of types for simple data stores:

$$DS \stackrel{\text{def}}{=} \{\text{put}\langle \underline{u}{:}\text{type}, u \rangle [\text{empty}][\text{full}], \text{get}\langle Back\,[\text{one}] \rangle [\text{full}][\text{empty}]\}$$

$$Back \stackrel{\text{def}}{=} \{\text{back}\langle \underline{u}{:}\text{type}, u \rangle [\text{one}][\,] \}$$

$$DS' \stackrel{\text{def}}{=} \{\text{put}\langle \underline{u}{:}\text{type}, u \rangle [\,][\text{full}], \text{get}\langle Back\,[\text{one}] \rangle [\text{full}][\,] \}$$

$$DS'' \stackrel{\text{def}}{=} \{\text{put}\langle \underline{u}{:}\text{type}, u \rangle [\text{ok}][\text{full}], \text{get}\langle Back\,[\text{one}] \rangle [\text{ok}, \text{full}][\,], \text{delete}\langle\rangle [\text{ok}^{\infty|0}][\,] \}$$

An object of type $DS[\text{empty}^{50}]$ accepts a message "put" whenever one of the fifty buffer slots is empty, and "get" if a slot is full; the empty slot becomes full, and the full slot empty. The type $DS[\text{empty}^{\infty}]$ is, in some sense, equivalent to $DS'[]$: An infinite data store of this type always accepts "put", but "get" only when there is a full slot. Instances of $DS'[\text{full}^{\infty}]$ accept "put" and "get" in arbitrary ordering. A data store of the type $DS''[\text{ok}^{\infty|0}]$ allows users to explicitly delete the store by sending "delete". An infinite number of tokens "ok" is present in the state as long as the data store is alive; the tokens are removed when "delete" is accepted. Because of $\text{ok}^{\infty|0}$ in the in-set, "delete" can be sent only if there is no other reference to the data store with a token "ok" in the type mark's state.

The expressiveness can be shown by a further example: An object of type

$$\{a_1[x^{1|0}][x^{1|0}, y], a_2[x^{1|0}, y][x^{2|0}, y, z], a_2[x^{2|0}][x^{2|0}, z],$$
$$a_3[x^{2|0}, y][x^{3|0}], a_3[x^{3|0}, y][x^{3|0}], a_4[x^{3|0}, y^{0|0}, z][x^{3|0}, y^{0|0}]\}[x^{1|0}, y^{0|0}]$$

accepts messages sequences of the form $a_1^n a_2^m a_3^n a_4^m$ (with $m, n > 0$). Using context-free grammars it is not possible to specify words of this form.

3   Type Equivalence and Subtyping

Some further notation is needed in the definition of the type equivalence and subtyping relations. $\mathit{free}(\tilde{e})$ denotes the set of all names occurring free in at least one expression in $\tilde{e}$. $\mathit{tok}(\tilde{r})$ denotes the set $\{x \mid \exists_{p \neq 0} x^p \in \{\tilde{r}\} \vee \exists_{p,q} x^{p|q} \in \{\tilde{r}\}\}$. $[\tilde{e}/\tilde{u}]\tilde{g}$ denotes the simultaneous substitution of e_i for all free occurrences of u_i $(i = 1, \ldots, |\tilde{e}|)$ in $\tilde{g}$, where $|\tilde{e}| = |\tilde{u}|$ and all names $\tilde{u}$ are pairwise different.

A symbol x is relevant as token for $\tilde{\alpha}$ if x^p (with $p \neq 0$) or $x^{p|q}$ occurs in the in-set of at least one message descriptor in $\tilde{\alpha}$; $\mathit{relev}(\tilde{\alpha})$ denotes the set of all symbols relevant for $\tilde{\alpha}$: $\mathit{relev}(a_1[\tilde{r}_1][\tilde{s}_1], \ldots, a_n[\tilde{r}_n][\tilde{s}_n]) = \mathit{tok}(\tilde{r}_1, \ldots, \tilde{r}_n)$.

The exact number of available tokens x is known for $\{\tilde{\alpha}\}[\tilde{r}]$ if $x \in \mathit{relev}(\tilde{\alpha})$ and an $x^{p|q}$ occurs in $[\tilde{r}]$; $\mathit{exact}(\{\tilde{\alpha}\}[\tilde{r}])$ denotes the set of all such symbols.

The state descriptor removing relation $\overset{\text{rem } x}{\longrightarrow}$ on message descriptors is defined by $a[\tilde{r}][\tilde{s}] \overset{\text{rem } x}{\longrightarrow} a[\tilde{r}'][\tilde{s}']$, where $[\tilde{r}] = [\tilde{r}']$ if there is an $x^{p|0} \in \{\tilde{r}\}$, and $[\tilde{s}] = [\tilde{s}']$ if there is an $x^{p|0} \in \{\tilde{s}\}$; otherwise $[\tilde{r}']$ and $[\tilde{s}']$ are constructed by removing all state descriptors x^q from $[\tilde{r}]$ and $[\tilde{s}]$, respectively.

A subtyping environment contains subtyping assumptions of the form $u \leq v$.

Structural equivalence $\equiv$ on types and their constituents is the least congruence closed under renaming of bound names (α-conversion) and these rules:

$$(p + p') + q \equiv p + (p' + q) \qquad p + q \equiv q + p \qquad p + 0 \equiv p$$
$$p + p' \equiv q \ (p + p' = q) \qquad p + \infty \equiv \infty \qquad [\tilde{r}, t, \tilde{s}] \equiv [\tilde{r}, \tilde{s}, t]$$
$$[\tilde{r}, x^{p_1|q+1}, x^{p_2|q+1}] \equiv [\tilde{r}, x^{p_1+p_2|q}] \qquad [\tilde{r}, x^0] \equiv [\tilde{r}] \qquad [\tilde{r}, x^p, x^q] \equiv [\tilde{r}, x^{p+q}]$$
$$[\tilde{r}, x^{p_1}, x^{p_2|q}] \equiv [\tilde{r}, x^{p_1+p_2|q}] \qquad \{\tilde{\alpha}, \beta, \beta\} \equiv \{\tilde{\alpha}, \beta\} \quad \{\tilde{\alpha}, \beta, \tilde{\gamma}\} \equiv \{\tilde{\alpha}, \tilde{\gamma}, \beta\}$$

Definition 1. *Type equivalence $\Pi \vdash \sigma \cong \tau$ (or $\sigma \cong \tau$ for empty Π) is the least reflexive, transitive and symmetric relation closed under these rules:*

$$\Pi \vdash \sigma \cong \tau \quad (\sigma \equiv \tau) \tag{equiv$_\cong$}$$

$$\Pi \vdash (\rho\|\sigma)\|\tau \cong \rho\|(\sigma\|\tau) \tag{assoc$_\cong$}$$

$$\Pi \vdash \sigma\|\tau \cong \tau\|\sigma \tag{commut$_\cong$}$$

$$\Pi \vdash \tau\|\{\}[] \cong \tau \tag{empty$_\cong$}$$

$$\Pi \vdash *\underline{u}\{\tilde{\alpha}\}[\tilde{r}] \cong \{[*\underline{u}\{\tilde{\alpha}\}[\tilde{r}]/u]\tilde{\alpha}\}[\tilde{r}] \tag{rec$_\cong$}$$

$$\Pi \vdash \{\tilde{\alpha}\}[\tilde{r}]\|\{\tilde{\gamma}\}[\tilde{s}] \cong \{\tilde{\alpha},\tilde{\gamma}\}[\tilde{r},\tilde{s}] \quad (1) \tag{comb$_\cong$}$$

$$\Pi \vdash *\underline{u}\{\tilde{\alpha}\}[\tilde{r},s] \cong *\underline{u}\{\tilde{\alpha}\}[\tilde{r}] \quad (tok(s) \cap relev(\tilde{\alpha}) = \emptyset) \tag{state$_\cong$}$$

$$\Pi \vdash *\underline{u}\{\tilde{\alpha},\beta\}[\tilde{r}] \cong *\underline{u}\{\tilde{\alpha}\}[\tilde{r}] \quad (\beta=a[\tilde{s},x^{p|0}][\tilde{t}]; x\notin exact(\{\tilde{\alpha}\}[\tilde{r}])) \tag{descr$_\cong$}$$

$$\Pi \vdash *\underline{u}\{\tilde{\alpha}\}[\tilde{r},x^\infty] \cong *\underline{u}\{\tilde{\gamma}\}[\tilde{r}] \quad (x\in relev(\tilde{\alpha}); x\notin tok(\tilde{r}); \tilde{\alpha} \xrightarrow{rem\ x} \tilde{\gamma}) \tag{avail$_\cong$}$$

$$\Pi \vdash *\underline{u}\{\tilde{\alpha},\tilde{\gamma}\}[\tilde{r}] \cong *\underline{u}\{\tilde{\alpha}\}[\tilde{r}] \quad (2) \tag{red$_\cong$}$$

(1) $tok(\tilde{r}) \subseteq relev(\tilde{\alpha}); \forall_x (x \in relev(\tilde{\gamma}) \wedge \exists_{p,q} x^{p|q} \in \{\tilde{r}\}) \Rightarrow \exists_{p,q} x^{p|q} \in \{\tilde{s}\};$
$tok(\tilde{s}) \subseteq relev(\tilde{\gamma}); \forall_x (x \in relev(\tilde{\alpha}) \wedge \exists_{p,q} x^{p,q} \in \{\tilde{s}\}) \Rightarrow \exists_{p,q} x^{p,q} \in \{\tilde{r}\}$

(2) $\forall_{\gamma\in\{\tilde{\gamma}\}}\exists_{\alpha\in\{\tilde{\alpha}\}} \alpha \succeq \gamma$, *where* $x\langle\underline{\tilde{u}}{:}\tilde{\mu},\tilde{\sigma}\rangle[\tilde{s}][\tilde{t},\tilde{t}'] \succeq x\langle\underline{\tilde{u}}{:}\tilde{\mu},\tilde{\tau}\rangle[\tilde{s},\tilde{s}',\tilde{s}''][\tilde{s}'',\tilde{t},\tilde{t}'']$
if $\Pi \vdash \tilde{\tau} \leq \tilde{\sigma}$ *and* $tok(\tilde{s}',\tilde{t}',\tilde{t}'') \cap exact(\{\tilde{\alpha},\tilde{\gamma}\}[\tilde{r}]) = \emptyset = tok(\tilde{t}'') \cap relev(\tilde{\alpha})$

Type combination is an associative, commutative operation with $\{\}[]$ as neutral element. Two types $\{\tilde{\alpha}\}[\tilde{r}]$ and $\{\tilde{\gamma}\}[\tilde{s}]$ are combined to $\{\tilde{\alpha},\tilde{\gamma}\}[\tilde{r},\tilde{s}]$ as specified by (comb$_\cong$). The side-conditions ensure that (1) all symbols irrelevant as tokens have been removed by applying (state$_\cong$), and (2) if a state descriptor $x^{p|q}$ occurs in $[\tilde{r},\tilde{s}]$, state descriptors of this form occur in both $[\tilde{r}]$ and $[\tilde{s}]$ if x is relevant there.

Rule (descr$_\cong$) removes message descriptors if the acceptability of such messages depends on an exact number of tokens, but this number is not known.

State descriptors that specify always available tokens can be removed from states, in-sets and out-sets simultaneously by using (avail$_\cong$).

Rule (red$_\cong$) allows us to remove redundant message descriptors. A message descriptor is redundant if there is another message descriptor that can be used wherever the redundant one can be used.

Definition 2. *Subtyping $\Pi \vdash \sigma \leq \tau$ (or $\sigma \leq \tau$ if Π is empty) is the least reflexive and transitive relation closed under these rules:*

$$\Pi \cup \{u \leq v\} \vdash u \leq v \tag{assmp$_\leq$}$$

$$\frac{\exists_\rho \Pi \vdash \sigma \cong \tau\|\rho}{\Pi \vdash \sigma \leq \tau} \tag{sub$_\leq$}$$

$$\frac{\Pi \vdash \pi \leq \rho \quad \Pi \vdash \sigma \leq \tau}{\Pi \vdash \pi\|\sigma \leq \rho\|\tau} \tag{comb$_\leq$}$$

$$\frac{\Pi \cup \{u \leq v\} \vdash \{\tilde{\alpha}\}[\tilde{r}] \leq \{\tilde{\gamma}\}[\tilde{s}]}{\Pi \vdash *\underline{u}\{\tilde{\alpha}\}[\tilde{r}] \leq *\underline{v}\{\tilde{\gamma}\}[\tilde{s}]} \quad (u \notin free(\tilde{\gamma},\tilde{s},\Pi); v \notin free(\tilde{\alpha},\tilde{r},\Pi)) \tag{rec$_\leq$}$$

Rule (sub$_\leq$) states that σ is a subtype of τ if there is a type ρ such that σ and the combination of τ and ρ are equivalent; ρ specifies the difference between σ and τ. As a special case with $\rho = \{\}[]$, (sub$_\leq$) states that $\Pi \vdash \sigma \cong \tau$ implies $\Pi \vdash \sigma \leq \tau$. With $\rho = \sigma$, $\Pi \vdash \sigma \leq \{\}[]$ follows from (sub$_\leq$). Rules like (rec$_\leq$) were shown to be useful as definition of subtyping for recursive types [1].

Type equivalence and subtyping are decidable relations. A proof can be found in an extended version of the present paper [16].

Using the above examples, it is easy to derive $DS[\mathrm{empty}^\infty] \cong DS'[]$ as well as $DS[\mathrm{empty}^{10}]\|DS[\mathrm{empty}^{40}] \cong DS[\mathrm{empty}^{50}]$ and $DS[\mathrm{empty}^{50}]\|DS'[] \cong DS'[]$ from the rules. Hence, $DS'[] \leq DS[\mathrm{empty}^{50}]$ and $DS[\mathrm{empty}^{50}] \leq DS[\mathrm{empty}^{10}]$ hold. However, $DS''[\mathrm{ok}^{\infty|0}] \leq DS'[]$ does not hold. But, $DS''[\mathrm{ok}^{\infty|0}]$ is a subtype of the type $\{\mathrm{put}\langle\underline{u}{:}\mathrm{type}, u\rangle[\mathrm{ok}][\mathrm{full}], \mathrm{get}\langle Back\,[\mathrm{one}]\rangle[\mathrm{ok}, \mathrm{full}][]\}[\mathrm{ok}^{\infty|0}]$. Although this type specifies the same set of acceptable message sequences as $DS'[]$ (see below), the types are not equivalent.

4 Message Signature Sequences

Definition 3. *A type* $\{\tilde{\alpha}\}[\tilde{r}]$ *with* $\alpha_i = x_i\langle\tilde{\underline{u}}_i{:}\tilde{\mu}_i, \tilde{\tau}_i\rangle[\tilde{s}_i][\tilde{t}_i]$ *is deterministic if for all* $i, j = 1..|\tilde{\alpha}|$ $(i \neq j)$ *with* $x_i = x_j$ *and* $|\tilde{\mu}_i, \tilde{\tau}_i| = |\tilde{\mu}_j, \tilde{\tau}_j|$ *implies one of these:*

- $\exists_{x,p}\,(\exists_{\tilde{t}'}\,[\tilde{s}_i] \equiv [x^{p|0}, \tilde{t}'] \wedge p \not\equiv \infty \wedge \forall_q x^q \notin \{\tilde{t}'\}) \Rightarrow \exists_{\tilde{s}'}\,[\tilde{s}_j] \equiv [x^{p+1}, \tilde{s}']$;
- *or* $\exists_{x,p}\,(\exists_{\tilde{t}'}\,[\tilde{s}_j] \equiv [x^{p|0}, \tilde{t}'] \wedge p \not\equiv \infty \wedge \forall_q x^q \notin \{\tilde{t}'\}) \Rightarrow \exists_{\tilde{s}'}\,[\tilde{s}_i] \equiv [x^{p+1}, \tilde{s}']$.

The type of each object shall be deterministic to ensure that, when a message is accepted, this type is updated in the same way as the corresponding type mark was updated when the message was sent. Deterministic types are free of redundant message descriptors. All examples of types of the form $\{\tilde{\alpha}\}[\tilde{r}]$ shown above are deterministic. Types used as type marks need not be deterministic, but they are supertypes of deterministic types. Subtyping keeps all important properties of deterministic types. Especially, each type of the form $\sigma\|\tau$ in a type-consistent system is equivalent to a type of the form $\{\tilde{\alpha}\}[\tilde{r}]$. Hence, in the rest of this paper, we consider mainly types of this form.

Definition 4. *A message signature a is active with follow-state* $[\tilde{s}]$ *in a type* $\{\tilde{\alpha}\}[\tilde{r}]$ *if* $a[\tilde{s}] \in act(\{\tilde{\alpha}\}[\tilde{r}])$, *where act is defined by:*

$$act(\{\tilde{\alpha}\}[\tilde{r}]) = \{a[\tilde{t}, \tilde{t}'] \mid a[\tilde{s}][\tilde{t}] \in \{\tilde{\alpha}\}; \; [\tilde{r}] \equiv [\tilde{s}, \tilde{t}']; \; \forall_{x,p}\, x^{p|0}{\in}\{\tilde{s}\} \Rightarrow x{\notin}tok(\tilde{t}');$$
$$\forall_{x,p}(x^\infty{\in}\{\tilde{r}\} \wedge x^{p|0}{\notin}\{\tilde{s}\}) \Rightarrow x^\infty{\in}\{\tilde{t}'\} \}$$

If a message signature a is active with a follow-state $[\tilde{s}]$ in a type $\{\tilde{\alpha}\}[\tilde{r}]$, an object of this type accepts a (single) message of signature a. When the message is accepted, the type is updated to $\{\tilde{\alpha}\}[\tilde{s}]$. The signatures of the messages acceptable next are active in $\{\tilde{\alpha}\}[\tilde{s}]$. Type marks are updated in the same way when sending messages. A message descriptor $a[\tilde{s}][\tilde{t}]$ is active in a type $\{\tilde{\alpha}\}[\tilde{r}]$ with follow-state $[\tilde{r}']$ if $a[\tilde{s}][\tilde{t}] \in \{\tilde{\alpha}\}$ and $a[\tilde{r}'] \in act(\{a[\tilde{s}][\tilde{t}]\}[\tilde{r}])$.

Definition 5. *The set seq(τ) of all message signature sequences conforming to a type τ of the form {ᾱ}[r̃] is constructed inductively:*

$$S_0 = \{\langle\rangle[\tilde{r}]\}$$
$$S_{i+1} = \{\langle\tilde{a}, x\langle\underline{\tilde{u}}:\tilde{\mu}, \tilde{\sigma}\rangle\rangle[\tilde{s}] \mid \langle\tilde{a}\rangle[\tilde{t}] \in S_i;\ x\langle\underline{\tilde{v}}:\tilde{\mu}, \tilde{\tau}\rangle[\tilde{s}] \in act(\{\tilde{a}\}[\tilde{t}]);$$
$$\tilde{\sigma} \leq [\tilde{u}/\tilde{v}]\tilde{\tau};\ \{\tilde{u}\} \not\subseteq free(\tilde{\tau}) \setminus \{\tilde{v}\} \} \qquad (i \geq 0)$$
$$seq(\tau) = \{\langle\tilde{a}\rangle \mid \langle\tilde{a}\rangle[\tilde{s}] \in \bigcup_{i \geq 0} S_i\}$$

Theorem 1. *Let σ and τ be types of the form {ᾱ}[r̃]. Then, σ ≅ τ implies seq(σ) = seq(τ), and σ ≤ τ implies seq(τ) ⊆ seq(σ).*

Proof. First, the proof of $seq(\sigma) = seq(\tau)$ if $\sigma \cong \tau$ is sketched. Most type equivalence rules do not influence whether a message signature is active. Only (red$_\cong$) is a bit more difficult. Let σ be of the form $\{\tilde{\alpha}\}[\tilde{r}]$ and τ of the form $\{\tilde{\gamma}\}[\tilde{s}]$; and let (red$_\cong$) be applied to σ and τ such that $[\tilde{r}] \cong [\tilde{s}, \tilde{t}]$ for some $\tilde{t}$ with $tok(t) \cap relev(\tilde{\gamma}) = \emptyset$, and $\{\tilde{\alpha}\} \cong \{\tilde{\gamma}, \tilde{\beta}\}$ for some $\tilde{\beta}$; the message descriptors $\tilde{\beta}$ are removed. Hence, $seq(\tau) \subseteq seq(\sigma)$. A message descriptor is deleted only if there remains a more general message descriptor being active whenever the removed one was active. This implies $seq(\sigma) \subseteq seq(\tau)$ and $seq(\sigma) = seq(\tau)$.

If there is a type $\{\tilde{\alpha}\}[\tilde{r}]$ with $\{\tilde{\alpha}\}[\tilde{r}] \cong \sigma\|\tau$, then $\langle\tilde{a}\rangle \otimes \langle\tilde{c}\rangle \subseteq seq(\{\tilde{\alpha}\}[\tilde{r}])$ for each $\langle\tilde{a}\rangle \in seq(\sigma)$ and $\langle\tilde{c}\rangle \in seq(\tau)$, where $\langle\tilde{a}\rangle \otimes \langle\tilde{c}\rangle$ is the set of all arbitrary interleavings of $\langle\tilde{a}\rangle$ and $\langle\tilde{c}\rangle$. This is easy to see from the definition of seq.

Let $\sigma \leq \tau$. There is a type ρ with $\sigma \cong \tau\|\rho$ according to (sub$_\leq$) and, therefore, $\langle\tilde{a}\rangle \otimes \langle\rangle \subseteq seq(\sigma)$ for all $\langle\tilde{a}\rangle \in seq(\tau)$. Hence, $seq(\tau) \subseteq seq(\sigma)$. ☐

The reverses of Theorem 1 do not hold: $seq(\sigma) = seq(\tau)$ does not imply $\sigma \cong \tau$, and $seq(\tau) \subseteq seq(\sigma)$ does not imply $\sigma \leq \tau$. A more accurate characterization of $\cong$ and $\leq$ concerning message signature sequences considers possible extensions of types. If $\sigma \cong \tau$ (or $\sigma \leq \tau$) holds, $\sigma\|\rho \cong \tau\|\rho$ (or $\sigma\|\rho \leq \tau\|\rho$) are also expected to hold for each type ρ combinable with σ and τ. The next theorems show soundness and completeness of $\cong$ and $\leq$: $seq(\sigma\|\rho) = seq(\tau\|\rho)$ (or $seq(\tau\|\rho) \subseteq seq(\sigma\|\rho)$) for all appropriate ρ is equivalent to $\sigma \cong \tau$ (or $\sigma \leq \tau$).

Definition 6. *Two types σ and τ of the form {ᾱ}[r̃] are extensibility-equivalent (denoted by σ ≃ τ) if and only if for each type ρ: $seq(\{\tilde{\beta}\}[\tilde{s}]) = seq(\{\tilde{\gamma}\}[\tilde{t}])$ for all $\tilde{\beta}$, $\tilde{\gamma}$, $\tilde{s}$ and $\tilde{t}$ with $\sigma\|\rho \cong \{\tilde{\beta}\}[\tilde{s}]$ and $\tau\|\rho \cong \{\tilde{\gamma}\}[\tilde{t}]$.*

Theorem 2. *Let σ and τ be types of the form {ᾱ}[r̃]. Then, σ ≅ τ ⟷ σ ≃ τ.*

Proof. $\sigma \cong \tau \Rightarrow \sigma \simeq \tau$ follows from Theorem 1. The other direction can be shown by a case analysis on all reasons for $\sigma \not\cong \tau$ if $seq(\sigma) = seq(\tau)$. A type ρ not satisfying the conditions of Def. 6 is constructed for each reason [16]. ☐

Definition 7. *For two types σ and τ of the form {ᾱ}[r̃], σ is extensibility-substitutable for τ (denoted by σ ≲ τ) if and only if for each type ρ: $seq(\{\tilde{\gamma}\}[\tilde{t}]) \subseteq seq(\{\tilde{\beta}\}[\tilde{s}])$ for all $\tilde{\beta}$, $\tilde{\gamma}$, $\tilde{s}$ and $\tilde{t}$ with $\sigma\|\rho \cong \{\tilde{\beta}\}[\tilde{s}]$ and $\tau\|\rho \cong \{\tilde{\gamma}\}[\tilde{t}]$.*

Theorem 3. *Let σ and τ be types of the form $\{\tilde{\alpha}\}[\tilde{r}]$. Then, $\sigma \leq \tau \Leftrightarrow \sigma \lesssim \tau$.*

Proof. $\sigma \leq \tau \Rightarrow \sigma \lesssim \tau$ follows from Theorem 1. The other direction can be shown by a case analysis on all reasons for $\sigma \not\leq \tau$ if $seq(\tau) \subset seq(\sigma)$. A type ρ not satisfying the conditions of Def. 7 is constructed for each reason [16].   □

5   Related Work

Most work on types in concurrent systems is based on Milner's π-calculus [8, 7] and similar calculi. The problems of inferring most general types [4, 19] and subtyping [11, 12, 18, 3, 17] were considered. But these type systems cannot represent message sequences and ensure statically that all sent messages are acceptable.

A large amount of work based on "path expressions" and process algebra shows that reasoning about the order of messages in concurrent systems is quite difficult. Nierstrasz [11] proposes "regular types" and "request substitutability" as foundations of subtyping. His very general results are not concrete enough to develop a static type system from them. A similar definition of subtyping was given by Bowman et al. [2]. The proposal of Nielson and Nielson [10] can deal with constraints on the ordering of messages. Their type system cannot ensure that all sent messages are understood. But, subtyping is supported so that instances of subtypes preserve the properties expressed in supertypes.

The work of Liskov and Wing on behavioral subtyping [6] shows the importance of "constraints" in subtyping: Assertions on methods are not sufficient to specify an object's behavior at the presence of subtyping and aliases.

Process types as presented in [14, 15] improve previous work on process types represented as expressions in a process calculus [13]. The present work increases the expressiveness of process types by allowing the acceptability of messages to depend on exact numbers of available tokens. Other than in previous work, the type equivalence relation and subtyping relation are shown to be equivalent to an extensibility-equivalence relation and an extensibility-substitutability relation, respectively. The object calculus and type checking rules presented in [14, 15] can be used together with non-regular process types.

The proposal of Najm and Nimour [9] has a similar purpose as process types. However, in their approach at each time only one client is allowed to interact with a server through an interface specifying acceptable message changes. Process types do not have this restriction because of type splitting.

6   Conclusions

The expressiveness of process types need not be restricted to regular sets of acceptable message sequences. Type equivalence and subtyping are decidable, sound and complete for non-regular process types, provided that these relations conform to an extensibility criterion.

Acknowledgments

Many thanks to Christof Peter for his useful comments. This work was supported by the FWF (Fonds zur Förderung wissenschaftlicher Forschung) under project number P12703-INF (Static Process Types for Active Objects).

References

[1] R. M. Amadio and L. Cardelli. Subtyping recursive types. In *Conference Record of the 18th Symposium on Principles of Programming Languages*, pages 104–118. ACM, 1991.

[2] H. Bowman, C. Briscoe-Smith, J. Derrick, and B. Strulo. On behavioural subtyping in LOTOS. In *Proceedings FMOODS '97*, Canterbury, United Kingdom, July 1997.

[3] J.-L. Colaco, M. Pantel, and P. Salle. A set-constraint-based analysis of actors. In *Proceedings FMOODS '97*, Canterbury, United Kingdom, July 1997. Chapman & Hall.

[4] S. J. Gay. A sort inference algorithm for the polyadic π-calculus. In *Conference Record of the 20th Symposium on Principles of Programming Languages*, Jan. 1993.

[5] J. E. Hopcroft and J. D. Ullman. *Formal Languages and their Relation to Automata*. Addison-Wesley, 1969.

[6] B. H. Liskov and J. M. Wing. A behavioral notion of subtyping. *ACM Transactions on Programming Languages and Systems*, 16(6):1811–1841, Nov. 1994.

[7] R. Milner. The polyadic π-calculus: A tutorial. Technical Report ECS-LFCS-91-180, Dept. of Comp. Sci., Edinburgh University, 1991.

[8] R. Milner, J. Parrow, and D. Walker. A calculus of mobile processes (parts I and II). *Information and Computation*, 100:1–77, 1992.

[9] E. Najm and A. Nimour. A calculus of object bindings. In *Proceedings FMOODS '97*, Canterbury, United Kingdom, July 1997.

[10] F. Nielson and H. R. Nielson. From CML to process algebras. In *Proceedings CONCUR'93*, number 715 in Lecture Notes in Computer Science, pages 493–508. Springer-Verlag, 1993.

[11] O. Nierstrasz. Regular types for active objects. *ACM SIGPLAN Notices*, 28(10):1–15, Oct. 1993. Proceedings OOPSLA'93.

[12] B. Pierce and D. Sangiorgi. Typing and subtyping for mobile processes. In *Proceedings LICS'93*, 1993.

[13] F. Puntigam. Types for active objects based on trace semantics. In E. N. et al., editor, *Proceedings FMOODS '96*, Paris, France, Mar. 1996. IFIP WG 6.1, Chapman & Hall.

[14] F. Puntigam. Coordination requirements expressed in types for active objects. In M. Aksit and S. Matsuoka, editors, *Proceedings ECOOP '97*, number 1241 in Lecture Notes in Computer Science, Jyväskylä, Finland, June 1997. Springer-Verlag.

[15] F. Puntigam. Dynamic type information in process types. In D. Pritchard and J. Reeve, editors, *Proceedings EuroPar '98*, number 1470 in Lecture Notes in Computer Science, Southampton, England, Sept. 1998. Springer-Verlag.

[16] F. Puntigam. Non-regular process types. Technical report, Institut für Computersprachen, Technische Universität Wien, Vienna, Austria, 1999.

[17] A. Ravara and V. T. Vasconcelos. Behavioural types for a calculus of concurrent objects. In *Proceedings Euro-Par '97*, Lecture Notes in Computer Science. Springer-Verlag, 1997.

[18] V. T. Vasconcelos. Typed concurrent objects. In *Proceedings ECOOP'94*, number 821 in Lecture Notes in Computer Science, pages 100–117. Springer-Verlag, 1994.

[19] V. T. Vasconcelos and K. Honda. Principal typing schemes in a polyadic pi-calculus. In *Proceedings CONCUR'93*, July 1993.

Decision Procedure for Temporal Logic of Concurrent Objects

Jean-Paul Bahsoun, Rami El-Baïda, and Hugues-Olivier Yar

Institut de Recherche en Informatique de Toulouse
118, route de Narbonne
31062 Toulouse cedex, France
{bahsoun, baida, yar}@irit.fr

Abstract. We present a tableau method for the temporal logic of concurrent objects. This method, which combines tableaux and model checking, is based on the semantics of systems written in an object-oriented programming model and a three level logic: *action* (Hoare triple-like notation), *object* (temporal logic allowing stuttering-states), and *system* (linear temporal logic taking into account communication axioms) [BMS95]. The particularity of this method is that it ensures, once a formula proved valid, that every execution of our programming model satisfies this formula.

1 Introduction

During the past decade, the concurrent object-oriented approach to software development has found wide acceptance. However, a formal method is one of the most suitable tools ensuring that a software system developed in such approaches behaves as expected. Among formal methods that integrate the object concept, only a few are devoted to verification of software systems [Ame90, Mes93, AH87] or take advantage of object oriented structuration at the formal level [FM95].

In this paper, our goal is to give an algorithmic method to prove properties, expressed in (propositional) linear temporal logic, of concurrent object-oriented systems. We suppose that these systems are written in the programming model presented in [BMS93]. We transform the system into an appropriate temporal logic (taking into account the semantics of the system and especially communication between objects) called Temporal Logic of Concurrent Objects (TLCO) [BMS95]. By combining model checking and tableaux, reasoning will be reduced to a simple one in a variant of linear temporal logic, instead of reasoning on the system and its properties. In fact, proving that a property is valid in the propositional part of TLCO will automatically show that our model always satisfies this property.

The rest of the paper starts with the presentation of the programming model for concurrent objects [BMS93]. Section 3 presents the logical framework (based on a three level logic: action, local and system) to formally design object systems and prove their properties [BMS95]. Finally the tableau method of the propositional part of TLCO is developed in section 4.

P. Amestoy et al. (Eds.): Euro-Par'99, LNCS 1685, pp. 1344–1352, 1999.
© Springer-Verlag Berlin Heidelberg 1999

2   A Concurrent Object-Oriented Language to Model Reactive Systems

The basis of this language is an object. With every object we associate : a unique identity, a set of objects or variables that describe the private state of the object, and a set of actions that describe the object behavior. This set is subdivided into two disjoint subsets of methods and reflexes. Methods are executed in response to requests from other objects, while reflexes are executed spontaneously by the object. A guard is associated with every action which must be true before the action is executed. The body of an action is expressed by multiple (and possibly conditional) assignments, and primitives to send messages.

Figure 1 contains the definition of a simple client-server system. For more information on this approach see [BMS93, BMS95]

```
class server                                class client(s: server)

    control                                     control
        inUse : boolean := false                    owner, waiting : boolean := false

    method  Enter(c:client; m:method)           reflex  Request
        when  not inUse                             when  (not owner) and (not waiting)
        ==> inUse := true                           ==> waiting := true
        do send-asy   (c, "m")                      do send-asy   (s, "Enter", me, "Receive")

    method  Leave                               method  Receive
        when  inUse                                 ==> (owner, waiting) := (true, false)
        ==> inUse := false
                                                reflex  Return
end  server                                         when  owner
                                                    ==> owner := false
                                                    do send-asy   (s, "Leave")

                                            end  client
```

Fig. 1. Client and Server classes

3   A Temporal Logic for Concurrent Objects (TLCO)

The TLCO formalism for proving properties of objects and object systems, reflects the two mechanisms we had to develop their structural definition by inheritance and parallel composition. Unlike many traditional temporal logic [Lam91, MP92], TLCO is compositional w.r.t. parallel composition, and sub-classes inherit their ancestors' temporal properties [BFS97]. To do this, three levels of reasoning about systems built from concurrently executing objects, each focusing on one particular aspect or view of object systems, has been proposed.

The *action level* is concerned with the local effects of single methods or reflexes offered by an object and ignores all communication related aspects of actions. The execution of a method or reflex transforms the object's local state. Therefore, a simple formalism (Hoare triple-like notation) based on pre- and post-conditions of methods and reflexes is sufficient at this level.

At the *object level*, we reason about behaviors of individual objects of a given class, dealing with both safety and liveness properties. The proof rules take advantage of the encapsulation of an object's private state which may only change by executing methods or reflexes defined for the object. In contrast to the action level, initialization and fairness conditions are taken into account. Aspects of communication and interaction with other objects are not dealt with except in the form of environment assumptions.

Finally, the *system level* models the top-level view of the entire system by an external observer. In particular, it cannot refer to the internal state of any object. Only the execution of actions by objects and the transmission of messages are visible. Typical properties of interest include synchronization and liveness involving several objects.

Three different formal languages and proof rules, one for each aspect of reasoning has been defined. These languages have been related by "interface rules". The following sections present briefly the object level and its local logic as well as the system level and its system logic. For more detail on the TLCO formalism see [BMS95].

3.1   The Object Level

At this level, we reason about behaviors of individual objects of a given class. A "local" temporal logic, $\mathcal{L}_a$ for each object a in the model, capable of expressing properties about infinite state sequences corresponding to the runs of the object is defined. The semantics of $\mathcal{L}_a$ differs from that of conventional linear-time temporal logic in that it takes into account of "Stuttering-States" [Lam91] which correspond to environment steps that pass completely unobserved by a. The syntax, semantics and an axiom system for the local temporal logic $\mathcal{L}_a$ associated to an object a, are given in [BMS95]. A tableau-method for its propositional part is also given in [Ser95].

Notation: $\alpha(t_1, \ldots, t_n)$ and $!\alpha(t_1, \ldots, t_n)$ are atomic formulas, representing the execution of the action α with parameters $t_1, \ldots, t_n$ and an outstanding request to execute the method α with parameters $t_1, \ldots, t_n$.

Example 1. The safety property: Enter(c) $\rightarrow \bigcirc_s(\neg$Enter($c'$) unless Leave) states that any two execution of Enter by the server have to be separated by an execution of Leave.

3.2   The System Level

Properties at the system level concern the interaction of objects. Private states of individual objects are invisible. A "system" temporal logic [BMS95]$\mathcal{L}$ to express system properties, ensuring a smooth interface between local and system reasoning, is employed. Connectives are the same as in linear temporal logic, but there is atomic formulas related to the programming model. These formulas are: a, $a.\alpha(t_1, \ldots, t_n)$ and $a.!\alpha(t_1, \ldots, t_n)$ (where a is an object of the system and α is an action of this object). The semantics of these formulas is: the object a is active, a is active and executing the action α, and there is an incoming request to a demanding the execution of the method α.

3.3   Formal Reasoning in System Temporal Logic $\mathcal{L}$

Such systems have been studied in [MP92, Gab92], among others. We only extend classical axioms to take into account communications produced by objects of the system. We suppose that messages are modeled by sets which means that identical messages are merged and thus some messages are lost. Under this hypothesis, the whole system is finite and decidable. In the following, we model asynchronous communication between objects.

First, we need formulas specifying that every sent message will arrive at its destination. For every statement $send(x, !\beta)$ appearing in the definition of some method or reflex α of an object a, we add an axiom "com1". The second group of axioms expresses that messages are consumed by executing the requested method. Such axioms take the form "com2" where the disjunction is over all those actions of objects which may cause the sending of a message $!\alpha$ to the object a (and since is the strong version of the operator backto). Moreover, we assume that there are no unprocessed requests at the beginning, formalized by the axioms "com3".

$$
\begin{array}{ll}
a.\alpha \to \bigcirc_b \Diamond b.!\beta & \text{(com1)} \\
\bigcirc_a a.!\alpha \to \bigcirc_a(\neg a.\alpha \text{ since } (b_1.\beta_1 \vee \ldots \vee b_n.\beta_n)) & \text{(com2)} \\
\text{first} \to \neg a.!\alpha & \text{(com3)}
\end{array}
$$

Example 2. Consider a system consisting of one **server** object and two client objects, say $\mathcal{A} = \{s, c1, c2\}$. An appropriate interleaving of local reasoning and applications of the communication axioms allows to prove

$$
\bigwedge_{i=1}^{2}(ci.Receive \to \bigwedge_{j \neq i} \neg cj Receive \text{ unless } ci.Return),
$$

i.e. mutual exclusive access of the clients to the shared resource.

4   A Tableau-Method for TLCO

The tableau methods developed for $\mathcal{L}_a$ and $\mathcal{L}$ are extensions of the tableau method for the propositional part of classical temporal logic presented in [Wo85]. Logical connectives are treated as in classical logic (see [Sm68]) and the temporal connectives are decomposed in a requirement on current state and requirements on the rest of the sequence. In favor of this decomposition, the tableau construction, which is an oriented graph, is made in by seeking, state by state, a model for the treated formula.

In the rest, we present a tableau method for the propositional part of the system logic $\mathcal{L}$. Our method uses some local properties that can be demonstrated in tableau methods for local logic $\mathcal{L}_a$ presented in [Ser95].

4.1   A Tableau-Method for the System Logic $\mathcal{L}$ Formulas

In this section, we present a tableau method (and a decision procedure) for the propositional part of the system logic $\mathcal{L}$. This logic, though being a linear temporal logic, depends on the semantic of the system (precisely on the semantic of each object in the system). Therefore a classical tableau method is not sufficient for proving the satisfiability of system formulas. In the *Client-Server* system for example, the (mutual exclusive) property (see previous page) taken as a formula in a classical linear temporal logic, can be said to be not valid, if we assume a state in which $c1.Receive$ and $c2.Receive$ are true simultaneously. However such state is not accessible by the *Client-Server* system.

The question we should set is thus: "A state where actions α and β are both executed is accessible or not by the system?". Therefore, for every state s_i, actions that each object can execute in s_i (actions whose preconditions are true in s_i), as well as in s_{i+1} (reflexes and methods with requests) should be known. Knowing that private states of objects are invisible at the system level, our solution was to combine model checking and tableau method. Thus, in the graph construction, we look only for states reachable by the system. For that, we use properties of local logic $\mathcal{L}_a$.

By local tableau methods, we look for satisfiable formulas of the forms:

$$\text{first} \rightarrow O_a a.\alpha \quad (\alpha \in \text{Act}(a))$$
$$a.\alpha \rightarrow O_a a.\beta \quad (\beta \in \text{Act}(a))$$

where $\text{Act}(a)$ is the set of objects in the system.

The first type of formulas gives us, for each object a, actions whose preconditions are true in the initial state of a and thus can be executed in the next state (in which a becomes active). The second type of formulas indicates the actions whose preconditions become true after the execution of an action α by the object a. We then note:

$$First = \{a.\alpha \mid a \in \mathcal{A}, \alpha \in \text{Act}(a) \text{ and first} \rightarrow O_a a.\alpha \text{ is satisfiable in } \mathcal{L}_a\}$$
$$Nextt_{a.\alpha} = \{a.\beta \mid \beta \in \text{Act}(a) \text{ and } a.\alpha \rightarrow O_a a.\beta \text{ is satisfiable in } \mathcal{L}_a\}$$

By using these two sets we can know, in each state s_i, the set of actions the system can execute in the state s_{i+1}, and thus, we can test all possible executions of the system. In the rest, we propose a tableau method for the propositional part of the system logic $\mathcal{L}$.

Node of the Graph. To test the satisfiability of a formula f, the tableau method constructs an oriented graph. Each node n is a quadruple (T_n, E_n, Act, Exe) of sets. T_n is a set of formulas that should be true in the node n. E_n is a set of labels (actions that are executed in n). Act and Exe are sets of actions that can be executed in the next state, only defined in special nodes called *states*. Once a formula f is treated, we mark this formula with the symbol $*$ (we obtain f^*) in order to not treat twice a formula.

A node containing only elementary ($\bigcirc f$, atomic or the negation of atomic formulas) or marked formulas is called *pre-state*. A *state* is a pre-state in which Act_n and Exe_n are calculated.

Only the graph states correspond to the states of a temporal structure. The other nodes help to determine the mark (T_n) and label (E_n) of their successors.

Decomposition Rules. Their major role is to decompose a non elementary formula into elementary formulas. For each non elementary formula f, they associate a set Σ of sets S_i of formulas f_j as shown in figure 2. Other decomposition rules for other temporal operators such as: $\Diamond(f) \rightarrow \{\{f\}, \{\bigcirc(\Diamond f)\}\}$ can also be established.

D1	$\neg\neg f$	$\rightarrow$	$\{\{f\}\}$
D2	$\neg\bigcirc f$	$\rightarrow$	$\{\{\bigcirc\neg f\}\}$
D3	$f_1 \wedge f_2$	$\rightarrow$	$\{\{f_1, f_2\}\}$
D4	$f_1 \vee f_2)$	$\rightarrow$	$\{\{f_1\}, \{f_2\}\}$
D5	$\Box f$	$\rightarrow$	$\{\{f, \bigcirc\Box f\}\}$
D6	$\neg\Box f$	$\rightarrow$	$\{\{\neg f\}, \{\bigcirc\neg\Box f\}\}$
D7	f_1 unless f_2	$\rightarrow$	$\{\{f_2\}, \{f_1, \bigcirc(f_1 \text{ unless } f_2)\}\}$
D8	$\neg(f_1 \text{ unless } f_2)$	$\rightarrow$	$\{\{\neg f_1, \neg f_2\}, \{\neg f_2, \bigcirc\neg(f_1 \text{ unless } f_2)\}\}$

Fig. 2. *: Decomposition rules*

The Graph Construction. Let's start with some preliminary notations and their intuitive means.

We note *F-State(n)* (father-state) the nearest ancestor node of n which is a state if it exists ($p(n) = F\text{-}State(n)$). We also define these sets of formulas :

- $F^{Next} = \{f \mid \bigcirc f \in F\}$ where F is a set of formulas.
- $Act = First$ (the initial one)
- $Exe = \{a.\alpha \mid a \in A,\ \alpha \in Refl(a) \text{ and } a.\alpha \in Act\} \cup \{a.\alpha \mid a.\alpha \in Act \text{ and } a.!\alpha \in T_n\}$
- $Next = 2^{Exe} - \{E \mid E \in 2^{Exe} \text{ and } \exists a.\alpha, a.\beta \in E \text{ with } \alpha \neq \beta\}$

- $Idle = \{a.\tau \mid \exists a.!\alpha \in T_n, a.!\alpha \notin T_{p(n)}\}$ and $\not\exists\beta \in Act(a) \mid a.\beta \in E_n\}$

The formula $\bigcirc f$ is satisfied in a state s_i if and only if f is satisfied in the state s_{i+1}. Other elementary formulas are atomic or the negation of atomic formulas, and their values in the state s_i do not depend on their values in the state s_{i+1}. For that reason we have defined F^{Next}.

Act, at the beginning, is the set of actions whose preconditions are true in the initial state, and *Exe* contains always actions of *Act* that can be executed in the next state (i.e. methods having requests and/or reflexes in *Act*). *Next* is then the set of actions that can be executed simultaneously (by the system), and *Idle* indicate the arrival of new requests.

Decision Procedure for $\mathcal{L}$. If f is the formula we want to test its satisfiability, the initial node is marked by the formula $\Diamond f$ and has the label $a.\tau$ for every $a \in \mathcal{A}$.

In the algorithm, "create a son node of n", signifies that this son is effectively created only if an identical node does not already exist. If so, we only add an arrow leading to this node. Two nodes are said to be identical if they are marked by same formulas and labels, and if Act and Exe of their fathers are equals.

The graph construction is made by repeating the following stages:

- Stage 1: If a node $n = (T_n, E_n)$, contains a non elementary formula f non marked, and if the decomposition rule for f is $f \rightarrow \{S_i\}$, then for each S_i, create a son node $((T_n - \{f\}) \cup S_i \cup \{f^*\}, E_n)$ where f^* is f marked by the symbol $*$.
- Stage 2: if a node $n = (T_n, E_n)$, is a pre-state, and $p(n)$ does not exist, then create the son node $(T_n, E_n, First, Exe)$. If $p(n)$ exists, then create the son node $(T_n, E_n \cup Idle, Act, Exe)$ where: $Act = (Act(p(n)) - \bigcup_{a.\alpha \in E_n} (\{a.\alpha \mid \alpha \in Act(a)\} - Nextt_{a.\alpha})) \cup \bigcup_{a.\alpha \in E_n} (Nextt_{a.\alpha})$

- Stage 3: If a node n is a state (T_n, E_n, Act, Exe) then, for each set G in $Next$, create the son node of n $T_n^{Next} \cup G \cup Com \cup Mes, G$ where:
  - $Com = \{\bigcirc \Diamond b.!\beta \mid a.\alpha \in G \text{ et } a.\alpha \rightarrow \bigcirc_b \Diamond b.!\beta\}$
  - $Mes = \{a.!\alpha \mid a.!\alpha \in T_n \text{ et } a.\alpha \notin E_n\}$

The first stage, repeated enough, on a starting node and his descendant, gives us nodes containing only elementary or marked formulas (pre-state).

Let us justify the stage 2. If in the node, the system is still inactive, the initial set Act is the set $First$ (actions whose preconditions are true in the initial state of each object). Exe is then the set of reflexes in Act (no requests at the beginning). If not (which means there are actions that have been executed, then for each action $a.\alpha$ executed in this node (i.e. appearing in E_n), the effect of this action on other actions (of the same object) is known through the set $Nextt_{a.\alpha}$. We should then remove, from $Act(p(n))$ of n's father, all the actions that their preconditions become false after the execution of $a.\alpha$ (those that do not figure in $Nextt_{a.\alpha}$) and add those of $Nextt_{a.\alpha}$. Exe is always the set of reflexes and methods having requests in Act. On the other hand, if a new request $a.!\alpha$ is obtained, during the decomposition of a formula, and if the object a is inactive in this state (i.e. $\forall \beta \in Meth(a), a.\beta \notin E_n$), then we should add $a.\tau$ to E_n. And that is what $Idle$ does.

As for the third stage, recall that we look for satisfying T_n formulas in a state s_i. Elementary formula $\bigcirc f$ is satisfied in s_i, iff the formula f is satisfied in s_{i+1}. Through the graph construction, we are sure that atomic formulas (a or $a.\alpha$) are satisfied in this state, if it is not the case, this node in eliminated by the elimination rules E3 or E5. Atomic formulas $a.!\alpha$ with $\alpha \in Meth(a)$, once becoming true, remain true till the execution of the method $a.\alpha$. Therefore we have added sets T_n^{Next} and Mes. Communication produced by the execution of an action, are added through the set Com. The actions that the system can execute in the next state are also added (the set G), and that is done for every possible combination of actions.

Finally the algorithm is: Repeat the first stage as much as possible. Leafs nodes of the tree obtained from the initial node are pre-states. We then apply the stage 2 on all pre-states in order to obtain states. Finally, each state gives, by applying the stage 3, one or several nodes (we repeat this until we obtain terminal states or states that do not contain $\bigcirc f$).

For deciding the satisfiability of the initial formula, we eliminate the nodes of the graph constructed using the elimination rules (figure 3). The decision procedure is terminated when all the unsatisfied nodes are eliminated. If the initial node is eliminated, then the formula $\Diamond f$ (the one of the initial node) is unsatisfied in the initial state of all temporal structure.

E1 The node n contains a formula f and its negation $\neg f$.
E2 The node n contains $a.\alpha$ and $a.\beta$ where $\alpha, \beta \in Act(a)$, and $\alpha \neq \beta$.
E3 The node n is marked by $a.\alpha$, and $a.\alpha \notin E_n$.
E4 The node n is marked by $T_n \mid a.\alpha \in T_n$, $\alpha \in Meth(a)$ and $a.!\alpha \notin T_n$.
E5 The node contains an elementary formula a, and for every action $\alpha \in \mathcal{O}_a \cup \tau$, we have $a.\alpha \notin E_n$.
E6 All the successors of the node are eliminated.
E7 The node n is a state which contains $a.!\alpha$, and there exists no path coming from this state and leading to a node containing $a.\alpha$
E8 The node n is a state and contains a formula of the form $\neg\Box f$ or $\neg(f_1$ unless $f_2)$ which is not s atisfied in that state. A formula $\neg\Box f$ (resp. $\neg(f_1$ unless $f_2)$) is said to be satisfied in a state n, if there exists, in the graph, a road coming from n and leading to a node containing the formula $\neg f$ (resp. $\neg f_1$).

Fig. 3. Elimination rules

5   Conclusion

In this paper, we have presented a method for verifying properties of concurrent object-oriented systems (written in TLCO which take into consideration the two mechanisms: *inheritance* and *parallel composition* [BFS97]) along with a decision procedure for algorithmic verification of such formulas. Recall that soundness and completeness of the axioms of the propositional part of local logic $\mathcal{L}_a$ have been proved in [Ser95]. Our method also uses results of the tableau method for local logic presented in that thesis. With these methods, we compute the sets of actions whose preconditions become true after the execution of each action in the system along with those whose preconditions are true in the initial state of the system (i.e. initial state of each object in the system).

Once computed, we can use them in the proof of any property of the system. Our tableau method will search for every possible execution of the system leading to a state verifying the property.

In future work, we shall investigate whether the axioms for the system part $\mathcal{L}$ are also sound and complete, and study the complexity of the decision procedure in order to improve further implementation of our method.

References

[AH87]   G. Agha, C. Hewitt: *Concurrent Programming Using Actors*. In: Object-Oriented Concurrent Programming, A. Yonezawa, M. Tokoro (eds), MIT Press, pp. 37–53 (1987)

[Ame90]   P. America: *Designing an Object-Oriented Programming Language with Behavioural Subtyping* in Foundations of Object-Oriented Languages (1990), LNCS 489, pp. 60-90

[BFS97]   J. P. Bahsoun, P. Fares, C. Servières: *Multilevel Proof System for Concurrent Object-Oriented Systems* 2de France-Japan workshop on Object Based Parallel and distributed Computing October 1997. To appear in HERMES 1999

[BMS93]   J. P. Bahsoun, S. Merz, C. Servières *A Framework for Programming and Formalizing Concurrent Objects* In proceedings of ACM Sigsoft 93

[BMS95]   J. P. Bahsoun, S. Merz, C. Servières *A Framework for formalizing and proving concurrent objects* In Parallélisme, répartition et réseaux. Hermès, 1995

[FM95]   J.Fiadeiro, T.Maibaum *Verifying for Reuse: foundations of object-oriented system verification*, in Theory and Formal Methods 1994. ICL Press, 1995

[Gab92]   D. M. Gabbay: *Temporal Logic: Mathematical Foundations*. Tech. Report MPI-I-92-213, Max-Planck-Institut Saarbrücken (1992)

[Lam91]   L. Lamport: *The Temporal Logic of Actions*. Research Report 79, DEC Systems Research Center, December 1991

[Mes93]   J. Meseguer: *Solving the Inheritance Anomaly in Concurrent Object-Oriented Programming*, ECOOP 93, LNCS 707

[MP92]   Z. Manna, A. Pnueli: *The Temporal Logic of Reactive and Concurrent Systems — Specification*. New York etc.: Springer-Verlag (1992)

[MW92]   J. Meseguer, T. Winkler *Parallel Programming in Maude*. In "Research Directions in High-Level Parallel Programming Languages" J.B. Banatre, D. LeMetayer (Eds) Springer Verlag 1992

[Ser95]   C.Servières: *Modélisation et Vérification Orientées Objet pour les Systèmes Réactifs*, Ph.D. Thesis, Université Paul Sabatier, November 1995.

[Sm68]   R.M. Smullyan: *First Order Logic*, Springer-Verlag Berlin 1968.

[Wo85]   P. Wolper: *The tableau method for temporal logic: an overview*. Logique et Analyse 110-111 numéro spécial, Nauwelaerts Printing, 1985.

Aliasing Models for Object Migration*

Uwe Nestmann[1], Hans Hüttel[1], Josva Kleist[1], and Massimo Merro[2]

[1] BRICS, Aalborg University, Denmark
[2] INRIA, Sophia-Antipolis, France

Abstract. In Obliq, a lexically scoped, distributed, object-oriented programming language, object migration was suggested as the creation of a copy of an object's state at the target site, followed by turning the object itself into an alias, also called *surrogate*, for the remote copy. We consider the creation of object surrogates as an abstraction of the above-mentioned style of migration. We introduce Øjeblik, a distribution-free subset of Obliq, and provide three different configuration-style semantics, which only differ in the respective *aliasing model*. We show that two of the semantics, one of which matches Obliq's implementation, render migration unsafe, while our new proposal for a third semantics is provably safe. Our work suggests a straightforward repair of Obliq's aliasing model such that it allows programs to safely migrate objects.

1 From Migration to Surrogation

Øjeblik is an object language that is not only inspired by Obliq [Car95], but rather represents its concurrent core. Obliq is a lexically scoped, distributed, object-oriented programming language. Lexical scoping in distributed settings renders program analysis easier since the binding of variables is determined by their location in the program text, not by the site at which execution takes place.

It is advantageous, for example for efficiency improvements, to be able to migrate an object from one site to another. In Obliq, object migration is carried out by creating a copy of an object—that is required to be protected and serialized—at the target site and then modifying the original (local) object such that it forwards future requests to the new (remote) object. Unless sites may fail, lexical scoping permits us to safely ignore the aspects of distribution, so we concentrate on just the concurrent aspects: *surrogation* of an object a can be described as creating a copy b of a and then turning a itself into a proxy for b, i.e., which forwards future request for methods of a to b.

2 Øjeblik, a Language for Serialized Concurrent Objects

In this section, we present Øjeblik as an untyped language, although types can be added in a straightforward manner. For the sake of simplicity, compared to

---

* This paper is a revised part of another paper called *Migration = Cloning ; Aliasing* [HKMN99] that appeared in the *Informal Proceedings of FOOL 6*.

P. Amestoy et al. (Eds.): Euro-Par'99, LNCS 1685, pp. 1353–1368, 1999.
© Springer-Verlag Berlin Heidelberg 1999

Obliq, we omit ground values, data operations, and procedures, we restrict field selection to method invocations, we restrict multiple cloning to single cloning, we omit flexibility of object attributes, we replace field aliasing with object aliasing, we omit explicit distribution, and we omit exceptions and advanced synchronization, so we get a feasible, but still non-trivial language. The set $\mathcal{L}$ of Øjeblik-expressions is generated as follows, where the l range over method *labels*.

$$
\begin{array}{llll}
a, b & ::= & \mathbb{O} & \text{object} \\
& | & a.l\langle a_1 .. a_n \rangle & \text{method invocation} \\
& | & a.l{\Leftarrow}m & \text{method update} \\
& | & a.\mathsf{alias}\langle b \rangle & \text{object aliasing} \\
& | & a.\mathsf{clone} & \text{shallow copy} \\
& | & a.\mathsf{surrogate} & \text{object surrogation} \\
& | & s, x, y, z & \text{variables} \\
& | & \mathsf{let}\, x = a\, \mathsf{in}\, b & \text{local definition} \\
& | & \mathsf{fork}\langle a \rangle & \text{thread creation} \\
& | & \mathsf{join}\langle a \rangle & \text{thread destruction} \\
\mathbb{O} & ::= & [l_j{=}m_j]_{j \in J} & \text{object record} \\
m & ::= & \varsigma(s, \tilde{x})b & \text{method}
\end{array}
$$

An object $[l_j{=}m_j]_{j \in J}$ consists of a finite collection of named methods $l_j{=}m_j$, more generally called fields, for pairwise distinct labels l_j.

In a method $\varsigma(s, \tilde{x})b$, the letter ς denotes a binder for the self variable s and argument variables $\tilde{x}$ (a tuple $x_1 .. x_n$) within the body b. Method invocation supplies a number of actual parameters: usually, $a.l\langle a_1 .. a_n \rangle$ with field l containing the method $\varsigma(s, \tilde{x})b$ results in the body b with the self variable s replaced by the enclosing object a, and the formal parameters $\tilde{x}$ replaced by the actual parameters $a_1 .. a_n$ of the invocation. The expression $a.l{\Leftarrow}m$ updates the content of the named field l in a with method m and evaluates to the modified object.

Every object in Øjeblik comes equipped with special methods for cloning and aliasing and surrogate, which cannot be overwritten by the update operation. The operation $a.\mathsf{clone}$ creates an object with the same fields as the original object and initializes the fields to the same entries as in the original object. The operation $a.\mathsf{alias}\langle b \rangle$ replaces the object a with a pointer to b, regardless of whether a is already a pointer or still an object record. Thus, like cloning (and surrogation), the aliasing operation itself is not subject to aliasing. The operation $a.\mathsf{surrogate}$ represents the abstraction of migration: by calling it, object a is turned into a local proxy for a remote copy of itself, which is implemented by providing a uniform method $\mathsf{surrogate}{=}\varsigma(s)s.\mathsf{alias}\langle s.\mathsf{clone}\rangle$. Note some crucial properties of such surrogation: like standard methods, surrogation is forwarded by aliased objects, and is permitted for external requests. These properties are justified since surrogation mimics migration (although without resorting to explicit distribution), so an object should be surrogatable more than once. In this realm, double-surrogation $a.\mathsf{surrogate};\, a.\mathsf{surrogate}$ (where ; denotes sequential composition, defined below) should obviously be equivalent to a.surrogate.surrogate. Without forwarding, the migration of an already migrated object would mistakenly migrate the proxy.

As usual, an expression let $x = a$ in b (only non-recursive) first evaluates a, binding the result to x, and then evaluates b within the scope of the new binding. Moreover, $a; b$ abbreviates let $x = a$ in b, where x does not occur free in b.

While objects represent persistent stateful structural entities, computational activity takes place within so-called *threads*. Apart from the main thread that is initially started up with the the execution of a term, new separate threads can be created by the fork command. The term fork$\langle a \rangle$ returns a new thread identifier to denote the thread evaluating a. The result of a forked computation is grabbed by the join command. If a evaluates to a thread identifier, then join$\langle a \rangle$ potentially blocks until that thread finishes and returns the thread's result, or blocks forever, if a join on thread a was already performed earlier.

Self-Infliction, Serialization, Protection The *current method* of a thread is the last method invoked in it that has not yet completed. The *current self* of a thread is the self of its current method. An Øjeblik operation is *self-inflicted* if it addresses the current self; an operation is *external* if it is not self-inflicted.

In concurrent object-based settings, the invariant that at most one thread at a time may be active within an object is often called *serialization*. The simplest way to ensure serialization is to associate with an object a *mutex* that is locked when a thread enters the object and released when the thread exits the object. However, this approach is too restrictive—for instance, it prevents recursion. Based on the notion of *thread*, so-called *reentrant* mutexes, as in Java, can be used to allow an operation to re-enter an object under the assumption that this operation belongs to the same thread as the operation that is currently active in the object. In Obliq, however, the more cautious idea of *self-serialization* requires, based on the above notion of self-infliction, that the mutex is always acquired for external operations, but never for self-inflicted ones. Note that this concept allows a method to recursively call its siblings through self, but it excludes the kind of inter-object mutual recursion, where a method in an object a calls a method in another object b, which then tries to 'call back' another method in a.

Based on self-infliction, objects are protected against external modifications: for protected objects, updates, cloning, and aliasing, are only allowed, if these operations are self-inflicted. In Øjeblik, all objects are protected *and* serialized.

3   Operational Semantics for Øjeblik

We give a transition semantics for Øjeblik terms that closely follows the one sketched by Talcott [Tal96b]. Her semantics addresses a larger subset of Obliq than we do with Øjeblik, in particular including distribution concepts, but nevertheless excludes, for example, migration and join.

3.1   Basic Concepts

The semantics performs changes on run-time configurations, which are mappings from references $\mathcal{R}$ to run-time entities. More precisely, a configuration $\mathfrak{C}$ maps task references $t \in \mathcal{R}_T$ to tasks T, and object references $o \in \mathcal{R}_\mathcal{O}$ to objects $\mathcal{O}$ (see below). We write $t, o \in \mathfrak{C}$, if t, o are defined in $\mathfrak{C}$, and $\uparrow$ for undefined references.

Run-Time Entities A run-time expression a is generated from the extended Øjeblik grammar as in Figure 1, where we introduce references as *values* v, as well as an additional construct wait, whose meaning will become clear from its use later on. Let us refer to this extended set of terms as $\mathcal{L}_\mathcal{R}$. A run-time object $O \in \mathcal{O}$ is either an object record $\mathbb{O}$ or a pointer $\gg o$ to an object reference $o \in \mathcal{R}_\mathcal{O}$. A run-time task T is a triple $\langle f, s, a \rangle \in \mathcal{R}_T \times \mathcal{R}_\mathcal{O} \times \mathcal{L}_\mathcal{R}$ that refers to a *parent* f, a current *self* s, and a run-time Øjeblik expression a, which remains to be evaluated. By the partial functions $s_\mathfrak{C}(t)$ and $f_\mathfrak{C}(t)$, we refer to the—also possibly undefined—current self and parent of the task associated with reference t in $\mathfrak{C}$.

Alias chains The partial function $\mathrm{ali}_\mathfrak{C} : \mathcal{R}_\mathcal{O} \rightharpoonup \mathcal{R}_\mathcal{O}^* \cup \{\uparrow\}$ with

$$
\mathrm{ali}_\mathfrak{C}(o) \stackrel{\mathrm{def}}{=} \begin{cases} \uparrow & \text{if } \mathfrak{C}(o) = \uparrow \\ o & \text{if } \mathfrak{C}(o) = \mathbb{O} \\ o \cdot \mathrm{ali}_\mathfrak{C}(o') & \text{if } \mathfrak{C}(o) = \gg o' \end{cases}
$$

computes the *alias chain*, starting at reference o, where $\cdot$ denotes concatenation of strings of references, possibly ending with $\uparrow$. This computation only terminates, if there are no cycles in the chain. We denote the endpoint of an alias chain by $\mathrm{end}(\mathrm{ali}_\mathfrak{C}(o))$; usually, if it exists, it is associated with an object record $\mathbb{O}$. We write $\hat{o} \in \mathrm{ali}_\mathfrak{C}(o)$ if $\hat{o}$ occurs in the string representing o's alias chain.

Threads Since tasks, in general, refer explicitly to their parent, we can build up task chains, which start in tasks that have no parent assigned. Such chains precisely correspond to Øjeblik threads, as formalized in [HKMN99]. In the current paper, we only use threads implicitly (cf. rule (FORK) in Figure 3).

Self-Infliction According to the need to test for the either self-inflicted or external character of operations on objects, we introduce some suitable notation. An object o is *idle* in $\mathfrak{C}$, if it is not the current self of any task in $\mathfrak{C}$; an object o is *available* for task t in $\mathfrak{C}$, if o is idle or the same as the current self of t.

$$
\begin{aligned}
\mathrm{Idle}_\mathfrak{C}(o) &\stackrel{\mathrm{def}}{=} \bigwedge_{t \in \mathcal{R}_T \cap \mathfrak{C}} (o \neq s_\mathfrak{C}(t)) \\
\mathrm{Avail}_\mathfrak{C}(o, t) &\stackrel{\mathrm{def}}{=} \mathrm{Idle}_\mathfrak{C}(o) \vee (o = s_\mathfrak{C}(t))
\end{aligned}
$$

Evaluation Figure 1 contains grammars for generating *redexes* r and *evaluation contexts* $e[\cdot]$, which we use to control the evaluation of (the expression part of) run-time tasks. The contexts are designed such that evaluation proceeds leftmost-innermost. A simple algorithm can compute for every run-time expression $a \notin \mathcal{R}$ a *unique* pair of a redex r and a context $e[\cdot]$ such that $a = e[r]$.

$$a, b ::= \ldots \mid \text{wait} \mid v \qquad\qquad v ::= o \mid t$$

$$r ::= \mathbb{O} \mid \text{wait} \mid o.\text{l}\Leftarrow m \qquad e[\cdot] ::= [\cdot] \mid e[\cdot].\text{l}\Leftarrow m$$
$$\mid o.\text{l}\langle \tilde{v} \rangle \qquad\qquad\qquad \mid e[\cdot].\text{l}\langle \tilde{a} \rangle \qquad\qquad \mid o.\text{l}\langle \tilde{v}, e[\cdot], \tilde{a} \rangle$$
$$\mid o.\text{alias}\langle o' \rangle \qquad\qquad \mid e[\cdot].\text{alias}\langle b \rangle \qquad \mid o.\text{alias}\langle e[\cdot] \rangle$$
$$\mid o.\text{clone} \quad\mid o.\text{surrogate} \qquad \mid e[\cdot].\text{clone} \qquad\quad \mid e[\cdot].\text{surrogate}$$
$$\mid \text{let } x = v \text{ in } b \qquad\qquad\qquad \mid \text{let } x = e[\cdot] \text{ in } b$$
$$\mid \text{fork}\langle a \rangle \quad\mid \text{join}\langle t \rangle \qquad\qquad \mid \text{join}\langle e[\cdot] \rangle$$

Fig. 1. Syntax of Øjeblik run-time expressions

Behaviors The semantics $[\![\, a\,]\!]$ of a *closed* Øjeblik term $a \in \mathcal{L}$ is given by assigning $\{t_m := \langle \uparrow, \uparrow, a \rangle\}$ as its initial configuration. Note that the task associated with t_m represents the start of a *main* thread. The behavior of configurations is generated from (subsets of) the rules in Figures 2–8. In each case we pick some task and object references in a particular configuration $\mathfrak{C}$, which, under the respective conditions may enable a transition to take place in $\mathfrak{C}$. In the premises, note that the expressions of tasks are always in unique context-redex decomposed form. In the conclusions of the rules, the notation $\mathfrak{C}\{t := T, o := O\}$ means that the mapping $\mathfrak{C}$ is either extended or overwritten with the association of task reference t with task T, and object reference o with run-time object O.

The rules in Figure 2 describe the local activity in a single task t in a straightforward manner; recall that let is not recursive and that the value v is either a task or an object reference whose actual run-time entity is accessible through $\mathfrak{C}$.

$$\frac{\mathfrak{C}(t) = \langle f, s, e[\text{let } x = v \text{ in } b] \rangle}{\mathfrak{C} \to \mathfrak{C}\{t := \langle f, s, e[b\{^v/_x\}] \rangle\}} \quad (\text{Let})$$

$$\frac{\mathfrak{C}(t) = \langle f, s, e[\mathbb{O}] \rangle \qquad \mathfrak{C}(o) = \uparrow}{\mathfrak{C} \to \mathfrak{C}\{t := \langle f, s, e[o] \rangle, o := \mathbb{O}\}} \quad (\text{New})$$

Fig. 2. Local transitions

The rules in Figure 3 exhibit the interplay of fork and join: in rule (FORK), a new task t' is spawned off, which runs the expression a without current self. In rule (JOIN), the parent referrring to his child t' is returned a value v. Note that forked tasks do not know their parent, so they represent initial tasks of threads.

$$\frac{\mathfrak{C}(t) = \langle\, f, s, e[\mathsf{fork}\langle a\rangle]\,\rangle \qquad \mathfrak{C}(t') = \uparrow}{\mathfrak{C} \;\to\; \mathfrak{C}\{t := \langle\, f, s, e[t']\,\rangle, t' := \langle\, \uparrow, \uparrow, a\,\rangle\}} \quad (\textsc{Fork})$$

$$\frac{\mathfrak{C}(t) = \langle\, f, s, e[\mathsf{join}\langle t'\rangle]\,\rangle \qquad \mathfrak{C}(t') = \langle\, \uparrow, \uparrow, v\,\rangle}{\mathfrak{C} \;\to\; \mathfrak{C}\{t := \langle\, f, s, e[v]\,\rangle, t' := \uparrow\}} \quad (\textsc{Join})$$

Fig. 3. Concurrency transitions

Most of the other rules of the operational semantics—used for describing protected operations (Ali)/(Cln)/(Upd) and invocations (Inv)/(Ret) on objects—crucially depend on how aliased objects should behave. Here, we start splitting up (in Figures 4–8) our presentation into different variants corresponding to different aliasing models, as introduced in the next subsection.

$$\frac{\mathfrak{C}(t) = \langle\, f, s, e[o.\mathsf{alias}\langle o'\rangle]\,\rangle \qquad s = o}{\mathfrak{C} \;\to\; \mathfrak{C}\{t := \langle\, f, s, e[o']\,\rangle, s := \gg o'\}} \quad (\textsc{C/R-Ali})$$

$$\frac{\mathfrak{C}(t) = \langle\, f, s, e[o.\mathsf{clone}]\,\rangle \qquad s = o \qquad \mathfrak{C}(o') = \uparrow}{\mathfrak{C} \;\to\; \mathfrak{C}\{t := \langle\, f, s, e[o']\,\rangle, o' := \mathfrak{C}(s)\}} \quad (\textsc{C/R-Cln})$$

$$\frac{\begin{array}{cc}\mathfrak{C}(t) = \langle\, f, s, e[o.\mathsf{l}\!\Leftarrow\!m]\,\rangle & s = o \\ \mathfrak{C}(s) = [\mathsf{l}_j = m_j]_{j \in J} & \mathsf{l} = \mathsf{l}_k \text{ for } k \in J\end{array}}{\mathfrak{C} \;\to\; \mathfrak{C}\{\, t := \langle\, f, s, e[s]\,\rangle, s := [\mathsf{l}_k = m, \mathsf{l}_{j \neq k} = m_j]_{j \in J}\}} \quad (\textsc{C-Upd})$$

Fig. 4. Protected transitions in the conservative (and relaxed) models

3.2   Aliasing Models for Øjeblik

With respect to the semantics of the alias operation, there is some freedom on how to precisely model serialization and protection in the aliased object. There are several possible variants, some of which we list in the order of strength:

- a *conservative model* (C) that keeps protection and serialization unchanged,
- a *relaxed model* (R) that keeps protection, but relaxes serialization,
- a *forwarder model* (F) that relaxes even protection to some extent, and
- an *ignorant model* that ignores protection and serialization completely.

According to Cardelli's intuitive semantics [Car95, Car98], where object aliasing is derived from field aliasing, Obliq adopts the conservative model. A reasonable explanation is that, there, the aliased object still exists unchanged, but only accesses to its fields are redirected. In the remainder of this section, we provide a formal semantics for each of the above aliasing models, except for the ignorant

$$\frac{\mathfrak{C}(t) = \langle\, f, s, e[o.\mathbf{l}\langle\tilde{v}\rangle]\,\rangle \qquad \mathrm{Avail}_{\mathfrak{C}}(o,t) \qquad \mathfrak{C}(t') = \uparrow}{\mathfrak{C}(o) = [\mathbf{l}_j = \varsigma(s_j,\tilde{x})b_j]_{j\in J} \qquad\qquad\qquad 1 = \mathbf{l}_k \ \text{for} \ k\in J} \quad (\text{C-Inv}_1)$$
$$\mathfrak{C} \ \rightarrow \ \mathfrak{C}\{\,t := \langle\, f, s, e[\mathsf{wait}]\,\rangle, t' := \langle\, t, o, b_k\{^{o\tilde{v}}/_{s_j\tilde{x}}\}\,\rangle\,\}$$

$$\frac{\mathfrak{C}(t) = \langle\, f, s, e[o.\mathbf{l}\langle\tilde{v}\rangle]\,\rangle \qquad \mathrm{Idle}_{\mathfrak{C}}(o) \qquad \mathfrak{C}(t') = \uparrow}{\mathfrak{C}(o) = \gg o'}{\mathfrak{C} \ \rightarrow \ \mathfrak{C}\{\,t := \langle\, f, s, e[\mathsf{wait}]\,\rangle, t' := \langle\, t, o, o'.\mathbf{l}\langle\tilde{v}\rangle\,\rangle\,\}} \quad (\text{C-Inv}_2)$$

$$\frac{\mathfrak{C}(t) = \langle\, f, s, e[\mathsf{wait}]\,\rangle \qquad \mathfrak{C}(t') = \langle\, t, s', v\,\rangle}{\mathfrak{C} \ \rightarrow \ \mathfrak{C}\{\,t := \langle\, f, s, e[v]\,\rangle, t' := \uparrow\}} \quad (\text{Ret})$$

Fig. 5. Method transitions in the conservative model

model, which extradites object surrogates fully to the will of its environment. In the following, the inclusion of a rule to the semantics of only a particular aliasing model is indicated by prefixing its name with either of C, R, or F.

The Conservative Model Each aliasing node is protected. Thus, the rules in Figure 4, all of which address protected operations on object o—which can only happen if they are self-inflicted—require the premise $o = s$. Note that for simplicity we do not generate run-time errors (as in Obliq) for invalid access to protected operations; instead, the calls to such operations just block forever.

The rules $(\text{C-Inv}_1)/(\text{Ret})$ in Figure 5 formalize the protocol of synchronous method invocation, where each call to a non-aliased object creates a subtask t' with the current task t as its parent—note the difference to fork-created subtasks. The parent is blocked, by means of wait, until the subtask t' returns a result to its parent t in rule (Ret). The use of $\mathrm{Avail}_{\mathfrak{C}}(o,t)$ captures self-infliction.

Each aliasing node is serialized. Thus, a method call on an aliased object o, as in rule (C-Inv_2), also creates a sub-task t' with current self o, which essentially means that o's mutex is now locked by this task, which is otherwise just forwarding the call to the reference o' mentioned in the configuration.

The Relaxed Model [Tal96b] While protection is kept in aliased objects, serialization is removed. Talcott proposes a scheme for method calls in Obliq, which is optimized in the sense that it directly addresses the endpoint of a chain,

$$\frac{\mathfrak{C}(t) = \langle\, f, s, e[o.\mathbf{l}\langle\tilde{v}\rangle]\,\rangle \qquad \mathrm{end}(\mathrm{ali}_{\mathfrak{C}}(o)) = \hat{o} \qquad \mathrm{Avail}_{\mathfrak{C}}(\hat{o},t)}{\mathfrak{C}(\hat{o}) = [\mathbf{l}_j = \varsigma(s_j,\tilde{x})b_j]_{j\in J} \qquad 1 = \mathbf{l}_k \ \text{for} \ k\in J \qquad \mathfrak{C}(t') = \uparrow}{\mathfrak{C} \ \rightarrow \ \mathfrak{C}\{\,t := \langle\, f, s, e[\mathsf{wait}]\,\rangle, t' := \langle\, t, \hat{o}, b_k\{^{\hat{o}\tilde{v}}/_{s_j\tilde{x}}\}\,\rangle\,\}} \quad (\text{R/F-Inv})$$

Fig. 6. Method invocation in the relaxed and forwarder models

$$\frac{\mathfrak{C}(t) = \langle\, f, s, e[o.\mathsf{l} \Leftarrow m]\,\rangle \qquad \mathsf{end}(\mathsf{ali}_{\mathfrak{C}}(o)) = s}{\mathfrak{C}(s) = [\mathsf{l}_j = m_j]_{j \in J} \qquad \mathsf{l} = \mathsf{l}_k \ \text{for} \ k \in J} \quad \mathfrak{C} \ \to \ \mathfrak{C}\{\, t := \langle\, f, s, e[s]\,\rangle, s := [\mathsf{l}_k = m, \mathsf{l}_{j \neq k} = m_j]_{j \in J}\}}{} \quad \text{(R/F-UPD)}$$

Fig. 7. Update transitions in the relaxed and forwarder models

$$\frac{\mathfrak{C}(t) = \langle\, f, s, e[o.\mathsf{alias}\langle o'\rangle]\,\rangle \qquad s \in \mathsf{ali}_{\mathfrak{C}}(o)}{\mathfrak{C} \ \to \ \mathfrak{C}\{\, t := \langle\, f, s, e[o']\,\rangle, s := \gg o'\}} \quad \text{(F-ALI)}$$

$$\frac{\mathfrak{C}(t) = \langle\, f, s, e[o.\mathsf{clone}]\,\rangle \qquad s \in \mathsf{ali}_{\mathfrak{C}}(o) \qquad \mathfrak{C}(o') = \uparrow}{\mathfrak{C} \ \to \ \mathfrak{C}\{\, t := \langle\, f, s, e[o']\,\rangle, o' := \mathfrak{C}(s)\}} \quad \text{(F-CLN)}$$

Fig. 8. Protected transitions in the forwarder model

if it exists, and instantly creates a sub-task there. In rule (R/F-INV) of Figure 6, this scheme is formalized in terms of the functions end and $\mathsf{ali}_{\mathfrak{C}}$. In particular, no tasks are created in intermediate nodes between o and $\hat{o}$, and the availability of the endpoint is checked with respect to the current self s of the calling task t.

Although aliased objects are protected, updates on aliased objects are modeled as in rule (R/F-UPD): if an update on o is originating from the current endpoint of its alias chain ($s = \hat{o}$), then it is forwarded there to take effect, there. Otherwise, the caller is blocked. Actually, this peculiar behavior may be interpreted as a relaxation of protection by allowing update requests to re-enter a protected object, as long as they have only passed predecessors of the latter.

The Forwarder Model We 'learn' from the peculiar modeling of update transitions in the relaxed model, and generalize it to also apply to cloning and aliasing: the rules in Figure 8 replace the former tests $s = o$ for immediate self-infliction by a test for self-infliction on successors in the alias chain starting from the entry in the chain. In effect, this behavior prescribes an implementation of an aliased object as a pure *forwarder* for external and a partial forwarder for internal requests, as we proposed in terms of a translation into a π-calculus [HKMN99].

3.3   Surrogation

Since surrogation in Øjeblik is represented as a uniform method that is, except for updates, treated like standard methods, also the semantics of surrogation in the three aliasing models is exactly like the (INV)-rules for standard methods.

$$
\begin{array}{rll}
C[\cdot] ::= & [\cdot] & \mid \; [\, 1_k{=}\varsigma(s,\tilde{x})C[\cdot]\,,\; 1_{j\neq k}{=}m_{j\neq k}\,]_{j\in J} \\
& \mid \; C[\cdot].1\langle\,\tilde{a}\,\rangle & \mid \; a.1\langle\,\tilde{a},C[\cdot],\tilde{a}\,\rangle \\
& \mid \; C[\cdot].1{\Leftarrow}m & \mid \; a.1{\Leftarrow}\varsigma(s,\tilde{x})C[\cdot] \\
& \mid \; C[\cdot].\mathsf{alias}\langle b\rangle & \mid \; a.\mathsf{alias}\langle C[\cdot]\rangle \\
& \mid \; C[\cdot].\mathsf{clone} & \mid \; C[\cdot].\mathsf{surrogate} \\
& \mid \; \mathsf{let}\; x = C[\cdot]\; \mathsf{in}\; b & \mid \; \mathsf{let}\; x = a\; \mathsf{in}\; C[\cdot] \\
& \mid \; \mathsf{fork}\langle C[\cdot]\rangle & \mid \; \mathsf{join}\langle C[\cdot]\rangle
\end{array}
$$

Fig. 9. Øjeblik contexts

3.4   Theory

Based on a may-variant[1] of convergence [Mor68], we define a contextual notion of equivalence uniformly for the three semantics of Øjeblik.

Definition 1 (Convergence). *A closed term $a \in \mathcal{L}$ converges, written $a{\Downarrow}$, if there is a configuration $\mathfrak{C}$ with $[\![\, a\,]\!] \to^{*} \mathfrak{C}$ and $\mathfrak{C}(t_{\mathrm{m}}) = \langle\,\uparrow,\uparrow,v\,\rangle$ for some v.*

This notion of convergence does not mean that the whole computation of a terminates, but rather that the main thread t_{m} does so: the evaluation of a *may* converge to a value v and *can* be reached in finite time within t_{m}. Note that there might be forked, but unjoined computations around, possibly running forever.

In order to define a contextual notion of program equivalence [GHL97], we also need a general notion of program context that differs from the notion of evaluation context given earlier. More specifically, according to Figure 9, a *context $C[\cdot]$* has a single hole $[\cdot]$ that may be filled with an Øjeblik term.

Definition 2 (Equivalence). *Two terms $a, b \in \mathcal{L}$ are equivalent, written $a \cong b$, if for all closing contexts $C[\cdot]$ (s.t. $C[a]$ and $C[b]$ are closed): $C[a] \Downarrow$ iff $C[b] \Downarrow$.*

4   On the Safety of Surrogation

The following statement formalizes the idea that *an object before and after surrogation should behave the same in all possible contexts* as an equation.

Guess 1 (Safety) *Let $a \in \mathcal{L}$ be an Øjeblik term. Then $a \cong a.\mathsf{surrogate}$.*

It turns out that this guess is rather naive, and indeed wrong with all three semantics, so in the remainder of this section we narrow down the above equation such that it becomes true, and provably so. The following discussion also generalizes to *migration* in a distributed lexically-scoped setting, like Obliq.

---

[1] In the context of a concurrent language with fork, threads may nondeterministically affect the outcome and convergence of evaluation. So, with respect to our goal of reasoning about surrogation, we regard *must*-variants of convergence as too strong.

The simplest case of Guess 1 is where a is an Øjeblik object $\mathbb{O}$. In this case the surrogation is surely safe in all three semantics, because (1) the process of surrogation is carried out correctly since only the surrogation thread can interact with the object $\mathbb{O}$, i.e., there cannot be any interference with another thread or activity, and (2) every interaction with $\mathbb{O}$ is mimicked identically by $\mathbb{O}$.surrogate, which suffices since after surrogation nobody has access to the previous $\mathbb{O}$.

In the general case, however, neither of the two above arguments holds. In order to simplify the discussion, assuming that $a \cong \mathsf{let}\, x = a\, \mathsf{in}\, x$ and the fact that the notion of equivalence takes all Øjeblik contexts into account, Guess 1 can be reduced to the problem of surrogation on variables: $x \cong x.\mathsf{surrogate}$.

A closer look at Øjeblik examples, as we will have in the remainder of this section, shows that for the safety of surrogation it is crucial to distinguish, whether or not the call $x.\mathsf{surrogate}$ within a given context $C[\cdot]$ is "external for" x, or whether it is self-inflicted. Note that, unfortunately, this problem is undecidable, as already observed for Obliq [Car95]—only at run-time may we observe which case applies. Intuitively, this means that we must therefore execute a term $C[x]$ until that particular access to x appears at the top-level for evaluation.

4.1 External Surrogation

We provide two single-threaded, distinguishing examples that exhibit unsafe surrogations in the conservative and relaxed models, while the forwarder model ensures safety. As an abbreviation, we use the method definitions l=id and k=Ω to denote l=$\varsigma(s)s$ and k=$\varsigma(s)s.$k, respectively, where $[\,\mathsf{l=id}\,].\mathsf{l} \Downarrow$ and $[\,\mathsf{k=}\Omega\,].\mathsf{k} \Uparrow$.

The two examples have a very similar structure, only differing in the object $\mathbb{O}$ that (at least) serves single-parameter methods on label l.

$$C[\cdot] := \mathsf{let}\, x = \mathbb{O}\, \mathsf{in}\, \mathsf{let}\, y = [\cdot]\, \mathsf{in}\, y.\mathsf{l}\langle x\rangle$$

Here, we have a kind of "diagonalization" in mind, in that we intend to apply the context $C[\cdot]$ to, on the one hand a variable x, and on the other hand its surrogated counterpart $x.\mathsf{surrogate}$, which may or may not lead to calling an object at variable y with a reference to itself as a parameter.

In Figure 10, we show the initial steps according to our operational semantics: we indicate the applied transition rules, as well as the respective uniquely defined redex for each following step. Here, we were not specific about the aliasing model on which the semantics we follow is based on since all models, so far, coincide. Note the four additional steps needed for performing the surrogation in between the two applications of rule (LET). Note further the different resulting states X and S, where either o or o' is called, respectively, from within the initial task with o as parameter.

Countering to the conservative model The following context distinguishes a variable x and its surrogated counterpart $x.\mathsf{surrogate}$ in the conservative alias model:

$$C_1[\cdot] := C[\cdot] \text{ with } \mathbb{O} \text{ instantiated as } \mathbb{O}_1 := [\,\mathsf{k=id}\,, \mathsf{l=}\varsigma(s,z)z.\mathsf{k}\,] \qquad (1)$$

$$[\![\, C[x]\,]\!] = \{t_m := \langle\, \uparrow, \uparrow, \text{let } x = \mathbb{O} \text{ in let } y = x \text{ in } y.l\langle x \rangle \,\rangle\}$$
$$\xrightarrow{\text{(NEW)}} \{t_m := \langle\, \uparrow, \uparrow, \text{let } x = o \text{ in let } y = x \text{ in } y.l\langle x \rangle \,\rangle, o := \mathbb{O}\}$$
$$\xrightarrow{\text{(LET)}} \{t_m := \langle\, \uparrow, \uparrow, \underline{\text{let } y = o \text{ in } y.l\langle o \rangle} \,\rangle, o := \mathbb{O}\}$$
$$\xrightarrow{\text{(LET)}} \{t_m := \langle\, \uparrow, \uparrow, \underline{o.l\langle o \rangle} \,\rangle, o := \mathbb{O}\} \overset{\text{def}}{=} X$$

$$[\![\, C[x.\text{surrogate}]\,]\!]$$
$$= \{t_m := \langle\, \uparrow, \uparrow, \text{let } x = \mathbb{O} \text{ in let } y = x.\text{surrogate in } y.l\langle x \rangle \,\rangle\}$$
$$\xrightarrow{\text{(NEW)}} \{t_m := \langle\, \uparrow, \uparrow, \text{let } x = o \text{ in let } y = x.\text{surrogate in } y.l\langle x \rangle \,\rangle, o := \mathbb{O}\}$$
$$\xrightarrow{\text{(LET)}} \{t_m := \langle\, \uparrow, \uparrow, \text{let } y = o.\text{surrogate in } y.l\langle o \rangle \,\rangle, o := \mathbb{O}\}$$
$$\xrightarrow{\text{(INV)}} \{t_m := \langle\, \uparrow, \uparrow, \text{let } y = \text{wait in } y.l\langle o \rangle \,\rangle, o := \mathbb{O}, t_1 := \langle t_m, o, o.\text{alias}\langle o.\text{clone} \rangle \,\rangle\}$$
$$\xrightarrow{\text{(CLN)}} \{t_m := \langle\, \uparrow, \uparrow, \text{let } y = \text{wait in } y.l\langle o \rangle \,\rangle, o := \mathbb{O}, t_1 := \langle t_m, o, o.\text{alias}\langle o' \rangle \,\rangle, o' := \mathbb{O}\}$$
$$\xrightarrow{\text{(ALI)}} \{t_m := \langle\, \uparrow, \uparrow, \text{let } y = \underline{\text{wait}} \text{ in } y.l\langle o \rangle \,\rangle, o := \gg o', t_1 := \langle t_m, o, \underline{o'} \,\rangle, o' := \mathbb{O}\}$$
$$\xrightarrow{\text{(RET)}} \{t_m := \langle\, \uparrow, \uparrow, \underline{\text{let } y = o' \text{ in } y.l\langle o \rangle} \,\rangle, o := \gg o', o' := \mathbb{O}\}$$
$$\xrightarrow{\text{(LET)}} \{t_m := \langle\, \uparrow, \uparrow, \underline{o'.l\langle o \rangle} \,\rangle, o := \gg o', o' := \mathbb{O}\} \overset{\text{def}}{=} S$$

Fig. 10. Initial executions for the counterexample

Let X_1 and S_1 denote the respective derivatives of $C_1[x]$ and $C_1[x.\text{surrogate}]$ in Figure 10. Then, we can trace some more reductions now specific to $\mathbb{O}_1$:

$$X_1 \xrightarrow{\text{(INV)}} \{t_m := \langle\, \uparrow, \uparrow, \text{wait} \,\rangle, t' := \langle t_m, o, \underline{o.k} \rangle, o := \mathbb{O}_1\}$$
$$\xrightarrow{\text{(INV)}} \{t_m := \langle\, \uparrow, \uparrow, \text{wait} \,\rangle, t' := \langle t_m, o, \underline{\text{wait}} \rangle, t'' := \langle t', o, \underline{o} \rangle, o := \mathbb{O}_1\}$$
$$\xrightarrow{\text{(RET)}} \{t_m := \langle\, \uparrow, \uparrow, \underline{\text{wait}} \,\rangle, t' := \langle t_m, o, \underline{o} \rangle, o := \mathbb{O}_1\}$$
$$\xrightarrow{\text{(RET)}} \{t_m := \langle\, \uparrow, \uparrow, o \,\rangle, o := \mathbb{O}_1\} \overset{\text{def}}{=} X_1' \not\rightarrow$$

Here we were not specific about the semantics we follow when applying rule (INV) since the behavior is the same in either case (C-INV$_1$) or (R/F-INV) and leads to convergence ($X_1 \Downarrow$ and thus $C_1[x] \Downarrow$).

In contrast, for the execution of S_1, the various semantics exhibit different behaviors (which we indicate by subscripts) after the first common invocation step:

$$S_1 \xrightarrow{\text{(INV)}} \{t_m := \langle\, \uparrow, \uparrow, \text{wait} \,\rangle, t' := \langle t_m, o', \underline{o.k} \rangle, o \gg o' := \mathbb{O}_1\} \overset{\text{def}}{=} S_1'$$

(where we abbreviate alias chains $\{o := \gg o', o' := \mathbb{O}_1\}$ with $\{o \gg o' := \mathbb{O}_1\}$). While the conservative model blocks after

$$S_1' \xrightarrow{\text{(INV}_2)}_{\mathsf{C}} \{t_m := \ldots, t' := \langle t_m, o', \text{wait} \rangle, t_i := \langle t', o, \underline{o'.k} \rangle, o \gg o' := \mathbb{O}_1\}$$

because the addressed object o' is already inhabited by task t', the relaxed and forwarder model exhibit further reductions, like X_1, although yielding value o':

$$S_1' \xrightarrow[\text{(INV)}]{}_{R/F} \{t_m := \ldots, t' := \langle t_m, o', \underline{\text{wait}} \rangle, t'' := \langle t', o', \underline{o'} \rangle, o \gg o' := \mathbb{O}_1 \}$$
$$\xrightarrow[\text{(RET)}]{}_{R/F} \{t_m := \langle \uparrow, \uparrow, \underline{\text{wait}} \rangle, t' := \langle t_m, o', \underline{o'} \rangle, o \gg o' := \mathbb{O}_1 \}$$
$$\xrightarrow[\text{(RET)}]{}_{R/F} \{t_m := \langle \uparrow, \uparrow, o' \rangle, o \gg o' := \mathbb{O}_1 \} \overset{\text{def}}{=} S_1'' \not\rightarrow$$

The reason is simply that in the latter modeling, the intermediate task t_i with parent o is not created at all, but instead the forwarding to o', which represents the endpoint of the alias chain when starting at o, is done immediately, while still coming from parent o', such that it is accepted as self-inflicted.

To summarize, we observe that in all three aliasing models, $C_1[x]$ converges since the critical call $z.k$ is transformed, at run-time, into $o.k$ with parent o, i.e. into a self-inflicted call. In contrast, for $C_1[x.\text{surrogate}]$, this call arises at run-time with parent o', by then the surrogation target of object o, and comes back to o' as a forwarded request. In the relaxed and forwarder model, the whole term converges in accordance with $C_1[x]$, but the conservative model blocks this call, as it is not self-inflicted due to the intermediate parent o created for task t_i.

Countering to the relaxed model The following context distinguishes a variable x and its surrogated counterpart $x.\text{surrogate}$ in the relaxed aliasing model:

$$C_2[\cdot] := C[\cdot] \text{ with } \mathbb{O} \text{ instantiated as } \mathbb{O}_2 := [\, l = \varsigma(s, z) z.\text{clone} \,] \tag{2}$$

Let X_2 and S_2 denote the respective derivatives of $C_2[x]$ and $C_2[x.\text{surrogate}]$ in Figure 10. As before, we now trace some more reductions of X_2 and S_2:

$$X_2 \xrightarrow[\text{(INV)}]{} \{t_m := \langle \uparrow, \uparrow, \text{wait} \rangle, t' := \langle t_m, o, \underline{o.\text{clone}} \rangle, o := \mathbb{O}_2 \}$$
$$\xrightarrow[\text{(CLN)}]{} \{t_m := \langle \uparrow, \uparrow, \underline{\text{wait}} \rangle, t' := \langle t_m, o, \underline{o'} \rangle, o := \mathbb{O}_2, o' := \mathbb{O}_2 \}$$
$$\xrightarrow[\text{(RET)}]{} \{t_m := \langle \uparrow, \uparrow, o' \rangle, o := \mathbb{O}_2, o' := \mathbb{O}_2 \} \overset{\text{def}}{=} X_2' \not\rightarrow$$

Immediately, we observe the well-behavior of X_2' in all aliasing models since the cloning requested within task t' is self-inflicted and leads to convergence ($X_2 \Downarrow$ and thus $C_2[x] \Downarrow$).

In contrast, the behavior of the corresponding S_2 after the first step

$$S_2 \xrightarrow[\text{(INV)}]{} \{t_m := \langle \uparrow, \uparrow, \text{wait} \rangle, t' := \langle t_m, o', \underline{o.\text{clone}} \rangle, o \gg o' := \mathbb{O}_2 \} \overset{\text{def}}{=} S_2'$$

is quite different for the respective models, due to the external cloning request. In the conservative model, we immediately get $S_2' \not\rightarrow_C$. However, also in the relaxed model, we get $S_2' \not\rightarrow_R$; the reason is that forwarding, which helped us in dealing with Example (1), was only proposed for invocations and updates.

This is exactly the motivation for our forwarder model, which remedies the above situation and leads to convergence ($S_2' \Downarrow$ and thus $C_2[x.\text{surrogate}] \Downarrow$):

$$S_2' \xrightarrow[\text{(CLN)}]{}_F \{t_m := \langle \uparrow, \uparrow, \underline{\text{wait}} \rangle, t' := \langle t_m, o', \underline{o''} \rangle, o \gg o' := \mathbb{O}_2, o'' := \mathbb{O}_2 \}$$
$$\xrightarrow[\text{(RET)}]{}_F \{t_m := \langle \uparrow, \uparrow, o'' \rangle, o \gg o' := \mathbb{O}_2, o'' := \mathbb{O}_2 \} \overset{\text{def}}{=} S_2'' \not\rightarrow$$

To summarize, again in all three aliasing models, $C_2[x]$ converges since the cloning on x becomes self-inflicted at run-time. However, among our three candidate semantics only the forwarder model handles these requests properly for $C_2[x.\mathsf{surrogate}]$, when becoming external at run-time.

4.2   Self-Inflicted Surrogation

A particular class of examples is represented by objects that perform surrogation in a self-inflicted way such that they afterwards—as long as the current method is active—still can perform self-inflicted operations on the surrogated object. Due to these examples surrogation is *not completely* safe, even not in the forwarder model. We classify two different sets of examples depending on whether they exhibit problems with access to a self-surrogated source or the target thereof.

Target problems An immediate source of problems is due to the incorrect external use of the surrogation target by means of protected operations, e.g., cloning:

$$C_3[\cdot] := [\,k{=}\varsigma(s)\mathsf{let}\, y = [\cdot]\,\mathsf{in}\, y.\mathsf{clone}\,].k \qquad (3)$$

yields $C_3[s]\Downarrow$ and $C_3[s.\mathsf{surrogate}]\Downarrow\!\!\!\!/\,$. In $C_3[s]$ the cloning of y is allowed, while in $C_3[s.\mathsf{surrogate}]$ the corresponding call is blocked due to a run-time error.

Source problems Another problem arises by externally sending a request to the surrogation target, but now via the surrogation source, e.g., for updates:

$$C_4[\cdot] := [\,k{=}\varsigma(s)\mathsf{let}\, y = [\cdot]\,\mathsf{in}\, s.k{\Leftarrow}\mathsf{id}\,].k \qquad (4)$$

yields $C_4[s]\Downarrow$ and $C_4[s.\mathsf{surrogate}]\Downarrow\!\!\!\!/\,$. By sending an update to itself after having surrogated, as in $C_4[s.\mathsf{surrogate}]$, the update of y is blocked, while without previous surrogation the call succeeds and the whole term converges.

The next (and final) example is intended to exhibit the effect of re-aliasing:

$$C_5[\cdot] := \mathsf{let}\, x = [\,1{=}\Omega\,,\mathsf{k}{=}\Omega\,]\,\mathsf{in} \atop [\,1{=}\mathsf{id}\,,\mathsf{k}{=}\varsigma(s)\mathsf{let}\, y = [\cdot]\,\mathsf{in}\, s.\mathsf{alias}\langle x\rangle; y.1\,].k \qquad (5)$$

yields $C_5[s]\Downarrow\!\!\!\!/\,$ and $C_5[s.\mathsf{surrogate}]\Downarrow$. Whereas $C_5[s]$ diverges since the alias call to s also affects y in that case, the counterpart $C_5[s.\mathsf{surrogate}]$ converges since the re-aliasing of s does not affect the target y.

Problems? It is the programmer of an object who is responsible for potential problems caused by self-inflicted surrogation. The above-mentioned examples can be avoided if, in the current method, the self-variable s is neither copied (to prevent from target problems), nor used after surrogation in a self-inflicted way for calls to state-changing methods (to prevent from source problems). The safest way is to use the self-variable in the current method for just surrogation.

4.3   A Provable Safety Theorem

According to the above remarks, we concentrate our efforts on reasoning about $x \cong x$.surrogate for the case that the surrogate operation is external to x, because we conjecture that the forwarder model is good enough for that purpose. According to Definition 1, this means:

Theorem 1 (Safety). *Let* $x \in \mathcal{L}$. *Then for all* $C[\cdot]$ *with* $[\cdot]$ *"external for"* x: $C[x]\Downarrow$ *iff* $C[x.\text{surrogate}]\Downarrow$, *where transition is as specified in the forwarder model.*

In previous work on the Imperative Object Calculus (IOC) [AC96a], equivalence between IOC terms was defined in a contextual way [GHL97], similar to Definition 1. In many cases, it is simpler to use a semantics by translation into π-calculus to establish the equivalence between terms [KS98]. The main advantage is the large number of equivalences and algebraic laws that can be used to reason about expressions. Since IOC is (almost) a concurrency-free subset of Øjeblik, we chose a similar path for establishing the safety of external surrogation.

Proof (Overview). A proof of a restricted version of Theorem 1, where only a subset of inductively defined contexts is taken into account, is found in [HKMN99].

In a forthcoming paper [MHKN99], we show how to carry out the lengthy proof using a refined translation into π-calculus that allows us to prove a much stronger result: the translation of a variable and the translation of its surrogated counterpart are *barbed congruent* [MS98] under the assumption that the surrogation call is external at run-time.

5   Conclusion

We have shown, by means of examples, that object surrogation in Øjeblik, and consequently object migration in Obliq, is not transparent. We have verified, using the Obliq interpreter [Car], that these examples indeed expose problems with surrogation/migration in Obliq. Most of the migration problems were actually discovered, when trying to prove the safety of surrogation using π-calculus translations that implemented, first Obliq's semantics, then Talcott's semantics. Experimenting with these failed attempts led us to our final semantics for Øjeblik. The major "improvement" suggested by it is the *forwarder model* for the treatment of aliasing, which seems to be necessary in order to ensure transparency of at least external surrogation.

The forwarder model may be understood as restricting the concept of reentrant mutexes mentioned in § 2 to *pre-entrant mutexes*: any request is allowed to re-enter an object, if it has only traveled through predecessors of the object since its last visit. A repaired version of Obliq, by adopting the forwarder model, employs object aliasing rather as a primitive operation than merely derived from field aliasing. By that, the appropriate amount of serialization and protection for aliased objects can be implemented.

The major lesson learned from the work presented in this paper, is that *concurrent objects need formal analysis*—not because one necessarily should prove properties, but because formal analysis is a good debugging tool.

Current and future work A programmer should be provided with syntactic criteria for those cases of self-inflicted surrogation that are safe. Such criteria could then be used to extend Theorem 1.

Since we are using a semantics by translation into π-calculus for the proofs, work on the preservation of types and an operational correspondence (to a refined operational semantics for the forwarder model) is required and under way.

It would be interesting to exploit the operational semantics of the forwarder model, or suitable variants of it, directly for carrying out safety proofs. Interaction semantics [Tal96a] seems to be a sensible candidate for such a project.

As noted in Section 1, our positive results will only hold unless sites may fail. Naturally, one should investigate to make precise what safety of migration in a faulty setting could mean and in what sense it would be satisfied.

Finally, we will use our semantics to show further properties about Øjeblik. For instance, we consider showing that $\mathsf{join}(\mathsf{fork}(a)){\cong}a$ under certain conditions, but also expect some laws of Moggi's computational λ-calculus [Mog88] to hold.

Related work Apart from Talcott [Tal96b], closest to our work are two *concurrent object calculi*, one by Gordon and Hankin [GH98], like ours based on Abadi and Cardelli's object calculus [AC96b], and another one by Di Blasio and Fisher [DF96], based on Fisher and Mitchell's object calculus [FM94]. However, no account on object migration has been addressed in their work.

Acknowledgments We are grateful for discussions with and remarks from Luca Cardelli, Davide Sangiorgi, Carolyn Talcott, and ν-klubben at BRICS Aalborg.

References

[AC96a]   M. Abadi and L. Cardelli. An Imperative Object Calculus. *Theory and Practice of Object Systems*, 1(13):151–166, 1996.

[AC96b]   M. Abadi and L. Cardelli. *A Theory of Objects*. Monographs in Computer Science. Springer, 1996.

[Car]   L. Cardelli. `obliq-std.exe` — Binaries for Windows NT. `http://www.luca.demon.co.uk/Obliq/Obliq.html`.

[Car95]   L. Cardelli. A Language with Distributed Scope. *Computing Systems*, 8(1):27–59, 1995. Short version in *Proceedings of POPL '95*. A preliminary version appeared as Report 122, Digital Systems Research, June 1994.

[Car98]   L. Cardelli. On the Semantics of Obliq. Personal Communication, 1998.

[DF96]   P. Di Blasio and K. Fisher. A Concurrent Object Calculus. In U. Montanari and V. Sassone, eds, *Proceedings of CONCUR '96*, volume 1119 of *LNCS*, pages 655–670. Springer, 1996. An extended version appeared as Stanford University Technical Note STAN-CS-TN-96-36, 1996.

[FM94]   K. Fisher and J. Mitchell. Notes on Typed Object-Oriented Programming. In M. Hagiya and J. C. Mitchell, eds, *Proceedings of TACS '94*, volume 789 of *LNCS*, pages 844–885. Springer, 1994.

[GH98]   A. D. Gordon and P. D. Hankin. A Concurrent Object Calculus: Reduction and Typing. In U. Nestmann and B. C. Pierce, eds, *Proceedings of HLCL '98*, volume 16.3 of *ENTCS*. Elsevier Science Publishers, 1998.

[GHL97]   A. D. Gordon, P. D. Hankin and S. B. Lassen. Compilation and Equivalence of Imperative Objects. In S. Ramesh and G. Sivakumar, eds, *Proceedings of FSTTCS '97*, volume 1346 of *LNCS*, pages 74–87. Springer, Dec. 1997. Full version available as Technical Report 429, University of Cambridge Computer Laboratory, June 1997.

[HKMN99] H. Hüttel, J. Kleist, M. Merro and U. Nestmann. Migration = Cloning ; Aliasing (Preliminary Version). In *Informal Proceedings of the Sixth International Workshop on Foundations of Object-Oriented Languages (FOOL 6, San Antonio, Texas, USA)*. Sponsored by ACM/SIGPLAN, 1999. Available from http://www.cs.williams.edu/~kim/FOOL/sched6.html.

[KS98]   J. Kleist and D. Sangiorgi. Imperative Objects and Mobile Processes. In D. Gries and W.-P. de Roever, eds, *Proceedings of PROCOMET '98*, pages 285–303. International Federation for Information Processing (IFIP), Chapman & Hall, 1998.

[MHKN99] M. Merro, H. Hüttel, J. Kleist and U. Nestmann. Mobile Objects As Mobile Processes. Draft. Available from Massimo.Merro@sophia.inria.fr, 1999.

[Mog88]   E. Moggi. Computational Lambda Calculus and Monads. Technical Report ECS-LFCS-88-66, LFCS, University of Edinburgh, Oct. 1988.

[Mor68]   J.-H. Morris. *Lambda Calculus Models of Programming Languages*. PhD thesis, MIT, 1968.

[MS98]   M. Merro and D. Sangiorgi. On Asynchrony in Name-Passing Calculi. In K. G. Larsen, S. Skyum and G. Winskel, eds, *Proceedings of ICALP '98*, volume 1443 of *LNCS*, pages 856–867. Springer, July 1998.

[Tal96a]   C. L. Talcott. Interaction Semantics for Components of Distributed Systems. In *1st IFIP Workshop on Formal Methods for Open Object-based Distributed Systems, FMOODS'96*, 1996. http://www-formal.stanford.edu/MT/96fmoods.ps.Z.

[Tal96b]   C. L. Talcott. Obliq semantics notes. Unpublished note. Available from clt@cs.stanford.edu, Jan. 1996.

Dynamic Extension of CORBA Servers

Marco Catunda, Noemi Rodriguez, and Roberto Ierusalimschy

Departamento de Informática, PUC-Rio
22453-900, Rio de Janeiro, Brazil
catunda,noemi,roberto@inf.puc-rio.br

Abstract. This paper describes LuaDSI, a system for implementing CORBA servers with the Lua scripting language. An object written in LuaDSI can be dynamically modified and extended without stopping its service. We also describe LuaRep, an extension to Lua which allows clients to have transparent access to CORBAs interface repository. In conjunction with LuaDSI, LuaRep allows new CORBA services to be dynamically defined and installed.

1 Introduction

The wide acceptance of the CORBA architecture has shown the importance of componentware as a current trend in software development. Currently, most CORBA bindings direct their design to support servers written in a statically compiled language, such as C++, using a statically compiled skeleton; although CORBA supports a dynamic interface for servers (DSI, the Dynamic Skeleton Interface), this interface is very low level and difficult to use. Static implementations are quite acceptable and even desirable in many cases, but they have shortcomings that can be serious in some contexts. First, static implementations make rapid prototyping, which is another important trend in current software development, difficult. Second, they make remote updates to servers very hard. Third, the updating of an existing service implemented as a static server usually requires its interruption for some period of time.

A scripting language brings an interesting design alternative to this static nature of CORBA components. Servers implemented with a scripting (and interpreted) language can be dynamically modified and extended without compiling or linking phases, and so, without interrupting their services. With an interpreted language, it is easy to send code across a network, which allows the system to do remote or interactive modifications and extensions to the server. An interpreted language greatly improves the support for rapid prototyping, as we can load and test new design alternatives for a system in a quick and simple way.

The OMG recognizes the relevance of scripting languages to manipulate CORBA components [6]. Although the primary use of such scripts is for CORBA clients, they must be able to handle events generated by the components. Usually, the callbacks to handle events are specified as methods of an object, a *listener*, that is passed to the component. Therefore, even a scripting language

P. Amestoy et al. (Eds.): Euro-Par'99, LNCS 1685, pp. 1369–1376, 1999.
© Springer-Verlag Berlin Heidelberg 1999

used only for writing client code should support the creation of server objects, to act as listeners.

Our work on dynamically extensible servers has been preceded by a study of the flexibility which a binding to an interpreted language can bring to a CORBA client. For that task we chose the scripting language Lua [4, 2]; Lua was developed at PUC-Rio in 1994, and since then it has been used in hundreds of places, both in Brazil and abroad; so it was a natural choice for us. We then implemented a binding between CORBA and Lua, called *LuaOrb* [5]. LuaOrb, which is based on CORBA's Dynamic Invocation Interface, allows clients to access any available CORBA server, independently of stubs; servers are represented by proxy Lua objects, and operations on these proxies map dynamically to remote method calls.

In this work we describe LuaDSI, a system which allows CORBA servers to be implemented in the language Lua. A server written in LuaDSI can be dynamically modified and extended without stopping its service; we can also change its interface without having to create and link new skeletons to the server. All these modifications are made through a CORBA interface, so that you can modify a dynamic server from any remote system with CORBA access; as a particular example, you can manage a server from a remote console running LuaOrb.

LuaDSI hides the complexity of using the DSI, CORBAs Dynamic Skeleton Interface. LuaDSI dynamically receives all calls to a server object, unmarshals the arguments according to their descriptions in the Repository Interface, calls the appropriate Lua method to handle the call, and marshals any results back to the ORB.

In this work we also describe LuaRep, an extension to Lua which allows clients to have transparent access to CORBAs interface repository. The Interface Repository, which contains interface definitions, acts as a regular CORBA object, and can be dynamically queried and updated. LuaRep facilitates these operations and, in conjunction with LuaDSI, allows new services to be dynamically defined and installed. If you need only to modify a server implementation, LuaDSI is enough. If you also have to modify the server interface, then you can use LuaRep for that. With both systems, you can have a console wherein you have complete control over a server.

2   The Dynamic Skeleton Interface and LuaDSI

The Dynamic Skeleton Interface is an interface for writing object implementations that do not have compile time knowledge of the type of the object they are implementing [7]. The basic idea of the DSI is to implement all calls to a particular object (which we will call a *dynamic server*) by invocations to a single upcall routine, the *Dynamic Implementation Routine* (DIR). This routine is responsible for unmarshalling the arguments and for dispatching the call to the appropriate code. To our knowledge, the most frequent application of DSI

```
class ServerRequest {
  public:
  Identifier op_name();
  OperationDef_ptr op_def();
  Context_ptr ctx();
  void params (NVList_ptr parameters);
  void result (Any *value);
  void exception (Any *value);
}
```

Fig. 1. ServerRequest interface

```
interface Hello {
  void Print (in string Hello)
}
```

Fig. 2. Hello interface.

has been to implement bridges between different ORBs [11]. In this context, the dynamic server acts as a proxy for an object in some other ORB.

To make a request to a dynamic server, the ORB invokes the corresponding DIR with a single argument, a ServerRequest object. Figure 1 shows the ServerRequest interface for C++: The op_name attribute identifies the method being invoked. Method params gets the list of parameters. The other methods are related to setting the invocation results and exception signalling, and to retrieving the context information specified in IDL for the operation.

To illustrate the use of the DSI in C++, we present a simple IDL interface, in Fig. 2, and a dynamic skeleton implementation for it, in Fig. 3. The C++ implementation of a dynamic server must inherit from class DynamicImplementation. This class implements the DIR through a method invoke, which must be reimplemented in each dynamic server implementation. In our example, the implementation first checks whether the called operation is Print (line 4); it should be, in any type correct invocation. Then it creates and initializes an NVList (lines 5–9), to get the operation arguments with correct types from the ORB (line 10). Finally, the code gets the only argument from the list (lines 11–12), converts it to a C++ string (lines 13–14), and prints it (line 15). Because C++ is statically typed and has no garbage collection, while the DSI is inherently dynamic, the code is quite complex.

Since DSI allows a server to implement requests to objects about which it has no compile time knowledge, it is natural to consider this mechanism as a basis for the dynamic extension of servers. This section describes LuaDSI, a facility which uses DSI to allow remote dynamic creation and updating of CORBA servers written in Lua. LuaDSI offers an IDL interface for server update. Servers with this interface are implemented as DSI servers, and are collections of Lua

```
1      class ImplHello: CORBA::DynamicImplementation {
2        public:
3        void invoke (CORBA::ServerRequest request) {
4          if (strcmp(request->op_name(), "Print") == 0) {
5            CORBA::NVList nv;
6            orb->create_list(0, nv);
7            CORBA::NamedValue *namedValue = nv->add(ARG_IN);
8            CORBA::Any *any = namedValue->value();
9            any->replace(CORBA::_tc_string, 0);
10           request->params(nv);
11           namedValue = nv->item(0);
12           any = namedValue->value();
13           char *str;
14           (*any) >> str;
15           printf("%s", str);
16         }
17       }
18     }
```

Fig. 3. A Dynamic Implementation Routine in C++.

```
interface LuaDSIObject {
  readonly attribute Object obj;
  void InstallImplementation (in string opname, in string luaCode);
}

interface ServerLua {
  LuaDSIObject Instance (in CORBA::InterfaceDef intf);
}
```

Fig. 4. ServerLua interface.

objects, each representing an instance of some IDL interface. A dynamically extensible server maps each request coming from the ORB (except requests for object instantiation) to an operation on the corresponding Lua object.

A LuaDSI dynamic object is encapsulated inside a LuaDSIObject interface, presented in Fig. 4. They are created through the interface ServerLua. To create a new instance at a dynamically extensible server, a client first invokes method Instance. The single parameter of this method is a reference to an interface definition in the interface repository (see Sect. 3). This reference is used by Instance to retrieve information about the attributes and methods of the new object. The new object is returned as the attribute obj inside a new LuaDSIObject. The manager client can then invoke the method InstallImplementation to install or modify each of the objects methods.

To illustrate the use of LuaDSI, we will again use the Hello interface presented in Fig. 2. Figure 5 presents a Lua chunk that installs an instance of

```
-- gets the desired interface from the Interface Repository
interfaceDefHello = IR:lookup("Hello")
-- creates a new LuaDSIObject
dsiobj = serverlua:Instance(interfaceDefHello)
-- installs a "Print" method
dsiobj:IntallImplementation("Print", "function (s) print(s) end")
newobj = dsiobj.obj          -- gets the object
newobj:Print("Hello World!")   -- uses it
```

Fig. 5. Creation of a new server in Lua.

this object. We assume that the user already has the bindings to **serverLua** and to the Interface Repository. The presented sequence of commands can be interactively issued from a simple LuaOrb console.

3   The Interface Repository and LuaRep

The interface repository (IR), defined as a component of the ORB, provides dynamic access to object interfaces. The IR is itself a CORBA object, and can thus be accessed through method invocations. In general, these methods can be used by any program, allowing the user, for instance, to browse through available interfaces. However, the IR is specially important for the dynamic interfaces of CORBA, DII and DSI. DII allows programs to invoke CORBA servers for which they have no precompiled stub. In order to build dynamic invocations, the program must possess information about available methods and their parameters; the interface repository provides this information. On the server side, DSI allows a server to handle requests for which it has no precompiled skeleton. Again, the correct signature of these requests must be obtained from the interface repository.

The possibility of dynamically updating the IR extends the flexibility obtained with LuaDSI. It allows a manager client to install not only new implementations for existing interfaces, but also unforeseen services in the server, by first adding their definitions to the interface repository.

Eight interfaces are defined in CORBA for interaction with the IR. The **Repository** interface represents the root object, the IR itself. The **ModuleDef**, **InterfaceDef**, **AttributeDef**, **OperationDef**, **TypedefDef**, **ConstantDef**, and **ExceptionDef** interfaces represent the definitions of a module, interface, attribute, operation, typedef, constant and exception, respectively. All these interfaces inherit from the **IRObject** interface, which contains operations **destroy**, for destroying an object in the repository, and **def_kind**, which identifies an object.

Although these interfaces provide any CORBA client with the possibility of querying and updating the repository interface, using them in a conventional binding, such as the one to C++, is quite complex. The goal of the LuaRep library is to simplify access to the IR. LuaRep makes extensive use of Lua data

```
-- creates an object representing the Hello Interface
HelloInt= CORBA_createinterface{
              Print = CORBA_createoperation{
                      result = CORBA_void,
                      params = {CORBA_string}}}

-- binds to the IR
repository = OM_BindRepository{}
-- Installs Hello Interface in the repository
repository.Hello = HelloInt
```

Fig. 6. Lua code for defining interface Hello.

description facilities, so as to describe with Lua structures the desired interfaces. When such a structure is "assigned" to a field of a repository, LuaRep automatically installs the described interface in the repository.

Basic IDL types are represented by Lua constants, and structured IDL types are represented by Lua constructors; the names of these constants and constructors are formed by prefixing the IDL type name with the string 'CORBA_'. To make descriptions simpler, both parameter names and parameter modes are optional, with mode PARAM_IN as default. As an example, Fig. 6 contains the Lua code needed to publish the IDL interface Hello (Fig. 2):

To create a Lua object representing an IDL type, we use a constructor. The name of the constructor is again formed by prefixing the IDL type name, in this case with the string 'CORBA_create'. Execution of the first assignment in Fig. 6 creates a Lua object with the description on the Hello interface. Such object has a single field, named Print, to represent the single method of the interface. The method description is again given by a Lua object, with two fields, result and params, that represent the result type (CORBA_void) and the parameter types (a list with a single element, CORBA_string) of the method.

To navigate in a CORBA repository, the Lua program must first create a local reference to the repository object. This is done in the second assignment in Fig. 6; function OM_BindRepository returns an object that acts as a proxy for the IR. The fields of this proxy object are themselves objects that represent type definitions contained in the IR. Operations on this proxy object are transparently reflected on the repository itself. Thus, the last assignment in Fig. 6 updates the IR by creating a new "field" and adding to it the new Hello interface. After adding the Hello interface to the IR, the client is free to install an implementation for it in the extensible server.

The reflexivity implemented by LuaRep allows not only updates to the repository, but queries as well. Again, expressions written in Lua are automatically translated to operations on the IR. For instance, to know the type of the third argument of a method Foo from interface Iface, you just write

```
argtype = repository.Iface.Foo.params[3]
```

4   Final Remarks

Although the OMG's Request for Proposals for a CORBA Scripting Language [6] has generated several responses, none of them offers facilities similar to LuaDSI.

CorbaScript [8] supports the creation of server classes written in the language. This allows easy development of servers. However, you must write the complete class code, which must be type compatible with the IDL description, before it is bound to the ORB. You can neither use partial implementations for tests nor add new methods on the fly.

The Python proposal [3] also offers a conventional support for servers. Either you write the server as a complete class, as in CorbaScript, or you can use the raw DSI. For that, you must implement a function invoke, that must do an explicit dispatching and type conversions, like in our C++ example (Fig. 3).

The work presented in this paper is part of a larger project wherein we study the flexibility that can be brought to distributed componentware with the use of an interpreted language [5, 10]. The use of a standard CORBA mechanism, the DSI, allows a dynamically extensible server to be accessed by any client, either through stubs or through the DII. Although the CORBA specification cites support for "monitors that want to dynamically interpose on objects" [7] as one of the possible applications for DSI, we believe this work has the important function of showing how this can be done in practice.

When we started work on LuaDSI, our main goal was to dynamically implement and reimplement existing interfaces. Since the implementation of a dynamic skeleton must rely on queries to the interface repository, this led us to a deeper understanding of the IR, and to the idea of developing LuaRep. LuaRep made our working environment much more flexible than we had anticipated, since now the manager client can install and implement new interfaces. This can be done either interactively or through programmed scripts, and seems to be an important facility in the context of CORBA application management.

Currently, we are working on the application of the tools we described in network management. In the initial network management model [1], the network management application concentrated all statistics and tests, collecting only raw data from the management agents. However, this model is neither appropriate to the current size of networks nor necessary, since nowadays most of the managed devices have enough memory and CPU resources to carry on part of the management processing. Thus, it is interesting that the agents be programmed to include tests and statistics collection. However, since the management application typically retains the role of maintaining a global view, it should be able to modify agent behavior according to the global state of the network. Using the CORBA framework which we implemented, the management application would be able not only to change parameters and limits, but also to carry out complex modifications; as an example, the administrator using the management application could dynamically define new statistical functions to be applied by the agents on the raw data.

Although we have used the language Lua in this work, our approach can be applied to other languages with similar characteristics, such as Tcl or Python.

The LuaOrb system, which includes both LuaDSI and LuaRep, is available at http://www.tecgraf.puc-rio.br/~rcerq/luaorb/.

Acknowledgments

We would like to thank Renato Cerqueira, who implemented the main part of LuaOrb (the client side of the binding between Lua and CORBA); and LES (Laboratório de Engenharia de Software — PUC/Rio), which hosted the project. This work has been partially supported by CNPq (The Brazilian Research Council), as part of PROTEM III.

References

[1] J. Case, M. Fedor, M. Schoffstall, and J. Davin. A simple network management protocol, 1990. RFC 1157.

[2] Luiz H. Figueiredo, Roberto Ierusalimschy, and Waldemar Celes. Lua: An extensible embedded language. *Dr. Dobb's Journal*, 21(12):26–33, December 1996.

[3] GMD-Fokus. *CORBA Scripting with Python*, 1998. OMG Document: orbos/98-12-19.

[4] R. Ierusalimschy, L. H. de Figueiredo, and W. Celes. Lua—an extensible extension language. *Software: Practice and Experience*, 26(6):635–652, 1996.

[5] Roberto Ierusalimschy, Renato Cerqueira, and Noemi Rodriguez. Using reflexivity to interface with CORBA. In *IEEE International Conference on Computer Languages (ICCL'98)*, pages 39–46, Chicago, IL, May 1998. IEEE Computer Society.

[6] Object Management Group. *CORBA Scripting Language Request for Proposal*, 1997. OMG Document: orbos/96-06-12.

[7] Object Management Group. *The Common Object Request Broker: Architecture and Specification*, 1998. revision 2.2.

[8] Object-Oriented Concepts, Inc. *CORBA Scripting Language*, 1998. OMG Document: orbos/98-12-09.

[9] J. K. Ousterhout. *Tcl and the Tk Toolkit*. Addison-Wesley, 1994.

[10] Noemi Rodriguez, Roberto Ierusalimschy, and Renato Cerqueira. Dynamic configuration with CORBA components. In *4th International Conference on Configurable Distributed Systems (ICCDS'98)*, pages 27–34, Annapolis, MD, May 1998. IEEE Computer Society.

[11] N. M. S. Zuquello and E. R. M. Madeira. A mechanism to provide interoperability between orbs with relocation transparency. In *IEEE Third International Symposium on Autonomous Decentralized Systems (ISADS'97)*, pages 195–202, Berlin, Germany, April 1997. IEEE.

On the Concurrent Object Model of UML[*]

Iulian Ober, Ileana Stan

INPT-ENSEEIHT, 2, rue Camichel, 31000 Toulouse, France
Phone (+33) 5.61.19.29.39, Fax (+33) 5.61.40.84.52
{iulian.ober, ileana.stan}@enseeiht.fr

Abstract. Designing object models with concurrency features was a preoccupation of numerous researchers for the past two decades. UML, the standard object-oriented modeling language adopted by the OMG, does not tackle the issues of concurrency in its object model in a systematic way. In this paper we outline the position of the UML object model within the design space of concurrent object models described in [10], discuss the inconsistencies in the UML semantics and finally propose an object model, comprising features from UML and ATOM – a state-of-the-art concurrent object model.

1   Introduction

The problem of integrating the object-oriented concepts and concurrency concepts has been a preoccupation of many researchers for the past two decades. Several concurrent object-oriented languages and models were developed [2,3,5,8,9,11,14,15] and there was research work put into analyzing and classifying them [7,10].

On the other hand, UML [1,13,17] is the standard object-oriented modeling language adopted by the Object Management Group and is becoming the de-facto standard support for object-oriented analysis and design.

This paper aims to give a clear idea about the concurrency features that exist in the adopted version of UML [17]. In order to do so, we proceed in a systematic fashion, using the classification of concurrent object models described in [10] and outlined in section 2. From this perspective, in the third section we try to determine the position of UML and outline the weak points of the semantics of certain concurrency features of UML. Finally, in section 4 we describe a fully-functional object model that we have developed, which presents the same facilities as UML, notably state machines, but unambiguously integrates them with the concurrency features of a complex concurrent object model, ATOM, defined in [11,12].

The intended auditory of this paper is both the UML community, since the modeling of concurrency in UML seems not to enjoy the same level of attention as do other features of the language, and the concurrent object-oriented programming community since UML has an ever increasing importance in the spectrum of object-oriented modeling languages.

---

* work supported by CS VERILOG, http://www.verilogusa.com

P. Amestoy et al. (Eds.): Euro-Par'99, LNCS 1685, pp. 1377-1384, 1999.
© Springer-Verlag Berlin Heidelberg 1999

2   Classification of Concurrent Object Models

A well known classification of concurrent object models is given by Papathomas in
the chapter 3 of [10]. It uses three dimensions. On the first dimension, object models
are divided with respect to what combination of objects they support, into three categories: *orthogonal* (objects are independent of threads), *homogenous* (all objects are
active), *heterogeneous* (objects may be active or passive). The second dimension
captures the internal concurrency of objects: *internally sequential, quasi-concurrent*
or *concurrent*. Finally, the third dimension captures the available inter-object communication and synchronization mechanisms: *synchronous/asynchronous feature
calls, conditional/unconditional acceptance of incoming calls*, etc. For details on the
classification criteria, the reader is referred to the original work [10].

In the next section we will use this classification for systematically analyzing the
facilities of the UML concurrent object model.

3   The Concurrency Features of the UML Object Model

Our goal in this section is not to explain in detail the semantics of the UML concepts
that we are interested in. We rather outline the modeling of concurrency features and
the related anomalies. Therefore, understanding this section may require some general
knowledge about the main concepts of the UML meta-model and semantics, like
Class, Operation, Method, Action, StateMachine, Event.

UML is a modeling language thought to be used for analysis, design and documentation rather than implementation. Its designers wanted it as general as possible.
However, some decisions that have been taken in its definition, decisions conforming
to the cultures from which UML emerged (OMT, Booch, Objectory, etc.), restrict its
area of application. For example, it would be hard model a highly reflective CLOS
architecture in UML, simply because UML does not have a meta-class concept.

Correspondingly, if UML wants to conform to all possible models of concurrency,
it will have to resign form describing its own model. But as soon as it describes such
a model for concurrency, it should do that in a coherent and complete way. This is
why we allow ourselves to study the inconsistencies in the concurrent object model of
UML and to propose an alternative solution.

The modeling of concurrency is spread across the UML specification [17], for each
concept that is involved in it (*Class, Operation, Action* and *StateMachine*). In what
follows, we will try to group this information, discuss it and infer the implications
with respect to the position of UML in the design space sketched in section 2.

3.1   Active and Passive Objects

Classes in UML may be either *active* or *passive*, as described by the attribute *isActive*
of the meta-model class *Class* ([17a] pp. 16): "specifies whether an *Object* of the
Class maintains its own thread of control and runs concurrently with other active

Objects. If false, then the *Operations* run in the address space and under the control of the active *Object* that controls the caller" ([17a] pp. 21).

Therefore the model is *heterogeneous*. An UML *active* object has only one thread of control, so active objects are *internally sequential*. However, the UML semantics says nothing about the situation when multiple concurrent calls are made to the same active object (except for the case when the object has a state machine, case that we will see in more detail later). As only one message may be treated at a time, there will have to exist some mechanism for queuing the calls. Whether this queuing mechanism exists or how it acts, it is not said in [17a].

In the case of *passive* objects, each *Operation* is either *sequential*, *guarded* or *concurrent*, according to the *concurrency* attribute of the meta-model class *Operation* ([17a pp. 26]). For a certain *passive* object, only one *sequential* operation may be active at a time, but this is not enforced by the model. *Guarded* operations are also mutually exclusive, but they are protected by the model against concurrent calls. Finally, *concurrent* operations may be called concurrently at any time.

We note that the meaning of active/passive is not entirely the same as in [10]: the passive objects still have a degree of control over the invocations made towards them. This statement is further supported by the semantics of state machines.

3.2   Messages

The *object interaction mechanisms* in UML are *messages* and *signals*. Signals are discussed together with state machines in the next section. The action of passing a message is called *CallAction* in the UML meta-model. *CallAction* has an attribute *mode* ([17a] pp. 67,69) that may specify either "synchronous" or "asynchronous".

Thus, in terms of our classification, the communication means offered by the UML object model are *one-way message passing* and *RPC*. An UML *synchronous* call is what the classification calls a *(blocking) RPC*. An *asynchronous* call is what we call *one-way* message passing since there are no mechanisms for the caller to synchronize with the end of execution of the called method and to retrieve the result.

UML does not specify constraints with respect to the target of an asynchronous call. This is problematic, since passive objects execute their methods on the calling thread, if the calling thread "does not wait for the completion of the execution of the *Operation* but continues immediately", the called operation will have no thread left to execute on.

3.3   State Machines

The UML object model is different from any other object model studied in connection with the issues of concurrency, in that the behavior of an object in UML may also be specified entirely or partially through a statechart ([17a] pp. 97).

According to [17a], if an object has a state machine then all the requests sent to the object pass through its state-machine, whether they are *signals* or *method calls*. State machines in UML respond to both *SingalEvents* and *CallEvents*. The state machine processes events in a sequential way so that all events are processed following a queuing policy, even if they are *method calls*. One thing that is not clear in [17a] is whether the methods of an object are executed by its state machine or the state machine is manipulated by the methods.

For both active and passive objects, the state machine can respond to a *CallEvent* either by invoking the corresponding operation, by performing some *other actions* or by doing *nothing*. Although slightly unusual, this is a powerful feature that makes it possible to express *request scheduling* policies through state machines. One problem is that UML does not define a notation to distinguish between the three cases above.

Another problem that persists, despite the fact that passive objects with state machines go a long way back in the OOA&D, is that state machines are primarily designed to respond to asynchronous stimuli like *Signals*. Or a passive object cannot respond to such stimuli, not having its own thread of control.

3.4   UML Position in the Design Space of Concurrent Object Models

The following table summarizes the conclusions we have drawn about the position of UML in the design space of concurrent object models as systematized in [10]:

Classification criterion	UML position	Remarks
Object Model	*Heterogeneous*	The definitions of passive and active in UML are not the same as those used in the classification. Passive objects are *not unprotected* against concurrent calls
Internal Concurrency	–Active objects are *sequential* –Passive objects are *internally concurrent*	The methods of a passive object are split in three classes, *sequential*, *guarded* and *concurrent*.
Client/Server Interaction	–One way message passing through *signals* and *asynchronous calls* –RPC	No support for reply scheduling through *futures*.
Constructs for accepting requests	Activation conditions supported through *state machines*	States/transitions/guards provide a powerful mechanism for expressing activation conditions, but there is no semantics for the inheritance of a *state machine*

The position of UML within the design space of concurrent object models is quite clear, despite the leaks in the specification that we saw in the earlier sections. Besides these leaks, there are some drawbacks in the approach taken by UML towards concurrency, among which:

- active objects are *sequential*
- *request scheduling policies* can be expressed only using statecharts, which makes it difficult to express very simple activation conditions and is prone to inheritance anomalies (especially since statechart inheritance is not formally defined in UML).
- *asynchronous request-reply* communication is not supported natively

A careful designer can avoid all these anomalies and UML is already used on a large industrial scale. Still, a cleaner semantics and more powerful primitives are desirable. It is not our goal to define a semantics in this paper, but merely to suggest improvements and clarifications to the model.

4   The ATOM-S Object Model

The drawbacks of the UML object model drove us in the research of a better combination of a *concurrent object model* and *state machines*. The result is the ATOM-S object model. Similar efforts are described in [4,16]. The essential difference is that we start from a concurrent object model, ATOM [11,12], proven to solve many of the classical problems of object-oriented concurrency, such as inheritance anomalies [14].

Our proposal represents an extension of UML. This requires to correct the anomalies that exist in UML and to use the UML extension mechanisms (stereotypes and tagged values) to model the concepts added by ATOM-S. We have implemented a fully functional prototype of this model in the Python [18] programming language.

4.1   Basic Notions of ATOM

This section is a brief presentation of ATOM and may be skipped by the readers familiar with it. More details on ATOM can be found in [11,12].

ATOM has only *quasi-concurrent active objects*. [11] presents this choice as a good compromise between expression power and protection of the integrity against concurrent calls.   Active objects communicate through two-ways (*request-reply*) messages, which can be *synchronous* (*blocking* or *non-blocking RPC*) or *asynchronous*.

For asynchronous messages ATOM uses advanced features for reply scheduling. An object can issue a requests towards another object and later retrieve the result or wait for this result to be produced. Delegation of the response by a method to another method is also possible and the reply is routed automatically to the initial requester.

Request scheduling is based on activation conditions. A specific mechanism associates actions to specific events, so one can write an action to be executed when a method call is received, when the execution begins or when it ends. [11] shows that this feature solves certain inheritance anomalies.

ATOM introduces the notion of abstract state, denoting a named predicate over the attributes of an object used to describe activation conditions. Abstract states are used for coordinating a set of objects. Each object has a notifier service, to which another object can subscribe, to be announced when the object reaches a certain state.

4.2   ATOM-S

ATOM-S does not make modifications to the features of ATOM, instead it makes an essential extension by integrating state machines. ATOM-S uses *passive objects*, because they exist in UML and they may be necessary when all the active object overhead is not needed. Passive objects do not own threads of control, but execute their methods on the caller threads. Therefore they can not respond to asynchronous calls and, unlike in UML, they *cannot have a state machine*. This difference is justi-

fied by the fact that state machines should be used to describe the response of objects
to asynchronous *signals*. This also solves the issue raised in the end of section 3.3.

```
1  class ThreadScheduler (ActiveObjectSupport):
2     methods = ['InsertThread','Schedule','EndThread',
              'AlertAdmin','RecoveryProc']
3     events = ['Recover']
4     conditions = {
5         'InsertThread':'not self.inState('Overloaded')'
6         'Schedule':'not self.inState('Overloaded')'
7     }
8     def InsertThread(self, thread, priority):
9         """ ... the bodies of the methods are omitted"""
10    statechart = {
```

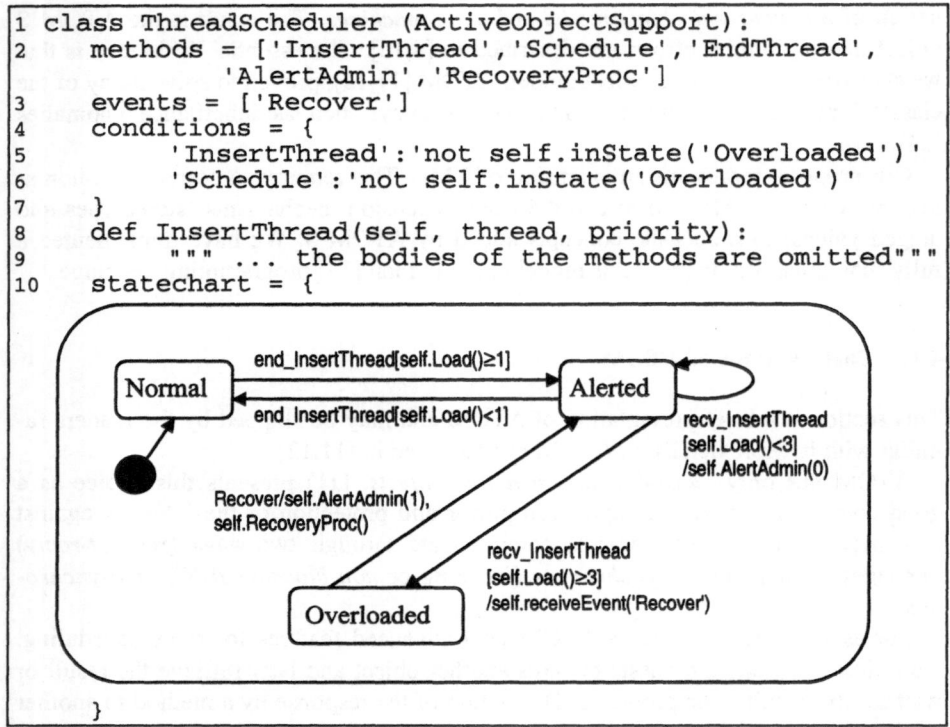

Fig. 1. A simple thread scheduler class in ATOM-S

As in UML, the methods of a passive object can be: *sequential* (raise an exception
when called concurrently), *guarded* (mutually exclusive and protected through semaphores) and *concurrent* (not protected and free to be called concurrently).

In ATOM-S *active objects* are *quasi-concurrent*. This represents an extension of
the UML internally sequential model: an executing method can explicitly yield the
control for example while it is waiting for a return from another object. ATOM-S
uses all the communication and synchronization mechanisms of ATOM.

To exemplify the features of ATOM-S, we use a simple example: a thread scheduler class whose code is partially given in Fig. 1 (the bodies of the methods are omitted). The example is written in our extension of Python, except for the statechart,
which has a Python form but it is more readable in a graphic Statechart [4] format.

Besides attributes and methods, an *active* object may have a *state machine* that
specifies the *reactive* part of its behavior and responds to *asynchronous one-way
stimuli*, called *signals*, that may carry parameters. The signals to which a class responds are part of the class signature, just like the methods (Fig. 1 line 3).

As in UML, the state machine may specify the complete behavior of an object or only its protocol, case in which its states are used as *activation conditions* for the object methods. In Fig.1 the activation conditions of `InsertThread` and `Schedule` are based on state machine: the methods cannot execute while the object is in the `Overloaded` state (lines 5,6).

To solve the ambiguity that exists in UML about the way *CallEvents* are processed (see section 3.3), method invocations sent towards an object do not pass through its state machine and only *signals* are queued and processed one-by-one by the machine. Thus, a *method invocation* will always result in the *execution of the method*.

The operational semantics of state machines in ATOM-S is based on the fact that the state machine of an active object *runs quasi-concurrently* with its methods. At object creation time a default thread is created to manage the state machine (if there is a machine). This thread *keeps track* of the current state of the machine, *retrieves signals* from the queue and *executes the transitions and actions* of the state machine. Depending on a priority policy chosen for the object, the state machine thread may yield the control between the processing of consecutive events or only when the event queue is empty.

The state machine of an object is notified when a message call is received, when a method starts executing or finishes. Note that, like in ATOM, the moment when the method is received may not coincide with the moment when it starts executing. The following signals are sent by default by the underlying model to the state machine: recv_m (when the method m is called on that object), begin_m (when the method m starts executing) and end_m (when the method m ends its execution). The introduction of these implicit messages augments the power of expressing *request scheduling policies* through the state machine in the same way the receipt, pre- and post-actions augment the power of expression of ATOM activation conditions (see [11]). In Fig.1 these implicit signals `recv_InsertThread` and `end_InsertThread` trigger transitions in the state machine.

Our model clarifies the relationship between the operations of a class and its state-chart, which is vague in UML. As an extension of the ATOM model, the ATOM-S solves the classical problems of concurrency. In addition, it has the expressing power of statecharts and means for combining the two ways of modeling behavior.

5   Conclusions

The substantial research work that has been done in the field of concurrent object-oriented programming, although scientifically recognized, has a long way ahead towards the integration in industrial standards, languages and environments. On the other side, UML, which is a popular standard language for analysis, design and documentation, has largely ignored the problem of concurrency in its model. It is therefore an open door through which research results in concurrent object oriented programming languages can penetrate to the industrial world.

This paper is only the beginning of a process of correcting the problems and identifying general solutions for the integration of concurrency in UML. We have pointed out some weaknesses of UML face to concurrency. To overpass these problems we have presented in the fourth section a concurrent model, which can be regarded either as an extension of UML or as a design pattern. The model we developed has about the same power of expression as UML. It benefits from the results of research in concurrent object oriented programming, embodied in the ATOM model, and takes an approach as close to UML as possible and does it in a clean, unambiguous way. The prototype that we developed (freely available from the authors of this paper) is an extensible tool, that can be used for experimenting with our model as well as with other variations on this theme.

References

1. Booch, G., Jacobson, I., Rumbaugh, J.: The Unified Modeling Language User Guide. Addison-Wesley, (1998)
2. Caromel, D: Towards a Method of Object Oriented Concurrent Programming, Communications of the ACM, vol. 36, no. 9, pp. 90-102, (1993)
3. Gehani, D., Roome, W.D: The Concurrent C Programming Language, Prentice Hall 1989
4. Harel, D., Gery, E.: Executable Object Modeling with Statecharts IEEE Computer July, pp.31-42, (1997)
5. Kafura, D.G., Lee, K.H.: Inheritance in Actor based concurrent object-oriented languages Proceedings of ECOOP'89, pag131-145, Cambridge University Press (1989)
6. Matsuoka, S., Watanabe, T., Yonezawa, A.: Hybrid group reflective architecture for object-oriented languages Proceedings of ECOOP/OOPSLA'90 Workshop on Reflection and Metalevel Architectures in Object Oriented Programming, (1990)
7. Matsuoka, S., Yonezawa, A.: Analysis of inheritance anomaly in OO concurrent programming language, Research directions in Concurrent OO programming, MIT Press, (Ed. G.Agha, P.Wegner, A.Yonezawa), pg 107-150, (1993)
8. Meyer, B.: Object Oriented software construction, Second edition, Prentice Hall, (1997)
9. Nierstrasz, O.: Active Objects in Hybrid. Proc OOPSLA'87, SIGPLAN Notices v.22, no. 12, pp. 243-253, (1987)
10. Papathomas, M.: Language Design Rationale and Semantic Framework for Concurrent Object-Oriented Programming. PhD Thesis. Université de Genève, (1992)
11. Papathomas, M.: ATOM: An Active Object Model for Enhancing Reuse in the Development of Concurrent Software, RR 963-I-LSR-2, LSR-IMAG, Grenoble, (1996)
12. Papathomas, M., Andersen, A.: Concurrent Object Oriented Programming in Python with ATOM 1st Conference on Python, (1996)
13. Rumbaugh, J., Jacobson, I., Booch, G.: The Unified Modeling Language Reference Manual., Addison-Wesley, (1998)
14. Yonezawa, A.: ABCL: An Object-Oriented Concurrent System. Computer Systems Series. Ed. Akinori Yonezawa, MIT Press (1990)
15. Tomlinson, V. Singh: Inheritance and synchronization with Enabled-Sets. Proceedings of OOPSLA'89, ACM Press, October (1989)
16. Selic, B.: "An Efficient Object-Oriented Variation of the Statecharts Formalism for Distributed Real-Time Systems", Paper submitted to CHDL '93: IFIP Conference on Hardware Description Languages and Their Applications, April 26-28, 1993, Ottawa, Canada.
17. UML Proposal to the OMG, OMG documents ad/97-08-02 to ad/97-08-09
    17a. UML Semantics, OMG document ad/97-08-04
18. The Python Language website: http://www.python.org

Object Oriented Design for Reusable Parallel Linear Algebra Software

Eric Noulard[1,2] and Nahid Emad[2]

[1] Société ADULIS – 3, rue René Cassin – F-91742 Massy Cedex – France
E.Noulard@adulis.fr
[2] Laboratoire PRiSM – UVSQ – Bât. Descartes – F-78035 Versailles Cedex – France
Nahid.Emad@prism.uvsq.fr

Abstract. Maintaining and reusing parallel numerical applications is not an easy task. We propose an OO design which enables very good code reuse for both sequential and parallel linear algebra applications. A linear algebra class library called LAKe is implemented using our design method. We show how the same code is used to implement both the sequential and the parallel version of the iterative methods implemented in LAKe. We show that polymorphism is insufficient to achieve our goal and that both genericity and polymorphism are needed. We propose a new design pattern as a part of the solution. Some numerical experiments validate our approach and show that efficiency is not sacrified.
Keywords: OO design, parallel code reuse, Krylov methods.

Introduction

In the area of numerical computing many people would like to use parallel machines in order to solve large problems. Parallel machines like modest sized SMPs or workstations clusters are becoming more and more affordable, but no easy way to program these architectures is known today. Our main goal is to evaluate the object oriented design as a mean to reuse most of the sequential and parallel software components. We focus on the domain of linear algebra and particularly one the Krylov subspace methods. They solve either eigenproblems [5] or linear systems [6] and are good candidates to code reuse and parallelization.

In section 1, we present the block Arnoldi method and recall the usual way to parallelize such a method. We list the elementary neeeded operations for either a sequential or a parallel implementation. In section 2, we first develop our goals in terms of code reuse and then present different design solutions We show the limit of polymorphism and dynamic binding as a reuse scheme when compared to genericity. Finally section 3 presents some numerical experiments.

1 The Block Arnoldi Method

The Block Arnoldi method is a projection method that computes some eigenelements (u, λ) satisfying $Au = \lambda u$, of a large non-symmetric sparse matrix A of

P. Amestoy et al. (Eds.): Euro-Par'99, LNCS 1685, pp. 1385–1392, 1999.
© Springer-Verlag Berlin Heidelberg 1999

size $n \times n$ where n is large. The very first step consists in reducing the projected matrix into an upper block Hessenberg matrix H_m of size $ms \times ms$ through the *Block Arnoldi Process*. For further details see [5].

1.1   Parallelizing Arnoldi Method

Our target parallel machines are distributed memory architectures. In this context, the classical way to parallelize Krylov subspace iterative methods is to distribute the large vectors and/or matrices and replicate the small ones on the processors. We first decompose and distribute all the matrices of size n, the matrix A, the Krylov subspace basis $V_m = [V_1, \cdots V_m]$ of size $n \times m \cdot s$ and possibly the temporary variable of size $m \cdot s$ we call W. The matrix H_m and all the m-sized matrices are replicated.

1.2   Necessary Elementary Operations

We list below the basic operations used by all the Krylov subspace methods including the block Arnoldi one. We have for $m, n, p \in \mathbb{N}$:

1. SAXPY: $Y = \alpha X + \beta Y$ with $Y, A, X \in \mathbb{R}^{m \times n}$ and $\alpha, \beta \in \mathbb{R}$. The matrices may be distributed.
2. Product: $Y = \alpha A \cdot X + \beta Y$ with $Y \in \mathbb{R}^{m \times p}$, $A \in \mathbb{R}^{m \times n}$, $X \in \mathbb{R}^{n \times p}$ and $\alpha, \beta \in \mathbb{R}$. The matrices may be distributed.
3. Sparse product: $Y = \alpha A \cdot X$ with $X, Y \in \mathbb{R}^{n \times p}$ and $A \in \mathbb{R}^{n \times n}$, A is **sparse**. These matrices may be distributed and A may be available only as a function to do matrix product.
4. subranging: $Y = A(i_1 : i_2, j_1 : j_2)$, with $A \in \mathbb{R}^{m \times n}$ and $Y \in \mathbb{R}^{(i_2 - i_1 + 1) \times (j_2 - j_1 + 1)}$. The matrices may be distributed.
5. point addressing: $A(i, j) = \alpha$ with $A \in \mathbb{R}^{m \times n}$ and $\alpha \in \mathbb{R}$. This operation is authorized only on full matrix.
6. allocate, deallocate and (re-)distribute $A \in \mathbb{R}^{m \times n}$.

2   Designing a Reusable Software

The Krylov subspace methods and Block Arnoldi one led us to design a class library called LAKe (**Linear Algebra Kernels**). The main goal of this library is the use of the *same code for the sequential and parallel* version of the iterative methods. In that way we only maintain a single code which is not cluttered with unreadable parallel code. We first present the sequential design of LAKe, then we explain why polymorphism and dynamic binding are not sufficient to make this feature possible. We finally demonstrate how genericity is the key of the solution. We point out that our design is the first one to reach such a reuse goal.

2.1   LAKe Architecture

The LAKe architecture is presented in figure 1. Each box represents a class, whose features have been omitted for the sake of clarity. Plain arrows stand for inheritance or the **is-a** relation [4, p. 811]. Dashed arrows represent the **client** relation. A class A is a **client** of another class B if it uses at least an object of type B. The client relation is *dynamic* if the relation is established at runtime and it is *static* if it may be established at compile time. Polymorphism

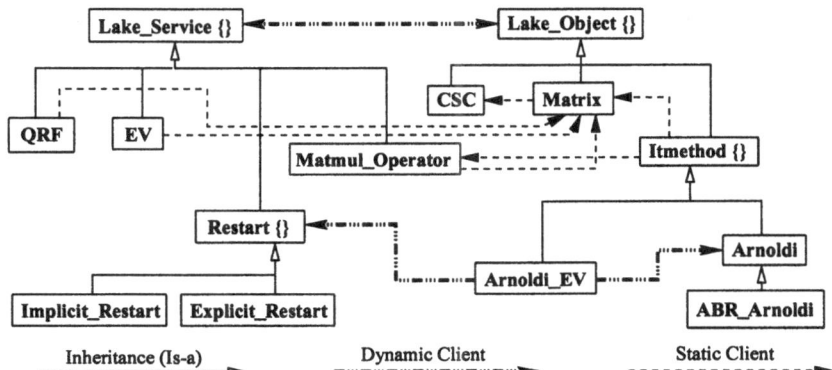

Fig. 1. The LAKe architecture

[4, p. 28] is the handling of different objects which share some parts of their interface. Polymorphism associated with dynamic binding [4, p. 29] enables the polymorphically handled object to act differently at runtime.

2.2   A Weakness of Polymorphism: Contravariance

Polymorphism seems an obvious way to parallelize iterative methods without touching the code. We only need to build a DMatrix class that is derived from Matrix which redefines the needed features in order to have a parallel implementation. Our iterative methods will then use the distributed matrix class DMatrix polymorphically. We will now show why polymorphism is not sufficient to reach our aim. The problem in object-oriented language is the implicit assumption that inheritance defines a subtype relation. Contravariance comes out when trying to subtype functions and the correct rule for function subtyping is given in definition 2.2.

Definition 2.1 (Subtype). *A type T' is a* subtype *of type T, also noted $T' \leq T$ iff every function that expects an argument of type T may take an argument of type T'.*

Definition 2.2 (Contravariance). *Let $TA \to TR$ be the type of a function taking an argument of type TA and returning an argument of type TR. The subtype rule for functions is: $TA' \to TR' \leq TA \to TR$ iff $TR' \leq TR$ and $TA \leq TA'$. We say that the outputs type of a function is covariant since it varies in the same way the type of the function does, but the type of the input arguments is contravariant since the subtype relation is inverted.*

The contravariance problem arises in the Block Arnoldi method when performing the algebraic operation $H_{ij} = V_i^H W$ on the distributed matrices V_i and W. Let TA be Matrix and TB be DMatrix. The operation is performed by a call to the method TA::tmatmul(TA*,TA*) which has been redefined in the distributed matrix class as TB::tmatmul(TB*,TB*). At this point the wrong method is called because the subtype relation on functions implies that TB::tmatmul(TB*,TB*) is **not** a subtype of TA::tmatmul(TA*,TA*) since the inputs arguments must be contravariant. The proper method redefinition is TB::tmatmul(TA*, TA*). Thus the type of the arguments must be checked dynamically in order to verify what they really are. This process is called *dispatch* of the arguments: single, double and multiple dispatch when doing it for one, two or more arguments. The *multiple dispatch* problem is an old OO problem and has been solved in the past. It is generally not integrated in OO langages since it is costly. It was noticed and solved, *in the same linear algebra context* by F. Guidec in [3, pp. 96–99] for the Paladin linear algebra library. We propose an improved solution as a design pattern called *Service Pattern*. It has two advantages over the Paladin solution: the dispatch of an argument is only done when it changes and the dispatch may be done for several operations using the same arguments.

2.3   Service Pattern Solution

The *Service Pattern* reifies the method which must dispatch its argument: the Matrix::matmul method becomes a Matmul_Operator service class. The *Service Pattern* (inspired from the *Visitor Pattern* [2, p. 331]) represents a set of operations which register (or connect) their arguments one by one, a status determine which operation can be called. As an example the figure 2 shows the DMatmul_Operator service which performs the task $Y = A \cdot X$ on distributed matrix. The advantages of the *Service Pattern* are that the related operations

```
// register A as "Y", B as "A" and C as "X"
Matmul.connect(A,"Y");  Matmul.connect(B,"A");  Matmul.connect(C,"X");
Matmul.matmul();        // compute A = B * C
Matmul.disconnect();
```

Fig. 2. Examples of a LAKe Service

are grouped together in an object offering a complete service, and that the arguments are dispatched only when needed. The participants of the pattern are:

a Service class, a base Object class and as many descendants of Object as needed. The Object class has no requirement other than having type information provided for it. In the following parm_name stands for the name of the argument of the service, typically ''A'', ''X'', ''Y'' in fig. 2 which represent the role the arguments play for the service Matmul_Operator. The Service class must provide:

- one redefinable method connect(Object* O, char* parm_name) which finds the dynamic type of O and calls the specialized method corresponding to the specified parm_name.
- one method connect_PARM_NAME(DType* O) for each parm_name which makes sense for the service and for each dynamic type DType accepted for this parameter. The call to such a method will register the object into the service.
- one redefinable method disconnect(char* par_name) which unregisters the specified parameter(s).
- one method do_task() for each computational task offered by the service.

The preceding technique works for all the operations listed in §1.2 but the memory allocation. We explain the reason and the solution in the next section.

2.4 The Need of Genericity

Some classes or functions of LAKe must be able to allocate matrices whose size depends on input parameters of these classes or functions. As an example the Arnoldi class must be able to allocate the matrices H, V and W. In a sequential context this is not a problem since the Matrix used by Arnoldi must have a method to create any rectangular matrix. In a parallel context, those matrices may be distributed and the Arnoldi class has no way to do this since it must not know if the matrices it uses are distributed or not. There is a simple solution to this problem: make all distributed object parameters of the Arnoldi class. Finally to create an Arnoldi object, all the matrices must be passed as parameters. We have made a big step backwards since *our class has the same structure as a Fortran subroutine: it requires input/output parameters and workspace!*

Remark 2.1 (Classes or Functions). It may seem a better choice to make the Arnoldi class a function as it is done in IML++ [1] or ITL [7]. We are convinced this is not when our goal is reusing the code implementing the Arnoldi algorithm. In fact, if Arnoldi were a function, every Arnoldi client would allocate the H, V and W matrices before using the function. In the end every related iterative method would become functions each of them requiring several preallocated matrices arguments including workspace like W. This approach breaks encapsulation since clients of Arnoldi should provide Arnoldi's private data and workspace. This is against reuse too since no client can polymorphically use a specialized Arnoldi function.

Another solution is to use the *Service Pattern* to design a Matrix_Allocator service and pass it as parameters of Arnoldi, but it would make the code of Arnoldi uglier which is just what we want to avoid. The concept that solves all the issues is *genericity*.

Definition 2.3 (Genericity). *Genericity is the ability to parameterize a class with a type. We note the generic class A<TB> where the class A is parameterized by the formal generic type parameter TB.*

A classical example of the use of genericity is the `Container<TElem>` which defines a `Container` whose elements are of type `TElem`. While writing the code of the methods of the generic class we implicitly assume that the formal generic parameter has some properties like having the +, and * operators. If an actual type or class `AT` fullfills the constraints of the formal generic parameter `T` we say that `AT` conforms to `T`.

The solution to the distributed allocation problem is to parameterize the matrix class with an opaque type `TShape`. The formal generic parameter `TShape` encapsulates the information needed to create a matrix. The generic `Matrix<TShape>` class will have two methods: `Matrix<TShape>::create(TShape S)`, whose purpose is to create a matrix knowing its shape and `TShape Matrix<TShape>::shape()` which returns its actual shape. To have a complete solution to our problem we finally need to define a set of operations on shape: *, +, subrange, expansion... corresponding to the needed operations on matrices. Now, at compile time when we instantiate a shaped matrix we know the exact type of its shape. This last point is important since we need to know the exact shape of the matrix in order to implement `Matrix<TShape>::create(TShape S)`. Now if we want to create W whose shape is the product of matmul operator shape and x_0 shape, we write: `W.create((SMatmul.shape())*(x0.shape()))`. If we want to create V whose shape is the shape of x_0 expanded along the columns $m+1$ times we write: `V.create((x0.shape()).expand(1,max_it()+1))` Operations on shapes fix the rules for distributing the result of distributed operations on matrices. For example the result of the product of a column-wise distributed matrix by a row-wise distributed matrix should be a duplicated matrix. Shapes unify guard conditions for matrix operations. When doing $Y = A \cdot X$ we should have `Y.shape() == A.shape()*X.shape()`.

For solving the contravariance problem genericity helps too. If we suppose `DMatrix` conforms to `TMatrix`, the instantiated class:
`Arnoldi_EV<Matmul_Operator<DMatrix>,DMatrix>` is a parallel iterative method which does not suffer from the contravariance problem. In fact the compiler is able to decide at *compile time* the right method it must call. The classes `Matrix` and `DMatrix` may even be unrelated (no inheritance relation) and this would work in the same way, since the conformance relation is weaker.

Remark 2.2 (Other generic libraries). IML++[1] and ITL[7] both define generic iterative methods. LAKe handles issues which are unresolved in those libraries:

1. they have not been used with distributed matrix classes.
2. the iterative methods are implemented as generic functions and not classes. This means that polymorphically reusing an Arnoldi process was not a goal of those libraries.
3. the functions implementing iterative methods cannot handle the allocation of a distributed variable.

The generic LAKe library fullfills its requirements. The code of the iterative methods hierarchy is *strictly the same* when used with parallel or sequential matrices. Iterative methods are really building blocks which may hold and allocate their own data distributed or not. We reuse most of the code of the sequential matrix to implement the distributed one. We must note that the Service/Object hierarchy and the Service Pattern are still usefull for implementing the *dynamic client* relation. Moreover we can point out a methodology to choose between Service Pattern and generic approach: identify the dynamic and static client relation.

3   Numerical Experiments

We have implemented the LAKe library in C++ and used MPI through OOMPI [8] for the parallel classes. We used block Arnoldi method in order to find the 10 eigenvalues of largest modulus. Iterations were stopped whenever the residual associated with the ritz pair was less than $1e - 6$. The first matrix (CRY2500) is taken from the matrix market (CRYSTAL set of the NEP collection) and has 2500 rows and 12349 entries. The second matrix (RAEFSKY3) has 21200 rows and 1488768 entries. Numerical experiments were done on the CRAY T3E of IDRIS[1].

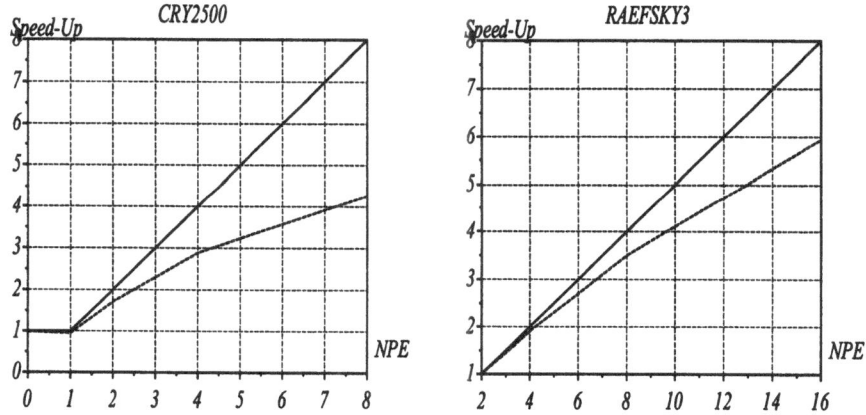

Fig. 3. Speed-up

Speed-up are shown in figure 3. The solid line curve corresponds to theoretical speed-up and the dashed curve to measured speed-up. The speed-up correspond-

[1] Institut du Developpement et des Ressources en Informatique Scientifique, CNRS, Orsay, France

ing to RAEFSKY3 begins with 2 processors since the code is unable to be run on one processor. For CRY2500 a number of processors $NPE = 0$ corresponds to the sequential code and $NPE \geq 1$ corresponds to the parallel one. The speed-up are good as soon as the number of processors is not too large in comparison with the size of the matrix. We note that the sequential and parallel code used for CRY2500 are derived from the *same* generic code. This means that for a data set that fits on a workstation we do not need to run the parallel version on one processor but we instantiate the sequential version. A raw comparison with a Fortran 77 code implementing the method in a *non-generic* way showed that the fortran code was 2 times faster than the generic C++ one. We must note that the comparison is not fair since the the F77 code is far from implementing every feature implemented in the C++ version.

Conclusion

We have presented how a coupled object-oriented and generic design enables the development of the *same* code for the sequential or parallel version of our linear algebra application. This is a key to parallel software maintenance and reuse. The basic idea is to parameterize the class which will become parallel by its abstract data type. We think the shaped matrix mecanism may be illustrative enough to give insight for other parallel applications. Experiments have shown that the same code is working for both sequential and parallel version, with promising scalability. We pointed out that both genericity and polymorphism are useful. A good perspective is a design methodology which explains how to choose between generic and polymorphic components in order to build reusable and extendable software for both sequential and parallel applications.

References

[1] Jack Dongarra, Andrew Lumsdaine, Roldan Pozo, and Karin. A. Remington. *Iterative Methods Library*, April 1996. Reference Guide.
[2] Erich Gamma, Richard Helm, Ralph Johnson, and John Vlissides. *Design Patterns: Elements of Reusable Object-Oriented Software*. Addison-Wesley, Reading, 1995.
[3] Frédéric Guidec. *Un Cadre Conceptuel pour la Programmation par Objets des Architectures Parallèles Distribuées: Application à l'Algèbre Linéaire*. PhD thesis, Université de Rennes 1, Rennes, France, Juin 1995. PhD thesis edited by IRISA.
[4] Bertrand Meyer. *Object-Oriented Software Construction*. Prentice Hall, 1997.
[5] Yousef Saad. *Numerical Methods For Large Eigenvalue Problems*. Manchester University Press, 1991.
[6] Yousef Saad. *Iterative Methods for Sparse Linear Systems*. PWS Publishing Company, New York, 1996.
[7] Jeremy G. Siek, Andrew Lumsdaine, and Lie Quan Lee. Generic programming for high performance numerical linear algebra. In *SIAM Workshop on Interoperable OO Sci. Computing*, 1998.
[8] Jeffrey M. Squyres, Brian C. McCandless, and Andrew Lumsdaine. *Object Oriented MPI (OOMPI): A C++ Class Library for MPI*, 1998. http://www.cse.nd.edu/~lsc/research/oompi.

Topic 18
Global Environment Modelling

Michel Déqué

Local Chair

This topic adresses problems linked to the numerical simulation of the atmosphere or the ocean. Environmental problems at the global scale can be solved with highly complex simulation models. The parallel computing offers a solution to increase widely the number of phenomena and interactions which occur in the atmosphere, ocean, cryosphere and biosphere, and to accelerate the production of real-time forecasts.

In the paper on the PALM project, a general structure for a modular implementation of a data assimilation system is presented. In this system, an assimilation algorithm is split up into elementary units such as the observation operator, the computation of the correlation matrix of observational errors, the forecast model, etc. This approach allows to separate the physical part of the problem from the algebraic part. PALM performs the algebra, ensures the synchronization of the units and drives the communication of the fields exchanged by the units. Thanks to the interface specifications, each physical unit can be implemented as an independent executable code and can be easily replaced or reutilised for other configurations.

The paper on the Princeton ocean model illustrates the use of message passing in parallelizing the model. Domain decomposition techniques are used for the horizontal discretization whereas in the vertical direction each column per grid point is computed. The required interprocessor communication is implemented on the PVM message passing library. Three different data partitioning schemes are studied. It is shown that the checkerboard partitioning produces the best results when the number of processors becomes larger than 15. Otherwise row blocked striped partitioning is to be preferred.

The other contributions are devoted to atmospheric models. In the paper on modular Fortran 90 implementation of a parallel atmospheric GCM, software engineering aspects of a parallel atmospheric General Circulation Model are presented. This GCM is currently nearing completion at the NASA Data Assimilation Office. The code is scientifically closely related to the current Goddard Earth Observing System (GEOS 2.x) GCM. The new software design, however, is a radical departure from the previous Fortran 77 code, using Fortran 90 derived data types and modules to compartmentalize the implementation. In order to reduce turnaround time needed to speed scientific development, the GEOS 3.x implementation uses a message-passing paradigm. This allows the performance to scale consistently to a large number of processors on both an Origin 2000 and a Cray T3E. The Message-Passing Interface (MPI) interface is used for portability. The modularity and performance of the parallel GEOS 3.x code

P. Amestoy et al. (Eds.): Euro-Par'99, LNCS 1685, pp. 1393–1394, 1999.
© Springer-Verlag Berlin Heidelberg 1999

allow faster development of scientific improvements such as variable resolution grids and a new land surface model.

The other papers concern limited area models. The first one presents the implementation of the limited area numerical weather prediction model Aladin in distributed memory. The technical challenges for the limited area model Aladin are various: Aladin is the French operational mesoscale model, run on the Meteo-France VPP-700E on 4 processors twice a day. Also, the model is used in operational or pre-operational mode in several European countries, plus Morocco, and therefore the model has to be portable and computationally efficient on many different platforms. The general choices for the porting of Aladin in a parallel environment are discussed, with an emphasis to Aladin-specific aspects, the general choices being closely related to those of Arpege. Finally, the question of applications to workstation clusters is addressed.

The next two papers are devoted to the mesoscale model Meso-NH. A parallel distributed fast 3D poisson solver is presented. The fast Poisson solvers based on FFT computations are among the fastest techniques to solve Poisson problems on uniform grids. In this paper, two parallel distributed implementations are described in the context of the atmospheric simulation code Meso-NH. The first parallel implementation consists in implementing data movement between each computational step so that no elementary computational routine implements communication. The second approach intends to reduce the global data movement and requires to parallelize one step of the fast Poisson solver algorithm. Experimental results are given on a 128 node Cray T3E to illustrate the advantages and the drawbacks of each approach.

The next paper presents the parallelization of the French meteorological mesoscale model MésoNH. This model, developed by Météo-France and the Laboratoire d'Aérologie, offers self-nesting possibilities: simultaneous interaction of several simulations working on grids with different resolutions. The parallelisation has been carried out with the help of CERFACS. The paper presents the message passing layer, the memory distribution strategy, and the serial code.

Another strategy of memory distribution is presented in the last paper on the porting of a limited area numerical weather forecasting model on a scalable shared memory parallel computer. This paper describes the porting of a weather forecast limited area model from a vector machine to a recent SMP-multiprocessor, the SGI Origin 2000, and evaluates the performances obtained. As one can see, the different fields, data assimilation, ocean and atmosphere modelling, global and local model are present in this topic. The different techniques, PVM, MPI, shared memory are employed to parallelize these big models and reduce the ellapsed time by multiple computers.

The Parallelization of the Princeton Ocean Model

L.A. Boukas[1], N.Th. Mimikou[1], N.M. Missirlis[1]*, G.L. Mellor[2], A. Lascaratos[3], and G. Korres[3]

[1] Department of Informatics, University of Athens,
Panepistimiopolis 157 10, Athens, Greece.
[2] Princeton University,
Princeton, NJ 08544–0710, USA.
[3] Department of Applied Physics, University of Athens,
Panepistimiopolis 157 84, Athens, Greece.

Abstract. In this paper we present the parallel implementation of the Princeton Ocean Model (POM) using message passing. Domain decomposition techniques are used for the horizontal discretization whereas in the vertical direction each column per grid point is computed. The required interprocessor communication was implemented on the PVM message passing library. Three different data partitioning schemes are studied. It is found that the checkerboard partitioning produces the best results when the number of processors becomes larger than 15, otherwise row block striped partitioning is to be preferred.

Keywords: Princeton Ocean Model, domain decomposition, message passing

1 Introduction

The use of parallelism as a vehicle to advance Earth's simulations is considered as one of the most complex problems facing the scientific community today and is one of the Grand Challenges of the next century.

Climate models and in particular ocean models have followed a similar development with atmospheric models and in fact they have been coupled to simulate the Earth's global climate dynamics. However, there is a high demand for high-resolution and inclusion of more complex physical processes in order to make accurate predictions for future climate conditions. For meeting the required demands a huge amount of computations is needed that can be carried out in realistic time using parallel computers.

In previous work [1, 12, 5] we have parallelized the atmospheric model eta which is in operational use at the Greek National Meteorological Service. In the present paper we consider the parallelization of the Princeton Ocean Model (POM) [4]. Many other groups are developing parallel ocean models. A cluster

---

* Author for correspondence: Email: `nmis@di.uoa.gr`

P. Amestoy et al. (Eds.): Euro-Par'99, LNCS 1685, pp. 1395–1402, 1999.
© Springer-Verlag Berlin Heidelberg 1999

version of the Ocean Circulation Model (OCM) based on a model of the Geophysical Fluid Dynamics Laboratory (GFDL) for the CEDAR was developed in [6, 7, 8]. In [2] and [20] simple shallow–sea models were parallelized on different parallel machines. A parallel version of POP (Parallel Ocean Program) was developed in [19] for the CM2 and CM–5 machines. Additional work towards to implementation of POP on parallel computers with recent architectures and couple it with sea ice, atmospheric and land surface models is described in [11]. Another global model developed at GFDL was the Modular Ocean Model (MOM). The code of this model was structured for vector machines. A two dimensional version of MOM code was developed in [16] and is known as the Modular Ocean Model Array (MOMA) code. The Ocean Circulation and Climate Advanced Modelling (OCCAM) is a message passing implementation of the MOMA code [10]. The ocean general circulation model, OPYC which uses implicit time stepping was recently parallelized on the CRAY T3D [15]. HOPE is an ocean model which was also parallelized for distributed memory processors using the PARMACS message passing package [9]. In [3, 17, 18] the multipurpose isopycnic model MICOM was parallelized following two different approaches.

In this paper POM is briefly presented in section 2. In section 3 we discuss the design issues for the parallelization of POM using domain decomposition. Implementation is described in section 4, performance is presented in section 5 and finally future plans conserning continuation and optimizing of the current work is discussed in section 6.

2   The POM Model

The POM model was developed by A. Blumberg and G. Mellor [4]. The model is a three dimensional ocean model incorporating a turbulence closure model to provide a realistic parameterization of the vertical mixing processes [14]. The model uses a sigma coordinate system which simulates well the bottom topography. The prognostic variables are the three components of velocity, temperature, salinity, turbulence kinetic energy and turbulence macroscale.

The equations which form the basis of the circulation model are the traditional hydrodynamic equations for conservation of mass, momentum, temperature and salinity coupled by an equation of state. The governing equations together with their boundary conditions are solved by finite difference techniques. The horizontal grid uses curvilinear orthogonal coordinates on an Arakawa staggered C grid. The horizontal time differencing is explicit whereas the vertical differencing is implicit. The latter eliminates time constraints for the vertical coordinate and permits the use of fine vertical resolution in the surface and bottom layers. The model has a free surface and a mode splitting technique in time has been adopted for computational efficiency. The barotropic (external) mode portion of the primitive equations is two–dimensional and uses a short time step based on the CFL condition and the external wave speed. The baroclinic (internal) mode is three dimensional and uses a long time step based on the CFL condition and the internal wave speed.

3 Parallelization of POM

In this section we describe how we have implemented POM using message passing. The most common technique to introduce parallelism in environmental models using finite differences for the discretization of the involved system of partial differential equations is domain decomposition. The basic idea is to decompose the computational domain into subdomains and assign each subdomain to a different processor. Each processor solves by a separate computing process the problem using the original code. In this way, there is only a single code to be maintained for both sequential and parallel computing platforms. In addition, the sequential code is reused entirely in the parallelization. To keep the computations consistent with the sequential code inter–processor communication is needed.

The computations in the horizontal mesh use explicit finite differences. This means that each grid point in the next time level is computed using only values from grid points in the previous time levels. So, the computations for each grid point in the advanced time level are independent and can be carried out in parallel by assigning a group of these points to each of the available processors. As the computations in the vertical direction use implicit schemes the previous approach cannont be applied since in the present case the values of the grid points in the next time level are computed via the solution of a linear system using direct methods (e.g. Gaussian elimination). Although there exist methods for the parallel solution of such systems, we postponed this approach for future investigation and decided not to introduce parallelism in the vertical direction.

3.1 Mapping Schemes and Communication Patterns

The discretization of the involved partial differential equations defines a rectangular mesh of grid points, where the value of each point is computed independently. There are a number of different ways to partition a rectangular domain [13]. However, for simplicity of algorithm implementation, the domain is usually divided into subdomains with simple and regular geometries.

In the present study we consider basicaly two ways to partition a rectangular two dimensional domain: block striped and checkerboard partitioning. In the block striped partitioning each processor is assigned a continuous set of rows or columns. In a rowwise block striping of an $N \times M$ mesh on p processors (labeled P_0, P_1, ..., P_{p-1}) processor P_i contains rows with indices $(N/p)i$, $(N/p)i + 1$, ..., $(N/p)(i+1) - 1$. In the checkerboard partitioning the mesh is partitioned into smaller rectangular blocks of equal size that are distributed among the processors.

A checkerboard partitioning splits both rows and columns of the mesh, so no processor is assigned any complete row or column. A checkerboard–partitioned rectangular mesh maps naturally onto a two–dimensional rectangular mesh of processors. The $N \times M$ mesh maps onto an $r \times q = p$ processor mesh by dividing it into blocks of size $N/r \times M/q$. The computations of each grid column are independent of each other and as we do not consider any parallelization in the vertical direction they do not cause any additional implications to the aforementioned partitioning techniques. Due to the structure of the computational stencil processors have to communicate their boundary points to the eight neighbouring processors after each scan of the domain. Therefore, each processor contains all the grid points corresponding to its domain, as well as those grid points which form its artificial boundary (in our case two rows/columns in each side). The communication pattern is carried out concurrently in all local processors. To avoid communication between diagonal processors which contain only a common corner, the communication is carried out in two phases. In the first phase the column boundaries are exchanged with the left and right neighbour and in the second phase the row boundaries are exchanged with the upper and lower neighbour. Note, that the aforementioned phases may be performed in any order.

4   Parallel Implementation

The main objective during the parallelization of the POM model was to produce a portable, easy to use optimised code that would execute on both shared and distributed memory architectures via interfaces that support message passing. This was achieved by gathering all the machine and interface dependent routines and grouping optimised versions of them in a number of files that were finally used to form a library serving for all parts of the parallel program discussed in this section, from the initial data partitioning phase to the communication patterns.

During the execution of the parallel POM model two types of communication arise:

- *local communication* used to update the halo points whenever neccessary.
- *global communication* at each external time step to check whether the CFL condition is satisfied.

4.1   Data Partitioning

Prior to the execution of the parallel program, a data partitioning scheme, as described in section 3, is used to allocate the data to the processors while maintaining a reasonably even distribution of the work to be performed. In this part, all the data that will be used as input to the parallel program are decomposed with consecutive calls to the library routines *par_split_2d* and *par_split_3d* for (x, y)–decomposition of 2d and 3d matrices, respectively. The z–direction is allocated to the same processor. For each processor that takes part in the data mapping, special routines are called in order to decide its position on the mesh, its neighbours and the amount of kernel and hallo points that it will hold.

For the checkerboard partitioning, the structure of the computational stencil requires two rows/columns artificial boundary for the computation of the points on the perimeter of each subdomain.

4.2   Local Communication

For the purposes of message passing, a subroutine *par_exchange* handles all the steps involved in the communication. Each message is formed by internal elements of the sending processor, then it is packed in a vector array and is finally transmited to the receiving processor, which will unpack the message and use it to update its halo points. According to the processor the message will be sent to, and considering the fact that there is an overlapping of message so as to avoid communication with the diagonal processors, we have the following four variants of the Send subroutine:

Send_up, Send_down, Send_left, Send_right

Similarly, according to the processor from which the message has been sent, we have the following variants of the Receive subroutine:

Recv_down, Recv_up, Recv_right, Recv_left

Considering the fact that the proposed communication scheme consists of four Send calls, that can be performed in pairs concurrently, and four blocking Receive calls, the body of subroutine Exchange is either: call Send_up, call Send_down, call Recv_down, call Recv_up, call Send_left, call Send_right, call Recv_right, call Recv_left or call Send_left, call Send_right, call Recv_right, call Recv_left, call Send_up, call Send_down, call Recv_down, call Recv_up with additional two alternative forms.

Communication, expressed in calls of subroutine *par_exchange*, is chosen to be performed at the beginning of the subprogram that requires it so as to avoid repeated and unnecessary updating of the same halo points within loops, unless this is critical to the correctness of the results.

Moreover, communication is restricted to 2d and 3d arrays and is handled differently by each processor, in the sense that the amount of halo points exchanged by each processor depends on its position in the 2d–mesh, as processors laying on the perimeter of the mesh have fewer "valid" halo points than the internal ones. With the term "valid", we refer to those halo points that are essential for the execution, in contrast to the "dummy" halo points assigned to the processors from the side that they lack a neighbour, for reasons of consistency as well as security regarding pointer referencings.

5   Performance

The aforementioned data partitioning and message passing techniques have been successfully tested and validated in both shared and distributed memory architectures. For the purposes of this paper, runs were made on the distributed memory Parsytec CC platform with 2, 4, 6, 8, 10, 12, 14 and 16 processors which was the platform used for purposes of development and testing of the results. The 128MB RAM restriction discouraged the execution of the sequential program. Thus, all results assume time for two processors to be the basis for the scalability study.

The programming model that was selected for the implementation was PVM as being our initial development environment while an optimised MPI version is currently being tested and refined.

The data sets used for our experiments simulate the Mediterranean basin and contain 30 layers of 360×120 of grid points each, that is a grid spacing of $0.125°$. Figures 1 and 2 present our numerical results of the model as they were estimated for a 30–days 3–d calculation run taking 1440 steps. Times are measured in seconds in all cases. From the results of Fig. 1 and 2 we note that the best

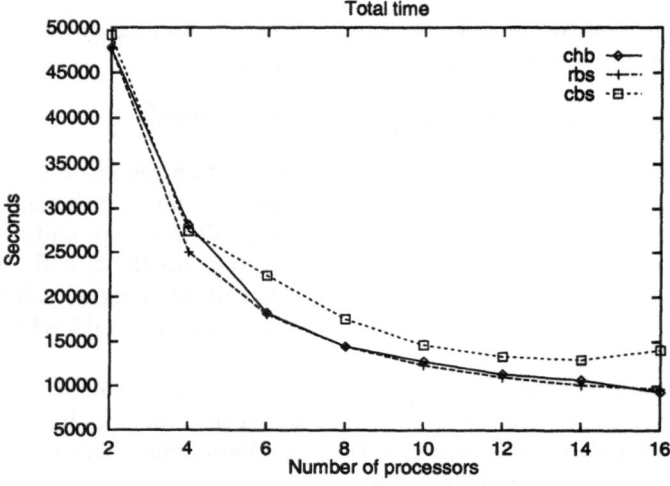

Fig. 1. Parallel run time

performance for the POM model is achieved for the checkerboard partitioning when the number of processors is greater than 15, otherwise block row striped partitioning is slightly better. This serves as an indication towards selecting, when real–time problems are studied, of that grid size–number of processors combination which would offer the

The achieved results 5.1 and 0.64 for the relative speedup (estimated true speedup 10.2) and efficiency, respectively, for 16 processors show also a satisfactory scalability behaviour of the model.

6   Future Work

Even though for a first approach the results are quite encouraging, they are amenable to improvements. More work has to be done, covering multiple areas not only of parallelization but also in conjuction to the parallel POM model with

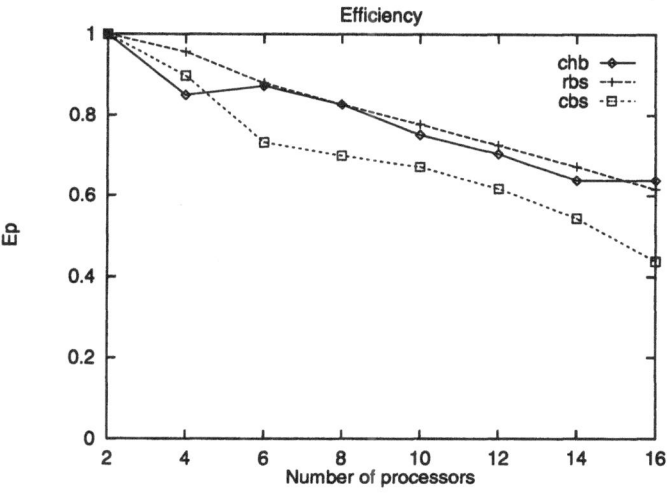

Fig. 2. Efficiency

other relative applications in order to provide a wider package of applications concerning enviroment.

References

[1] Argyropoulos, N.E., Boukas, L.A., Mimikou, N.Th., Missirlis, N.M., Papageorgiou, J.G.: A Distributed Implementation of the Numerical Weather Prediction Eta Model. Presented at the IASTED International Conference Parallel and Distributed Systems, Euro–PDS'97, June 9–11, 1997, Barcelona, Spain and appeared in the Proceedings of the IASTED Conference on Parallel and Distributed Computing and Networks, IASTED/Acta Press (also accepted, after selection, to be published in extended form in the IASTED Journal for Parallel and Distributed Systems) (1997) 301–304 .

[2] Ashworth, M.: Parallel Processing in Environmental Modeling. Parallel Supercomputing in Atmospheric Sciences, G–R. Hoffmann and T. Kauranne eds., World Scientific, Singapore (1993) 1–25.

[3] Bleck, R., Dean, S., O'Keefe, M., Sawdey, A.: A comparison of data parallel and message–passing versions of the Miami Isopycnic Coordinate Ocean Model (MICOM). Parallel Computing, **21** no. 10 (1995) 1695–1720.

[4] Blumberg, A.F., Mellor, G.L.: A coastal ocean numerical model. Mathematical Modelling of Estharine Physics, Proc. Int. Symp., Hamburg, August 24–26, 1978, edited by Sundermann and K.–P. Holz, Springer–Verlag, Berlin (1980) 203–214.

[5] Boukas, L.A., Mimikou, N.Th., Missirlis, N.M.: A parallel implementation of the Eta model. Abstracts of the Symposium on Regional Weather Prediction On Parallel Computer Enviroments Athens, Greece (1997).

[6] DeRose, L., Gallivan K., Gallopoulos, E.: 3–d land avoidance and load balancing in regional ocean simulation. Proc. 1996 Int'l. Conf. Parallel Processing **2** (1996) 158–165.

[7] DeRose, L., Gallivan K., Gallopoulos, E.: Status Report: Parallel Ocean Circulation Modeling on CEDAR. Parallel Supercomputing in Atmospheric Science. G.–R. Hoffmann and T. Kauranne eds., World Scientific,Singapore (1993) 157–172.

[8] DeRose, L, Gallivan, K., Gallopoulos, E.: Experiments with an ocean circulation model on CEDAR. Proc. 1992 ACM International Conference on Supercomputing (1992) 397–408.

[9] Gülzow, V., Kleese, K.: About the parallelization of the HOPE Ocean Model. Proceedings of the Sixth ECMWF Workshop on the Use of Parallel Processors in Meteorology, Reading, England, November 1994. Proceedings published by World Scientific Publishers, Coming of Age, edited by G–R. Hoffmann and N. Kreitz (1995) 505–511.

[10] Gwilliam, C.S.: The OCCAM Global Ocean Model. Proceedings of the Sixth ECMWF Workshop on the Use of Parallel Processors in Meteorology, Reading, England, November 1994. Proceedings published by World Scientific Publishers, Coming of Age, edited by G–R. Hoffmann and N. Kreitz (1995) 446–454.

[11] Jones, P.W.: The Los Alamos parallel ocean program (POP) and coupled model on MPP and clustered SMP architectures. Making its Mark, G–R. Hoffmann and N. Kreitz, World Scientific, Singapore (1997) 226–238.

[12] Kallos, G., Nickovic, S., Jovic, D., Kakaliagou, O., Papadopoulos, A., Missirlis, N., Boukas, L., Mimikou, N.: The Eta Model Operational Forecasting System and its Parallel Implementation. Proceedings of the First Workshop on Large–Scale Scientific Computations, eds. M. Griebel, O. Iliev, S.D. Margenov, P.S. Vassilevski, Varna, Bulgaria (1997) 176–188.

[13] Kumar, V., et al: Introduction to Parallel Computing: Design and Analysis of Algorithms. The Benjamin/Cummings Publ. (1994).

[14] Mellor, G.L., Yamada, T.: Development of a turbulence closure model for geophysical fluid problems. Rev. Geophys. Space Phys. 10 No. 4 (1982) 851–875.

[15] Oberhuber, J.M., Ketelsen, K.: Parallelization of an OGCM on the CRAY T3D. Proceedings of the Sixth ECMWF Workshop on the Use of Parallel Processors in Meteorology, Reading, England, November 1994. Proceedings published by World Scientific Publishers, Coming of Age, edited by G–R. Hoffmann and N. Kreitz (1995) 494–504.

[16] Pacanowski, R.C., Dixon, K., Rosati, A.: The GFDL Modular Ocean Model Users Guide. Technical Report 2 GFDL Ocean Group (1990).

[17] Sawdey, A., O'Keefe, M., Bleck, R., Numrich, R.W.: The design, implementation and performance of a parallel ocean circulation Model. Proceedings of the Sixth ECMWF Workshop on the Use of Parallel Processors in Meteorology, Reading, England, November 1994. Proceedings published by World Scientific Publishers, Coming of Age, edited by G–R. Hoffmann and N. Kreitz (1995) 523–550.

[18] Sawdey, A.C., O'Keefe, M.T., Jones, W.B.: A General Programming Model for Developing Scalable Ocean Circulation Applications. Making its Mark, G–R. Hoffmann and N. Kreitz, World Scientific, Singapore (1997) 209–225.

[19] Smith, R.D., Dukowicz, J.K., Malone, R.C.: Physica D 60 (1992) 38.

[20] Wait, R., Harding, T.J.: Numerical Software for 3D Hydrodynamic Modeling using Transputer Array. Parallel Supercomputing in Atmospheric Sciences, G–R. Hoffmann and T. Kauranne eds., World Scientific, Singapore (1993) 453–464.

Modular Fortran 90 Implementation of a Parallel Atmospheric General Circulation Model

William Sawyer[1,2], Lawrence Takacs[1], Andrea Molod[1], and Robert Lucchesi[1]

[1] NASA Goddard Space Flight Center, Data Assimilation Office
Code 910.3, Greenbelt MD, 20771, USA
{wsawyer,ltakacs,amolod,rlucchesi}@dao.gsfc.nasa.gov
http://dao.gsfc.nasa.gov
[2] Department of Meteorology, University of Maryland at College Park
College Park MD, 20742-2425, USA

Abstract. In this paper we present software engineering aspects of a parallel atmospheric General Circulation Model (GCM) currently nearing completion at the NASA Data Assimilation Office. The code is scientifically closely related to the current Goddard Earth Observing System (GEOS 2.x) GCM. The new software design, however, is a radical departure from the previous Fortran 77 code, using Fortran 90 derived data types and modules to compartmentalize the implementation.

In order to reduce turnaround time needed to facilitate scientific development, the GEOS 3.x implementation uses a message-passing paradigm. This allows the performance to scale consistently to a large number of processors on both an Origin 2000 and a Cray T3E. The Message-Passing Interface (MPI) is used for portability.

The modularity and performance of the parallel GEOS 3.x code allow faster development of scientific improvements such as variable resolution grids and a new land surface model which are key to keeping the GCM competitive with other leading models.

1 Introduction

The goal of the Data Assimilation Office (DAO) is to produce accurate gridded datasets of atmospheric fields by assimilating a range of observations along with physically consistent model forecasts. These datasets are produced by the Goddard Earth Observing System (GEOS) Data Assimilation System (DAS) and used by the climate research community. The key components of GEOS DAS are the gridpoint-based atmospheric General Circulation Model (GCM) [1] and a data analysis system [2]. They are computationally intensive and have traditionally run on large Cray PVP (C90 and J90) and SGI Origin 2000 systems on up to 32 processors by "multitasking" key loops.

Recent improvements in the GCM — the increase in the horizontal resolution to $1^o \times 1^o$, variable resolution grids to acquire better scientific results in local areas, and a new land surface model [3] — make the multitasking parallelism insufficient. In addition, the transition in the supercomputing industry to more

P. Amestoy et al. (Eds.): Euro-Par'99, LNCS 1685, pp. 1403–1410, 1999.
© Springer-Verlag Berlin Heidelberg 1999

economical commodity multiprocessors and the high cost of maintaining Cray PVP systems indicate that new parallelism paradigms must be utilized.

The message-passing implementation of the GCM has been realized in a series of prototypes. The authors undertook an incremental restructuring of the GEOS GCM to meet the need for increased flexibility and performance. Fortran 90 constructs were introduced to improve its readability and modularity. Common blocks were removed in favor of derived data structures which are manipulated in an object-oriented fashion. A high-level communication library was written to obviate the need for explicit MPI [4] calls in the scientific code. All this was achieved without requiring extensive changes to the computationally intensive, low-level legacy code which represents years of development.

Similar efforts to revamp large legacy modeling codes can be seen in the earth science community. For example, MM90 [5] is a re-engineered Fortran 90, message-passing implementation of the well-known MM5 Mesoscale Model. PALM [6] is the back-bone of an oceanic data assimilation software which allows parallel execution on distributed memory systems. It is designed to use Fortran 90 and the MPI-2 standard.

Section 2 of this paper discusses the Fortran 90 design of the GEOS 3.x GCM, which is now in the final implementation stage before validation and production. Section 3 discusses issues in the parallel implementation. Scientific and performance results are presented in Section 4.

2   Fortran 90 Modular Design of the GCM

The GCM employs a finite-difference algorithm to integrate the primitive equations. It uses a geographic spherical coordinate grid with horizontal spacing of 1° latitude by 1° longitude and a 48-level vertical sigma coordinate. Physical processes are modeled using land-surface, moisture, turbulence, and longwave and shortwave radiative transfer schemes [1]. The dynamical core [7] of the GCM computes the time tendencies due to advection of winds, temperatures, surface pressure, and an arbitrary number of tracers.

The existing GCM's scientific validity has been well documented [3, 8]. Less has been written about the recent software engineering improvements. Most common blocks have been replaced by F90 derived types making the code much more modular and flexible. Each of the internal models for the earth, ocean, chemistry, land and atmosphere has a corresponding Fortran 90 module which defines the model's intrinsic *state*, and provides facilities to dynamically create and destroy states. For example, the derived data type for the dynamics state used in the atmospheric model is:

```
type    dynamics_vars_type
  real , pointer ::    p(:,:)            ! Surface pressure
  real , pointer ::    u(:,:,:)          ! East-west winds
  real , pointer ::    v(:,:,:)          ! North-south winds
  real , pointer ::    t(:,:,:)          ! Temperature
  real , pointer :: pkht(:,:,:)          ! Geopotential heights
  real , pointer ::    q(:,:,:,:)        ! Tracers
endtype dynamics_vars_type
```

```
type    dynamics_grid_type
  type ( LatticeType ) Lattice              ! Lattice description
  type (  DecompType ) Decomposition        ! Decomposition
  integer im,jm,lm                          ! Local grid sizes
  real    lam_np, phi_np, lam_0             ! Rotation parameters
  real, pointer :: dlam(:), dphi(:)         ! Stretched grid increments
  real, pointer :: sige(:), sig(:), dsig(:) ! Sigma coordinates
endtype dynamics_grid_type

type    dynamics_state_type
  type ( dynamics_grid_type ) grid          ! Grid description
  type ( dynamics_vars_type ) vars_n        ! Variables this timestep
  type ( dynamics_vars_type ) vars_nm1      ! Variables last timestep
endtype dynamics_state_type
```

This data structure includes a description of the grid on which prognostic fields are defined, as well as the fields themselves at the current and previous time steps. Arrays dlam and dphi specify the (possibly non-uniform) grid spacing. The computational poles can be relocated on the geophysical globe by specifying rotation parameters lam_np, phi_np and lam_0. Finally, information concerning the decomposition of the grid for the message-passing version is held in Lattice and Decomposition. Similar data structures have been defined for the chemistry, earth, ocean, and land states.

Not only can there be multiple instances of states, they can have different dimensions. For example, states with $1^o \times 1^o$ and $2^o \times 2.5^o$ resolution can co-exist, expanding the possibilities for multi-resolution modeling. The GCM can also support variable resolution *stretched grids*, which can improve the forecast's quality in local areas of interest [8]. The software for stretched grids has been generalized in such a way that it can also support interpolation between uniform grids of different resolution or even from the stretched grid directly to an unstructured grid of observation locations — a feature which could be used in GEOS DAS in the future.

Revisions have been made incrementally to support the new land surface model, in which a "tile-space" is defined — a packed, one-dimensional depiction of all surface tiles, each having one or several surface types.

The Fortran 90 design implemented in the GCM is similar to the one proposed in [9], without the "mediation layer" employed there. The parallel GCM is also related to the one described in [10], although the former is used in the framework of data assimilation instead of simulation. Furthermore, prognostic and diagnostic data required by the DAO's clients require a parallel I/O concept to remove a potential performance bottleneck. Finally, the need to support a stratospheric model incurs numerical instabilities near the poles which do not play a role in the tropospheric model used by Suárez, et al. These instabilities require that the computational grid be transformed or "rotated" [1] such that they disappear.

3   Parallelization Issues

In order to support the message-passing parallelization of GEOS 3.x in relatively opaque fashion, a Parallel Library for Grid Manipulations, or PILGRIM [11] was designed and implemented. PILGRIM supports the basic operations needed in regular and irregular grid applications. It is similar to RSL [12], the GMD Communications Library [13] and PLUMP [14], but is most like NNT [15]. The major differences are its extensive use of Fortran 90 features, its object-oriented design, its use of MPI for communication, and its compactness with less than 10,000 lines of code.

Due to the requirement to adopt large amounts of legacy code, it is not possible to replace existing arrays with a special distributed type, such as those available for distributed vectors in the PETSc library [16]. Instead, the existing arrays change meaning to denote the *local* data segments in the parallel version. Every such distributed array is associated with decomposition type which defines how it fits into the global array. The PILGRIM decomposition module was designed to support several different types of grids, both regular and irregular.

PILGRIM offers several high-level communication primitives to gather and scatter distributed data, perform asynchronous and synchronous some-to-some block transfer as well as collective summation and the exchange of boundary (or "ghost") regions. With a sparse linear algebra module, PILGRIM provides facilities to create, destroy and initialize distributed sparse matrices as well as to perform sparse matrix-vector and matrix-matrix multiplications.[1] Finally, a module for redistributing data allows mappings between two decompositions to be defined, resulting in a redistribution type. The redistribution can then be performed on given input and output arrays on the local processor:

```
CALL RedistributeCreate( DecompA, DecompB, InterDecomp )
CALL RedistributePerform( InterDecomp, LocalIn, LocalOut )
```

The GCM's physical parameterizations operate on 3-D arrays with only vertical dependencies and are generally independent of the geographical location. If data are distributed properly, the physics can run in parallel without communication, with each processor working on its local region. The land surface model is also in this class of parameterizations, although many of its argument are distributed 1-D tile-space arrays.

There are several different types of data decompositions employed in the GEOS GCM, the premier being a horizontal "checkerboard" decomposition of 2-D arrays, or, correspondingly, a "column" distribution of 3-D arrays, which are both quite regular. The 1-D tile-space decomposition, on the other hand is irregular, as indicated in Figure 1. The proper distribution of these arrays makes use of PILGRIM's support for both regular and irregular grids. For example, if a global version of a tile-space array is required on one processor for output, it can be gathered with the following PILGRIM primitive:

```
CALL ParGather( COMMUNICATOR, ROOT, LocalTileSpaceArray,
   1                TileSpaceDecomposition, GlobalTileSpaceArray )
```

---

[1] Possible extensions, such as iterative solvers, are in the discussion stage, but are not needed in GEOS GCM.

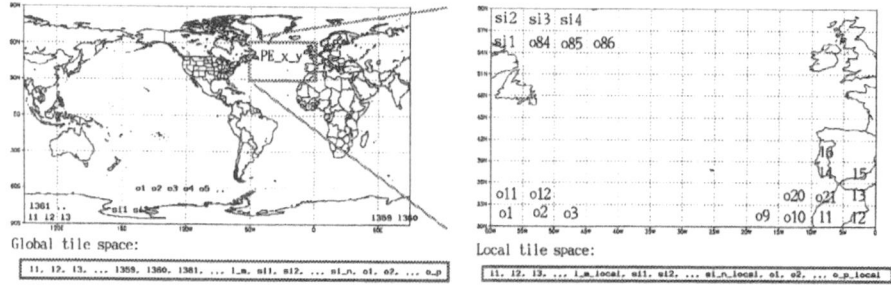

Fig. 1. A global tile-space array (left) consists of one long vector of *tiles*, first all the land tiles, of which there can be several types for each tile, then all sea-ice tiles, and finally all ocean tiles. Each processor PE$_{x,y}$ has a local version (right) of this tile-space array, ordered again by land, sea-ice, and ocean. The mapping of data from the global tile-space to the local is irregular. It depends on the geographical coordinates of each tile and on the processor where the tile resides.

In contrast to the physical parameterizations, the dynamics has dependencies in all three dimensions. Due to the checkerboard decomposition of the underlying arrays, horizontal communication must be performed. This is done by the *ghosting* technique which requires the arrays' boundary regions to be exchanged periodically. Boundary region transfers are grouped over all levels to reduce the total number of messages and thus the aggregate message latency. The dynamical core makes use of PILGRIM's ghosting facilities to bring boundary regions up to date. In the following example, the column face USB(IM+1,1:JM,:) is copied from USB(1,1:JM,:) on the neighboring processor directly to the east.

```
ALLOCATE(USB(0:IM+1,0:JM+1,LM))
  :
CALL ParGhost( East, IM, JM, LM, USB )
```

The dynamics includes a filter for the high latitudes where instabilities tend to occur. The high-latitude filter creates dependencies across entire latitudes, which is unfortunate since they are generally distributed over several processors. The filter is, however, computationally intensive. Therefore, with proper overlapping of the gather operation with the intrinsic dense matrix-matrix multiplication local to the processor, adequate performance can be obtained.

The "glue" between different portions of the GCM consists of couplers which operate on different local grids. For example, land surface couplers translate between tile-space and gridded data. Other couplers include the conversion between C- and A-grids and the conversion between a stretched and rotated grid used in the dynamics and the uniform grids used in the physical parameterizations (see [11]). Most of these routines have horizontal dependencies, work on data with a checkerboard decomposition, and therefore require communication.

The input and output of data to the GEOS Data Assimilation System (GEOS DAS) are the responsibility of the GEOS Parallel I/O Subsystem (GPIOS). Efficient parallel I/O is critical if a distributed-memory version of GEOS DAS is to achieve strategic performance goals. The assimilation of thirty days per day requires daily input of more than 1.5 gigabytes and produces 25 to 35 gigabytes of gridded output data in HDF format. I/O in the GEOS DAS is complicated by the need to translate data from the internal computational grid to a grid appropriate for users. These transformations can be computationally expensive and require significant communication when done in a distributed domain.

The input of boundary conditions and the pre-existing atmospheric condition is not considered a bottleneck and is performed on one processor. The information is read on to one processor and either broadcasted or distributed, depending on the variable's nature. On the other hand, there are several processors designated for output. The number of output processors can be increased at run-time to allow flexibility in the definition of output streams and memory management on machines with limited memory nodes. In addition, the output processors are in an MPI group distinct from the other compute processors and are dedicated to I/O and related functions. They act as a cache for output data, allowing computation in the GCM to continue once the MPI transfer of data from compute to I/O processors is complete. One advantage of such an approach is that it allows data transformations in preparation for output to occur locally on I/O processors without slowing the forward progression of the assimilation system.

4   Results

Software improvements to the GEOS GCM have allowed the DAO to increase model resolution. Figure 2 indicates that simulations at $1^{\circ} \times 1^{\circ}$ resolution can resolve significant weather characteristics where the previous $2^{\circ} \times 2.5^{\circ}$ could not.

The increased performance requirements have been met by better scalability of the message-passing over the older multitasking code. This allows the GCM to be run on large machine configurations, increasing throughput substantially. Benchmarks in [11, 17] indicate that the parallelization of individual GCM components such as the computational pole rotation, satisfies the performance objectives. Figure 3 illustrates the good scalability and performance of the physics and dynamics on the Cray T3E and SGI Origin 2000.

Acknowledgments

The work of Will Sawyer was partially funded by the High Performance Computing and Communications Initiative (HPCC) Earth and Space Science (ESS) program.

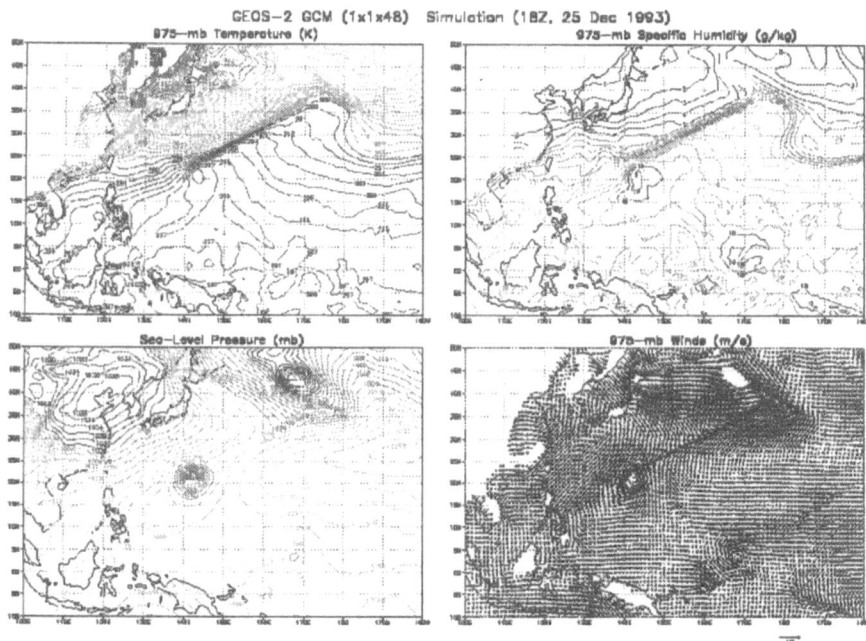

Fig. 2. By either increasing the global resolution or stretching the grid in appropriate regions, important meteorological phenomena can be discerned. Here, in a one-month simulation, a low pressure area (lower left) has developed in the Pacific Ocean southeast of Japan. This causes a storm weather front to form (lower right). The storm line can be seen in the high temperature gradients (upper left) and humidity (upper right).

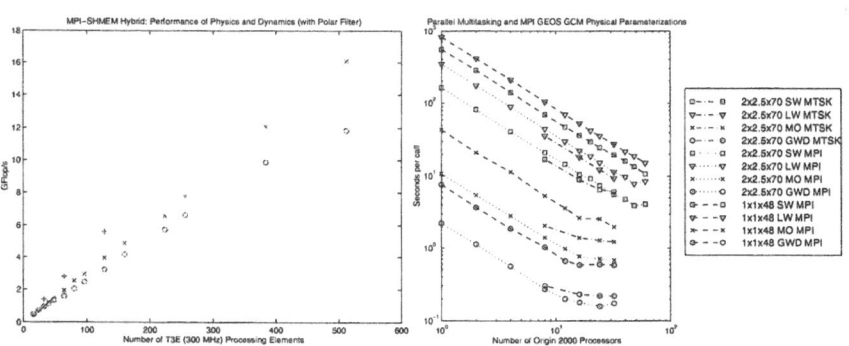

Fig. 3. The graph at left depicts the GFlop/s performance of the dynamics, which performs nearest neighbor communication using MPI and SHMEM. It scales only slightly less well for both $2° \times 2.5° \times 70$ (o) and $1° \times 1° \times 43$ (x) resolutions than the ensemble physical parameterizations (+). On the right, the absolute timings (multitasking and MPI) are given for the individual physical parameterizations (gravity wave drag, moisture, shortwave and longwave radiation) on a 64-processor SGI Origin 2000.

References

[1] L. L. Takacs, A. Molod, and T. Wang. Documentation of the Goddard Earth Observing System (GEOS) General Circulation Model — Version 1. Technical Memorandum 104606, NASA Code 910.3, 1994.

[2] A. da Silva and J. Guo. Documentation of the Physical-space Statistical Analysis System (PSAS). DAO Office Note 96-02, NASA Code 910.3, 1996.

[3] M. Bosilovich, P. Houser, A. Molod, and S. Nebuda. A Comparison of FIFE Observation with GEOS Assimilated Data Including a Heterogeneous LSM. In *Proceedings of the Conference on Hydrology*, 1999. Submitted.

[4] MPI Forum. MPI: A Message-Passing Interface Standard. *International Journal of Supercomputer Applications*, 8(3&4):157–416, 1994.

[5] J. Michalakes. MM90: A Scalable Parallel Implementation of the Penn State / NCAR Mesoscale Model (MM5). *Parallel Computing*, 23(14):2173–2186, December 1997. Also ANL/MCS-P659-0597.

[6] Le Groupe PALM. Etude de faisabilité du projet PALM. Technical Report TR/CMGC/98/50, CERFACS, 1998.

[7] M. J. Suárez and L. L. Takacs. Documentation of the ARIES/GEOS Dynamical Core: Version 2. NASA Technical Memorandum 104606, NASA Code 910.3, 1995.

[8] M. Fox-Rabinovitz and G. Stenchikov and M. Suárez and L. L. Takacs. A Finite-Difference GCM Dynamical Core With a Variable Resolution Stretched Grid. *Mon. Wea. Rev.*, 125:2943–2968, 1997.

[9] J. Michalakes, J. Dudhia, D. Gill, J. Klemp, and W. Skamarock. Design of a Next-generation Weather Research and Forecast Model. In *Proceedings of the Eighth ECMWF Workshop on the Use of Parallel Processors in Meteorology*. ECMWF, 1998. Also ANL/MCS-P735-1198.

[10] D. S. Schaffer and M. J. Suárez. Next Stop: Teraflop; The Parallelization of an Atmospheric General Circulation Model. In *High Performance Computing Symposium 98 Proceedings*, April 1998.

[11] W. Sawyer, P. Lyster, A. da Silva, and L. Takacs. Parallel Grid Manipulations in Earth Science Calculations. In *3rd International Meeting on Vector and Parallel Processing*. Springer-Verlag, 1998. Lecture Notes in Computer Science 1573.

[12] J. Michalakes. RSL: A Parallel Runtime System Library for Regular Grid Finite Difference Models Using Multiple Nests. Technical Report ANL/MCS-TM-197, ANL, 1994.

[13] R. Hempel and H. Rizdorf. The GMD Communications Library for Grid-Oriented Problems. Technical Report 589, GMD, 1991.

[14] O. Bröker, V. Deshpande, P. Messmer, and W. Sawyer. Parallel Library for Unstructured Mesh Problems. Technical Report CSCS-TR-96-15, CSCS, 1996.

[15] B. Rodriguez, L. Hart, and T. Henderson. A Library for the Portable Parallelization of Operational Weather Forecasting Models. In *Proceedings of the Sixth ECMWF Workshop on the Use of Parallel Processors in Meteorology*, pages 22–27. ECMWF, 1995.

[16] L. C. McInnes and B. F. Smith. PETSc 2.0: A Case Study of using MPI to Develop Numerical Software Libraries. In *1995 MPI Developers' Conference*, June 1995.

[17] W. Sawyer, R. Lucchesi, P. Lyster, L. Takacs, J. Larson, A. Molod, S. Nebuda, and C. Pabon-Ortiz. Parallelization of the DAO Atmospheric General Circulation Model. In *Applied Parallel Computing, 4th International Workshop, PARA'98*. Springer-Verlag, 1998. Lecture Notes in Computer Science 1541.

Implementation of the Limited-Area Numerical Weather Prediction Model Aladin in Distributed Memory

Claude Fischer[1], Jean-François Estrade[2], and Jure Jerman[3]

[1] Météo-France/CNRM/GMAP/EXT, 42 avenue Gustave Coriolis,
31057 TOULOUSE CEDEX (France)
claude.fischer@meteo.fr
[2] Météo-France/SCEM/TTI
[3] Slovenian Hydrometeorological Institute, Ljubljana
jure.jerman@rzs-hm.si

Abstract. The technical challenges for the limited area model Aladin are various: Aladin is the French operational mesoscale model, run on the Météo-France VPP-700E on 4 processors twice a day. Also, the model is used in operational or pre-operational mode in several European countries, plus Morocco, and therefore the model has to be portable and computationally efficient on several platforms. In the presentation, the general choices for the porting of Aladin in a parallel environment will be discussed, with an emphasis to Aladin-specific aspects, the general choices being closely related to those of Arpège. Finally, the question of applications to workstation clusters will be addressed.

1 Introduction

The numerical weather prediction model Aladin is the current operational mesoscale model at Météo-France. It is the result of a fruitful international cooperation started in 1991. Presently, 13 countries are involved in the project. Aladin is built on the basis of the Arpège-IFS libraries (the global model commonly developped by Météo-France and the European Center for Medium-range Weather Forecasts, ECMWF). As a consequence, the technical choices for Aladin often derive directly from Arpège-IFS.

Presently, Aladin runs in 3 operational configurations at Météo-France: creation of domain specific climatological data (mono-processor only), interpolation of model fields between two different domains (spherical to plane or plane to plane geometry), forecast mode (hydrostatic, semi-lagrangian two-time level, Staniforth and Côté, 1991[11], with a boundary relaxation towards the global model solution).

Besides these operational tools, a number of development configurations have to be maintained: non-hydrostatic version (Bubnova *et al*, 1995[4]), sensitivity computations, 3D-VAR.

P. Amestoy et al. (Eds.): Euro-Par'99, LNCS 1685, pp. 1411–1416, 1999.
© Springer-Verlag Berlin Heidelberg 1999

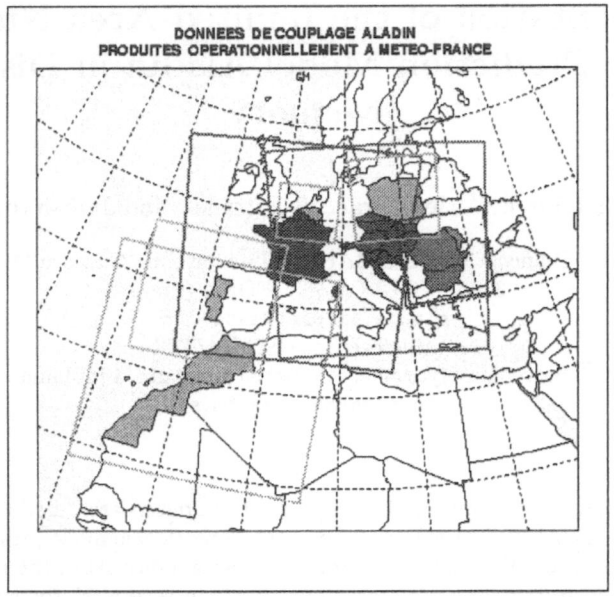

Fig. 1. Limited area domains of those Aladin models which are immediatly coupled with Météo-France originating data

Portability of the code is a major issue in Aladin. Indeed, our partners run the model operationally on a wide panel of machines: NEC-SX4 vector computer (3 procs), SGI-Origin (6 procs in Hungary), SUN, HP, DEC workstations, Cray J90 computers (12 procs in Belgium). For the time being, Météo-France is the only center which operates in a distributed memory, multi-processor environment. On the NEC-SX4, the model runs on 2 processors in multitask mode. In Bruxelles and Casablanca, Aladin runs with Cray multitask facilities. Hungary plans to install the distributed memory, message passing version of Aladin on its SGI computer.

Furthermore, a cascade of nested, coupled (pre-)operational models does now exist, such as Arpège → Aladin-France → Aladin-Belgium or Arpège → Aladin-LACE → Aladin-country (in the latter, the Limited Area Central European Aladin run in the Prague computing center on a NEC-SX4 provides boundary conditions for the national central european versions, twice a day). Therefore, there is a strong need for efficiency driven by scheduling constraints and the delivery of the coupling data in the national numerical prediction centers. Figure 1 shows those model applications and domains which are coupled immediately with Toulouse forecast data (either Arpège or Aladin-France).

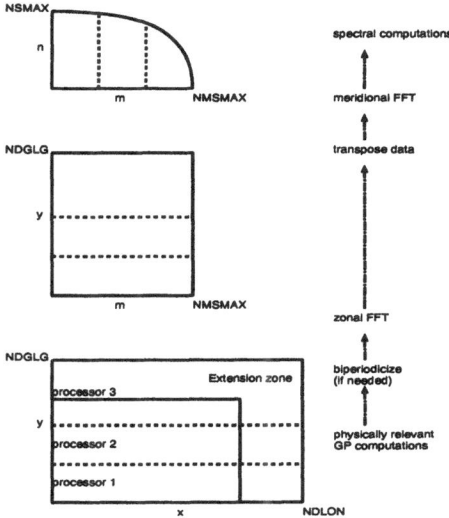

Fig. 2. Overview of the main computational steps
in distributed memory. This picture is valid both
for forecast mode and for geometrical interpolations
(grid projections).

2   Parallel Aladin

The set of operational models and configurations of Arpège and Aladin have
been ported to the distributed memory environment at Météo-France (CNRM-
GMAP and SCEM-TTI, see Estrade and Birman, 1994[7], Radi and Estrade,
1995[9]) following the first developments for Arpège-IFS at ECMWF (see Dent,
1992[6], Barros et al, 1994[1]). The global model Arpège uses the spectral method
with a succession of Fourier and Legendre transforms to switch between the
gridpoint collocation grid (where the physical parametrizations are applied and
semi-lagrangian trajectories are estimated) and the spherical harmonic space
(which is the space where semi-implicit corrections are performed, for example).
These general settings are kept in Aladin. One main difference arises from the
impact of the limited area constraint: the spectral space is spanned by a bi-
Fourier basis of functions and gridpoint data are "bi-periodicized" to obtain
smooth, periodic fields in both horizontal directions (Radnoti et al, 1995[10]).
Despite these mathematical differences, the main technical frame for distributed
memory environments is the same in Aladin and in Arpège (Barros et al, 1994[2]):

- domain splitting and data transpositions (see figure 2 and Barros and Kauranne, 1992[3], Barros et al, 1994[1]).
- control of the distribution in the "setup" part of the model.
- definition of the level of parallelism in the model algorithm.

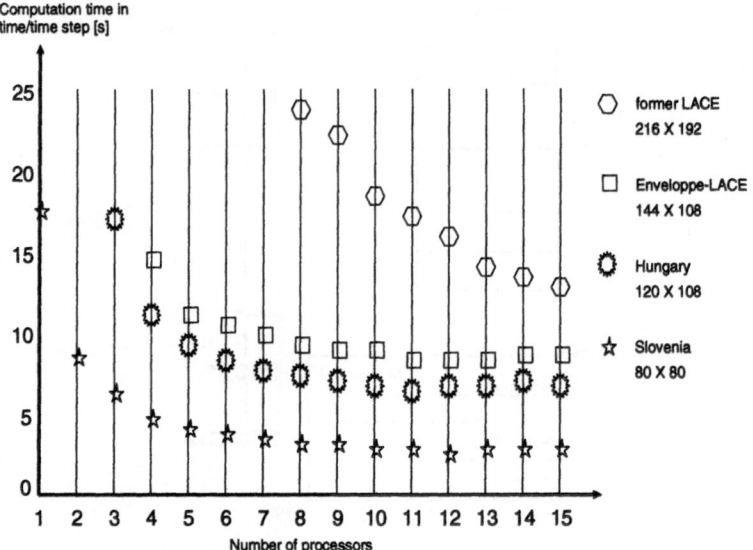

Fig. 3. Total computational time (divided by duration of one timestep) for a range of number of nodes. Results are obtained on the homogeneous workstation cluster (with DEC-α processors) presented in §3. 4 operational gridpoint domains have been displayed.

– choice for the MPI library and its overhead MPE (ECMWF, see Dent, 1997[5]).

Specificities in Aladin arise for example from the existence of the so-called extension zone. This area is an extra gridpoint domain in which the physical fields are biperiodicized before the transformation to bi-Fourier space (see figure 2). In forecast mode, the number of computations in this area is very small (especially, the physics are not activated) and some load imbalance is possible in the integration.

3   Workstation Cluster

This part presents the first results, obtained in a development framework, for the porting and evaluation of Aladin on a cluster of workstations. The work was mainly performed by our Slovenian colleague Jure Jerman. The distributed memory version of the model has been installed on a set of workstations using MPI and touching only 3 routines of the whole code. The forecast mode has so far been validated.

The cluster consists of a set of "off-the-shelf" stations which are associated to a homogeneous hardware configuration. Every processor runs its own operating system (21 nodes run LINUX, 1 node runs Digital Unix). The processors are Alpha SX with 533 MHz clock rate and 128 Mbytes of memory per node. In this research environment, 4 Gbytes of disk were available per node. The total theoretical power reaches 25 Gflops.

Figure 3 shows the total computing time per timestep for a range of number of processors and for 4 different operational Aladin domains. The scalability of the Aladin model is reasonably good as long as the size of the domain allows for an efficient data distribution. As soon as the domain becomes too small, domain splitting becomes useless. Eventually, when a very small number of latitudes is treated by one processor in gridpoint, then the amount of communications increases because each processor needs to exchange halo data with many other processors (for the semi-lagrangian trajectories, the backward trajectory computation can lead easily to origin points outside of the band of latitudes treated by one processor, then the halo data should contain the surrounding data to interpolate on the local node, see Barros *et al*, 1994[1], or Estrade *et al*, 1996[8]).

Figure 3 demonstrates the possibilities for an efficient implementation of Aladin on a homogeneous workstation cluster. Technical problems are still under investigation. Thus, some deadlocks have appeared for specific numbers of processors. Also, problems of hang-offs inside **mpe_send** were encountered: their dependence to the parameters for the communications in message passing will be checked.

Present plans contain a pre-operational implementation of Aladin on a 5-node cluster and tests with new Linux and LAM MPI implementations. Also, in future, the cluster installation might be a basis for more validation of distributed memory features in Aladin.

4   State of the Art and Future Evolution

Aladin runs operationally on 4 processors on the Fujitsu VPP-700E of Météo-France. Some figures concerning code performances will be presented. The optimization of Aladin on the VPP benefits at the same time from some work already performed on Arpège and to specific studies for Aladin. For the near future, the present strategy based on a one-dimensional data decomposition in every computational space will be kept. Future issues might be either a second level data distribution or macro-tasking features, but time is not ripe for an in-depth discussion of these items. Also, several development configurations still have to be ported to message passing mode (tangent linear and adjoint versions of the model, for example).

References

[1] S.R.M. Barros, D. Dent, L. Isaksen, G. Robinson, and F.G. Wollenweber. The message passing version of ECMWF's weather forecast model. In W. Gentzsch and U. Harms, editors, *High-performance computing and networking - Lecture notes in computer science 796*, pages 299–304. Springer Verlag, 1994.

[2] S.R.M. Barros, D. Dent, Isaksen L., and G. Robinson. The IFS model: overview and parallel strategies. In G.-R. Hoffmann and N. Kreitz, editors, *Coming of age - Proceedings of the 6th ECMWF workshop on the use of parallel processors in Meteorology*, pages 303–318. World Scientific, 1994.

[3] S.R.M. Barros and T. Kauranne. On the parallelization of global spectral eulerian shallow-water models. In G.-R. Hoffmann and T. Kauranne, editors, *Parallel supercomputing in atmospheric science - Proceedings of the 5th ECMWF workshop on the use of parallel processors in Meteorology*, pages 36–43. World Scientific, 1992.

[4] R. Bubnova, G. Hello, Bénard P., and J.-F. Geleyn. Integration of the fully elastic equations cast in the hydrostatic pressure terrain-following coordinate in the framework of the ARPEGE-ALADIN NWP system. *Mon. Wea. Rev.*, 123:515–535, 1995.

[5] D. Dent. A parallel production weather prediction system. In G. Kallos, V. Kotroni, and K. Lagouvardos, editors, *Proceedings of the symposium on regional weather prediction on parallel computer environments*, pages 169–173. University of Athens, 1997.

[6] D.W. Dent. Parallelisation of the IFS model. In G.-R. Hoffmann and T. Kauranne, editors, *Parallel supercomputing in atmospheric science - Proceedings of the 5th ECMWF workshop on the use of parallel processors in Meteorology*, pages 73–87. World Scientific, 1992.

[7] J.F. Estrade and D. Birman. Adapting parallel IFS-Arpège to Météo-France implementation. In G.-R. Hoffmann and N. Kreitz, editors, *Coming of age - Proceedings of the 6th ECMWF workshop on the use of parallel processors in Meteorology*, pages 206–222. World Scientific, 1994.

[8] J.F. Estrade and L. Dragulanescu. Advances in Météo-France models parallelisation. In G.-R. Hoffmann and N. Kreitz, editors, *Making its mark - Proceedings of the 7th ECMWF workshop on the use of parallel processors in Meteorology*, pages 101–112. World Scientific, 1996.

[9] B. Radi and J.F. Estrade. Preliminary results on the parallelization of Arpège-IFS with the implementation of the full Météo-France physics. In B. Hertzberger and G. Serazzi, editors, *High-performance computing and networking - Lecture notes in computer science 919*, pages 164–169. Springer Verlag, 1995.

[10] G. Radnoti, R. Ajjaji, Bubnova R., M. Caian, E. Cordoneanu, K. von der Emde, J.-D. Gril, J. Hoffman, A. Horanyi, S. Issara, V. Ivanovici, M. Janousek, A. Joly, P. LeMoigne, and S. Malardel. The spectral limited area model ARPEGE-ALADIN. In WMO, editor, *PWPR report series n. 7 - WMO TD n. 699*, pages 111–118, 1995.

[11] Andrew Staniforth and Jean Côté. Semi-lagrangian integration schemes for atmospheric models- a review. *Mon. Wea. Rev.*, 119:2206–2223, 1991.

Parallelization of the French Meteorological Mesoscale Model MésoNH

Patrick Jabouille[1], Ronan Guivarch[4], Philippe Kloos[12], Didier Gazen[3],
Nicolas Gicquel[1], Luc Giraud[2], Nicole Asencio[1], Veronique Ducrocq[1],
Juan Escobar[3], Jean-Luc Redelsperger[1], Joël Stein[1], and Jean-Pierre Pinty[3]

[1] Centre National de Recherches Météorologiques (CNRM)
42 av G Coriolis, 31057 Toulouse Cedex 01, France
[2] Centre Européen de Recherche et de Formation Avancée en Calcul Scientifique,
42 av G Coriolis, 31057 Toulouse Cedex 01, France
[3] Laboratoire d'Aérologie, Observatoire Midi-Pyrénées,
14 av E Belin, 31400 Toulouse, France
[4] ENSEEIHT-LIMA-IRIT, 2 rue C Camichel, 31071 Toulouse Cedex 7, France

Abstract. Numerical simulation of the atmospheric motions requires
the most powerful machines and a high performance computer technology. The French meteorological model MésoNH is designed as a research
tool for small and mesoscale atmospheric processes. This paper describes
softwares and techniques used for implementing this numerical model
on parallel processor computers and presents first results and performances.. . .

1 Introduction

The MésoNH atmospheric simulation system is a joint effort of Centre National de Recherches météorologiques (CNRM Météo-France) and Laboratoire
d'aérologie (LA).

It comprises several elements : a numerical model with a comprehensive physical package, an ensemble of facilities to prepare initial states, post-processing
and graphical tools to visualize the results.

MésoNH is a gridpoint limited area model. To be able to use a horizontal
resolution lower than 5 km, the model is nonhydrostatic, i.e. that it uses a
complete prognostic equation for the vertical speed. This assumption implies to
solve an elliptic equation to determine the pressure.

The model solves the equations governing the atmospheric state evolution
on a computational grid. This grid is non orthogonal to maintain the vertical
coordinate and take into account the topography (more details about the Mésonh
equations and discretization can be found in [1] or by visiting our web site :
http://www.aero.obs-mip.fr/mesonh/).

In addition to the dynamic part of the flow, the model takes account of
many physical parameterizations (turbulence, radiations, surface processes, mi-
crophysics ...). It allows for the transport and diffusion of passive scalars, to be
coupled with a chemical module.

P. Amestoy et al. (Eds.): Euro-Par'99, LNCS 1685, pp. 1417–1422, 1999.
© Springer-Verlag Berlin Heidelberg 1999

The model is able to perform several simultaneous simulations on a nested grid. This technique (called grid nesting) allows to focus on specific regions described by a higher spatial resolution, preserve a correct representation of large scale flow with a moderate size memory occupation.

The code is entirely written in FORTRAN 90 and in a first stage it has been developed to be run on a mono processor computer.

In order to improve the spatial resolution and the representation of physical processes, a parallel implementation of the MésoNH model is become very crucial. To achieve this goal, the CERFACS laboratory expert on efficient algorithms for solving large-scale scientific problem has joined CNRM and LA to develop tools and techniques used for the model parallelization.

2   Interface Routines

The work of parallelization began by a precise specification of requirements related to communications between processors to run the model on a parallel machine.

- Decomposition of a model on n processors
- Decomposition of m horizontally nested models on n processors
- Parallelization is achieved to be as much as possible transparently to users not aware of parallelization technics
- Full compatibility of model running on 1 processor with the same code
- Portability on any computers owning the MPI library
- Routines allowing I/O flow along a transparent way for users

To meet these goals, an interface library named ComLib has been developed. It contains all routines necessary to parallelize the Méso-NH model and is based on the standard library MPI (Message Passing Interface).

Current development is focusing on the parallelization of multiple nested models which requires additional interface routines to perform exchanges data between parent and child models, both decomposed over processors. This important part of ComLib are not described in the paper.

The Méso-NH ComLib package has been designed to enable an easy implementation of the Méso-NH code on parallel distributed memory computers. This package is implemented in Fortran 90 on top of the MPI (Message Passing Interface) library. Its main purpose is to provide the Méso-NH developers, who are not necessarily expert in parallel computing, with all the required communication capabilities while hiding to him the low level message passing paradigm details.

The implemented coarse grain parallelism exploits a parameterizable 2D decomposition in the x-y direction of the 3D physical domain (the vertical dimension is not decomposed). The automatic partitioning of the 3D domains generates by ComLib produces as many non-overlapping vertical rectangular beams as processors requested by the user. Each beam is then assigned to one processor of the parallel target computer. Some overlapping data-structure referred to as "halo" in the sequel are required; this enables to reuse all the computational routines existing in the sequential version of the Méso-NH code. Because most of the operators are discretized using 5 point stencil in the horizontal plane the "halo" have a width of only one cell. However this width may be easily increased.

The ComLib library is based on data structure organized in an object-oriented style, with Fortran 90 user defined types. There are two top-level types

- the configuration of all the processors i.e. the way the domain is splitted into subdomains corresponding to a processor
- the communications to be performed by the current processor (send and receive operations) for each kind of communication (halo updates, transpositions, grid-nesting...).

These types are recursive, (which means that they contain pointers to variables of the same type), in order to treat the model nesting.

The definition of the characteristics of the 2D decomposition and the communication informations is entirely managed the ComLib package and its related representation is stored in private data structure of the package.

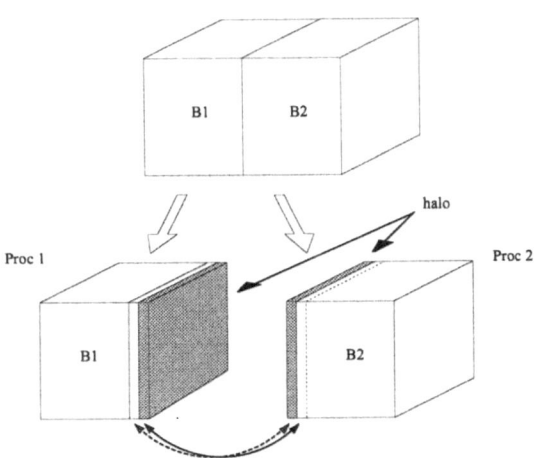

Fig. 1. Example of decomposition

In Figure 1 we depict an example of decomposition suitable for a parallel execution of Méso-NH on a two processor computer. Beam B_1 and B_2 are respectively allocated to processor 1 and 2; the variables associated with internal nodes in B_i can be updated simultaneously provided that data in the halo (shadowed area in Figure 1) have been updated correctly prior to the computation.

It is up-to the user to define the list of fields that should be kept updated in the halo region through specific subroutines to manipulate list of fields , as well as when this list of field should be updated by calling an other subroutine.

The distribution depicted on figure 2a does not work to solve the elliptic pressure system because a fast Poisson solver based on FFT computations is used. Two other types of distribution are used (figure 2b and c) respectively called x-slice and y-slice. Communication routines have been implemented that move a field between these different decompositions, then it is possible to perform the FFT for each horizontal direction. The parallel implementation of the pressure solver is presented in an other paper [2].

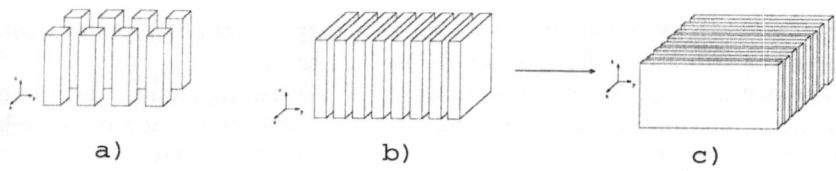

a)   b)   c)

Fig. 2. the different distributions used in the model

The IO routines have been modified in order to carry out at the same time the reading and the distribution of the input fields and the gathering and the writing of the output fields, in a transparent way for the user.

3   MésoNH Parallelization

The parallelization of the code has been achieved in modifying some routines in the temporal loop to take into account the data domain decomposition. Three types of routines can be distinguished: routines to communicate fields between processors, routines allowing the treatment of physical boundary conditions and routines doing global operations.

- The main processor communications in a time step are located at the end of temporal loop to refresh the halo of all prognostic variables (wind, temperature, humidity ...). Because of the only one width of the halo area, other supplementary communications for a few fields are required to obtain a correct computation in the inner subdomain (communication of surface fluxes for example). These minor communications should be removed by increasing the width of the halo.
- Modifications in some Méso-NH routines have been brought to distinguish processor located on physical border. Special ComLib functions are used to do this.

- Global operations using the FORTRAN 90 functions SUM(A) or MIN(A), where A is an array, have been replaced by equivalent parallel ComLib function involving processor communications.

It can be noticed that the modifications in the fortran code are very limited and correspond to the call of a few ComLib routines. So, it will be easy for a Mésonh user to add his own source modification without problem concerning the parallelization.

4   First Results and Performance Analysis

To validate the parallel implementation of MésoNH, a numerical simulation has been performed two times using one and four processors. The meteorological situation corresponds to a large convective area located over France. The grid is composed by $100 \times 100 \times 30$ points which represent an horizontal domain of 2000 by 2000 km. The time step of the numerical integration is 30s. The simulation has been made on Fujitsu VPP-700 of Météo-France.

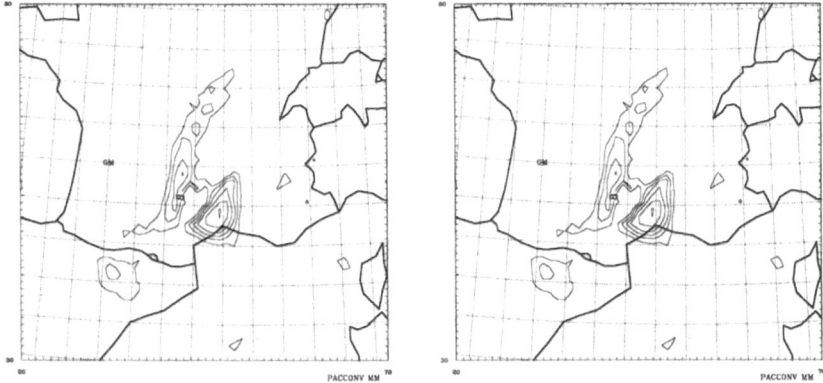

Fig. 3. cumulative convective rainfall during 3 hours (contour interval is 4 mm) for one and four processors simulation

The simulation lasts three hours and the complete physical package (turbulence, radiation, convective scheme, microphysic...) is used. The integrated convective rainfall is compared for the two simulations in figure3. No difference can be observed in these two pictures and a precise computation shows that the maximum numerical discrepancy between the two simulations is about 1%. This discrepancy is due to the compiler optimization in the pressure solver part. Simulations made over a few time steps without use of pressure solver do not reveal significant discrepancy.

For the three hours simulation, the speed-up on 4 processors is equal to 2. This moderate result can be partially explained by the decrease of vectorization efficiency. By using 4 processors, each processor performs its computations on a $50 \times 50 \times 30$ size domain. This size is rather short for a vectorial calculator such VPP700. A comparison between a sequential simulation on a 100×100 domain and a 4 processors parallel simulation on a 200×200 domain has been made. In this configuration, the amount of data per processor is the same. The efficiency of parallelization becomes 0.65. The optimization of the parallel version is still in progress.

A complete study of the parallel performance including different platforms and using more processors will be shown during the presentation.

5 Future Evolution

Some developments are required to finish the parallelization of the whole Méso-NH atmospheric simulation system : the implementation of processor communications for the grid nesting technique, the implementation of a distributed method for preparation of initial files, the use of a station network...

The final goal is to provide a research tool able to produced very fine simulations (using a huge number of grid point and including advanced parameterizations) to help the understanding of complex mesoscale phenomena.

References

[1] Lafore, J.P., Stein, J., Asencion, N., Bougeault, P., Ducrocq, V., Duron, J., Fischer, C., Hereil, P., Mascart, P., Pinty, J.P., Redelsperger, J.L, Richard, E., and Vila-Guerau de Arellano, J.: The Méso-NH atmospheric simulation system. Part I: adiabatic formulation and control simulations. Annales Geophysicae,**16** (1988) 90-109.

[2] Giraud, L., Guivarch, R., and Stein, J.: Parallel distributed fast 3D Poisson solver for Meso-scale atmospheric simulation. in this proceedings

The PALM Project: MPMD Paradigm for an Oceanic Data Assimilation Software

A. Fouilloux[1] and A. Piacentini[1]

Global Change and Climate Modelling Team - CERFACS
42 Av. Coriolis - 31057 Toulouse - France

Abstract. The PALM project aims to provide a general structure for a modular implementation of a data assimilation system. In this system, an assimilation algorithm is split up into elementary "units" such as the observation operator, the computation of the correlation matrix of observational errors, the forecast model, etc. This approach allows to separate the physical part of the problem from the algebraic part. PALM performs the algebra, ensures the synchronization of the units and drives the communication of the fields exchanged by the units. Thanks to the interface specifications, each physical unit can be implemented as an independent executable code and can be easily replaced or reutilised for other configurations.

1 Introduction

The MERCATOR project aims to implement an oceanic forecasting system with assimilation of satellite-born and *in situ* data.

This system will be used as a research tool and for operational purposes. This double aim imposes the full modularity of the whole system and the best tuning of the performances on all the actual reference plate-forms. Modularity, performances and portability are three aspects hardly ever compatible: the best trade-off is looked for.

The core of a forecast system with data assimilation is the assimilation cycle: this is a large scale optimisation problem which makes use of very different computational components, such as the forecasting model, the observation operator mapping the state of the model onto the data space, the statistics, the minimiser and so on. Each component is best implemented as a stand alone code developed and tested by a separate scientific team. The effort due to the merging of these heterogeneous components in an efficient code is the main hindrance for the research in data assimilation.

The philosophy of the PALM software, the back-bone of the MERCATOR project, is to keep the components as independent as possible and to make them work together in a MPMD distributed system. The definition of algorithm independent interfaces will ensure the full modularity of the scheme.

This document presents the motivation for such an approach on a simple example and the implications that the users requirements have on the design of

P. Amestoy et al. (Eds.): Euro-Par'99, LNCS 1685, pp. 1423–1430, 1999.
© Springer-Verlag Berlin Heidelberg 1999

the code. The PALM philosophy is explained in detail in the central sections. The technical choices are then sketched in section 5.

2   MPMD for Data Assimilation

Let's consider as a simple example the computation of the gradient of the cost function J in the 3D-Var assimilation scheme

$$\nabla J(\mathbf{x}) = P^{-1}(\mathbf{x} - \mathbf{x}^b) + \mathbf{H}_{|\mathbf{x}}^T R^{-1} \left[H(\mathbf{x}) - \mathbf{y}^0 \right]$$

where $\mathbf{x}$ is the optimal initial condition to be found by minimization of J, $\mathbf{x}^b$ is the first guess previously forecast, $\mathbf{y}^0$ are the observations, H is the operator going from the model space to the space of the observed variables and P, R are the variance/covariance matrices of the forecast and the observation errors.

Some remarks determine the choice of the MPMD paradigm.

- A huge amount of data has to be exchanged between operators. The size of $\mathbf{x}$ can be up to 10^9 8 bytes REAL for the high resolution models. The operators have therefore to be implemented as distributed codes.
- Therefore two levels of parallelism coexist: between tasks and inside each task

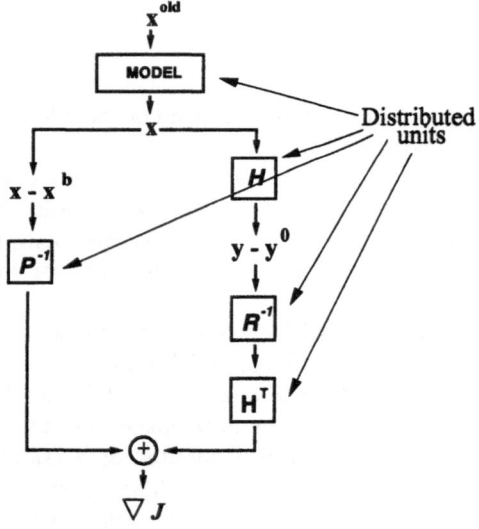

- Different units are developed in different laboratories.
- Different assimilation methods reuse the same units in different order.
- Research in this field often consists in testing different versions of one unit without changing the rest of the system.
- Units often are legacy codes.

The MPMD paradigm allows to compose an assimilation scheme "à la carte", using existing units. This is the idea of the PALM project.

3   Implications

- PALM aims to become a common tool for all the potential users (see the experience with the coupler OASIS developed at CERFACS) ⇒ <u>portability</u>
- PALM is a research tool ⇒ <u>flexibility</u>
- PALM is an operational tool <u>as well</u> ⇒ <u>performances</u>
- Data assimilation is a machine dimensioning problem ⇒ the technical solutions retained for PALM can affect the requirements. In particular, the MPMD paradigm implies an under-exploitation of the machine ⇒ <u>economical interest</u> in the research of the best trade-off.

4   The PALM Philosophy

4.1   The Concepts

The main idea of PALM is that the physics of the problem are IN the units.

PALM handles the algorithm and the linear algebra. For this purpose it has to perform the following tasks

0. definition of the parallel strategy for the algorithm and the linear algebra
1. process management: spawning and monitoring
2. process synchronisation: on a "on event" basis
3. communication management: addressing and monitoring
4. memory buffer management
5. distributed linear algebra: strategy and computations
6. interaction with the operational system's supervisor
7. statistics

The meaning of these tasks will be clarified in the next simple examples.

To plug the units in PALM an interface layer will be added to the units. This interface has to be independent of the actual algorithm to allow portability and reuse of the units.

4.2   A Simple Case

The function of the tasks will be illustrated on a simple case in which the driver spawns a first unit, waits for the unit to notify that a new result is available, spawns the next unit, waits for the second unit to notify that it is ready to receive, sends the addresses to each unit to the other one and waits for the first unit to finish after the direct transmission of the large amount of data to the second one.

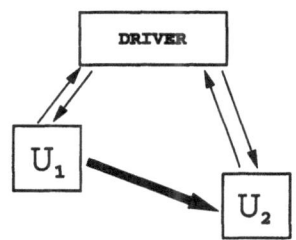

The sequence of tasks is shown in the next figure, where the bold numbers refer to the tasks.

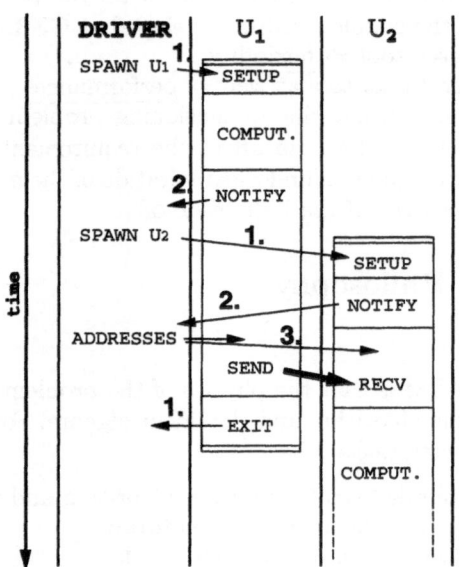

These actions have to be performed for each branch of a scheme like the one in section 2. Moreover, the units can be distributed. It implies the management of parallel communications between single processes of the units.

4.3   Delayed Communications

In the slighter complicated case in which the result of a unit will not be used by the immediately following unit, it is necessary to store the temporary result in a memory buffer external to the units.

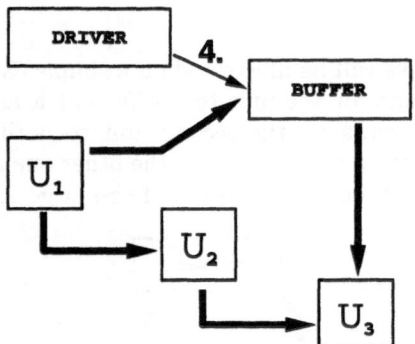

This buffer can be useful also for other purposes as, for instance, the pre-loading of data files used by several units.

4.4   Linear Algebra

Let's illustrate the issues connected to the treatment of the linear algebra on one branch of the example of section 2. Compute $R^{-1}\left[H(\mathbf{x}) - \mathbf{y}^0\right]$ with $\mathbf{y}^0$ in the buffer. Let $\mathbf{d}$ be $H(\mathbf{x}) - \mathbf{y}^0$

In the non distributed case the straightforward strategy is to perform the linear algebra operation (–) in the buffer and to send the result to the following unit.

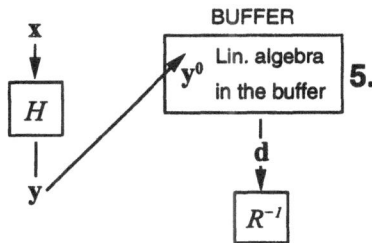

In the distributed case two strategies are possible. The task 0 deals with the choice of the most effective strategy.

a. H and R^{-1} **do not** have the same distribution.

In this case the same strategy than in the non distributed case can be used.

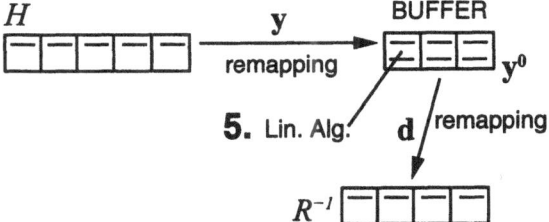

In this case the advantage is that the two units do not need to run together, but on the other hand a big buffer is needed and two remappings are necessary.

b. H and R^{-1} **have** the same distribution.

In this case it is better to perform the linear algebra in the interface layer.

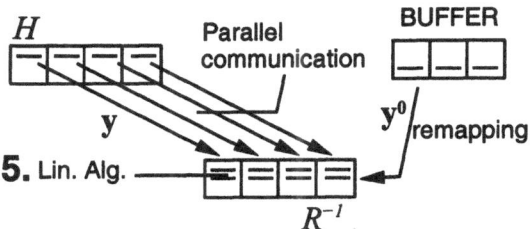

The advantage is that H can send y to R^{-1} with a parallel communication processor to processor. In this case only one remapping is needed. The two units have to run concurrently.

Obviously, more complicated linear algebra operations are to be performed, requiring the use of an efficient distributed linear algebra library.

5   Functional Decomposition

The design of PALM can be summarised through a top-down functional decomposition on four levels:

1. The highest level corresponds to the <u>actions</u> of the driver. In few words the driver gets as input from PREPALM some configuration files providing the execution parameters, the description of the spaces and of the objects, the sequence of the units, the parallelism between tasks, the distribution of the units, the communication patterns, the algebra to perform and any other relevant information. The driver acts as a finite state automaton evolving following the execution of the algorithm and performing a set of predefined tasks. The monitoring of the execution is accessible to the user through an ergonomic interface, possibly graphical.
2. The second level corresponds to the <u>tasks</u> of the driver. They have been grouped in four categories:
   (a) Treatment of the objects: this category includes the algebra and the control of coherence of the objects
   (b) Exchange of messages, objects and structures between units, driver and memory buffer.
   (c) Sequencing of units inside a parallel branch of the execution tree: in particular, the spawning and the synchronisation of the units are the object of this category.
   (d) Scheduling of actions as a function of resources and of the synchronisation between parallel branches.
   Moreover, the statistics collection can be considered a transversal task belonging to all the categories.
3. The third level corresponds to the <u>tools</u> which realise the tasks. They are of two kinds:
   (a) The *public* tools which can be used either in the driver or in the units interface layer:
       i. The algebra and more generally all the computational tools
       ii. The events manager and in particular the notification mechanism
       iii. The communication tools
   (b) The tools which are *private* to the driver
       i. The memory buffer
       ii. The process manager and monitor
4. The last level is the <u>implementation</u> level and corresponds to the technical solutions which can be adopted for the implementation of the tools of the third level.

The decomposition can be summarised with the following diagram:

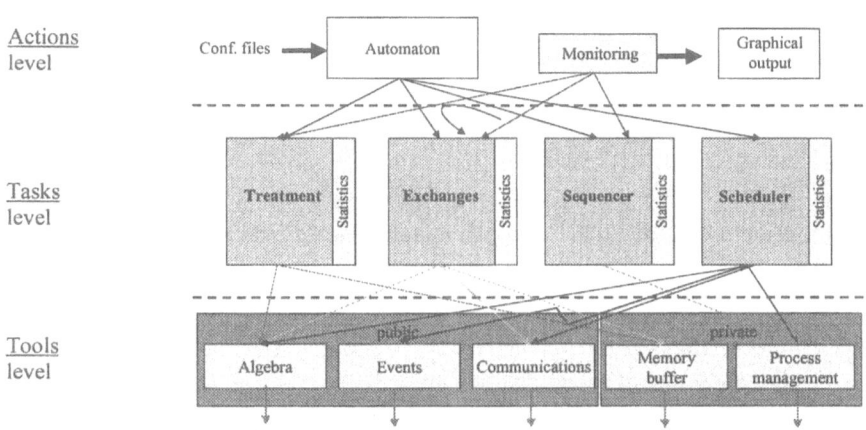

6   Performance Issues

The main drawback of the MPMD approach seems to be the overhead introduced by the message passing communications and the process manager. In order to assess the overhead, a simulator of the performances of an assimilation suite has been implemented [8]. This simulator allows to compare the PALM implementation of a suite and a standard Unix/Fortran implementation. Two operational suites were used as case study: the 3D-Var of Météo-France and the sub-optimal scheme SOFA applied to the global oceanic model OPA. The results of the simulations show that in the sequential distributed configuration, following the level of optimisation, the overhead is between 1.6% and 2.4% of the elapsed time of the Fortran implementation. This overhead is acceptable. Furthermore, a PALM implementation exploiting the parallelism among tasks leads to a gain of elapsed time.

7   Conclusions

This short overview points out some major advantages of the MPMD paradigm for data assimilation. It can open the way for new tools, like the PALM software, which can interest a large community of users, not only oceanographers but also atmospheric and geophysical scientists. The technical issues concerned by the MPMD approach are state-of-the- art in parallel computing. Some of the implementation choices described here can evolve following the most recent developments in the message passing libraries and in the communication techniques implemented by constructors on the main supercomputers.

References

[1] D. Beaucourt and Caremoli C. Calcium: a new tool for code coupling. In *Proceedings of Fall CUG 94, 10–15 October 1994*, 1994.

[2] C. Cassou, P. Noyret, E. Sevault, O. Thual, L. Terray, D. Beaucourt, and M. Imbard. Distributed ocean-atmosphere modelling and sensitivity to the coupling flux precision: the cathode project. *Monthly Weather Review*, pages 1035–1053, April 1998.

[3] P. Courtier, J.N. Thépaut, and A. Hollingsworth. A strategy for operational implementation of 4d-var, using an incremental approach. *Quart. J. Roy. Meteor. Soc.*, 120:1367–1387, 1994.

[4] J. Derber and A. Rosati. A global oceanic data assimilation system. *J. Phys. Oceanogr.*, 19:1333–1347, 1989.

[5] K. Ide, P. Courtier, M. Ghil, and A.C. Lorenc. Unified notation for data assimilation: operational, sequential and variational. *J. Meteorol. Soc. Jap.*, 75 1B:181–189, 1997.

[6] T. Kentemich, H.C. Hoppe, and U. Keller. MPI–2 enhancements for the MERCATOR project. Technical Report: Internal Confidential, Revision 1.00, PALLAS, Germany, 1998.

[7] Le groupe PALM. A coding norm for PALM. Technical Report GLOBC Team Log Book lb050598_*, CERFACS, Toulouse, 1998.

[8] Le groupe PALM. Etude de faisabilité du projet PALM. Technical Report TR/CMGC/98-50, CERFACS, Toulouse, 1998.

[9] Le groupe PALM. A source code documentation system: Prodoc. Technical Report GLOBC Team Log Book lb011299_*, CERFACS, Toulouse, 1999.

[10] A. Piacentini. Some remarks on the use of the CRAY T3E network PVM version for the implementation of a MPMD data assimilation code. Technical Report: TRACS Visit Final Report 08/98, EPCC, Edinburgh, 1998.

[11] W. Sawyer, L. Takacs, A. da Silva, and P. Lyster. Parallel grid manipulations in earth science calculations. DAO Office Note.

[12] E. Sevault and L. Terray. The coupling library for interfacing models (clim): User guide and reference manual. Technical Report TR/CMGC/95-47, CERFACS, Toulouse, 1995.

A Parallel Distributed Fast 3D Poisson Solver for Méso-NH

Luc Giraud[1], Ronan Guivarch[2], and Joël Stein[3]

[1] Centre Européen de Recherche et de Formation Avancée en Calcul Scientifique,
42, Avenue Gaspard Coriolis, 31057 Toulouse Cedex 01, France
giraud@cerfacs.fr

[2] ENSEEIHT-LIMA-IRIT, 2 rue Charles Camichel, 31071 Toulouse Cedex 7, France
guivarch@enseeiht.fr

[3] Centre National de Recherche en Météorologie,
42, Avenue Gaspard Coriolis, 31057 Toulouse Cedex 01, France
Joel.Stein@meteo.fr

Abstract. The fast Poisson solvers based on FFT computations are among the fastest techniques to solve Poisson problems on uniform grids. In this paper, we present a parallel distributed implementation of a 3D fast Poisson solver in the context of the atmospheric simulation code Meso-NH [3]. This parallel implementation consists in implementing data movement between each computational step so that no elementary computational routine implements communication. Experimental results are given on a 128 node Cray T3E to illustrate the advantages of this method.

1 Introduction

The solution of Poisson problem occurs in many applications, like computational fluid dynamics, quantum chemistry, ... The availability of a fast and robust parallel solvers then becomes a key component to perform intensive simulations on large and complex problems. For Poisson problems defined on regular grids, the fast solvers based on FFT (Fast Fourier Transform) techniques are amongst the fastest methods. In this paper, we describe a parallel implementation of a fast Poisson solver based on FFT that has been developed in the context of the atmospheric simulation code Meso-NH. In this code, the fast Poisson solver acts as preconditioner for the solution of the pressure equation (see [3]). Basically, in order to solve the following Poisson-type equation (1), the sequential algorithm can be summarized as follow:

$$\begin{cases} F(u) = \frac{\partial^2 u}{\partial x^2} + \frac{\partial^2 u}{\partial y^2} + \frac{\partial}{\partial z}(g(z)\frac{\partial u}{\partial z}) = f \\ \text{Boundary conditions} \end{cases} \qquad (1)$$

on an uniform 3D grid the solver implements 5 steps:

1. Perform a FFT on each horizontal line of the grid in the x-direction; that is, diagonalize the $\frac{\partial^2 u}{\partial x^2}$ part of the operator. This computational step is referred to as FFT_x in the sequel of this paper.

P. Amestoy et al. (Eds.): Euro-Par'99, LNCS 1685, pp. 1431–1434, 1999.
© Springer-Verlag Berlin Heidelberg 1999

2. Repeat this procedure for the y-direction (referred to as FFT_y in the sequel of this paper). At this stage the transformation of the operator F reduces to a set of tridiagonal systems, each linear system is associated with one vertical grid line.
3. Solve a tridiagonal system for each vertical line of the discretization.
4. Perform a FFT inverse on each horizontal line of the grid in the y-direction (referred to as FFT_y^{-1}).
5. Finally, perform a FFT inverse on each horizontal line of the grid in the x-direction (referred to as FFT_x^{-1}).

In Section 2 we present the parallel distributed implementation of the code we have considered in our study and present in Section 3 the performance observed on a Cray T3E.

2   Parallel Implementation

In the framework of the Meso-NH code, the exploited parallelism is based on a partitioning of the physical domain into sub-domains. Because a lot of physical phenomena simulated by the code take place in the vertical direction, only 2D decomposition in x and y direction are considered. That is the original physical domain is decomposed in N_x pieces in the x-direction and in N_y pieces in the y-direction. This partitioning defined a set of vertical beams and is referred to as a $N_x \times N_y$ decomposition (see Fig 1).

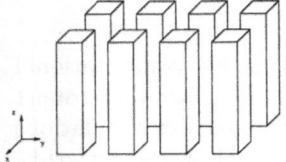
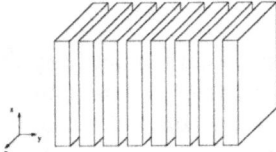
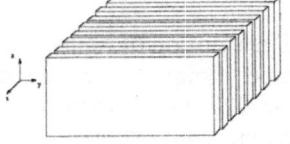

Fig. 1. 2×4 decomposition

Fig. 2. x-slice decomposition

Fig. 3. y-slice decomposition

In our parallel approach, rather than parallelizing any of the computational kernel of sequential algorithm, we have implemented communication between each of its computational step so that we can reuse concurrently the sequential code on subsets of data. In this respect, communication routines have been implemented that move the data first from a vertical $N_x \times N_y$ beam decomposition (2D decomposition depicted in Figure 1) into $(N_x \times N_y)$ vertical slices aligned to the x-direction (1D decomposition depicted in Figure 2), refer to as x-slice decomposition so that each process can perform the FFT_x on the x-slice it stores. The next step consists in transposing the data to move them from a x-slice decomposition to vertical slices aligned to the y-direction, refer to as y-slice decomposition (see Figure 3). With this data partitioning each processor performs on the data it stores the FFT_y on each horizontal line, the solution of the tridiagonal systems associated with the vertical lines and the FFT_y^{-1} again on each horizontal line. Before performing the FFT_x^{-1}, the data are transposed back in a x-slice decomposition; finally the data are moved from the x-slice decomposition to the original beam decomposition.

Because this method requires to move the data from a vertical beam decomposition to x-slices and y-slices, it is referred to as the X-Y variant and its algorithm can be summarized as follows:

Algorithm 1 (X-Y variant)

> **Step 1** Move data from a beam to a x-slice decomposition
> **Step 2** for $i=1:n_y^{local} \times n_z$; $FFT_x(i)$; end
> **Step 3** Transpose the data from a x-slice to a y-slice decomposition
> **Step 4** for $i=1:n_x^{local} \times n_z$; $FFT_y(i)$; end
> **Step 5** for $i=1:n_x^{local} \times n_y$; Solve_tridiag(i) ; end
> **Step 6** for $i=1:n_x^{local} \times n_z$; $FFT_y^{-1}(i)$; end
> **Step 7** Transpose the data from a y-slice to a x-slice decomposition
> **Step 8** for $i=1:n_y^{local} \times n_z$; $FFT_x^{-1}(i)$; end
> **Step 9** Move data from a x-slice to a beam decomposition

where n_x^{local} (respectively n_y^{local}) denotes the number of points in the x-direction (resp. in the y-direction) allocated to the processor.

3 Experimental Results

The experiments reported have been observed on a 128 node Cray-T3E located at CEA Grenoble and correspond to the solution of a problem defined on a $74 \times 74 \times 38$ grid with MPI library.

We present the scaled speed-ups observed with the X-Y approach described above. The aim of the scaled speed-up is to study the behavior of the parallel code when the amount of data per processor is kept constant while the number of processors is increased. In particular this shows how the code would behave on huge problems when numerous processors will be used [2].

The experiments reported correspond to $64 \times 64 \times 150$ local problem size. In our experiments we have only scaled the x and y direction of the global problem as considering too many grid points in the z-direction might not make sense for the physical simulation usually performed with Meso-NH.

In this particular study we defined the scaled speed-up as $S_p = \frac{p * T_4}{T_p}$, where T_4 is the elapsed time to solve a problem of global size resp. $128 \times 128 \times 150$ on 4 processors and a 2×2 decomposition and T_p is the elapsed time to solve a problem on p processors.

For algorithms where amount of computation is proportional to the amount of stored data, the upper bound of the scaled speed-up is $S_p \leq p$. Closer S_p is from p better the parallel implementation is.

nb of processors	scaled speed-up
4 × 2 decomposition	6.52
2 × 4 decomposition	7.46
4 × 4 decomposition	13.13
8 × 4 decomposition	20.60
4 × 8 decomposition	24.25
8 × 8 decomposition	43.32
16 × 8 decomposition	62.74
8 × 16 decomposition	71.35

Table 1: Scaled speed-up for a
64 × 64 × 150 grid per processor

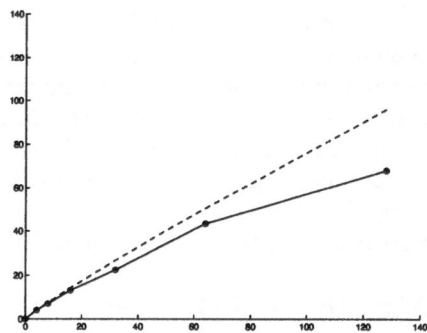

Fig. 4. Optimal scaled speed up and experimental scale speed up.

The optimal scaled speed up is not linear because of the $\mathcal{O}(n \log(n))$ complexity of the FFT computations. In this respect the optimal speed-up depicted by a dashed line in Figure 4 is not the usual $y = x$ straight line. In this figure, it may be seen that the observed scaled speed-ups are close to the optimal one for a moderate number of processors.

4   Concluding Remarks

In this work we have proposed an approache to parallelize a Fast Poisson solver on distributed memory multiprocessor. The performances of this approach have been studied on a 128 node Cray T3E and its scalability has been investigated. It appears that this approach exhibits reasonable parallel performance and is appropriated in the framework of Meso-NH. For more details and results, we refer to [1]

It can be added than the parallelization of the whole model Meso-NH is currently performed. This parallel code is dedicated to run on many platforms as the supercomputers Fujitsu VPP-700 of Météo-France or Cray-T3E as well as station networks. Some experiments are still in progress on the VPP-700 in order to gain a better understanding of the performance of the parallel distributed fast 3D Poisson solver.

References

[1] Giraud, L., Guivarch, R.,Stein,J.: Parallel Distributed Fast 3D Poisson Solver for Meso-scale Atmospheric Simulation, CERFACS Technical Report, in preparation.
[2] Gustafson, J., Montry, G., and, Benner, R.: Development of Parallel Methods for a 1024-Processor Hypercube, SIAM J. Sci. Stat. Comput., **9**, 609-638, 1988.
[3] Lafore, J. P., Stein, J., Asencio, N., Bougeault, P., Ducrocq, V., Duron, J., Fischer, C., Hereil, P., Mascart, P., Pinty, J.P., Redelsperger, J.L., Richard, E., and Vila-Guerau de Arellano, J.: The Meso-NH Atmospheric Simulation System, I, Adiabatic formulation and control simulations, Annales Geophysicae, **16**, 90-109, 1998.

Porting a Limited Area Numerical Weather Forecasting Model on a Scalable Shared Memory Parallel Computer

Roberto Ansaloni[1], Paolo Malfetti[2], and Tiziana Paccagnella[3]

[1] Silicon Graphics SRL, Milano, Italy, Roberto.Ansaloni@sgi.com
[2] CINECA, Casalecchio di Reno, Bologna, Italy, Malfetti@cineca.it
[3] ARPA, Bologna, Italy, T.Paccagnella@smr.arpa.emr.it

Abstract. As new parallel machines become popular commercially, it is important to understand if they are adequate for performing and scaling well for a range of applications. This paper describes the porting of a weather forecast limited area model from a vector machine to a recent SMP-multiprocessors and evaluates the performances obtained.

1 Introduction

Recently, ccNUMA multiprocessors have been shown to deliver good performance on many applications. Weather forecast limited area models run commonly on vector machines or on MPP systems. It is important to know that the ccNUMA systems perform and scale well even for the above mentioned models, and a relatively simple parallelization effort due to the programming model available on these system is needed.

In Section 2 we begin by briefly describing the limited area model used and, in Section 3, the memory and communication architecture of the SGI Origin 2000 (Origin). In Section 4 we address the problem of porting the code from a vector machine, like the Cray C90 (C90), to a scalable shared memory processor machine (SMP), like the Origin, while in Section 5 we evaluate the performances of the code on SMP multiprocessor. Finally, Section 6 briefly explains the operational suite and Section 7 concludes the paper.

2 Model Description

LAMBO, Limited Area Model BOlogna, is a grid-point primitive equations model, based on the 1989 and 1993 versions of the ETA model, used at the National Centre for Environmental Prediction of Washington [2].

LAMBO is running operationally since 1993 at Agenzia Regionale Prevenzione Ambiente - Servizio Meteorologico Regionale (S.M.R.).

As already mentioned before, LAMBO is a grid-point, primitive equations limited-area model: in such models the only basic approximation, well justified by the scale analysis of the vertical component of the momentum equation, is the

P. Amestoy et al. (Eds.): Euro-Par'99, LNCS 1685, pp. 1435–1438, 1999.
© Springer-Verlag Berlin Heidelberg 1999

hydrostatic approximation, which assumes pressure at any point simply equal to the weight of the unit cross-section column of air above that point. Atmospheric motion is predicted by applying the principles of conservation of momentum, energy and mass and using the law of ideal gases. Such set of differential equations constitutes the initial and boundary value problem, the solution of which provides the future state of the atmosphere.

The equations of motion are solved in practice using finite difference methods and all model variables are defined on the so-called Arakawa E-type grid.

Particular numerical schemes were developed to integrate on the E-grid the part of the equations related to adiabatic processes [1].

LAMBO has a full complement of physical parameterizations. Parameterizations are modules of the model code which attempt to represent the effects on model prognostic (dependent) variables of those processes which cannot be explicitly resolved during model integration. Such processes include in general moist processes, vertical turbulent exchanges, radiative exchanges, lateral diffusion, surface exchanges of moisture, heat and momentum and so on.

3   LAMBO on the SGI Origin 2000

A serial version of LAMBO has been running in production on CINECA C90 since September 1993. The purpose of this work was to implement a parallel version of LAMBO on a Origin, following few basic criteria:

• the parallel version of the code had to run on the 16 Origin processors at least in the same time as the serial C90 version

• the code modifications had to be kept at a minimum level in order to retain code readability and portability

The Origin version has been written in a shared memory programming model, by far the most natural and efficient way to implement parallel code on Origin systems. Parallelism has been achieved by exploiting the auto-parallelizing compiler features and by the insertion of SGI parallel directives.

The migration of the code to the parallel Origin system has been articulated in four major steps: porting, single processor tuning, parallelization, performance analysis.

The porting process has been really straightforward: the only important issue was related to the numerical precision required, due to the different default variable size on C90 (64 bits) and on Origin (32 bits).

In order to efficiently run the LAMBO vector code on the cache-based Origin architecture, aggressive optimization compiler flags had to be turned on, in particular for the loop nesting and cache prefetching analysis.

The major problem arose when considering the code parallelization. The original version of LAMBO made a large use of EQUIVALENCEd variables, to save memory and increase code readability, but the presence of an EQUIVALENCEd variable in a loop inhibits its parallelization. Thus, in order to achieve a significant level of parallelism, it has been necessary to remove most of the EQUIVALENCE statements, thus reducing the code readability for the authors.

Different parallelization schemes have been applied to different subroutines, always choosing the best approach according to the algorithm implemented.

Another advantage given by the programming model chosen, is the possibility to adopt an incremental code parallelization approach.

Eventually it turned out that 10 subroutines have been manually parallelized by the insertion of compiler directives and 6 have been automatically parallelized by the compiler. In the case of the radiation package, the parallelization has been achieved at a higher level, by parallelizing the main loop in the driver routine which calls the other radiation routines.

4   Parallel Performance

In order to evaluate the parallel performance, the results are compared with those foreseen by Amdhal's law which represents the parallel execution time T(p) as a function of the number of processor p and of the parallel fraction fp of the serial time (see Table ??). S(p), the speed-up function definition, is used to evaluate the parallel performance; in the ideal case, when all the code is perfectly parallel (fp=1) , the speed-up function is the linear function S(p)=p. Since only the routines that account for the 96% of the total execution time were considered for parallelization, the Amdhal's curve corresponding to fp=0.95 should be considered. Due to imperfect load balance, cache misses and data contention between processors that fraction is positioned between 90% and 95%. LAMBO performance is in good agreement with those predicted by the Amdhal's law model as shown in Figure 1.

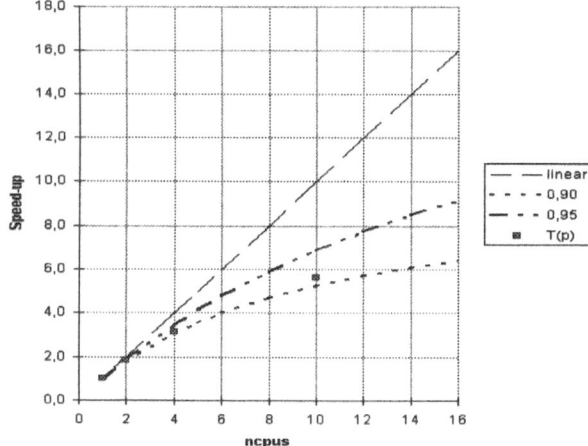

Fig. 1. Linear speedup, theoretical speedup for f_p

Pes	1	2	4	6	8	10	12	14	16
T(p)	275	150	87	66	56	50	46	43	40

Table 1. Parallel execution times on Origin

5   The Operational Suite

LAMBO runs daily at two different horizontal and vertical resolutions. The first run, 40 km horizontal resolution with 20 vertical sigma levels, is performed taking initial and boundary conditions from European Center for Medium range Weather Forecast (E.C.M.W.F.) 00GMT analysis and forecast. The second run, 20 km horizontal resolution with 32 vertical sigma levels, is nested in the previous run using a one way nesting procedure developed at S.M.R.. Both the runs are computed on integration domains covering all the Italian territory. E.C.M.W.F. products required by LAMBO are obtained through a link with the Servizio Meteorologico Aeronautica Militare, the National Weather Service.

After one month of testing phase, LAMBO is operative on CINECA Origin since the 1st July 1998, and, on average over a six months period, on 10 Origin processors, the first run takes about 5' while the second takes about 32'. This should be compared with the 10' and 50', respectively required by the previous C90 runs.

6   Conclusions

In the recent years it has been a common perception that only vector processor machines were adequate for running limited area weather forecast models. Few recent models were developed on MPP systems, others were ported using the message passing paradigm. We have shown that a limited area model on a scalable shared memory parallel computer can give good performances, providing satisfactory operational forecasts.

Moreover it should be stressed that these results have been obtained by a relatively simple parallelization effort due to the programming model available on the Origin system: a much higher effort would have been necessary in the case of a message passing implementation.

References

[1] Janjic, Z., 1979: Forward-backward scheme modified to prevent two-grid interval noise and its application in sigma coordinate model. Contr. Atm. Phys., 52, 69-84.

[2] Mesinger F., and Z. I. Janjic, S. Nickovic, D. Gavrilov and D. G. Deaven, 1988: The step-mountain coordinate: Model description and performance for cases of Alpine lee cyclogenesis and for a case of Appalachian redevelopment. Mon. Wea. Rev., 116, 1493-1518.

Topic 22
High Performance Data Mining
and Knowledge Discovery

David Skillicorn and Domenico Talia

Co-chairmen

Many, perhaps most, organizations use computers when they interact with their customers. As a result, and almost by accident, many organizations have accumulated huge amounts of data about such interactions. Over the past five to ten years, they have increasingly tried to use this data for commercial advantage. This process began by accumulating transaction data into *data warehouses*, where it could be made available for decision support and retrospective analysis. The effectiveness of such analysis largely depends on the ability of individuals to induce queries that will reveal key facts about the organization and its customers.

Increasingly, both the volume of data and its complexity have taken the problem beyond the ability of any individual to analyze. *Data mining* is the automated analysis of large volumes of data, looking for the relationships and knowledge that are implicit in large volumes of data and are 'interesting' in the sense of impacting an organization's practice. Research and development work in the area of knowledge discovery and data mining concerns the study and definition of techniques, methods, and tools for the extraction of novel, useful, and implicit patterns from data. It builds on machine learning, database technology, and statistics, but is distinguished by problems of scale: the data involved is so large that most applications tend to use conceptually straightforward, but carefully optimized, algorithms.

There is a natural confluence between parallel computation and data mining. For researchers in parallel computation, data mining is an application area that is growing in importance, and that introduces interesting new problems (irregularity, data representation and storage, multiple parallelization strategies, symbolic computation) that have not been so critical in scientific and numerical computing. For organizations who want to use data mining in their day to day work, parallel computation offers increased performance, which in turn may translate into commercial advantage. When data mining tools are implemented on high-performance parallel computers, they can analyze massive databases in a reasonable time. Faster processing also means that users can experiment with more models to understand complex data. High performance makes it practical for users to analyze greater quantities of data. Larger databases, in turn, yield improved predictions.

Data mining, even sequentially, is not yet mature, and many of the existing applications are relatively unsophisticated. Nevertheless, it seemed useful to explore the fledgling projects that are looking at the connections between parallel

P. Amestoy et al. (Eds.): Euro-Par'99, LNCS 1685, pp. 1439–1440, 1999.
© Springer-Verlag Berlin Heidelberg 1999

computing and data mining. This track has assembled a small number of papers describe such research experiences.

The first paper "Mining of Association Rules in Very Large Databases: A Structured Parallel Approach" by Becuzzi, Coppola, and Vanneschi, presents a case study implementing the Apriori parallel association rule algorithm using the skeleton-based language SkIE. The paper is as much about parallel software engineering as it is about data mining. It demonstrates the effectiveness of the skeleton approach as a software construction technique for a real problem. The development of a parallel program from a a sequential one, and its subsequent tuning, are shown to be straightforward and inexpensive. The second paper "Parallel k-means Clustering for Large Data Sets" by Stoffel and Belkoniene presents a parallel version of the k-means clustering algorithm where data points are distributed across processors. Experiments on a cluster of 32 PCs are discussed. This implementation shows interesting scalability properties of the h-means version of the algorithm on distributed-memory parallel computers. The third paper "Performance Analysis for Parallel Generalized Association Rule Mining on a Large Scale PC Cluster", by Shintani, Oguchi and Kitsuregawa, addresses the parallel computation of association rules with an item hierarchy. The authors present a performance evaluation, on a large scale PC cluster, of algorithms for mining generalized association rules. The performance results show that the algorithms are effective for handling skew on highly parallel computers. The last paper "Inducing Load Balancing and Efficient Data Distribution Prior to Association Rule Discovery in a Parallel Environment" by Manning and Keane, proposes a statistical method to achieve efficient data partitioning in parallel data mining. In particular, the authors suggest a database redistribution based on principal component analysis (PCA) to achieve load balancing and reduction in candidate set duplication in a parallel algorithm for mining association rules.

The four papers presented in this session give some idea of the issues and approaches to using parallel computation to implement scalable data mining algorithms. Parallel data mining is a long way from mature status. We hope the results presented here will stimulate interest in the confluence of parallel computing and data mining, and will lead to more work in this important area.

We would like to thank the two Vice-Chairs, Vipin Kumar and Hannu Toivonen, who organized the track and reviewed papers with us. We would also like to thank the external referees who helped us review all of the submissions. Their work tends to be unnoticed, but is not unappreciated.

Mining of Association Rules in Very Large Databases: A Structured Parallel Approach*

P. Becuzzi, M. Coppola, and M. Vanneschi

Dipartimento di Informatica, Università degli Studi di Pisa, Italy
{becuzzi,coppola,vannesch}@di.unipi.it

Abstract Newer and newer parallel architectures being developed raise a strong demand for high-level and programmer-friendly parallel tools. We show some results regarding mining of association rules, a well-known Data Mining algorithm, which we ported from sequential to parallel within the PQE2000/SkIE environment. The main goals achieved are the low effort spent in parallelizing the code, the machine independence of the application produced, source code portability and performance portability. Here we report test results for the same parallel program on three different architectures.

1 Introduction

During the last years the problem of extracting valuable knowledge from bigger and bigger amounts of automatically collected data has become of greater industrial relevance. In the field of Data Mining new techniques and tools are being developed [3, 7, 11] to cope with the growing size of modern databases, and parallel computing is widespreadly seen as the main way to High Performance Data Mining. This sharpens the demand of software engineering techniques and environments for programming parallel computers: the industrial need for short time-to-market of application software conflicts with the typically very long development time on today's parallel platforms.

The aim of the PQE2000 project [10] is to transfer the research results in the field of parallel architectures, structured parallel programming tools and strategic applications to the industry. Knowledge Data Discovery is definitely one field in which the technological transfer is needed.

This work is part of a joint project for High Performance Tools for Tax Fraud Detection, carried on by the PQE2000 project, the Italian Ministry of Finance and SOGEI SpA [2]. The first step in developing such an applications is the selection of a set of Data Mining kernels as ordinary and parallelization case studies. In this paper we show some results regarding the classic problem of mining of association rules, a well-known Data Mining algorithm. We ported

---

* Computing resources for this research were provided both by C3P center of C.N.R. (STRIDE project, co-founded by the EC) and by Italian M.U.R.S.T. ("Design Methodologies and Tools of High Performance Systems for Distributed Applications" – MOSAICO project).

P. Amestoy et al. (Eds.): Euro-Par'99, LNCS 1685, pp. 1441–1450, 1999.
© Springer-Verlag Berlin Heidelberg 1999

from sequential to parallel the code for finding the *frequent sets* of a database of transactions, working within the PQE2000/SkIE environment.

The main goals achieved are the low effort spent in parallelizing the code, the machine independence of the application produced, source code portability and performance portability. Here we report test results for the same parallel program on three different architectures.

The next section contains the basic definition of our subproblem and an overview of the sequential and parallel algorithms involved. Section three defines our target, the quick development of a portable application, introduces to the SkIE programming environment and to the parallel systems that we used. Section four describes the structure of the prototype we have realized, discussing various design issues. Their impact on performance is explained through section five. The last section evaluates the experiment with respect to the initial target, and presents future research and development directions.

2   The Frequent Itemsets Problem

The problem of discovering simple association rules in databases has been proposed back in 1993. We'll refer to [3], chapter 12, for exact definitions and a thorough theoretical background regarding the generation of association rules. It's enough to say that the rules are easily produced once the subproblem of finding *frequent sets* in the data is solved. It's the most heavy sub-task, most of the efforts in the literature being addressed to its efficient solution.

Given a set $\mathcal{I} = \{i_1, \ldots, i_m\}$ of items, we will call *transaction* a pair T of a unique identifier and a set of items; a *k-itemset* will be a simple set of k items. Our database $\mathcal{D}$ will be a set of transactions. Itemsets can contain each other, and be contained in a transaction. The parameters of the problem are the set of attributes $\mathcal{I}$, the database $\mathcal{D}$ and a fixed real number $0 < s < 1$, the *minimum support*. The support $S(X)$ of an itemset $X \subseteq \mathcal{I}$ is the fraction of transactions in $\mathcal{D}$ that contain that itemset; if $S(X) > s$, then we say that X is *frequent*, or has (at least) minimum support. We want to find all itemsets that are frequent in the given database $\mathcal{D}$, and compute their support counts.

The itemsets inherit a natural order by inclusion from being subsets of the set of attributes $\mathcal{I}$. They form a lattice structure, and since the minimum support property for a set implies the same holds for all its subsets, the frequent sets are the union of a class of sub-lattices: we want to discover this structure. Several approaches are viable, most of them are reviewed in [11, 12] together with a theoretical background. The one we have chosen for our experiment is the bottom-up exploration done by the Apriori algorithm. The key idea is that since supersets of non-frequent itemsets are non frequent we can build up the set of frequent sets levelwise, instead of exploring the whole power set of $\mathcal{I}$.

From the set L_k of frequent k-itemsets we build a candidate set of $(k+1)$-itemsets C_{k+1}, using as heuristic filter the property that for each element $x \in C_{k+1}$, all k-itemsets in x belong to L_k. The size of C_{k+1} can be much less than that of $(\mathcal{I})^{k+1}$. A linear scan of the database is required at each pass to extract

L_k from C_k, and as much passes as the length of the biggest frequent itemset. Pruning candidates using information gathered during the previous pass is not effective at the second one: $C_2 = L_1 \times L_1$ is often roughly the same as $\mathcal{I} \times \mathcal{I}$. Of course the algorithm requires temporary data structures holding information about L_k and C_k: so the amount of memory needed can be quite large.

We regard the parallel solutions of Apriori as belonging to one of three main classes outlined in [1]. **Count Distribution** solutions replicate the candidate set at each node, and partition the database among the processors. The itemset counting is distributed, global communications are needed at each step of Apriori. **Data Distribution** parallelizations spread the candidate set over the nodes, regardless of its structure. Partitioning a big database requires "local" transaction data to travel to most or all of the processors at each pass. **Candidate Distribution** programs partition the candidate set trying exploit its structure, so that the database can be accordingly partitioned in order to minimize or avoid subsequent communications, and/or load imbalancing.

The three approaches deal in different ways with the problem of keeping in the local memory part of the transaction data and the intermediate counts. A sequential algorithm that addresses the same problem is the **Partition** scheme.

It was introduced in [7] to allow efficient use of the main memory, reducing the I/O burden due to making repeated scans of the data in the main loop of Apriori. Here the computation is divided in two consecutive *phases*, and the database is partitioned.

Phase I computes frequent itemset information $\{\forall k, {}^j L_k\}$ separately for each slice j of the data, working with the minimum support specified. It's easy to prove that a globally frequent k-itemset must belong to at least one ${}^j L_k$. Hence the family of sets $\mathcal{H} = \{k > 0 \,|\, \cup_j ({}^j L_k)\}$ must contain all of the frequent k-itemsets, and is an upper approximation of the solution of the problem.

Phase II has a simpler structure and lower computational load. It extracts the solution computing itemset counts with a linear scan of the whole data.

Parallel implementations of Partition are akin to Count parallelizations, as they scale up almost linearly with DB size, but cannot easily handle very low minimum support (very large L_k and C_k sets) because of the memory requirements. Some previous results about this kind of parallelization of Apriori can be found in [6].

The statistical properties of the data can vary among partitions: too much different ${}^j L_k$ lead to a very coarse approximation of the true global results. The problem is known as *data skew*, and increases memory usage and computational load of phase II. Sequential Partition usually avoids this pitfall, since to minimize the overall workload the partitions in phase I are kept as big as possible. The parallel implementation however has to find a tradeoff between the total amount of computation and the degree of exploitable parallelism, bounded by the number of partitions. We just mention that data skew can be solved by hashing the database unless the partition size is too small with respect to data size and minimum support.

3   Target and Environment of the Experimentation

Parallelizing a non trivial, data-intensive sequential code is a severe test of the efficiency of a programming environment. Our main requirements were:
1) no limits imposed on database size to be dealt with; 2) high reuse of sequential code, with more efforts in modularization than in writing ad-hoc code; 3) ease of integration with other algorithms and within more complex application; 4) to exploit as much as possible the portability of code and performance; 5) to shorten as much as possible the development time of the program.

Some of the conditions were easily fulfilled by working with SkIECL, the coordination language of our programming environment. It allows very simple exploitation of parallelism between code modules, as well as seamless substitution of sequential modules with semantically equivalent, parallel ones.

We started from an "off-the-shelf" Apriori implementation of average performance. The original structure has then been changed into a partition scheme. The choice of a parallel partition scheme comes up from the two main facts that Parallel Partition schemes scale well with respect to increasing database size and processor number, and that they can be easily combined with other parallelization schemes by substituting the Apriori module in the first phase. This is even easier when using a structured coordination language.

We didn't spent much time in tuning the algorithm, the main effort in modifying the sequential code was to make it fully modular, so to allow the different processes to share the same high-level library code for the operations, just exchanging data when needed. Having this done, it was straightforward to select the functions to parallelize, insert the corresponding modules into parallel constructs and starting to run tests.

The structure and features of the SkIE programming environment can't be described in full here. We refer to some of the literature about the environment and the compilation techniques [9, 10], just outlining here the basis of the approach and the technical details that are strictly necessary to the analysis of our prototype. SkIECL is a coordination language, designed to allow easy parallel composition of high-level modules written in different languages. A parallel application is developed by integrating (possibly existing) program modules in a SkIECL framework. Parallelism can be expressed only by means of *parallel constructs* or *skeletons*, which are machine-independent and fully compositional. Every communication detail is handled by the compiler, as well as some forms of compile-time optimisation that are based on profiling data and expected performance models.

Our sequential code is mostly C++ routines that call some class methods defined in the sequential version of the application. The classes were compiled to a standard linked library, while the code of the main algorithm was encapsulated using some of the SkIECL constructs. Apart from the seq construct, that simply defines the interfaces of a sequential module, we used the constructs pipe (expresses the pipeline composition of the contained modules), farm (functional replication of the contained *worker* construct over a stream of independent tasks, with load-balancing) and comp (used to express functional composition of two pipe skeletons without parallel execution). The first two skeleton are tailored to stream parallelism, while the comp in its general form is designed to express data parallel computations.

The very beginning of the development relied upon an heterogeneous cluster of LINUX machines over ordinary Ethernet. We have then been able to make extensive tests over three parallel architectures, all of them using MPICH, of those supported by the SkIECL compiler:

CS-2 — a QSW Computing Surface 2 with 24 processing nodes, (2 ×130Mhz Hyper-SPARC processors, 128Mb RAM) with a proprietary fat-tree network. This machine is in fact the MIMD portion of one of the PQE2000 prototypes.

LINUX — a cluster of LINUX CPUs made up of 10 Pentium II (266Mhz, 128 Mb RAM), that uses Fast-Ethernet and a crossbar hub as communication network.

SMP — a SMP workstation (4 Ultra-SPARC 250Mhz, 256Mb RAM).

4   Structure of the Prototype

When designing a parallel version of the Partition scheme, keeping the data inside the local memory of each processor turns out to reduce the global communications to just two: the first one at the end of the first phase, to globally compute the approximation of the result set, the second one at the end of the algorithm, to exchange counting information for the itemsets.

We see in Fig.1 the overall structure of the prototype. It is a comp of the two Partition phases, which do not compute in parallel, each one realized as a pipe of modules, with the slower ones functionally replicated inside farm constructs. The sequential modules 1–3 realize Phase I of the Partition algorithm, modules 4–6 Phase II. There you see also support processes (d,c) automatically generated by the compiler to act as data dispatchers and collectors inside each farm construct.

The database is read by process 1, and flows as a stream of partitions through the pipeline of following stages. Module 2, which is functionally replicated, applies Apriori to each partition in its incoming stream, producing a stream of hash-tree structures holding local results. Module 3 collapses this stream, computing the upper approximation $\mathcal{H} = \{k > 0| \cup_j (^j L_k)\}$ of the result.

Phase II starts with reading again the database and producing a stream of partitions; the verification work is done in module 5, also replicated by a farm, and $\mathcal{H}$ is sent to all the worker processes as initialization data. Module 6 collects the stream of itemset support counts and sums them to produce the final results. It also produces the association rules using the sequential algorithm.

In principle there is no need to exploit the same degree of parallelism or use the same partition size in both phases, but in our case using a different number of workers for the two farm is useless. Even centralized reading of the data inside modules 1 and 4 is not required by the algorithm. In the beginning, we decided to do no preprocessing or off-line partitioning of the data on local disks.

In our view, repeatedly mining the data (like in meta-learning algorithms) would be carried out through some data-mart support, with data filtering parameters that may change each run. Having a single input point for the whole

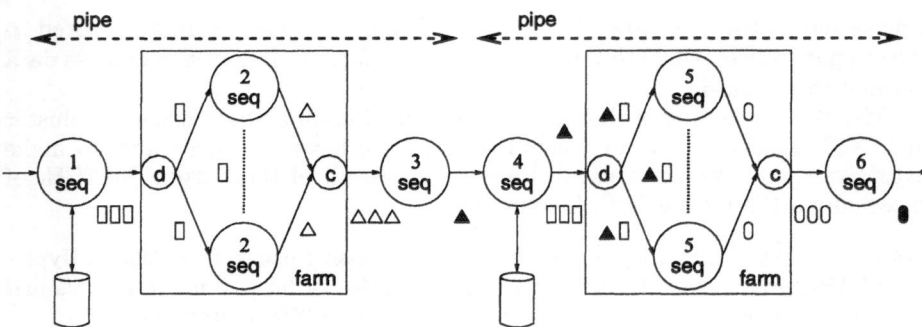

Fig.1. Block structure of the prototype

algorithm makes allowance for filter and interface functions, and is not necessarily a performance loss if the data come from a separate database server.

We know that the exploitable parallelism degree in a farm is bounded by the ratio between communication/scheduling time and computation time of the worker nodes. Since some of the data structures may become quite large, we must check that the communication network does not congest.

This is not a sharp bound for systems with a high performance interconnection network (CS2), or with communication happening efficiently in memory (small SMP like the one we used). The test results on these architectures are measured with the centralized file read module that we have just described.

However, on the LINUX cluster the many software layers between the MPI API and the Ethernet, and this being a slower device, can create a bottleneck. We changed our code to distribute only partition references, and load the data in modules 2 and 5 from local disks. The data has not been partitioned: each node has a copy of the whole file, and the load distribution remains dynamic. This kind of code changes are made easy and quick by the high level parallel language, requiring only limited editing of the source code and its SkIE interface definitions. The net effect on the LINUX cluster has been to cut down the completion times nearly by one third, down to the figures shown in the following. We are now planning to use the very same code on the CS-2 at first to exploit its local disks, then to develop an interface to the parallel file system.

5   Performance Analysis

The test files used to measure the performance of the program were generated with the Quest synthetic data generation code, used in the literature as a common reference. An explanation of the underlying data model can be found in [3]. In this paper we show results obtained from files having average transaction length of 20 items, and frequent itemsets of average size of 6. These parameters provide a heavy combinatorial load to the Apriori module, hitting the weak point of simple count/partition parallel schemes. Our test databases are made up of 1, 4 and 12 million transactions, and are about 90, 360 and 1100 Mbytes in size.

Results over the CS-2 are shown in Fig. 2. We must keep in mind that direct comparison between CPUs built with different technologies is not fair. For our purposes, we can say that the three architectures used are ordered by the computation/communication time ratio, with the LINUX (the fastest CPUs and the slowest network) having the lowest value, and the SMP the highest.

In Fig.3b the fall of the speed-up curves for the 90Mbyte file at 2% shows that raising the parallelism degrees the sequential workload (reading the data and computing the set of rules) begins dominating the completion time.

The CS2 and SMP are less vulnerable to this effect, as is clear by the comparison of the two tests with centralized reading (the ones with a *), and from the first tests on the SMP, Fig.3a. The SMP results are amazingly close to ideal speedup limit for four processors: the high speed of inter-node communication can partly explain the figures, but more tests are needed to evaluate the influence of various other factors, like the interaction between the database size and the amount of memory used for disk buffering.

In the general setting, we can reach substantially better results by removing the two sequential bottlenecks of reading the data, and using bigger databases.

Our view is confirmed by the results: we can see the effect of removing the centralization in Fig.3b for 90 and 360 Mbytes files at 2%. Fig.4 shows a smooth behaviour of the program, and that the speedup increases with the computational load (more data and/or lower support).

Apart from the results taken at 0.5%, we note that the program reaches an asyntotic behaviour, because the speedup does not change significantly (Fig.4b). The 0.5% minimum support value is not directly comparable to the other results: the speedup is even higher, but we had to remove the sequential generation of association rules. The intermediate data were becoming too big, and the sequential part too slow, meaning that rule generation needs to be parallelized as well at this point.

We point out that the limit on the exploitable parallelism comes from the nature of the problem: for too small partitions no Partition scheme is effective. We ran tests and verified that very small partitions slow down the program as expected, but then the performance is almost unaffected until the data structures exceed the main memory size.

We conclude that communications and I/O do not impose severe restrictions on this parallelization, and more performance improvement can come from substitution of the algorithm inside module 2 and from the introduction of some degree of parallelism in the association rule generation of module 6.

6   Evaluation of the Results and Future Work

To evaluate the re-engineering costs of the parallelization, consider first that the original sequential C++ code was about 2900 lines; the modular version is about 3550 lines, with code mainly added to dump the status of some objects.

The main file of the final prototype consists of 350 lines of parallel code (SkIECL and C++) replacing the original sequential main file of 220 lines.

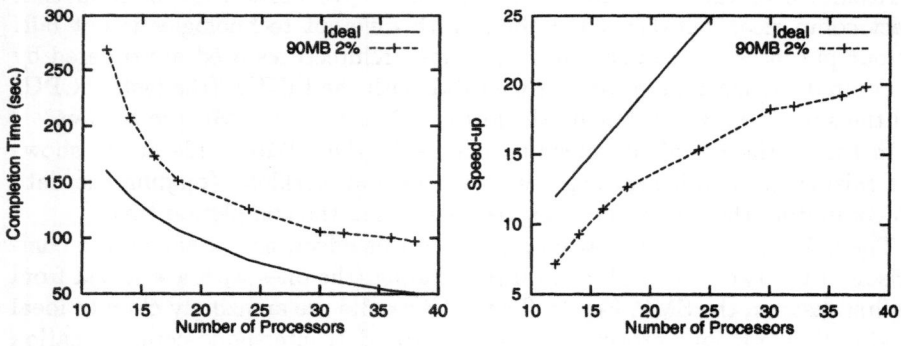

Fig.2. CS-2 parallel completion time and speed-up, T20.I6.D1M

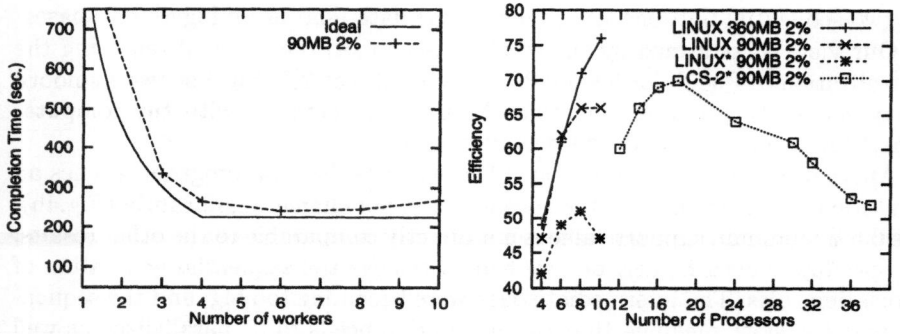

Fig.3. (a) SMP completion time versus workers, up to 4 processors; (b) program efficiency over LINUX and CS-2 with support set at 2%.

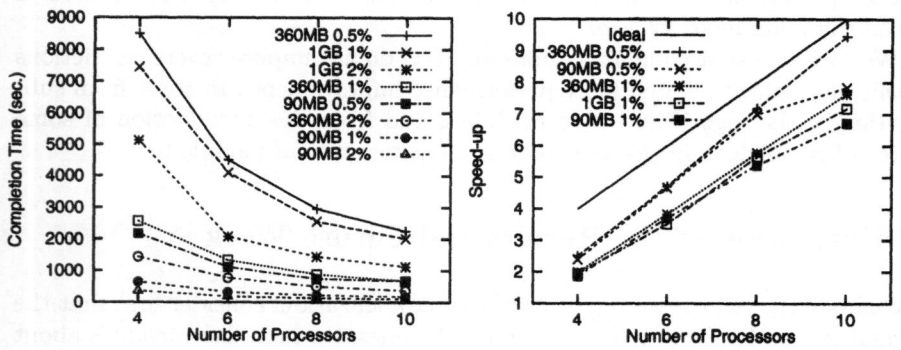

Fig.4. LINUX parallel completion time and speed-up, T20.I6.D1M / 4M / 12M

Relative to the original code size, the percentage of reused sequential code is over 90%, while new code has been written that is 22% (sequential part) and 12% (parallel). The overall size of the source code has grown by less than 25%. The amount of time used for development and testing is up to now about three man-months, with two months spent on the sequential code, and one more to test and try the different parallel solutions.

Devised a global structure for the program, we started with the parallelization of the slowest portion of the code (phase I), moving to more complex solutions (farm in phase II, distributed I/O) as early as it was needed to improve performance. This process is incremental, because the effort spent in parallelizing one section is never wasted by changes in other modules.

We have reached the goal of quickly producing a compact, reasonably efficient parallel code that can be both practically used and further improved along the same path. No bound has been forced on the size of the database. The limit of the developed application is only the size of the intermediate data, as every Partition or Count parallelization of Apriori, but the underlying structured model allows us to substitute Apriori with a more scalable and/or parallel algorithm.

Compared to the small effort required by parallelizing the code, the results are good: we succeeded in parallelizing an existing high-level application code with no deep insight of its structure, achieving good speedup and scalability over the original program.

There are results in the field which outperform our sequential code, like in [7], where the Partition algorithm is introduced. Nevertheless, if we compare with [12] we discover that our prototype on all the three machines outperforms the algorithm presented on dataset with the same structure. The CS-2 and LINUX have a better completion time, while the SMP can reach nearly the same performance using 4 processor instead of 32. Notably, the sequential algorithm on the CS-2 is nearly 3 times slower than the sequential Par-MaxClique, but the parallel version is nearly twice as fast when using 32 processors.

From the test results is clear that the same source code is of practical use from small, low-cost cluster of PC's to real parallel machines. This saving of programming efforts has no comparison within the literature, with the parallel programs being architecture-tailored (not portable) or written using communication libraries. See for instance how much attention is due to implementation details, like communication cost analysis or synchronizations, in [1] or [8]. Besides, the good results maintained across different architectures are also an evidence that structured parallel environments can promote restructuring of existing programs instead of development of new ones from scratch.

We plan to complete the test set for the algorithm with much larger databases, and to improve the behaviour with respect to decreasing minimum support for the itemsets. Most of the optimisations devised could benefit from the availability of a virtual shared memory support, to scale-up the size of data structures beyond the limit of a single node memory. We could tune up the sequential Apriori code; a better solution would be to completely replace it with another algorithm. New, more efficient algorithms exist that cut down the average com-

putational complexity of finding the frequent itemsets, for instance see [4] and
[11]. The newer algorithms, which are closely related to the Candidate Distribution, as well as the ones that exploit only Data Distribution allow us to handle
very large candidate sets, and lower the minimum support for the search. They
can be used together with the partition approach to process huge databases, and
can also be mixed with a Count Distribution scheme just like in [5].

References

[1] R. Agrawal and J.C. Shafer. Parallel mining of association rules: Design, imple-
    mentation and experience. *IEEE Transactions on Knowledge and Data Engineer-
    ing*, 8(6), December 1996. IBM Research Report RJ 10004, January 1996.

[2] P. Becuzzi, M. Coppola, D. Laforenza, S. Ruggieri, D. Talia, and M. Vanneschi.
    Data analysis and data mining with parallel architectures: Techniques and ex-
    periments. Technical report, project "Parallel Intelligent Systems for Tax Fraud
    Detection", December 1998.

[3] U. M. Fayyad, G.Piatetsky-Shapiro, P. Smyth, and R. Uthurusamy, editors. *Ad-
    vances in Knowledge Discovery and Data Mining*. AAAI press / MIT press, 1996.

[4] D. Gunopulos, H. Mannila, R. Khardon, and H. Toivonen. Data mining, hyper-
    graph transversals, and machine learning (ext. abstract). In *PODS '97. Proc. of
    the 16th ACM Symposium on Principles of Database Systems, May 1997, Tucson,
    Arizona*, pages 209–216, New York, 1997. ACM Press.

[5] E.H. Han, G. Karypis, and V. Kumar. Scalable parallel data mining for association
    rules. In *Proc. of the ACM SIGMOD Int. Conf. on Management of Data*, volume
    26,2 of *SIGMOD Record*, pages 277–288, New York, May13–15 1997. ACM Press.

[6] Andreas Mueller. Fast sequential and parallel algorithms for association rule
    mining: A comparison. Technical Report CS-TR-3515, Dept. of Computer Science,
    Univ. of Maryland, College Park, MD, August 1995.

[7] A. Savasere, E. Omiecinski, and S. Navathe. An efficient algorithm for mining
    association rules in large databases. In U. Dayal, P.M.D. Gray, and S. Nishio,
    editors, *VLDB '95: Proc. of the 21st Int. Conf. on Very Large Data Bases, Zurich,
    Switzerland*, pages 432–444, Los Altos, CA, 1995. Morgan Kaufmann Publishers.

[8] T. Shintani and M. Kitsuregawa. Hash based parallel algorithms for mining asso-
    ciation rules. In *PDIS '96: 4th Int. Conf. on Parallel and Distributed Information
    Systems*, pages 19–30, Los Alamitos, Ca., USA, December 1996. IEEE Computer
    Society Press.

[9] M. Vanneschi. Heterogeneous HPC Environments. In David Pritchard and Jeff
    Reeve, editors, *Euro-Par '98 Parallel Processing*, volume 1470 of *LNCS*, pages
    21–34, Southampton, UK, September 1998. ACM / IFIR, Springer-Verlag.

[10] M. Vanneschi. PQE2000: HPC tools for industrial applications. *IEEE Concur-
     rency: Parallel, Distributed & Mobile Computing*, 6(4):68–73, Oct-Dec 1998.

[11] M.J. Zaki. *Scalable Data Mining for Rules*. PhD thesis, University of Rochester,
     Rochester, New York, 1998.

[12] M.J. Zaki, S.Parthasarathy, and M.Ogihara. Parallel algorithms for discovery of
     association rules. In *Data Mining and Knowledge Discovery*, volume 1. Kluwer
     Academic Publishers, 1997.

Parallel k/h-Means Clustering for Large Data Sets

Kilian Stoffel and Abdelkader Belkoniene

Université de Neuchâtel
Groupe Informatique
Pierre-à-Mazel 7
CH-2000 Neuchâtel
(Switzerland)
{Kilian.Stoffel;Abdelkader.Belkoniene}@seco.unine.ch
Phone: ++41 32 718 1376, Fax: ++41 32 718 1231

Abstract. This paper describes the realization of a parallel version of the k/h-means clustering algorithm. This is one of the basic algorithms used in a wide range of data mining tasks. We show how a database can be distributed and how the algorithm can be applied to this distributed database. The tests conducted on a network of 32 PCs showed for large data sets a nearly ideal speedup.

1 Introduction

Clustering, the process of grouping similar objects, is a well known and a well studied problem. Some of early work has been done in statistics (e.g. [2][7]). In more recent years clustering was identified as a key technique in data mining tasks [3]. This fundamental operation can be applied to many common tasks such as unsupervised classification, segmentation and dissection. We are focusing here on one specific algorithm for clustering namely k/h-means clustering [1]. The original version of the k/h-means algorithm was designed for numerical data [4][6][5].

Our contribution in this paper is the development of a parallel version of the k/h-means algorithm. We present a realization of this algorithm on top of a distributed object store installed on a simple PC network.

2 The k/h-Means Algorithm

The k/h-means belongs to the class of the partitional or non-hierarchical clustering algorithms. The goal of the algorithm is, starting from a set of objects $\mathcal{X}$ and an integer number K, to find a partitioning of the objects in $\mathcal{X}$ into K clusters. If M denotes the number of objects in the database then $K \leq M$. The partitioning is optimized in the sense that the distance between the objects in one cluster is minimized. This optimization is often called "whiting group sum squared error minimization".

P. Amestoy et al. (Eds.): Euro-Par'99, LNCS 1685, pp. 1451–1454, 1999.
© Springer-Verlag Berlin Heidelberg 1999

More formally this problem can be defined in the following way:
Given $\mathcal{X}$ a set of M objects. Let $X_i = [x_{i1}, x_{i2}, ..., x_{in}]$ denote a vector that represents the values of ith object over the n attributes, let $\mathcal{P}_0(M, K)$ be a partition matrix, and let C_l, $l = 1, K$ be the clusters of the partition. Each of the M objects lies in one of the K clusters. The mean of the K clusters are defined by $\bar{X}_l = \frac{1}{n_l} \sum_{i \in C_l} X_i$, $l = 1, K$ where n_l stands for the number of elements in cluster C_l. The error for a given cluster C_l is given by: $\epsilon_l = \sum_{i \in C_l} \delta(X_i, \bar{X}_l)$ where δ represents any distance function. Then the error for a partition $\mathcal{P}_0(M, K)$ can be defined as: $\epsilon[\mathcal{P}_0(M, K)] = \sum_{l=1}^{K} \epsilon_l$ The overall goal is to minimize $\epsilon[\mathcal{P}_0(M, K)]$. The large number of all possible partitions makes it impracticable to search for an optimal partition $\mathcal{P}^*(M, K)$ such that $\epsilon[\mathcal{P}^*(M, K)]$ is minimal. Instead, it is necessary to use a local optimization technique.

Let $\mathcal{P}_1(M, K)$ be a partition in the neighborhood of $\mathcal{P}_0(M, K)$ obtained by moving the object k from the cluster C_r of the partition $\mathcal{P}_0(M, K)$ to the cluster C_j of the partition $\mathcal{P}_1(M, K)$. If for δ we use the squared Euclidean distance, then the relation between $\epsilon[\mathcal{P}_0(M, K)]$ and $\epsilon[\mathcal{P}_1(M, K)]$ is given by:
$\epsilon[\mathcal{P}_1(M, K)] = \epsilon[\mathcal{P}_0(M, K)] + \frac{n_j}{n_j+1}\delta(\bar{X}_j, X_k) - \frac{n_r}{n_r-1}\delta(\bar{X}_r, X_k)$. $\epsilon[\mathcal{P}_1(M, K)]$ decreases if $\frac{n_j}{n_j+1}\delta(\bar{X}_j, X_k) - \frac{n_r}{n_r-1}\delta(\bar{X}_r, X_k) < 0$ or $\frac{n_j}{n_j+1}\delta(\bar{X}_j, X_k) < \frac{n_r}{n_r-1}\delta(\bar{X}_r, X_k)$

3   The Parallel k/h-Means Algorithms

In the previous section we describe the original k-means algorithm. This algorithm can be translated into the pseudo code version given in Figure 1. This is the version we will use to present the reflection to be made in order to parallelize the algorithm given above. However, another formulation of the k-means algorithm exists. This version is sometimes also called h-means . Very often people do not distinguish between the two algorithms. However, we think it is important to make this difference, even though the conceptual difference in the algorithms is very small. The only difference is the place in the algorithm where the means are recalculated. Instead of recalculating them after each migration of an object, all objects are migrated if a better cluster assignment can be found, and then the new mean values are calculated (see Figure 1). The theoretical properties of the two algorithms are different. The k-means algorithm has better convergence properties and is less likely to get stuck in a local minimum. However in a wide range of tests we conducted, no major differences could be measured.

We will now show how the k-means algorithm can be parallelized. We will first start with the mainLoop given in Figure 1. At first glance this loop seems to be ideal for the parallelization of an object based data distribution. However, the parallelization of the loop as it stands is very difficult. The problem lies in the link between the two function calls getNearestMean and recalculateMean. Once an object is selected to be processed all three function calls inside the main loop have to be finished before another object can be considered. Therefore an easy parallelization of the mainLoop is not possible. The h-means algorithm given in Figure 1 does not have this problems and is therefore much easier to parallelize. Every processor can, independently of all other processors, find the

```
function K-MeanMainLoop {                      function H-MeanMainLoop {
    assign each object randomly to one cluster;    assign each object randomly to one cluster;
    do {                                           do {
        for each object t in the database {            for each object t in the database {
            nC = getNearestMean(t);                        nC = getNearestMean(t);
            insertIntoCluster(t,nC);                       insertIntoCluster(t,nC);
            recalculateMeans(t,nC);                    }
        }                                              recalculateMeans();
    } while at least one t changes its cluster     } while at least one t changes its cluster
}                                              }
```

Fig. 1. Pseudo code of the k-means and h-means algorithm.

nearest clusters for all local objects. Once the cluster membership for every object is defined, the mean value and the membership values have to be globally update. This modified version has exactly the same properties as the sequential algorithm given in Figure 1.

4   Experiments

In order to test the performance we conducted a series of tests. The environment we used consists of a network of 32 PC connected through a 10 MBits Ethernet. This is a very simple environment, but fairly realistic for many of the application environments we are focusing on.

In order to get real timings we used some reference data sets from the *Machine Learning Database Repository at UC Irvine* as well as artificially generated ones. To present the timings here we use a database with the following characteristics: 20 continuous attributes, 100'000 objects in 20 clusters.

If we execute the program "as it is" on one PC the average execution time for the clustering is 528.7 seconds. This time can not be directly compared to timings for the parallel version running in an environment of two and more nodes. Every node could keep its part of the data in memory. In order to get a usable reference time we added memory to one PC to allow it to keep the whole database in memory. The execution time under these conditions goes down to 343.5 seconds. This time can now be used as reference time for the parallel timings.

In order to measure the execution time in a parallel/distributed setting, first we have to distribute the data to all the processors participating in the clustering. In the real world scenario this distribution would be given. Here we installed the data that was processed by each node on its local disk. This distribution is done by the underlying object store [8]. The time for this operation is not included in the timings presented here.

The first measures of interest are of course the overall execution times[1] of the application. For the one node configuration we measured the timings for the standard machine (528.7 secs) and the machine with the augmented memory (343.5). With an increasing number of nodes the speedup slowly degrades, but

---

[1] The timings are the mean values over 100 runs.

even with 32 nodes we still have 90% efficiency. This is reasonable with respect to the slow interconnection network we are using. The increase in time consumption we are measuring after adding nodes is not only related to the increase in communication overhead, but is also related to the variations in the execution times of the different processors. These variations are relatively high and are very hard to control largely because of OS limitations (Windows 95) that do not allow us to control many of the parameters we would like to.

The previously presented results are representative for all the other tests we conducted. Because of space constraints we did not add other similar results. Overall, the results are very satisfactory for the given environment.

5   Conclusion and Future Work

We have presented a parallel version of the k/h-means clustering algorithm. The algorithm is designed to be used on large distributed data sets. Even on a very simple distributed computing environment, namely a PC cluster on a 10 MBits Ethernet, we are able to achieve about 90% efficiency for a configuration up to 32 processors. These results show that parallel k/h-means is scalable and thus enlarges its field of application to clustering tasks where it would be the preferred algorithm, but the task's computational complexity previously made it impossible

The basic algorithm presented here can be used in conjunction with broad variety of other settings. We are currently using this algorithm in a system that handles a wide range of classification problems. The parallel algorithm presented was extended to handle, not only continuous, but also categorical data. We are currently working on a version that would allow us to combine both types of data.

References

[1] M.R. Anderberg. *Cluster Analysis for Applications*. Academic Press, 1973.
[2] John A. Hatigan. *Clustering Algorithms*. John Wiley and Sons, 1975.
[3] W. Kloesgen and J.M. Zytkow. Knowledge discovery in database terminology. *Advances in Knowledge Discovery and Data Mining*, pages 573–592, 1996.
[4] J.B. MacQueen. Some methods for classification and analysis of multivariate observations. *Proceedings of the 5th Berkeley Symposium on Mathematical Statistics and Probability*, 1967.
[5] C.F. Olson. Parallel algorithms for hierarchical clustering. *Parallel Computing*, 21, 1995.
[6] E.M. Rasmussen and P. Willett. Efficiency of hierarchical agglomerative clustering using the icl distributed array oricessor. *Journal of Documentation*, 45(1), 1989.
[7] Helmuth Spaeth. *Cluster Analysis Algorithms*. John Wiley and Sons, 1980.
[8] Kilian Stoffel. Pattern matching in time series. Technical Report University of Neuchâtel, September 1998.

Performance Analysis for Parallel Generalized Association Rule Mining on a Large Scale PC Cluster

Takahiko Shintani, Masato Oguchi, and Masaru Kitsuregawa

Institute of Industrial Science, The University of Tokyo
7-22-1 Roppongi, Minato-ku, Tokyo, 106 Japan
{shintani,oguchi,kitsure}@tkl.iis.u-tokyo.ac.jp

Abstract. One of the most important problems in data mining is discovery of association rules in large database. We had proposed parallel algorithms for mining generalized association rules with classification hierarchy. In this paper, we implemented the proposed algorithms on a large scale PC cluster which consists of one hundred PCs interconnected by an ATM switch, and analyzed the performance of our algorithms using a large amount of transaction dataset. Performance evaluations show our parallel algorithms are effective for handling skew for such large scale parallel systems.

1 Introduction

Association rule mining is one of the most important problems in data mining. Association rule is the rule about what items are bought together within the transaction. Usually, the classification hierarchy over the data items is available. Users are interested in generalized association rules that span different levels of the hierarchy, since sometimes more interesting rules can be derived by taking the hierarchy into account[1, 2]. In our previous study, we proposed parallel algorithms for mining generalized association rules and evaluated their performance on 16-node shared-nothing parallel machine[3].

Recently, PC (Personal computer) clusters have become a hot research topic in the field of parallel and distributed computing. PC performance is increasing incredibly rapidly these days and the price of PCs remains inexpensive compared with that of workstation. The PC clusters are very promising platform for massively parallel processing. We developed a PC cluster system consisting of 100 PCs for parallel processing.

In this paper, we implement the parallel algorithms for mining generalized association rules proposed in [3] on a cluster of 100 PCs interconnected with an ATM network, and analyze the performance of our algorithms using a large amount of transaction dataset. In [5], the parallel algorithms for mining flat association rules are experimented on 128 processor system. However, the transaction data is not read from actual disk in the experiments. In that experiments, the small transactions are kept in the buffer, and the transactions are read from

P. Amestoy et al. (Eds.): Euro-Par'99, LNCS 1685, pp. 1455–1459, 1999.
© Springer-Verlag Berlin Heidelberg 1999

the buffer instead of the actual disks. Moreover, the size of transaction data is 50MBytes. On the other hand, the transactions are read from the actual disk and used a large amount of transactions (1GBytes).

2   Parallel Generalized Association Rule Mining

Here, we introduce some basic concepts of generalized association rules, using the formalism presented in [1]. Let $\mathcal{T}$ be a classification hierarchy on the items, which organize relationships of items in a tree form. In $\mathcal{T}$, a node is item and an edge represents *is-a* relationship. Let $\mathcal{D} = \{t_1, t_2, \ldots, t_n\}(t_i \subseteq \mathcal{I})$ be a set of transactions, where each transaction is constructed with set of items and unique identifier. We say a transaction t *contains* a set of items X, if X is in t or is an ancestor of some item in t. A k-itemset is the set of k items. The itemset X has *support* s in the transaction set $\mathcal{D}$, if $s\%$ of transactions in $\mathcal{D}$ contain X, here we denotes $s = sup(X)$. An *generalized association rules with classification hierarchy* is an implication of the form $X \Rightarrow Y$, where $X, Y \subset \mathcal{I}$, $X \cap Y = \phi$ and no item in Y is an ancestor of any item in X. Each rule has two measures of value, *support* and *confidence*. The *support* of the rule $X \Rightarrow Y$ is $sup(X \cup Y)$. The *confidence* c of the rule $X \Rightarrow Y$ in the transaction set $\mathcal{D}$ means $c\%$ of transactions in $\mathcal{D}$ that contain X also contain Y, which can be written as the ratio $sup(X \cup Y)/sup(X)$. The procedure to find the rules is as follows: (1)Find all itemsets that have support above user-specified minimum support. These itemsets are called the *large itemsets*. (2)For each large itemsets, derive all rules that have more than user-specified minimum confidence.

2.1   Parallel Algorithms

In this section, we describe our parallel algorithms for finding all large itemsets on shared-nothing environment proposed in [3].

Non Partitioned Generalized association rule Mining : NPGM

NPGM copies the candidate itemsets over all the nodes. Each node can work independently. The procedure to find large k-itemsets is as follows: (1)Each node generates the candidate k-itemsets (C_k) using the large $(k-1)$-itemsets. If k is 2, delete the candidates that contains an items and its ancestor. (2)Each node reads the transaction database from its local disk, generates extended transaction t' by adding all ancestors of the items in a transaction t that are present in $\mathcal{T}$ to t'. Increment the support count of all candidates in that are contained in t'. (3)After reading all the transaction data, all node's support count are gathered into the coordinator node and checked to determine whether the minimum support condition is satisfied or not.

Hash Partitioned Generalized association rule Mining : HPGM

HPGM partitions the candidate itemsets among the nodes using a hash function like in the hash join, which eliminate broadcasting. The procedure is as follows: (1) Each node generates the candidate in the same way as NPGM. Apply the hash function to the candidates in C_k and determine the destination node

ID. If the ID is its own, hold it into main memory(C_k^n). (2) Each node generates the extended transaction t' in the same way as NPGM. Generate k-itemsets from t' and apply the same hash function used in phase 1. Derive the destination node ID and send the k-itemset to it. For the itemsets received from other nodes and those locally generated whose ID equals the node's own ID, the support is checked. If hit, increment its support count value. (3) After reading all the transaction data, each node can determine individually whether each candidate in C_k^n satisfy minimum support or not. Each node send L_k^n to the coordinator node, where all the large k-itemsets $\mathcal{L}_k := \bigcup_n L_k^n$ are derived.

Hierarchical HPGM : H-HPGM

H-HPGM partitions the candidate itemsets among the nodes taking the classification hierarchy into account so that all the candidate itemsets whose root items are identical be allocated to the identical node, which eliminates communication of the ancestor items. Thus the communication overhead can be reduced significantly compared with original HPGM. The procedure is as follows: (1) Each node generates the candidate itemsets in the same way as NPGM. Apply the hash function to the candidate itemsets in $\mathcal{C}_k$. Here each item of the candidate itemset is replaced by its root items, then hash function is applied and destination node ID is determined. If the ID is its own, hold it into the hash table(C_k^n). (2) Each node reads the transaction database from its local disk and generates extended transaction t' by replacing the item in t with the large item in its ancestors which is closest to the bottom, if there are small items. For each node n, select all items in t' whose root item is allocated to n-th node and send them to n-th node. For the itemsets received from other nodes and those locally generated whose root item is allocated to own, generate k-itemset from the itemsets and increment the support count value of this k-itemset and its all ancestor candidates. (3) Same as in HPGM.

H-HPGM with Fine Grain Duplicate: H-HPGM-FGD

In the case the size of the candidate itemsets is smaller than available system memory, H-HPGM-FGD utilizes the remaining free space. H-HPGM-FGD detects the frequently occurring itemsets which consists of the any level items. It duplicates them and their all ancestor itemsets over all the nodes and counts the support count locally for those itemsets like in NPGM. The procedure is as follows: (1)Count up the number of candidates allocated to each node by generating the k-itemsets using $\mathcal{L}_{k-1}$. Sort the large items based on their count support value. Choose the first most frequently occurring candidate itemsets, and insert them and their all ancestor candidates to C_k^D so that free space be occupied as much as possible. Delete the candidates in C_k^D from $\mathcal{C}_k$. Hold the candidates in $\mathcal{C}_k$ if its root itemset's hashed value is corresponding to its own node ID(C_k^n). (2) Each node generates the extended transaction t' in the same way as H-HPGM. Increment the support count of C_k^D that are contained in t' and send the itemsets in the same way as H-HPGM. (3) C_k^D are checked the support condition in the same way as NPGM, and C_k^n are checked in the same way as H-HPGM.

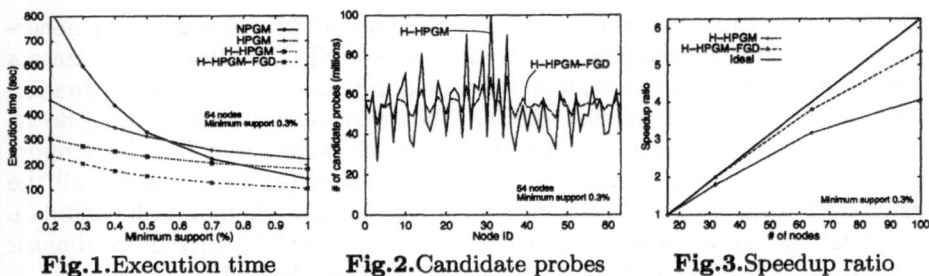

Fig.1. Execution time   **Fig.2.**Candidate probes   **Fig.3.**Speedup ratio

3   Performance Analysis

We implemented all the above algorithms in PC cluster system. Our PC cluster system consists of one hundred commodity PCs, connected with ATM switch[4]. Each PC has Intel Pentium Pro 200MHz CPU, 4.3GB SCSI hard disk and 64MB main memory.

To evaluate the performance , synthetic dataset emulating retail transactions is used. The generation procedure is described in [1]. The following are the parameters of the generated synthetic transaction data: (1)the number of items is 50,000, the number of roots is 100, the number of levels is 4–5, fanout is 5, (2)the total number of transactions is 20,000,000(1GBytes), the average size of transactions is 5, and (3)the number of potentially large itemsets is 10,000. The transaction database is evenly spread over the node's local disks.

We show the execution time at pass 2 of all parallel algorithms varying the minimum support in Figure 1. The execution time of all the algorithms increases when the minimum support becomes small. When the minimum support is small, the candidate partitioned methods can attain good performance. H-HPGM-FGD significantly outperform other algorithms.

Next, the workload distribution of H-HPGM and H-HPGM-FGD is examined. Figure 2 shows the number of candidate probes to increment the support count in each node at pass 2. In H-HPGM, the distribution of the number of probes is largely fractured, since the candidate itemsets are partitioned in the unit of hierarchy of the candidate itemsets. H-HPGM-FGD detects the frequently occurring candidate itemsets and duplicate them. The support counting process for these duplicated candidate itemsets can be locally processed, which can effectively balance the load among the nodes.

Figure 3 shows the speedup ratio with varying the number of nodes used 16, 32, 64 and 100. The curves are normalized by the execution time of 16 nodes system. H-HPGM-FGD attains higher linearity than H-HPGM. Since H-HPGM duplicates no candidate itemsets, the workload skew degrades the linearity. The skew handling methods detect the frequently occurring candidate itemsets and duplicate them so that the remaining free memory space can be utilized as much as possible. In Figure 3, H-HPGM-FGD achieves good performance on one hundred nodes system.

4   Conclusions

In this paper, we implemented parallel algorithms for mining generalized association rules on large scale PC cluster system and analyzed their performance.

We examined the effectiveness of parallel algorithms on large scale parallel computer system using the large amount of transaction dataset. Our system is consisted with one hundred of PCs. As far as the authors know, there has no research on parallel data mining over such large scale systems using a large amount of transaction dataset. Through several experiments, we showed H-HPGM-FGD could attain sufficiently high performance and achieve good workload distribution on one hundred PC cluster system.

References

[1] R.Srikant, R.Agrawal: Mining generalized association rules. Proc. of VLDB (1995)
[2] J.Han, Y.Fu: Discovery of multi-level association rules from large databases. Proc. of VLDB (1995)
[3] T.Shintani, M.Kitsuregawa: Parallel algorithms for mining generalized association rules with classification hierarchy. Proc. of ACM SIGMOD (1998)
[4] T.Tamura, M.Oguchi, M.kitsuregawa: Parallel database processing on a 100 node PC cluster: cases for decision support query processing and data mining. Proc. of SC97:High Performance Networking and Computing (1997)
[5] E.-H.Han, G.Karypis, Vipin Kumar: Scalable parallel data mining for association rules. Proc. of ACM SIGMOD (1997)

Inducing Load Balancing and Efficient Data Distribution Prior to Association Rule Discovery in a Parallel Environment

Anna M. Manning and John A. Keane

Department of Computation, UMIST, Manchester, M60 1QD, UK
anna@crimson1.demon.co.uk, jak@sna.co.umist.co.uk

Abstract. Many association rule algorithms operate in a parallel environment where the database is divided up among a number of processors, a procedure which is usually carried out indiscriminately. The nature of the database partitioning can affect both the number of candidate sets produced and the workload at each processor. This paper demonstrates that Principal Component Analysis can be used successfully to help arrange the records of a database among processors so that efficient load balancing is enabled and candidate set duplication minimised.

1 Introduction

The discovery of association rules is an important example of data mining [1]. We consider an algorithm called Distributed Mining of Association rules (DMA) [5] which is outlined briefly in section 2. DMA performs better the greater the inter-site record pattern variation as this can result in fewer duplicate candidate itemsets which in turn can reduce the size and number of messages passed between sites [4, 5]. It is possible to induce non-uniform record distribution before applying DMA by considering a centralised database which has been partitioned across the nodes of a parallel system as this gives scope for selecting the destination for data records; an efficient method that provides good predictions of attribute patterns within the database must be employed to achieve this. Our work investigates the application of Principal Component Analysis (PCA) [6] for this purpose where PCA is a statistical tool which can identify significant variation in data, representing it in the form of linear functions (Principal components, PCs) of the data variables.

2 The DMA Algorithm

The DMA algorithm [5] is based on the Apriori Algorithm [3] and distinguishes between itemsets that are frequent at any one site (*locally frequent*) and those that are frequent across the whole database (*globally frequent*). An itemset that is both locally and globally frequent is termed *heavy*. DMA finds heavy itemsets as follows::

P. Amestoy et al. (Eds.): Euro-Par'99, LNCS 1685, pp. 1460–1463, 1999.
© Springer-Verlag Berlin Heidelberg 1999

(1) For k=1 the candidate sets at each site consist of the single attributes. For k>1 the candidate sets at each site are generated from the heavy itemsets found at the same site during the previous iteration .

(2) Local support counts for the candidate sets are computed at each partition together with the supports of the candidate sets at the other sites.

(3) Candidate sets that are not locally frequent at their site are pruned.

(4) Each site gathers support counts for all of its locally frequent itemsets from each of the other sites and computes its heavy k-itemsets.

(5) The heavy k-itemsets are broadcast to all other sites.

3   Proposed Method for Record Distribution

(1) Find (or estimate) the variance/covariance matrix Z for the attribute column means in the database and apply PCA to Z. Choose the required number of processors, S and associate each with one of the S PCs with the highest eigenvalues.

(2) Arrange the attribute weights (taking absolute values) contained within each PC in descending order, associating each with a numerical label that represents its original position in the list. Assess the number of attributes that are most significant in generating the variance for each of the S PCs (with the help of a graph if necessary) and calculate the average. Compose rules for each processor that aim to select records which contain the prominent attributes of its allotted PC. Intra-processor variance can be further enhanced by attempting to force zero support for at least one attribute at each, the exact number of attributes considered depending on support count levels and attribute correlations. If the 3 most prominent attributes for a particular PC are 3, 10, 7 and support levels and correlations are high the rules will aim to maximise the number of records at this processor containing attribute numbers 10, &and to eliminate those containing 3 as follows:

   1 .   add any record with a 1 for attribute3 and a 0 for attribute 3
   2 .   add any record with a 1 for attribute10 and a 0 for attribute 3
   3.   add any record with a 0 for attribute 3
   4.   add any record

During the first pass of DMA the records are taken one at a time and compared against the rules at each processor beginning with the one associated with the weakest PC and progressing to the next strongest in order to encourage load balancing. If the record does not match the first rule at any processor the second rules are considered and so on until a match is found. The local support contribution of the matched record is then added to the total stored at its allotted processor.

At the end of pass 1 the data will have been redistributed and the 1-itemset supports for each processor calculated without an additional pass over the data. The DMA algorithm can then proceed as normal with pass 2. Binary databases were randomly generated to test the above method using three levels of sparsity; these were referred to as sparse data (approx. 25% 1s), middle data (approx.

50% 1s) and dense data (approx. 75% 1s). Each had 60,000 records with 10 attributes and were referred to as *raw* data. Each raw database was distributed over 3, 4 and 6 processors using the above PCA process to form *new* data as well as being divided over 3, 4 and 6 processors using a straightforward slicing process for comparison purposes.

4   Results

The average reduction in numbers of candidate sets together with it value as a percentage of the total are shown in Figure 2 with the details for the case of middle data over 4 processors in Figure 1. The general reduction in candidate sets, irrespective of data sparsity and processor numbers can be explained by the more varied nature of the supports for the itemsets in the new data which is directly attributable to the requirement that each PC must maximise its own variance independently of each of its predecessors.

σ	5%	10%	15%	20%	25%	30%	35%	40%	45%	50%
# Raw candidate sets	627	375	165	165	58	45	45	45	45	6
# New candidate sets	442	259	132	120	44	36	36	36	36	3
Improvement	185	116	33	45	14	9	9	9	9	3
% Change	29.5	30.9	20	27.3	24.1	20.0	20.0	20.0	20.0	50.0

Fig. 1. Candidate set reduction for middle data over 4 processors

Candidate set reduction is greatest for the sparse data no matter how many processors are used as the difference in frequent itemsets for the new and raw data is significant for a greater proportion of support thresholds.

3 processors			4 processors			6 processors		
sparse	middle	dense	sparse	middle	dense	sparse	middle	dense
21.6 sets	29.5 sets	15.5 sets	26.4 sets	43.2 sets	28.3 sets	24.8 sets	41.7 sets	27.9 sets
34.0%	17.2%	2.0%	45.6%	26.2%	11.9%	42.4%	26.2%	12.5%

Fig. 2. Average reduction in candidate set numbers

Raising the number of processors from 3 to 4 or from 3 to 6 improves the results for all levels of sparsity. With fewer processors there are less potential destinations for records and it would appear that the effect of the PCA process is diluted by those records which do not match any rules of high priority but which need to be placed somewhere.

5   Conclusions and Further Work

The PCA data redistribution technique involves two payoffs for the application of DMA to a centralised database: load balancing and candidate set reduction. Both payoffs are useful for other parallel algorithms that are based on Apriori [3], e.g. Count Distribution Algorithm [2]. Candidate Distribution Algorithm [2], benefits from the prior knowledge about attribute relationships as candidate sets can be allocated to processors without the need for the study of connected graphs. Records matching these candidate sets can also be placed at the corresponding processors before the algorithm is applied which results in the sending of fewer data pages during execution.

Further work is required to test the PCA process on wider variations of sparsity, larger numbers of attributes, larger databases and on other parallel algorithms.

References

[1] R.Agrawal, T.Imielinski and A.Swami, 'Mining association rules between sets of items in large databases', SIGMOD 93, pp 207-216, 1993.

[2] R.Agrawal and J.C. Shafer, Parallel mining of association rules, IEEE Transactions on Knowledge and Data Engineering, 8(6), pp962-969, 1996.

[3] R.Agrawal and R.Srikant, 'Fast algorithms for mining association rules in large databases', VLDB-94, pp 487-499, 1994.

[4] D.W.Cheung and Y.Xiao, Effect of Data Skewness in Parallel Mining of Association Rules, PAKDD-98, pp 48-60, 1998.

[5] D.W.Cheung, V.T.Ng, A.W.Fu and U.Fu, 'Efficient Mining of Association Rules in Distributed Databases', IEEE Transactions on Knowledge and Data Engineering Vol.8 No.6 pp 911-922, 1996.

[6] I.T.Jolliffe, 'Principal Component Analysis', New York : Springer (Springer series in statistics), 1986.

Topic 23
Symbolic Computation

Mike Dewar

Local chair

The field of symbolic computation covers a wide range of areas including computer algebra, theorem proving and logic programming. Applications which arise in these areas often lead to extremely large and long computations, so it is natural to look to parallelism as a means to deal with this. However the environments used for symbolic computation are very different from those used for numerical computation, and thus it is not always easy to benefit directly from advances in what might be regarded as main-stream high performance computing. Typically these are interactive systems with their own high-level language, and where many issues which can have a critical impact on performance (such as memory management) are handled automatically. Although attempts have been made to produce parallel versions of such systems these have not been very successful.

In this session we have four papers, each of which addresses some of the issues outlined above. In their paper *A Library for Parallel Modular Arithmetic*, Bradford & Power outline a design for a library implementing a novel kind of parallel arithmetic. Gautier & Mannhart, in *Parallelism in Aldor — the communication library Π^{it} for parallel, distributed computation*, describe an implementation which uses standard message-passing libraries (such as MPI) with an existing symbolic language. Both these systems provide environments in which to implement and experiment with more sophisticated algorithms.

The papers by Calegarion & Dutra (*Performance Evaluation of Or-Parallel Logic Programming Systems on Distributed Shared-Memory Architectures*) and Matooane & Norman (*A Parallel Symbolic Computation Environment: Structures and Mechanics*), look in detail at the behaviour of particular symbolic computations on existing hardware. Such detailed work provides valuable insights to those who wish to combine parallelism and symbolic computation.

P. Amestoy et al. (Eds.): Euro-Par'99, LNCS 1685, pp. 1465–1465, 1999.
© Springer-Verlag Berlin Heidelberg 1999

Parallelism in Aldor — The Communication Library Π^{it} for Parallel, Distributed Computation

Thierry Gautier[1] and Niklaus Mannhart[2]

[1] INRIA, project APACHE,
100 rue des Mathématiques, B.P. 53, F-39041 Grenoble Cedex 9, France
Thierry.Gautier@inrialpes.fr
http://www-apache.imag.fr
[2] Institute for Scientific Computing
ETH Zentrum, IFW D25.2, 8092 Zürich, Switzerland
mannhart@inf.ethz.ch
http://www.inf.ethz.ch/personal/mannhart

Abstract. In this paper, we present the design of Π^{it}, an ALDOR library to express parallel programs. ALDOR is a general purpose programming language designed for computer algebra and Π^{it} provides an ALDOR low-level interface that interacts with hardware or system tools in order to express parallelism. Additionally, Π^{it} provides an API that hides any low-level details such as sending messages, creating threads and provides an interface for data parallelism. This paper presents our design decisions and our implementation as well as examples of how easy ALDOR programmers can implement parallel algorithms in a high-level abstract way with Π^{it}.

1 Introduction

During the last decade, an important thread of research in the area of computer algebra has addressed questions of parallelism: parallel analysis of the complexity of problems, development of parallel algorithms and the design/implementation of parallel systems.

Currently few of these systems are in use for general purpose computer algebra problems. Even if main challenges (data management, tasks scheduling and interface design) in the implementation of a computer algebra system have solutions, most of these systems have been depreciated due to the high costs of development necessary to ensure portability of sources with good efficient programs at runtime.

This paper deals with the design of Π^{it}, an ALDOR library for parallel computation. By reducing the number of functions needed for portability and by using standard tools (MPI and a POSIX thread kernel, ALDOR) we believe that the library is portable across different architectures. Moreover, by using a high-level programming interface with a scheduling algorithm, we ensure portability and efficiency.

P. Amestoy et al. (Eds.): Euro-Par'99, LNCS 1685, pp. 1466–1475, 1999.
© Springer-Verlag Berlin Heidelberg 1999

One key point in implementing a parallel computer algebra system is the ability to reuse a sequential computer algebra kernel that will be plugged in to a library supporting parallelism (thread kernel, data communication). The key point of this method is to have an efficient integration for communication of data exchange without a costly overhead due to conversion of representation. Moreover, due to the irregularity of computer algebra programs and for portability on various architecture, a parallel computer algebra system should be able to:

- *manage parallelism* such as task creation, communication, shared memory etc.
- have a *dynamic scheduler* as most of the task dependencies are only known at runtime.
- automatically free non-used objects with a *garbage collector*.

In the last ten years, several experimental systems has been developed in order to experiment with parallel computer algebra algorithms on various parallel architectures. PARSAC-2[16] is a parallel version of the SAC-2[12] Computer Algebra system, based on S-threads for shared memory systems and a fork-join paradigm. A distributed version called DTS (Distributed Thread System)[2] has also been implemented. A first version of the C++ library for computer algebra called GIVARO [15, 14] has been developed using the fork-join paradigm. Currently GIVARO relies on a new version of ATHAPASCAN-1 [13] that permits easy expression of synchronizations between tasks by putting annotation of access made on objects into a (virtual) shared memory (that enables at runtime the unfolding of the data flow graph). Moreover, using ATHAPASCAN-1, the scheduling is a plug-in that could be specify by annotation without any change to the source code.

Sugarbush[6] is a parallel symbolic computation system based on the computer algebra system Maple and communicates a textual representation through the tuple space provided by Linda. Some applications based on the system have been reported [5]. DSC[9, 4] is an environment for distributed symbolic computation. A DSC program distributes code and the corresponding input data file over the net. DSC is concerned more with the dynamic load balancing mechanism than with the efficiency of basic parallel constructs (communication, process creation). Nevertheless, some experiments with coarse grain applications such as sparse linear solving, have shown good results using network of workstations [8].

There are many other parallel systems such as ||Maple||[20] (parallelism is expressed using Horn clauses and a prototype has been implemented on shared memory machine), PACLIB/pD[19] (parallel programming using fork-join paradigm and message passing on shared or distributed memory architecture) and MuPAD[17] (paradigm based on message passing and global object – remote read/write – implemented on a distributed memory architecture). An overview can be found in [18].

The goal of Π^{it} is to extend, as a library, the general purpose language ALDOR for parallel programs. The library should be *portable* and *efficient* on the machine it runs on. It should provide a basis for a high-level programming interface with a dynamic load balancing mechanism. Currently, we propose such an interface based on a data parallelism paradigm.

2   Design of Π^{it}

The target parallel architectures for the Π^{it} library are networks of workstations. One workstation could be a symmetric multi-processors (SMP) machine. The library is written in the programming language ALDOR and C. The interface between ALDOR and the parallel primitives provided by the operating system or a library are written in C. This *runtime* system ensures the portability and the efficiency of the whole library. It is viewed as a minimal parallel extension to FOAM [22, 1] (First Order Abstract Machine: target machine of the ALDOR compiler) and defines a set of functions that are callable by ALDOR. The interface can be implemented in a straight-forward fashion on top of existing libraries for communication and thread management such as ATHAPASCAN [7] and MPI[10] plus a thread package.

The Π^{it} library itself is based on three layers:

level 0 contains the C-ALDOR interface. It contains the basic categories and domains for low-level access to the C runtime system. We refer to this interface as the low-level interface of Π^{it}. Its role is to be a portable low-level library for parallel programming in ALDOR.

level 1 offers both the control and the data parallel programming paradigms with explicit mapping of tasks among processors. This layer is composed of a set of ALDOR domain and category definitions and it is based on the low level interface of Π^{it}. Communication by stream and (virtual) shared memory, thread creation and synchronization are part of this interface.

level API provides a parallel application programming interface that hides low level details such as synchronization and mapping of tasks of previous levels. This interface is based on data parallelism paradigm and a general scheduler that maps the created tasks to nodes automatically.

2.1   Level 0: Foam Extensions to Aldor

The C *runtime* of the Π^{it} is a set of new definitions of the FOAM functions and types for parallel computations, with an interface that respects the ALDOR C calling convention [21]. The level 0 – ALDOR interface maps the C *runtime* of Π^{it} interface. It defines the ALDOR domain "Π^{it} abstract machine" in the same way as the definition of the FOAM interface by the ALDOR domain "machine".

System environment composed of the definition of a set of variables: The number of processors and a unique identifier (an integer) for each *node* (a node is an address space (unix process)).

Data format of primitive FOAM data types for communications. The format of data is close to the definition of the MPI[10] or ATHAPASCAN-0b[3] data type. FOAM basic types have a format definition in Π^{it} and complex data types in

ALDOR are explicitly (see section 2.2) translated to a set of basic format in a recursive manner.

Message interface is responsible for communicating between nodes using a *port* number. A message is a continuous memory region of basic data types or a buffer. To send or receive data, a message is posted and a request is returned in order to test the completion of the communication, i.e. it is an asynchronous operation and the caller is not blocked. After the completion of the operation the data could be (re-)used. Messages sent to a node and port identifiers are buffered.

Light-weight processes or threads, for low cost management of new thread of control with synchronization objects. The chosen interface is close to a subset[1] (thread creation, mutex and condition variables) of POSIX 1003.1c (PThread) without functions to control or change attributes of scheduling policies.

Remote service requests (close to active messages in [11]) to invoke functions remotely. Basically a remote service request is a message and a handler (a function). The given function is called with the data packed in the message. The function is executed like an interruption of the C runtime support. It cannot block but it can post messages. This feature is used to implement one-sided communications (remote read/write, remote creation of thread).

Our implementation[2] of the C runtime is based on top of ATHAPASCAN-0b[3][3] which is a portable C library that offers a distributed thread system which communicates via a network. Moreover, our choice in the set of functions for the portability of the whole library permits us to remove about 50% of the number[4] of functions provided by ATHAPASCAN-0.

Unfortunately due to limitations on the current implementation of the FOAM-C library, the current implementation has one and only one thread of computation that executes an ALDOR function. We hope to change this implementation in the near future when a thread safe version of the ALDOR runtime system is available.

2.2 Level 1 - Communication Interface

Objects that can be sent over the net are variables of domains of the communicable category (ComType). Most of the basic domains in ALDOR library have been extended to be of this type of category. For example, the integer domain Integer is extended to be of category ComType with

```
extend Integer: ComType == add {
    write(s:OutStream, v:%):() == {...};
    read(s:InStream): % == {...};
}
```

---

[1] As defined in ATHAPASCAN-0b.

[2] Note we also have an other implementation directly based on MPI plus a Posix thread kernel.

[3] This library is based on MPI and a Posix Thread kernel

[4] Approximately 40 functions in level 0 should be compared to 100 functions in ATHAPASCAN.

Now, the programmer can use an integer value as arguments to functions which could to be called remotely. Hence he or she does not have to change existing code with a new type, i.e. `ComInteger`.

In addition, the level 0 − ALDOR provides a stream interface and a distributed shared memory (without management of copies) built on top of the message interface and the remote service request. With respect to messages in the level-0 layer, identification of streams is done by using a node and a port number that is globally unique. Operations onto stream are blocking.

A *distributed shared memory* domain provides remote write/read operations and lock/unlock onto shared ALDOR objects. The consistency of shared objects is up to the responsibility of the user of these functions. At the creation, the shared object could be allocated locally or on a remote node. A shared object is a reference into a (virtual) shared memory. A read or write operation of a shared object located on a remote processor is a communication: a copy is made into or from a local ALDOR variable. Due to a strong interaction[5] between the high-level parallel programming interface and the distributed garbage collector of unused objects. In this level-1 layer all shared reference should be explicitly destroyed because there exists no distributed garbage collector.

The following example illustrates the use of streams to communicate a shared object (the reference) followed by the read operation to get its value. This portion of code is supposed to run on a node and communicate with node 0 (per definition the master node).

```
-- read a reference to a shared object from a new input stream
-- connected to node 0
inS : InStream := create( 0, portid);
ref:  Shared(String) := read(inS);
close!(inS);
-- Take an exclusive access on the shared object
lock(ref);
-- read the value pointed by the reference to the shared object
val: String := read(ref);
<Do an update on ref ..>
-- Unlock the access to ref
unlock(ref);
```

2.3   Level 1 - Thread Creation

The Π^{it} level 1 parallel programming interface is also based on the fork/join scheme of computations. A thread of control can fork a *task* and wait until its completion. Several threads inside a node share the memory of the (Unix) process.

The following example shows the use of fork/join on local node.

```
-- function f
f(arg:Integer, res: Shared(Integer) ): ()== {
   ...write(res, value);...
}
```

---

[5] For instance, the scope of an object is given by the semantics at the high-level constructs.

```
-- initialization of the the join variable
jv:JoinVar := create();
r1, r2: Shared(Integer); r1 = ...; r2 = ...
fork(jv, f, 2);  -- span first task  -> f(2)
fork(jv, f, 4);  -- span second task -> f(4)
...              -- do some more work
-- wait for the completion of f(2) and f(4)
wait jv;
destroy!(jv);
destroy!(r1); destroy!(r2);
```

Note, the join variable jv can be used as a synchronization variable for many fork operations. The result of a function should be passed as a reference to an object in its arguments.

The remote creation of thread is a communication. The Π^{it} level 1 library does not implement it, but it provides the basic mechanism as a remote service request. A service is a function that must be registered to the Π^{it} library. A remote service request is composed of the identifier of the function and its effective parameters. This request is packed into a buffer, before a message is sent over the network. At the receiving node, the message is unpacked and the Π^{it} runtime system calls the associated function with its effective parameters.

2.4   Level API

The common idea behind parallel programming paradigms is to abstract the need of writing explicit synchronizations. Using a message passing interface (e.g. PVM, MPI, BSP), the descriptions of synchronizations are given at the level of remote data accesses.

Within the *fork/join* paradigm, the computations are split into tasks following the dependencies provided by the user between data accesses. Therefore, synchronization is expressed at the level of call of functions rather than at the level of data accesses. The *functional parallelism* paradigm is based on the fact that a purely functional program does not make side-effects and hence is relatively easy to split into several tasks that can be executed in parallel. The computation of synchronizations could be achieved automatically by a compiler. Nevertheless, some experiences have shown that it is difficult to have good efficiency by using automatic parallelization of programs. Therefore, code annotations have been defined in order to specify a good *grain of parallelism* for both data and computation. For instance, using high-order functions, standard control structures are abstract and could have efficient parallel implementations such as *data parallel* evaluation over collection (e.g. parallel map, parallel reduction).

name	definition	example
pmap	$(G_T, T \to S) \to G_S$	$([a,b,c,d], f) \mapsto [f(a), f(b), f(c), f(d)]$
preduce	$(G_T, (T,T) \to T, T) \mapsto T$	$([a,b,c,d], \oplus, n) \mapsto [a \oplus b \oplus c \oplus d]$
combine	$(G_T, G_S) \mapsto G_{T \times S}$	$([a,b,c], [1,2,3,4]) \mapsto [(a,1), (b,2), (c,3)]$

Table 1. Parallel collection functions in Π^{it}

In order to support data parallelism, Π^{it} provides an interface with parallel map *pmap*, parallel reduce *preduce* and *combine*. Table 1 shows the types of of pmap, preduce and combine. The mapping of $f(a), f(b), \ldots$ can be executed in parallel, assuming that the order of execution of applying f to the arguments $a, b, \ldots$ are independent. Hence f is not allowed to update any shared resources. Parallel reduce *preduce* works in a similar way: no order of the reduction is assumed. Parallel map as well as parallel reduce are implemented with ALDOR generators [21].

```
square(i:Integer):Integer == i * i;
-- register square
rsquare := register(square);
...
-- create our list 1, 2, 3, ... 10
gin:Generator(Integer) := generator(1..10);
-- apply rsaquare to 1, 2, 3, ... 10
gout:Generator(Integer) := pmap(rsquare)(gin);
for res in gout repeat {
    print << gout ;
}
```

In the above example, `gin` is a generator containing the elements $1, 2, \ldots 10$. The function `pmap(rsquare)` returns a function that has to be applied with a generator. As soon as we apply the created `gin` to the returned function, 10 threads are created and enqueued in the the ready task list. The function immediately returns an output generator containing the result. The results are printed in the for loop. If any result is ready, then it will be returned (i.e. in the above example printed), otherwise the task is suspended until a result has been calculated. Note, the order of execution is random (in the above example a possible output could be $81, 16, 4, 1, 64, 49, 100, 36, 25, 9$). The function preduce works in the same way. Again, the parallel/distributed execution is hidden from the user.

We are still experimenting with this layer in order to have a user friendly interface and an efficient implementation.

3   Experimentation of Level-0

Two kind of experiments were done in order to estimate the impact of our design and implementation relative to the maximum performance possible. Firstly, experiments were run to measure the loss of performance (latency, bandwidth) between the implementation of our low-layer C interface between underlying library of communication and the library of thread management. The second set of experiments show the impact of ALDOR to C calls that increase the latency.

Experiments were done with a IBM SP1.x[6] running AIX3.2.5 at the IMAG located in Grenoble, France. The C interface was based on top of ATHAPASCAN-0b which is based on MPI-F, the IBM implementation of MPI, and a Posix thread kernel. On this parallel machine the theoretical bandwidth is 40 MBytes/s, and

---

[6] Processors are Power1 and interfaces to high-speed network are TB2.

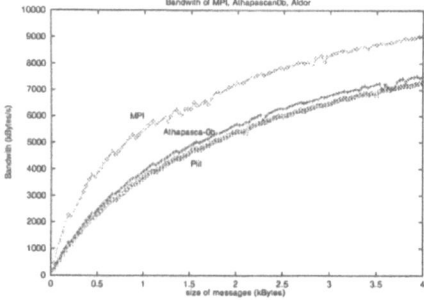
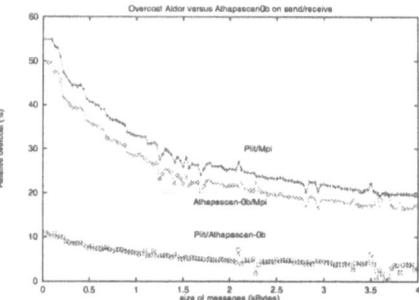

Fig. 1. Bandwidth on SP1.x at the three different layers of the design of Π^{it} (=Piit): MPI, ATHAPASCAN-0b and Π^{it} (at the ALDOR level). Small size messages.

Fig. 2. Over-cost due to Π^{it} (=Piit) compared to ATHAPASCAN-0b. Small size messages.

the practical obtainable is about 28 MBytes/s using MPI-F. Hence, we decided to compare our result using Π^{it} with both ATHAPASCAN and MPI-F.

For small sized messages (from 32 bytes to 4 kBytes), figures 1 and 2 shows the bandwidth and the over-cost of MPI, ATHAPASCAN-0b and Π^{it}. For 32 bytes size messages, the over-cost is 50% for ATHAPASCAN-0b versus MPI-F and 55% for Π^{it} versus MPI-F. This is explained by the increase of latency of implementation (more function calls to effectively attack communication primitives).

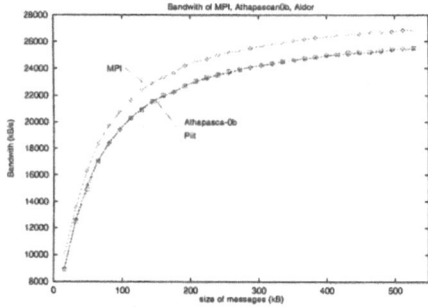
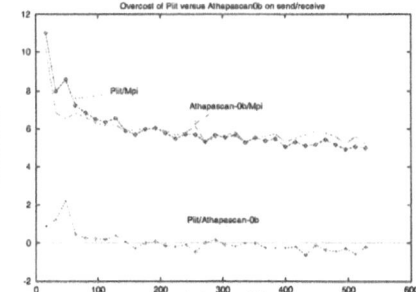

Fig. 3. Bandwidth on SP1.x at the three different layers of the design of Π^{it}: MPI, ATHAPASCAN-0b and Π^{it} (=Piit) (at the ALDOR level). Large size messages.

Fig. 4. Over-cost due to Π^{it} (=Piit) compared to ATHAPASCAN-0b. Large size messages.

Figure 3 shows medium-size messages (from 16kB to 512kB) for the three different layers used by our Π^{it} implementation. The program is a ping/pong data exchange. The Π^{it} program uses the low-level interface in ALDOR, which is a link to our low-level C interface based on ATHAPASCAN-0b. For $16kB$ size, bandwidths are $9.7MB/s$ with MPI-F, $8.8MB/s$ with ATHAPASCAN-0b and $8.7MB$ with Π^{it}. At the larger size $512kB$, this values are respectively about $27MB$ with MPI-F and 24.8 for both ATHAPASCAN-0b and Π^{it}.

Relative differences are showed in figure 4. Most of the loss in performance versus MPI-F comes at the ATHAPASCAN-0b layer: ATHAPASCAN-0b's bandwidth is less than those of MPI of about 11% for $16kB$ messages, and less than 6% for $512kB$ messages. These results are to be compared with less than 2% of loss in performance with Π^{it} versus ATHAPASCAN-0b.

Important results are: less than 20% of loss of performance of communication versus MPI-F are achieved for message of size 4 kBytes, and for big messages (more than 16kBytes) the loss of performance is at least 6%.

4   Conclusions and Future Work

We have defined a minimal interface for a portable library. We implemented the communication library Π^{it} for parallel, distributed programming in ALDOR. 95% of the library is written in ALDOR itself, 5% is written in the C language. Level 0 is inspired from ATHAPASCAN with an interface to fewer functions. The implementation of that interface has shown good efficiency of communication. On top of this level we defined a set of categories and domains for a natural integration in ALDOR. This layer implements only a few parallel programming paradigms, i.e. message passing, remote service requests, threads, shared memory. In order to overcome the difficulties that ALDOR programmers would run into by using level 1, we implemented an API based on data parallel programming model. Hence, the API user does not have to care about low-level details. Level 0 and 1 has been implemented and documented.

In the future we will test our API level for efficiency and easy of use before we are releasing the library. Target applications come from linear algebra (Gaussian elimination, rank computation,...) used in the $\sum^{it}$ library. These experimentations will permit us to tune a few schedulers on two main architectures: a network of workstations and a parallel computer with high-performance network such as the IBM-SP.

References

[1] Peter A. Broadbery. FOAM – An Intermediate Language Description. unpublished draft version.

[2] Tilmann Bubeck, Martin Hiller, Wolfgang Küchlin, and Wolfgang Rosenstiel. Distributed Symbolic Computation with DTS. In Afonso Ferreira and Jose Rolim, editors, *Proceedings of Parallel Algorithms for Irregularly Structured Problems*, LNCS 980. Springer, Sep 1995.

[3] A. Carissimi and M. Pasin. Athapascan: An experience on Mixing MPI Communications and Threads. In Vassil Alexandreov and Jack Dongarra, editors, *Proceedings of 5th European PVM/MPI Users' Group Meeting*, LNCS 1497, pages 137–144, Liverpool, UK, sep 1998. Springer Verlag.

[4] K. C. Chan, A. Díaz, and E. Kaltofen. A distributed apporach to problem solving in Maple. In R. Lopez, editor, *Maple Summer Workshop and Sympsium*, Maple V: Mathematics and Application, pages 13–21, Boston, 1994. Brikhäuser Verlag.

[5] Bruce Char and Jeremy Johnson. Some Experiments with Parallel Bignum Arithmetic. In Hoon Hong, editor, *Parallel Symbolic Computation*, PASCO, pages 94–103, Sep 1994.

[6] Burce Char. Progress report on a system for general-purpose parallel symbolic alg ebraic computation. In *International Symposium on Symbolic and Algebraic Computation*, ISSAC, 1990.

[7] M. Christaller, M.-R. Castaneda Retiz, and T. Gautier. Control Parallelism on top of PVM: The ATHAPASCAN Environment. In Jack Dongarra, Marc Gengler, Bernard Tourancheau, and Xavier Vigouroux, editors, *Proc. Second European PVM User's Group Meeting*, pages 71–76, Ecole Nationale Suprieure, Lyon, France, 1995 1995. Hermes.

[8] A. Diaz, M. Hitz, A. Lobo, and T. Valente. Process Scheduling in DSC and the Large Sparse Linear Systems Challenge. *Journal of Symbolic Computation*, 11(1-000), 1994.

[9] A. Diaz, E. Kaltofen, K. Schmitz, and T. Valente. DSC A System for Distributed Symbolic Computation. In S. M. Watt, editor, *International Symposium on Symbolic and Algebraic Computation*, ISSAC, pages 323–332, 1991.

[10] The MPI forum. MPI: A Message Passing Interface. Technical report, University of Tenessee, Knoxville, 1993.

[11] I. Foster, C. Kesselman, R. Olson, and S. Tuecke. Nexus: An Interoperability toolkit for parallel and distributed computer systems. Technical Report ANL/MCS-TM-189, Argonne National Laboratory, 1993.

[12] R. Loos G. E. Collins. ALDES/SAC-2 Now Available. *ACM SIGSAM*, 1982.

[13] Franois Galile, Jean-Louis Roch, Gerson G. H. Cavalheiro, and Ma htias Doreille. A General Modular Specification for Distributed Schedulers. In *Pact'98*, Paris, France., October 1998.

[14] T. Gautier. *Calcul Formel et Parallélisme : Conception du Système PAC et Applications au Calcul dans les Extensions Algébriques*. PhD thesis, Institut National Polytechnique de Grenoble, 1996.

[15] T. Gautier and J.L. Roch. PAC++ system and parallel algebraic numbers computations. In Hoon Hong, editor, *PASCO'94*, volume 5 of *Lecture Notes Series on Computing*, pages 145–153. World Scientific Publishing Co. Pte. Ltd., 1994.

[16] Wolfgang Küchlin. PARSAC-2: A Parallel SAC-2 Based on Threads. *Lecture Notes in Computer Science*, LNCS 508:341–353, 1990.

[17] H. Naundorf. Parallelism in MuPAD. In *First IMACS Conference on Applications of Computer Algebra*, Albuquerque, NM USA, 1995.

[18] J.-L. Roch and G. Villard. Parallel computer algebra. Tutorial of ISSAC'97, Preprint IMAG Grenoble, France, July 1997. http://www-lmc.imag.fr/~gvillard/BIBLIOGRAPHIE/POSTSCRIPT/tutorial.ps.

[19] Wolfgang Schreiner. A Para-Functional Programming Interface for a Parallel Computer Algebra Package. *Journal of Symbolic Computation*, 11:1–22, 1996.

[20] K. Siegl. ||Maple|| - A system for parallel symbolic computations. In *Parallel Systems Fair at the 7th International Parallel Processing Symposium*, Newport Beach, CA, April 1993.

[21] S.M. Watt et al. *Axiom Library Compiler User Guide*. NAG Ltd, 1994.

[22] Stephen M. Watt et al. FOAM: A First Order Abstract Machine, V 0.35. Technical report, IBM Thomas J. Watson Research Center, RC 19528, 1994.

A Library for Parallel Modular Arithmetic

David Power and Russell Bradford

Department of Mathematical Sciences
University of Bath
Claverton Down
Bath
England BA2 7AY.
djp@maths.bath.ac.uk, rjb@maths.bath.ac.uk

Abstract. This paper describes a library of platform independent functions for performing modular arithmetic on a range of parallel hardware. It is based around an approximate Chinese remainder reconstruction which allows the most significant bits of the stored number to be calculated without the cost of a full reconstruction. We describe how this can be used to calculate the length of a modular number, and also its applications to comparison and division.

1 Introduction

In a modular representation of integers an integer is stored as a list of residues to a list of co-prime moduli. This representation is well suited to parallel computing, and we shall discuss how the basic arithmetic operations can be performed on such a representation.

To perform addition, subtraction or multiplication in a modular representation is a straightforward task, as the result can be calculated for each modulus independently. This results in a low parallel complexity, as no communication is needed.

Tasks that are less straightforward are the calculation of number *length* (the number of moduli required to represent the number) and division with quotient and remainder. To enable these two tasks to be performed, two basic algorithms are used. These are an *approximate Chinese remainder reconstruction* and the reconstruction of a number to a small modulus (a previously unpublished result). Both of these methods enable information to be gathered which concerns the total number without the expense of a full reconstruction.

We also outline a proposed library for modular arithmetic. This library is to follow the style of the Gnu Multiple Precision (GMP) library [7], in providing a platform independent set of functions. The library is to run on a variety of parallel platforms, including the MasPar MP-1, shared memory machines and clusters of workstations. It is hoped that the implementation of this library will provide both a contrast between competing algorithms, and a contrast between synchronous and asynchronous programming styles.

P. Amestoy et al. (Eds.): Euro-Par'99, LNCS 1685, pp. 1476–1483, 1999.
© Springer-Verlag Berlin Heidelberg 1999

2   Modular Representation of Integers

In a modular representation a number is stored as a sequence of residues. Each residue corresponds to a different modulus, each of which is relatively prime to all of the others. If the moduli are $p_0, p_1, \ldots, p_{n-1}$ then an integer X would be represented as $\langle x_0, x_1, \ldots, x_{n-1} \rangle$ where $x_i \equiv X \pmod{p_i}$, $0 \leq x_i < p_i$. The product of all the moduli is $M = p_0 p_1 \ldots p_{n-1}$.

The Chinese Remainder Theorem (CRT) states that the system of simultaneous congruences,

$$X \equiv x_i \bmod p_i, 0 \leq i < n - 1 \tag{1}$$

has a unique solution $\bmod \ p_0 p_1 \ldots p_{n-1}$, assuming $\gcd(p_i, p_j) = 1$, for $i \neq j$.

We can obtain this unique solution using

$$\text{If } Y \equiv \sum_{i=0}^{n-1} M_i y_i \pmod{M} \tag{2}$$

where $M_i = M/p_i$ and $y_i \equiv M_i^{-1} x_i \pmod{p_i}$, then X is the least non-negative value of the residue class Y.

2.1   An Approximate Chinese Remainder Reconstruction

To reconstruct a number using the summation of Equation 2 involves numbers of magnitude M, which is much larger than a machine word size. This can be inconvenient and time consuming, particularly if all we require is some information about X, say its magnitude, and not its precise value. In such cases it is enough to compute an *approximation* to X. In an *approximate Chinese Remainder Reconstruction* each term is calculated to a limited floating point precision reducing both the amount of work done and the accuracy of the reconstruction.

Taking Equation 2 as a start point an approximation of X can be derived as follows.

$$\text{If } \sum_{i=0}^{n-1} M_i y_i \equiv X \pmod{M}, \text{ then}$$

$$\sum_{i=0}^{n-1} M_i y_i = kM + X, \text{for some } k \in \mathcal{Z}, \text{ so}$$

$$\sum_{i=0}^{n-1} \frac{y_i}{p_i} = k + \frac{X}{M}, 0 \leq \frac{X}{M} < 1 \tag{3}$$

Our approximate reconstruction is X/M, which is conveniently bounded by 1, and is found using only floating point computations. This value, X/M, together with k, is enough for many of our later algorithms.

Thus to calculate the approximate reconstruction the fractions y_i/p_i need to be summed and the integer part k needs to be subtracted. The accuracy of this summation depends on the accuracy of the floating point approximations of the fractions y_i/p_i. If each fraction is calculated to d bits of precision then using normal rounding the maximum error for each fraction is $\pm 2^{-(d+1)}$, and as there are n fractions to be summed the maximum total error is bounded by $\pm n2^{-(d+1)}$.

In [10] a more detailed error analysis is undertaken, this gives the the upper and lower bounds to be

$$\pm 2^{-d} \sum_{i=0}^{n-1} \frac{p_i - 1}{2p_i}. \tag{4}$$

which is very close to the simple bound of $\pm n2^{-(d+1)}$ when the p_i are large.

In order for this reconstruction to work, we must ensure the error bound is small enough to identify a single value of k. That is, the value X/M is more than $\pm n2^{-(d+1)}$ distant from an integer. We consider the case of being too close later when we look at reconstructing to fewer moduli.

2.2   Reconstructing to a Small Modulus

While an approximate reconstruction gives information about the most significant digits of a number, a *reconstruction to a small modulus* gives precise information about a number without the cost of a full reconstruction.

It is assumed that $\gcd(m, M) = 1$, and that m is sufficiently small to allow calculations mod m to be performed in unit time. Rearrange Equation 3, and take modulo m:

$$X = \sum_{i=0}^{n-1} \frac{M}{p_i} y_i - kM$$

$$X \equiv M \left(\left(\sum_{i=0}^{n-1} p_i^{-1} y_i \right) - k \right) \quad (\text{mod } m) \tag{5}$$

The value of k is known from Equation 3, and the value of $M \bmod m$ can be found by taking the product of all $p_i \bmod m$. The value of $p_i^{-1} \bmod m$ can either be calculated using an extended Euclidean GCD, or can be looked up from tables. (Note: for the value of k to be constructed correctly, we must have X at least twice the error bound less than M.)

Both of these reconstruction methods can be completed sequentially in time proportional to the number of moduli being used. This is significantly faster than a full reconstruction, which will have a time complexity of $O(n^2)$.

More interestingly, both of the above algorithms are well suited to parallel computation as they take a constant amount of calculation for each modulus, followed by one or more reductions. Each reduction will take $\log_2 N$ steps where N is the number of processors. If $N = n$ this will give each of the above algorithms a time complexity $O(\log n)$.

3   Calculating Length

In this section we will consider the problem of calculating the *length* of a number X, as this is needed when comparing numbers, performing division, and choosing the number of primes needed to represent a number.

The length will be called l and will obey the following inequality:

$$p_0 p_1 \ldots p_{l-2} \leq X < p_0 p_1 \ldots p_{l-2} p_{l-1} \tag{6}$$

3.1   Repeated Estimate

A method suited to numbers close to M, or ones in which the size is known to lie between an upper and lower bound is to use repeated approximate Chinese remainder reconstructions. If a number X has a length which is significantly smaller than n, then the value of X/M will be close to zero. If the approximated value of X/M is small enough it will be safe to remove one of the moduli and repeat the approximate reconstruction. By repeating this process the value of X/M will eventually become significantly large enough to calculate the value of l.

If X/M is less than or equal to $1/p_{n-1}$ then the value of l is less than n as $X < p_0 p_1 \ldots p_{n-2}$. Taking a general case with n' moduli the condition to allow the removal of a modulus is

$$\frac{X}{M'} < \frac{1}{p_{n'-1}} \tag{7}$$

Assuming relatively large moduli the error bound of Equation 4 will approach $\pm n' 2^{-(d+1)}$. We can be certain that a further modulus can be removed if the approximate reconstruction of X/M', here denoted as $\mathrm{Approx}(X/M')$, obeys the following inequality:

$$-n' 2^{-(d+1)} < \mathrm{Approx}\left(\frac{X}{M'}\right) < \frac{1}{p_{n'-1}} - n' 2^{-(d+1)} \tag{8}$$

This assumes that the value of X/M' is less than $1 - 2n' 2^{-(d+1)}$ else it would be possible for such a number to obey the above inequality while being very large rather than very small. To avoid this case it is necessary to start with the condition that $X/M < 1 - 2n 2^{-(d+1)}$ and also to tighten Equation 8 to the following:

$$-n' 2^{-(d+1)} < \mathrm{Approx}\left(\frac{X}{M'}\right) < \frac{1}{p_{n'-1}} - n' 2^{-(d+1)} - \frac{2(n'-1)2^{-(d+1)}}{p_{n'-1}} \tag{9}$$

This new bound will ensure that the condition $X/M < 1 - 2n 2^{-(d+1)}$ is maintained through each step as

$$\frac{X}{M'} \left(\frac{1}{p_{n'-1}} - \frac{2(n'-1)2^{-(d+1)}}{p_{n'-1}} \right) = \frac{X}{M'/p_{n'-1}} \left(1 - 2(n'-1)2^{-(d+1)} \right) \quad (10)$$

To be certain that the value of X needs all $n'-1$ moduli the following must hold:

$$\text{Approx} \left(\frac{X}{M'} \right) > \frac{1}{p_{n'-1}} + n'2^{-(d+1)} \quad (11)$$

However this leaves all the values of X that are approximately equal to $M'/p_{n'-1}$, in these cases it is simplest to return a value which is possibly one greater than the real l.

This repeated approximate reconstruction is good for numbers with lengths close to n, or alternatively numbers close to a known bound. The number of steps required will be $n - l$, each of these steps will involve an approximate reconstruction which sequentially takes $O(n)$ steps, giving a sequential complexity of $O((l - n)n)$. In parallel using $N = n$ processors, the complexity of a single approximate reconstruction is $O(\log n)$ this will lead to a total complexity of $O((l - n) \log n)$.

4   Division

Division with quotient and remainder is the hardest of the basic arithmetic operations to perform in the residue representation. Let us suppose we wish to divide A by B as $A = qB + r$, when both A and B are given as a list of residues.

If we happen to know we have an exact division, there is a fairly straightforward technique we can use. If $A = qB$, then $A \equiv qB \bmod p_i$ for each p_i. If $B \not\equiv 0 \bmod p_i$, then $A \equiv B^{-1}A \bmod p_i$, If $B \equiv 0 \bmod p_i$, that is to say $p_i \mid B$, then the modulus is removed from the calculation, and can subsequently be reconstructed using the method described in section 2.2.

For the division $A = qB + r$ when B is a small integer we can first find r by taking $A \bmod B$ as in section 2.2. Then subtracting r from A we can get q by exact division as before.

When B is large the process is much more difficult. While several different algorithms have been proposed [9, 3, 8, 13], only [9] proposes an algorithm for general division on a parallel computer.

General division is performed using a modified version of Knuth's classical "$\hat{q}$" division algorithm presented in [11]. An approximate reconstruction is used to calculate the most significant digits of the numbers in each step. As in the original algorithm the divisor is normalised so that it is greater than half of the total modulus needed for its reconstruction. This makes the estimation of each quotient digit boundedly accurate. Unlike the original algorithm no correction step is employed. Instead if a quotient digit is over estimated the result is allowed to become negative since we don't compute q directly. Instead we reduce A by multiples of B until r remains, and then we can compute q by exact division.

5   Comparison

If two numbers are equal, then their values will be the same in each and every modulus. This can be tested using a single globalor operation.

To compare two non-equal numbers, their approximate reconstructions can be be used. If the two numbers are too close together to distinguish (closer than the error bounds), then we can subtract one from the another and the sign of the result can be found.

To find the sign of the result, a modified *length* algorithm can be used. The difference as approximated will be close to a whole number, as the two original numbers were within error bounds of each other. If the approximate reconstruction is greater than the approximation error, then the result is positive. If it is less than minus the approximation error then the result is negative. If it is too close to zero, then a prime can be removed and a new approximation made.

6   A Modular Arithmetic Library

We are constructing a library of functions to allow arbitrary length integer arithmetic on a wide range of platforms with the underlying representation being modular, such as has been discussed throughout this paper. The library provides a common set of datatypes and functions for each platform so that the same program can be run on each platform. We provide basic arithmetic functions, and later will implement a limited range of more involved functions such as divisibility checks and GCDs.

6.1   Platforms

The library runs on a range of parallel and sequential hardware. There are four basic versions of the library being written in MPI, AFAPI, MPL and sequential C. The library runs on a variety of platforms including a MasPar MP-1, a cluster of four Linux boxes communicating using TTL_PAPERS, four Linux boxes communicating using Ethernet, a Solaris shared memory computer and a single sequential Linux computer.

The AFAPI library (Aggregate Function API [5]) provides a library of architecture independent function designed for parallel or distributed sytems. It was originally designed to be used with the Purdue Adapter for Parallel Execution and Rapid Synchronisation (hardware to achieve very fast synchronisation between machines), TTL_PAPERS [6, 4]. We also use the AFAPI library on a shared memory computer.

The MasPar MP-1 is a SIMD array processor, has 1024 processors and is programmed in a language called MPL [12], which is based on C. The processors are connected in a 32 by 32 grid which theoretically should lead to poor performance when performing global reductions. However, in practice a 32 bit modulus operation (using a division) takes longer than reduction using 32 bit addition due to hardware support for reduction operations.

6.2   Datatypes

The library will use one main datatype, that of an arbitrary length integer. This datatype is to contain six fields as shown in Table 1.

Field	Type	Description
Sign	int	True if number is negative
Array	unsigned int*	Pointer to an array of residues
Num_res	int	The number of residues
Length	int	The number of moduli needed for reconstruction
Approx	double	The approx reconstruction using Length moduli
Size	int	The size of the moduli in bits

Table 1. Modular Number Datatype

We shall use 16 bit moduli as the default as they will allow the storing of inverses in tables. This is useful as the calculation of inverses is a costly but very common operation. However as a number becomes larger it may not be possible to represent it using 16 bit moduli. This is because there are just 6542 primes less than 2^{16}, and the product of these has 28,305 decimal digits. If a number larger than this is needed then larger primes will have to be used. At this point we can switch to 32 bit moduli and calculate inverses on the fly using the extended Euclidean algorithm. It is unlikely that primes above 32 bits would ever be needed as there are approximately 203 million of them [1].

The purpose of the Length and Approx fields is to allow fast comparison between numbers, and to detect overflow. Keeping track of these at all times will involve many approximate reconstructions, but this is a relatively quick operation. The number of residues can change with the size of the number. The process of adding more moduli (*base extension*) is achieved by using a Mixed Radix Conversion (MRC) followed by individual prime reconstructions. More details of this method can be seen in [13].

The basic functions provide by the library will include addition, subtraction, multiplication, division, remainder, comparison, equality, divisibility and GCD. In addition there will be more implementation specific functions such as length, increase number of residues and change from 16 bit to 32 bit moduli. Addition, subtraction and multiplication are straightforward, as these can be performed on each modulus individually. For division and comparison it will be necessary to keep track of the length and approximate reconstruction of each number but as these are already known for the operands, this will only require one approximate reconstruction in the majority of cases.

7   Conclusion

We have described a modular representation for arbitrary length integer arithmetic that has advantages both in performing the basic operations of multiplication and addition and also in the easy partitioning of data for parallel computation.

We have shown an algorithm for approximating a number stored in modular form, this approximate reconstruction is used throughout the report and allows all basic arithmetic operations to be performed without having to reconstruct the number fully using the Chinese Remainder Theorem.

We have the design of a portable library of arbitrary length integer functions which is to be in the style of GMP. It will work on a wide range of platforms including a MasPar MP-1, a cluster of PCs, a shared memory multiprocessor and a single PC workstation.

The library will provide a full range of arithmetic operations along with limited more complex functions such as GCD. We intend later to implement univariate polynomial arithmetic using the library, as this is one application where large numbers often arise [2].

References

[1] D.M. Burton. *Elementary Number Theory*. Wm. C. Brown, 1989.

[2] J.H. Davenport, Y. Siret, and E. Tournier. *Computer Algebra*. Academic Press, 1988.

[3] G.I. Davida and B. Litow. Fast parallel arithmetic via modular representation. *SIAM J. Comput*, 20(4):756–765, 1991.

[4] H.G. Dietz, T.M. Chung, T. Mattox, and T. Muhammad. Purdue adapter for parallel execution and rapid synchronistaion: The TTL_PAPERS design. Technical report, School of Electrical Engineering, Purdue University, 1995.

[5] H.G. Dietz, T. Mattox, and G. Krishnamurthy. The aggregate function API: It's not just for PAPERS anymore. Technical report, School of Electrical Engineering, Purdue University, 1997.

[6] H.G. Dietz, T. Muhammad, and T. Mattox. TTL implementation of purdue's adapter for parallel execution and rapid synchronistaion. Technical report, School of Electrical Engineering, Purdue University, 1994.

[7] T. Granlung. *The Gnu Multiple Precision Arithmetic Library*. TMG Datakonsult, 2.0.2 edition, June 1996.

[8] M.A. Hitz and E. Kaltofen. Integer division in residue number systems. *IEEE Transactions on Computers*, 44(8):983–989, 1995.

[9] C.Y. Hung and B. Parhami. An approximate sign detection method for residue numbers and its application to RNS division. *Computers Math. Applic.*, 27(4):22–35, 1994.

[10] C.Y. Hung and B. Parhami. Error analysis of approximate chinese-remainder theorem decoding. *IEEE Transactions on Computers*, 44(11):1344–1348, 1994.

[11] D.E. Knuth. *The Art of Computer Programming*, volume 2. Addison Wesley, 3rd edition, 1998.

[12] Maspar Computer Corporation. *Maspar Parallel Application Language (MPL), User Guide*, 9302-0101-a5 edition, July 1993.

[13] K.C. Posch and R. Posch. Modulo reduction in residue number systems. *IEEE Transactions on Parallel and Distributed Systems*, 6(5):449–454, 1995.

Performance Evaluation of Or-Parallel Logic Programming Systems on Distributed Shared-Memory Architectures*

Vanusa Menditi Calegario[1] and Inês de Castro Dutra[2]

[1] Empresa Municipal de Informática S/A, Rio de Janeiro, Brazil
[2] COPPE – Systems Engineering and Computer Science
Federal University of Rio de Janeiro, Brazil
{vanusa,ines}@cos.ufrj.br

Abstract. In this work we investigate how Distributed Shared Memory (DSM) architectures affect performance of or-parallel logic programming systems and how this performance approaches that of conventional C systems. Our work concentrates on basic performance, scalability, and programmability. We use execution-driven simulation of a hardware DSM (DASH) to investigate the access patterns and caching behaviour exhibited by parallel C programs and by Aurora, a parallel logic programming system capable of exploiting implicit parallelism in Prolog programs. Aurora was originally written to run on bus-based shared-memory platforms.

1 Introduction

Distributed shared-memory architectures have been object of research by many computer science groups. Research goes broadly from hardware based coherence protocols to DSM software protocols on networks of workstations passing through high technology inter-connection networks that reduce network latency. In this work we investigate how DSM architectures affect performance of parallel logic programming systems and compare to results obtained for parallel conventional systems.

We use execution-driven simulation of a hardware DSM (DASH [8]) to investigate the access patterns and caching behaviour exhibited by parallel C programs and by Aurora [9], a parallel logic programming system capable of exploiting implicit parallelism in Prolog programs. Aurora was written originally to run on bus-based shared-memory platforms. The simulator allows us to vary several machine settings. Our work concentrates on the study of Prolog and parallel C programs under different cache coherence protocols regarding basic performance, scalability, and programmability.

Our work is the first to study the differences between parallel logic programming systems and parallel C-compatible systems. In contrast, previous works such as [13], [10], [5], [6], [4], [12], [11], [3] and [7] have also studied the performance of logic programming systems in other contexts.

The paper is organised as follows. Section 2 briefly describes our work methodology. Section 3 describes the applications and algorithms used and results obtained. Finally, section 4 concludes this work and draws our next steps.

---

* Research supported by CNPq and Capes, Brazilian Research Councils

P. Amestoy et al. (Eds.): Euro-Par'99, LNCS 1685, pp. 1484–1491, 1999.
© Springer-Verlag Berlin Heidelberg 1999

2 Methodology

In our experiments we used a detailed on-line, execution-driven simulator that simu-lates a 32-node, DASH-like [8], directly-connected multiprocessor to execute Aurora, an or-parallel system, and C parallel applications. In order to keep caches coherent we used write-invalidate (WI), write-update (WU) and dynamic hybrid protocols. Detailed information about the simulator, Aurora and the protocols can be found at [2].

3 Applications and Results

We concentrated our studies on some applications commonly used as benchmarks for Prolog or C systems. Our first benchmark is a typical scientific application that com-putes matrix multiplication. We used 150x150 matrices. The second benchmark is a classical combinatorial problem in the artificial intelligence area that consists of plac-ing N queens on an N x N chessboard so that no two queens attack each other. We used a board of size 10. Our third benchmark is the classical travelling salesperson problem with a 16-node graph. Finally, the fourth benchmark is a floor plan design that consists of arranging several components on a surface in order to minimise the total area oc-cupied by the components. The number of components to be arranged is 60, with an arrangement of 14 objects on the surface.

The Prolog and C source codes for the programs can be found at the following URL: http://www.cos.ufrj.br/~vanusa/benchs.html.

Table 1. SIMULATED RESULTS - AURORA AND C ELAPSED SIMULATED CYCLES - 1 PROCESSOR

Application	Aurora	C	A/C
Matrix 150x150	1365311633	60522671	22.55
10-Queens	787832881	206354254	3.81
TSP 16	6228451159	5560316620	1.12
Floor Plan	8835176394	578872490	15.26

Table 1 shows the sequential execution times for the C and Aurora Prolog versions of the programs. Times are given in number of simulated cycles. The column A/C shows the ratio between Aurora and C execution times. We can observe that the C programs run from 1.12 to about 22 times faster than Aurora.

From the programs shown in table 1, matrix, a typical scientific application, is the one that has the worst results, as expected. The other applications are at most 15 times slower in Prolog than in C. This can be considered a good result, since Aurora does not use a state-of-the-art sequential engine that generates native code, and since the algo-rithms can be improved. According to our previous work, modern sequential engines can achieve significant speedups [3].

However, more important than showing that Prolog systems can perform better than conventional ones, is to show that with no effort from the Prolog programmer, we can obtain parallelism from the Prolog applications on DSM architectures and can outperform parallelised C applications. Therefore, in this section we also show results for Aurora and parallel C running the four applications with 2, 4, 8, 16, 24 and 32 processors using four cache coherence protocols (WI, WU, Hybrid 1 and Hybrid 2). It is important to notice that base times to compute speedups are different for C and Aurora. Figures were built in order to show how Aurora and C individually manage to efficiently exploit all parallelism available in the applications.

Matrix Multiplication: The matrix multiplication program written in Prolog has or-parallelism that stems from the choicepoints created for the calculation of each element of the output matrix. The matrices are represented as facts in the database in Prolog. In C the first matrix is private to each processor while the second and the output matrices are shared. Parallelism is implicitly and efficiently exploited by Aurora which makes the system achieve slightly better speedups than parallel C up to 24 processors as can be seen in table 2. As we increase the number of processors, Aurora does not scale well up. When we observe the execution in more detail, we find out that some processors in the 32-processor simulation stay blocked much longer than the 24-processor simulation. This indicates that there is contention to access a lock with processors trying to grab work always from the same worker. This also indicates that the or-scheduler version we use for Aurora is not scalable due to the way workers try to fetch work from other workers. In contrast, the C parallel program scales well up due to the static load distribution of the matrices among the processors. It is important to notice that the total amount of parallel work is the same as the sequential work. Therefore, the speedups correspond to actual gains in performance for efficiently utilising the resources.

Table 2. MATRIX SPEEDUPS, INVALIDATE PROTOCOL

System	2	4	8	16	24	32
Aurora	1.88	3.56	6.41	10.87	14.23	15.67
C	1.59	2.91	5.54	10.21	14.10	18.95

It is interesting to notice that while the parallel C program has a miss rate of 2.49% with a smaller total of references to shared memory ($\sim$6M) for one processor, Aurora with a much larger total of references ($\sim$142M) has a very low miss rate running this application (0.46%) for one processor. Note that these misses correspond only to read misses and that only shared data is cached. The simulator does not cache instructions. This indicates that Aurora running the matrix application has better locality than C. This is mainly due to the design of the abstract machine used for Aurora that is based on the WAM [1] and it is optimised for locality.

Most of the C misses in this application come from eviction misses. This happens because of the memory organisation (one-way associative) and cache size (64Kbytes) that does not fit one entire matrix. The cold miss rate in the C program increases as we increase the number of processors, because the second matrix needs to be loaded by every processor.

Contrasting with the very low miss rates obtained for the matrix application, the network traffic for Aurora is much larger than for the parallel C matrix program. This high network traffic for Aurora is easily explained by the high number of references to shared memory. Even so, Aurora using dynamic scheduling achieves good parallel performance that is comparable to C, which uses a static distribution of work. This application has almost no sharing misses, and no coherence network traffic. The speedups were not affected by the cache coherence protocols and the same occurred to miss rates. As this application is very regularly distributed, a change in the protocol is hardly noticed.

N-queens: In figure 1 we can see the speedups for the 10-queens problem. This problem is very difficult to parallelise efficiently in the procedural program. In our C implementation, ten boards are initially created by the parent process. The parent process starts working with the first board created and puts the other boards on a shared task queue. Each idle process takes one board from this task queue, tries to place a queen, continues working on its own board and generates new boards to put on the task queue. The program in C uses a forward checking algorithm in order to prune the search space. The Prolog program uses a generate-and-test algorithm that is much less efficient than forward checking.

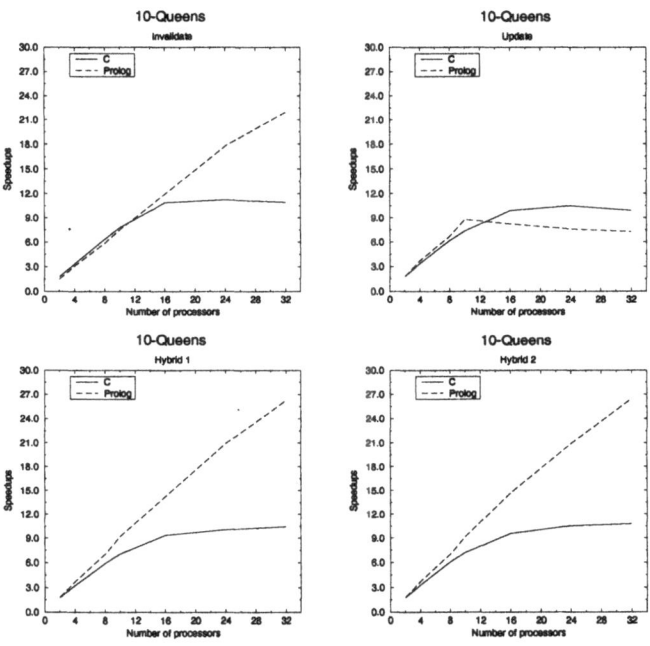

Fig. 1. QUEENS SPEEDUPS, FOUR PROTOCOLS

Because of the way boards are distributed through the processors, we can observe a very high rate of cold misses for the C parallel application. Boards that are created by one processor will cause multiplication of cold misses by other processors that grab those boards from the task queue. This problem could be minimised with the utilisation of local task queues per processor and some kind of affinity scheduling. This application can only sustain parallelism in C up to 16 processors because of the way the work is distributed among the processors. The overhead of accessing the global task queue is larger than the work granularity when we reach 16 processors, which cause a degradation in parallel performance for C. In contrast, Aurora manages to maintain performance and scales quite well. This program demonstrates that without effort from the end-programmer, we can obtain a very good performance when exploiting parallelism in logic programming.

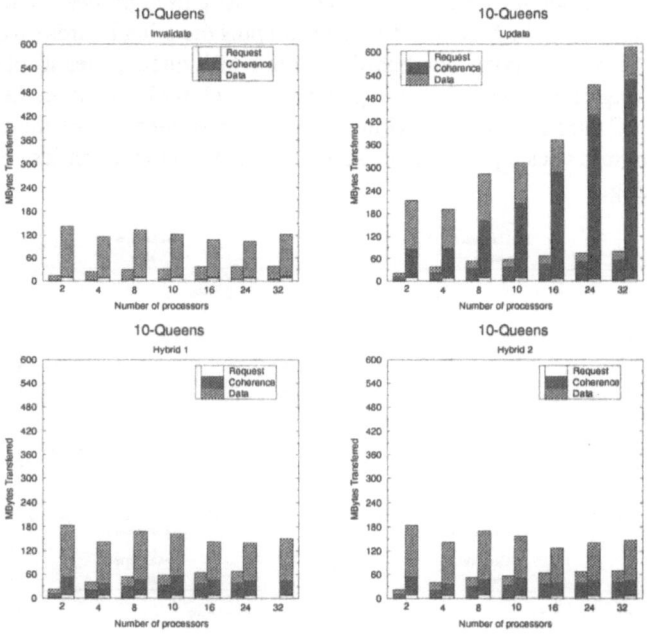

Fig. 2. QUEENS NETWORK TRAFFIC, FOUR PROTOCOLS, C (LEFT) AND AURORA (RIGHT)

Figure 1 also shows that the Aurora queens version can benefit from the hybrid protocols. They manage to reduce the network traffic caused by WU as can be seen in figure 2. Hyb2 yields the best performance for the Aurora version with an improvement of 20.79% over the invalidate protocol for 32 processors. Aurora manages to keep performance achieving efficiency of 81%, on 32 processors, with the hybrid protocols.

TSP: Figure 3 shows the speedup curves for the tsp benchmark. The performance of the C parallel program is quite good while Aurora is not scalable. It is important to notice that the parallel C application we use is well-tuned for DSM architectures.

The parallel C program shows a very low read miss rate when compared to the Prolog program. Similarly, the network traffic for this application in C is very low. This shows that this application is very well tuned for the DSM architecture we are simulating.

In figure 3, no hybrid protocol causes significant impact on the C parallel application, since this application is already well-tuned for the architecture we simulate. In contrast, the Prolog application benefits from the WU and hybrid protocols with an improvement in performance of 42% related to the invalidate protocol with 32 processors.

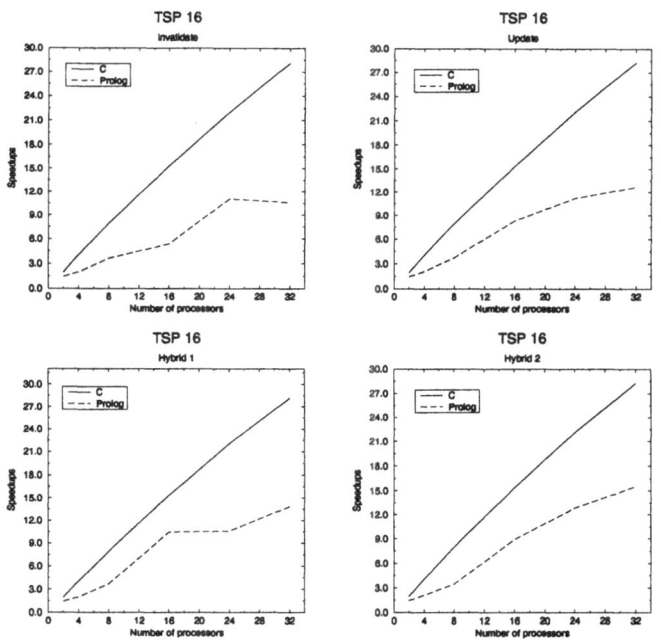

Fig. 3. TSP SPEEDUPS, FOUR PROTOCOLS

Floor Plan: The floor plan program in C and in Prolog were written using a branch-and-bound algorithm. These algorithms when run in parallel can produce superlinear speedups. Our speedup results indeed show that we have superlinear speedups in C and in Prolog as shown in figure 4.

The Prolog program achieves spectacular speedups while the C program fails to maintain performance after 8 processors. This is due to the fact that the Prolog program reduces dramatically the search space when running in parallel. Miss rates and network traffic are very much influenced by the search done to find a solution, for each set of processors. As can be also seen in figure 4, the C application has very poor performance when using any hybrid protocol. Speedups are not improved related to the WI protocol. Aurora can benefit from the hybrid protocols, managing to achieve best speedups of

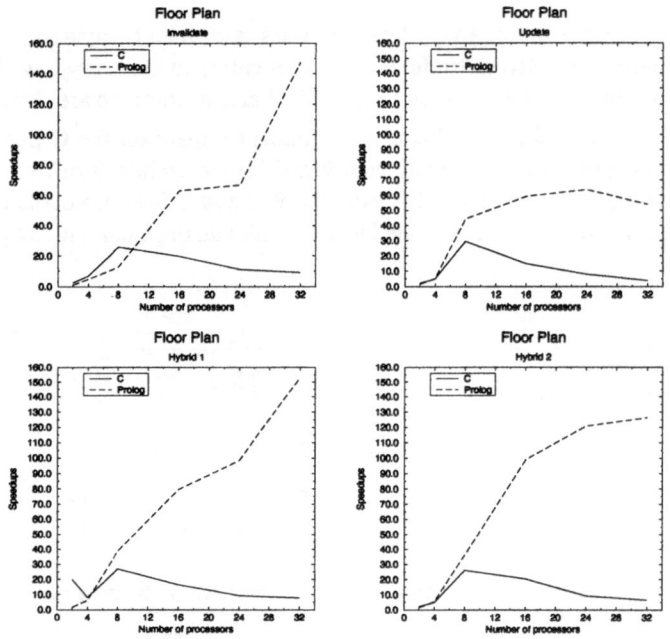

Fig. 4. FLOOR PLAN SPEEDUPS, FOUR PROTOCOLS

150 on 32 processors, because of the reduction in search space compared to the one-processor result.

4   Conclusions and Future Work

Our results indicate that or-parallel logic programming systems can perform well on DSM architectures. More important than being able to efficiently exploiting parallelism, is the ability to do that transparently with no effort from the Prolog programmer, and with a minimum effort from the parallel programmer, i.e., migrating parallel logic programming systems originally written for shared-memory machines to distributed shared-memory architectures. Regarding the impact of different cache coherence protocols on the Aurora performance we can conclude that they do not significantly affect extremely regular applications such as our typical scientific application, matrix multiplication, but they can significantly affect irregular ones. Particularly, hybrid protocols can improve performance up to 42% from an invalidate protocol, as our tsp application has shown.

Results also indicate that optimisations such as improvements on the abstract engine, data organisation and scheduling can make parallel logic programming systems competitive and more attractive than parallel C systems for some applications, specially if we consider that parallel logic programming systems exploit implicit parallelism. Our results also show that although Aurora has a huge number of references to shared memory, in general, it manages to maintain good reference locality while parallel C programs with very low number of references need a good programming style for the kind of architecture we are studying. The main problems encountered on our experiments are related to scheduling issues. Our next steps will be to use different scheduling strategies and algorithm restructuring both for the parallel C programs and for Aurora in order to improve performance. Another further step will be to evaluate other real Prolog and C symbolic applications and to test different input parameters for the applications.

Acknowledgments

We would like to thank Ricardo Bianchini and Vítor Santos Costa. Vanusa would like to thank her bosses Sérgio Fontoura, Sérgio Abramovitch, Vânia Pintos and Volnei Marques da Costa for allowing her to carry out this research work.

References

[1] H. Aït-Kaci. *Warren's Abstract Machine — A Tutorial Reconstruction*. MIT Press, 1991.
[2] V. M. Calegario and I. C. Dutra. Parallel Conventional Systems versus Parallel Logic Programming Systems on Distributed Shared Memory Architectures. Technical Report ES-489/98, COPPE/Systems Engineering and Computer Science, Dezembro 1998.
[3] V. M. Calegario and I. C. Dutra. Performance Comparison between Conventional and Logic Programming Systems. Technical Report ES-478/98, COPPE/Systems Engineering and Computer Science, Setembro 1998.
[4] V. S. Costa and R. Bianchini. Optimising Parallel Logic Programming Systems for Scalable Machines. In *Proceedings of the EUROPAR'98*, pages 831–841, Sep 1998.
[5] V. S. Costa, R. Bianchini, and I. C. Dutra. Evaluating the Impact of Coherence Protocols on Parallel Logic Programming Systems. In *Proceedings of the 5th EUROMICRO Workshop on Parallel and Distributed Processing*, pages 376–381, 1997.
[6] V. S. Costa, R. Bianchini, and I. C. Dutra. Parallel Logic Programming Systems on Scalable Multiprocessors. In *Proceedings of the 2nd International Symposium on Parallel Symbolic Computation, PASCO'97*, pages 58–67, July 1997.
[7] V. W. Freeh. A Comparison of Implicit and Explicit Parallel Programming. *Journal of Parallel and Distributed Computing*, 1(34):50–65, 1996.
[8] D. Lenoski, J. Laudon, T. Joe, D. Nakahira, L. Stevens, A. Gupta, and J. Hennessy. The DASH Prototype: Logic Overhead and Performance. *IEEE Transactions on Parallel and Distributed Systems*, 4(1):41–61, Jan 1993.
[9] E. Lusk, D. H. D. Warren, S. Haridi, et al. The Aurora Or-Parallel Prolog System. *New Generation Computing*, 7(2,3):243–271, 1990.
[10] S. Raina, D. H. D. Warren, and J. Cownie. Parallel Prolog on a Scalable Multiprocessor. In Peter Kacsuk and Michael J. Wise, editors, *Implementations of Distributed Prolog*, pages 27–44. Wiley, 1992.
[11] P. V. Roy and A. M. Despain. High-Performance Logic Programming with the Aquarius Prolog Compiler. *IEEE Computer*, 25(1):54–68, January 1992.
[12] M. G. Silva, I. C. Dutra, R. Bianchini, and V. S. Costa. The Influence of Computer Architectural Parameters on Parallel Logic Programming Systems. In *Workshop on Practical Aspects of Declarative Languages (PADL'99)*, pages 122–136, January 1999.
[13] E. Tick. *Memory Performance of Prolog Architectures*. Kluwer Academic Publishers, Norwell, MA 02061, 1987.

A Parallel Symbolic Computation Environment: Structures and Mechanics

'Mantŝika Matooane and Arthur Norman

University of Cambridge Computer Laboratory,
New Museums Site, Pembroke Street,
Cambridge CB2 3QG, United Kingdom,
mam32@cam.ac.uk, acn1@cam.ac.uk

Abstract. We describe a set of representations for polynomials and sparse matrices suited for use with fine-grain parallelism on a distributed memory multiprocessor system. Our aim is to support use of supercomputers with this style of architecture to perform computations that would exceed the main memory capacity of more traditional computers: although such systems have very high performance communication networks it is still essential to avoid letting any one part of the network become a bottleneck. We use randomised data placement both to avoid hot-spots in the communication patterns and to balance (in a probabilistic sense) the memory load placed upon each processing element. The expected application areas for such a system will be those where intermediate expression swell means that the huge primary memory available on MPP systems will be needed if the smaller final result is to be successfully computed.

1 Introduction

We study very fine grained parallelism in computing with small numbers of huge polynomials. The fine granularity is a result of the need to distribute each polynomial across several processing elements (PE's) taking advantage of the memory available and enabling concurrent computation on multiple arguments on each processing element. Granularity and communication cost are inversely related [9], therefore the fine granularity in our algorithms incurs large communication costs.

The emphasis in this paper, on the low level details of data structures and the communication flow, is motivated by the need to compare the efficiency of different representations in parallel and sequential implementations. Secondly, we would like to compare any differences between symbolic and other algorithms regarding network use [1].

To ensure that each polynomial is uniformly distributed we follow the universal hashing mechanism introduced in CABAL [7]. In this paper we elaborate on the strategy and evaluate it performance and progress compared to the earlier version. We also monitor the traffic generated by two computations: multiplication and solution of large sparse systems of equations. An analysis of the

P. Amestoy et al. (Eds.): Euro-Par'99, LNCS 1685, pp. 1492–1495, 1999.
© Springer-Verlag Berlin Heidelberg 1999

communication patterns generated is conducted. These experiments reveal the need to address the bandwidth requirements while attempting to balance peak memory demand.

The parallel environment consists of processing elements with tightly-coupled memory, and cooperation through message passing. Although we have been able to carry out small scale testing and debugging on networks of workstations, the target for serious applications is a Hitachi SR2201[4] computer with 256 nodes, each with 256 Mbytes of memory giving a total 64 Gbytes. We note that this amount of memory should eventually allow us to address problems that are significantly beyond the capabilities of large individual workstations.

The speed of the communication network enables us to consider algorithms that generate large numbers of messages. However, this does not remove the need for bandwidth and synchronization consideration in the algorithm.

For the work described here the message passing uses MPI[12, 8] rather than PVM, which had been used in an earlier exploration of these ideas. This change was to fit in better with the preferred and best supported software environment on the target computer, and reflects the growing maturity of MPI [2] as a widely available low-level substrate for parallel processing.

2 Polynomial and Matrix Representations

The data distribution problem for numerical matrix algorithms is a static one and is solved in various ways: block distribution, cyclic, block-cyclic, cyclic-cyclic [5, 6]. The choice between these is made on the basis of how the distribution interacts with the algorithm being performed, and an attempt is normally made to minimize communication requirements.

Symbolic computations often exhibit intermediate expression swell [13]. This recurring hazard requires that we implement dynamic distribution which balances storage requirements on each processing element during computation. The growth of terms is not easy to predict and therefore a policy such as block distribution may result in heavy storage use on a single processor and possibly overflow.

We count the need to distribute parts of large intermediate results across many separate memory modules as the driving force in this work.

The basic strategy used is to represent polynomials in a distributed format, using a fixed collection of indeterminates so that the associated exponents could be kept in packed vectors of bytes. No a-priori ordering is imposed on the terms making up a polynomial[3]. The justification for this is that the asymptotic cost of multiplication for polynomials where the terms are kept sorted is $O(n^2 \log n)$ and for very large formula the logarithmic factor would be especially important. A system-wide hash table supports the unordered merge operations involved in polynomial arithmetic and allows polynomial multiplication of n-term polynomials to complete in expected time $O(n^2)$. Whenever a monomial is created a hash value was computed for it; part of this hash value is used to determine which PE the monomial belongs in, while the rest determines its placement in the hash table maintained within that PE. This scheme results is every polynomial having its terms pseudo-randomly spread among all the PE's . The hashing leads to even distribution so the memory load will be balanced, although of course there could be pathological cases where hashing attempted to map almost all terms to just one of the PE's. Each PE holds copies of a few index values (as well as the hash table) that provide access to the head of each polynomial stored there.

Our application area is symbolic linear algebra, and specifically the calculation of determinants of sparse symbolic matrices [11, 10]. Because the structures needed to represent the sparse matrices are fairly compact we keep a copy of them in each PE so that this information at least does not need repeated broadcast.

The most communication-intensive part of our computation occurs when two large polynomials need to be multiplied. Every possible pair of terms will have to be combined, and the resulting product term will then have to be sent to its proper PE for storage. We arrange that each PE in turn acts in an active state, broadcasting the terms that it holds. Some degree of global synchronization (and hence cost) is needed in the handshakes that control which PE's take this active role and which remain passive. To improve overlap between communication and computation we make use of the MPI non-blocking persistent communication and through that arrange our own buffering of messages.

There are two communication levels; one-to-all communication of the current work terms, and asynchronous one-to-one communication of storage messages that have hashed to a particular PE. All messages are the same size, and bandwidth requirements are dependent on the frequency and number of messages to each PE. While the communication of current term from master to slaves may be implemented in a broadcast with synchronization side-effect, the hashed terms are random 'events'. The second type of message requires anticipation of an asynchronous event and a decision of when to handle it. We expect requests for memory balancing to be handled as soon as possible to free the receive buffer for a possible next message from the same source.

3   Evaluation and Conclusions

We have been measuring the communication traffic generated by our randomized distribution algorithm. To verify the fine-grain patterns in which messages interleave and also to assist with debugging we have a version of the code where each processor writes a time stamped message to its output file indicating a sent message, and its destination. All messages generated are the same size therefore bandwidth requirements are dependent only on the frequency and number of messages to each destination. This testing framework also makes it easy to insert arbitrary extra delays into selected parts of the processing to explode deadlock and buffer overflow potential. Sample traces will be included as part of the conference presentation of this work, as will be timings for the system running on the SR2201 and showing that although move to a distributed memory system leads to an significant initial overhead compared with the costs of running on a single processor, for large polynomial calculations our scheme can show speed-up as well as the ability to perform calculations that would otherwise be too big to complete.

We have described structures that support efficient parallel algorithms for symbolic computation on distributed memory architectures. The hashing mechanisms generate communication patterns that counter intermediate expression swell, but introduce some extra synchronization and event-handling requirements. Recent work on solution of sparse systems of equations indicates their viability and they can be refined further for matrix computations. The structures and mechanisms discussed here are also amenable to vector operations and future work would explore their combination with pseudo-vector capabilities of the Hitachi SR2201 in manipulation of symbolic data.

References

[1] DINDA, P. A., GARCIA, B. M., AND LEUNG, D.-S. The measured network traffic of compiler-parallelized programs. Tech. rep., Carnegie Mellon University School of Computer Science, 1998. Techincal Report CMU-CS-98-144.

[2] GEIST, G., KOHL, J., AND PAPADOPOULOS, P. Pvm vs mpi: A comparison of features. In *Calculateurs Paralleles* (1996).

[3] GUSTAVSON, F., AND YUN, D. Y. Y. Arithmetic complexity of unordered sparse polynomials. In *SYMSAC'76* (August 1976), R. D. Jenks, Ed., SIGSAM, ACM, pp. 149–153.

[4] HITACHI LTD. *A tutorial for parallel programming*, 1995.

[5] KUMAR, V., GRAMA, A., GUPTA, A., AND KARYPIS, G. *Introduction to Parallel Computing:Design and Analysis of Algorithms*. Benjamin/Cummings Publishing Company, Inc., Redwood City, California, 1994.

[6] MILLER, R., AND STOUT, Q. F. *Parallel Algorithms for Regular Architectures: Meshes and Pyramids*. The MIT Press, Cambridge, Massachusetts, 1996.

[7] NORMAN, A., AND FITCH, J. Cabal: Polynomial and power series algebra on a parallel computer. In *Proceedings of PASCO 1997* (1997).

[8] PACHECO, P. S. *Parallel Programming with MPI*. Morgan Kaufmann Publishers, Inc., San Francisco, California, 1997.

[9] SASAKI, T., AND KANADA, Y. Parallelism in algebraic computaion and parallel algorithms for symbolic linear systems. In *Proceedings of 1981 ACM Symposium on Symbolic and Algebraic Computation* (1981), P. S. Wang, Ed., pp. 160–167.

[10] SMIT, J. New recursive minor expansion algorithms: A presentation in a comparative context. In *Proceedings of International Symposium on Symbolic and Algebraic Manipulation* (1979), E. W. Ng, Ed., Lecture Notes in Computer Science 72, Springer-Verlag, pp. 74–87.

[11] SMIT, J. A cancellation free algorithm, with factoring capabilities, for the efficient solution of large sets of equations. In *Proceedings of the 1981 ACM Symposium on Symbolic and Algebraic Computation* (1981), P. S. Wang, Ed.

[12] SNIR, M., OTTO, S. W., HUSS-LEDERMAN, S., WALKER, D. W., AND DONGARRA, J. *MPI:The Complete Reference*. The MIT Press, Cambridge, Massachusetts, 1996.

[13] TOBEY, R. Experience with formac algorithm design. In *Communications of the ACM* (1966), pp. 589–597.

Index of Authors

Åge Øye, Geir, 586

Abu-Ghazaleh, Nael B., 1204
Aggarwal, Sudhir, 505
Ahr, Dino, 98
Ait Saadi, Karima, 1018
Alexandre, Frédéric, 935
Almeida, F., 877
d'Almeida, Filomena D., 608
Alonso, P., 1073
Altman, Erik R., 1269
Anile, Marcello A., 663
Ansaloni, Roberto, 1435
Antoniu, Gabriel, 117
Antonoiu, Gheorghe, 823
Arantes, Luciana Bezerra, 815
Arbenz, Peter, 1078
Asano, Takashi, 914
Asencio, Nicole, 1417
August, David I., 1301
Avresky, D. R., 515
Ayache, Alain, 939

Bäcker, Andreas, 98
Böszörményi, László, 258
Béchennec, J-L., 659
Béchennec, Jean-Luc, 262, 757
Bagherzadeh, Nader, 727
Bahsoun, Jean-Paul, 1344
Baldoni, Roberto, 450
Bampis, Evripidis, 369
Baraka, Noria, 1018
Baron, C., 511
Barrado, Cristina, 213
Barreteau, Michel, 1171
Baryamureeba, Venansius, 1044
Basermann, A., 613
Baum, M., 613
Becker, Ch., 643
Becuzzi, P, 1441
Beivide, R., 1222
Belkoniene, Abdelkader, 1451
Benkner, Siegfried, 1155
Benner, Peter, 1120
Berzins, Martin, 651

Bietenholz, Wolfgang, 1040
Biham, Eli, 777
Bilardi, Gianfranco, 543
Blake, R.J., 295
Blayo, Eric, 303
Bobbio, B., 1166
Bodin, François, 125, 1171, 1265
Boeres, Cristina, 340
Boniface, Yann, 935
Borensztejn, Patricia, 213
Botti, O., 1166
Bougé, Luc, 831
Boukas, L.A., 1395
Boukerche, Azzedine, 562
Bradford, Russell, 1476
Brahimi, Zahia, 1018
Brandes, Thomas, 833
Brent, Richard P., 1
Brorsson, Mats, 888
Brun, Olivier, 291
Byrd, R., 578

Calegario, Vanusa Menditi, 1484
Campo, Renato, 1153
Cantoni, Virginio, 939
Cappello, Franck, 790
Carroll, L., 511
Carvalho, Luiz M., 1032
Casanova, Henri, 30
Cassinari, F., 1166
Castro, Luís Fernando, 899
Catthoor, Francky, 668
Catunda, Marco, 1369
Chalot, F., 595
Chamski, Zbigniew, 1171
Chapman, Barbara, 125, 373
Charles, Henri-Pierre, 1171
Chassin de Kergommeaux, J., 154
Chaussumier, Frédérique, 570
Chehayeb, Nassouh A., 1199
Chevalier, G., 595
Chi, Chi-Hung, 1243
Choi, Jong Hyuk, 753
Choudhary, Alok, 430
Ciciani, Bruno, 165, 961

Clark, A.F., 995
Claver, José M., 1104
Cleary, Andrew, 1078
Cocco, F., 1176
Cohen, Albert, 375
Colajanni, M., 165
Collard, Jean-François, 383
Collet, Jacques, 745
Combs, Candice Bechem, 1322
Combs, Jonathan, 1322
Connors, Daniel A., 1301
Coppola, M., 1441
Cordsen, Joerg, 188
Corrêa, Ricardo C., 272
Correia, Manuel E., 899
Cortina, R., 1073
Cosnard, Michel, 523
Cotofana, Sorin, 708
Courtier, Philippe, 23
Crivelli, S., 578
Crop, Jason B., 409
Crozier, Kevin M., 1301

Déqué, Michel, 1393
Dalton, Niall J., 866
Danckaert, Koen, 668
Daoudi, El Mostafa, 686
Darling, Gordon, 691
Das, Sajal K., 562
De Florio, V., 1166
Debreu, Laurent, 303
Deconinck, G., 1166
Desbiens, Jocelyn, 785
Desprez, Frédéric, 570
Desprez, Frédéric, 89
Dewar, Mike, 1465
Diaz, Michel, 769
Dillon, E., 113
Dimitrelos, Dimitris, 159
Dinh, Q.V., 595
Doallo, Ramón, 183, 229
Donatelli, S., 1166
Dongarra, Jack J., 30, 1078
Dowers, Steve, 691
Downton, A.C., 995
Drach-Temam, Nathalie, 659, 757
Driesen, Karel, 1312
Drira, Khalil, 769
Ducrocq, Veronique, 1417
Dutra, Inê de Castro, 1484

Ebcioğlu, Kemal, 1269
Eicker, Norbert, 1040
Eisenbeis, Christine, 1171
El-Baïda, Rami, 1344
Elling, Volker, 203
Elsässer, Robert, 280
Emad, Nahid, 1385
Ercan, M. Fikret, 317
Erhard, W., 1005
Es-sqalli, T., 113
Escobar, Juan, 1417
Eskow, E., 578
Espadas, D., 173
Estrade, Jean-François, 1411

FEAST Group, 643
Feil, Manfred, 1013
Fenwick, Jr., James B., 1255
Ferreira, Afonso, 272
Ferretti, Marco, 977
Fesquet, Laurent, 745
Fey, D., 1005
Fezzani, Djamel, 785
Feßler, Armin, 459
Filali, Mamoun, 831
Filho, Eliseu M.C., 727
Fimmel, Dirk, 401
Fingberg, Jochen, 313
Fischer, Claude, 1411
Fleury, E., 113
Fleury, M., 995
Folliot, Bertil, 815
Fotis, S., 1195
Fouilloux, A., 1423
Fraguela, Basilio B., 229
Freitag, Burkhard, 449
Friedman, Roy, 777
Frommer, Andreas, 280, 1040
Fung, Yu-Fai, 317

Garcia, F., 877
Garcia, Jean-Marie, 291
Gauchard, David, 291
Gautier, Thierry, 1466
Gazen, Didier, 1417
Geffroy, J.C., 511
Geib, Jean-Marc, 271
Genius, Daniela, 1260
Gentzsch, Wolfgang, 561

Geoffray, Patrick, 633
Geoghegan, S., 515
Geyer, Cláudio F.R., 899
Ghose, Kanad, 440, 505
Ghosh, Abhrajit, 505
Gicquel, Nicolas, 1417
Giraud, Luc, 595, 1032, 1153, 1417, 1431
Giroudeau, Rodolphe, 369
Gittings, Bruce M., 691
Glover, Fred, 533
Gołębiewski, M., 613
Goldman, David, 505
Gonzalez, J., 877
Gouëzec, Frédéric, 769
Gregorio, J.A., 1222
Großer, B., 1088
Gschwind, Michael, 1269
Gubitoso, Marco Dimas, 188
Guerra, Concettina, 939
Guinand, Frédéric, 350
Guivarch, Ronan, 1417, 1431
Gupta, Anshul, 1096
Gupta, Manish, 668
Gurd, John, 1171
Gustavson, Fred, 1096
Gutiérrez, E., 422
Guyard, J., 113
Guyon, Marc, 603

Höhn, T., 1005
Hölzle, Urs, 1312
Hénon, Pascal, 1059
Hadad, Erez, 777
Haines, Matthew, 413
Halatsis, Constantine, 159
Hameurlain, Kader, 449
Hamma, Beidi, 1199
Head-Gordon, T., 578
Hegland, Markus, 1078
Hempel, R., 613
Hendrickson, Bruce, 271
Hernández, Vicente, 1104
Herrmann, Christoph A., 930
Hertzberger, Louis O., 217
Hill, L., 125
Hlayhel, Wissam, 745
Hoogerbrugge, Jan, 1171
Hu, Ping, 1171
Hu, Weiwu, 909
Hu, Y. Charlie, 248

Hu, Y.F., 295
Hwu, Wen-mei W., 1301
Hüttel, Hans, 1353

Iavernaro, Felice, 1136
Ierusalimschy, Roberto, 1369
Imbard, Maurice, 603
Itzkovitz, Ayal, 777
Izaguirre, Javier G., 1187, 1199
Izu, C., 1222

Jaâra, El Miloud, 686
Jabouille, Patrick, 1417
Jacobi, Christian, 477
Jalby, William, 1171
Jerman, Jure, 1411
Jesshope, Chris, 695
Jiménez, José M., 1199
Jimenéz, José M., 1187
Jin, Guohua, 248
John, Lizy Kurian, 239, 266
Jonker, Pieter, 939

Kähäri, Andreas, 1124
König, Jean-Claude, 369
Kacsuk, P., 90
Kalinov, Alexey, 1024
Kalinovsky, Vladislav, 777
Kandemir, Mahmut, 430
Karaivanov, Alexander, 1096
Karypis, George, 322
Kasche, B., 1005
Keane, John A., 1195, 1460
Kilian, S., 643
Kim, MyungHo, 30
Kisuki, Toru, 1171
Kitsuregawa, Masaru, 1455
Klein, A., 1166
Kleist, Josva, 1353
Kleyman, Sergey, 777
Kloos, Philippe, 1417
Knijnenburg, Peter M.W, 1171
Knoop, Jens, 391
Korres, G., 1395
Kovács, J., 90
Královič, R., 1231
Kramer, William T.C., 61
Kranzlmüller, D., 154
Kshemkalyani, Ajay D., 795
Kufner, H., 1166

Kulkarni, Chidamber, 668
Kumar, Vipin, 322
Kurdahi, Fadi, 727
Kutil, Rade, 1013

Löwe, Welf, 332
Labarta, Jesus, 213
Laccetti, G., 1144
Lafage, Thierry, 1265
Laitenberger, Jan, 930
Lamoureux, Michael G., 525
Lang, B., 1088
Langlais, Michel, 677
Lapegna, M., 1144
Larzábal, Alberto, 1187
Lascaratos, A., 1395
Lastovetsky, Alexey, 1024
Latu, Guillaume, 677
Laure, Erwin, 413, 925
Lauwereins, R., 1166
Leair, Mark, 418
Lebeck, Alvin R., 1251
Lee, Ming-hau, 727
Lefèvre, Laurent, 633
Lefebvre, Vincent, 375
Lefer, Wilfrid, 624
Legat, Jean-Didier, 735
Lelait, Sylvain, 1260
Lenders, Patrick M., 1049
Lengauer, Christian, 930
Lichtenau, Cédric, 477
Limousin, C., 262
Lippert, Thomas, 1040
Llorente, I.M., 173
Loi, Michel, 570
Lombardi, Luca, 987
Lonsdale, Guy, 313, 1155
Lovas, R., 90
Lu, Guangming, 727
Lucas, Robert F., 61
Lucchesi, Robert, 1403

Märtens, Holger, 469
Méhaut, Jean-François, 139
Macq, Benoit, 735
Madec, Gurvan, 603
Madoń, Dominik, 716
Maerten, Bart, 313
Maguire, K.C.F., 295
Malfetti, Paolo, 1435

Malony, Allen D., 135
Mamalis, Basilis, 482
Mancini, Stéphane, 940
Manneback, Pierre, 271
Mannhart, Niklaus, 1466
Manning, Anna M., 1460
Martín, María J., 1068
Martín, Unai, 1187
Martin, Bruno, 557
Matessi, Andrea, 987
Matey, Luis M., 1187
Matooane, Mantŝika, 1492
Mazzia, Francesca, 1136
McColl, Bill, 831
Medeke, Björn, 1040
Mehrotra, Piyush, 413
Meier, René, 519
Melin, Emmanuel, 552
Mellor, G.L., 1395
Merker, Renate, 401
Merlin, J., 125
Merro, Massimo, 1353
Miles, Douglas, 418
Mimikou, N.Th., 1395
Mineter, Michael J., 691
Miron, David J., 1049
Missirlis, N.M., 1395
Molod, Andrea, 1403
Monien, Burkhard, 280
Monnier, Stefan, 716
Monteil, Thierry, 291
Morales, D., 877
Moreno, Rafael A., 1291
Morrison, John P., 866
Moschella, Francesco, 663
Motet, Gilles, 487
Moukrim, Aziz, 350
Mounié, Grégory, 303
Mourlas, Costas, 497
Mulholland, Connor, 691

Nascimento, Aline, 340
Nash, Jonathan, 651
Nestmann, Uwe, 1353
Ni, Sze-Yao, 1209
Nisbet, Andy, 1171
Nixon, Paddy, 519
Norman, Arthur, 1492
Norris, Cindy, 1255
Noulard, Eric, 1385

O'Boyle, Michael F.P, 1171
Oğuz, Ceyda, 317
Ober, Iulian, 1377
Oguchi, Masato, 1455

Pacalet, Renaud, 940
Paccagnella, Tiziana, 1435
Padiou, Gérard , 767
Pancake, Cherri M., 44
Pantel, Marc, 1333
Papatheodorou, Theodore S., 358
Pardines, Inmaculada, 1068
Park, Cheol Ho, 753
Park, Daeyeon, 753
Park, Kyu Ho, 753
Paruolo, G., 1176
Pavlov, Velisar, 1149
Pechanek, Gerald G., 761
Pedersen, Jan B., 108
Perez, Christian, 117
Pessolano, Francesco, 1296
Petrosino, Alfredo, 969
Pfannenstiel, W., 445
Pham, CongDuc, 633
Piacentini, A., 1423
Pierson, Jean-Marc, 624
Pietracaprina, Andrea, 543
Pimentel, Andy D., 217
Pinty, Jean-Pierre, 1417
Pinuel, Luis, 1291
Pitsianis, Nikos, 761
Plank, James S., 30
Planquelle, Benoît, 139
Plata, O., 422
Polychronopoulos, Eleftherios D., 358
Powell, David, 487
Power, David, 1476
Preis, Robert, 280
Prieto, M., 173
Proença, Alberto José, 845
Prylli, Loïc, 633
Pucci, Geppino, 543
Puente, V., 1222
Puiatti, Jean-Michel, 1301
Puntigam, Franz, 1334

Quaglia, Francesco, 165, 450
Quintana-Ortí, Enrique S., 1104, 1120
Quintana-Ortí, Gregorio, 1120
Quisquater, Jean-Jacques, 735

Radhakrishnan, Ramesh, 239
Raffin, Bruno, 552
Rajasekaran, Umesh Kumar V, 1204
Rajopadhye, Sanjay, 698
Ramanujam, J., 430
Ramet, Pierre, 1059
Ranilla, J., 1073
Rau-Chaplin, Andrew, 525
Ravindran, Binoy, 489
Raymond, David R., 1251
Raynal, Michel, 450, 806
Rebello, Vinod E. F., 340
Rebeuf, Xavier, 552
Redelsperger, Jean-Luc, 1417
Reeve, Jeff, 193
Reme, Hilde, 586
Revol, Nathalie, 139
Reymann, Olivier, 633
Richard, Olivier, 790
Rinaudo, Salvatore, 663
Risset, Tanguy, 698
Ritzdorf, H., 613
Rivera, Francisco F., 1068
Rizzo, Davide, 977
Roda, J., 877
Rodman, Andreas, 888
Rodriguez, C., 877
Rodriguez, Noemi, 1369
Rohou, Erven, 1171, 1265
Roman, Jean, 271, 677, 1059
Romero, Milton, 961
Roose, Dirk, 313
Roux, Francois-Xavier, 603
Rovan, B., 1231
Ružička, P., 1231

Sébot Julien, J., 659
Sánchez, Eduardo, 716
Sagastizábal, Claudia A., 1112
Sainrat, Pascal, 1241
Sallé, Patrick, 1333
Salvi, Giuseppe, 969
Sanlaville, Eric, 350
Santos Costa, Vítor , 899
Sassa, Masataka, 914
Sathaye, Sumedh, 1269
Sato, Toshinori, 1281
Savarese, Daniel, 78
Sawyer, William, 1403

Schaller, Christian, 930
Schaubschläger, Ch., 154
Schek, Hans-Jörg, 459
Schilling, Klaus, 1040
Schimmler, Manfred, 950
Schiper, André, 767
Schloegel, Kirk, 322
Schmidt, Bertil, 950
Schnabel, R., 578
Schuster, Assaf, 777
Schuster, Vincent, 418
Schwan, Karsten, 203
Scott, R. I., 1195
Selwood, Paul, 651
Sens, Pierre, 815
Seznec, André, 1171, 1265
Shafer, Stephen, 440
Sharma, Girindra D., 1204
Shen, John Paul, 1322
Sheu, Jang-Ping, 1209
Shi, Weisong, 909
Shintani, Takahiko, 1455
Silan, Patrick, 677
Silva, Fernando, 899
Simon, Horst D., 61
Singh, Hartej, 727
Singhal, Mukesh, 795
Sips, Henk, 831
Skillicorn, David, 1439
Sloan, Terence M., 691
Sobral, João , 845
Solodov, Mikhail V., 1112
Sottile, Matthew J., 135
Spirakis, Paul, 482
Srimani, Pradip K., 823
Stöhr, Elena A., 1171
Stan, Ileana, 1377
Stathis, Pyrrhos, 708
Steihaug, Trond, 1044
Stein, Joël, 1417, 1431
Sterling, Thomas, 78
Stoffel, Kilian, 1451
Sulatycke, Peter, 505
Sundberg, Samuel, 1124

Taddei, F., 1176
Tadonki, Claude, 698
Takacs, Lawrence, 1403
Talia, Domenico, 1439
Talla, Deependra, 266

Tampakas, Basil, 482
Tang, Zhimin, 909
Thiaudière, Yan, 1199
Thomas, Bruno, 1187
Thottethodi, Mithuna S., 1251
Thulasiraman, Krishnaiyan, 533
Thurner, E., 1166
Tirado, Francisco, 173, 1291
Tondu, B., 511
Toulouse, Michel, 533
Tourancheau, Bernard, 633
Touriño, Juan, 183
Träff, J.L., 613
Treffers, Menno, 1171
Trystram, Denis, 303
Tseng, Yu-Chee, 1209
Turek, S., 643

Uhl, Andreas, 1013

Valero, Mateo, 1241
van der Mark, Paul, 1171
Vanderstraeten, Denis, 1128
Vanneschi, M., 1441
Vargas, Patrícia Kayser, 899
Vartanian, Alexis, 262, 659, 757
Vasconcelos, Paulo B., 608
Vasek, Pavel, 505
Vassiliadis, Stamatis, 708, 761
Verhulst, E., 1166
Verians, Xavier, 735
Vialle, Stéphane, 935
Viland, G., 125
Vincent, Jean-Marc, 163
Virot, Bernard, 552
Vitenberg, Roman, 777
Vogel, David R., 505

Wagner, Alan, 108
Wakita, Ken, 914
Wang, San-Yuan, 1209
Ward, Paul A.S., 144
Wasniewski, Jerzy, 1096
Westrelin, Roland, 633
Wickner, Andreas, 258
Wijshoff, Harry A.G, 1171
Wilde, Doran K., 409
Wilsey, Philip A., 1204
Winstanley, Noel, 858
Wolf, Harald, 258

Wolfe, Michael, 418
Wollenweber, F., 125

Yalamov, Plamen, 1096, 1149
Yang, Chia-Lin, 1251
Yar, Hugues-Olivier, 1344
Yuan, Jun-Li, 1243

Zanghirati, G., 1176
Zapata, Emilio L., 229, 422
Zavanella, Andrea, 853
Zhang, Yin, 1044
Zima, Hans, 413, 1155
Zimmermann, Wolf, 332

Lecture Notes in Computer Science

For information about Vols. 1–1584
please contact your bookseller or Springer-Verlag

Vol. 1587: J. Pieprzyk, R. Safavi-Naini, J. Seberry (Eds.), Information Security and Privacy. Proceedings, 1999. XI, 327 pages. 1999.

Vol. 1589: J.L. Fiadeiro (Ed.), Recent Trends in Algebraic Development Techniques. Proceedings, 1998. X, 341 pages. 1999.

Vol. 1590: P. Atzeni, A. Mendelzon, G. Mecca (Eds.), The World Wide Web and Databases. Proceedings, 1998. VIII, 213 pages. 1999.

Vol. 1592: J. Stern (Ed.), Advances in Cryptology – EUROCRYPT '99. Proceedings, 1999. XII, 475 pages. 1999.

Vol. 1593: P. Sloot, M. Bubak, A. Hoekstra, B. Hertzberger (Eds.), High-Performance Computing and Networking. Proceedings, 1999. XXIII, 1318 pages. 1999.

Vol. 1594: P. Ciancarini, A.L. Wolf (Eds.), Coordination Languages and Models. Proceedings, 1999. IX, 420 pages. 1999.

Vol. 1595: K. Hammond, T. Davie, C. Clack (Eds.), Implementation of Functional Languages. Proceedings, 1998. X, 247 pages. 1999.

Vol. 1596: R. Poli, H.-M. Voigt, S. Cagnoni, D. Corne, G.D. Smith, T.C. Fogarty (Eds.), Evolutionary Image Analysis, Signal Processing and Telecommunications. Proceedings, 1999. X, 225 pages. 1999.

Vol. 1597: H. Zuidweg, M. Campolargo, J. Delgado, A. Mullery (Eds.), Intelligence in Services and Networks. Proceedings, 1999. XII, 552 pages. 1999.

Vol. 1598: R. Poli, P. Nordin, W.B. Langdon, T.C. Fogarty (Eds.), Genetic Programming. Proceedings, 1999. X, 283 pages. 1999.

Vol. 1599: T. Ishida (Ed.), Multiagent Platforms. Proceedings, 1998. VIII, 187 pages. 1999. (Subseries LNAI).

Vol. 1600: M. Wooldridge, M. Veloso (Eds.), Artificial Intelligence Today. VIII, 489 pages. 1999. (Subseries LNAI).

Vol. 1601: J.-P. Katoen (Ed.), Formal Methods for Real-Time and Probabilistic Systems. Proceedings, 1999. X, 355 pages. 1999.

Vol. 1602: A. Sivasubramaniam, M. Lauria (Eds.), Network-Based Parallel Computing. Proceedings, 1999. VIII, 225 pages. 1999.

Vol. 1603: J. Vitek, C.D. Jensen (Eds.), Secure Internet Programming. X, 501 pages. 1999.

Vol. 1604: M. Asada, H. Kitano (Eds.), RoboCup-98: Robot Soccer World Cup II. XI, 509 pages. 1999. (Subseries LNAI).

Vol. 1605: J. Billington, M. Diaz, G. Rozenberg (Eds.), Application of Petri Nets to Communication Networks. IX, 303 pages. 1999.

Vol. 1606: J. Mira, J.V. Sánchez-Andrés (Eds.), Foundations and Tools for Neural Modeling. Proceedings, Vol. I, 1999. XXIII, 865 pages. 1999.

Vol. 1607: J. Mira, J.V. Sánchez-Andrés (Eds.), Engineering Applications of Bio-Inspired Artificial Neural Networks. Proceedings, Vol. II, 1999. XXIII, 907 pages. 1999.

Vol. 1608: S. Doaitse Swierstra, P.R. Henriques, J.N. Oliveira (Eds.), Advanced Functional Programming. Proceedings, 1998. XII, 289 pages. 1999.

Vol. 1609: Z. W. Raś, A. Skowron (Eds.), Foundations of Intelligent Systems. Proceedings, 1999. XII, 676 pages. 1999. (Subseries LNAI).

Vol. 1610: G. Cornuéjols, R.E. Burkard, G.J. Woeginger (Eds.), Integer Programming and Combinatorial Optimization. Proceedings, 1999. IX, 453 pages. 1999.

Vol. 1611: I. Imam, Y. Kodratoff, A. El-Dessouki, M. Ali (Eds.), Multiple Approaches to Intelligent Systems. Proceedings, 1999. XIX, 899 pages. 1999. (Subseries LNAI).

Vol. 1612: R. Bergmann, S. Breen, M. Göker, M. Manago, S. Wess, Developing Industrial Case-Based Reasoning Applications. XX, 188 pages. 1999. (Subseries LNAI).

Vol. 1613: A. Kuba, M. Šámal, A. Todd-Pokropek (Eds.), Information Processing in Medical Imaging. Proceedings, 1999. XVII, 508 pages. 1999.

Vol. 1614: D.P. Huijsmans, A.W.M. Smeulders (Eds.), Visual Information and Information Systems. Proceedings, 1999. XVII, 827 pages. 1999.

Vol. 1615: C. Polychronopoulos, K. Joe, A. Fukuda, S. Tomita (Eds.), High Performance Computing. Proceedings, 1999. XIV, 408 pages. 1999.

Vol. 1616: P. Cointe (Ed.), Meta-Level Architectures and Reflection. Proceedings, 1999. XI, 273 pages. 1999.

Vol. 1617: N.V. Murray (Ed.), Automated Reasoning with Analytic Tableaux and Related Methods. Proceedings, 1999. X, 325 pages. 1999. (Subseries LNAI).

Vol. 1618: J. Bézivin, P.-A. Muller (Eds.), The Unified Modeling Language. Proceedings, 1998. IX, 443 pages. 1999.

Vol. 1619: M.T. Goodrich, C.C. McGeoch (Eds.), Algorithm Engineering and Experimentation. Proceedings, 1999. VIII, 349 pages. 1999.

Vol. 1620: W. Horn, Y. Shahar, G. Lindberg, S. Andreassen, J. Wyatt (Eds.), Artificial Intelligence in Medicine. Proceedings, 1999. XIII, 454 pages. 1999. (Subseries LNAI).

Vol. 1621: D. Fensel, R. Studer (Eds.), Knowledge Acquisition Modeling and Management. Proceedings, 1999. XI, 404 pages. 1999. (Subseries LNAI).

Vol. 1622: M. González Harbour, J.A. de la Puente (Eds.), Reliable Software Technologies – Ada-Europe'99. Proceedings, 1999. XIII, 451 pages. 1999.

Vol. 1623: T. Reinartz, Focusing Solutions for Data Mining. XV, 309 pages. 1999. (Subseries LNAI).

Vol. 1625: B. Reusch (Ed.), Computational Intelligence. Proceedings, 1999. XIV, 710 pages. 1999.

Vol. 1626: M. Jarke, A. Oberweis (Eds.), Advanced Information Systems Engineering. Proceedings, 1999. XIV, 478 pages. 1999.

Vol. 1627: T. Asano, H. Imai, D.T. Lee, S.-i. Nakano, T. Tokuyama (Eds.), Computing and Combinatorics. Proceedings, 1999. XIV, 494 pages. 1999.

Col. 1628: R. Guerraoui (Ed.), ECOOP'99 - Object-Oriented Programming. Proceedings, 1999. XIII, 529 pages. 1999.

Vol. 1629: H. Leopold, N. García (Eds.), Multimedia Applications, Services and Techniques - ECMAST'99. Proceedings, 1999. XV, 574 pages. 1999.

Vol. 1631: P. Narendran, M. Rusinowitch (Eds.), Rewriting Techniques and Applications. Proceedings, 1999. XI, 397 pages. 1999.

Vol. 1632: H. Ganzinger (Ed.), Automated Deduction – Cade-16. Proceedings, 1999. XIV, 429 pages. 1999. (Subseries LNAI).

Vol. 1633: N. Halbwachs, D. Peled (Eds.), Computer Aided Verification. Proceedings, 1999. XII, 506 pages. 1999.

Vol. 1634: S. Džeroski, P. Flach (Eds.), Inductive Logic Programming. Proceedings, 1999. VIII, 303 pages. 1999. (Subseries LNAI).

Vol. 1636: L. Knudsen (Ed.), Fast Software Encryption. Proceedings, 1999. VIII, 317 pages. 1999.

Vol. 1637: J.P. Walser, Integer Optimization by Local Search. XIX, 137 pages. 1999. (Subseries LNAI).

Vol. 1638: A. Hunter, S. Parsons (Eds.), Symbolic and Quantitative·Approaches to Reasoning and Uncertainty. Proceedings, 1999. IX, 397 pages. 1999. (Subseries LNAI).

Vol. 1639: S. Donatelli, J. Kleijn (Eds.), Application and Theory of Petri Nets 1999. Proceedings, 1999. VIII, 425 pages. 1999.

Vol. 1640: W. Tepfenhart, W. Cyre (Eds.), Conceptual Structures: Standards and Practices. Proceedings, 1999. XII, 515 pages. 1999. (Subseries LNAI).

Vol. 1642: D.J. Hand, J.N. Kok, M.R. Berthold (Eds.), Advances in Intelligent Data Analysis. Proceedings, 1999. XII, 538 pages. 1999.

Vol. 1643: J. Nešetřil (Ed.), Algorithms – ESA '99. Proceedings, 1999. XII, 552 pages. 1999.

Vol. 1644: J. Wiedermann, P. van Emde Boas, M. Nielsen (Eds.), Automata, Languages, and Programming. Proceedings, 1999. XIV, 720 pages. 1999.

Vol. 1645: M. Crochemore, M. Paterson (Eds.), Combinatorial Pattern Matching. Proceedings, 1999. VIII, 295 pages. 1999.

Vol. 1647: F.J. Garijo, M. Boman (Eds.), Multi-Agent System Engineering. Proceedings, 1999. X, 233 pages. 1999. (Subseries LNAI).

Vol. 1648: M. Franklin (Ed.), Financial Cryptography. Proceedings, 1999. VIII, 269 pages. 1999.

Vol. 1649: R.Y. Pinter, S. Tsur (Eds.), Next Generation Information Technologies and Systems. Proceedings, 1999. IX, 327 pages. 1999.

Vol. 1650: K.-D. Althoff, R. Bergmann, L.K. Branting (Eds.), Case-Based Reasoning Research and Development. Proceedings, 1999. XII, 598 pages. 1999. (Subseries LNAI).

Vol. 1651: R.H. Güting, D. Papadias, F. Lochovsky (Eds.), Advances in Spatial Databases. Proceedings, 1999. XI, 371 pages. 1999.

Vol. 1652: M. Klusch, O.M. Shehory, G. Weiss (Eds.), Cooperative Information Agents III. Proceedings, 1999. XI, 404 pages. 1999. (Subseries LNAI).

Vol. 1653: S. Covaci (Ed.), Active Networks. Proceedings, 1999. XIII, 346 pages. 1999.

Vol. 1654: E.R. Hancock, M. Pelillo (Eds.), Energy Minimization Methods in Computer Vision and Pattern Recognition. Proceedings, 1999. IX, 331 pages. 1999.

Vol. 1656: S. Chatterjee, J.F. Prins, L. Carter, J. Ferrante, Z. Li, D. Sehr, P.-C. Yew (Eds.), Languages and Compilers for Parallel Computing. Proceedings, 1998. XI, 384 pages. 1999.

Vol. 1661: C. Freksa, D.M. Mark (Eds.), Spatial Information Theory. Proceedings, 1999. XIII, 477 pages. 1999.

Vol. 1662: V. Malyshkin (Ed.), Parallel Computing Technologies. Proceedings, 1999. XIX, 510 pages. 1999.

Vol. 1663: F. Dehne, A. Gupta. J.-R. Sack, R. Tamassia (Eds.), Algorithms and Data Structures. Proceedings, 1999. IX, 366 pages. 1999.

Vol. 1664: J.C.M. Baeten, S. Mauw (Eds.), CONCUR'99. Concurrency Theory. Proceedings, 1999. XI, 573 pages. 1999.

Vol. 1666: M. Wiener (Ed.), Advances in Cryptology – CRYPTO '99. Proceedings, 1999. XII, 639 pages. 1999.

Vol. 1668: J.S. Vitter, C.D. Zaroliagis (Eds.), Algorithm Engineering. Proceedings, 1999. VIII, 361 pages. 1999.

Vol. 1671: D. Hochbaum, K. Jansen, J.D.P. Rolim, A. Sinclair (Eds.), Randomization, Approximation, and Combinatorial Optimization. Proceedings, 1999. IX, 289 pages. 1999.

Vol. 1678: M.H. Böhlen, C.S. Jensen, M.O. Scholl (Eds.), Spatio-Temporal Database Management. Proceedings, 1999. X, 243 pages. 1999.

Vol. 1684: G. Ciobanu, G. Păun (Eds.), Fundamentals of Computation Theory. Proceedings, 1999. XI, 570 pages. 1999.

Vol. 1685: P. Amestoy, P. Berger, M. Daydé, I. Duff, V. Frayssé, L. Giraud, D. Ruiz (Eds.), Euro-Par'99. Parallel Processing. Proceedings, 1999. XXXII, 1503 pages. 1999.

Vol. 1688: P. Bouquet, L. Serafini, P. Brézillon, M. Benerecetti, F. Castellani (Eds.), Modeling and Using Context. Proceedings, 1999. XII, 528 pages. 1999. (Subseries LNAI).

Vol. 1689: F. Solina, A. Leonardis (Eds.), Computer Analysis of Images and Patterns. Proceedings, 1999. XIV, 650 pages. 1999.